Aircraft Design: A Conceptual Approach

Fourth Edition

Aircraft Design: A Conceptual Approach

Fourth Edition

Daniel P. Raymer
Conceptual Research Corporation
Playa del Rey, California

EDUCATION SERIES
Joseph A. Schetz
Series Editor-in-Chief
Virginia Polytechnic Institute and State University
Blacksburg, Virginia

Published by the
American Institute of Aeronautics and Astronautics, Inc.
1801 Alexander Bell Drive, Reston, Virginia 20191-4344

American Institute of Aeronautics and Astronautics, Inc., Reston, Virginia

2 3 4 5 6 7 8 9 10

Library of Congress Cataloging-in-Publication Data

Raymer, Daniel P.
Aircraft design: a conceptual approach/Daniel P. Raymer. – 4th ed.
p. cm. – (AIAA education series)
Includes bibliographical references and index.
ISBN 1-56347-829-3 (alk. paper)
1. Airplanes–Design and construction. I. Title. II. Series.
TL671.2.R29 2006
629.134'1–dc22 2006004706

This book is dedicated to all who taught me, especially Lester Hendrix, Richard Hibma, Louis Hecq, Harry Scott, Richard Child, George Owl, Robert Maier, Ed McGachan, Doug Robinson, Steve White, Harvey Hoge, Michael Robinson, George Palmer, Henry Yang, Robert Swaim, C. T. Sun, David Schmidt, Bruce Reese, William Heiser, and Gordon Raymer (test pilot, aeronautical engineer, and my father).

Thanks also to Rockwell North American Aircraft Operations, SAAB Aircraft, and Lockheed Martin for permission to use various illustrations. All other artwork is original, in the public domain, or copyrighted by AIAA.

Foreword

We are delighted to present the Fourth Edition of *Aircraft Design: A Conceptual Approach* by Daniel Raymer. The first three editions have been very well received and widely used in the aerospace community. This new edition has updated the material and expanded the coverage, and we anticipate that it will be equally well and even better received. The current volume has 23 chapters and six appendices in more than 800 pages. A key feature of this and the earlier volumes is the encyclopedic scope of the coverage. In addition, every topic necessary to the understanding of aircraft design such as aerodynamics, structures, stability and control, propulsion, etc. is discussed from the point-of-view of the designer, not the specialist in a given topic area.

Daniel Raymer is uniquely qualified to write this book because of his broad expertise in the area. His command of the material is excellent, and he is able to organize and present it in a very clear manner. Clearly, his extensive industrial experience provides the background and perspective needed for the author of a successful design textbook.

The AIAA Education Series aims to cover a very broad range of topics in the general aerospace field, including basic theory, applications and design. A complete list of titles can be found at www.aiaa.org. The philosophy of the series is to develop textbooks that can be used in a university setting, instructional materials for continuing education and professional development courses, and resources that can serve as the basis for independent study or as working references. Suggestions for new topics or authors are always welcome.

Joseph A. Schetz
Editor-in-Chief
AIAA Education Series

Author with display model of his Advanced Supercruise Fighter Concept (Ref. 13). Photo courtesy of Rockwell International North American Aircraft Operations.

Table of Contents

Learjet (USAF C21-A) (U.S. Air Force photo).

Preface

There are two equally important aspects of aircraft design: *design layout* and *design analysis*. These very different activities attract different types of people. Some love playing with numbers and computers, whereas others can't stop doodling on every piece of paper within reach.

This book was written to fill the need for a texbook in which both aircraft analysis and design layout are covered equally and the interactions between them are explored in a manner consistent with industry practice.

Aircraft design obviously depends on the reliable calculation of numbers; however, in the end the only thing that actually gets built is the drawing. Its creation is not a trivial task of drafting based upon the analysis results, but rather it is a key element of the overall design process and ultimately determines the performance, weight, and cost of the aircraft. Bluntly stated, if you don't have a good drawing, you don't have an aircraft design.

It is difficult to visualize and draw a new aircraft that has a streamlined aerodynamic shape and an efficient internal layout and yet satisfies an incredible number of real-world constraints and design specifications. This is a rare talent that takes years to cultivate. Although to some extent good designers are "born, not made," the proven methods and best practices of aircraft configuration layout can be taught, and are covered here in the first half of this book.

It is also true that a nice aircraft drawing is nothing without the analytical results to support it, and it will be a much nicer drawing if clever optimization methods are employed to make it better. So, a good designer or design team must find an appropriate balance between layout and analysis. To provide such balance, the second half of this book covers analysis and optimization methods that will tell you if the design works, if it meets its design requirements, and how you can make it better in the next drawing.

One of the most important lessons that a student of aircraft design must absorb is that design is an iterative process. One does not draw a neat-looking concept, analyze its range and performance, and stop there. You must use design judgment and computational analysis to identify all possible areas for improvement and optimization of your concept, and then draw it again . . . and again, and maybe again!

The specific analysis techniques presented in the book are simplified to permit the student to experience the whole design process in a single course. No textbook can contain the methods actually used in industry, which tend to be proprietary and highly computerized. However, the methods presented here are sufficient, and give reasonable results for most categories of aircraft. In fact, they are good enough to be used to check the results of the sophisticated computerized methods, and if they are far apart, the computer results are probably wrong!

The Aircraft Conceptual Design Web site at *www.aircraftdesign.com* includes examination questions for the book, advice to students and would-be inventors, sample aircraft design layouts, free design software, tips for the use of the design software RDS-Student (based on methods in this book), and information on aircraft design short courses. All are welcome!

I, and the AIAA, would like to thank the many people who have offered constructive suggestions for this new edition, as well as the 30,000 students and working engineers who have made this book an AIAA best seller. Writing this book has been an educating and humbling experience. It is my sincere wish that it helps aspiring aircraft designers to "learn the ropes" more quickly.

Daniel P. Raymer
March 2006

Author's Note Concerning Use of Metric Weight Units

Metric (SI) units are more universal and technically consistent than British "Imperial" units (fps) and also reduce the possibility of stupid errors in aircraft calculations. However, one must still decide exactly which metric unit multipliers to use. Should masses be defined in grams or in kilograms? Should times be in seconds or in hours, or used as needed to make the numbers "nice." These decisions change the numbers, and unfortunately, different companies use slightly different combinations of unit multipliers and times. To maximize consistency with prior literature, the metric unit terms used in *Jane's All the World's Aircraft* (Ref. 1) and in Stinton's *The Design of the Aeroplane* (Ref. 28) were employed in this book. Values in this book are presented first in British units, and then in metric units enclosed in braces { }.

A key issue and the source of much confusion is the treatment of "weight" in metric units. Weight by definition is a force, not a mass. However, pilots and working engineers describing the weight of the Airbus A340 would say 126,000 kg, not 1,235,682 kN. What those pilots and engineers really mean is, "the Airbus exerts a weight force equivalent to that exerted by a 126,000 kg mass in a 1-g gravitational field." This book follows this common practice—don't let it confuse you! When doing an analysis such as calculating lift force and equating it to weight, the weight of 126,000 kg must first be converted to proper force units (Newtons) by multiplying by the 1-g acceleration constant ($g = 9.807\ \text{m/s}^2$).

This verbal equating of weight with force in a 1-g gravitational field is carried over to the definitions of ratios such as wing loading (kg/m^2) and power loading (kg/kW). Because of this, the values of these ratios as given in the tables are not technically correct when applied to the various equations that use them. The mass terms must be converted to force by multiplying by g. Thus, a wing loading given in "pilot talk" as $586\ \text{kg/m}^2$ must be converted to $5746.9\ \text{N/m}^2$ to apply in equations relating lift to weight (for example, see Table 5.5).

The values given for thrust-to-weight ratio (T/W) do not require conversion. In traditional British-unit practice, the thrust is given in lbs-force, and the weight is given in lbs-mass (exerted force assuming a 1-g field), so that the ratio is nondimensional and the same as the desired SI units of Newtons/Newton. A T/W greater than one means the aircraft can accelerate straight up, regardless of the units in which it was designed!

X-15 rollout (U.S. Air Force photo).

1
Design—A Separate Discipline

1.1 What is Design?

Aircraft design is a separate discipline of aeronautical engineering—different from the analytical disciplines such as aerodynamics, structures, controls, and propulsion. An aircraft designer needs to be well versed in these and many other specialties, but will actually spend little time performing such analysis in all but the smallest companies. Instead, the designer's time is spent doing something called "design," creating the geometric description of a thing to be built.

To the uninitiated, "design" looks a lot like "drafting" (or in the modern world, "computer-aided drafting"). The designer's product is a drawing, and the designer spends the day hunched over a drafting table or computer terminal. However, the designer's real work is mostly mental.

If the designer is talented, there is a lot more than meets the eye on the drawing. A good aircraft design seems to miraculously glide through subsequent evaluations by specialists without major changes being required. Somehow, the landing gear fits, the fuel tanks are near the center of gravity, the structural members are simple and lightweight, the overall arrangement provides good aerodynamics, the engines install in a simple and clean fashion, and a host of similar detail seems to fall into place.

This is no accident, but rather the product of a lot of knowledge and hard work by the designer. This book was written primarily to provide the basic tools and concepts required to produce good designs that will survive detailed analysis with minimal changes.

Other key players participate in the design process. Design is not just the actual layout, but also the analytical processes used to determine what should be designed and how the design should be modified to better meet the requirements. In a small company, this may be done by the same individuals who do the layout design. In the larger companies, aircraft analysis is done by the sizing and performance specialists with the assistance of experts in aerodynamics, weights, propulsion, stability, and other technical specialties.

1.2 Introduction to the Book

This book describes the process used to develop a credible aircraft conceptual design from a given set of requirements. As a part of the AIAA Education Series, the book is written primarily for the college student. Every effort has been made to achieve a self-contained book.

In an aircraft company, the designer can ask a functional specialist for a reasonable initial tire size, inlet capture area, weight savings due to the use of composites, or similar estimates. Such specialists are not available at most universities. This book thus gives various "rule-of-thumb" approximations for initial estimation of design parameters.

The book has 23 chapters, and approximately follows the actual design sequence. Chapters 2 and 3 provide an overall introduction to the design process. Chapter 2 discusses how the conceptual design process works and how it fits into the overall process of aircraft development. Chapter 3 presents a "first-pass" design procedure to familiarize the reader with the essential concepts of design, including design layout, analysis, takeoff-weight estimation, and trade studies.

In Chapters 4–11 the techniques for the development of the initial configuration layout are presented. These include the conceptual sketch, initial sizing, wing geometry selection, lofting, inboard layout, and integration of propulsion, crew station, payload/passenger compartment, fuel system, and landing gear. Considerations for observability, producibility, and supportability are also discussed. While the text implies that the design is done on a drafting board, it should be understood that in major aircraft companies today most aircraft design work is done on a computer-aided design system. However, the same basic design techniques are used whether on a drafting table or computer scope.

Chapters 12–19 address the analysis, sizing, and optimization of the design layout. Various chapters discuss aerodynamics, weights, installed propulsion characteristics, stability and control, performance, cost, and sizing. Optimization based upon design requirements is introduced in a section on trade studies.

These methods are simplified to allow rapid design analysis by students. Simplified analysis methods allow the student more time to experience the all-important optimization and iteration process.

The next three chapters discuss the design of flight vehicles that are in some way different from "normal" vehicles. Chapter 20 covers vertical flight, including helicopters and vertical takeoff jets. Chapter 21 introduces the extremes of flight—very fast to very slow—with subchapters on spacecraft, hypersonics, and airships. In Chapter 22, a number of unconventional designs are discussed, including flying wings, canard pushers, joined wings, and aysmmetric aircraft. This material builds upon the methods for conventional aircraft design, but introduces additional considerations that affect the design layout and analysis.

The last chapter, 23, contains two complete design project examples that use the methods presented in the previous chapters. These are provided instead of numerous example calculations throughout the text to illustrate how the different aspects of design fit together as a whole.

The appendices contain information useful in conceptual design, such as conversion tables, atmosphere tables, and data on airfoils and engines. Also included is a summary of the current civil and military design requirements and specifications, which have been taken primarily from Federal Aviation Regulations (FAR) and Military Specifications (Mil-Specs).

2
Overview of the Design Process

2.1 Introduction

Those involved in design can never quite agree as to just where the design process begins. The designer thinks it starts with a new airplane concept. The sizing specialist knows that nothing can begin until an initial estimate of the weight is made. The customer, civilian or military, feels that the design begins with requirements.

They are all correct. Actually, design is an iterative effort, as shown in the "Design Wheel" of Fig. 2.1. Requirements are set by prior design trade studies. Concepts are developed to meet requirements. Design analysis frequently points toward new concepts and technologies, which can initiate a whole new design effort. However a particular design is begun, all of these activities are equally important in producing a good aircraft concept.

2.2 Phases of Aircraft Design

Conceptual Design

Aircraft design can be broken into three major phases, as depicted in Fig. 2.2. Conceptual design is the primary focus of this book. It is in conceptual design that the basic questions of configuration arrangement, size and weight, and performance are answered. Conceptual design is characterized by a large number of design alternatives and trade studies, and a continuous, evolutionary change to the aircraft concepts under consideration.

The critical question is, "Can *any* affordable aircraft be built that meets the requirements?" If not, the customer may wish to revise or relax the requirements. In conceptual design, the design requirements are used to guide and evaluate the development of the overall aircraft configuration arrangement. This design arrangement includes wing and tail overall geometry (areas, sweeps, etc.), fuselage shape and internal locations of crew, payload, passengers, and equipment, engine installation, landing gear, and other design features. The level of detail in configuration design is not very deep (for example, the landing gear may be shown only as a circle for the tire and a "stick" for the gear leg), but the interactions among all the different components are so crucial that it requires years of experience to create a good conceptual design.

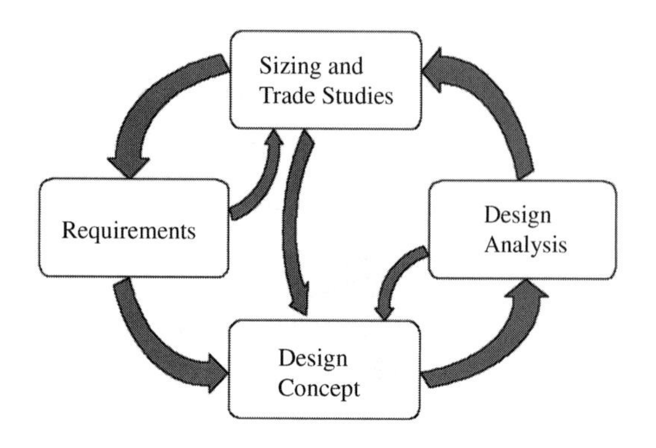

Fig. 2.1 The design wheel.

A key aspect of conceptual design is that it is a very fluid process, and the design layout is always being changed, both to incorporate new things learned about the design and to evaluate potential improvements to the design. Trade studies and an ever-increasing level of analysis sophistication cause the design to evolve on almost a week-by-week basis, and changes can be made in every aspect of the design including wing geometry, tail arrangement, and even the number of engines. Furthermore, during conceptual design a number of alternative designs are studied to determine which design approach is preferred. If we think the design requirements point to a canard, we may first design a concept with that arrangement, but the wise designer will also design several aft-tail concepts, and perhaps a tailless one, and let the numbers (not opinion, prejudice, or preconceived notions) make the final selection.

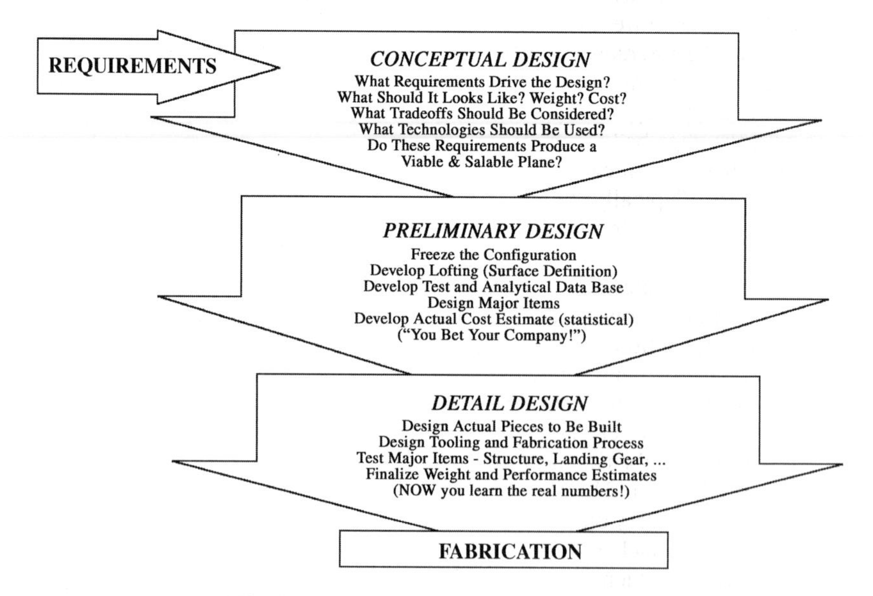

Fig. 2.2 Three phases of aircraft design.

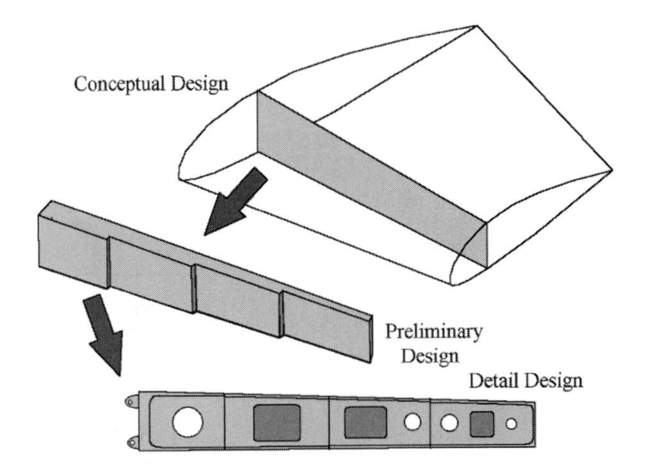

Fig. 2.3 Design phases: front wing spar.

As you go through conceptual, preliminary, and detail design, the level of detail of the design steadily increases. Figure 2.3 illustrates this for a typical piece of aircraft geometry, the front wing spar.

The top of figure 2.3 depicts the design of a front wing spar in the amount of detail typical of conceptual design. It is "designed" as nothing more than a flat plate from root to tip at the desired location of the spar. In other words, the designer draws a line in top view from root to tip, and the spar is assumed to be the depth of the wing airfoils at that location. While this seems crude, keep in mind that the entire aircraft arrangement is being determined at this stage of design, and the interactions between components are more important than the exact geometry of any one part. This simple definition answers the key questions for the initial conceptual layout: How big can the wing box, wing fuel tank, and leading-edge flaps be?

Conceptual design can take as little as a week (done poorly!) or as much as several years. Typically one should plan on six months to properly study the requirements, technologies, and configuration alternatives and to downselect to a best concept.

Note that computer-aided design (CAD) tools used during conceptual design should be tailored toward the fluid environment of conceptual design. A key ability is a collection of tools that permit one to rapidly develop a notional design concept (in approximately one day), and to continuously revise design concepts and perform geometric trade studies. CAD capabilities for rapidly locating rivets or cutter paths are worthless at this point, but a CAD capability to change the wing's sweep and automatically revise the geometry of the front spar accordingly would be of tremendous use, since the wing sweep will probably be changed after every optimization study or wind-tunnel test. Unfortunately, few companies have such CAD tools for conceptual design because of the very limited market for them.

The steps of conceptual design are described later in more detail.

Preliminary Design

Preliminary design can be said to begin when the major changes are over. The big questions such as whether to use a canard or an aft tail have been resolved. The configuration arrangement can be expected to remain about as shown on current drawings, although minor revisions may occur. Preliminary design is characterized by a maturation of the selected design approach (no more quick studies to determine if a canard might be better). The design evolves over a period of many months, with an ever-increasing level of understanding of the design, an ever-increasing level of design and analysis detail, and an ever-increasing level of confidence that the design will work. At some point the company believes it has sufficient information to "freeze" the design, forbidding further changes to the overall design arrangement. This schedule milestone is crucial, because it allows other designers to begin serious development of structure and subsystems without fear that their work will be invalidated by later changes to the overall design configuration.

During preliminary design the specialists in areas such as structures, landing gear, and control systems will design and analyze their portion of the aircraft. Testing is initiated in areas such as aerodynamics, propulsion, structures, and stability and control. A mockup may be constructed at this point, either physically or electronically using a modern CAD system.

A key activity during preliminary design is "lofting." Lofting is the mathematical modeling of the outside skin of the aircraft with sufficient accuracy to ensure proper fit between its different parts, even if they are designed by different designers and possibly fabricated in different locations. Lofting originated in shipyards and was originally done with long flexible rulers called "splines." This work was done in a loft over the shipyard, hence the name.

The ultimate objective during preliminary design is to ready the company for the detail design stage, also called full-scale development. Thus, the end of preliminary design usually involves a full-scale development proposal. In today's environment, this can result in a situation jokingly referred to as "you-bet-your-company." The possible loss on an overrun contract or from lack of sales can exceed the net worth of the company! Preliminary design must establish confidence that the airplane can be built on time and at the estimated cost.

In preliminary design, our example wing spar's overall geometry as defined in the conceptual design phase is refined, including the actual shaping of the spar's cross section (middle of Fig. 2.3). Fairly sophisticated methods are used to perform a structural analysis of the overall spar, with the objective of determining the thickness (or number of composite plies) required to handle the expected loads. The spar is only one element of the overall structure of the aircraft that will be defined in preliminary design, and extensive analysis (and sometimes testing) will be done of the whole structural concept to assess and optimize the overall concept.

Note that the spar design in the preliminary design phase is still not "buildable." Full consideration has not yet been given to attachments, cutouts, access panels, flanges, manufacturing limitations, fuel sealing, and other "real-world" details. These are the subject of detail design (bottom of Fig. 2.3), and are typically considered only after the aircraft structural concept as a whole has been validated during the preliminary design phase.

Preliminary design should take somewhere between a few months (done poorly) and perhaps two years for a complicated, high technology design such as a supersonic transport or stealth fighter.

CAD tools in preliminary design should include the conceptual design capabilities of rapidly reshaping the overall configuration, but must also permit definition of the entire design in production-quality surface definition. Since the number of designers working on the geometry will grow beyond a handful, some means of managing access to the geometry to avoid chaos is essential.

Detail Design

Assuming a favorable decision for entering full-scale development, the detail design phase begins in which the actual pieces to be fabricated are designed. This last and most expensive part of the design process is characterized by a large number of designers (sometimes thousands) preparing detailed drawings or CAD files with actual fabrication geometries and dimensions. While in conceptual design, the designers are concerned about such top-level issues as the number of engines required or the sweep of the wing; in detail design the designers are concerned about such things as the exact radius of the corner of a pocket cutout on a flap track, and the locations and dimensions of the holes that must be drilled for fasteners.

Furthermore, thousands of "little pieces" not considered during preliminary design must be designed during the detail design phase. These include flap tracks, brackets, structural clips, doors, avionics racks, and similar components. Every single piece of the aircraft's structure and its hydraulic, electrical, pneumatic, fuel, and other systems must be designed in the detail design phase.

For example, during conceptual and preliminary design the wing box will be designed and analyzed as a whole. During detail design, that whole will be broken down into individual ribs, spars, and skins, each of which must be separately designed and analyzed.

Another important part of detail design is called production design. Specialists determine how the airplane will be fabricated, starting with the smallest and simplest subassemblies and building up to the final assembly process. Production designers frequently wish to modify the design for ease of manufacture; that modification can have a major impact on performance or weight. Compromises are inevitable, but the design must still meet the original requirements.

It is interesting to note that in the former Soviet Union, the production design was done by a design bureau separate from the conceptual and preliminary design, resulting in superior producibility at some expense in performance and weight.

During detail design, the testing effort intensifies. Actual structure of the aircraft is fabricated and tested. Control laws for the flight control system are tested on an "iron-bird" simulator, a detailed working model of the actuators and flight control surfaces. Flight simulators are developed and flown by both company and customer test-pilots.

The further along a design progresses, the more people are involved. In fact, most of the engineers who go to work for a major aerospace company will work in preliminary or detail design.

CAD tools for use during detail design are very well developed, and programs such as UNIGRAPHICS and CATIA have numerous tools to assist in design of "little pieces," as just described, and in definition of typical production features such as cutouts, pockets, radii, and holes. It is the CAD database developed in detail design that is actually passed to computer-aided manufacturing machinery.

Detail design ends with fabrication of the first aircraft. Often to meet a schedule the fabrication of some parts of the aircraft must begin before the entire detail design effort is completed. Sometimes this leads to changes in already-fabricated parts or tools, at enormous expense!

Usually the prototypes are built on "soft" or temporary tooling, and are often built with fabrication processes different from those envisioned for the production run. While initially cheaper, this may not be a good idea in the long run. It may be as important to test the tooling as it is to test the prototype aircraft. Using production tooling, as was done on the Mitsubishi F-2 (derivative of F-16), will uncover production problems earlier and should reduce the total program cost, even if the initial costs are higher.

Production begins with the design and fabrication of the production tooling. Historically this has been a massive and expensive undertaking, with hundreds or thousands of expensive jigs and fixtures being built. Once production begins, problems are often uncovered and the tooling and production processes must be modified. An ongoing trend for cost reduction is to use CAM technologies and innovative design concepts to minimize such hard tooling.

2.3 Aircraft Conceptual Design Process

Figure 2.4 depicts the conceptual design process in greater detail. Conceptual design will usually begin with either a specific set of design requirements established by the prospective customer or a company-generated guess as to what future customers may need. Design requirements include parameters such as the aircraft range and payload, takeoff and landing distances, and maneuverability and speed requirements.

The design requirements also include a vast set of civil or military design specifications that must be met. These include landing sink-speed, stall speed, structural design limits, pilots' outside vision angles, reserve fuel, and many others.

Sometimes a design will begin as an innovative idea rather than as a response to a given requirement. The flying wings pioneered by John Northrop were not conceived in response to a specific Army Air Corps requirement at that time, but instead were the product of one man's idea of the "better airplane." Northrop pursued this idea for years before building a flying wing to suit a particular military requirement.

Before a design can be started, a decision must be made as to what technologies will be incorporated. If a design is to be built in the near future, it must use only currently available technologies as well as existing engines and avionics. If it is being designed to be built in the more distant future, then an estimate of the technological state of the art must be made to determine which emerging technologies will be ready for use at that time.

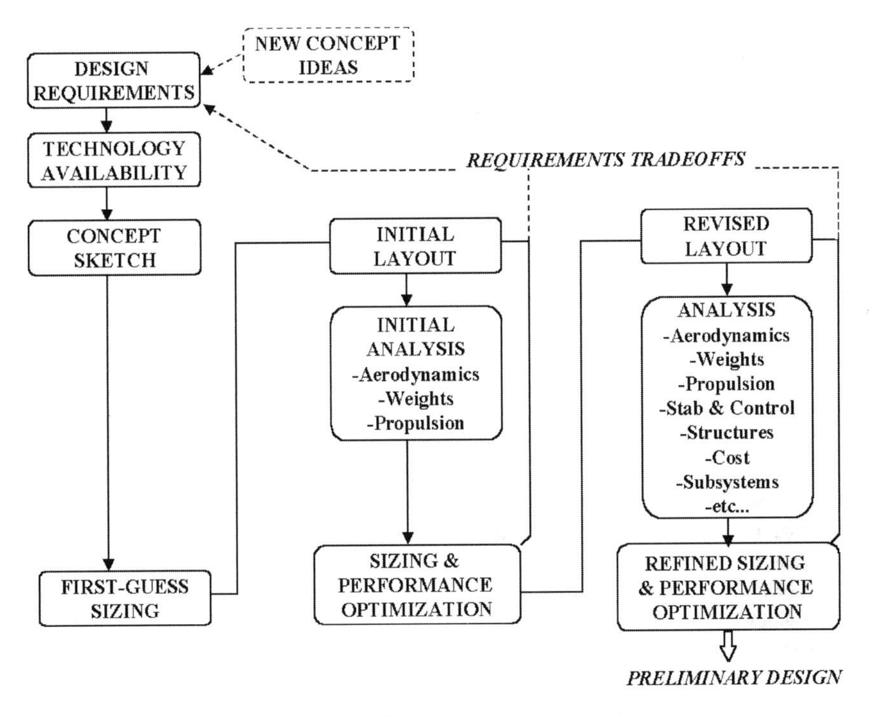

Fig. 2.4 Aircraft conceptual design process.

For example, an aircraft with all-electric actuators has yet to enter production as of 2005, but that technology poses no great risk based on successful flight demonstration. On the other hand, active laminar flow control by suction pumps shows great payoff analytically and in recent flight tests, but would be considered by many to be too risky to incorporate into a new transport jet in the near future.

An overly optimistic estimate of the technology availability will yield a lighter, cheaper aircraft to perform a given mission, but will also result in a higher development risk. Conversely, use of only "yesterday's technology" will result in a heavy and underperforming airplane that nobody will buy!

The actual design effort usually begins with a conceptual sketch (Fig. 2.5). This is the "back of a napkin" drawing of aerospace legend, and gives a rough indication of what the design may look like. A good conceptual sketch will include the approximate wing and tail geometries, the fuselage shape, and the internal locations of the major components such as the engine, cockpit, payload/passenger compartment, landing gear, and fuel tanks.

The conceptual sketch can be used to estimate aerodynamics and weight fractions by comparison to previous designs. These estimates are used to make a first estimate of the required total weight and fuel weight to perform the design mission, by a process called "sizing." The conceptual sketch may not be needed for initial sizing if the design resembles previous ones.

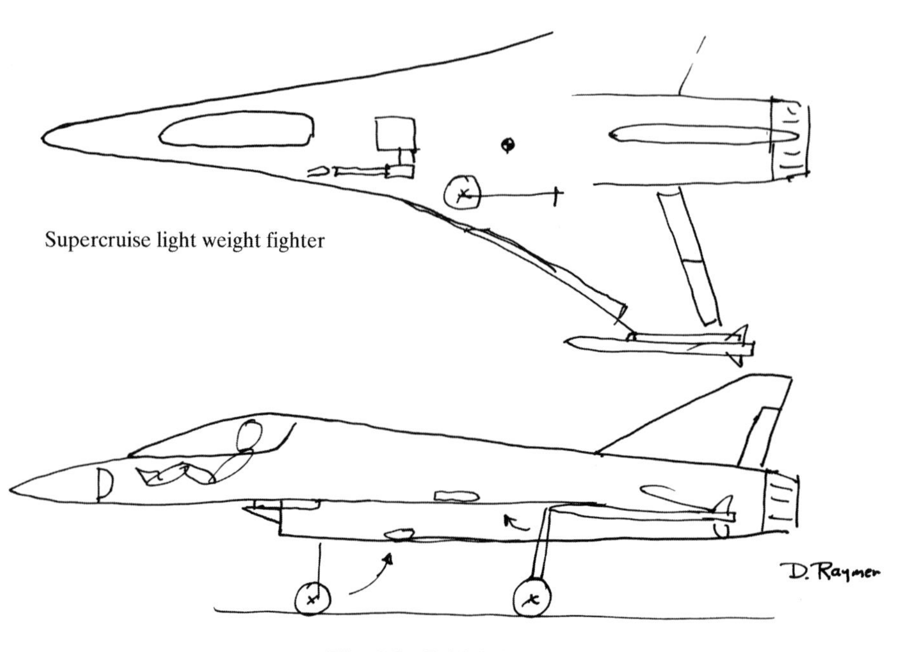

Fig. 2.5 Initial sketch.

The "first-order" sizing provides the information needed to develop an initial design layout (Fig. 2.6). This is a three-view drawing complete with the more important internal arrangement details, including typically the landing gear, payload or passenger compartment, engines and inlet ducts, fuel tanks, cockpit, major avionics, and any other internal components that are large enough to affect the overall shaping of the aircraft. Enough cross sections are shown to verify that everything fits.

On a drafting table, the three-view layout is done in some convenient scale such as 1/10, 1/20, 1/40, or 1/100 (depending upon the size of the airplane and the available paper). On a computer-aided design system, the design work is usually done in full scale (numerically).

This initial layout is analyzed to determine if it really will perform the mission as indicated by the first-order sizing. Actual aerodynamics, weights, and installed propulsion characteristics are analyzed and subsequently used to do a detailed sizing calculation. Furthermore, the performance capabilities of the design are calculated and compared to the requirements just mentioned. Optimization techniques are used to find the lightest or lowest-cost aircraft that will both perform the design mission and meet all performance requirements.

The results of this optimization include a better estimate of the required total weight and fuel weight to meet the mission. The results also include required revisions to the engine and wing sizes. This frequently requires a new or revised design layout in which the designer incorporates these changes and any others suggested by the effort to date.

In industry, designs are typically given a project and drawing number such as D645-5. The first drawing of a project is called the "Dash One" (−1), and an

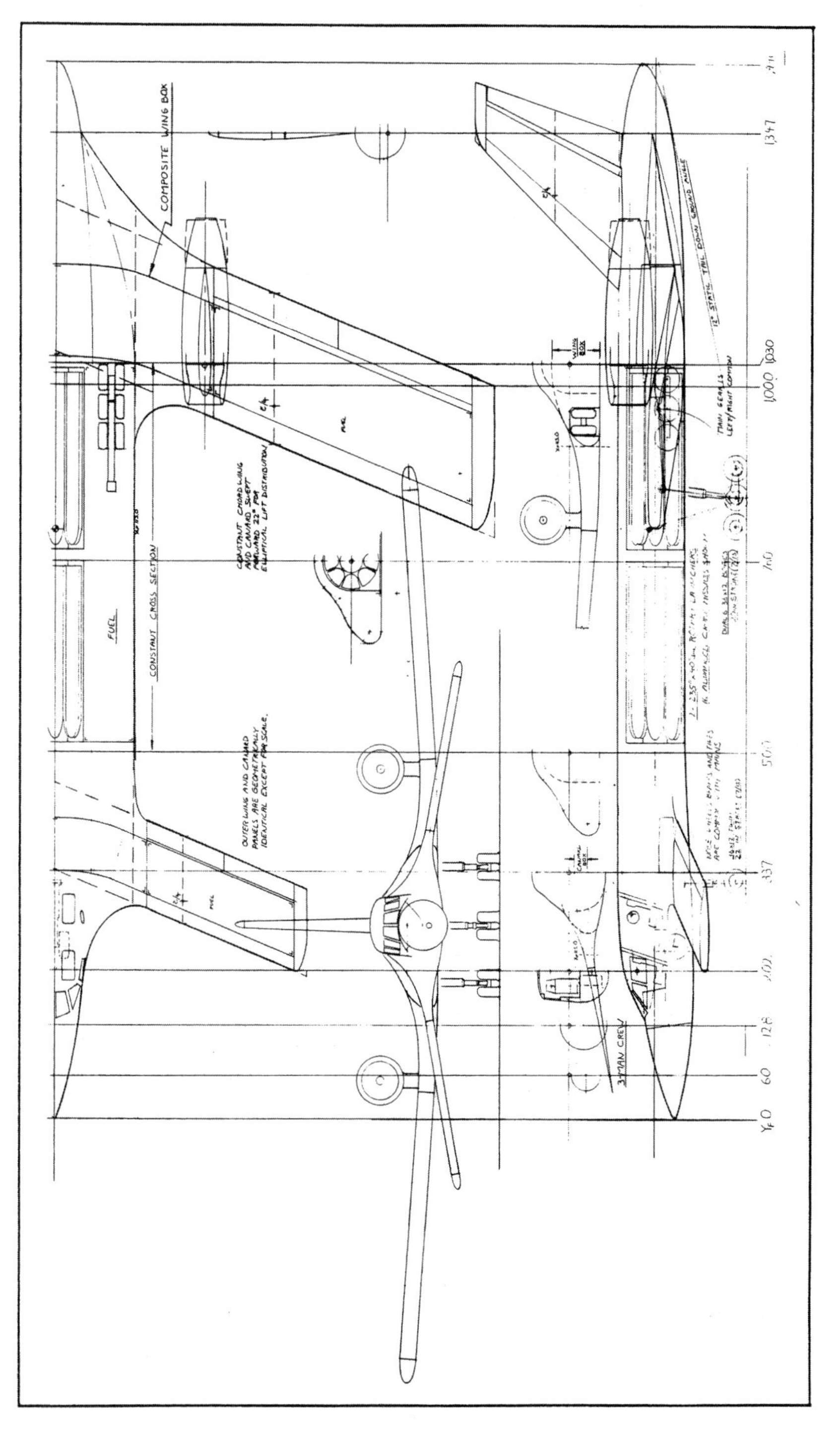

Fig. 2.6 Configuration layout.

important thing to realize is this: *you never build the Dash One*. The Dash One is a tool for making the Dash Two, which is a tool for making the Dash Three, which is . . . and so on. A Dash-50 is not unheard of before the design that will be built is locked in.

The revised drawing, after some number of iterations, is then examined in detail by an ever-expanding group of specialists, each of whom ensures that the design meets the requirements of that specialty.

For example, controls experts will perform a six-degrees-of-freedom analysis to ensure that the designer's estimate for the size of the control surfaces is adequate for control under all conditions required by design specifications. If not, they will instruct the designer as to how much each control surface must be expanded. If a larger aileron is required, the designer must ensure that it can be incorporated into the design without adversely affecting something else, such as the flaps or the landing gear.

The end product of all this will be an aircraft design that can be confidently passed to the preliminary design phase, as previously discussed. While further changes should be expected during preliminary design, major revisions will not occur if the conceptual design effort has been successful.

2.4 Integrated Product Development and Aircraft Design

Increasingly, aircraft design is being done in what is now called an Integrated Product Development (IPD) environment, and the design work is being accomplished by Integrated Product Teams (IPT). The USAF Materiel Command Guide on Integrated Product Development, a major impetus for the adoption of these methods by U.S. aircraft contractors, defines the IPD as a "philosophy that systematically employs a teaming of functional disciplines to integrate and concurrently apply all necessary processes to produce an effective and efficient product that satisfies customer's needs" (Ref. 111). It goes on to call for a cultural change to focus on, in order of importance, the customer, product, process, constraints, and finally, the organizational structure.

IPD refutes the traditional hierarchical structure of large, bureaucratic engineering organizations and calls for decision making to be pushed down to the lowest possible level. Multidisciplinary IPTs bring together design, engineering, production, and operations personnel along with customer representatives to define and develop new products, up to and including entire new aircraft. IPTs are to be established for the creation of one particular product, and are not to become self-perpetuating permanent organizations. They are to be collocated as much as possible to maximize communication between team members.

In many ways, IPTs are like the old "Project" side of a matrix-management structure, and the best of the old Projects were run almost exactly as a best-practice IPT is run today. A well-run project quickly found a collocated home of its own, and brought together a diverse group of specialists to accomplish the project. It communicated with customers early and often and single-mindedly focused on the creation of the best product. However, many advanced design projects had difficulties justifying a collocated home because the functional heads of the different disciplines preferred to keep their people together. Also, obtaining a large enough budget to include the production and operations experts was always a struggle (management would say "we'll bring them in later, when we need them!"). Furthermore, design micromanagement from above by people who weren't involved in or aware of all the tradeoffs and constraints often demoralized the team and deoptimized the design. IPTs and the IPD environment make it very clear that those problems must be fixed, and the IPT way of doing business is almost universally accepted in industry today.

Kelly Johnson, the legendary leader of the Lockheed "Skunk Works" who developed such revolutionary aircraft as the F-104 and SR-71, was a firm advocate of a "strong but small" project office, emphasizing the authority of the project manager and team to get the job done without micromanagement from above. However, Johnson warns "there is a tendency today, which I hate to see, toward design by committee—reviews and recommendations, conferences and consultations, by those not directly doing the job. Nothing very stupid will result, but nothing brilliant either. And it's in the brilliant concept that a major advance is achieved" (Ref. 112).

Care must be taken that an IPT doesn't substitute for, or tie the hands of, an experienced aircraft designer doing the layout portions of the conceptual design process. We don't have a team vote on whether the wing will flutter off—the best technical expert makes that judgment. Similarly, an experienced aircraft configuration designer should have final say over the configuration arrangement (within the goals and constraints set by the customer, the management, and the IPT itself). But, within the IPT environment, that aircraft designer can learn from the collected knowledge of the other members of the team, and create the best possible design accordingly. And, with the collocated team working toward a common and understood goal, the all-important design iterations and trade studies can be done more quickly and with greater creativity.

Concurrent engineering is an important part of the IPT environment. Historically, product development has been done serially. First, Advanced Design created the concept and took it through conceptual and preliminary design. Then they "threw it over a wall" to a large detail design functional organization, which completed their task and threw it over another wall to the manufacturing people, who usually said, "How can I build this stupid thing?!"

With concurrent engineering, detail design and production personnel are brought in at the earliest stages of design, in an IPT environment. Benefits include reduced manufacturing cost and better product quality, with fewer required engineering changes in production. There will be an increase in up-front costs, but in the long run, those are trivial compared to the benefits.

In an extreme form of concurrent engineering, the designer trying to develop a new aircraft concept would see, on the next CAD scope, a production designer trying to develop tooling for the aircraft that hasn't been designed yet! This is actually done in the automotive industry, where the parts and overall geometries from one car to the next don't change very much. They can design and order, say, a die for stamping fenders and know that the actual fender shaping will be enough like the last one that the die can be revised at the last minute to the desired contours. Aircraft aren't cars, and wings aren't fenders, but still the presence of detail design and production personnel from the earliest stages of design is beneficial.

Raymer's reverse installation vectored engine thrust ("RIVET") supersonic VSTOL concept.

3
Sizing from a Conceptual Sketch

3.1 Introduction

"Sizing" is the most important calculation in aircraft design—more so than drag, or stress, or even cost (well, maybe not cost). Sizing literally determines the size of the aircraft, specifically the weight that the aircraft must possess to perform its intended mission carrying its intended payload.

To the rest of the aircraft community—pilots, detail design engineers, mechanics, military officers—our process of aircraft sizing seems backwards. Most people would assume that we draw a new aircraft design, and then determine how far it goes. We do it the other way around. We know how far it goes—it goes as far as the requirements say it goes. What we do not know, and will find out by calculation, is how big to draw it. That process is called sizing.

There are many levels of aircraft sizing procedure. The simplest level just adopts past history. For example, if you need an immediate estimate of the takeoff weight of an airplane to replace the Air Force F-15 fighter, use 44,500 lb. That is the design weight of the F-15 and is probably a good number to start with.

To get the "right" answer takes several years, many people, and lots of money. Design requirements must be rigorously analyzed and then used to develop a number of candidate designs, each of which must be designed, analyzed, sized, optimized, and redesigned any number of times. The best of our candidates, sized to its minimum weight to perform the required mission, yields the right answer—we presume.

Analysis techniques include all manner of computer code as well as correlations to wind-tunnel and other tests. Even with this extreme level of design sophistication, the actual airplane when flown will never exactly match predictions.

In between these extremes of sizing procedure lie the methods used for most conceptual design activities. As an introduction to the design process, this chapter presents a quick sizing method, which will allow you to estimate required takeoff weight from a conceptual sketch and a sizing mission.

The sizing method presented in this chapter is most accurate when used for missions that do not include any combat or payload drops. Although admittedly simplified, this method introduces all of the essential features of the most sophisticated sizing methods used by the major aerospace manufacturers. In a later chapter, the concepts introduced here will be expanded to a sizing method capable of handling all types of missions and with greater accuracy.

3.2 Takeoff-Weight Buildup

"Design takeoff gross weight" is the total weight of the aircraft as it begins the mission for which it was designed. This is not necessarily the same as the "maximum takeoff weight." Many military aircraft can be overloaded beyond design weight but will suffer a reduced maneuverability. Unless specifically mentioned, takeoff gross weight, or W_0, is assumed to be the design weight.

Design takeoff gross weight can be broken into crew weight, payload (or passenger) weight, fuel weight, and the remaining (or "empty") weight. The empty weight includes the structure, engines, landing gear, fixed equipment, avionics, and anything else not considered a part of crew, payload, or fuel. Equation (3.1) summarizes the takeoff-weight buildup.

$$W_0 = W_{\text{crew}} + W_{\text{payload}} + W_{\text{fuel}} + W_{\text{empty}} \tag{3.1}$$

The crew and payload weights are both known since they are given in the design requirements. The only unknowns are the fuel weight and empty weight. However, they are both dependent on the total aircraft weight. Thus an iterative process must be used for aircraft sizing.

To simplify the calculation, both fuel and empty weights can be expressed as fractions of the total takeoff weight, that is, (W_f/W_0) and (W_e/W_0). Thus Eq. (3.1) becomes

$$W_0 = W_{\text{crew}} + W_{\text{payload}} + \left(\frac{W_f}{W_0}\right)W_0 + \left(\frac{W_e}{W_0}\right)W_0 \tag{3.2}$$

This can be solved for W_0 as follows:

$$W_0 - \left(\frac{W_f}{W_0}\right)W_0 - \left(\frac{W_e}{W_0}\right)W_0 = W_{\text{crew}} + W_{\text{payload}} \tag{3.3}$$

$$W_0 = \frac{W_{\text{crew}} + W_{\text{payload}}}{1 - (W_f/W_0) - (W_e/W_0)} \tag{3.4}$$

Now W_0 can be determined if (W_f/W_0) and (W_e/W_0) can be estimated. These are described next.

3.3 Empty-Weight Estimation

After the aircraft has been drawn, the actual empty weight will be calculated by estimating and summing the weights of all of the components of the aircraft. For now it can be estimated as a fraction (W_e/W_0) using simpler methods. The empty-weight fraction (W_e/W_0) can be estimated statistically from historical trends as shown in Fig. 3.1, developed by the author from data taken from Ref. 1 and other sources. Empty-weight fractions vary from about 0.3 to 0.7, and diminish with increasing total aircraft weight.

As can be seen, the type of aircraft also has a strong effect, with flying boats having the highest empty-weight fractions and long-range military aircraft having the lowest. Flying boats are heavy because they need to carry extra

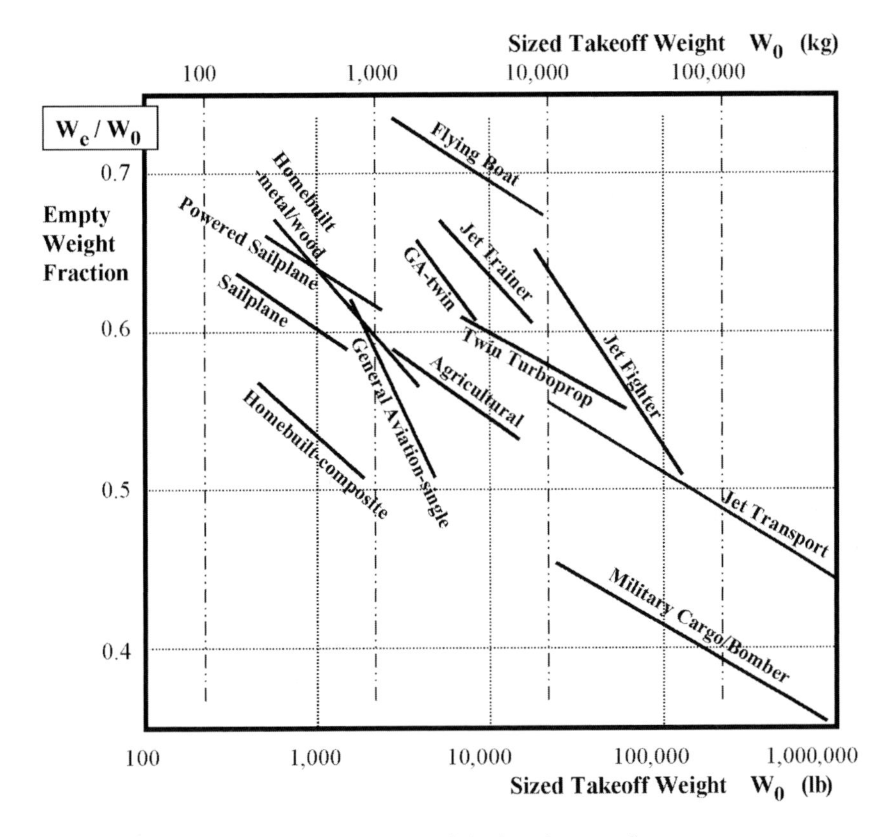

Fig. 3.1 Empty-weight fraction trends.

weight for what amounts to a boat hull. Notice also that different types of aircraft exhibit different slopes to the trend lines of empty-weight fraction vs takeoff weight.

Table 3.1 presents statistical curve-fit equations for the trends shown in Fig. 3.1. Note that these are all exponential equations based upon takeoff gross weight (pounds or kilograms). The exponents are small negative numbers, which indicates that the empty-weight fractions decrease with increasing takeoff weight, as shown by the trend lines in Fig. 3.1. The differences in exponents for different types of aircraft reflect the different slopes of their trend lines and imply that some types of aircraft are more sensitive in sizing than others.

A variable-sweep wing is heavier than a fixed wing, and is accounted for at this initial stage of design by multiplying the empty-weight fraction as determined from the equations in Table 3.1 by about 1.04.

Advanced composite materials such as graphite–epoxy are replacing aluminum in a number of new designs. There have not yet been enough composite aircraft flown to develop statistical equations. Based on a number of design studies, the empty-weight fraction for other types of composite aircraft can be estimated at this stage by multiplying the statistical empty-weight fraction by 0.95.

Table 3.1 Empty weight fraction vs W_0

$W_e/W_0 = AW_0^C K_{vs}$	A	{A-metric}	C
Sailplane—unpowered	0.86	{0.83}	−0.05
Sailplane—powered	0.91	{0.88}	−0.05
Homebuilt—metal/wood	1.19	{1.11}	−0.09
Homebuilt—composite	1.15	{1.07}	−0.09
General aviation—single engine	2.36	{2.05}	−0.18
General aviation—twin engine	1.51	{1.4}	−0.10
Agricultural aircraft	0.74	{0.72}	−0.03
Twin turboprop	0.96	{0.92}	−0.05
Flying boat	1.09	{1.05}	−0.05
Jet trainer	1.59	{1.47}	−0.10
Jet fighter	2.34	{2.11}	−0.13
Military cargo/bomber	0.93	{0.88}	−0.07
Jet transport	1.02	{0.97}	−0.06

K_{vs} = variable sweep constant = 1.04 if variable sweep
= 1.00 if fixed sweep

It is possible to improve on these numbers. The round-the-world Rutan GlobalFlyer has an empty weight fraction below 18%—but is little more than a flying fuel tank, designed and optimized solely for that mission and highly impractical for any normal application.

3.4 Fuel-Fraction Estimation

We also need to estimate the fuel available to perform the mission. Simple statistical methods will not work—we need to "fly" the aircraft over its required mission. Only part of the aircraft's fuel supply is available for performing the mission ("mission fuel"). The other fuel includes "reserve fuel" as required by civil or military design specifications (mostly to allow for degradation of engine performance) and also includes "trapped fuel," which is the fuel that cannot be pumped out of the tanks.

The required amount of mission fuel depends upon the mission to be flown, the aerodynamics of the aircraft, and the engine's fuel consumption. The aircraft weight during the mission affects the drag, so that the fuel used is a function of the aircraft weight.

As a first approximation, the fuel used can be considered to be proportional to the aircraft weight, so that the fuel fraction (W_f/W_0) is approximately independent of aircraft weight. Fuel fraction can be estimated based on the mission to be flown using approximations of the fuel consumption and aerodynamics.

Mission Profiles

Typical mission profiles for various types of aircraft are shown in Fig. 3.2. The Simple Cruise mission is used for many transport and general-aviation

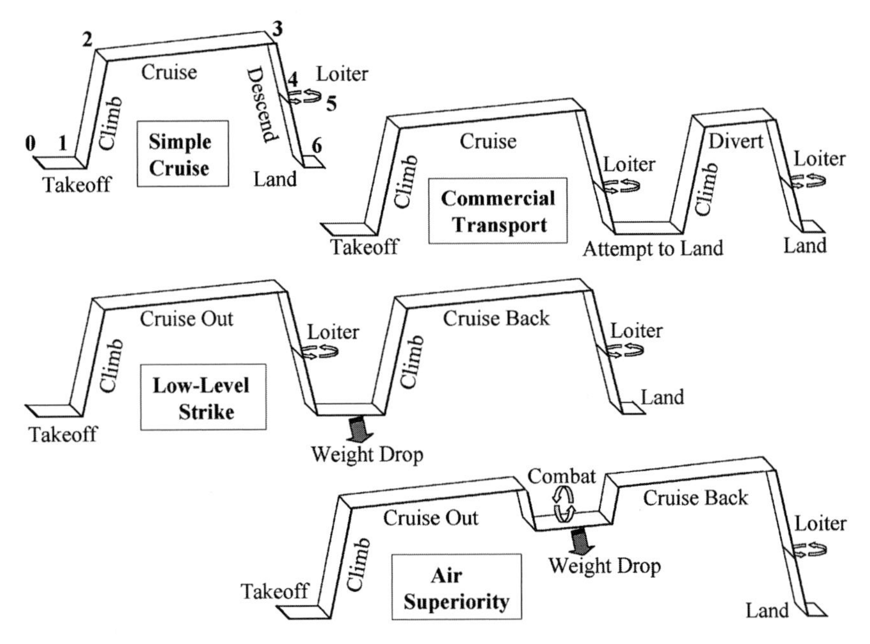

Fig. 3.2 Typical mission profiles for sizing.

designs, including homebuilts. The aircraft is sized to provide some required cruise range.

For safety you would be wise to carry extra fuel in case your intended airport is closed, so a loiter of typically 20–30 min (at 10,000 ft {3048 m}) is added. Alternatively, additional range could be included, representing the distance to the nearest other airport or some fixed number of minutes of flight at cruise speed. [The FAA requires 30 min of additional cruise fuel for daytime flights under visual flight rules (VFR), and 45 min of fuel at night or under instrument conditions (IFR).] Under commercial IFR regulations you also need fuel to fly to an alternate airport after loitering and attempting to land at your intended destination.

The Low-Level Strike mission includes "dash" segments that must be flown at just a few hundred feet off the ground. This is to improve the survivability of the aircraft as it approaches its target. Unfortunately, the aerodynamic efficiency of an aircraft, expressed as "lift-to-drag ratio" (L/D), is greatly reduced during low-level, high-speed flight, as is the engine efficiency. The aircraft may burn almost as much fuel during the low-level dash segment as it burns in the much-longer cruise segment.

The typical Air Superiority mission includes a cruise out, a combat consisting of either a certain number of turns or a certain number of minutes at maximum power, a weapons drop, a cruise back, and a loiter. The weapons drop refers to the firing of gun and missiles, and is often left out of the sizing analysis to ensure that the aircraft has enough fuel to return safely if the weapons are not used. Note that the second cruise segment is identical to the first, indicating that the aircraft must return to its base at the end of the mission.

Many military missions include aerial refueling. The aircraft meets up with a tanker aircraft such as an Air Force KC-135 and receives some quantity of fuel. This enables the aircraft to achieve far more range, but adds to the overall operating cost because a fleet of tanker aircraft must be dedicated to supporting the bombers. Analytically, this "resets the clock." The onloaded fuel brings the aircraft weight up to or even greater than the takeoff weight, so that the post-refuel segments are treated as an entire separate mission.

In addition to the mission profile, requirements will be established for a number of performance parameters such as takeoff distance, maneuverability, and climb rates. These are ignored in the simplified sizing method of this chapter, but will be discussed in detail later.

Mission Segment Weight Fractions

For analysis, the various mission segments, or "legs," are numbered, with zero denoting the start of the mission. Mission leg "one" is usually engine warmup and takeoff for first-order sizing estimation. The remaining legs are sequentially numbered.

For example, in the simple cruise mission the legs could be numbered as 1) warmup and takeoff, 2) climb, 3) cruise, 4) loiter, and 5) land (see the example mission at the end of this chapter).

In a similar fashion, the aircraft weight at each part of the mission can be numbered. Thus, W_0 is the beginning weight ("takeoff gross weight").

For the simple cruise mission, W_1 would be the weight at the end of the first mission segment, which is the warmup and takeoff. W_2 would be the aircraft weight at the end of the climb. W_3 would be the weight after cruise, and W_4 after loiter. Finally, W_5 would be the weight at the end of the landing segment, which is also the end of the total mission.

During each mission segment, the aircraft loses weight by burning fuel. (Remember that our simple sizing method doesn't permit missions involving a payload drop.) The aircraft weight at the end of a mission segment divided by its weight at the beginning of that segment is called the "mission segment weight fraction." This will be the basis for estimating the required fuel fraction for initial sizing.

For any mission segment "i," the mission segment weight fraction can be expressed as (W_i/W_{i-1}). If these weight fractions can be estimated for all of the mission legs, they can be multiplied together to find the ratio of the aircraft weight at the end of the total mission, W_x (assuming "x" segments altogether) divided by the initial weight, W_0. This ratio, W_x/W_0, can then be used to calculate the total fuel fraction required.

These mission segment weight fractions can be estimated by a variety of methods. For our simplified form of initial sizing, the types of mission leg will be limited to warmup and takeoff, climb, cruise, loiter, and land. As previously mentioned, mission legs involving combat, payload drop, and refuel are not permitted in this simplified sizing method but will be discussed in a later chapter.

The warmup, takeoff, and landing weight fractions can be estimated historically. Table 3.2 gives typical historical values for initial sizing. These values

Table 3.2 Historical mission segment weight fractions

Mission segment	(W_i/W_{i-1})
Warmup and takeoff	0.970
Climb	0.985
Landing	0.995

can vary somewhat depending on aircraft type, but the averaged values given in the table are reasonable for initial sizing.

In our simple sizing method we ignore descent, assuming that the cruise ends with a descent and that the distance traveled during descent is part of the cruise range.

Cruise-segment mission weight fractions can be found using the Breguet range equation (derived in Chapter 17):

$$R = \frac{V}{C}\frac{L}{D}\ell n\frac{W_{i-1}}{W_i} \tag{3.5}$$

or

$$\frac{W_i}{W_{i-1}} = \exp\frac{-RC}{V(L/D)} \tag{3.6}$$

where

R = range (ft or m)
C = specific fuel consumption (see following section)
V = velocity (ft/s or m/s)
L/D = lift-to-drag ratio

Loiter weight fractions are found from the endurance equation (also derived in Chapter 17):

$$E = \frac{L/D}{C}\ell n\frac{W_{i-1}}{W_i} \tag{3.7}$$

or

$$\frac{W_i}{W_{i-1}} = \exp\frac{-EC}{L/D} \tag{3.8}$$

where E = endurance or loiter time.

(Note: It is very important to use consistent units! Convert all values to feet-lb-s, or to m-k-s. Also note that C and L/D vary with speed and altitude. Furthermore, C varies with throttle setting, and L/D varies with aircraft weight. These will be discussed in detail in later chapters.)

Specific Fuel Consumption

Specific fuel consumption ("SFC" or simply "C") is the rate of fuel consumption divided by the resulting thrust. For jet engines, specific fuel consumption is measured in fuel mass flow per hour per unit thrust force. In British (fps) units, SFC is in pounds of fuel per hour, per pound of thrust. We jokingly "cancel" the pounds yielding 1/hr as the units! In 22 metric terms we use the more reasonable mg/Ns. Figure 3.3 shows SFC vs Mach number.

Propeller engine SFC is normally given as C_{bhp}, the pounds of fuel per hour to produce one horsepower at the propeller shaft (or one "brake horsepower": bhp = 550 ft-lb/s). In metric, power SFC is given in mg/W-s (mg/J, or in μg/J to make "nice" numbers).

A propeller thrust SFC equivalent to the jet-engine SFC can be calculated. The engine produces thrust via the propeller, which has an efficiency η_p defined as thrust power output per horsepower input [Eq. (3.9)]. The 550 term converts horsepower to power in British units, and assumes that V is in feet per second.

$$\eta_p = \frac{TV}{P} = \frac{TV}{550 \text{ hp}} \text{ (fps units)} \tag{3.9}$$

Equation (3.10) shows the derivation of the equivalent-thrust SFC for a propeller-driven aircraft. Note that for a propeller aircraft, the thrust and the SFC are a function of the flight velocity. For a typical aircraft with a propeller efficiency of about 0.8, one horsepower equals one pound of thrust at about

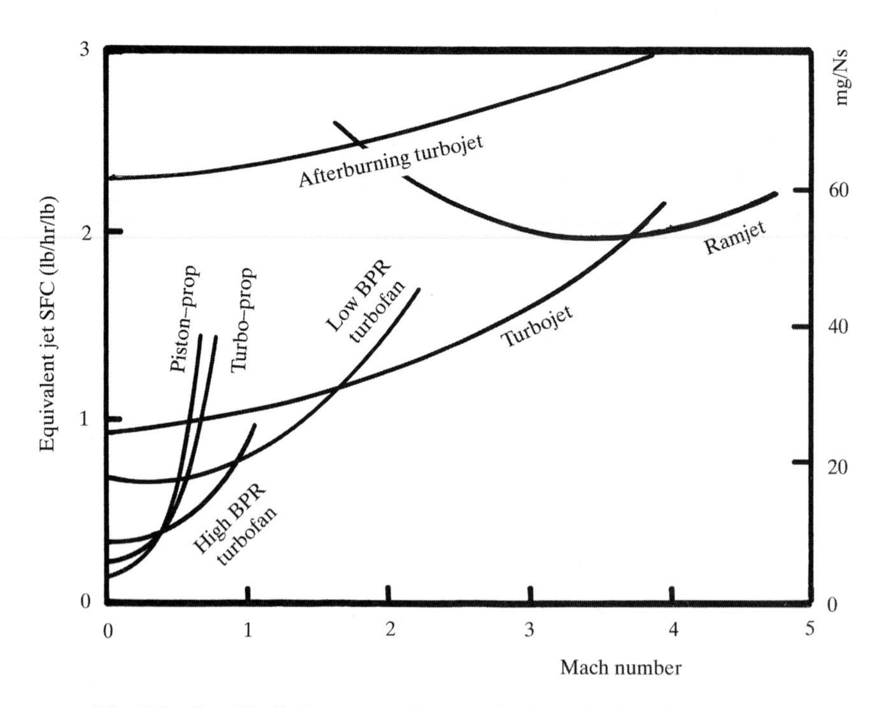

Fig. 3.3 Specific fuel consumption trends (at typical cruise altitudes).

Table 3.3 Specific fuel consumption, C

Typical jet SFCs: 1/hr {mg/Ns}	Cruise	Loiter
Pure turbojet	0.9 {25.5}	0.8 {22.7}
Low-bypass turbofan	0.8 {22.7}	0.7 {19.8}
High-bypass turbofan	0.5 {14.1}	0.4 {11.3}

440 ft/s, or about 260 knots {484 km/hr}.

$$C = \frac{W_f/\text{time}}{\text{thrust}} = C_{\text{power}} \frac{V}{\eta_p} = C_{\text{bhp}} \frac{V}{550\ \eta_p} (\text{ fps units}) \tag{3.10}$$

Table 3.3 provides typical SFC values for jet engines, while Table 3.4 provides typical C_{bhp} values for propeller engines. Typically one can assume $\eta_p = 0.8$ except for a fixed pitch propeller during loiter, where $\eta_p = 0.7$. These can be used for rough initial sizing. In later chapters more detailed procedures for calculating these values, which change as a function of altitude, velocity, and power setting, will be presented.

L/D Estimation

The remaining unknown in both range and loiter equations is the L/D, or lift-to-drag ratio, which is a measure of the design's overall aerodynamic efficiency. Unlike the parameters just estimated, the L/D is highly dependent upon the configuration arrangement. At subsonic speeds L/D is most directly affected by two aspects of the design: wing span and wetted area.

In level flight, the lift is known. It must equal the aircraft weight. Thus, L/D is solely dependent upon drag.

The drag at subsonic speeds is composed of two parts. "Induced" drag is the drag caused by the generation of lift. This is primarily a function of the wing span.

"Zero-lift," or "parasite" drag is the drag that is not related to lift. This is primarily skin-friction drag, and as such is directly proportional to the total surface area of the aircraft exposed ("wetted") to the air.

Table 3.4 Propeller specific fuel consumption, C_{bhp}

Propeller: $C = C_{\text{power}}\ V/\eta_p = C_{\text{bhp}}\ V/(550\eta_p)$ Typical C_{bhp}: lb/hr/bhp {mg/W-s}	Cruise	Loiter
Piston-prop (fixed pitch)	0.4 {0.068}	0.5 {0.085}
Piston-prop (variable pitch)	0.4 {0.068}	0.5 {0.085}
Turboprop	0.5 {0.085}	0.6 {0.101}

The "aspect ratio" of the wing has historically been used as the primary indicator of wing efficiency. Aspect ratio is defined as the square of the wing span divided by the wing reference area. For a rectangular wing the aspect ratio is simply the span divided by chord.

Aspect ratios range from under 1 for reentry lifting bodies to over 30 for sailplanes. Typical values range between 3 and 8. For initial design purposes, aspect ratio can be selected from historical data. For final determination of the best aspect ratio, a trade study as discussed in Chapter 19 should be conducted.

Aspect ratio could be used to estimate subsonic L/D, but for one major problem. The parasite drag is not a function of just the wing area, as expressed by aspect ratio, but also of the aircraft's total wetted area.

Figure 3.4 shows two widely different aircraft concepts, both designed to perform the same mission of strategic bombing. The Boeing B-47 features a conventional approach. With its aspect ratio of over 9, it is not surprising that it attains an L/D of over 17. On the other hand, the AVRO Vulcan bomber has an aspect ratio of only 3, yet it attains almost exactly the same L/D.

The explanation for this curious outcome lies in the actual drivers of L/D as already discussed. Both aircraft have about the same wing span, and both have about the same wetted areas, so both have about the same L/D. The aspect ratio of the B-47 is higher not because of a greater wing span, but because of a smaller wing area. However, this reduced wing area is offset by the wetted area of the fuselage and tails.

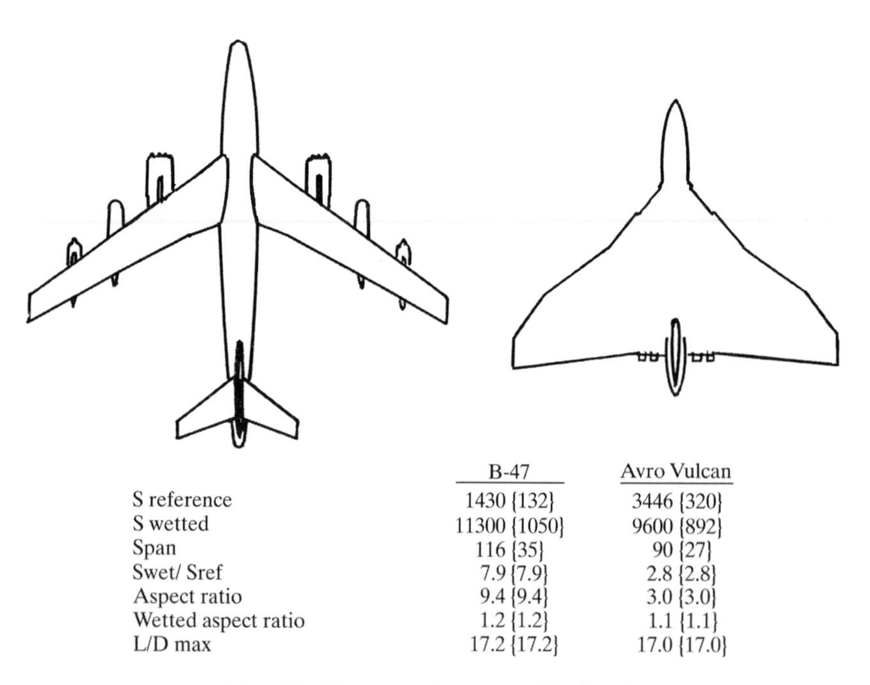

	B-47	Avro Vulcan
S reference	1430 {132}	3446 {320}
S wetted	11300 {1050}	9600 {892}
Span	116 {35}	90 {27}
Swet/ Sref	7.9 {7.9}	2.8 {2.8}
Aspect ratio	9.4 {9.4}	3.0 {3.0}
Wetted aspect ratio	1.2 {1.2}	1.1 {1.1}
L/D max	17.2 {17.2}	17.0 {17.0}

Fig. 3.4 Does aspect ratio predict drag?

This is illustrated by the ratios of wetted area to wing reference area (S_{wet}/S_{ref}). While the AVRO design has a total wetted area of less than three times the wing area, the Boeing design has a wetted area of eight times the wing area.

This wetted-area ratio can be used, along with aspect ratio, as a more reliable early estimate of L/D. Wetted-area ratio is clearly dependent on the actual configuration layout. Figure 3.5 shows a spectrum of design approaches and the resulting wetted-area ratios.

As just stated, L/D depends primarily on the wing span and the wetted area. This suggests a new parameter, the "Wetted Aspect Ratio," which is defined as the wing span squared divided by the total aircraft wetted area. This is very similar to the aspect ratio except that it considers total wetted area instead of wing reference area.

Figure 3.6 plots maximum L/D for a number of aircraft vs the wetted aspect ratio, and shows clear trend lines for jet, prop, and fixed-gear prop aircraft. Note that the wetted aspect ratio can be shown to equal the wing geometric aspect ratio divided by the wetted-area ratio, S_{wet}/S_{ref}.

The trend lines of Fig. 3.6 extend far to the right for high-aspect-ratio designs such as sailplanes and high-altitude aircraft like the Boeing Condor. At a wetted aspect ratio of 10, an L/D of about 45 can be expected!

It should be clear at this point that the designer has control over the L/D. The designer picks the aspect ratio and determines the configuration arrangement, which in turn determines the wetted-area ratio.

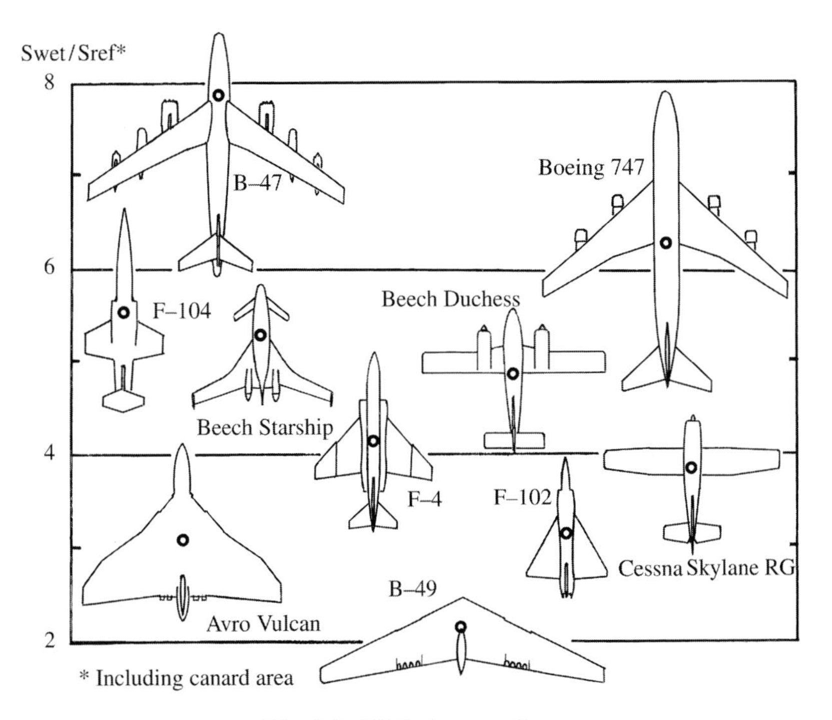

Fig. 3.5 Wetted area ratios.

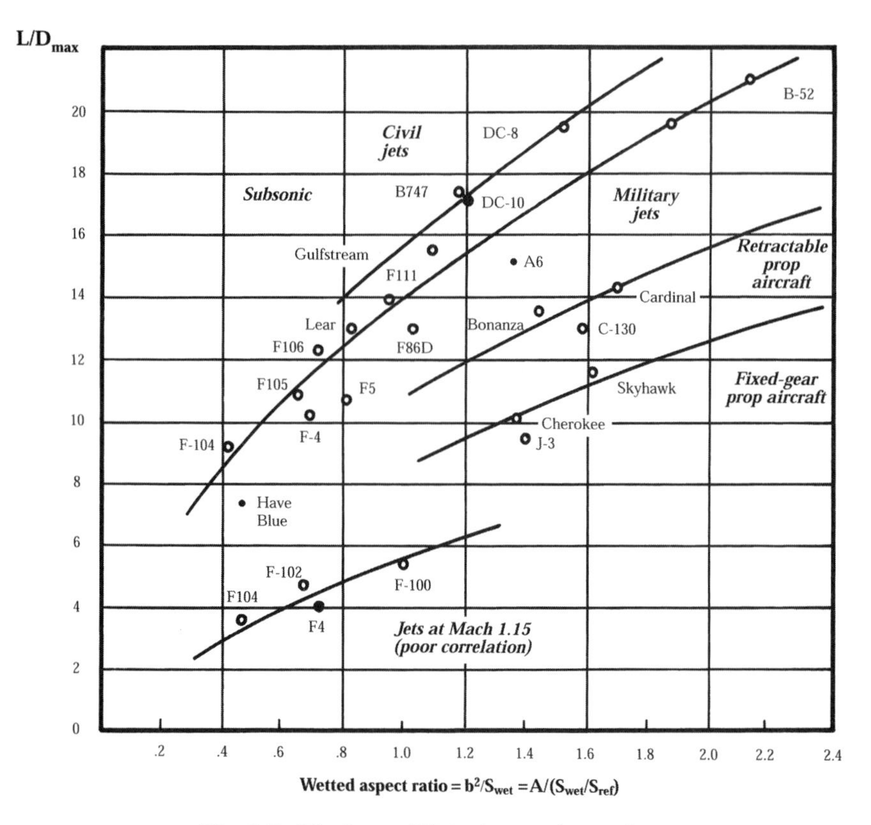

Fig. 3.6 Maximum lift-to-drag ratio trends.

However, the designer must strike a compromise between the desire for a high L/D and the conflicting desire for low weight. The statistical equations just provided for estimating the empty-weight fraction are based on "normal" designs. If the aspect ratio selected is much higher than that of other aircraft in its class, the empty-weight fraction would be higher than estimated by these simple statistical equations.

L/D can now be estimated from a conceptual sketch. This is the crude, "back of a napkin" drawing mentioned earlier. On the conceptual sketch the designer arranges the major components of the aircraft, including wings, tails, fuselage, engines, payload or passenger compartment, landing gear, fuel tanks, and others as needed.

From the sketch the wetted-area ratio can be "eyeball-estimated" using Fig. 3.5 for guidance. The wetted aspect ratio can then be calculated as the wing aspect ratio divided by the wetted-area ratio. Figure 3.6 can then be used to estimate the maximum L/D.

Note that the L/D usually can be estimated without a sketch by an experienced designer. The wetted aspect ratio method is provided primarily for the student, but can be useful for quickly evaluating novel concepts.

Drag varies with altitude and velocity. For any altitude there is a velocity that maximizes L/D. To maximize cruise or loiter efficiency the aircraft should fly at approximately the velocity for maximum L/D.

For reasons which will be derived later, the most efficient loiter for a jet aircraft occurs exactly at the velocity for maximum L/D, but the most efficient loiter speed for a propeller aircraft occurs at a slower velocity that yields an L/D of 86.6% of the maximum L/D.

Similarily, the most efficient cruise velocity for a propeller aircraft occurs at the velocity yielding maximum L/D, whereas the most efficient cruise for a jet aircraft occurs at a slightly higher velocity yielding an L/D of 86.6% of the maximum L/D:

	Cruise	Loiter
Jet	$0.866\ L/D_{max}$	L/D_{max}
Prop	L/D_{max}	$0.866\ L/D_{max}$

For initial sizing, these percentages can be multiplied times the maximum L/D as estimated using Fig. 3.6 to determine the L/D for cruise and loiter.

Fuel-Fraction Estimation

Using historical values from Table 3.2 and the equations for cruise and loiter segments, the mission-segment weight fractions can now be estimated. By multiplying them together, the total mission weight fraction, W_x/W_0, can be calculated.

Since this simplified sizing method does not allow mission segments involving payload drops, all weight lost during the mission must be due to fuel usage. The mission fuel fraction must therefore be equal to $(1 - W_x/W_0)$. If you assume, typically, a 6% allowance for reserve and trapped fuel, the total fuel fraction can be estimated as in Eq. (3.11):

$$\frac{W_f}{W_0} = 1.06\left(1 - \frac{W_x}{W_0}\right) \tag{3.11}$$

3.5 Takeoff-Weight Calculation

Using the fuel fraction found with Eq. (3.11) and the statistical empty-weight equation selected from Table 3.1, the takeoff gross weight can be found iteratively from Eq. (3.4). This is done by guessing the takeoff gross weight, calculating the statistical empty-weight fraction, and then calculating the takeoff gross weight. If the result doesn't match the guess value, a value between the two is used as the next guess. This will usually converge in just a few iterations. This first-order sizing process is diagrammed in Fig. 3.7.

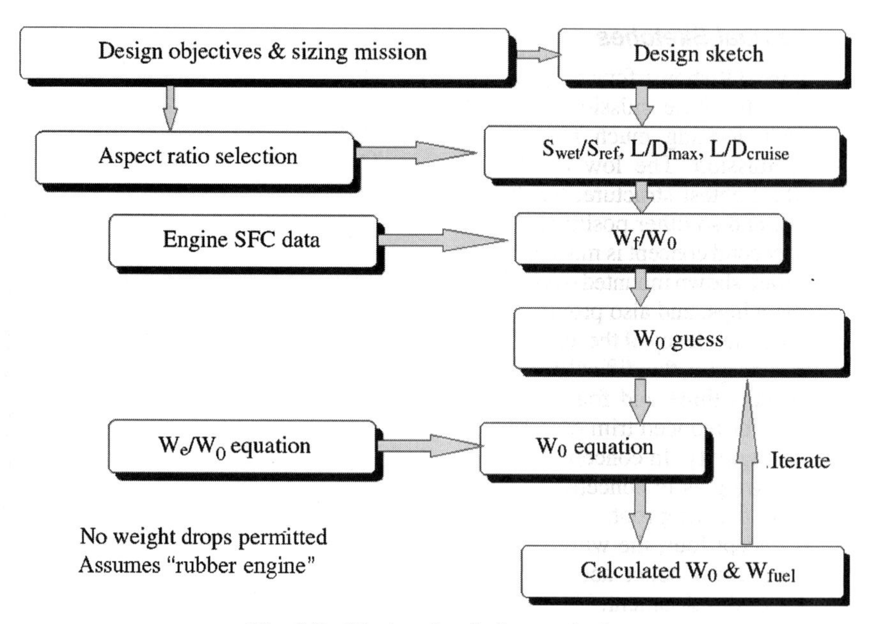

Fig. 3.7 First-order design method.

3.6 Design Example: ASW Aircraft

ASW Requirements

Figure 3.8 illustrates the mission requirement for a hypothetical antisubmarine warfare (ASW) aircraft. The key requirement is the ability to loiter for 3 hr at a distance of 1500 n miles {2778 km} from the takeoff point. While loitering on-station, this type of aircraft uses sophisticated electronic equipment to detect and track submarines. For the sizing example, this equipment is assumed to weigh 10,000 lb {4536 kg}. Also, a four-man crew is required, totalling 800 lb {363 kg}. The aircraft must cruise at 0.6 Mach number.

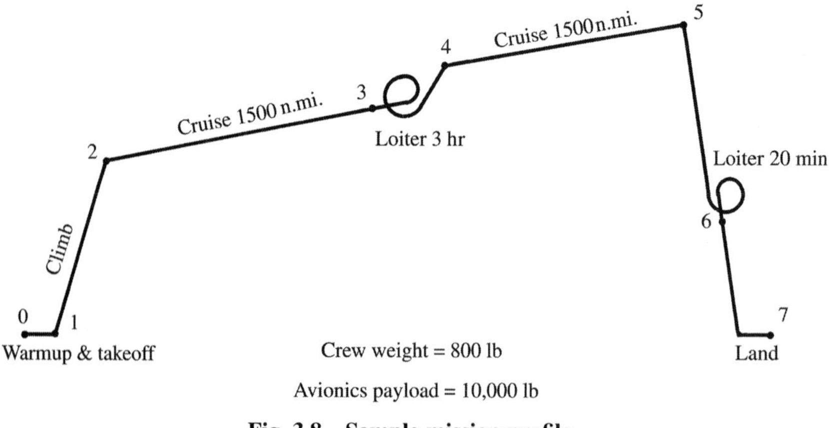

Fig. 3.8 Sample mission profile.

Conceptual Sketches

Figure 3.9 shows four conceptual approaches considered by the designer in response to these mission requirements. Concept one is the conventional approach, looking much like the Lockheed S-3A that currently performs a similar mission. The low horizontal tail position shown in solid line would offer the lightest structure, but may place the tail in the exhaust stream of the engines, and so other positions for the horizontal tail are shown in dotted lines.

The second concept is much like the first except for the engine location. Here the engines are shown mounted over the wing. This provides extra lift due to the exhaust over the wings, and also provides greater ground clearance for the engines, which reduces the tendency of the jet engines to suck up debris. However, the disadvantage of this concept is the difficulty in reaching the engines for maintenance work.

Concepts three and four explore the canarded approach. Canards offer the potential for reduced trim drag and may provide a wider allowable range for the center of gravity. In concept three, the wing is low and the engines are mounted over the wing as in concept two. This would allow the main landing gear to be stowed in the wing root.

In concept four, the wing is high with the engines mounted below. This last approach offers better access to the engines, and for this reason was selected for further development.

Figure 3.10 is a conceptual sketch prepared, in more detail, for the selected concept. Note the locations indicated for the landing-gear stowage, crew station, and fuel tanks.

This points out a common problem with canard aircraft, the fuel tank locations. The fuel tanks should be placed so that the fuel is evenly distributed about the aircraft center of gravity (estimated location shown by the circle with two quarters shaded). This is necessary so that the aircraft when loaded has nearly the same center of gravity as when its fuel is almost gone. However, the wing is located aft of the center of gravity whenever a canard is used, so that the fuel located in the wing is also aft of the center of gravity.

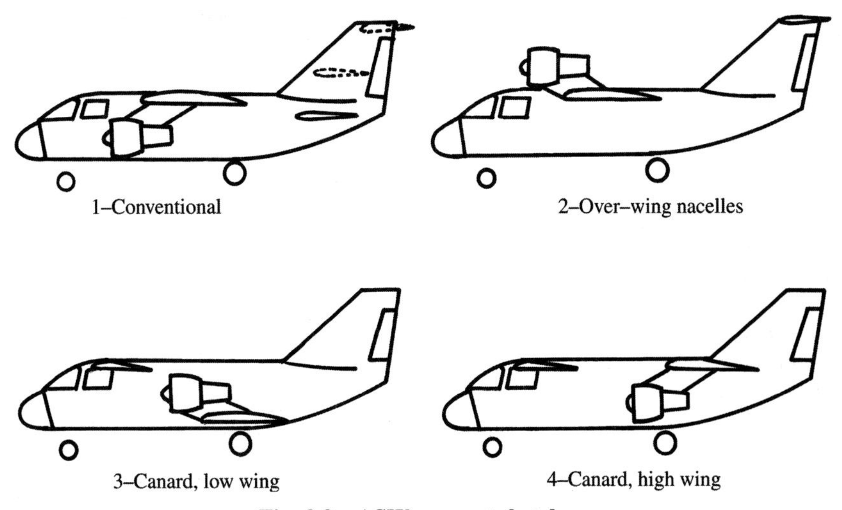

Fig. 3.9 ASW concept sketches.

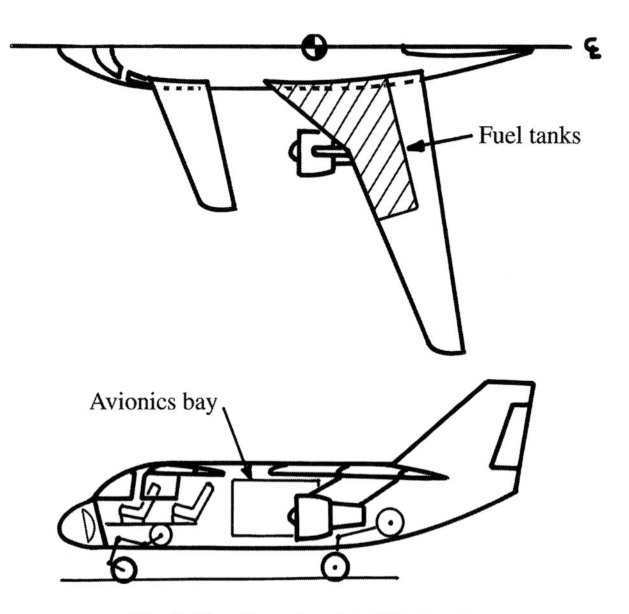

Fig. 3.10 Completed ASW sketch.

One solution to this problem would be to add fuel tanks in the fuselage, forward of the center of gravity. This would increase the risk of fire in the fuselage during an accident, and is forbidden in commercial aircraft. Although this example is a military aircraft, fire safety should always be considered.

Another solution, shown on the sketch, is to add a wing strake full of fuel. This solution is seen on the Beech Starship among others. The strakes do add to the aircraft wetted area, which reduces cruise aerodynamic efficiency.

This example serves to illustrate an extremely important principle of aircraft design, namely, that there is no such thing as a free lunch! All aircraft design entails a series of tradeoffs. The canard offers lower trim drag, but in this case may require a higher wetted area. The only true way to determine whether a canard is a good idea for this or any aircraft is to design several aircraft, one with and one without a canard. This type of trade study comprises the majority of the design effort during the conceptual design process.

L/D Estimation

For initial sizing, a wing aspect ratio of about 10 was selected. With the area of the wing and canard both included, this is equivalent to a combined aspect ratio of about 7.

Comparing the sketch of Fig. 3.10 to the examples of Fig. 3.5, it would appear that the wetted area ratio ($S_{\text{wet}}/S_{\text{ref}}$) is about 5.5. This yields a wetted aspect ratio of 1.27 (i.e., $7/5.5$).

For a wetted aspect ratio of 1.27, Fig. 3.6 indicates that a maximum lift-to-drag ratio of about 16 would be expected. This value, obtained from an initial sketch and the selected aspect ratio, can now be used for initial sizing.

Since this is a jet aircraft, the maximum L/D is used for loiter calculations. For cruise, a value of 0.866 times the maximum L/D, or about 13.9 is used.

Takeoff-Weight Sizing

From Table 3.3, initial values for SFC are obtained. For a subsonic aircraft the best SFC values are obtained with high-bypass turbofans, which have typical values of about 0.5 for cruise and 0.4 for loiter.

Table 3.1 does not provide an equation for statistically estimating the empty weight fraction of an antisubmarine aircraft. However, such an aircraft is basically designed for subsonic cruise efficiency so that the equation for military cargo/bomber can be used. The extensive ASW avionics would not be included in that equation, and so it is treated as a separate payload weight.

Box 3.1 gives the calculations for sizing this example. Note the effort to ensure consistent dimensions, including the conversion of cruise velocity (Mach 0.6) to

Box 3.1 ASW sizing calculations

Mission Segment Weight Fractions (British Units)

1) Warmup and takeoff $W_1/W_0 = 0.97$ (Table 2)

2) Climb $W_2/W_1 = 0.985$ (Table 2)

3) Cruise

$$R = 1500 \text{ n.mi.} = 9{,}114{,}000 \text{ ft}$$
$$C = 0.5 \text{ 1/hr} = 0.0001389 \text{ 1/s}$$
$$V = 0.6\text{M} \times (994.8 \text{ ft/s}) = 596.9 \text{ ft/s}$$
$$L/D = 16 \times 0.866 = 13.9$$
$$W_3/W_2 = e^{\{-RC/VL/D\}} = e^{-0.153} = 0.858$$

4) Loiter

$$E = 3 \text{ hr} = 10{,}800 \text{ s}$$
$$C = 0.4 \text{ 1/hr} = 0.0001111 \text{ 1/s}$$
$$L/D = 16$$
$$W_4/W_3 = e^{\{-EC/L/D\}} = e^{-0.075} = 0.9277$$

5) Cruise (same as 3) $W_5/W_4 = 0.858$

6) Loiter

$$E = \tfrac{1}{3} \text{ hr} = 1200 \text{ s}$$
$$C = 0.0001111 \text{ 1/s}$$
$$L/D = 16$$
$$W_6/W_5 = e^{-0.0083} = 0.9917$$

7) Land $W_7/W_6 = 0.995$ (Table 2)

$$W_7/W_0 = (0.97)(0.985)(0.858)(0.9277)(0.852)(0.9917)(0.995) = 0.6441$$
$$W_f/W_0 = 1.06(1 - 0.635) = 0.3773$$
$$W_e/W_0 = 0.93\ W_0^{-0.07} \quad \text{(Table 1)}$$

$$W_0 = \frac{10{,}800}{1 - 0.3773 - \dfrac{W_e}{W_0}}$$

W_0, guess	W_e/W_0	W_e	W_0, calculated
50,000	0.4361	21,803	57,863
60,000	0.4305	25,832	56,198
56,000	0.4326	24,227	56,814
56,500	0.4324	24,428	56,733
56,700	0.4322	24,508	56,702

ft/s by assuming a typical cruise altitude of 30,000 ft {9144 m}. At this altitude the speed of sound (see Appendix B) is 994.8 ft/s {303.2 m/s}.

The calculations in Box 3.1 indicate a takeoff gross weight of 56,702 lb {25,720 kg}. Although these calculations are based upon crude estimates of aerodynamics, weights, and propulsion parameters, it is interesting to note that the actual takeoff gross weight of the Lockheed S-3A, as quoted in Ref. 1, is 52,539 lb {23,831 kg}. While strict accuracy should not be expected, this simple sizing method will usually yield an answer in the "right ballpark."

Figure 3.11 illustrates an alternative way to size the aircraft, by a graphical method. Here a number of guesses of W_0 that bound the likely solution have been made. Rather than attempt to iterate to the correct answer as just done, we simply graph these answers with W_0 guess on the horizontal axis and W_0 calculated on the vertical axis. A 45-deg line from the origin represents where the guess equals the calculated value, so that the intersection of this line with the line of the answers is the solution.

An Excel™ spreadsheet of this sizing example illustrating both methods is available at the author's website, *www.aircraftdesign.com*, and is free to purchasers of this book.

Trade Studies

An important part of conceptual design is the evaluation and refinement, with the customer, of the design requirements. In the ASW design example, the required range of 1500 n miles (each way) is probably less than the customer would really like. A "range trade" can be calculated to determine the increase in design takeoff gross weight if the required range is increased.

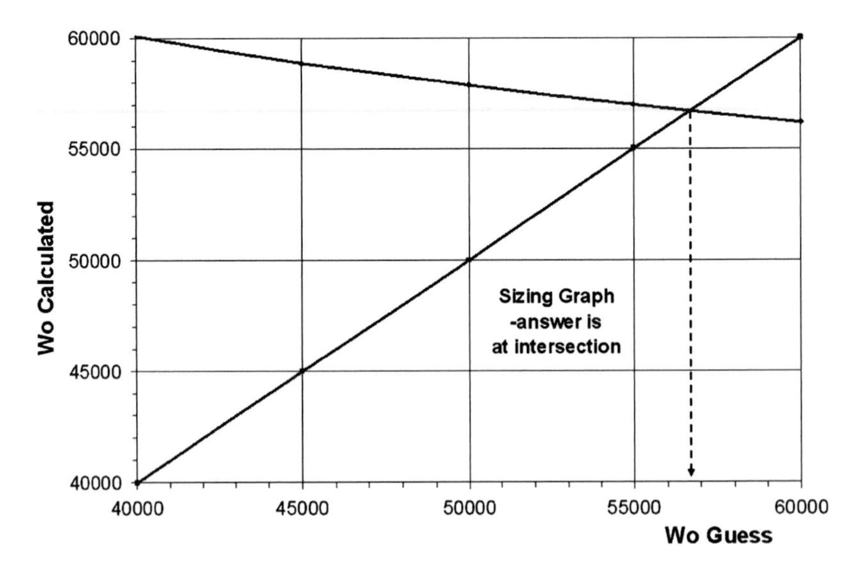

Fig. 3.11 Graphical sizing method for ASW example.

This is done by recalculating the weight fractions for the cruise mission segments, using arbitrarily selected ranges. For example, instead of the required 1500 n miles, we will calculate the cruise weight fractions using 1000 and 2000 n miles, and will size the aircraft separately for each of those ranges. These calculations are shown in Box 3.2, and the results are plotted in Fig. 3.12.

Box 3.2 Range trade

1000 n miles Range

$$W_3/W_2 = W_5/W_4 = e^{-0.1020} = 0.9030$$
$$W_7/W_0 = 0.7132$$
$$W_f/W_0 = 1.06\,(1 - 0.7132) = 0.3040$$

$$W_0 = \frac{10{,}800}{1 - 0.3040 - \dfrac{W_e}{W_0}}$$

W_0, guess	W_e/W_0	W_e	W_0, calculated
50,000	0.4361	21,803	41,544
40,000	0.4429	17,717	42,670
42,000	0.4414	18,540	42,417
42,400	0.4411	18,704	42,369
42,370	0.4412	18,692	42,372

2000 n miles Range

$$W_3/W_2 = W_5/W_4 = e^{-0.2040} = 0.8154$$
$$W_7/W_0 = 0.5816$$
$$W_f/W_0 = 0.4435$$

$$W_0 = \frac{10{,}800}{1 - 0.4435 - \dfrac{W_e}{W_0}}$$

W_0, guess	W_e/W_0	W_e	W_0, calculated
50,000	0.4361	21,803	89,671
80,000	0.4220	33,756	80,265
80,200	0.4219	33,835	80,221
80,210	0.4219	33,839	80,219
80,218	0.4219	33,842	80,217

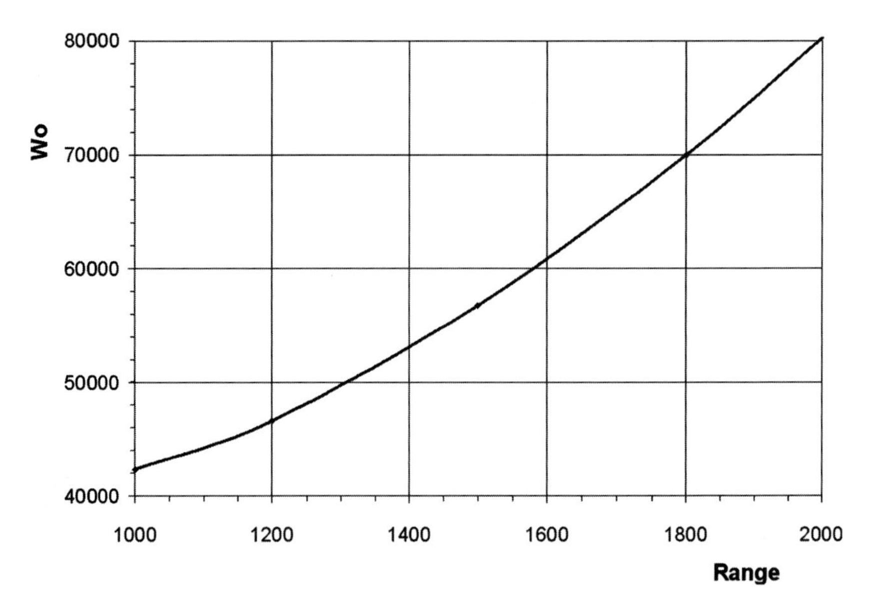

Fig. 3.12 Range trade.

Box 3.3 Payload trade

Payload = 5000 lb; $W_0 = \dfrac{5800}{1 - 0.3773 - \dfrac{W_e}{W_0}}$

W_0, guess	W_e/W_0	W_e	W_0, calculated
50,000	0.4361	21,803	31,074
32,000	0.4499	14,397	33,563
33,000	0.4489	14,815	33,376
33,300	0.4487	14,940	33,321
33,320	0.4486	14,949	33,318

Payload = 15,000 lb; $W_0 = \dfrac{15,800}{1 - 0.3773 - \dfrac{W_e}{W_0}}$

W_0, guess	W_e/W_0	W_e	W_0, calculated
50,000	0.4361	21,803	84,651
75,000	0.4239	31,790	79,456
78,000	0.4227	32,971	78,994
78,800	0.4224	33,285	78,875
78,865	0.4224	33,311	78,866

In a similar fashion, a "payload trade" can be made. The mission-segment weight fractions and fuel fraction are unchanged, but the numerator of the sizing equation, Eq. (3.4), is parametrically varied by assuming different payload weights. The given payload requirement is 10,000 lb of avionics equipment. Box 3.3 shows the sizing calculations assuming payload weights of 5000 and 20,000 lb. The results are plotted in Fig. 3.13.

The statistical empty-weight equation used here for sizing was based upon existing military cargo and bomber aircraft, which are all of aluminum construction. The preceding takeoff gross weight calculations have thus implicitly assumed that the new aircraft would also be built of aluminum.

To determine the effect of building the aircraft out of composite materials, the designer must adjust the empty-weight equation. As previously mentioned, this can be approximated in the early stages of design by taking 95% of the empty-weight fraction obtained for a metal aircraft. The calculations for resizing the aircraft using composite materials are shown in Box 3.4.

The use of composite materials reduces the takeoff gross weight from 56,702 lb {25,720 kg} to only 51,585 lb {23,399 kg} yet the aircraft can still perform the same mission. This is a 9% takeoff-weight savings, resulting from only a 5% empty-weight saving.

This result sounds erroneous, but is actually typical of the "leverage" effect of the sizing equation. Unfortunately, this works both ways. If the empty weight creeps up during the detail-design process, it will require a more-than-proportional increase in takeoff gross weight to maintain the capability to perform the sizing mission. Thus it is crucial that realistic estimates of empty weight be used during early conceptual design, and that the weight be strictly controlled during later stages of design.

There are many trade studies that could be conducted other than range, payload, and material. Methods for trade studies are discussed in detail in Chapter 19.

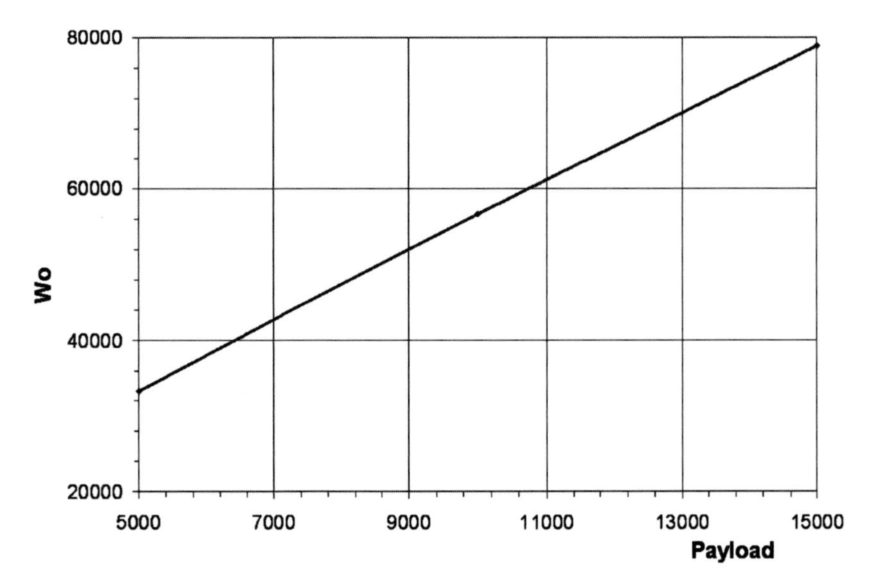

Fig. 3.13 Payload trade.

Box 3.4 Composite material trade

$$W_e/W_0 = (0.95)(0.93\ W_0^{-0.07}) = 0.8835\ W_0^{-0.07}$$

$$W_0 = \frac{10{,}800}{1 - 0.3773 - \dfrac{W_e}{W_0}}$$

W_0, guess	W_e/W_0	W_e	W_0, calculated
50,000	0.4143	20,713	51,810
51,000	0.4137	21,098	51,668
51,500	0.4134	21,291	51,598
51,550	0.4134	21,310	51,591
51,585	0.4134	21,323	51,587

The remainder of the book presents better methods for design, analysis, sizing, and trade studies, building on the concepts just given. In this chapter a conceptual sketch was made, but no guidance was provided as to how to make the sketch or why different features may be good or bad. Following chapters address these issues and illustrate how to develop a complete three-view drawing for analysis. Then more-sophisticated methods of analysis, sizing, and trade studies will be provided.

4
Airfoil and Geometry Selection

4.1 Introduction

Before the design layout can be started, values for a number of parameters must be chosen. These include the airfoil(s), the wing and tail geometries, wing loading, thrust-to-weight or horsepower-to-weight ratio, estimated takeoff gross weight and fuel weight, estimated wing, tail, and engine sizes, and the required fuselage size. These are discussed in the next three chapters.

This chapter covers selecting the airfoil and the wing and tail geometry. Chapter 5 addresses estimation of the required wing loading and thrust-to-weight ratio (horsepower-to-weight ratio for a propeller aircraft). Chapter 6 provides a more refined method for initial sizing than the quick method presented in the last chapter, and concludes with the use of the sizing results to calculate the required wing and tail area, engine size, and fuselage volume.

4.2 Airfoil Selection

The airfoil, in many respects, is the heart of the airplane. The airfoil affects the cruise speed, takeoff and landing distances, stall speed, handling qualities (especially near the stall), and overall aerodynamic efficiency during all phases of flight.

Much of the Wright Brothers' success can be traced to their development of airfoils using a wind tunnel of their own design, and the in-flight validation of those airfoils in their glider experiments of 1901–1902. The P-51 was regarded as the finest fighter of World War II in part because of its radical laminar-flow airfoil. More recently, the low-speed airfoils developed by Peter Lissaman contributed much to the success of the man-powered Gossamer Condor, and the airfoils designed by John Rontz were instrumental to the success of Burt Rutan's radical designs.

Airfoil Geometry

Figure 4.1 illustrates the key geometric parameters of an airfoil. The front of the airfoil is defined by a leading-edge radius that is tangent to the upper and lower surfaces. An airfoil designed to operate in supersonic flow will have a sharp or nearly sharp leading edge to prevent a drag-producing bow shock. (As discussed later, wing sweep may be used instead of a sharp leading edge to reduce the supersonic drag.)

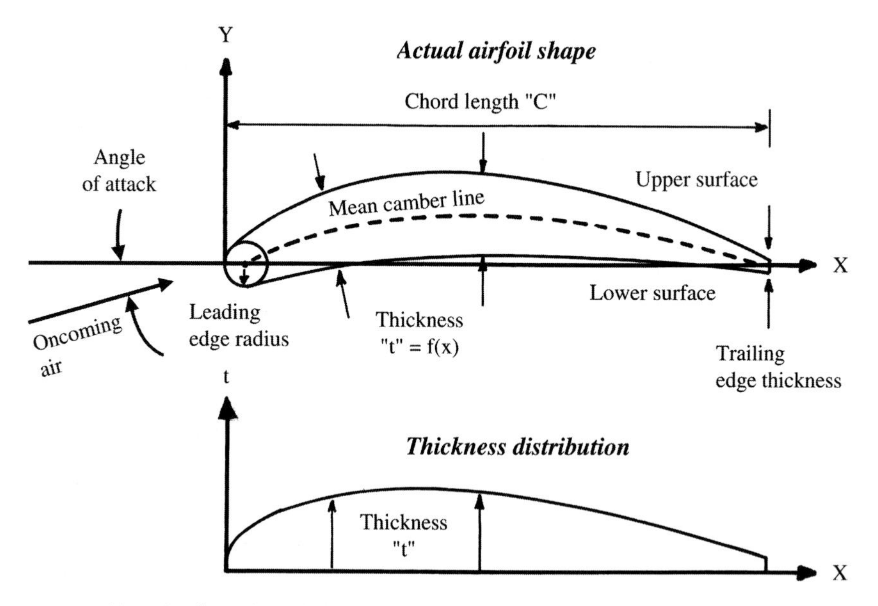

Fig. 4.1 Airfoil geometry.

The chord of the airfoil is the straight line from the leading edge to the trailing edge. It is very difficult to build a perfectly sharp trailing edge, and so most airfoils have a blunt trailing edge with some small finite thickness.

"Camber" refers to the curvature characteristic of most airfoils. The "mean camber line" is the line equidistant from the upper and lower surfaces. Total airfoil camber is defined as the maximum distance of the mean camber line from the chord line, expressed as a percent of the chord.

In earlier days, most airfoils had flat bottoms, and it was common to refer to the upper surface shape as the "camber." Later, as airfoils with curved bottoms came into usage, they were known as "double-cambered" airfoils. Also, an airfoil with a concave lower surface was known as an "under-cambered" airfoil. These terms are technically obsolete but are still in common usage.

The thickness distribution of the airfoil is the distance from the upper surface to the lower surface, measured perpendicular to the mean camber line, and is a function of the distance from the leading edge. The "airfoil thickness ratio" (t/c) refers to the maximum thickness of the airfoil divided by its chord.

For many aerodynamic calculations, it has been traditional to separate the airfoil into its thickness distribution and a zero-thickness camber line. The former provides the major influence on the profile drag, whereas the latter provides the major influence upon the lift and the drag due to lift.

When an airfoil is scaled in thickness, the camber line must remain unchanged, so that the scaled thickness distribution is added to the original camber line to produce the new, scaled airfoil. In a similar fashion, an airfoil which is to have its camber changed is broken into its camber line and thickness distribution. The camber line is scaled to produce the desired maximum camber; then the original thickness distribution is added to obtain the new airfoil. In this fashion, the airfoil can be reshaped to change either the profile drag or lift characteristics, without greatly affecting the other.

Airfoil Lift and Drag

An airfoil generates lift by changing the velocity of the air passing over and under itself. The airfoil angle of attack and/or camber causes the air over the top of the wing to travel faster than the air beneath the wing.

Bernoulli's equation shows that higher velocities produce lower pressures, so that the upper surface of the airfoil tends to be pulled upward by lower-than-ambient pressures while the lower surface of the airfoil tends to be pushed upward by higher-than-ambient pressures. The integrated differences in pressure between the top and bottom of the airfoil generate the net lifting force.*

Figure 4.2 shows typical pressure distributions for the upper and lower surfaces of a lifting airfoil at subsonic speeds. Note that the upper surface of the wing contributes about two-thirds of the total lift, so that a designer should avoid disturbing the top of the wing.

Figure 4.3a illustrates the flowfield around a typical airfoil as a number of airflow velocity vectors, with the vector length representing local velocity magnitude. In Fig. 4.3b, the freestream velocity vector is subtracted from each local velocity vector, leaving only the change in velocity vector caused by the presence of the airfoil. It can be seen that the effect of the airfoil is to introduce a change in airflow, which seems to circulate around the airfoil in a clockwise fashion if the airfoil nose is to the left.

This "circulation" is the theoretical basis for the classical calculation of lift and drag due to lift. The greater the circulation, the greater the lift. Circulation is usually represented by Γ and is shown as a circular flow direction as in Fig. 4.3c.

A flat board at an angle to the oncoming air will produce lift. However, the air going over the top of the flat "airfoil" will tend to separate from the surface, thus disturbing the flow and therefore reducing lift and greatly increasing drag (Fig. 4.4). Curving the airfoil (i.e., camber) allows the airflow to remain attached,

---

*There is another way of looking at lift—behind the wing there will be a downwash, geometrically caused by the airfoil angle of attack and camber. Thus, the wing has accelerated the air downwards requiring a force to have been applied to the air, and by application of Newton's laws this means that the air has applied an equal and opposite force to the wing. This downwash momentum in the air adds up to and equals the lift on the wing. People continue to have arguments over this distinction, often in the popular aviation magazines. Both ways of looking at lift are 100% correct. Lift equals the total downwash momentum imparted on the air, *and* lift equals the integrated vertical component of pressures on the wing. Which one truly "causes" the lift? Well, the only way a force is exerted on the wing is through pressures, and so this author leans towards that explanation—but it really does not matter.

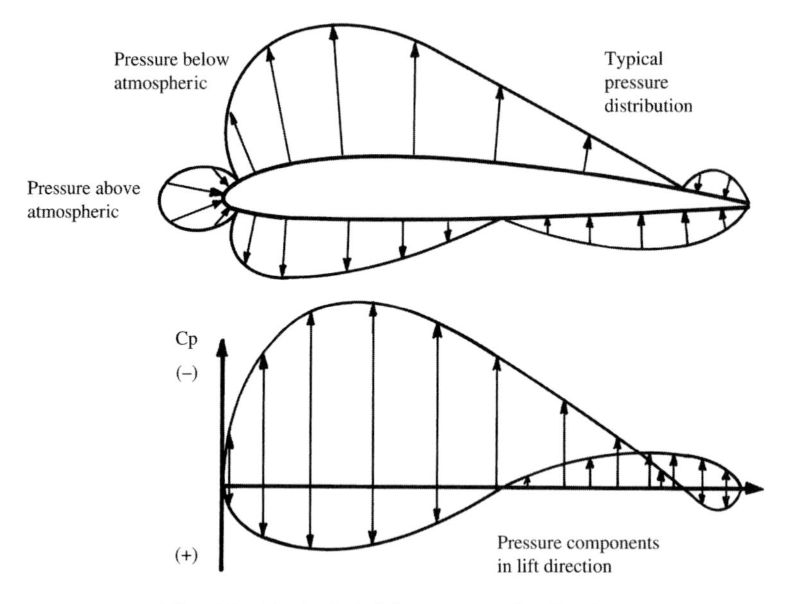

Fig. 4.2 Typical airfoil pressure distribution.

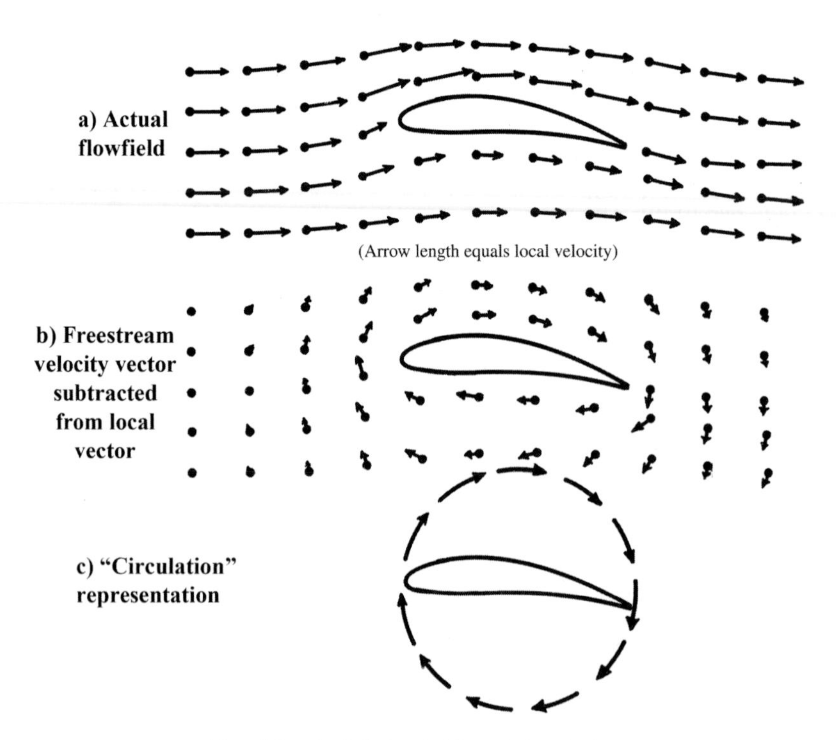

Fig. 4.3 Airfoil flowfield and circulation.

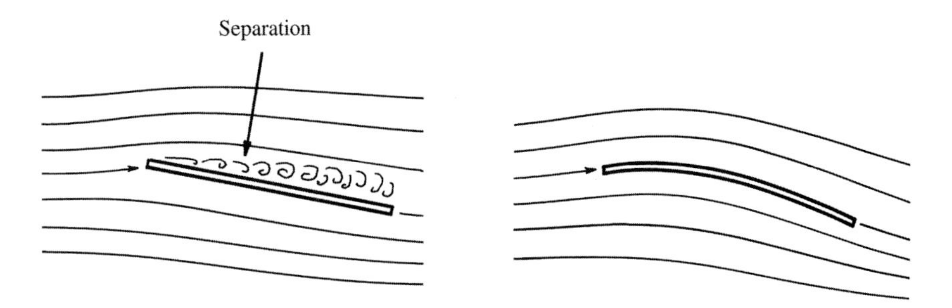

Fig. 4.4 Effect of camber on separation.

thus increasing lift and reducing drag. The camber also increases lift by increasing the circulation of the airflow.

In fact, an airfoil with camber will produce lift even at zero angle between the chord line and the oncoming air ("angle of attack"). For a cambered airfoil there is some negative angle at which no lift is produced, the "angle of zero lift." This negative angle is approximately equal (in degrees) to the percent camber of the airfoil.

Odd as it sounds, an airfoil in two-dimensional (2-D) flow does not experience any drag due to the creation of lift. The pressure forces produced in the generation of lift are at right angles to the oncoming air. All 2-D airfoil drag is produced by skin friction and pressure effects resulting from flow separation and shocks. It is only in three-dimensional (3-D) flow that drag due to lift is produced.

The airfoil section lift, drag, and pitching moment are defined in nondimensional form in Eqs. (4.1), (4.2), and (4.3). By definition, the lift force is perpendicular to the flight direction while the drag force is parallel to the flight direction. The pitching moment is usually negative when measured about the aerodynamic center, implying a nose-down moment. Note that 2-D airfoil characteristics are denoted by lowercase subscripts (i.e., C_ℓ), whereas the 3-D wing characteristics are denoted by uppercase subscripts (i.e., C_L).

$$\text{Section Lift Coefficient:} \quad C_\ell = \frac{\text{section lift}}{qc} \tag{4.1}$$

$$\text{Section Drag Coefficient:} \quad C_d = \frac{\text{section drag}}{qc} \tag{4.2}$$

$$\text{Section Moment Coefficient:} \quad C_m = \frac{\text{section moment}}{qc^2} \tag{4.3}$$

Where

c = chord length
q = dynamic pressure = $\rho V^2/2$
α = angle of attack
C_{ℓ_α} = slope of the lift curve = 2π (theoretical thin airfoil)

The point about which the pitching moment remains constant for any angle of attack is called the "aerodynamic center." The aerodynamic center is not the same as the airfoil's center of pressure (or lift). The center of pressure is usually behind the aerodynamic center. The location of the center of pressure varies with angle of attack for most airfoils.

Pitching moment is measured about some reference point, typically the quarter-chord point (25% of the chord length back from the leading edge). The pitching moment is almost independent of angle of attack about the quarter-chord for most airfoils at subsonic speeds (i.e., the aerodynamic center is usually at the quarter-chord point).

Lift, drag, and pitching-moment characteristics for a typical airfoil are shown in Fig. 4.5.

Airfoil characteristics are strongly affected by the "Reynolds number" at which they are operating. Reynolds number, the ratio between the dynamic and the viscous forces in a fluid, is equal to $(\rho Vl/\mu)$, where V is the velocity, l the length the fluid has traveled down the surface, ρ the fluid density, and μ the fluid viscosity coefficient. The Reynolds number influences whether the flow will be laminar or turbulent, and whether flow separation will occur. A typical aircraft wing operates at a Reynolds number of about ten million.

Figure 4.5 illustrates the so-called laminar bucket. For a "laminar" airfoil operating at the design Reynolds number, there is a range of lift coefficient for which the flow remains laminar over a substantial part of the airfoil. This causes a significant reduction of drag for a given lift coefficient.

However, this effect is very dependent upon the Reynolds number as well as the actual surface smoothness. For example, dirt, rain, or insect debris on the leading edge may cause the flow to become turbulent, causing an increase in drag to the dotted line shown in Fig. 4.5. This also can change the lift and pitching-moment characteristics.

In several canarded homebuilt designs with laminar airfoils, entering a light rainfall will cause the canard's airflow to become turbulent, reducing canard

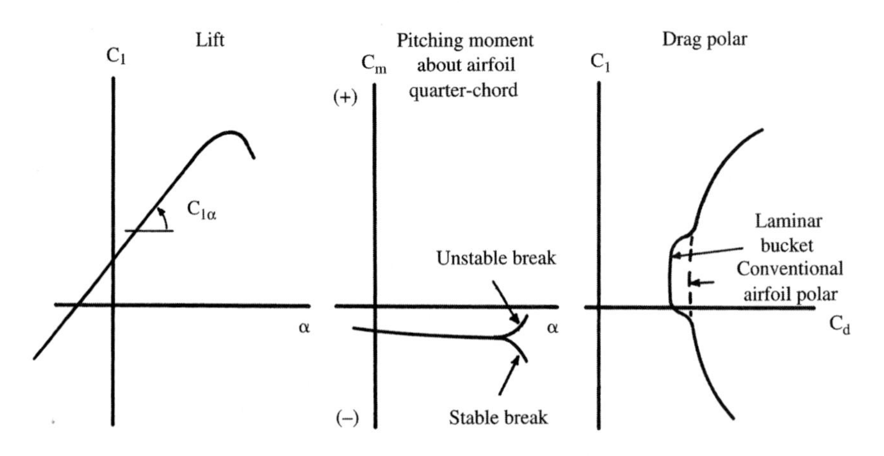

Fig. 4.5 Airfoil lift, drag, and pitching moment.

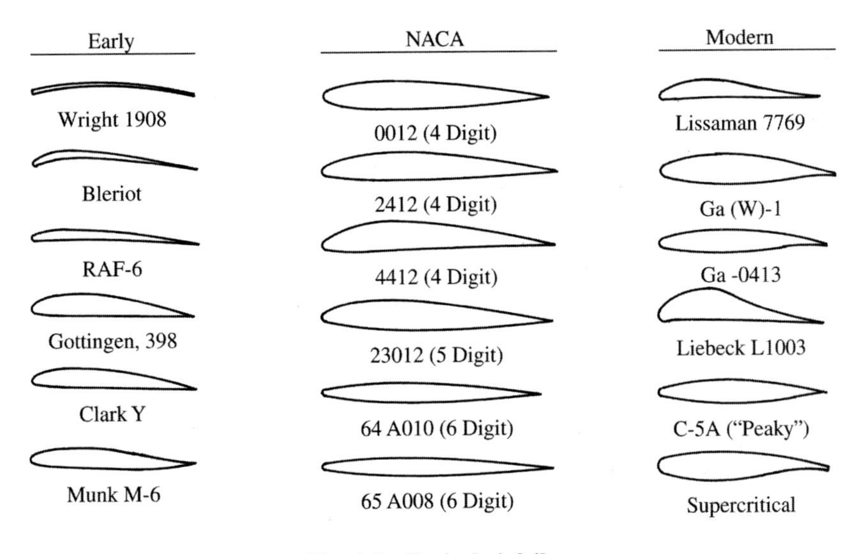

Fig. 4.6 Typical airfoils.

lift and causing the aircraft to pitch downward. Earlier, nonlaminar airfoils were designed assuming turbulent flow at all times and do not experience this effect.

Airfoil Families

A variety of airfoils is shown in Fig. 4.6. The early airfoils were developed mostly by trial and error. In the 1930s, the NACA developed a widely used family of mathematically defined airfoils called the "four-digit" airfoils. In these, the first digit defined the percent camber, the second defined the location of the maximum camber, and the last two digits defined the airfoil maximum thickness in percent of chord. While rarely used for wing design today, the uncambered four-digit airfoils are still commonly used for tail surfaces of subsonic aircraft.

The NACA five-digit airfoils were developed to allow shifting the position of maximum camber forward for greater maximum lift. The six-series airfoils were designed for increased laminar flow, and hence reduced drag. Six-series airfoils such as the 64A series are still widely used as a starting point for high-speed-wing design. The Mach 2 F-15 fighter uses the 64A airfoil modified with camber at the leading edge. Geometry and characteristics of these "classical" airfoils are summarized in Ref. 2, a must for every designer's library.

Airfoil Design

In the past, the designer would select an airfoil (or airfoils) from such a catalog. This selection would consider factors such as the airfoil drag during cruise, stall and pitching-moment characteristics, the thickness available for structure and fuel, and the ease of manufacture. With today's computational

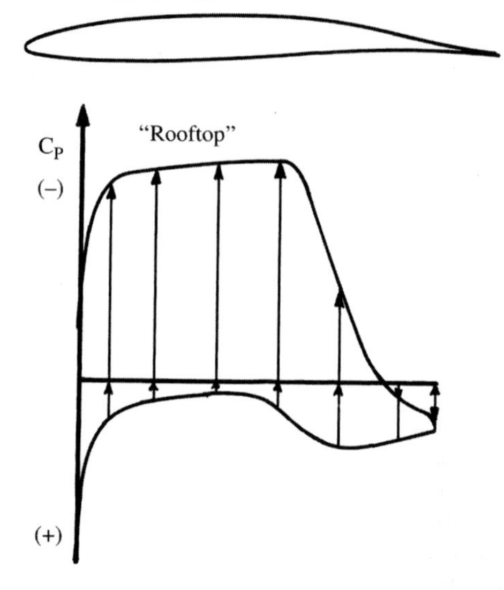

Fig. 4.7 Laminar airfoil.

airfoil design capabilities, it is becoming common for the airfoil shapes for a wing to be custom-designed.

Modern airfoil design is based upon inverse computational solutions for desired pressure (or velocity) distributions on the airfoil. Methods have been developed for designing an airfoil such that the pressure differential between the top and bottom of the airfoil quickly reaches a maximum value attainable without airflow separation. Toward the rear of the airfoil, various pressure recovery schemes are employed to prevent separation near the trailing edge.

These airfoil optimization techniques result in airfoils with substantial pressure differentials (lift) over a much greater percent of chord than a classical airfoil. This permits a reduced wing area (and wetted area) for a required amount of lift. Such airfoil design methods go well beyond the scope of this book.

Another consideration in modern airfoil design is the desire to maintain laminar flow over the greatest possible part of the airfoil. Laminar flow can be maintained by providing a negative pressure gradient, i.e., by having the pressure continuously drop from the leading edge to a position close to the trailing edge. This tends to "suck" the flow rearward, promoting laminar flow.

A good laminar-flow airfoil combined with smooth fabrication methods can produce a wing with laminar flow over about 50–70% of the wing. Figure 4.7 shows a typical laminar flow airfoil and its pressure distribution.

As an airfoil generates lift, the velocity of the air passing over its upper surface is increased. If the airplane is flying at just under the speed of sound, the faster air traveling over the upper surface will reach supersonic speeds causing a shock to exist on the upper surface, as shown in Fig. 4.8. The speed at which supersonic flow first appears on the airfoil is called the "critical Mach" (M_{crit}).

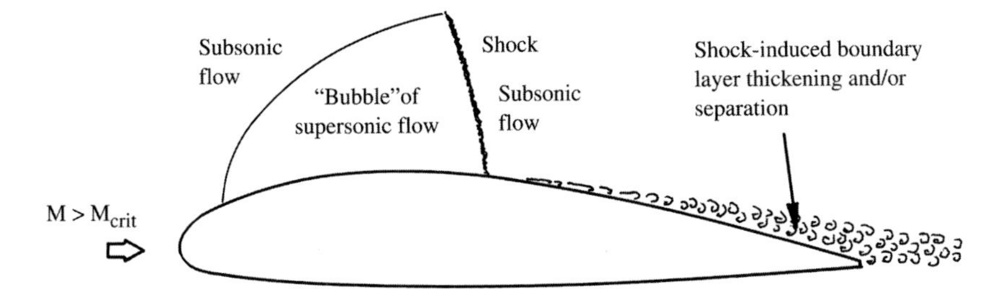

Fig. 4.8 Transonic effects.

The upper-surface shock creates a large increase in drag, along with a reduction in lift and a change in the pitching moment. The drag increase comes from the tendency of the rapid pressure rise across the shock to thicken or even separate the boundary layer.

A "supercritical" airfoil is one designed to minimize these effects. Modern computational methods allow design of airfoils in which the upper-surface shock is minimized or even eliminated by spreading the lift in the chordwise direction, thus reducing the upper surface velocity for a required total lift. This increases the critical Mach number.

Design Lift Coefficient

For early conceptual design work, the designer must frequently rely upon existing airfoils. From existing airfoils, the one should be selected that comes closest to having the desired characteristics.

The first consideration in initial airfoil selection is the "design lift coefficient." This is the lift coefficient at which the airfoil has the best L/D (shown in Fig. 4.9 as the point on the airfoil drag polar that is tangent to a line from the origin and closest to the vertical axis).

In subsonic flight a well-designed airfoil operating at its design lift coefficient has a drag coefficient that is little more than skin-friction drag. The aircraft should be designed so that it flies the design mission at or near the design lift coefficient to maximize the aerodynamic efficiency.

As a first approximation, it can be assumed that the wing lift coefficient, C_L, equals the airfoil lift coefficient, C_ℓ. In level flight the lift must equal the weight, so the required design lift coefficient can be found as follows:

$$W = L = qSC_L \cong qSC_\ell \tag{4.4}$$

$$C_\ell = \frac{1}{q}\left(\frac{W}{S}\right) \tag{4.5}$$

Dynamic pressure (q) is a function of velocity and altitude. By assuming a wing loading (W/S) as described later, the design lift coefficient can be calculated for the velocity and altitude of the design mission.

Note that the actual wing loading will decrease during the mission as fuel is burned. Thus, to stay at the design lift coefficient, the dynamic pressure

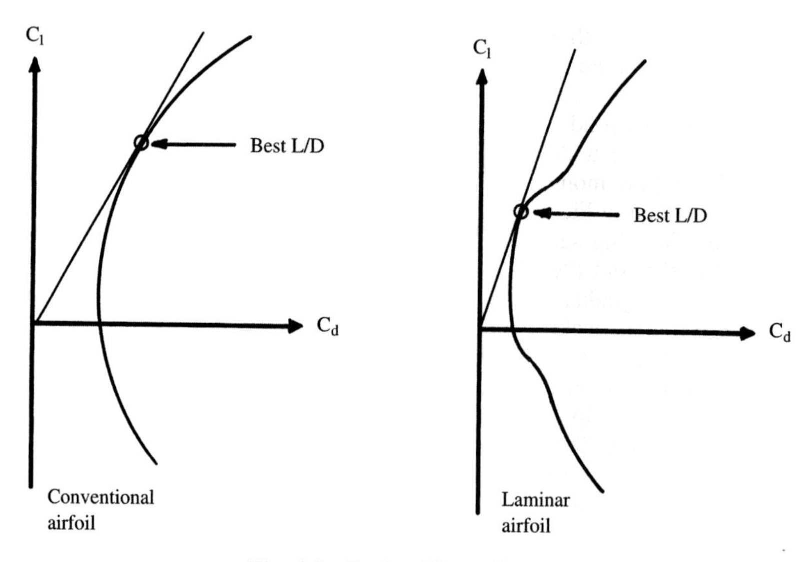

Fig. 4.9 Design lift coefficient.

must be steadily reduced during the mission by either slowing down, which is undesirable, or climbing to a higher altitude. This explains the "cruise-climb," which is followed by an aircraft trying to maximize range.

In actual practice, a design lift coefficient usually will be based upon past experience, and for most types of aircraft typically will be around 0.3–0.5. In early design studies the initial selection of the airfoil is often based simply upon prior experience or copied from some successful design.

Stall

Stall characteristics play an important role in airfoil selection. Some airfoils exhibit a gradual reduction in lift during a stall, whereas others show a violent loss of lift, accompanied by a rapid change in pitching moment. This difference reflects the existence of three entirely different types of airfoil stall.

"Fat" airfoils (round leading edge and t/c greater than about 14%) stall from the trailing edge. The turbulent boundary layer increases with angle of attack. At around 10 deg the boundary layer begins to separate, starting at the trailing edge and moving forward as the angle of attack is further increased. The loss of lift is gradual. The pitching moment changes only a small amount.

Thinner airfoils stall from the leading edge. If the airfoil is of moderate thickness (about 6–14%), the flow separates near the nose at a very small angle of attack, but immediately reattaches itself so that little effect is felt. At some higher angle of attack the flow fails to reattach, which almost immediately stalls the entire airfoil. This causes an abrupt change in lift and pitching moment.

Very thin airfoils exhibit another form of stall. As before, the flow separates from the nose at a small angle of attack and reattaches almost immediately.

However, for a very thin airfoil this "bubble" continues to stretch toward the trailing edge as the angle of attack is increased. At the angle of attack where the bubble stretches all the way to the trailing edge, the airfoil reaches its maximum lift. Beyond that angle of attack, the flow is separated over the whole airfoil, so that the stall occurs. The loss of lift is smooth, but large changes in pitching moment are experienced. The three types of stall characteristics are depicted in Fig. 4.10.

Twisting the wing such that the tip airfoils have a reduced angle of attack compared to the root ("washout") can cause the wing to stall first at the root. This provides a gradual stall even for a wing with a poorly stalling airfoil. Also, the turbulent wake off the stalled wingroot will vibrate the horizontal tail, notifying the pilot that a stall is imminent.

In a similar fashion, the designer may elect to use different airfoils at the root and tip, with a tip airfoil selected which stalls at a higher angle of attack than the root airfoil. This provides good flow over the ailerons for roll control at an angle of attack where the root is stalled.

If different airfoils are used at the root and tip, the designer must develop the intermediate airfoils by interpolation (discussed later). These intermediate airfoils will have section characteristics somewhere between those of the root and tip airfoils, and can also be estimated by interpolation. This interpolation of section characteristics does not work for modern supercritical or laminar-flow airfoils. Estimation of the section characteristics in those cases must be done computationally.

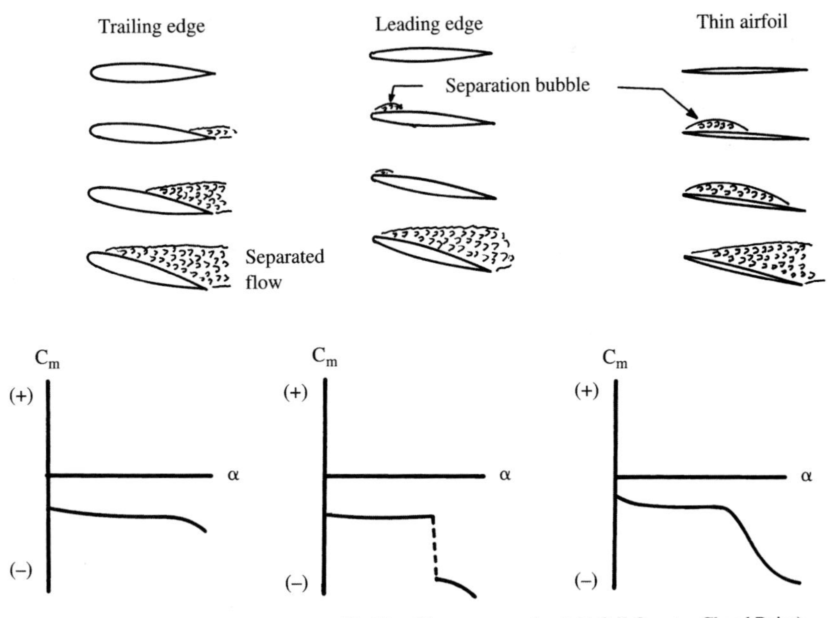

Fig. 4.10 Types of stall.

Stall characteristics for thinner airfoils can be improved with various leading-edge devices such as slots, slats, leading-edge flaps, Krueger flaps, and active methods (e.g., suction or blowing). These are discussed in the Aerodynamics chapter.

Wing stall is directly related to airfoil stall only for high-aspect-ratio, unswept wings. For lower-aspect-ratio or highly swept wings the 3-D effects dominate stall characteristics, and airfoil stall characteristics can be essentially ignored in airfoil selection.

Pitching moment must also be considered in airfoil selection. Horizontal tail or canard size is directly affected by the magnitude of the wing pitching moment to be balanced. Some of the supercritical airfoils use what is called "rear-loading" to increase lift without increasing the region of supersonic flow. This produces an excellent L/D, but can cause a large nose-down pitching moment. If this requires an excessive tail area, the total aircraft drag may be increased, not reduced.

For a stable tailless or flying-wing aircraft, the pitching moment must be near zero. This usually requires an "S"-shaped camber with the characteristic upward reflex at the trailing edge. Reflexed airfoils have poorer L/D than an airfoil designed without this constraint. This tends to reduce some of the benefit that flying wings experience due to their reduced wetted area. A computerized, "active" flight control system can remove the requirement for natural stability, and thus allow a nonreflexed airfoil.

Airfoil Thickness Ratio

Airfoil thickness ratio has a direct effect on drag, maximum lift, stall characteristics, and structural weight. Figure 4.11 illustrates the effect of thickness ratio on subsonic drag. The drag increases with increasing thickness due to increased separation.

Figure 4.12 shows the impact of thickness ratio on critical Mach number, the Mach number at which supersonic flow first appears over the wing. A supercritical airfoil tends to minimize shock formation and can be used to reduce drag for a given thickness ratio or to permit a thicker airfoil at the same drag level.

The thickness ratio affects the maximum lift and stall characteristics primarily by its effect on the nose shape. For a wing of fairly high aspect ratio and moderate sweep, a larger nose radius provides a higher stall angle and a greater maximum lift coefficient, as shown in Fig. 4.13.

The reverse is true for low-aspect-ratio, swept wings, such as a delta wing. Here, a sharper leading edge provides greater maximum lift due to the formation of vortices just behind the leading edge. These leading-edge vortices act to delay wing stall. This 3-D effect is discussed in the Aerodynamics chapter.

Thickness also affects the structural weight of the wing. Statistical equations for wing weight show that the wing structural weight varies approximately inversely with the square root of the thickness ratio. Halving the thickness ratio will increase wing weight by about 41%. The wing is typically about 15% of the total empty weight, so that halving the thickness ratio would increase empty weight by about 6%. When applied to the sizing equation, this can have a major impact.

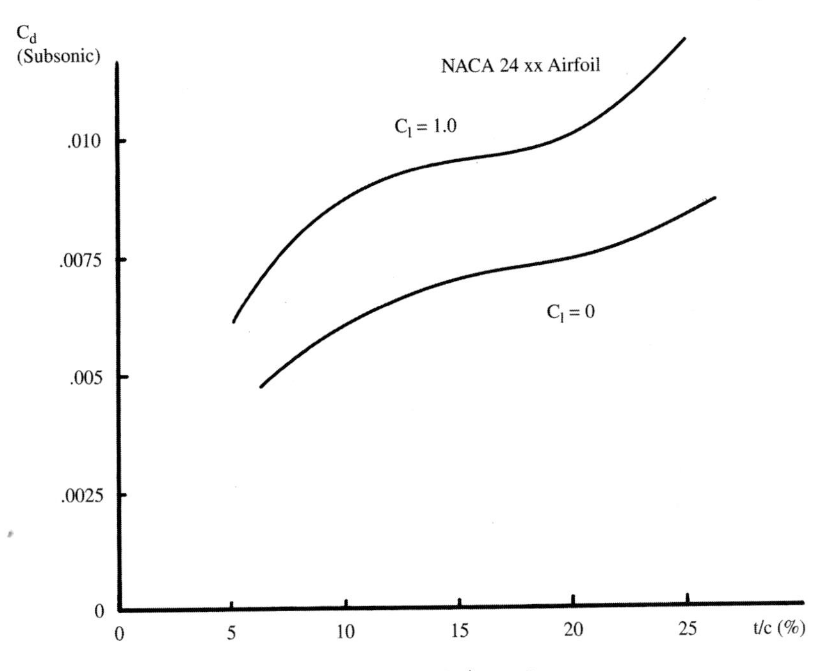

Fig. 4.11 Effect of t/c on drag.

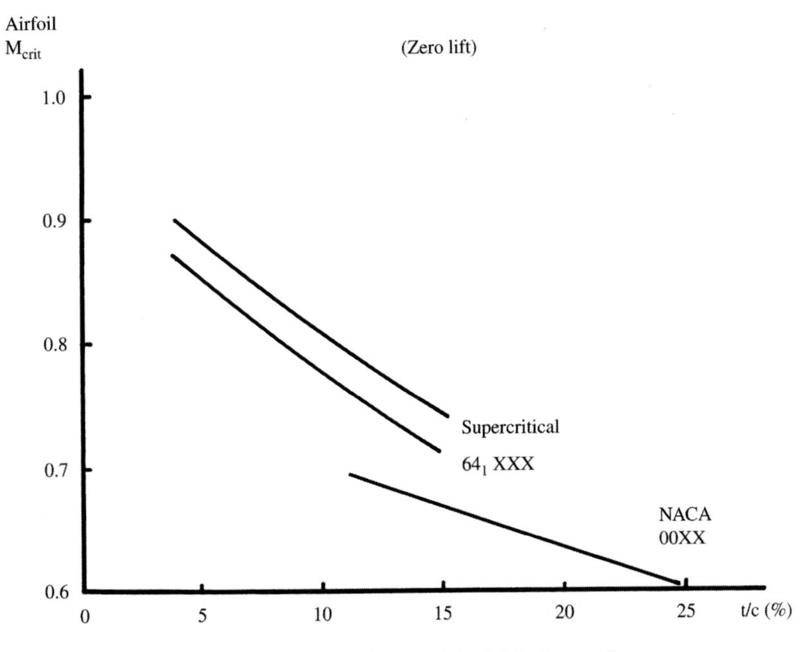

Fig. 4.12 Effect of t/c on critical Mach number.

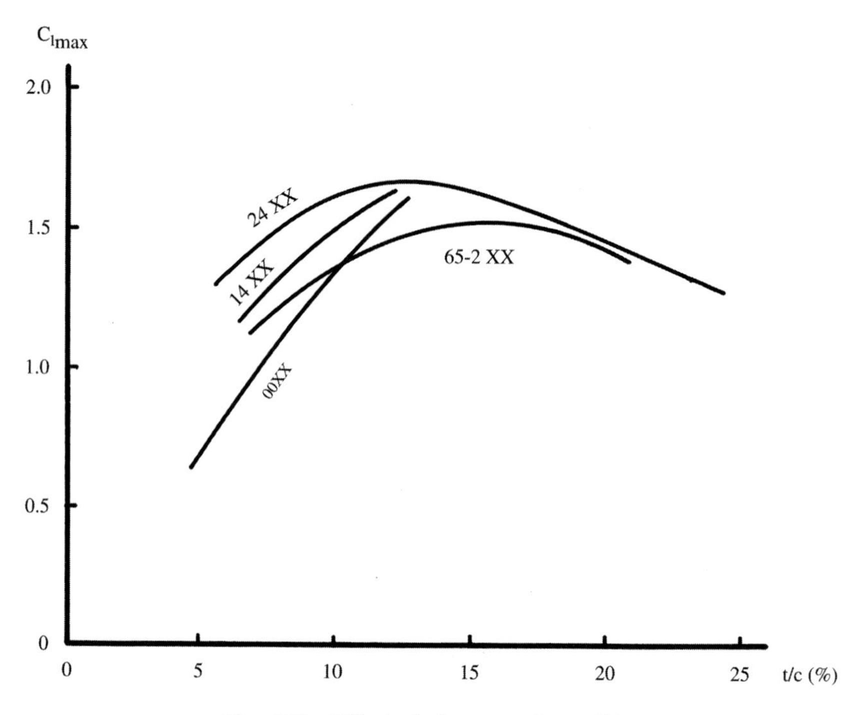

Fig. 4.13 Effect of t/c on maximum lift.

For initial selection of the thickness ratio, the historical trend shown in Fig. 4.14 can be used. Note that a supercritical airfoil would tend to be about 10% thicker (i.e., conventional airfoil thickness ratio times 1.1) than the historical trend.

Frequently the thickness is varied from root to tip. Due to fuselage effects, the root airfoil of a subsonic aircraft can be as much as 20–60% thicker than the tip airfoil without greatly affecting the drag. This is very beneficial, resulting in a structural weight reduction as well as more volume for fuel and landing gear. This thicker root airfoil should extend to no more than about 30% of the span.

Other Airfoil Considerations

Another important aspect of airfoil selection is the intended Reynolds number. Each airfoil is designed for a certain Reynolds number. Use of an airfoil at a greatly different Reynolds number (half an order of magnitude or so) can produce section characteristics much different from those expected.

This is especially true for the laminar-flow airfoils, and is most crucial when an airfoil is operated at a lower-than-design Reynolds number. In the past this has been a problem for homebuilt and sailplane designers, but there are now suitable airfoils designed especially for these lower Reynolds number aircraft.

The laminar airfoils require extremely smooth skins as well as exact control over the actual, as-manufactured shape. These can drive the cost up significantly.

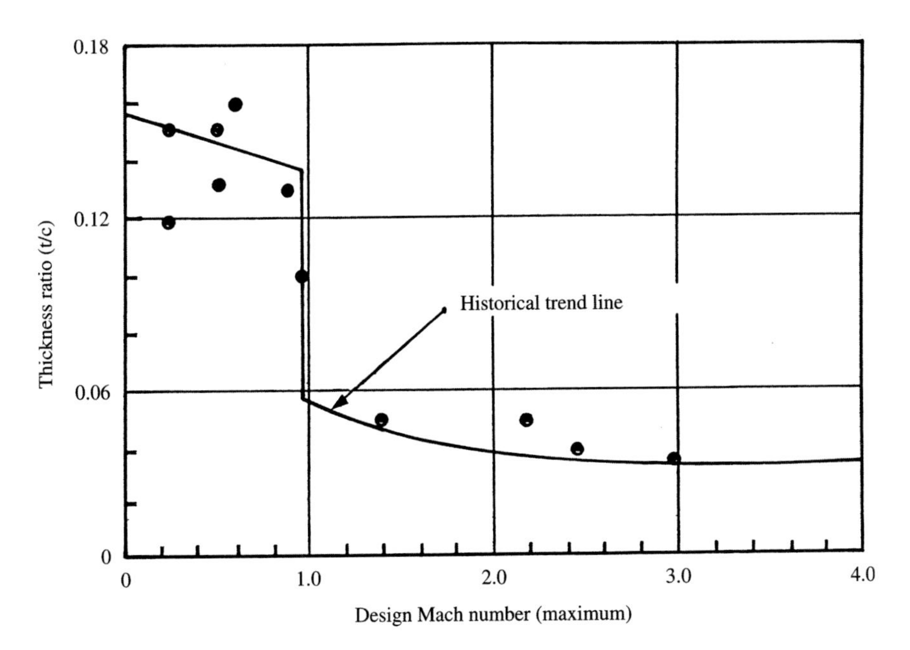

Fig. 4.14 Thickness ratio historical trend.

Also, the camouflage paints used on military aircraft are rough compared to bare metal or composite skins. This must be considered before selecting certain airfoils.

While an understanding of the factors important to airfoil selection is important, an aircraft designer should not spend too much time trying to pick exactly the "right" airfoil in early conceptual design. Later trade studies and analytical design tools will determine the desired airfoil characteristics and geometry. For early conceptual layout, the selected airfoil is important mostly for determining the thickness available for structure, landing gear, and fuel.

Appendix D provides geometry and section characteristics for a few airfoils useful in conceptual design. For swept-wing supersonic aircraft, the NACA 64A and 65A sections are good airfoils for initial design. The appendix describes a supercritical section suitable for transports and other high-subsonic aircraft, along with a typical modern NASA section for general aviation. A few specialized airfoils are provided for other applications.

The airfoils presented in Appendix D are not being recommended as the "best" sections for those applications, but rather as reasonable airfoils with which to start a conceptual design. Again, Ref. 2 is highly recommended.

4.3 Wing Geometry

The "reference" ("trapezoidal") wing is the basic wing geometry used to begin the layout. Figures 4.15 and 4.16 show the key geometric parameters of the reference wing.

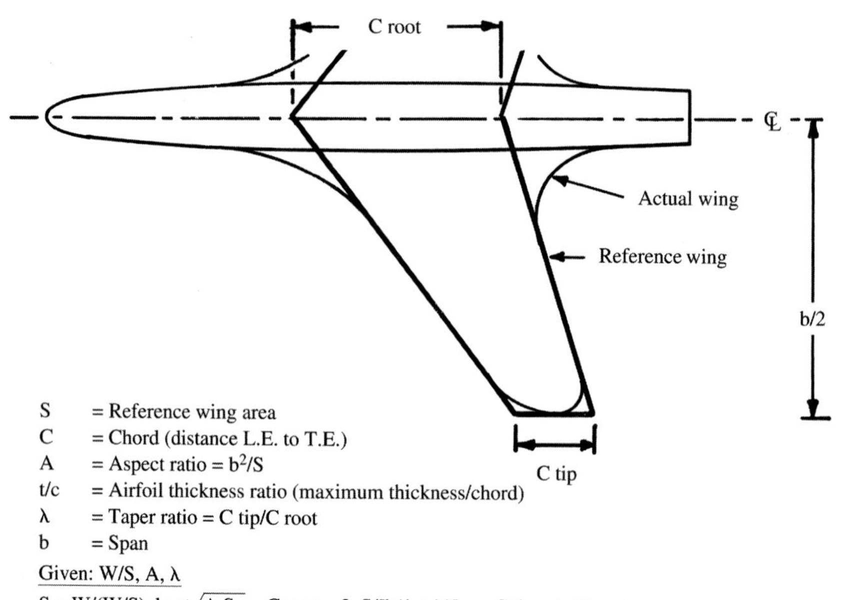

S = Reference wing area
C = Chord (distance L.E. to T.E.)
A = Aspect ratio = b^2/S
t/c = Airfoil thickness ratio (maximum thickness/chord)
λ = Taper ratio = C tip/C root
b = Span

Given: W/S, A, λ

$S = W/(W/S)$ $b = \sqrt{A \cdot S}$ $C\ root = 2 \cdot S/[b(1+\lambda)]$ $C\ tip = \lambda \cdot C\ root$

Fig. 4.15 Wing geometry.

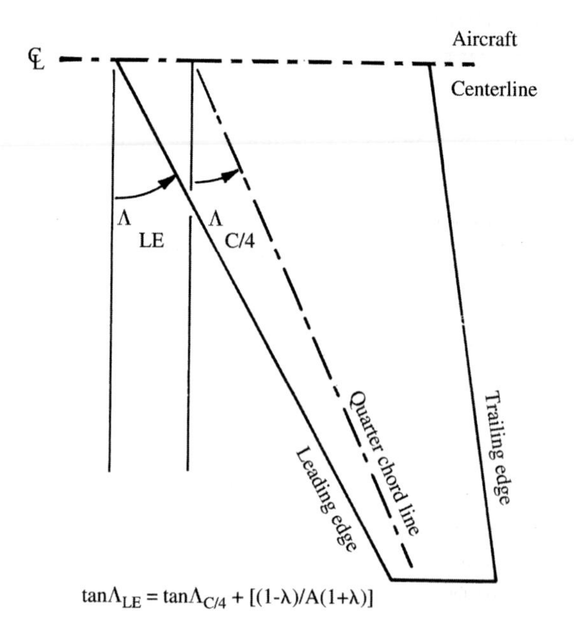

$$\tan\Lambda_{LE} = \tan\Lambda_{C/4} + [(1-\lambda)/A(1+\lambda)]$$

Fig. 4.16 Wing sweep Λ.

Note that the reference wing is fictitious, and extends through the fuselage to the aircraft centerline. Thus the reference wing area includes the part of the reference wing that sticks into the fuselage. For the reference wing, the root airfoil is the airfoil of the trapezoidal reference wing at the centerline of the aircraft, not where the actual wing connects to the fuselage.

There are two key sweep angles, as shown in Fig. 4.16. The leading-edge sweep is the angle of concern in supersonic flight. To reduce drag, it is common to sweep the leading edge behind the Mach cone. The sweep of the quarter-chord line is the sweep most related to subsonic flight. It is important to avoid confusing these two sweep angles. The equation at the bottom of Fig. 4.16 allows converting from one sweep angle to the other.

Airfoil pitching-moment data in subsonic flow are generally provided about the quarter-chord point, where the airfoil pitching moment is essentially constant with changing angle of attack (i.e., the "aerodynamic center"). In a similar fashion, such a point is defined for the complete trapezoidal wing and is based on the concept of the "mean aerodynamic chord." The mean aerodynamic chord (Fig. 4.17) is the chord $\bar{c}$ of an airfoil, located at some distance $\bar{Y}$ from the centerline.

The entire wing has its mean aerodynamic center at approximately the same percent location of the mean aerodynamic chord as that of the airfoil alone. In subsonic flow, this is at the quarter-chord point on the mean aerodynamic chord. In supersonic flow, the aerodynamic center moves back to about 40% of the mean aerodynamic chord. The designer uses the mean aerodynamic chord

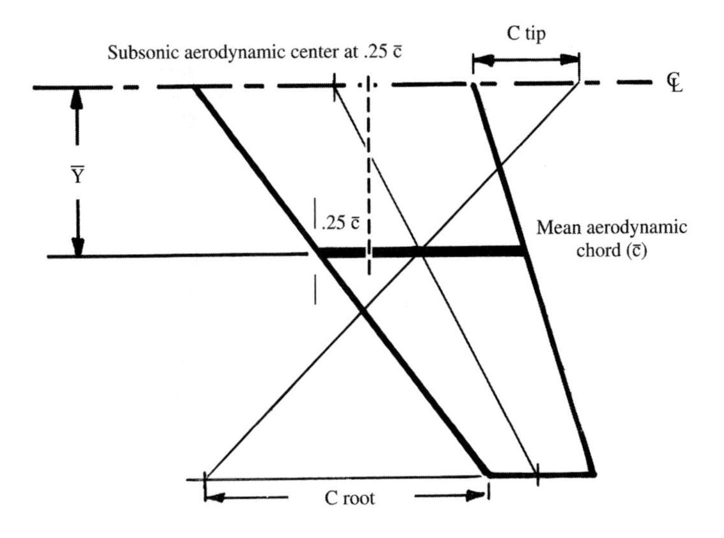

$\bar{c} = (2/3)\ C\ root\ (1+\lambda+\lambda^2)/(1+\lambda)$

$\bar{Y} = (b/6)[(1+2\lambda)/(1+\lambda)]$ (assuming lift is proportional to chord)

$\bar{Y}$ must be doubled for a vertical tail.

Fig. 4.17 Mean aerodynamic chord.

and the resulting aerodynamic center point to position the wing properly. Also, the mean aerodynamic chord will be important to stability calculations. Figure 4.17 illustrates a graphical method for finding the mean aerodynamic chord of a trapezoidal-wing planform.

The required reference wing area S can be determined only after the takeoff gross weight is determined. The shape of the reference wing is determined by its aspect ratio, taper ratio, and sweep.

Aspect Ratio

The first to investigate aspect ratio in detail were the Wright Brothers, using a wind tunnel they constructed. They found that a long, skinny wing (high aspect ratio) has less drag for a given lift than a short, fat wing (low aspect ratio). This is due to the 3-D effects.

As most early wings were rectangular in shape, the aspect ratio was initially defined as simply the span divided by the chord. For a tapered wing, the aspect ratio is defined as the span squared divided by the area (which defaults to the earlier definition for a wing with no taper).

When a wing is generating lift, it has a reduced pressure on the upper surface and an increased pressure on the lower surface. The air would like to "escape" from the bottom of the wing, moving to the top. This is not possible in 2-D flow unless the airfoil is leaky (a real problem with some fabric wing materials unless properly treated). However, for a real, 3-D wing, the air can escape around the wing tip (Fig 4.18).

Air escaping around the wing tip lowers the pressure difference between the upper and the lower surfaces. This reduces lift near the tip. Also, the air flowing around the tip flows in a circular path when seen from the front, and in effect pushes down on the wing. Strongest near the tip, this reduces the effective angle of attack of the wing airfoils. This circular, or "vortex," flow pattern continues downstream behind the wing.

A wing with a high aspect ratio has tips farther apart than an equal area wing with a low aspect ratio. Therefore, the amount of the wing affected by the tip

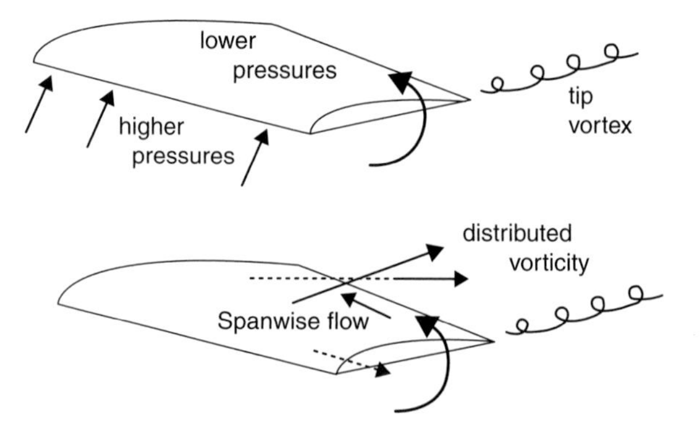

Fig. 4.18 "Escape" of air around the wing tip.

vortex is less for a high-aspect-ratio wing than for a low-aspect-ratio wing, and the strength of the tip vortex is reduced. Thus, the high-aspect-ratio wing does not experience as much of a loss of lift and increase of drag due to tip effects as a low-aspect-ratio wing of equal area.

(It is actually the wing span that determines the drag due to lift. A simple derivation based on equations in Chapter 12 will prove that the drag due to lift is proportional to the inverse of the square of the span. Aspect ratio per se has nothing to do with it. However, when looking at various options for the wing planform, the wing area is usually held constant unless widely different aircraft concepts are being evaluated. When wing area is held constant, the wing span varies by the square root of the aspect ratio, and so the drag due to lift becomes inversely proportional to aspect ratio.)

Another result of the air escaping around the wing tip is an outward flow beneath the wing and an inward flow above it, shown in the bottom of Fig. 4.18. This actually changes local flow directions and should be considered when orienting nacelles or stores on the wing.

As shown in Fig. 3.6, the maximum subsonic L/D of an aircraft increases approximately by the square root of an increase in aspect ratio (when wing area and S_{wet}/S_{ref} are held constant). On the other hand, the wing weight also increases with increasing aspect ratio, by about the same factor.

Another effect of changing aspect ratio is a change in stalling angle. Due to the reduced effective angle of attack at the tips, a lower-aspect-ratio wing will stall at a higher angle of attack than a higher-aspect-ratio wing (Fig. 4.19). This is one reason why tails tend to be of lower aspect ratio. Delaying tail stall until well after the wing stalls assures adequate control.

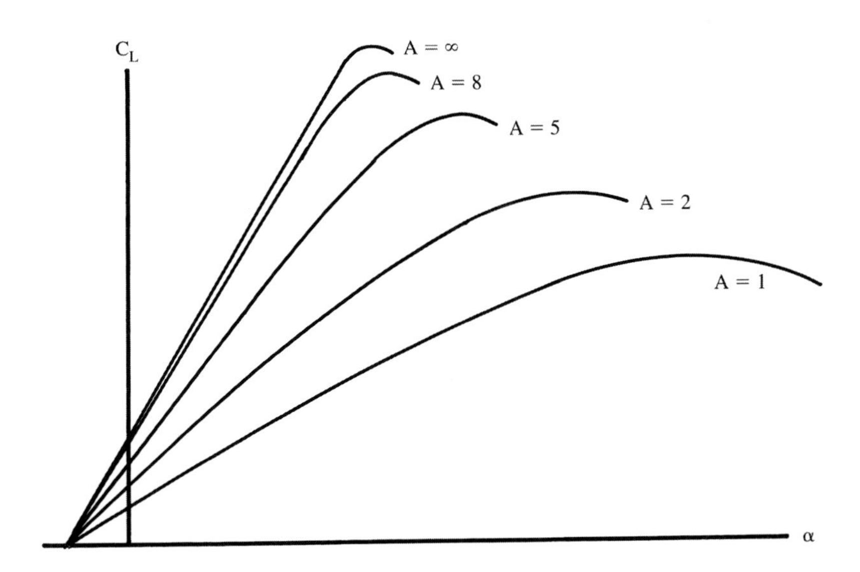

Fig. 4.19 Effect of aspect ratio on lift.

Conversely, a canard can be made to stall before the wing by making it a very high-aspect-ratio surface. This prevents the pilot from stalling the wing, and is seen in several canarded homebuilt designs.

Later in the design process, the aspect ratio will be determined by a trade study in which the aerodynamic advantages of a higher aspect ratio are balanced against the increased weight. For initial wing layout, the values and equations provided in Table 4.1 can be used. These were determined through statistical analysis of a number of aircraft, using data from Ref. 1.

Sailplane aspect ratio was found to be directly related to the desired glide ratio, which equals the L/D. Propeller aircraft showed no clear statistical trend, so that average values are presented. Jet aircraft show a strong trend of aspect ratio decreasing with increasing Mach number. This is probably due to drag-due-to-lift becoming relatively less important at higher speeds. Designers of high-speed aircraft thus use lower-aspect-ratio wings to save weight.

Note that, for statistical purposes, Table 4.1 uses an equivalent wing area that includes the canard area in defining the aspect ratio of an aircraft with a lifting canard. To determine the actual wing geometric aspect ratio, it is necessary to decide how to split the lifting area between the wing and canard. Typically, the canard will have about 10–25% of the total lifting area, and so the wing aspect ratio becomes the statistically determined aspect ratio divided by 0.9–0.75.

It is fairly common for the wing aspect ratio to be determined by a climb requirement, especially in the critical case of an engine failure for a multi-engine aircraft. If thrust is limited, a required rate of climb may be attained if the drag is reduced, which can be accomplished by an increase in aspect ratio. When

Table 4.1 Aspect ratio

Sailplane equivalent[a] aspect ratio $= 0.19 \, (\text{best } L/D)^{1.3}$

Propeller aircraft	Equivalent aspect ratio
Homebuilt	6.0
General aviation—single engine	7.6
General aviation—twin engine	7.8
Agricultural aircraft	7.5
Twin turboprop	9.2
Flying boat	8.0

Jet aircraft	Equivalent aspect ratio $= aM_{\max}^{C}$ a	C
Jet trainer	4.737	−0.979
Jet fighter (dogfighter)	5.416	−0.622
Jet fighter (other)	4.110	−0.622
Military cargo/bomber	5.570	−1.075
Jet transport	7.50 to 10	0

[a] Equivalent aspect ratio = wing span squared/(wing and canard areas)

the DC-10-20 was being developed, it was found that the weight increase over earlier models reduced the engine-out rate of climb below required values (Ref. 96). To fix this, the wing was extended by 10 ft {3 m}.

Consideration of the effect of aspect ratio on rate of climb requires more detailed analysis (given in Chapters 12 and 17). For now, the trends of Table 4.1 are reasonable.

Wing Sweep

Wing sweep is used primarily to reduce the adverse effects of transonic and supersonic flow. Theoretically, shock formation on a swept wing is determined not by the actual velocity of the air passing over the wing, but rather by the air velocity in a direction perpendicular to the leading edge of the wing. The distance from leading edge to trailing edge is shorter when measured perpendicular to the leading edge, so that its velocity appears slower, thus the shocks don't form! This result, first applied by the Germans during World War II, allows an increase in *critical Mach number* by the use of sweep.

At supersonic speeds the loss of lift associated with supersonic flow can be reduced by sweeping the wing leading edge aft of the Mach cone angle [arcsin(1/Mach no.)].

Figure 4.20 shows a historical trend line for wing leading-edge sweep vs Mach number. Note that sweep is defined aft of a line perpendicular to the flight direction, while the Mach angle is defined with respect to the flight direction. Thus, the

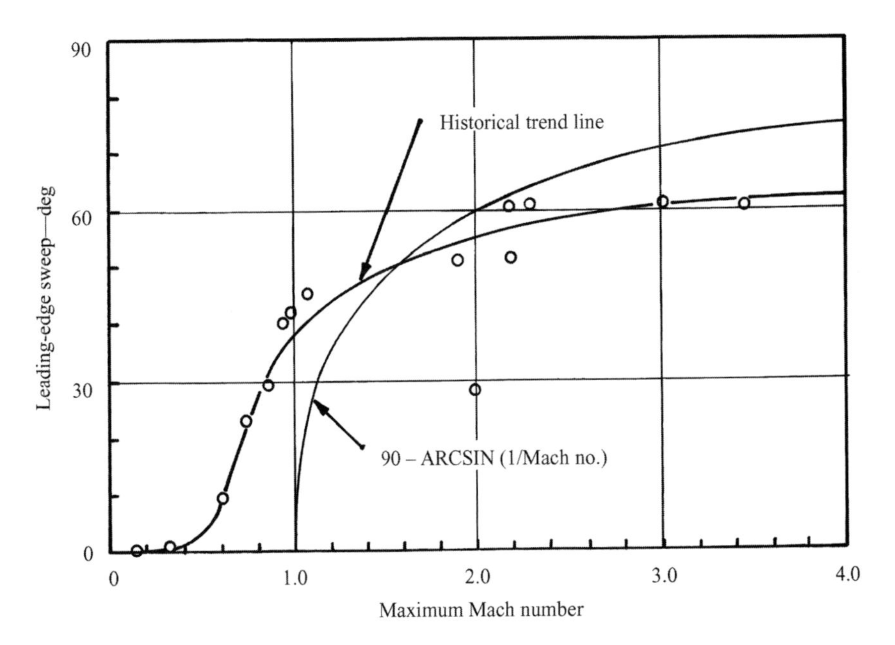

Fig. 4.20 Wing sweep historical trend.

line labeled "90-arcsin(1/Mach no.)" is the wing sweep required to place the wing leading edge exactly on the Mach cone.

The historical trend differs from this theoretical result for two reasons. In the high-speed range, it becomes structurally impractical to sweep the wing past the Mach cone. In this speed regime, over about Mach 2.5, it is necessary to use sharp or nearly sharp airfoils.

Selecting the wing sweep to equal the Mach-cone angle would indicate a zero sweep for speeds at or below Mach 1.0. However, in the transonic speed regime (roughly Mach 0.9–1.2) the desire for a high critical Mach number predominates. This requires subsonic airflow velocity over the airfoil (when measured perpendicular to the leading edge), and thus a swept wing.

The exact wing sweep required to provide the desired critical Mach number depends upon the selected airfoil(s), thickness ratio, taper ratio, and other factors. For initial wing layout the trend line of Fig. 4.20 is reasonable.

There is no theoretical difference between sweeping a wing aft and sweeping it forward. In the past, wings have been swept aft because of the structural divergence problem associated with forward sweep. With the use of composite materials, this can be avoided for a small weight penalty.

Also, there is no reason why one cannot sweep one wing aft and the other wing forward, creating an "oblique wing." This arrangement produces unusual control responses, but a computerized flight control system can easily provide normal handling qualities. The oblique wing also tends to have lower wave drag due to a better volume distribution (See Chapter 8, Sec. 8.2).

There are other reasons for sweeping a wing. For example, the fuselage layout may not otherwise allow locating the wing carry-through structure at the correct place for balancing the aircraft. Canarded aircraft with pusher engines are frequently tail-heavy, requiring wing sweep to move the aerodynamic center back far enough for balance. This is why most canard pushers have swept wings.

Wing sweep improves lateral stability. A swept wing has a natural dihedral effect. In fact, it is frequently necessary to use zero or negative dihedral on a swept wing to avoid excessive stability. Also, an aft-swept wing with some washout has additional pitch stability because the center of gravity must be moved forward for balance.

If an aircraft has its vertical tails at the wing tips, sweeping the wing will push the tails aft, increasing their effectiveness. This is also seen on many canard pusher aircraft.

Note on Fig. 4.20 the data point at Mach 2.0 and leading-edge sweep just under 30 deg. This is the Lockheed F-104, which used a different approach for reducing drag at supersonic speeds. The F-104 has a razor-sharp leading edge, so sharp that it was covered on the ground for the safety of line personnel. The F-104 also had a very thin wing, only 3.4% thick.

The wing sweep and aspect ratio together have a strong effect on the wing-alone pitchup characteristics. "Pitchup" is the highly undesirable tendency of some aircraft, upon reaching an angle of attack near stall, to suddenly and uncontrollably increase the angle of attack. The aircraft continues pitching up until it stalls and departs totally out of control. The F-16 fighter requires a computerized angle-of-attack limiter to prevent a severe pitchup problem at about 25-deg angle of attack.

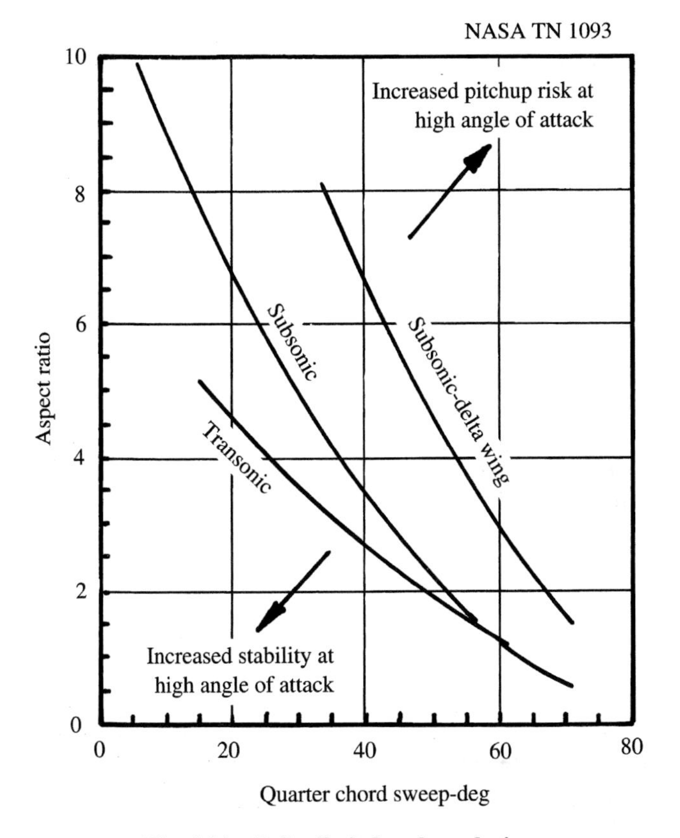

Fig. 4.21 Tail-off pitchup boundaries.

Figure 4.21 describes boundaries for pitchup avoidance for combinations of wing quarter-chord sweep angle and aspect ratio. Pitchup avoidance should be considered for military fighters, aerobatic aircraft, general-aviation aircraft, and trainers.

These boundaries may limit the allowable aspect ratio to a value less than that estimated earlier. However, Fig. 4.21 provides data for the wing alone. If a properly designed horizontal tail is used, the aspect ratio may be higher than that allowed by the graph. This is discussed later. Also, a large, all-moving canard such as that seen on the Grumman X-29 can be used to control a pitchup tendency. However, this requires a computerized flight control system.

For high-speed flight, a swept wing is desirable. For cruise as well as takeoff and landing, an unswept wing is desirable. A wing of variable sweep would offer the best of both worlds. Variable sweep was first flight-tested in the 1950s, and is now on several operational military aircraft including the F-111, F-14, B-1B, and the European Toronado and Soviet Backfire.

For design purposes, the planform for a variable-sweep aircraft should be developed in the unswept position, and then swept to the desired leading-edge angle for high-speed flight. The pivot position about which the wing is swept must be near the thickest part of the chord, between about the 30- and 40%-chord locations. Also, provisions must be made for smoothly fairing the wing root in both extended and fully swept positions.

Controlling the balance of a variable-sweep aircraft is a major design problem. When the wing swings aft, the aerodynamic center moves with it. The center of gravity also moves due to the wing movement, but not nearly as much as the aerodynamic center. To balance the aircraft, either fuel must be pumped to move the center of gravity, or the tail must provide a tremendous download (or both).

Yet another problem with the variable-sweep wing is the weight penalty associated with the pivot mechanism and less-than-optimal load paths. As shown in Table 3.1, variable sweep increases total empty weight roughly 4%. The detailed statistical weight equations of Chapter 15 show a 19% increase in the weight of the wing itself if it has variable sweep.

Taper Ratio

Wing taper ratio, λ, is the ratio between the tip chord and the centerline root chord. Most wings of low sweep have a taper ratio of about 0.4–0.5. Most swept wings have a taper ratio of about 0.2–0.3.

Taper affects the distribution of lift along the span of the wing. As proven by the Prandtl wing theory early in this century, minimum drag due to lift, or "induced" drag, occurs when the lift is distributed in an elliptical fashion. For an untwisted and unswept wing, this theoretically occurs when the wing planform is shaped as an ellipse, as shown in Fig. 4.22. This result was the basis of the graceful wing of the Supermarine Spitfire, a leading British fighter of World War II.

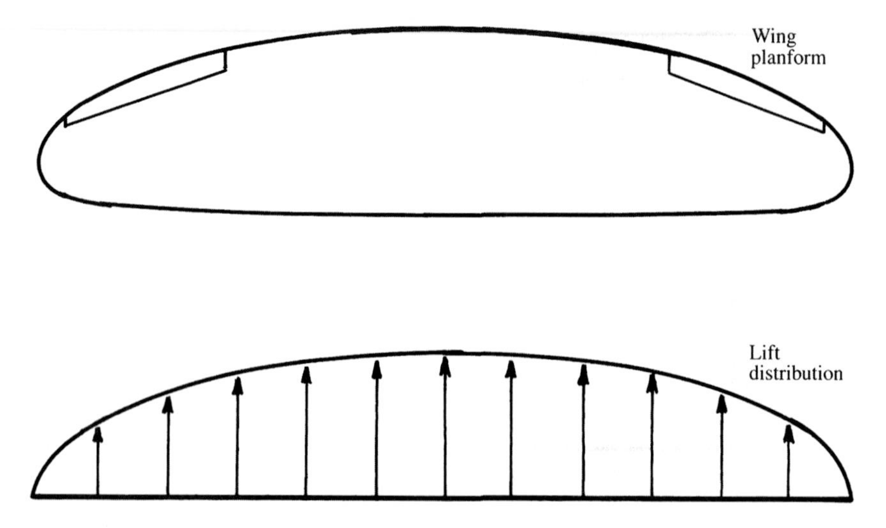

Fig. 4.22 Elliptical wing.

An elliptical wing planform is difficult and expensive to build. The easiest wing to build is the untapered ($\lambda = 1.0$) rectangular wing. However, the untapered wing has constant chord length along the span, and so has excessive chord toward the tip when compared to the ideal elliptical wing. This "loads up" the tip, causing the wing to generate more of its lift toward the tip than is ideal. The end result is that an untwisted rectangular wing has about 7% more drag due to lift than an elliptical wing of the same aspect ratio.

When a rectangular wing is tapered, the tip chords become shorter, alleviating the undesired effects of the constant-chord rectangular wing. In fact, a taper ratio of 0.45 almost completely eliminates those effects for an unswept wing, and produces a lift distribution very close to the elliptical ideal (Fig. 4.23). This results in a drag due to lift less than 1% higher than the ideal, elliptical wing. When the weight reduction from increased taper is taken into account, a taper ratio of about 0.4 is ideal for most unswept wings.

A wing swept aft tends to divert the air outboard, toward the tips. This loads up the tips, creating more lift outboard than for an equivalent unswept wing. To return the lift distribution to the desired elliptical lift distribution, it is necessary to increase the amount of taper (i.e., reduce the taper ratio, λ).

Figure 4.24 illustrates the results of NACA wind-tunnel tests to determine the taper ratio required to approximate the elliptical lift distribution for a swept untwisted wing. This figure can be used for a first approximation of the desired taper ratio for a swept wing. However, it should be noted that taper ratios much lower than 0.2 should be avoided for all but delta wings, as a very low taper ratio tends to promote tip stall.

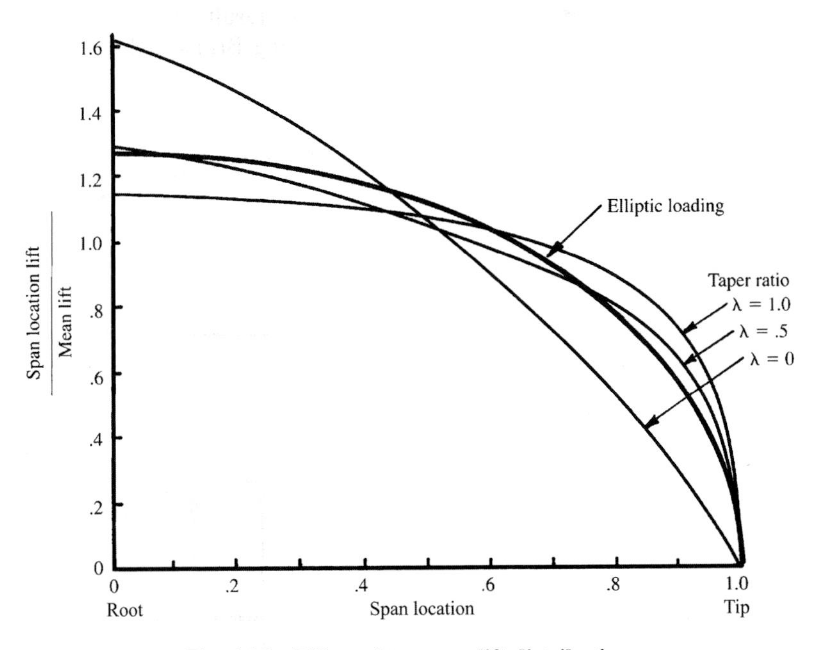

Fig. 4.23 Effect of taper on lift distribution.

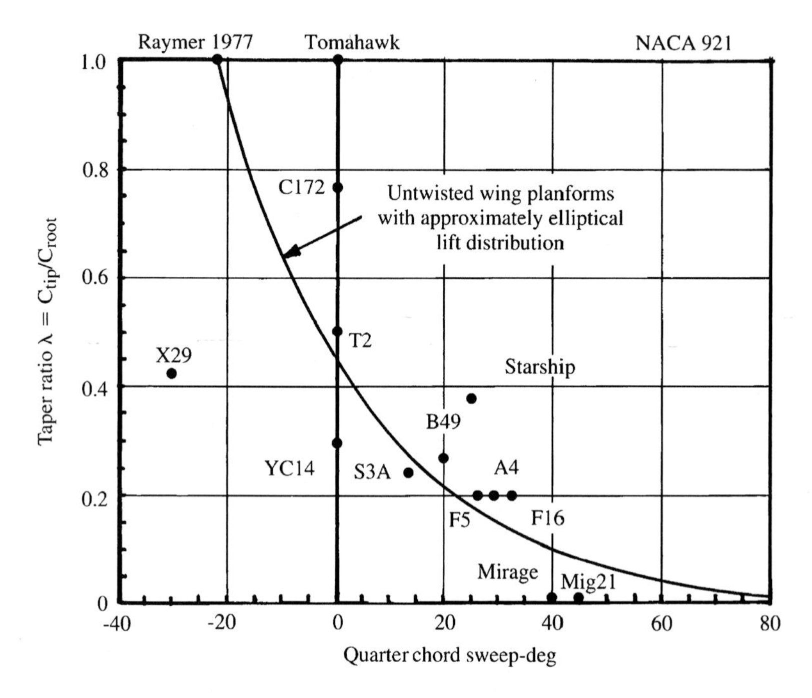

Fig. 4.24 Effect of sweep on desired taper ratio.

Figure 4.24 also indicates that an untwisted wing with no taper should have a forward sweep of 22 deg to approximate an elliptical lift distribution. This unusual planform was the basis of the design presented in the first section as Fig. 2.5. The intent was to provide an elliptical lift distribution with an easy-to-construct rectangular wing.

However, cost analysis indicated that the total reduction in manufacturing cost was small. Furthermore, the weight increase caused by the lack of wing thickness at the root when compared to a conventional, tapered wing caused this design to cost more than a regular design. (Well, at least it was an interesting trade study!)

The unusual Republic XF-91 of 1949 actually had reverse tapered wings. In other words, the wing tips had a greater chord than at the root, so that λ was greater than one! This was apparently intended to reduce wing-tip stalling at low speeds, and perhaps, to reduce wing-fuselage interference. Needless to say, it hasn't been attempted since!

Twist

Wing twist is used to prevent tip stall and to revise the lift distribution to approximate an ellipse. Typically, wings are twisted between 0 and 5 deg.

"Geometric twist" is the actual change in airfoil angle of incidence, usually measured with respect to the root airfoil. A wing whose tip airfoil is at a negative (nose-down) angle compared to the root airfoil is said to have "washout." A wing with washout will tend to stall at the root before the tip, which improves control during the stall and tends to reduce wing rock. If a wing has "linear twist," the twist angle changes in proportion to the distance from the root airfoil.

"Aerodynamic twist" is the angle between the zero-lift angle of an airfoil and the zero-lift angle of the root airfoil. If the identical airfoil is used from root to tip, the aerodynamic twist is the same as the geometric twist.

On the other hand, a wing with no geometric twist can have aerodynamic twist if, for example, the root airfoil is symmetric (zero-lift angle is zero), but the tip airfoil is highly cambered (zero-lift angle is nonzero). The total wing aerodynamic twist equals the wing geometric twist plus the root airfoil zero-lift angle, minus the tip airfoil zero-lift angle.

When wing twist is used to reshape the lift distribution, the change in lift at some chord station along the span is proportional to the ratio between the new airfoil angle of attack and the original one. Thus, the effect on lift distribution depends upon the original angle of attack of the wing, which in turn depends upon the lift coefficient at which the wing is flying.

In other words, any attempt to optimize the lift distribution by twisting the wing will be valid only at one lift coefficient. At other lift coefficients, the twisted wing will not get the whole benefit of the twist optimization. The more twist required to produce a good lift distribution at the design lift coefficient, the worse the wing will perform at other lift coefficients. It is for this reason that large amounts of twist (much over 5 deg) should be avoided.

It is very difficult to optimize twist for an arbitrary wing planform. A computerized solution is employed at large companies. For initial design purposes, historical data should be used. Typically, 3 deg of twist provides adequate stall characteristics.

Twist also changes the spanwise lift distribution because it changes the local angle of attack seen by each airfoil. This has an effect on the drag due to lift. If the optimum taper ratio is found as in Fig. 4.24 but washout is used to improve stall characteristics, the washout reduces lift at the tips so that the tip chord must be increased a bit and the root chord reduced. This means that the taper ratio as found in Fig. 4.24 must be increased, typically by about 0.1. For example, an unswept wing with typical twist should have a taper ratio of 0.55, not the 0.45 found from the figure. But as already described, structural weight effects can cause the optimum taper ratio to be a bit lower, perhaps 0.5.

Wing Incidence

The wing incidence angle is the pitch angle of the wing with respect to the fuselage. If the wing is untwisted, the incidence is simply the angle between the fuselage axis and the wing's airfoil chordlines. If the wing is twisted, the incidence is defined with respect to some arbitrarily chosen spanwise location of the wing, usually the mean aerodynamic chord or the root of the exposed wing where it intersects the fuselage. Frequently the incidence is given at the root and tip, which then defines the twist as the difference between the two.

Wing incidence angle is chosen to minimize drag at some operating condition, usually cruise. The incidence angle is chosen such that when the wing is at the correct angle of attack for the selected design condition, the fuselage is at the angle of attack for minimum total drag.

For a typical, circular straight fuselage, this is often a few degrees nose-up, allowing the fuselage to contribute to lift. For passenger aircraft, the incidence angle must be carefully chosen to ensure that the flight attendants do not have to push the food carts uphill!

Wing incidence angle is ultimately set using wind-tunnel data. For most initial design work, it can be assumed that general aviation and homebuilt aircraft will have an incidence of about 2 deg, transport aircraft about 1 deg, and military aircraft approximately zero. Later in the design process, aerodynamic calculations can be used to check the actual wing incidence angle required during the design condition.

These values are for untwisted wings. If the wing is twisted, the average incidence should equal these values.

A few aircraft have been built with a variable wing incidence angle. The wing aft-attachment is pivoted, and the forward attachment connects to a powerful actuator that pushes the front of the wing up for landing. This arrangement, seen on the Vought F8U Crusader aircraft, allows a short landing gear because the aircraft does not need to rotate to a high fuselage angle for additional lift during takeoff and landing. However, this arrangement is heavy and complicated, and has not been incorporated in a new design in several decades.

Dihedral

Wing dihedral is the angle of the wing with respect to the horizontal when seen from the front. Positive (tips higher) dihedral tends to roll the aircraft level whenever it is banked. This is frequently, and incorrectly, explained as the result of a greater projected area for the wing that is lowered.

Actually, the rolling moment is caused by a sideslip introduced by the bank angle. The aircraft "slides" toward the lowered wing, which increases its angle of attack and therefore its lift (Fig. 4.25). The resulting rolling moment is approximately proportional to the dihedral angle.

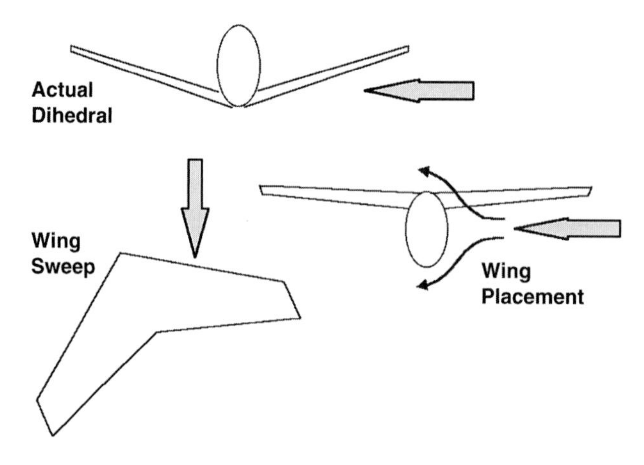

Fig. 4.25 Increased angle of attack and lift.

Table 4.2 Dihedral guidelines

	Wing position		
	Low	Mid	High
Unswept (civil)	5 to 7	2 to 4	0 to 2
Subsonic swept wing	3 to 7	−2 to 2	−5 to −2
Supersonic swept wing	0 to 5	−5 to 0	−5 to 0

Wing sweep also produces a rolling moment due to sideslip, caused by the change in relative sweep of the left and right wings. For an aft-swept wing, the rolling moment produced is negative and proportional to the sine of twice the sweep angle. This creates an effective dihedral that adds to any actual geometric dihedral.

Roughly speaking, 10 deg of sweep provides about 1 deg of effective dihedral. For a forward swept wing, the sweep angle produces a negative dihedral effect, requiring an increased geometric dihedral in order to retain natural rolling stability.

In addition, the position of the wing on the fuselage has an influence on the effective dihedral, with the greatest effect provided by a high wing. This is frequently, and incorrectly, explained as a pendulum effect.

Actually, the fuselage in sideslip pushes the air over and under itself. If the wing is high-mounted, the air being pushed over the top of the fuselage pushes up on the forward wing, providing an increased dihedral effect. The reverse is true for a low-mounted wing.

Due to the additive effects of sweep and wing position, many high-winged transports such as the Lockheed C-5 actually require a negative geometric dihedral angle to avoid an excess of effective dihedral. Excessive dihedral effect produces "Dutch roll," a repeated side-to-side motion involving yaw and roll. To counter a Dutch roll tendency, the vertical tail area must be increased, which increases weight and drag.

Unfortunately, as yet no simple technique for selecting dihedral angle takes all of these effects into account. Like so many parameters in initial design, the dihedral angle must be estimated from historical data and then revised following analysis of the design layout.

Table 4.2, developed by the author from data taken from Ref. 1, provides initial estimates of dihedral. For a wing in which the center section is flat and the outer sections alone have dihedral, a first approximation of the required dihedral for the outer panels is the one that places the wing tips as high as they would be for a wing with dihedral starting at the root.

Wing Vertical Location

The wing vertical location with respect to the fuselage is generally set by the real-world environment in which the aircraft will operate. For example, virtually all high-speed commercial transport aircraft are of low-wing design, yet military

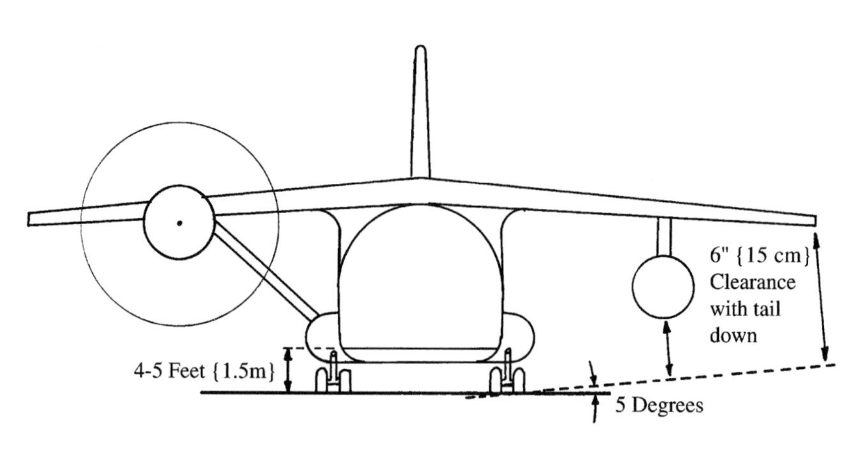

Fig. 4.26 High wing.

transport aircraft designed to similar mission profiles and payload weights are all of high-wing design. The reasons for this are discussed later.

The major benefit of a high wing is that it allows placing the fuselage closer to the ground (Fig. 4.26). For military transport aircraft such as the C-17, C-5, and C-141, this allows loading and unloading the cargo without special ground-handling gear. In fact, these aircraft place the floor of the cargo compartment about 4–5 ft {1.5 m} off the ground, which is the height of the cargo area of most trucks. If cargo is needed at a remote field lacking ground-handling gear, the trucks can be backed right up to the aircraft for loading.

With a high wing, jet engines or propellers will have sufficient ground clearance without excessive landing-gear length. Also, the wing tips of a swept high wing are not as likely to strike the ground when in a nose-high, rolled attitude. For these reasons, landing-gear weight is generally reduced for a high-wing aircraft.

For low-speed aircraft, external struts can be used to greatly lower wing weight. However, external struts add substantially to the drag. Since roughly two-thirds of the lift is contributed by the upper surface of the wing, it follows that less drag impact will be seen if the strut disturbs the airflow on the lower surface of the wing than if the strut is above the wing, as would be necesary for a strut-brace, low wing.

Another structural benefit occurs if the wing box is carried over the top of the fuselage rather than passing through it. When the wing box passes through the fuselage, the fuselage must be stiffened around the cut-out area. This adds weight to the fuselage. However, passing the wing box over the fuselage will increase drag due to the increase in frontal area.

For an aircraft designed with short takeoff and landing (STOL) requirements, a high wing offers several advantages. The high position allows room for the very large wing flaps needed for a high lift coefficient. The height of the wing above the ground tends to prevent "floating," where the ground effect increases lift as the aircraft approaches the ground. A floating tendency makes it difficult to touch down on the desired spot. Finally, most STOL designs are also intended to operate from unimproved fields. A high wing places the engines and propellers away from flying rocks and debris.

There are several disadvantages to the high-wing arrangement. While landing-gear weight tends to be lower than other arrangements, the fuselage weight is usually increased because it must be strengthened to support the landing-gear loads. In many cases an external blister is used to house the gear in the retracted position. This adds weight and drag. The fuselage is also usually flattened at the bottom to provide the desired cargo-floor height above ground. This flattened bottom is heavier than the optimal circular fuselage. If the top of the fuselage is circular, as shown in Fig. 4.26, a fairing is required at the wing-fuselage junction.

For small aircraft, the high wing arrangement can block the pilot's visibility in a turn, obscuring the direction toward which the aircraft is turning. Also, the high wing can block upward visibility in a climb. (A classic midair collision features a high-wing aircraft climbing into a low-wing one descending!) Many high-winged light aircraft have transparent panels in the roof to help the pilot see.

If the fuselage is roughly circular and fairings are not used, the midwing arrangement (Fig. 4.27) provides the lowest drag. High- and low-wing arrangements must use fairings to attain acceptable interference drag with a circular fuselage.

The midwing offers some of the ground clearance benefits of the high wing. Many fighter aircraft are midwinged to allow bombs and missiles to be carried under the wing. A high-wing arrangement would restrict the pilot's visibility to the rear—the key to survival of a fighter in combat.

The midwing arrangement is probably superior for aerobatic maneuverability. The dihedral usually required for adequate handling qualities in a low-wing design in normal flight will act in the wrong direction during inverted flight, making smooth aerobatic manuevers difficult. Also, the effective-dihedral contribution of either high or low wings will make it more difficult to perform high-sideslip maneuvers such as the knife-edge pass.

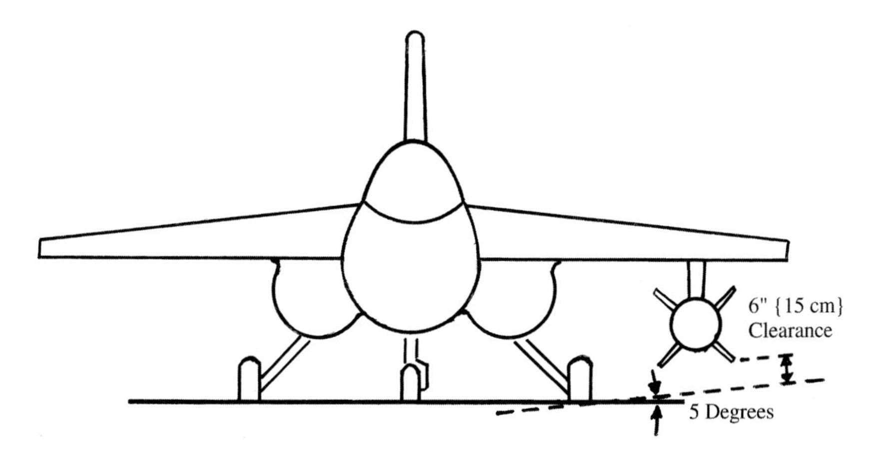

Fig. 4.27 Midwing.

Structural carrythrough presents the major problem with the midwing. As will be discussed in Chapter 8, the bending moment produced by the lift on the wing must be carried across the fuselage either by an extension of the wing box ("wing carrythrough box") or by a set of massive ring frames built into the fuselage.

The carrythrough box often proves lighter, but cannot be used in a midwing design that must carry cargo or passengers. (One exception to this, the German Hansa executive jet, uses a mild forward sweep to place the carrythrough box behind the passenger compartment.) A carrythrough box is also difficult to incorporate in a midwing fighter, in which most of the fuselage will be occupied by the jet engines and inlet ducts.

The major advantage of the low-wing approach (Fig. 4.28) comes in landing-gear stowage. With a low wing, the trunnion about which the gear is retracted can be attached directly to the wing box, which, being strong already, will not need much extra strengthening to absorb the gear loads. When retracted, the gear can be stowed in the wing itself, in the wing-fuselage fairing, or in the nacelle. This eliminates the external blister usually used with the high-wing approach.

To provide adequate engine and propeller clearance, the fuselage must be placed farther off the ground than for a high-wing aircraft. While this adds to the landing-gear weight, it also provides greater fuselage ground clearance. This reduces the aft-fuselage upsweep needed to attain the required takeoff angle of attack. The lesser aft-fuselage upsweep reduces drag.

While it is true that the low-wing arrangement requires special ground equipment for loading and unloading large airplanes, the high-speed commercial transports are only operated out of established airfields with a full complement of equipment. This is the main reason why military and commercial transports are so different.

Large transports have a fuselage diameter on the order of 20 ft {6 m}, which allows an uninterrupted passenger compartment above the wing carrythrough box. The wing carrythrough box usually passes through the fuselage for reduced drag, and splits the lower cargo compartment into two compartments.

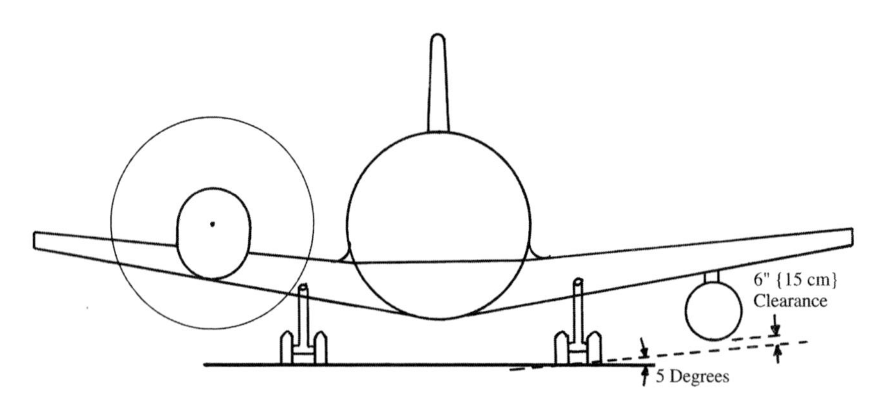

Fig. 4.28 Low wing.

This efficient internal fuselage layout is virtually standard for commercial transports.

If the center wing-panel of a low-wing aircraft lacks dihedral, a one-piece flap that passes under the fuselage can be used. This reduces complexity as well as the risk of asymmetric lift caused by the failure of one flap to extend. Also, the continuous flap will produce more lift and drag than an equal-area flap that is broken at the fuselage.

Several disadvantages of the low-wing approach have already been mentioned, including ground-clearance difficulties. Frequently low-wing aircraft will have dihedral angle set not by aerodynamics but by the angle required to avoid striking the wing tip on the ground during a bad landing. As was mentioned before, it may require an increase in vertical-tail size to avoid Dutch roll with an excessive dihedral angle.

Clearance also affects propellers. To minimize the landing-gear length, many low-wing aircraft have the propellers mounted substantially above the plane of the wing. This will usually increase the interference effects between the wing and propeller, and result in an increase in fuel consumption during cruise.

Wing Tips

Wing-tip shape has two effects upon subsonic aerodynamic performance. The tip shape affects the aircraft wetted area, but only to a small extent. A far more important effect is the influence the tip shape has upon the lateral spacing of the tip vortices. This is largely determined by the ease with which the higher-pressure air on the bottom of the wing can "escape" around the tip to the top of the wing.

A smoothly rounded tip (when seen nose-on) easily permits the air to flow around the tip. A tip with a sharp edge (when seen nose-on) makes it more difficult, thus reducing the induced drag. Most of the new low-drag wing tips use some form of sharp edge. In fact, even a simple cutoff tip offers less drag than a rounded-off tip, due to the sharp edges where the upper and lower surfaces end (Fig. 4.29).

The most widely used low-drag wing tip is the Hoerner wing tip (developed by S. Hoerner, Ref. 8). This is a sharp-edged wing tip with the upper surface continuing the upper surface of the wing. The lower surface is "undercut" and canted approximately 30 deg to the horizontal. The lower surface may also be "undercambered" (i.e., concave).

The "drooped" and "upswept" wing tips are similar to the Hoerner wing tip except that the tip is curved upward or downward to increase the effective span without increasing the actual span. This effect is similar to that employed by endplates, as discussed next.

The sweep of the wing tip also affects the drag. The tip vortex tends to be located approximately at the trailing edge of the wing tip, so an aft-swept wing tip, with a greater trailing-edge span, tends to have lower drag. However, the aft-swept wing tip tends to increase the wing torsional loads.

A cutoff, forward-swept wing tip is sometimes used for supersonic aircraft. The tip is cut off at an angle equal to the supersonic Mach-cone angle, because

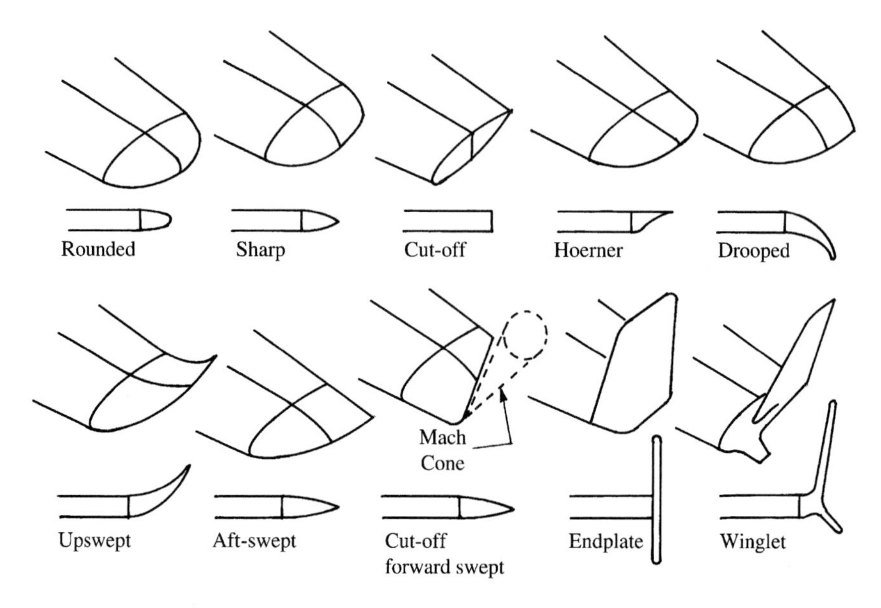

Fig. 4.29 Wing tips.

the area of the wing within the shock cone formed at the wing tip will contribute little to the lift. Also, this tip shape will reduce the torsional loads applied to the wing. The F-15 fighter uses such a cutoff tip for both wings and horizontal tails.

Induced drag is caused by the higher-pressure air at the bottom of the wing escaping around the wing tip to the top of the wing. An obvious way to prevent this would be to mount a vertical plate at the wing tip.

The endplate effect has been known almost since the dawn of flight, but has been seen rarely. The wetted area of the endplate itself creates drag. Also, an endplated wing has an effective span increase of only about 80% of the actual span increase caused by adding the endplates' height to the wing span. However, endplates can be useful when span must be limited.

An advanced version of the endplate can offer lower drag than an equal-area increase in wing span. The "winglet," designed by NASA's R. Whitcomb, gets an additional drag reduction by using the energy available in the tip vortex and can increase lift-to-drag ratio by up to 20%.

The winglet is cambered and twisted so that the rotating vortex flow at the wing tip creates a lift force on the winglet that has a forward component. This forward lift component acts as a "negative" drag, reducing the total wing drag.

A properly designed winglet can potentially provide an effective span increase up to double that bought by adding the winglets' height to the wing span. Winglets provide the greatest benefit when the wing-tip vortex is strong, so that a low-aspect-ratio wing will see more advantage from the use of winglets than an already-efficient high-aspect-ratio wing.

One problem with winglets is that they add weight behind the elastic axis of the wing, which can aggravate flutter tendencies. Also, the twist and camber of a winglet must be optimized for one velocity. At other than design speed, the winglet will provide less benefit.

For these and other reasons, winglets tend to be used more as add-on devices for existing wings requiring a little more efficiency without major redesign. When an all-new wing is being designed, it is usually better to rely upon increased aspect ratio to improve aerodynamic efficiency. This is not always true, and so a trade study should be conducted sometime during the conceptual design effort. Winglet design layout is presented in Chapter 7.

4.4 Biplane Wings

Biplanes dominated aviation for the first 30 years. The Wright Brothers were influenced by Octave Chanute, a noted architect and civil engineer who applied a structural concept used in bridge building to create lightweight biplane gliders. The early airfoils were thin and birdlike, requiring external bracing, and the biplane arrangement provided more structural efficiency than an externally braced monoplane.

With the thicker airfoils now in use, the biplane arrangement is mainly reserved for recreational purposes. However, it should be considered whenever low structural weight is more important to the design than aerodynamic efficiency, or when low speed is required without complicated high-lift devices or excessive wing span.

A biplane should theoretically produce exactly half the induced drag of a monoplane with equal span. Induced drag, or drag due to lift, is proportional to the square of the lift being generated. If that lift is split evenly between two wings, each wing should have only one-fourth of the drag of the original wing. Therefore, the total induced drag of a biplane should be two-fourths, or one-half of the value obtained with a monoplane of equal span.

Unfortunately, mutual-interference effects prevent the full benefit from being attained. Good design can yield on the order of a 30% reduction in drag due to lift for a biplane when compared to a monoplane of equal span. However, if the total wing area is held constant to provide the same wing loading for biplane and monoplane, and the monoplane has the same wing span as the biplane, then the aspect ratio of the two wings of the biplane must each be double the aspect ratio of the monoplane.

For a typical monoplane aspect ratio of seven, a biplane would need each wing to have an aspect ratio of 14 to maintain the same total wing area (and wetted area) while attaining the approximately 30% reduction in induced drag claimed above. Also, if the total wing area and span of a biplane and monoplane are identical, the biplane will have chord lengths half as long as the monoplane. Due to the Reynolds-number effect upon airfoil drag, an additional penalty will befall the biplane.

If a monoplane were designed with the same total wing area as the biplane but with an aspect ratio the same as each of the wings of the biplane, it would have a span 41% greater (square root of two, minus one) than the biplane. This would provide a net reduction in drag due to lift of about 31% when compared to the biplane of equal wing area (1−0.5/0.7). Thus, a biplane will actually provide a reduction in induced drag only if the aircraft's total span is limited for some reason to a value less than that desired for a monoplane.

Span can be limited for a number of reasons. For an aerobatic aircraft, a reduced span will increase the roll rate. For an aircraft flying at very low speeds, the wing area required to support the aircraft may require a wing span larger than practical from a structural viewpoint. Span can also be limited by the available hangar width. All of these reasons contributed to the prevalence of the biplane during World War I.

Biplane aerodynamic analysis using Prandtl's interference factor is described in Chapter 12. For initial design purposes, several key concepts should be considered. These are the "gap," "span ratio," "stagger," and "decalage."

Gap is the vertical distance between the two wings. If the gap were infinite, the theoretical result of a halving of the biplane induced drag when compared to an equal-span monoplane would be attained. However, structural weight and the drag of connecting struts generally limit the gap to a value approximately equal to the average chord length. A shorter gap will produce increasing interference between the two wings, raising the overall drag.

Span ratio is the ratio between the shorter wing and the longer wing. If both wings are the same length, the span ratio is one. When span is limited, the minimum induced drag is obtained from equal-length wings. As described, the only technical reason for using the biplane arrangement is the case where span is limited, and so the biplane with wings of unequal length should be rarely seen. However, a shorter lower wing has been used in the past to provide better ground clearance.

Stagger is the longitudinal offset of the two wings relative to each other. Positive stagger places the upper wing closer to the nose than the lower wing. Stagger has little or no effect upon drag, and is usually used to improve the visibility upward from a rear-located cockpit. Negative stagger was used in the beautiful Beech D-17 Staggerwing to improve visibility from an enclosed cabin cockpit and to reduce the pitching moment of the large flaps on the lower wing.

Decalage is the relative angle of incidence between the two wings of a biplane. Decalage is positive when the upper wing is set at a larger angle than the lower.* In early years much attention was paid to the selection of decalage to minimize induced drag while encouraging the forward wing to stall before the aft one, thus providing natural stall recovery. Most biplanes since World War I have been designed with zero decalage, although the Pitts Special, holder of numerous world aerobatic championships, has a positive decalage of 1.5 deg.

Much of the preceding discussions concerning the initial selection of wing geometry can be applied to biplane wings. Most biplanes have wing aspect ratios comparable to monoplanes of similar class (six to eight). As discussed, this yields induced drag levels much higher than obtained from a monoplane with similar wing loading. Taper ratios for biplanes can be selected as for a

---

* Some early sources define it the other way.

monoplane, although many biplanes have untapered wings for ease of manufacture.

One or both biplane wings can be swept to enhance stability, improve pilot visibility, or provide room for retractable landing gear. Biplanes typically have dihedral of about 2 deg. Aerobatic biplanes may apply dihedral only to the lower wing.

The mean aerodynamic chord of a biplane can be found as the weighted average of the mean chords of the two wings, weighted by the relative areas of the wings. The biplane aerodynamic center is at approximately 23% of the mean aerodynamic chord, rather than 25% as for a monoplane, due to the wing interference effects.

4.5 Tail Geometry and Arrangement

Tail Functions

Tails are little wings. Much of the previous discussion concerning wings can also be applied to tail surfaces. The major difference between a wing and a tail is that, while the wing is designed to routinely carry a substantial amount of lift, a tail is designed to operate normally at only a fraction of its lift potential. Any time in flight that a tail comes close to its maximum lift potential, and hence its stall angle, something is very wrong!

Tails provide for trim, stability, and control. Trim refers to the generation of a lift force that, by acting through some tail moment arm about the center of gravity, balances some other moment produced by the aircraft.

For the horizontal tail, trim primarily refers to the balancing of the moment created by the wing. An aft horizontal tail typically has a negative incidence angle of about 2–3 deg to balance the wing pitching moment. As the wing pitching moment varies under different flight conditions, the horizontal tail incidence is usually adjustable through a range of about 3 deg up and down.

Concerning the vertical tail, most aircraft are left–right symmetric, and so unbalanced aerodynamic yawing moments requiring trim are not created during normal flight. Propeller aircraft experience a yawing moment called "*p*-effect," which has several thrust-related causes. When the disk of the propeller is at an angle, such as during climb, the blade going downward has a higher angle of attack and is also at a slightly higher forward velocity. This condition produces higher thrust on the downward-moving side and hence a yawing moment away from that side. Also, the propeller tends to "drag" the air into a rotational corkscrew motion. The vertical tail is pushed on sideways by the rotating propwash causing a yawing moment, which adds to the *p*-effect. To counter *p*-effect many single-engine propeller airplanes have the vertical tail offset several degrees.

The vertical tails of multi-engine aircraft must be capable of providing sufficient trim in the event of an engine failure. This produces yawing both from lack of thrust on one side and the extra drag of the stopped or windmilling engine. For props, engine-out yaw is especially severe when the engine that is still running has its downward-traveling blade on the side away from the fuselage.

Some multi-engine aircraft have counter-rotating propellers to minimize the engine-out yawing.

The tails are also a key element of stability, acting much like the fins on an arrow to restore the aircraft from an upset in pitch or yaw. Although it is possible to design a stable aircraft without tails, such a design is usually penalized in some other area, as discussed in Chapter 20.

The other major function of the tail is control. The tail must be sized to provide adequate control power at all critical conditions. These critical conditions for the horizontal tail or canard typically include nosewheel liftoff, low-speed flight with flaps down, and transonic maneuvering. For the vertical tail, critical conditions typically include engine-out flight at low speeds, maximum roll rate, and spin recovery.

Note that control power depends upon the size and type of the movable surface as well as the overall size of the tail itself. For example, several airliners use double-hinged rudders to provide more engine-out control power without increasing the size of the vertical tail beyond what is required for Dutch-roll damping. Several fighters, including the YF-12 and the F-107, have used all-moving vertical tails instead of separate rudders to increase control power.

Preliminary methods for sizing tails are provided in Chapter 6, and stability and control analysis methods are provided in Chapter 16.

Tail Arrangement

Figure 4.30 illustrates some of the possible aft-tail arrangements. The first shown has become "conventional" for the simple reason that it works. For

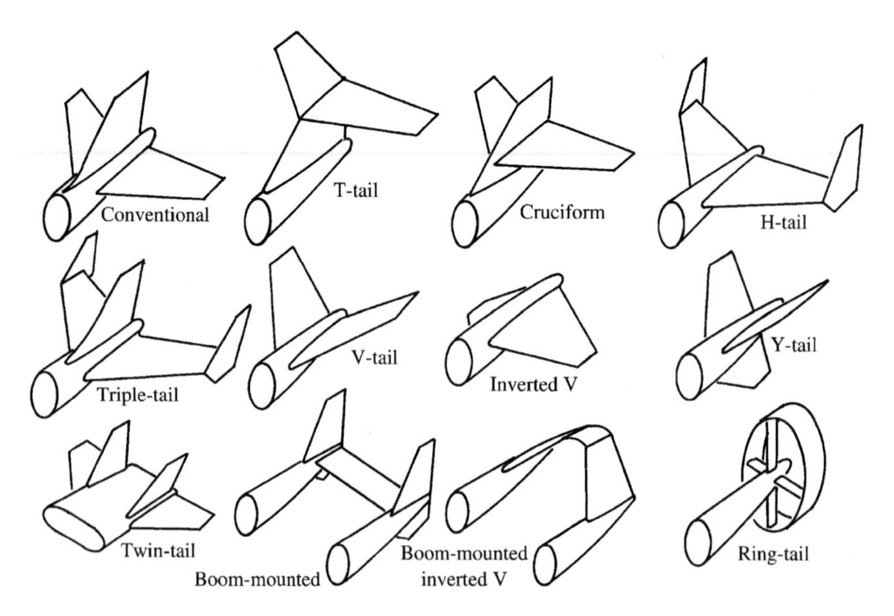

Fig. 4.30 Aft tail variations.

most aircraft designs, the conventional tail will usually provide adequate stability and control at the lightest weight. Probably 70% or more of the aircraft in service have such a tail arrangement. However, there are many reasons for considering others.

The "T-tail" is also widely used. A T-tail is inherently heavier than a conventional tail because the vertical tail must be strengthened to support the horizontal tail, but the T-tail provides compensating advantages in many cases.

Due to end-plate effect, the T-tail allows a smaller vertical tail. The T-tail lifts the horizontal tail clear of the wing wake and propwash, which makes it more efficient and hence allows reducing its size. This also reduces buffet on the horizontal tail, which reduces fatigue for both the structure and the pilot.

In jet transport aircraft such as the DC-9 and B-727, the T-tail allows the use of engines mounted in pods on the aft fuselage. Finally, the T-tail is considered stylish, which is not a trivial consideration.

The cruciform tail, a compromise between the conventional and T-tail arrangements, lifts the horizontal tail to avoid proximity to a jet exhaust (as on the B-1B), or to expose the lower part of the rudder to undisturbed air during high angle-of-attack conditions and spins. These goals can be accomplished with a T-tail, but the cruciform tail will impose less of a weight penalty. However, the cruciform tail will not provide a tail-area reduction due to endplate effect as will a T-tail.

The "H-tail" is used primarily to position the vertical tails in undisturbed air during high angle-of-attack conditions (as on the T-46) or to position the rudders in the propwash on a multiengine aircraft to enhance engine-out control. The H-tail is heavier than the conventional tail, but its endplate effect allows a smaller horizontal tail.

On the A-10, the H-tail serves to hide the hot engine nozzles from heat-seeking missiles when viewed from an angle off the rear of the aircraft. H-tails and the related triple-tails have also been used to lower the tail height to allow an aircraft such as the Lockheed Constellation to fit into existing hangars.

The "V-tail" (Fig. 4.31) is intended to reduce wetted area. With a V-tail, the horizontal and vertical tail forces are the result of horizontal and vertical projections of the force exerted upon the "*V*" surfaces. For some required horizontal and vertical tail area, the required *V* surface area would theoretically be found from the Pythagorean theorem, and the tail dihedral angle would be found as the arctangent of the ratio of required vertical and horizontal areas. The resulting wetted area of the *V* surfaces would clearly be less than for separate horizontal and vertical surfaces.

However, extensive NACA research (Ref. 3) has concluded that to obtain satisfactory stability and control, the *V* surfaces must be upsized to about the same total area as would be required for separate horizontal and vertical surfaces. Even without the advantage of reduced wetted area, V-tails offer reduced interference drag but at some penalty in control-actuation complexity, as the rudder and elevator control inputs must be blended in a "mixer" to provide the proper movement of the V-tail "ruddervators."

When the right rudder pedal of a V-tail aircraft is pressed, the right ruddervator deflects downward, and the left ruddervator deflects upward. The combined

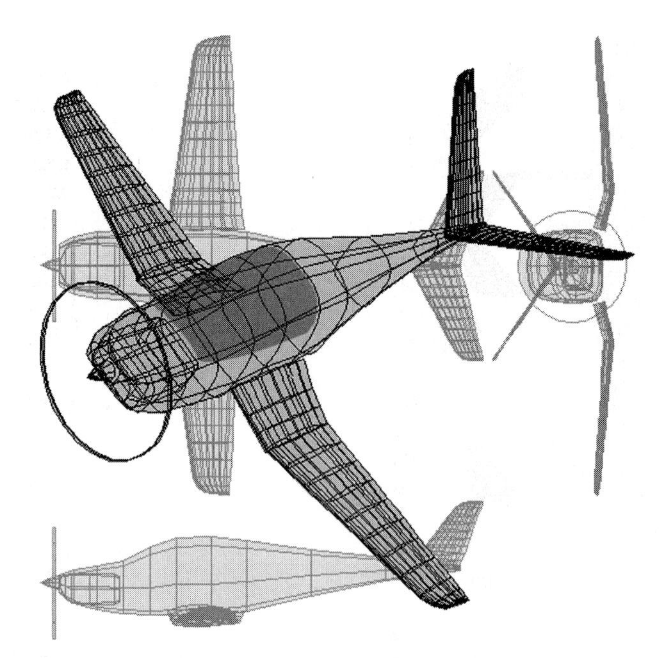

Fig. 4.31 Notional V-Tail Gullwing Homebuilt (D. Raymer 2005).

forces push the tail to the left, so the nose goes to the right as desired. However, the ruddervators also produce a rolling moment toward the left—in opposition to the desired direction of turn, an action called "adverse roll–yaw coupling."

The inverted V-tail shown in Fig. 4.32 avoids this problem, and instead produces a desirable "proverse roll–yaw coupling." The inverted V-tail is also said to reduce spiraling tendencies. This tail arrangement can cause difficulties in providing adequate ground clearance.

The "Y-tail" is similar to the V-tail, except that the dihedral angle is reduced and a third surface is mounted vertically beneath the *V*. This third surface contains the rudder, whereas the *V* surfaces provide only pitch control. This tail arrangement avoids the complexity of the ruddervators while reducing interference drag when compared to a conventional tail. An inverted Y-tail is used on the F-4, primarily to keep the horizontal surfaces out of the wing wake at high angles of attack.

Twin tails on the fuselage can position the rudders away from the aircraft centerline, which may become blanketed by the wing or forward fuselage at high angles of attack. Also, twin tails have been used simply to reduce the height required with a single tail. Twin tails are usually heavier than an equal-area centerline-mounted single tail, but are often more effective. Twin tails are seen on most large modern fighters such as the F-14, F-15, F-18, and MiG-25.

Boom-mounted tails have been used to allow pusher propellers or to allow location of a heavy jet engine near the center of gravity. Tail booms are typically

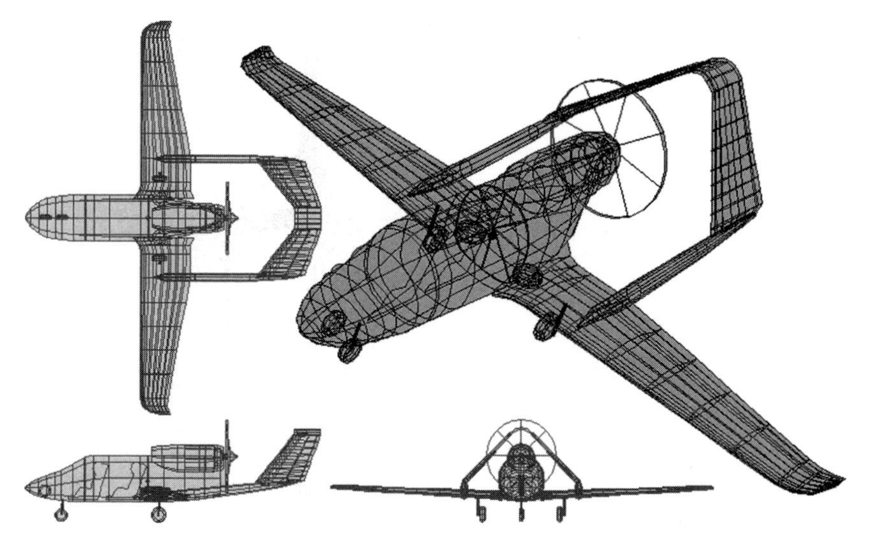

Fig. 4.32 Notional Inverted-V Pusher (D. Raymer 2005).

heavier than a conventional fuselage construction, but can be desirable in some applications.

Boom-mounted tails can have a midmounted horizontal tail or a high horizontal, as on the Cessna Skymaster. Also, the inverted V-tail arrangement can be used with tail booms. The unmanned NASA HiMat research aircraft used boom-mounted verticals with no connecting horizontal tail, instead relying on a canard for pitch control.

The "ring-tail" concept attempts to provide all tail contributions via an airfoil-sectioned ring attached to the aft fuselage, usually doubling as a propeller shroud. While conceptually appealing, the ring-tail has proven inadequate in application. The ring-tail JM-2 raceplane was ultimately converted to a T-tail.

The location of an aft horizontal tail with respect to the wing is critical to the stall characteristics of the aircraft. If the tail enters the wing wake during the stall, control will be lost and pitchup may be encountered. Several T-tailed aircraft encountered "deep stall" from which they could not be extricated. One T-Tail trainer was recently found to be three to seven times more likely to have a stall/spin accident than other similar trainers.

Figure 4.33 illustrates the boundaries of the acceptable locations for a horizontal tail to avoid this problem. Note that low tails are best for stall recovery. Also notice that a tail approximately in line with the wing is acceptable for a subsonic aircraft, but may cause problems at supersonic speeds due to the wake of the wing.

A T-tail requires a wing designed to avoid pitchup without a horizontal tail, as described by Fig. 4.21. This requires an aircraft stable enough to recover from a stall even when the tail is blanketed by the wing wake. Several general-aviation aircraft use this approach, which has the added benefit of a positive warning to

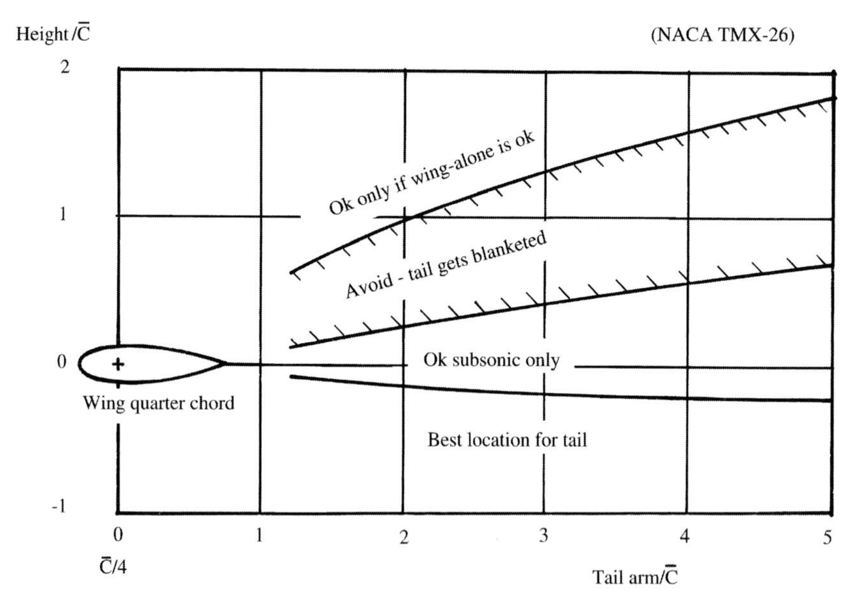

Fig. 4.33 Aft tail positioning.

the pilot of impending stall caused by buffeting on the tail as it enters the wing wake at high angle of attack.

Other possible tail arrangements are depicted in Fig. 4.34. Canards were used by the Wright Brothers but soon fell out of favor because of the inherent difficulty of providing sufficient stability. The early Wright airplanes were quite unstable and required a well-trained pilot with quick reflexes. Movie footage taken by passengers shows the Wright canards being continuously manipulated from almost full-up to full-down as the pilot responded to gusts.

The canard configuration has several advantages. One obvious advantage, and perhaps the reason it was used by the Wrights, is that it places the control surface in a region of undisturbed flow where its control response is sure and predictable. An aft tail is always flying in air that has been disturbed by its passage over the wing and fuselage and, in the case of the Wright Flyer, through the bracing wires and propellers.

There is an important canard advantage for an aircraft that relies on a manual flight control system—improved stall safety. The canard itself can be designed to stall before the wing so that the nose lowers before the plane can get into trouble. This is the main reason why Burt Rutan and others have used canards on home-built designs such as the VariEze. To make the canard stall first, the center of gravity of the aircraft is located far forward, and the canard surface is designed with a higher aspect ratio than the wing to encourage it to stall first. However, this results in other penalties as described next.

The canard configuration can be used to avoid pitch-up, wherein the nose comes up to a high angle and the pilot cannot put it back down. An all-moving

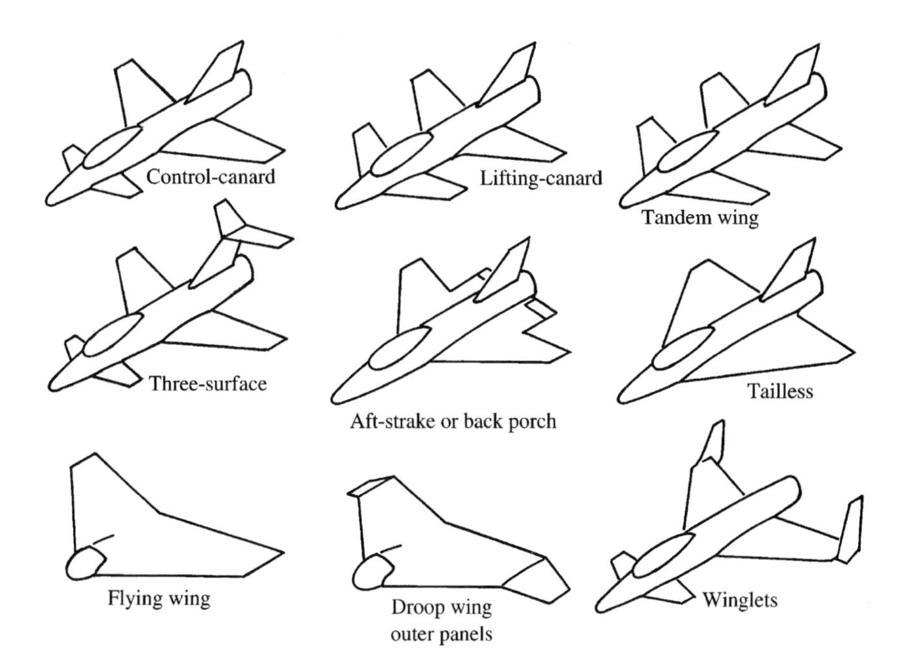

Fig. 4.34 Other tail configurations.

canard capable of downward deflections of 45 deg or more can be used to put the nose back down under almost any situation. This permits optimizing the wing's aspect ratio and sweep without compromising for pitch-up avoidance (see Fig. 4.21), but requires a sophisticated flight control system. Such an approach was used on the X-31, capable of flight at 90-deg angle of attack.

A subtle aerodynamic benefit can be obtained with a canard configuration. If both wing and canard are highly swept, the canard vortex can be made to interact with the leading-edge vortex on the wing, increasing its strength and therefore augmenting its lift. This beneficial interference is very geometry dependent and is difficult to predict. The SAAB Viggen and Rockwell HiMat both used this effect.

Canard advocates sometimes claim a lift and trim drag advantage for the canard configuration when compared to an aft tail. It is quite true that the canard's lift reduces the lift that must be produced by the wing. This permits a smaller wing and, all else being equal, would reduce the wing's drag due to lift. Traditional aft-tail designs frequently fly with a download on the tail to produce natural stability, which increases the amount of lift that the wing must produce and therefore increases its drag due to lift and increases trim drag.

However, a modern and sophisticated aft-tail aircraft is designed to a slight level of instability so that it normally flies with an upload, not a download on its tail. This is the very reason that computerized flight control systems with artificial stability were developed and put into production, first on the

F-16 over 30 years ago. Because modern high-performance canard designs also require artificial stability, it is misleading to compare an old "download-on-the-tail" design with the modern canard design. When comparing two noncomputerized designs, other canard disadvantages as described next tend to obliterate this supposed advantage.

There are actually two distinct classes of canard: the control-canard and the lifting-canard. In the control-canard, the wing carries most of the lift, and the canard is used primarily for control as is the case for an aft-tail design. Modern high-performance canard aircraft such as the Grippen, Typhoon, X-29, and X-31 are of this type.

On a control-canard aircraft, the canard is used to control the angle of attack of the wing and to balance out the pitching moment produced by deflection of the wing flaps. Such designs are highly unstable in pitch due to the forward canard, but are designed to be roughly neutral in stability with the canard removed. This implies that the canard normally operates at about zero angle of attack and thus carries little of the aircraft's weight. This instability requires a computerized flight control system that rapidly changes the angle of the canard, much as the pilot of an early Wright aircraft was forced to do manually.

In contrast, a lifting-canard aircraft uses both the wing and the canard to provide lift under normal flight conditions. A lifting-canard surface will usually have a higher aspect ratio and greater airfoil camber than a control-canard, to reduce the canard's drag due to lift. In other words, a lifting-canard must be a good wing as well as a tail.

In the extreme, the lifting-canard surface is as big as the wing—a tandem wing design. The supposed benefit of the tandem wing is a theoretical 50% reduction in the drag due to lift. This, also called induced drag, is a function of the square of the lift being produced. If the weight of the aircraft is evenly distributed to two wings, each wing would have only one-fourth of the induced drag of a single wing. Thus, the sum of the induced drags of the two wings should be one-half of the drag of a single wing—theoretically.

As with the supposed reduction in drag for biplane wings, this theoretical result is not seen in practice. First, for it to be even theoretically true the two wings would have to have the same span as the original, double-the-area wing. They would not be photographically scaled small versions of the original wing, but would have to have aspect ratios twice as high. The weight penalty is obvious, and therefore tandem wing designs usually have "normal" aspect ratios thus invalidating the one supposed benefit.

However, the fundamental problem is even simpler—the second wing must fly in the downwash of the first wing. This requires a higher angle of incidence on the second wing, but even more important, the direction of lift is turned. Lift is always perpendicular to the local flow direction, and, as can be seen in Fig. 4.35, that direction has been turned by the front wing. The lift of the back-wing therefore has a component to the rear that is a newly created drag term!

An even bigger problem is this: pitch stability requires that if the nose comes up, moments are created that push the nose back down. In a tandem wing design the front wing sees the full increase in angle of attack that results when the nose comes up and it generates additional lift. It also turns the flow,

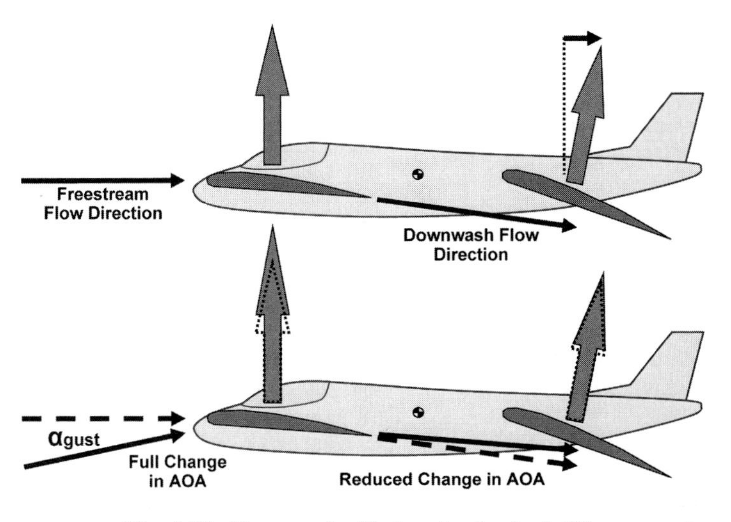

Fig. 4.35 Downwash effect on back wing's lift.

so that the back wing does not see the full increase in angle of attack, and so it does not produce nearly as much extra lift.

With the full extra lift in front and less extra lift in back, a nose-up moment is created that is the exact opposite of what is needed. The only way to get natural pitch stability in a tandem wing design is to locate the center of gravity substantially forward of the location that would provide for an even weight split. Thus, the aft wing is "lazy"—it carries much less than an equal share of the aircraft's weight. To create the required lift for flight, the total area of the two lifting surfaces must be increased substantially.

This problem is worse when flaps are to be used. Flaps on the back of the front wing are very near the center of gravity, and so they produce only a small nose-up pitching moment. Flaps on the back of the rear wing are far from the c.g., and so they produce a huge nose-down moment—it is literally impossible to balance the aircraft. For this reason tandem wing designs cannot normally use flaps on the back wing. Without flaps on all lifting surfaces, those surfaces must be made larger to meet the stall speed requirement.

Sometimes the tandem wing arrangement is useful for other reasons, such as to efficiently carry a large and bulky load from its ends. This author used the tandem wing arrangement to carry and air launch an ICBM in a never-built design done for Rockwell, and the recent Scaled Composites White Knight that carries SpaceShipOne between tandem wings.

To minimize the penalties of a tandem wing design, it is desirable to separate the two wings as far apart as possible, both horizontally and vertically. It is best if the front wing is low and the back wing is high, to maximize the vertical separation of the downwash from the back wing. However, watch out that the disturbed air shed when the front wing stalls does not go over the rear wing and make it stall too.

The lifting-canard is actually a tandem wing design with a similar front wing. It therefore suffers the same penalties caused by the downwash from the front wing (i.e., the canard). The center of gravity must be far forward for stability, the rear wing is lazy requiring a greater total area, and it is difficult to use flaps on the rear wing. To allow at least some flaps on the wing, several designs have resorted to the use of slotted canard flaps or even a canard with variable sweep (as on the Beech Starship).

The efficiency of the lifting-canard design is improved if the center of gravity can be moved farther to the rear, but this reduces stability and eventually requires a computerized flight control system. The farther back the center of gravity is placed, the greater the efficiency and the less of the lift is carried by the canard until finally it becomes a control-canard!

A three-surface arrangement includes both aft-tail and lifting-canard surfaces. This allows use of the canard for efficient trim and pitch control without the difficulty of incorporating wing flaps as seen on a canard-only configuration.

The three-surface aircraft theoretically offers minimum trim drag. A canard or aft tail, when generating lift for trim purposes, will change the aircraft total lift distribution, which increases total induced drag. On a three-surface configuration the canard and aft tail can act in opposite directions, thus cancelling out each other's effect upon the total lift distribution. For example, to generate a nose-up trim the canard can generate an upward lift force while the tail generates an equal downward lift force. The combined effect upon total lift distribution would then be zero.

However, this reduction in trim drag is a theoretical far-field effect and might not be fully realized in an actual design. The drawback of the three-surface arrangement is the additional weight, complexity, and interference drag associated with the extra surfaces.

The "back-porch" or "aft-strake" is a horizontal control surface that is incorporated into a faired extension of the wing or fuselage. This device, seen on the X-29, is mostly used to prevent pitchup but can also serve as a primary pitch control surface in some cases.

The tailless configuration offers the lowest weight and drag of any tail configuration, if it can be made to work. For a stable aircraft, the wing of a tailless aircraft must be reflexed or twisted to provide natural stability. This reduces the efficiency of the wing.

For an unstable aircraft with a computerized flight control system, this need not be done. In fact, an unstable, tailless aircraft can be designed to be "self-trimming," meaning that the wing trailing-edge flap angles required to balance the aircraft at different speeds and angles of attack can be designed to be almost exactly the optimal flap angles for maximum L/D.

This is very difficult to accomplish, and is very sensitive to the location of the center of gravity. In fact, all tailless designs are sensitive to center-of-gravity location, and are most successful in designs in which the expendable fuel and payload are located very close to the empty center of gravity.

The vertical tail can also be eliminated for reduced weight and drag. However, the fully tailless (flying-wing) design is probably the most difficult configuration to stabilize, either naturally or by computer. Fully tailless designs must rely exclusively upon wing control surfaces for control, unless vectored thrust is provided. Rudder control is usually provided by wing-tip-mounted drag devices.

Some fully tailless designs utilize drooped outer wing panels for stability and control enhancement. These act somewhat like an inverted V-tail and provide the desirable proverse roll-yaw coupling with rudder deflection.

Winglets or endplates mounted at the wing tips can be used in place of a vertical tail. This may provide the required vertical tail surface for free, since the effective increase in wing aspect ratio may more than compensate for the wetted area of the tail. To place these tip surfaces far enough aft to act like vertical tails requires either extreme wing sweep or a canard arrangement, or both.

Tail Arrangement for Spin Recovery

The vertical tail plays a key role in spin recovery. An aircraft in a spin is essentially falling vertically and rotating about a vertical axis, with the inside wing fully stalled. The aircraft is also typically at a large sideslip angle. To recover from the spin requires that the wing be unstalled, so the angle of attack must be reduced. However, first the rotation must be stopped and the sideslip angle reduced, or the aircraft will immediately enter another spin. This requires adequate rudder control even at the high angles of attack seen in the spin.

Figure 4.36 illustrates the effect of tail arrangement upon rudder control at high angles of attack. At high angles of attack the horizontal tail is stalled, producing a turbulent wake extending upward at approximately a 45-deg angle.

In the first example, the rudder lies entirely within the wake of the horizontal tail, so little rudder control is available. The second example shows the effect of moving the horizontal tail forward with respect to the vertical tail. This

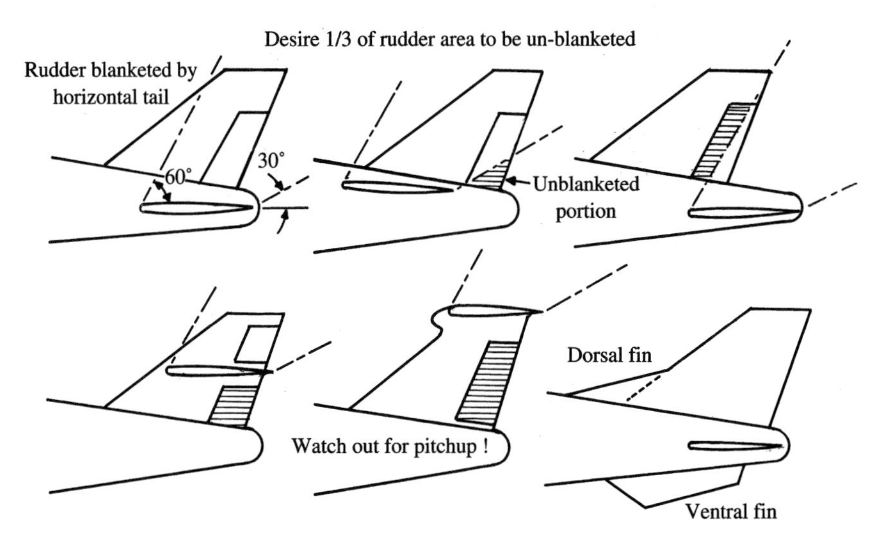

Fig. 4.36 Tail geometry for spin recovery.

"uncovers" part of the rudder, improving rudder control. The next example moves the horizontal tail aft with respect to the vertical tail, with the same result. As a rule of thumb, at least a third of the rudder should be out of the wake.

The next two examples show the effect of moving the horizontal tail upward. The T-tail arrangement completely uncovers the rudder, but can result in pitchup and loss of elevator control.

The last illustration in Fig. 4.36 shows the use of dorsal and ventral fins. The dorsal fin improves tail effectiveness at high angles of sideslip by creating a vortex that attaches to the vertical tail. This tends to prevent the high angles of sideslip seen in spins, and augments rudder control in the spin. The ventral tail also tends to prevent high sideslip, and has the extra advantage of being where it cannot be blanketed by the wing wake. Ventral tails are also used to avoid lateral instability in high-speed flight.

Tail Geometry

The surface areas required for all types of tails are directly proportional to the aircraft's wing area, so the tail areas cannot be selected until the initial estimate of aircraft takeoff gross weight has been made. The initial estimation of tail area is made using the "tail volume coefficient" method, which will be discussed in Chapter 6.

Other geometric parameters for the tails can be selected at this time. Tail aspect ratio and taper ratio show little variation over a wide range of aircraft types. Table 4.3 provides guidance for selection of tail aspect ratio and taper ratio. Note that T-tail aircraft have lower vertical-tail aspect ratios to reduce the weight impact of the horizontal tail's location on top of the vertical tail. Also, some general-aviation aircraft use untapered horizontal tails ($\lambda = 1.0$) to reduce manufacturing costs.

Leading-edge sweep of the horizontal tail is usually set to about 5 deg more than the wing sweep. This tends to make the tail stall after the wing, and also provides the tail with a higher critical Mach number than the wing, which avoids loss of elevator effectiveness due to shock formation. For low-speed aircraft, the horizontal tail sweep is frequently set to provide a straight hinge line for the elevator, which usually has the left and right sides connected to reduce flutter tendencies.

Table 4.3 Tail aspect ratio and taper ratio

	Horizontal tail		Vertical tail	
	A	λ	A	λ
Fighter	3–4	0.2–0.4	0.6–1.4	0.2–0.4
Sail plane	6–10	0.3–0.5	1.5–2.0	0.4–0.6
Others	3–5	0.3–0.6	1.3–2.0	0.3–0.6
T-Tail	—	—	0.7–1.2	0.6–1.0

Vertical-tail sweep varies between about 35 and 55 deg. For a low-speed aircraft, there is little reason for vertical-tail sweep beyond about 20 deg other than aesthetics. For a high-speed aircraft, vertical-tail sweep is used primarily to ensure that the tail's critical Mach number is higher than the wing's.

The exact planform of the tail surfaces is actually not very critical in the early stages of the design process. The tail geometries are revised during later analytical and wind-tunnel studies. For conceptual design, it is usually acceptable simply to draw tail surfaces that "look right," based upon prior experience and similar designs.

Tail thickness ratio is usually similar to the wing thickness ratio, as determined by the historical guidelines provided in the wing-geometry section. For a high-speed aircraft, the horizontal tail is frequently about 10% thinner than the wing to ensure that the tail has a higher critical Mach number.

Note that a lifting canard or tandem wing should be designed using the guidelines and procedures given for initial wing design, instead of the tail-design guidelines already described.

Takeoff of the supersonic–capable F86 (U.S. Air Force photo).

F-35 Joint Strike Fighter (U.S. Air Force photo).

5
Thrust-to-Weight Ratio and Wing Loading

5.1 Introduction

The thrust-to-weight ratio (T/W) and the wing loading (W/S) are the two most important parameters affecting aircraft performance. Optimization of these parameters forms a major part of the analytical design activities conducted after an initial design layout. The methods used for this optimization are described in Chapter 19.

However, it is essential that a credible estimate of the wing loading and thrust-to-weight ratio be made before the initial design layout is begun. Otherwise, the optimized aircraft may be so unlike the as-drawn aircraft that the design must be completely redone.

For example, if the wing loading used for the initial layout is very low, implying a large wing, the designer will have no trouble finding room for the landing gear and fuel tanks. If later optimization indicates the need for a much higher wing loading, the resulting smaller wing may no longer hold the landing gear and fuel. Although they could be put in the fuselage, this would increase the wetted area and therefore the drag, and so the optimization results would probably be no good.

Wing loading and thrust-to-weight ratio are interconnected for a number of performance calculations, such as takeoff distance, which is frequently a critical design driver. A requirement for short takeoff can be met by using a large wing (low W/S) with a relatively small engine (low T/W). While the small engine will cause the aircraft to accelerate slowly, it only needs to reach a moderate speed to lift off the ground.

On the other hand, the same takeoff distance could be met with a small wing (high W/S) provided that a large engine (high T/W) is also used. In this case, the aircraft must reach a high speed to lift off, but the large engine can rapidly accelerate the aircraft to that speed.

Because of this interconnection, it is frequently difficult to use historical data to independently select initial values for wing loading and thrust-to-weight ratio. Instead, the designer must guess at one of the parameters and use that guess to calculate the other parameter from the critical design requirements.

In many cases, the critical requirement for wing loading will be the stall speed during the approach for landing. Approach stall speed is independent of engine size, so the wing loading can be estimated based upon stall speed alone.

The estimated wing loading can then be used to calculate the T/W required to attain other performance drivers such as the single-engine rate of climb.

For less obvious cases, the designer must guess one parameter, calculate the other parameter to meet various performance requirements, then recheck the first parameter. In this book, the thrust-to-weight ratio appears as the first guess because that parameter better lends itself to a statistical approach, and also because it shows less variation within a given class of aircraft.

However, for certain aircraft the designer may wish to begin instead with the wing loading. In such cases, those equations presented next for calculating the wing loading can be solved for thrust-to-weight ratio instead.

5.2 Thrust-to-Weight Ratio

Thrust-to-Weight Definitions

T/W directly affects the performance of the aircraft. An aircraft with a higher T/W will accelerate more quickly, climb more rapidly, reach a higher maximum speed, and sustain higher turn rates. On the other hand, the larger engines will consume more fuel throughout the mission, which will drive up the aircraft's takeoff gross weight to perform the design mission.

T/W is not a constant. The weight of the aircraft varies during flight as fuel is burned. Also, the engine's thrust varies with altitude and velocity (as does the horsepower and propeller efficiency, η_p).

When designers speak of an aircraft's thrust-to-weight ratio, they generally refer to the T/W during sea-level static (zero-velocity), standard-day conditions at design takeoff weight and maximum throttle setting. Another commonly referred to T/W concerns combat conditions.

You can also calculate T/W at a partial-power setting. For example, during the approach to landing the throttle setting is near idle. The operating T/W at that point in the mission is probably less than 0.05.

It is very important to avoid confusing the takeoff T/W with the T/W at other conditions in the following calculations. If a required T/W is calculated at some other condition, it must be adjusted back to takeoff conditions for use in selecting the number and size of the engines. These T/W adjustments will be discussed later.

Power Loading and Horsepower-to-Weight

The term "thrust-to-weight" is associated with jet-engined aircraft. For propeller-powered aircraft, the equivalent term has classically been the "power loading," expressed as the weight of the aircraft divided by its horsepower (W/hp).

Power loading has an opposite connotation from T/W because a high power loading indicates a smaller engine. Power loadings typically range from 10–15 lb per horsepower for most aircraft. An aerobatic aircraft may have a power loading of about six. A few aircraft have been built with power loadings as low as three or four. One such overpowered airplane was the Pitts Sampson, a one-of-a-kind airshow airplane.

A propeller-powered aircraft produces thrust via the propeller, which has an efficiency η_p defined as the thrust output per horsepower provided by the engine. Using Eq. (3.9), an equivalent T/W for propellered aircraft can therefore be expressed as follows:

$$\frac{T}{W} = \left(\frac{\eta_p}{V}\right)\left(\frac{P}{W}\right) = \left(\frac{550\,\eta_p}{V}\right)\left(\frac{\text{hp}}{W}\right)(\text{fps units}) \tag{5.1}$$

Note that this equation includes the term P/W, the power-to-weight ratio. In British units this is simply the inverse of the classical power loading (W/hp). To avoid confusion when discussing requirements affecting both jet- and propeller-powered aircraft, this book refers to the power-to-weight ratio rather than the classical power loading. The reader should remember that the power loading can be determined simply by inverting the horsepower-to-weight ratio.

Also, to avoid excessive verbiage in the following discussions, the term "thrust-to-weight ratio" should be understood to include the power-to-weight ratio for propeller aircraft.

Statistical Estimation of T/W

Tables 5.1 and 5.2 provide typical values for T/W and P/W for different classes of aircraft. Table 5.2 also provides reciprocal values, i.e., power loadings, for propeller aircraft. These values are all at maximum power settings at sea level and zero velocity ("static").

At takeoff weights, a modern fighter plane approaches a T/W of 1.0, implying that the thrust is nearly equal to the weight. At combat conditions when some fuel has been burned off, these aircraft have T/W values exceeding 1, and are capable of accelerating while going straight up! The jet dogfighter T/W values are with afterburning engines, whereas the other jets typically do not have afterburning.

Thrust-to-weight ratio is closely related to maximum speed. Later in the design process, aerodynamic calculations of drag at the design maximum speed will be used, with other criteria, to establish the required T/W.

For now, Tables 5.3 and 5.4 provide curve-fit equations based upon maximum Mach number or velocity for different classes of aircraft. These can be used as a first estimate for T/W or P/W. The equations were developed by the author using data from Ref. 1, and should be considered valid only within the normal range of maximum speeds for each aircraft class.

Table 5.1 Thrust-to-weight ratio (T/W)[a]

Aircraft type	Typical installed T/W
Jet trainer	0.4
Jet fighter (dogfighter)	0.9
Jet fighter (other)	0.6
Military cargo/bomber	0.25
Jet transport (higher value for fewer engines)	0.25–0.4

[a]In mks units, the thrust force is found as T/W times mass times $g = 9.807$.

Table 5.2 Power-to-weight ratio

Aircraft type	Typical P/W hp/lb	{Watt/g}	Typical power loading (lb/hp)
Powered sailplane	0.04	{0.07}	25
Homebuilt	0.08	{0.13}	12
General aviation—single engine	0.07	{0.12}	14
General aviation—twin engine	0.17	{0.3}	6
Agricultural	0.09	{0.15}	11
Twin turboprop	0.20	{0.33}	5
Flying boat	0.10	{0.16}	10

Thrust Matching

For aircraft designed primarily for efficiency during cruise, a better initial estimate of the required T/W can be obtained by "thrust matching." This refers to the comparison of the selected engine's thrust available during cruise to the estimated aircraft drag.

In level unaccelerating flight, the thrust must equal the drag. Likewise, the weight must equal the lift (assuming that the thrust is aligned with the flight path). Thus, T/W must equal the inverse of L/D [Eq. (5.2)]:

$$\left(\frac{T}{W}\right)_{\text{cruise}} = \frac{1}{(L/D)_{\text{cruise}}} \tag{5.2}$$

L/D can be estimated in a variety of ways. Chapter 12 will discuss the detailed drag-buildup approach. For the first estimation of T/W, the method for L/D estimation presented in Chapter 3 is adequate.

Recall that this procedure for L/D estimation uses the selected aspect ratio and an estimated wetted-area ratio (Fig. 3.5) to determine the wetted aspect ratio. Figure 3.6 is then used to estimate the maximum L/D. For propeller aircraft, the cruise L/D is the same as the maximum L/D. For jet aircraft, the cruise L/D is 86.6% of the maximum L/D.

Table 5.3 T/W_0 vs M_{max}

$T/W_0 = a\, M_{max}^C$	a	C
Jet trainer	0.488	0.728
Jet fighter (dogfighter)	0.648	0.594
Jet fighter (other)	0.514	0.141
Military cargo/bomber	0.244	0.341
Jet transport	0.267	0.363

Table 5.4 P/W_0 vs V_{max} knots or {km/hr}

$P/W_0 = a\, V_{max}^C$: hp/lb or {watt/g}	a	C
Sailplane—powered	0.043 {0.071}	0
Homebuilt—metal/wood	0.005 {0.006}	0.57
Homebuilt—composite	0.004 {0.005}	0.57
General aviation—single engine	0.025 {0.036}	0.22
General aviation—twin engine	0.036 {0.048}	0.32
Agricultural aircraft	0.009 {0.010}	0.50
Twin turboprop	0.013 {0.016}	0.50
Flying boat	0.030 {0.043}	0.23

Note that this method assumes that the aircraft is cruising at approximately the optimum altitude for the as-yet-unknown wing loading. The method would be invalid if the aircraft were forced by the mission requirements to cruise at some other altitude, such as sea level.

When the wing loading has been selected, as described later in this chapter, the L/D at the actual cruise conditions should be calculated and used to recheck the initial estimate for T/W.

The thrust-to-weight ratio estimated using Eq. (5.2) is at cruise conditions, not takeoff. The aircraft will have burned off part of its fuel before beginning the cruise, and will burn off more as the cruise progresses.

Also, the thrust of the selected engine will be different at the cruise conditions than at sea-level, static conditions. These factors must be considered to arrive at the required takeoff T/W, used to size the engine.

The highest weight during cruise occurs at the beginning of the cruise. The weight of the aircraft at the beginning of the cruise is the takeoff weight less the fuel burned during takeoff and climb to cruise altitude. From Table 3.2, the typical mission weight fractions for these mission legs are 0.970 and 0.985, or 0.956 when multiplied together.

A typical aircraft will therefore have a weight at the beginning of cruise of about 0.956 times the takeoff weight. This value is used next to adjust the cruise T/W back to takeoff conditions.

Thrust during cruise is different from the takeoff value. Jet aircraft are normally designed to cruise at approximately the altitude at which the selected engine has the best (lowest) specific fuel consumption, typically 30,000–40,000 ft {approx. 10,000 m}. While SFC is improved at these altitudes, the thrust decreases. Also, the engine is sized using the thrust setting that produces the best SFC. This is usually 70–100% of the maximum continuous, nonafterburning thrust.

The cruise thrust at altitude is therefore less than the maximum takeoff thrust at sea level, and so the required cruise T/W must be adjusted to obtain the equivalent takeoff T/W.

Typically, a subsonic, high-bypass-ratio turbofan for a transport aircraft will have a cruise thrust of 20–25% of the takeoff thrust, while a low-bypass afterburning turbofan or turbojet will have a cruise thrust of 40–70% of the takeoff maximum value (see Fig. 5.1). Appendix A provides thrust and fuel-consumption data for several representative engines.

For a piston-powered, propeller-driven aircraft, the power available varies with the density of the air provided to the intake manifold. If the engine is not supercharged, then the power falls off with increasing altitude according to the density ratio, σ. For example, a nonsupercharged engine at 10,000 ft {3,048 m} will have about 73% of its sea-level power.

To prevent this power decrease, many piston engines use a supercharger to maintain the air provided to the manifold at essentially sea-level density up to the compression limit of the supercharger. Above this altitude, the power begins to drop off (see Fig. 5.2). Piston-powered aircraft typically cruise at about 75% of takeoff power.

For a turbine-powered, propeller-driven (turboprop) aircraft, the horsepower available increases somewhat with increasing speed, but the thrust drops off anyway due to the velocity effect on the propeller [Eq. (5.1)].

With a turboprop, there is an additional, residual thrust contribution from the turbine exhaust. It is customary to convert this thrust to its horsepower equivalent and add it to the actual horsepower, creating an "equivalent shaft horsepower (eshp)." For a typical turboprop engine installation, the cruise eshp is about 60–80% of the takeoff value.

The takeoff T/W required for cruise matching can now be approximated using Eq. (5.3). The ratio between initial cruise weight and takeoff weight

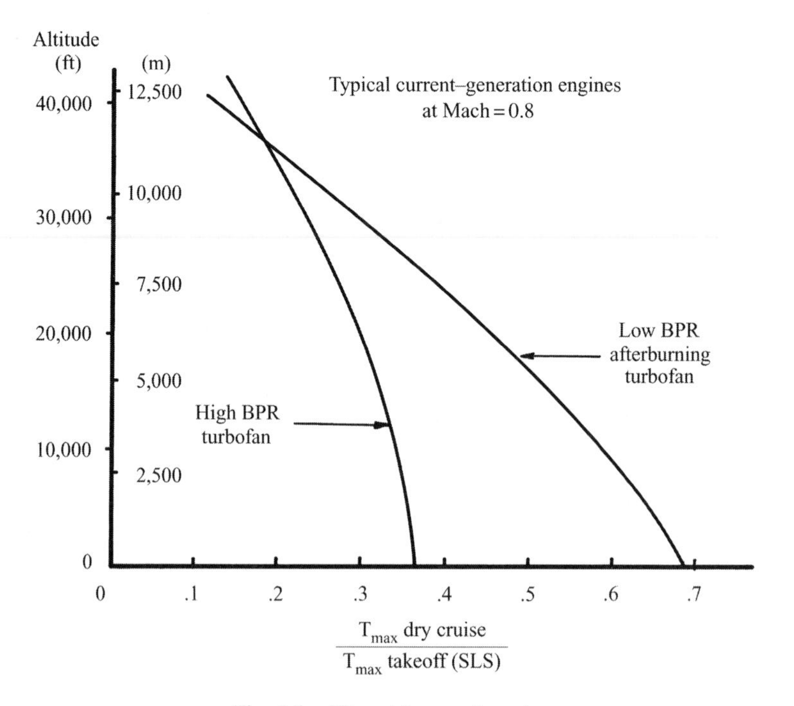

Fig. 5.1 Thrust lapse at cruise.

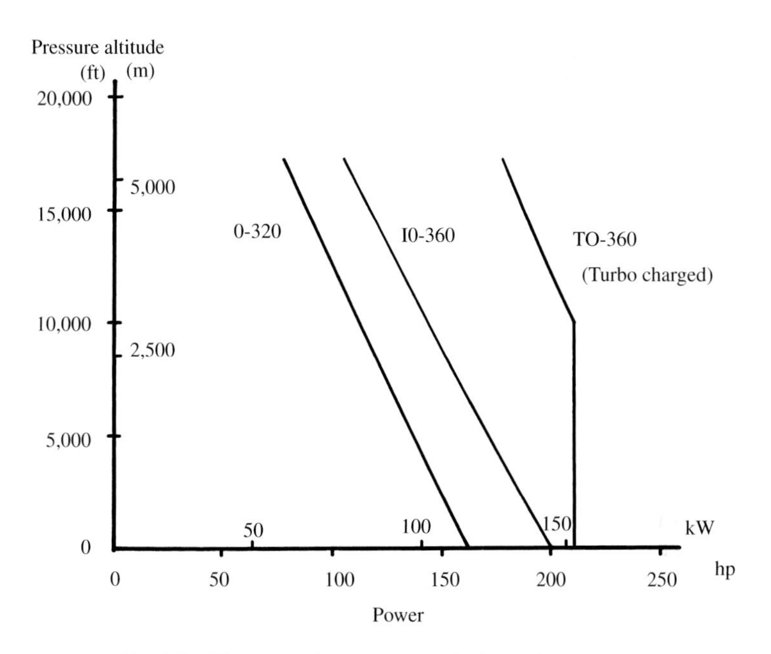

Fig. 5.2 Piston engine power variation with altitude.

was shown to be about 0.956. If a better estimate of this ratio is available, it should be used.

$$\left(\frac{T}{W}\right)_{\text{takeoff}} = \left(\frac{T}{W}\right)_{\text{cruise}} \left(\frac{W_{\text{cruise}}}{W_{\text{takeoff}}}\right) \left(\frac{T_{\text{takeoff}}}{T_{\text{cruise}}}\right) \tag{5.3}$$

The thrust ratio between takeoff and cruise conditions should be obtained from actual engine data if possible. Otherwise, data for a similar engine from Appendix A or some other source should be used.

For a propeller aircraft, the required takeoff P/W can be found by solving in Eq. (5.1) and adjusting weight and power back to takeoff conditions.

After an initial layout has been completed, actual aerodynamic calculations are made to compare the drag during cruise with the thrust available.

Thrust-to-weight ratio is often determined by a climb requirement rather than by cruise conditions. In fact, this leads to a very common problem. The T/W for climb can be so large that the engines must be throttled way back during cruise, and an aircraft engine running at only a fraction of its available power during cruise is usually very inefficient. This is especially true for jet engines.

T/W for a climb requirement can be found from a small adjustment to Eq. (5.2). As derived later in Chapter 17, the T/W for climb is the T/W for level flight, plus the extra thrust power required for the climb gradient, leading to Eq. (5.4). The aircraft's vertical velocity during the climb is generally specified in the design requirements or in military or civilian specifications (see Appendix

F.2). Note that the L/D for climb may be less than the L/D during cruising flight, especially during initial climb when the gear and flaps may still be down.

$$\left(\frac{T}{W}\right)_{\text{climb}} = \frac{1}{(L/D)_{\text{climb}}} + \frac{V_{\text{vertical}}}{V} \tag{5.4}$$

There are many other criteria that can set the thrust-to-weight ratio, such as takeoff distance and turning performance. These other criteria also involve the wing loading and are described in the next section.

For the first-pass estimate, the T/W (or P/W) should be selected as the higher of either the statistical value obtained from the appropriate equation in Tables 5.3 and 5.4, or the value obtained from the thrust matching as just described. After selection of the wing loading, as described next, the selected T/W should be rechecked against all requirements.

5.3 Wing Loading

The wing loading is the weight of the aircraft divided by the area of the reference (not exposed) wing. As with the thrust-to-weight ratio, the term "wing loading" normally refers to the takeoff wing loading, but can also refer to combat and other flight conditions.

Wing loading affects stall speed, climb rate, takeoff and landing distances, and turn performance. The wing loading determines the design lift coefficient, and impacts drag through its effect upon wetted area and wing span.

Wing loading has a strong effect upon sized aircraft takeoff gross weight. If the wing loading is reduced, the wing is larger. This may improve performance, but the additional drag and empty weight due to the larger wing will increase takeoff gross weight to perform the mission. The leverage effect of the sizing equation will require a more-than-proportional weight increase when factors such as drag and empty weight are increased. Table 5.5 provides representative wing loadings.

Table 5.5 Wing loading[a]

	Typical takeoff W/S	
Historical trends	lb/ft^2	$\{\text{kg/m}^2\}$
Sailplane	6	{30}
Homebuilt	11	{54}
General aviation—single engine	17	{83}
General aviation—twin engine	26	{127}
Twin turboprop	40	{195}
Jet trainer	50	{244}
Jet fighter	70	{342}
Jet transport/bomber	120	{586}

[a]In mks units, multiply metric values times $g = 9.807$ to use in equations.

Wing loading and thrust-to-weight ratio must be optimized together. Such optimization methods are presented in Chapter 19 using aerodynamic, weight, and propulsion data calculated from the initial design layout. The remainder of this chapter provides methods for initially estimating the wing loading to meet various requirements. These allow the designer to begin the layout with some assurance that the design will not require a complete revision after the aircraft is analyzed and sized.

This material generally assumes that an initial estimate of T/W has been made using the methods presented in the last section. However, most of the equations could also be used to solve for T/W if the wing loading is defined by some unique requirement (such as stall speed).

These methods estimate the wing loading required for various performance conditions. To ensure that the wing provides enough lift in all circumstances, the designer should select the lowest of the estimated wing loadings. However, if an unreasonably low wing loading value is driven by only one of these performance conditions, the designer should consider another way to meet that condition.

For example, if the wing loading required to meet a stall speed requirement is well below all other requirements, it may be better to equip the aircraft with a high-lift flap system. If takeoff distance or rate of climb require a very low wing loading, perhaps the thrust-to-weight ratio should be increased.

Stall Speed

The stall speed of an aircraft is directly determined by the wing loading and the maximum lift coefficient. Stall speed is a major contributor to flying safety, with a substantial number of fatal accidents each year due to "failure to maintain flying speed." Also, the approach speed, which is the most important factor in landing distance and also contributes to post-touchdown accidents, is defined by the stall speed.

Civil and military design specifications establish maximum allowable stall speeds for various classes of aircraft. In some cases the stall speed is explicitly stated. FAR 23 certified aircraft (under 12,500 lb TOGW {5,670 kg}) must stall at no more than 61 knots {113 km/hr} unless they are multi-engined and meet certain climb requirements (see Appendix F). While not stated in any design specifications, a stall speed of about 50 knots would be considered the upper limit for a civilian trainer or other aircraft to be operated by low-time pilots.

The approach speed is required to be a certain multiple of the stall speed. For civil applications, the approach speed must be at least 1.3 times the stall speed. For military applications, the multiple must be at least 1.2. Approach speed may be explicitly stated in the design requirements or will be selected based upon prior, similar aircraft. Then the required stall speed is found by division by 1.3 or 1.2.

Equation (5.5) states that lift equals weight in level flight, and that at stall speed, the aircraft is at maximum lift coefficient. Equation (5.6) solves for the required wing loading to attain a given stall speed with a certain maximum lift coefficient. The air density, ρ, is typically the sea-level standard value of 0.00238 slugs/cubic ft {1.23 kg/m^3} or sometimes the 5000-ft-altitude

{1524 m} hot-day value of 0.00189 {0.974} to ensure that the airplane can be flown into Denver during summer.

$$W = L = q_{\text{stall}} S C_{L_{\max}} = \frac{1}{2} \rho V_{\text{stall}}^2 S C_{L_{\max}} \tag{5.5}$$

$$W/S = \frac{1}{2} \rho V_{\text{stall}}^2 C_{L_{\max}} \tag{5.6}$$

The remaining unknown, the maximum lift coefficient, can be very difficult to estimate. Values range from about 1.2 to 1.5 for a plain wing with no flaps to as much as 5.0 for a wing with large flaps immersed in the propwash or jetwash.

The maximum lift coefficient for an aircraft designed for short takeoff and landing (STOL) applications will typically be about 3.0. For a regular transport aircraft with flaps and slats (leading-edge flaps with slots to improve airflow), the maximum lift coefficient is about 2.4. Other aircraft, with flaps on the inner part of the wing, will reach a lift coefficient of about 1.6–2.0.

Maximum lift coefficient depends upon the wing geometry, airfoil shape, flap geometry and span, leading-edge slot or slat geometry, Reynolds number, surface texture, and interference from other parts of the aircraft such as the fuselage, nacelles, or pylons. The trim force provided by the horizontal tail will increase or reduce the maximum lift, depending upon the direction of the trim force. If the propwash or jetwash impinges upon the wing or the flaps, it will also have a major influence upon maximum lift during power-on conditions.

Most aircraft use a different flap setting for takeoff and landing. During landing, the flaps will be deployed the maximum amount to provide the greatest lift and drag. However, for takeoff the maximum flap angle will probably cause more drag than desirable for rapid acceleration and climb, so the flaps will be deployed to about half the maximum angle. Therefore, the maximum lift coefficient for landing will be greater than for takeoff. Typically, the takeoff maximum lift coefficient is about 80% that of the landing value.

For a wing of fairly high aspect ratio (over about 5), the maximum lift coefficient will be approximately 90% of the airfoil maximum lift coefficient at the same Reynolds number, provided that the lift distribution is nearly elliptical. However, if partial-span flaps are used, their deflection will introduce a large, discontinuous twist into the wing geometry that changes the lift distribution, and thus the induced downwash, causing the effective angle of attack to vary at different span stations.

As a crude approximation, the designer can ignore this effect. Then the maximum lift can be estimated by determining the maximum angle of attack before some part of the wing stalls. Typically, the part of the wing with the flap deflected will stall first. Then, for that angle of attack the lift contributions of the flapped and unflapped sections can be summed, weighted by their areas (see Fig. 12.19 for definitions of flapped and un-flapped areas). This crude approximation for wings of a fairly high aspect ratio is given in Eq. (5.7):

$$C_{L_{\max}} \cong 0.9 \left\{ (C_{\ell_{\max}})_{\text{flapped}} \frac{S_{\text{flapped}}}{S_{\text{ref}}} + (C_{\ell})_{\text{unflapped}} \frac{S_{\text{unflapped}}}{S_{\text{ref}}} \right\} \tag{5.7}$$

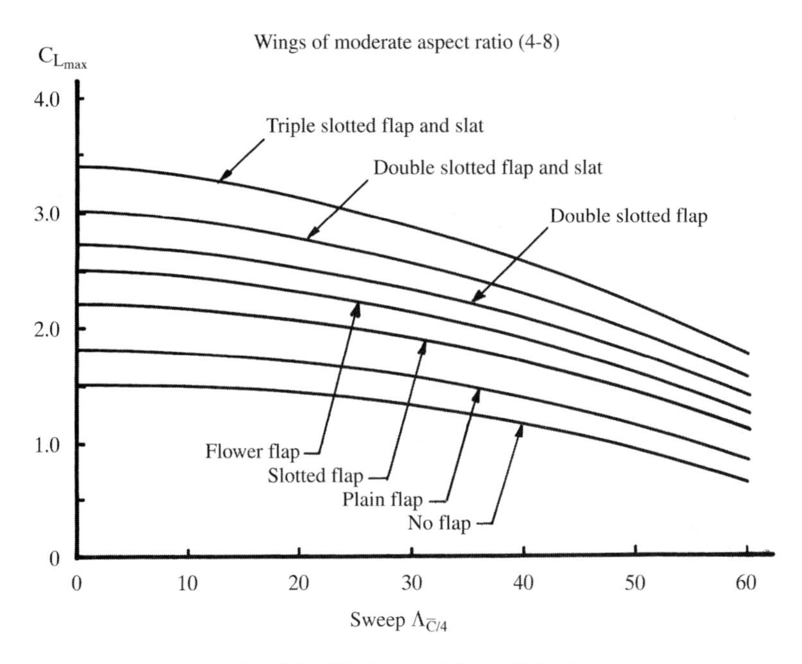

Fig. 5.3 Maximum lift coefficient.

where $C_{\ell_{\text{unflapped}}}$ is the lift coefficient of the unflapped airfoil at the angle of attack at which the flapped airfoil stalls.

For a better initial estimate of maximum lift, it is necessary to resort to test results and historical data. Figure 5.3 provides maximum-lift trends vs sweep angle for several classes of aircraft. Remember that the maximum lift using the takeoff flap setting will typically be about 80% of these landing maximum values. Maximum lift is discussed in more detail in Chapter 12.

Takeoff Distance

A number of different values are referred to as "takeoff distance." The "ground roll" is the actual distance traveled before the wheels leave the ground. The liftoff speed for a normal takeoff is 1.1 times the stall speed.

The "obstacle clearance distance" is the distance required from brake release until the aircraft has reached some specified altitude. This is usually 50 ft {15.24 m} for military or small civil aircraft and 35 ft {10.7 m} for commercial aircraft.

The "balanced field length" ("BFL") is the length of the field required for safety in the event of an engine failure at the worst possible time in a multi-engine aircraft. When the aircraft has just begun its ground roll, the pilot would have no trouble stopping it safely if one engine were to fail. As the speed increases, more distance would be required to stop after an engine failure. If the aircraft is nearly at liftoff speed and an engine fails, the pilot would be unable to stop safely and instead would continue the takeoff on the remaining engines.

The speed at which the distance to stop after an engine failure exactly equals the distance to continue the takeoff on the remaining engines is called the "decision speed." The balanced field length is the length required to take off and clear the specified obstacle when one engine fails exactly at decision speed. Note that use of reversed thrust is not permitted for calculation of balanced field length.

The Federal Aviation Administration specifies a field length requirement for FAR 25 certified aircraft called "FAR takeoff field length." This has a 35-ft (10.7-m) obstacle clearance requirement and requires that the aircraft meet the worst of either balanced field length as just described, or a value of 15% greater than the all-engines-operating obstacle clearance takeoff distance. FAR 23 certified aircraft are not required to meet a balanced field length requirement. See Appendix F for more information about FAR requirements or to obtain the full FARs (available online).

For military aircraft the balanced field length has a 50-ft (15.24-m) obstacle clearance requirement.

Both the wing loading and the thrust-to-weight ratio contribute to the takeoff distance. The following equations assume that the thrust-to-weight ratio has been selected and can be used to determine the required wing loading to attain some required takeoff distance. However, the equations could be solved for T/W if the wing loading is known.

Other factors contributing to the takeoff distance are the aircraft's aerodynamic drag and rolling resistance. Aerodynamic drag on the ground depends largely upon pilot technique. For example, if the pilot rotates (lifts the nose) too early, the extra drag may prevent the aircraft from accelerating to takeoff speed. This was a frequent cause of accidents in early jets, which were underpowered by today's standards.

The aircraft's rolling resistance, μ, is determined by the type of runway surface and by the type, number, inflation pressure, and arrangement of the tires. A thin, high-pressure tire operated over a soft dirt runway will have so much rolling resistance that the aircraft may be unable to move. A large, low pressure tire can operate over a softer runway surface but will have more aerodynamic drag if not retracted, or will take up more room if retracted. Values of μ for different runway surfaces are provided in the detailed takeoff analysis in Chapter 17.

In later stages of analysis the takeoff distance will be calculated by integrating the accelerations throughout the takeoff, considering the variations in thrust, rolling resistance, weight, drag, and lift. For initial estimation of the required wing loading, a statistical approach for estimation of takeoff distance can be used.

Figure 5.4, based upon data from Refs. 4 and 5, permits estimation of the takeoff ground roll, takeoff distance to clear a 50-ft {15.24 m} obstacle, and FAR balanced field length over a 35-ft {10.7 m} obstacle. For a military multi-engined aircraft, the balanced field length over a 50-ft {15.24 m} obstacle is approximately 5% greater than the FAR (35-ft {10.7 m}) balanced field value.

Note that a twin-engined aircraft has a greater FAR balanced field length than a three- or four-engined aircraft with the same total thrust. This occurs because the twin-engined aircraft loses half its thrust from a single engine failure, whereas the three- and four-engined aircraft lose a smaller percentage of their total thrust from a single engine failure.

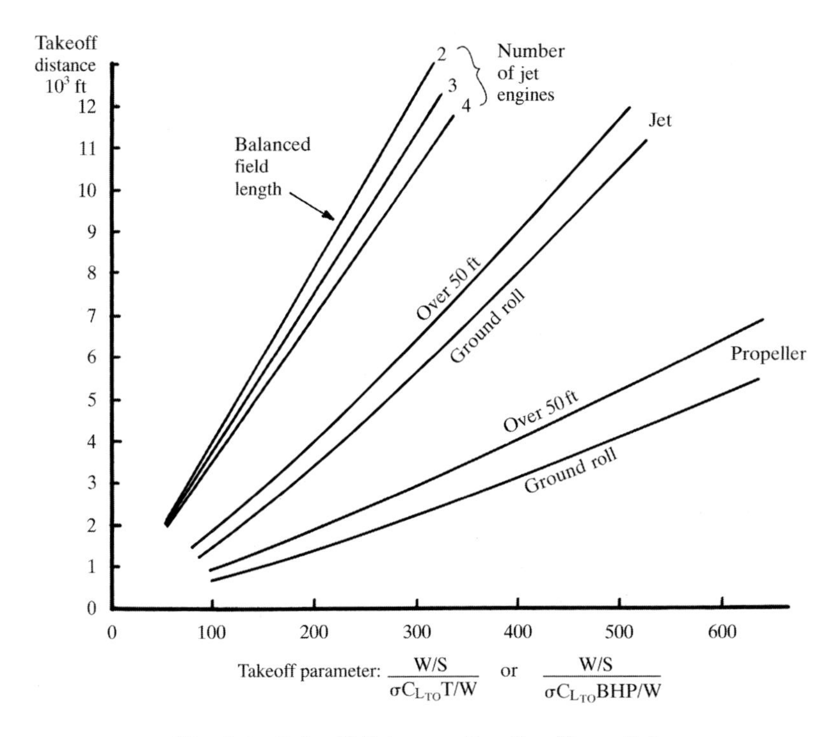

Fig. 5.4 Takeoff distance estimation (fps units).

The takeoff parameter (TOP) of Fig. 5.4 is the takeoff wing loading divided by the product of the density ratio (σ), takeoff lift coefficient, and takeoff thrust-to-weight (or horsepower-to-weight) ratio. The density ratio is simply the air density (ρ) at the takeoff altitude divided by the sea-level density.

The takeoff lift coefficient is the actual lift coefficient at takeoff, not the maximum lift coefficient at takeoff conditions as used for stall calculation. The aircraft takes off at about 1.1 times the stall speed so the takeoff lift coefficient is the maximum takeoff lift coefficient divided by 1.21 (1.1 squared). However, takeoff (and landing) lift coefficient may also be limited by the maximum tail-down angle permitted by the landing gear (typically not more than 15 deg).

To determine the required wing loading to meet a given takeoff distance requirement, the takeoff parameter is obtained from Fig. 5.4. Then the following expressions give the maximum allowable wing loading for the given takeoff distance:

$$\text{Prop: } (W/S) = (\text{TOP})\sigma C_{L_{\text{TO}}}(\text{hp}/W) \tag{5.8}$$

$$\text{Jet: } (W/S) = (\text{TOP})\sigma C_{L_{\text{TO}}}(T/W) \tag{5.9}$$

Catapult Takeoff

Most naval aircraft must be capable of operation from an aircraft carrier. For takeoff from a carrier, a catapult accelerates the aircraft to flying speed in a very short distance.

Catapults are steam-operated, and can produce a maximum force on the aircraft depending on the steam pressure used. Therefore, a light aircraft can be accelerated to a higher speed by the catapult than a heavy one. Figure 5.5 depicts the velocities attainable as a function of aircraft weight for three catapults in use by the U.S. Navy. Note that a rough guess of takeoff weight is required.

For a catapult takeoff, the airspeed as the aircraft leaves the catapult must exceed the stall speed by 10%. Airspeed is the sum of the catapult end speed (V_{end}) and the wind-over-deck of the carrier (V_{wod}), plus the velocity added by the engine's thrust, typically 3–10 knots, or 5–18 km/hr.

For aircraft launch operations the carrier will be turned into the wind which will produce a wind-over-deck on the order of 20–40 knots. However, the design specifications for a Navy aircraft frequently require launch capabilities with zero wind-over-deck or even a negative value, to enable aircraft launch while at anchor. Once the end speed is known, the maximum wing loading is defined by

$$\left(\frac{W}{S}\right)_{\text{takeoff}} = \frac{1}{2}\rho(V_{\text{end}} + V_{\text{wod}} + \Delta V_{\text{thrust}})^2 \frac{(C_{L_{\max}})_{\text{takeoff}}}{1.21} \tag{5.10}$$

where $\rho = 0.00219$ slug/ft^3 {1.13 kg/m^3} tropical day.

Sometimes the takeoff stall margin in this equation is defined not as a velocity margin of 10%, but as a lift coefficient margin of 15%. This changes the "1.21" term to a value of "1.18."

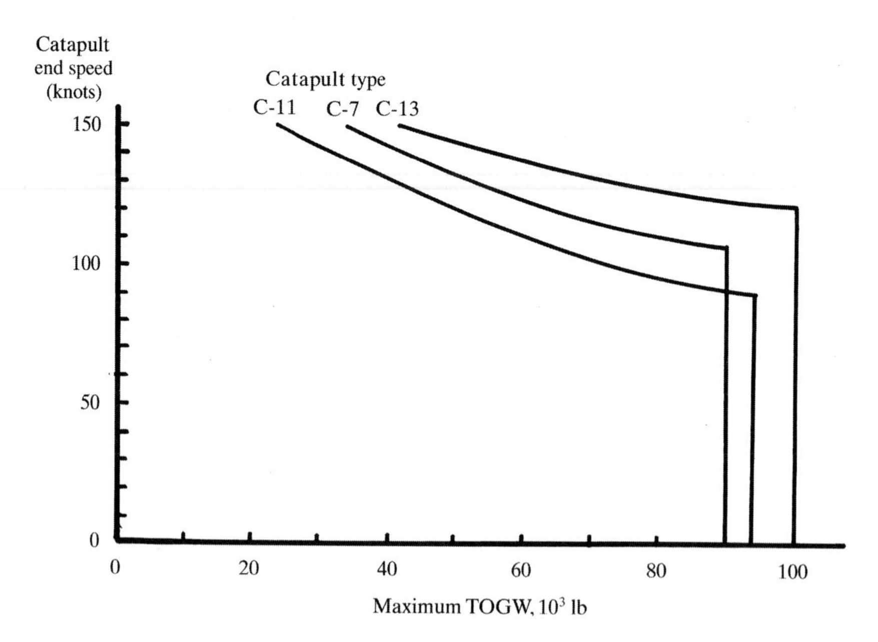

Fig. 5.5 Catapult end speeds.

Landing Distance

There are a number of different values referred to as the "landing distance." "Landing ground roll" is the actual distance the aircraft travels from the time the wheels first touch to the time the aircraft comes to a complete stop.

The "FAR 23 landing field length" includes clearing a 50-ft {15.24 m} obstacle while the aircraft is still at approach speed and on the approach glidepath (normally 3 deg). After crossing the obstacle, the pilot slows the aircraft to the touchdown speed of typically 1.15 times the stall speed. The obstacle-clearance distance roughly doubles the ground-roll distance alone.

The "FAR 25 landing field length" includes the 50-ft {15.24 m} obstacle clearance at approach speed, and also adds an arbitrary two-thirds to the total distance to allow a safety margin. The landing distance definition for military aircraft is normally specified in a request for proposals (RFP), but typically resembles the FAR 23 definition.

Landing distance is largely determined by wing loading. Wing loading affects the approach speed, which must be a certain multiple of stall speed (1.3 for civil aircraft, 1.2 for military aircraft). Approach speed determines the touchdown speed, which in turn defines the kinetic energy which must be dissipated to bring the aircraft to a halt. The kinetic energy, and hence the stopping distance, varies as the square of the touchdown speed.

In fact, a reasonable first-guess of the total landing distance in feet, including obstacle clearance, is approximately 0.3 times the square of the approach speed in knots (Ref. 5). This is approximately true for FAR 23 and military aircraft without thrust reversers, and FAR 25 aircraft with thrust reversers. While the FAR 25 aircraft have the additional requirement of a two-thirds distance increase, the thrust reversers used on most FAR 25 aircraft shorten the landing distance by about the same amount.

Equation (5.11) provides a better approximation of the landing distance, which can be used to estimate the maximum landing wing loading. The first term represents the ground roll to absorb the kinetic energy at touchdown speed. The constant term, S_a, represents the obstacle-clearance distance.

$$
\begin{aligned}
S_{\text{landing}} &= 80\left(\frac{W}{S}\right)\left(\frac{1}{\sigma C_{L_{\max}}}\right) + S_a\ (\text{ft}) \\
&= 5\left(\frac{W}{S}\right)\left(\frac{1}{\sigma C_{L_{\max}}}\right) + S_a\ (\text{m})
\end{aligned}
\tag{5.11}
$$

where

$\sigma =$ density ratio
$S_a =$ 1000 ft {305 m} (airliner-type, 3-deg glideslope)
$=$ 600 ft {183 m} (general aviation-type power-off approach)
$=$ 450 ft {137 m} (STOL, 7-deg glideslope)

For landing calculation with thrust reversers or reversible-pitch propellers, multiply the ground portion of the landing [first term in Eq. (5.11)] by 0.66. However, FAR and other requirements often specify that thrust reversers cannot be used to meet landing specifications for a simple reason—they may break, right when you need them the most.

For commercial (FAR 25) aircraft, multiply the total landing distance calculated with Eq. (5.11) by 1.67 to provide the required safety margin.

The landing wing loading must be converted to takeoff conditions by dividing by the ratio of landing weight to takeoff weight. This ratio is usually not based upon the calculated end-of-mission weight, but is instead based upon some arbitrary landing weight as specified in the design requirements.

For most propeller-powered aircraft and jet trainers, the aircraft must meet its landing requirement at or near the design takeoff weight, so the ratio is about 1.0. For most jet aircraft, the landing is typically calculated at a weight of about 0.85 times the takeoff weight. Military design requirements will frequently specify full payload and some percent of fuel remaining (usually 50%) for the landing.

Arrested Landing

Aircraft that land on Navy aircraft carriers are stopped by a cable-and-brake arrangement called "arresting gear." One of several cables strung across the flight deck is caught by a hook attached to the rear of the aircraft. The cable is attached at both ends to drum mechanisms which exert a drag upon the cable as it is pulled by the aircraft, thus stopping it in a very short distance.

For carrier-based aircraft, the approach speed (1.2 times the stall speed) is the same as the touchdown speed. Carrier pilots do not flare and slow down for landing. Instead, they are taught to fly the aircraft right into the deck, relying upon the arresting gear to stop the aircraft. By using this technique, the aircraft has enough speed to go around if the cables are missed.

The landing weight limits for three standard arresting gears are depicted in Fig. 5.6. This figure can be used to determine the allowable approach speed based upon a first-guess of the landing weight. The approach speed divided by 1.2 defines the stall speed, which can then be used to estimate the wing loading.

Wing Loading for Cruise

At this point we must bring in the use of two aerodynamic coefficients, C_{D_0} and "e." C_{D_0} is the zero lift drag coefficient, and equals approximately 0.015 for a jet aircraft, 0.02 for a clean propeller aircraft, and 0.03 for a dirty, fixed-gear propeller aircraft. The Oswald efficiency factor e is a measure of drag due to lift efficiency, and during cruise equals approximately 0.6 to 0.8 for a fighter and 0.8 for other aircraft. These coefficients are extensively discussed in Chapter 12. Chapter 12 also contains methods for estimation of C_{D_0} and e.

To maximize range during cruise, the wing loading should be selected to provide a high L/D at the cruise conditions. The following discussion provides methods for selecting wing loading to optimize cruise range.

A propeller aircraft, which loses thrust efficiency as speed goes up, gets the maximum range when flying at the speed for best L/D, while a jet aircraft maximizes range at a somewhat higher speed where the L/D is slightly reduced. The speed for best L/D can be shown to result in parasite drag equaling the induced drag (see Chapter 17). Therefore, to maximize range a propeller aircraft should

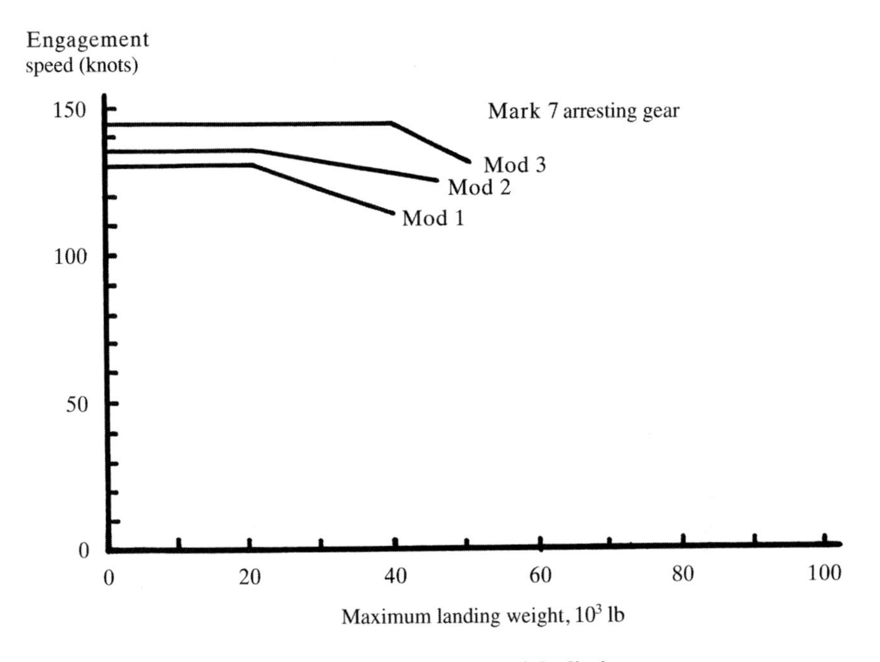

Fig. 5.6 Arresting gear weight limits.

fly such that

$$qSC_{D_0} = qS\frac{C_L^2}{\pi Ae} \tag{5.12}$$

During cruise, the lift equals the weight, so the lift coefficient equals the wing loading divided by the dynamic pressure. Substitution into Eq. (5.12) allows solution for the required wing loading to maximize L/D for a given flight condition. This result [Eq. (5.13)] is the wing loading for maximum range for a propeller aircraft.

$$\text{Maximum Prop Range: } W/S = q\sqrt{C_{D_0}/k} = q\sqrt{\pi AeC_{D_0}} \tag{5.13}$$

As the aircraft cruises, its weight reduces due to the fuel burned, so the wing loading also reduces during cruise. Optimizing the cruise efficiency while the wing loading is steadily declining requires reducing the dynamic pressure by the same percent [see Eq. (5.13)]. This can be done by reducing velocity, which is undesirable, or by climbing to obtain a lower air density. This range optimizing technique is known as a "cruise-climb."

A jet aircraft flying a cruise-climb will obtain maximum range by flying at a wing loading such that the parasite drag is three times the induced drag (see Chapter 12 for the derivation of this relationship). This yields the following formula for wing-loading selection for range optimization of jet aircraft:

$$\text{Maximum Jet Range: } W/S = q\sqrt{C_{D_0}/3k} = q\sqrt{\pi A e C_{D_0}/3} \tag{5.14}$$

Frequently an aircraft will not be allowed to use the cruise-climb technique to maximize range. Air traffic controllers prefer that aircraft maintain a single assigned altitude until given permission to climb or descend to another altitude. The pilot will attempt to obtain permission from the air traffic controllers to climb several times during the flight, thus forming the characteristic "stairstep climb schedule." This allows the thrust setting to be maintained approximately at the setting that minimizes fuel consumption.

Wing Loading for Loiter Endurance

Most aircraft will have some loiter requirement during the mission, typically 20 min of loiter before landing. Unless the loiter requirement is a substantial fraction of the total mission duration, it is better to optimize the wing loading for cruise.

Patrol aircraft such as the ASW design example of Chapter 3 are sometimes more concerned with time on station than with cruise efficiency. Other aircraft which may be concerned with loiter endurance are airborne command posts and intelligence-gathering aircraft.

For an aircraft that must be optimized for loiter, the wing loading should be selected to provide a high L/D. For jet aircraft, the best loiter occurs at maximum L/D, and so Eq. (5.12) [repeated next as Eq. (5.15)] should be used. For a propeller aircraft, loiter is optimized when the induced drag is three times the parasite drag, which yields Eq. (5.16). This also provides the wing loading for minimum power required.

$$\text{Maximum Jet Loiter: } W/S = q\sqrt{C_{D_0}/k} = q\sqrt{\pi A e C_{D_0}} \tag{5.15}$$

$$\text{Maximum Prop Loiter: } W/S = q\sqrt{3C_{D_0}/k} = q\sqrt{3\pi A e C_{D_0}} \tag{5.16}$$

These equations assume that the loiter velocity and altitude are known. If the loiter altitude is not specified, it should be selected for best specific fuel consumption at the loiter power setting. This is typically 30,000–40,000 ft {approximately 10,000 m} for a jet, and the limit altitude for the turbocharger for a piston-propeller aircraft. For a nonturbocharged engine, best loiter occurs at sea level.

Usually, the loiter velocity is not specified. Instead the designer must determine the best loiter velocity and select the wing loading accordingly. This requires cross plotting of wing loadings with the resulting L/D and specific fuel consumption for various velocities and altitudes. Such a procedure is too complex for initial design purposes.

For initial design purposes, it can be assumed that the best loiter velocity will be about 150–200 knots {about 325 km/hr} for turboprops and jets, and about 80–120 knots {about 180 km/hr} or piston-props. If altitude is not specified, the altitude for best fuel consumption should be selected.

The wing loading estimated from Eqs. (5.15) or (5.16) is the average during the loiter. This must be converted to takeoff conditions by dividing the loiter wing loading by the ratio of the average loiter weight to the takeoff weight. In the absence of better information, this ratio can be assumed to be about 0.85.

Remember that Eqs. (5.15) and (5.16) are to be used for designing an aircraft optimized solely for loiter. Optimizing for loiter alone is very rare in aircraft design. For most aircraft, the wing loading will be selected for best cruise or other requirements and the loiter capabilities will be a secondary consideration.

Instantaneous Turn

An aircraft designed for air-to-air dogfighting must be capable of high turn rate. This parameter, $d\psi/dt$ or $\dot{\psi}$, will determine the outcome of a dogfight if the aircraft and pilots are evenly matched otherwise. When air-to-air missiles are in use, the first aircraft to turn toward the other aircraft enough to launch a missile will probably win. In a guns-only dogfight, the aircraft with the higher turn rate will be able to maneuver behind the other. A turn rate superiority of 2 deg/s is considered significant.

There are two important turn rates. The "sustained" turn rate for some flight condition is the turn rate at which the thrust of the aircraft is just sufficient to maintain velocity and altitude in the turn. If the thrust acts approximately opposite to the flight direction, then the thrust must equal the drag for a sustained turn.

If the aircraft turns at a quicker rate, the drag becomes greater than the available thrust, so the aircraft begins to slow down or lose altitude. The "instantaneous" turn rate is the highest turn rate possible, ignoring the fact that the aircraft will slow down or lose altitude.

The "load factor," or "g-loading," during a turn is the acceleration due to lift expressed as a multiple of the standard acceleration due to gravity ($g = 32.2$ ft/s$^2 = 9.8$ m/s^2). Load factor ("n") is equal to the lift divided by the aircraft's weight.

Level, unturning flight implies a load factor of one ($n = 1$). In a level turn, the wing must provide 1-g lift in the vertical direction to hold up the aircraft, so the remaining "g's" available to turn the aircraft in the horizontal direction are equal to the square root of n squared minus 1 (see Fig. 17.4). Thus the radial acceleration in a level turn is g times the square root of $(n^2 - 1)$.

Turn rate is equal to the radial acceleration divided by the velocity. For a level turn, this results in Eq. (5.17). Note that this equation provides turn rate in radians per second, which must be multiplied by 57.3 to obtain degrees per second:

$$\dot{\psi} = \frac{g\sqrt{n^2 - 1}}{V} \tag{5.17}$$

where

$$n = \frac{qC_L}{W/S} \tag{5.18}$$

Instantaneous turn rate is limited only by the usable maximum lift, up to the speed at which the maximum lift exceeds the load-carrying capability of the wing structure. Typically, a fighter aircraft will be designed to an operational maximum load factor of 7.33 g, although newer fighters are being designed to 8 or 9 g. This g limit must be met at some specified combat weight.

The speed at which the maximum lift available exactly equals the allowable load factor is called the "corner speed," and provides the maximum turn rate for that aircraft at that altitude. In a dogfight, pilots try to get to corner speed as quickly as possible as it provides the best turn rate. Typically, a modern fighter has a corner speed of about 300–350 knots {550–650 km/hr} indicated airspeed (i.e., dynamic pressure) regardless of altitude.

Design specifications will usually require some maximum turn rate at some flight condition. Equation (5.17) can be solved for the load factor at the specified turn rate as follows:

$$n = \sqrt{\left(\frac{\dot{\psi}V}{g}\right)^2 + 1} \tag{5.19}$$

If this value of load factor is greater than the ultimate load factor specified in the design requirements, somebody has made a mistake. The required wing loading can be solved for in Eq. (5.18) as follows:

$$\frac{W}{S} = \frac{qC_{L_{\max}}}{n} \tag{5.20}$$

The only unknown is the maximum lift coefficient at combat conditions. This is not the same as the maximum lift coefficient for landing. During combat, use of full flap settings is not usually possible. Also, there is a Mach number effect which reduces maximum lift at higher speeds. Frequently the combat maximum usable lift will be limited by buffeting or controllability considerations.

For initial design purposes, a combat maximum lift coefficient of about 0.6–0.8 should be assumed for a fighter with only a simple trailing-edge flap for combat. For a fighter with a complex system of leading- and trailing-edge flaps which can be deployed during combat, a maximum usable lift coefficient of about 1.0–1.5 is attainable. Chapter 12 provides better methods of estimating the maximum lift coefficient.

Again, the resulting wing loading must be divided by the ratio of combat weight to takeoff weight to obtain the required takeoff wing loading. Usually the combat weight is specified as the aircraft design takeoff weight with any external fuel tanks dropped and 50% of the internal fuel gone. This is approximately 0.85 times the takeoff weight for most fighters.

The resulting wing loading is the maximum which will allow the required instantaneous turn.

Sustained Turn

The sustained turn rate is also important for success in combat. If two aircraft pass each other in opposite directions, it will take them about 10 seconds to complete 180-deg turns back towards the other. The aircraft will probably not be able to maintain speed while turning at the maximum instantaneous rate. If one of the aircraft slows down below corner speed during this time, it will be at a turn rate disadvantage to the other, which could prove fatal.

Sustained turn rate is usually expressed in terms of the maximum load factor at some flight condition that the aircraft can sustain without slowing or losing altitude. For example, the ability for sustaining 4 or 5 g at 0.9 Mach number at 30,000 ft {9144 m} is frequently specified. Equations (5.17) or (5.19) can be used to relate turn rate to load factor.

If speed is to be maintained, the thrust must equal the drag (assuming that the thrust axis is approximately aligned with the flight direction). The lift must equal the weight times the load factor, so we can write:

$$n = (T/W)(L/D) \tag{5.21}$$

Load factor in a sustained turn is maximized by maximizing the T/W and L/D. The highest L/D occurs when the induced drag equals the parasite drag, as expressed by Eq. (5.12). During a turn, the lift equals the weight times n, so the lift coefficient equals the wing loading times n divided by the dynamic pressure. Substitution into Eq. (5.12) yields:

$$W/S = \frac{q}{n}\sqrt{\pi A e C_{D_0}} \tag{5.22}$$

This equation gives the wing loading that maximizes the sustained turn rate at a given flight condition. Note that if n equals one, Eq. (5.22) is the same as Eq. (5.13), the wing loading for best L/D in level flight.

Equation (5.22) estimates the wing loading that maximizes the sustained turn rate regardless of thrust available. This equation will frequently give ridiculously low values of wing loading that will provide the required sustained turn rate using only a fraction of the available thrust.

The wing loading to exactly attain a required sustained load factor n using all of the available thrust can be determined by equating the thrust and drag, and using the fact that since lift equals weight times n, the lift coefficient during maneuver equals the wing loading times n, divided by the dynamic pressure. This yields Eq. (5.23):

$$T = qSC_{D_0} + qS\left(\frac{C_L^2}{\pi A e}\right) = qSC_{D_0} + \frac{n^2 W^2}{qS\pi A e} \tag{5.23}$$

or

$$\frac{T}{W} = \frac{qC_{D_0}}{W/S} + \frac{W}{S}\left(\frac{n^2}{q\pi A e}\right) \tag{5.24}$$

Equation (5.24) can be solved for W/S to yield the wing loading that exactly attains a required sustained load factor n [Eq. (5.25)]. Also, Eq. (5.24) can be used later to recheck the T/W after the wing loading is selected.

$$\frac{W}{S} = \frac{(T/W) \pm \sqrt{(T/W)^2 - (4n^2 C_{D_0}/\pi Ae)}}{2n^2/q\pi Ae} \tag{5.25}$$

The thrust-to-weight ratio for this calculation is at combat conditions, so the takeoff T/W must be adjusted to combat conditions by dividing by the ratio between combat and takeoff weight, and by multiplying by the ratio between combat thrust and takeoff thrust.

If the term within the square root in Eq. (5.25) becomes negative, there is no solution. This implies that, at a given load factor, the following must be satisfied regardless of the wing loading:

$$\frac{T}{W} \geq 2n\sqrt{\frac{C_{D_0}}{\pi Ae}} \tag{5.26}$$

It is very important to realize in these calculations that the efficiency factor e is itself a function of the lift coefficient at which the aircraft is operating. This is due to the separation effects at higher lift coefficients that increase drag above the parabolic drag polar values. At high angles of attack the effective e value may be reduced by 30% or more.

Unfortunately, the preceding equations for turning flight are very sensitive to the e value. If these equations yield W/S values far from historical values, the e value is probably unrealistic and the calculated W/S values should be ignored. Methods in Chapter 12 will better account for the separation effects.

Climb and Glide

Appendix F cites numerous climb requirements for FAR or military aircraft. These specify rate of climb for various combinations of factors such as engine-out, landing-gear position, and flap settings. While the details may vary, the method for selecting a wing loading to satisfy such requirements is the same.

Rate of climb is a vertical velocity, typically expressed in feet-per-minute (which must be converted to feet-per-second for the following calculations). Climb gradient, "G," is the ratio between vertical and horizontal distance traveled. As will be shown in Chapter 17, at normal climb angles the climb gradient equals the excess thrust divided by the weight, that is,

$$G = (T - D)/W \tag{5.27}$$

or

$$\frac{D}{W} = \frac{T}{W} - G \tag{5.28}$$

D/W can also be expressed as in Eq. (5.29), where in the final expression the lift coefficient is replaced by (W/qS).

$$\frac{D}{W} = \frac{qSC_{D_0} + qS(C_L^2/\pi Ae)}{W} = \frac{qC_{D_0}}{W/S} + \frac{W}{S}\frac{1}{q\pi Ae} \tag{5.29}$$

Equating Eqs. (5.28) with (5.29) and solving for wing loading yields:

$$\frac{W}{S} = \frac{[(T/W) - G] \pm \sqrt{[(T/W) - G]^2 - (4C_{D_0}/\pi Ae)}}{2/q\pi Ae} \tag{5.30}$$

Note the similarity to Eq. (5.25). Equation (5.30) is merely Eq. (5.25) for a load factor of 1.0, with the (T/W) term replaced by $[(T/W) - G]$. As before, T/W must be ratioed to the flight conditions and weight under consideration. The resulting W/S must then be ratioed to a takeoff-weight value.

The term within the square root symbol in Eq. (5.30) cannot go below zero, and so the following must be true regardless of the wing loading:

$$\frac{T}{W} \geq G + 2\sqrt{\frac{C_{D_0}}{\pi Ae}} \tag{5.31}$$

This equation says that no matter how "clean" your design is, the T/W must be greater than the desired climb gradient! [T/W for a propeller aircraft was defined in Eq. (5.1).]

Another implication of this equation is that a very "clean" aircraft that cruises at a high speed despite a very low T/W will probably climb poorly. A 200-mph airplane that flies on 20 hp can't be expected to climb as well as an airplane that requires 200 hp to reach 200 mph (unless the latter weighs 10 times as much).

C_{D_0} and e values for some of the climb conditions specified in Appendix F must include the effects of flaps and landing gear. Chapter 12 will provide methods for estimating these effects, but for now, approximations can be used.

For takeoff flap settings, C_{D_0} will increase by about 0.02 and e will decrease about 5%. For landing flap settings, C_{D_0} will increase by about 0.07 and e will decrease by about 10% relative to the no-flap value. Retractable landing gear in the down position will increase C_{D_0} by about 0.02 (Ref. 7).

Sometimes the rate of climb must also be calculated with one engine windmilling or stopped. The thrust loss due to a "dead" engine can be accounted for in the T/W. For example, if a three-engined aircraft loses one engine, the T/W becomes two-thirds of the original T/W.

The drag increase due to a windmilling or stopped engine will further reduce the climb rate. Chapter 12 provides methods for estimating this drag. For rough initial analysis, however, it can probably be ignored.

Equation (5.30) can also be used to establish the wing loading required to attain some specified glide angle, by setting T/W to zero and using a negative value of G (i.e., a glide is a climb in the negative direction). If a particular sink rate must be attained, the value of G to use is the sink rate divided by the forward velocity. Make sure that both are in the same units.

Maximum Ceiling

Equation (5.30) can be used to calculate the wing loading to attain some maximum ceiling, given the T/W at those conditions. The climb gradient G

can be set to zero to represent level flight at the desired altitude. Frequently a small residual climb capability, such as 100 ft/min {30.5 m/min} is required at maximum ceiling. This can be included in Eq. (5.30) by first solving for the climb gradient G (climb rate divided by forward velocity).

For a high-altitude aircraft such as an atmospheric research or reconnaissance plane, the low dynamic pressure available may determine the minimum possible wing loading. For example, at 100,000 ft {30,480 m} and 0.8 Mach number, the dynamic pressure is only 10 psf {0.5 kN/m^2}. Equation (5.13) [repeated below as Eq. (5.32)] can be used to determine the wing loading for minimum power.

$$W/S = q\sqrt{\pi A e C_{D_0}} \tag{5.32}$$

This may suggest a wing loading so low as to be impractical, and so should be compared with the wing loading required to fly at a given lift coefficient, that is:

$$W/S = qC_L \tag{5.33}$$

For efficiency during high-altitude cruise, the lift coefficient should be near the airfoil design lift coefficient. For a typical airfoil, this is about 0.5. For a high-altitude aircraft, new high-lift airfoils with design lift coefficients on the order of 0.95–1.0 can be used.

5.4 Selection of Thrust to Weight and Wing Loading

An initial estimate of the thrust-to-weight (or horsepower-to-weight) ratio was previously made. From the wing loadings just estimated, the lowest value should be selected to ensure that the wing is large enough for all flight conditions. Don't forget to convert all wing loadings to takeoff conditions prior to comparisons.

A low wing loading will always increase aircraft weight and cost. If a very low wing loading is driven by only one of the requirements, a change in design assumptions (such as a better high-lift system) may allow a higher wing loading.

Also, keep in mind that the wing loadings calculated by Eqs. (5.13–5.16), (5.22), and (5.32) are aerodynamic optimizations for only a portion of the mission. If these give wing loadings far lower than those in Table 5.5, they may be ignored.

When the best compromise for wing loading has been selected, the thrust-to-weight ratio should be rechecked to ensure that all requirements are still met. The equations in the last section which use T/W should be recalculated with the selected W/S and T/W. Only then can the next step of design, initial sizing, be initiated.

6
Initial Sizing

6.1 Introduction

Aircraft sizing is the process of determining the takeoff gross weight and fuel weight required for an aircraft concept to perform its design mission. Sizing was introduced in Chapter 3, in which a quick method based upon minimal information about the design was used to estimate the sizing parameters. That sizing method was limited to fairly simple design missions. This chapter presents a more refined method capable of dealing with most types of aircraft-sizing problems.

6.2 "Rubber" vs "Fixed-Size" Engines

An aircraft can be sized using some existing engine or a new design engine. The existing engine is fixed in size and thrust, and is referred to as a "fixed-size-engine" or "fixed-engine" ("fixed" refers to engine size).

The new design engine can be built in any size and thrust required, and is called a "rubber engine" because it can be "stretched" during the sizing process to provide any required amount of thrust.

Rubber-engine sizing is used during the early stages of an aircraft development program that is sufficiently important to warrant the development of an all-new engine. This is generally the case for a major military fighter or bomber program, and is sometimes the case for a transport-aircraft project such as the SST.

In these cases, the designer will use a rubber engine in the early stages of design, and then, with the customer, tell the engine company what characteristics the new engine should have. When the engine company finalizes the design for the new engine, it becomes fixed in size and thrust. The aircraft concept will then be finalized around this now-fixed engine.

Developing a new jet engine costs several billion dollars. Developing and certifying a new piston engine is also very expensive. Most aircraft projects do not rate development of a new engine, and so must rely on selecting the best of the existing engines. However, even projects that must use an existing engine may begin with a rubber-engine design study to determine what characteristics to look for in the selection of an existing engine.

The rubber engine can be scaled to any thrust so the thrust-to-weight ratio can be held to some desired value even as the aircraft weight is varied. The rubber-engine sizing approach allows the designer to size the aircraft to meet both performance and range goals, by solving for takeoff gross weight while holding the thrust-to-weight ratio required to meet the performance objectives. As the weight varies, the rubber-engine is scaled up or down as required.

This is not possible for fixed-engine aircraft sizing. When a fixed-size engine is used, either the mission range or the performance of the aircraft must become a fallout parameter.

For example, if a certain rate of climb must be attained, then the thrust-to-weight ratio cannot be allowed to fall to an extremely low value. If the calculation of the takeoff gross weight required for the desired range indicates that the weight is much higher than expected, then either the range must be reduced or the rate of climb must be relaxed.

A typical example of this is the would-be homebuilder who got a good buy on a Lycoming 0-320, and is designing an aircraft around that 150–hp engine. If the sizing results say that a bigger engine is required, the homebuilder will change the sizing requirements!

6.3 Rubber-Engine Sizing

Review of Sizing

Chapter 3 presented a quick method of sizing an aircraft using a configuration sketch and the selected aspect ratio. From this information a crude estimate of the maximum L/D was obtained. Using approximations of the specific fuel consumption, the changes in weight due to the fuel burned during cruise and loiter mission segments were estimated, expressed as the mission-segment weight fraction (W_i/W_{i-1}). Using these fractions and the approximate fractions for takeoff, climb, and landing that were provided in Table 3.2, the total mission weight fraction (W_x/W_0) was estimated.

For different classes of aircraft, statistical equations for the aircraft empty-weight fraction were provided in Table 3.1. Then, the takeoff weight was calculated using Eq. (3.4), repeated below as Eq. (6.1).

Since the empty weight was calculated using a guess of the takeoff weight, it was necessary to iterate toward a solution. This was done by calculating the empty-weight fraction from an initial guess of the takeoff weight and using Eq. (6.1) to calculate the resulting takeoff weight. If the calculated takeoff weight did not equal the initial guess, a new guess was made somewhere between the two.

$$W_0 = \frac{W_{\text{crew}} + W_{\text{payload}}}{1 - (W_f/W_0) - (W_e/W_0)} \tag{6.1}$$

where

$$\frac{W_f}{W_0} = 1.06\left(1 - \frac{W_x}{W_0}\right) \tag{6.2}$$

Equation (6.1) is limited in use to missions which do not have a sudden weight change, such as a payload drop. Also, in many cases Eq. (6.1) cannot be used for fixed-engine sizing.

Refined Sizing Equation

For missions with a payload drop or other sudden weight change, a slightly different sizing equation must be used. The takeoff weight is calculated by summing the crew weight, payload weight, fuel weight, and empty weight. This is shown in Eq. (6.3), which resembles Eq. (3.1) except that the payload now includes a fixed payload and a dropped payload. The empty weight is again expressed as an empty-weight fraction, but the fuel weight is determined directly.

$$W_0 = W_{\text{crew}} + W_{\text{fixed payload}} + W_{\text{dropped payload}} + W_{\text{fuel}} + W_{\text{empty}} \tag{6.3}$$

or

$$W_0 = W_{\text{crew}} + W_{\text{fixed payload}} + W_{\text{dropped payload}} + W_{\text{fuel}} + \left(\frac{We}{W_0}\right) W_0 \tag{6.4}$$

As before, an initial guess of the takeoff weight is used to determine a calculated takeoff weight, and the solution is iterated until the two are approximately equal to within a few percent. Refined methods for determining the empty-weight fraction and fuel used are discussed in the following.

Empty-Weight Fraction

The empty-weight fraction is estimated using improved statistical equations. Tables 6.1 and 6.2 were prepared using data from Ref. 1 to provide empty-weight equations which better reflect the weight impact of the major design variables. These are the aspect ratio, thrust-to-weight (or horsepower-to-weight) ratio, wing loading, and maximum speed.

The equations of Tables 6.1 and 6.2 result in a much better statistical fit, with only about half the standard deviation of the equations in Table 3.1. However, these equations should not be used to conduct design trade studies for one particular aircraft. That must be done using the component weight buildup methods in Chapter 15.

Table 6.1 Empty weight fraction vs W_0, A, T/W_0, W_0/S, and $M_{\max}$

$$W_e/W_0 = (a + bW_0^{C1}A^{C2}(T/W_0)^{C3}(W_0/S)^{C4}M_{\max}^{C5})K_{vs}$$

fps units	a	b	$C1$	$C2$	$C3$	$C4$	$C5$
Jet trainer	0	4.28	−0.10	0.10	0.20	−0.24	0.11
Jet fighter	−0.02	2.16	−0.10	0.20	0.04	−0.10	0.08
Military cargo/bomber	0.07	1.71	−0.10	0.10	0.06	−0.10	0.05
Jet transport	0.32	0.66	−0.13	0.30	0.06	−0.05	0.05

K_{VS} = variable sweep constant = 1.04 if variable sweep
= 1.00 if fixed sweep

Table 6.2 Empty weight fraction vs W_0, A, hp/W_0, W_0/S, and V_{max} (knots)

$$W_e/W_0 = a + bW_0^{C1}A^{C2}(\text{hp}/W_0)^{C3}(W_0/S)^{C4}V_{max}^{C5}$$

fps units	a	b	$C1$	$C2$	$C3$	$C4$	$C5$
Sailplane—unpowered	0	0.76	−0.05	0.14	0	−0.30	0.06
Sailplane—powered	0	1.21	−0.04	0.14	0.19	−0.20	0.05
Homebuilt—metal/wood	0	0.71	−0.10	0.05	0.10	−0.05	0.17
Homebuilt—composite	0	0.69	−0.10	0.05	0.10	−0.05	0.17
Gen. Av.—single engine	−0.25	1.18	−0.20	0.08	0.05	−0.05	0.27
Gen. Av.—twin engine	−0.90	1.36	−0.10	0.08	0.05	−0.05	0.20
Agricultural aircraft	0	1.67	−0.14	0.07	0.10	−0.10	0.11
Twin turboprop	0.37	0.09	−0.06	0.08	0.08	−0.05	0.30
Flying boat	0	0.42	−0.01	0.10	0.05	−0.12	0.18

Fuel Weight

The remaining unknown in Eq. (6.4) is the fuel weight. Previously this was estimated as a fuel fraction by determining the ratio between the weight at the end of the mission and the takeoff weight (W_x/W_0). As the only weight loss during the mission was due to fuel usage, the fuel fraction was found simply as $(1 - W_x/W_0)$. This cannot be assumed if the mission includes a weight drop.

If the mission includes a weight drop, it is necessary to calculate the weight of the fuel burned during every mission leg, and sum for the total mission fuel. The mission segment weight fractions (W_i/W_{i-1}) are calculated as before for all mission segments other than those that are weight drops. For each mission segment, the fuel burned is then equal to

$$W_{f_i} = \left(1 - \frac{W_i}{W_{i-1}}\right) W_{i-1} \tag{6.5}$$

The total mission fuel, W_{f_m}, then is equal to

$$W_{f_m} = \sum_1^x W_{f_i} \tag{6.6}$$

The total aircraft fuel includes the mission fuel as well as an allowance for reserve and trapped fuel. This reserve fuel allowance is usually 5%, and accounts for an engine with poorer-than-normal fuel consumption. An additional allowance of 1% for trapped (i.e., unusable) fuel is typical. Thus, the total aircraft fuel is

$$W_f = 1.06\left(\sum_1^x W_{f_i}\right) \tag{6.7}$$

The methods used for estimating the mission segment weight fractions are presented next. These are a combination of analytical and statistical methods, similar to the methods used in Chapter 3.

Engine Start, Taxi, and Takeoff

As before, the mission segment weight fraction for engine start, taxi, and takeoff is estimated historically. A reasonable estimate is

$$W_i/W_{i-1} = 0.97 - 0.99 \tag{6.8}$$

Climb and Accelerate

From data in Ref. 10, the weight fraction for an aircraft climbing and accelerating to cruise altitude and Mach number "M," (starting at Mach 0.1), will be approximately as follows:

$$\text{Subsonic:} \quad W_i/W_{i-1} = 1.0065 - 0.0325M \tag{6.9}$$

$$\text{Supersonic:} \quad W_i/W_{i-1} = 0.991 - 0.007M - 0.01M^2 \tag{6.10}$$

For an acceleration beginning at other than Mach 0.1, the weight fraction calculated by Eqs. (6.9) or (6.10) for the given ending Mach number should be divided by the weight fraction calculated for the beginning Mach number using Eqs. (6.9) or (6.10).

For example, acceleration from Mach 0.1–0.8 requires a weight fraction of about 0.9805, whereas acceleration from Mach 0.1–2.0 requires a weight fraction of 0.937. To accelerate from Mach 0.8–2.0 would require a weight fraction of (0.937/0.9805), or 0.956.

Cruise

Equation (3.6), repeated below as Eq. (6.11), is derived from the Breguet range equation for cruise as derived in Chapter 17. For propeller aircraft, the specific fuel consumption "C" is calculated from the propeller specific fuel consumption (C_p or C_{bhp}) using Eq. (3.10). Substitution of Eq. (3.10) into Eq. (6.11) yields Eq. (6.12).

$$\text{Jet:} \quad \frac{W_i}{W_{i-1}} = \exp \frac{-RC}{V(L/D)} \tag{6.11}$$

$$\text{Prop:} \quad \frac{W_i}{W_{i-1}} = \exp\left[\frac{-RC_{\text{power}}}{\eta_p(L/D)}\right] = \exp\left[\frac{-RC_{\text{bhp}}}{550\ \eta_p(L/D)}\right] (\text{fps units}) \tag{6.12}$$

where

R = range
C = specific fuel consumption
V = velocity
L/D = lift-to-drag ratio
η_p = propeller efficiency

During cruise and loiter, the lift equals the weight, so the L/D can be expressed as the inverse of the drag divided by the weight:

$$\frac{L}{D} = \frac{1}{\dfrac{qC_{D_0}}{W/S} + \dfrac{W}{S}\dfrac{1}{q\pi Ae}} \tag{6.13}$$

Note that the wing loading used in Eq. (6.13) and subsequent weight fraction equations is the actual wing loading at the condition being evaluated, not the takeoff wing loading.

Loiter

Repeating Eq. (3.8), the weight fraction for a loiter mission segment is:

$$\text{Jet:} \quad \frac{W_i}{W_{i-1}} = \exp\frac{-EC}{L/D} \tag{6.14}$$

where E = endurance or loiter time.

(Note—watch the units!) Substitution of Eq. (3.10) into Eq. (6.14) yields:

$$\text{Prop:} \quad \frac{W_i}{W_{i-1}} = \exp\left[\frac{-EVC_{\text{power}}}{\eta_p(L/D)}\right] = \exp\left[\frac{-EVC_{\text{bhp}}}{550\ \eta_p(L/D)}\right] (\text{fps units}) \tag{6.15}$$

Combat/Known-Time Fuel Burn

The combat mission leg is normally specified as either a time duration ("d") at maximum power (typically $d = 3$ min), or as a certain number of combat turns at maximum power at some altitude and Mach number. The weight of the fuel burned is equal to the product of thrust, specific fuel consumption, and duration of the combat, so the mission segment weight fraction is:

$$W_i/W_{i-1} = 1 - C(T/W)(d) \tag{6.16}$$

Note that the T/W is defined at combat weight and thrust, not at takeoff conditions. Again, watch the units, especially the time units.

If the combat is defined by some number of turns, the duration of combat (d) must be calculated. The time to complete "x" turns is the total number of radians to turn divided by the turn rate. When combined with Eq. (5.17), this yields

$$d = \frac{2\pi x}{\psi} = \frac{2\pi V x}{g\sqrt{n^2 - 1}} \tag{6.17}$$

The load factor "n" for a sustained combat turn is found by assuming that the thrust angle is approximately aligned with the flight direction, and so the thrust must equal the drag. The lift must equal the weight times the load factor n, which yields

$$n = (T/W)(L/D) \tag{6.18}$$

This is subject to the constraints of maximum structural load factor [Eq. (6.19)] and maximum available lift [Eq. (6.20)].

$$n \leq n_{\max} \tag{6.19}$$

$$n \leq \frac{qC_{L_{\max}}}{W/S} \tag{6.20}$$

The lift-to-drag ratio is found by including the load factor term into Eq. (6.13), which results in Eq. (6.21). The changes to e at combat conditions discussed in the last chapter should be used in Eq. (6.21).

$$\frac{L}{D} = \frac{1}{q\dfrac{C_{D_0}}{n(W/S)} + \dfrac{n(W/S)}{q\pi Ae}} \tag{6.21}$$

Descent for Landing

Descent is estimated historically:

$$W_i/W_{i-1} = 0.990 \text{ to } 0.995 \tag{6.22}$$

Landing and Taxi Back

Again, a historical approximation is used:

$$W_i/W_{i-1} = 0.992 \text{ to } 0.997 \tag{6.23}$$

Summary of Refined Sizing Method

The design and sizing method just presented, as summarized in Fig. 6.1, resembles in many respects the first-order method presented as Fig. 3.7, but makes use of more sophisticated analytical techniques and also permits sizing to missions that include weight drops.

From the design objectives and sizing mission, the wing geometry can be selected and an estimate of e obtained. A conceptual sketch or initial layout is used to estimate the wetted-area ratio, from which C_{D_0} is estimated. (Remember that e will be reduced during high-lift, combat conditions.)

The methods of the previous chapter are used to select initial values for thrust-to-weight (or horsepower-to-weight) ratio and wing loading. Then the methods of this chapter are used along with engine data to determine the mission-segment weight fractions for each leg of the design mission.

The iteration for takeoff gross weight (W_0) begins with an initial guess as to W_0, and then the aircraft weight is calculated throughout the mission. For each mission leg, the aircraft weight will be reduced by either the weight of fuel burned or the payload weight dropped. Also, the total fuel burned is summed throughout the mission. Equations (6.7) and (6.4) are then used, along with a statistical empty weight fraction estimation, to arrive at a calculated W_0.

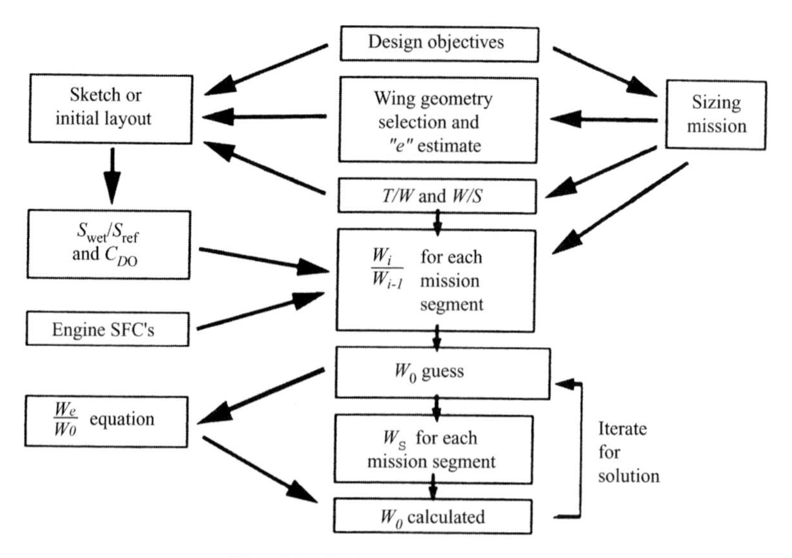

Fig. 6.1 Refined sizing method.

If this does not equal the guessed value for W_0, a new guess for W_0 is selected between the two values. Experience indicates that the solution will converge most rapidly if the new guess for W_0 is about three-fourths of the way from the initial guess to the calculated W_0 value.

This procedure is less complicated than it sounds. Examples can be found in Chapter 23.

(An alternative sizing method starts with a W_0 guess, and then subtracts the payload weight, crew weight, and calculated fuel weight to arrive at the "empty weight required" to perform the mission. This is compared to the statistical "empty weight available." If the empty weight required exceeds the empty weight available, then W_0 must be increased for the next iteration. This method is mathematically identical to the method just presented, but provides less obvious guidance as to the next value of W_0 to use for iteration.)

6.4 Fixed-Engine Sizing

The sizing procedure for the fixed-size engine is similar to the rubber-engine sizing, with several exceptions. These result from the fact that either the mission range or the performance must be considered a secondary parameter, and allowed to vary as the aircraft is sized.

If the range is allowed to vary, the sizing problem is very simple. The required thrust-to-weight ratio (T/W) is determined as in the last section to provide all required performance capabilities, using the known characteristics of the selected engine. Then the takeoff gross weight is determined as the total engine takeoff thrust divided by the required takeoff thrust-to-weight ratio.

$$W_0 = \frac{NT_{\text{per engine}}}{(T/W)} \tag{6.24}$$

where $N =$ number of engines.

With the takeoff weight known, the range capability can be determined from Eq. (6.4) using a modified iteration technique. The known takeoff weight is repeatedly used as the "guess" W_0, and the range for one or more cruise legs is varied until the calculated W_0 equals the known W_0.

This technique can also be used to vary mission parameters other than range. For example, a research aircraft may be sized for a certain radius (range out and back) with the number of minutes of test time as the variable parameter.

If some range requirement must be satisfied, then performance must be the secondary parameter. The takeoff gross weight will be set by fuel requirements, and the fixed-size engine may not necessarily provide the thrust-to-weight ratio desired for performance considerations.

In this case the takeoff gross weight can be solved by iteration of Eq. (6.4) as for the rubber-engine case, with one major exception. The thrust-to-weight ratio is now permitted to vary during the sizing iterations. Equation (6.16) cannot be used for determining a weight fraction for combat mission legs as it assumes a known T/W.

Instead, the fuel burned during combat by a fixed-size engine is treated as a weight drop. For a given engine, the fuel burned during a combat leg of duration d is simply the thrust times the specific fuel consumption times the duration:

$$W_f = CTd \tag{6.25}$$

The weight of fuel calculated by Eq. (6.25) is treated as a weight drop in the iterations to solve Eq. (6.4). Once the takeoff gross weight is determined, the resulting thrust-to-weight ratio must be used to determine the actual aircraft performance for the requirements evaluated in the last chapter. If the requirements are not met, then either your aircraft design is not very good or the requirements are too tough!

6.5 Geometry Sizing

Fuselage

Once the takeoff gross weight has been estimated, the fuselage, wing, and tails can be sized. Many methods exist to initially estimate the required fuselage size.

For certain types of aircraft, the fuselage size is determined strictly by "real-world constraints." For example, a large passenger aircraft devotes most of its length to the passenger compartment. Once the number of passengers is known and the number of seats across is selected, the fuselage length and diameter are essentially determined.

For initial guidance during fuselage layout and tail sizing, Table 6.3 provides statistical equations for fuselage length developed from data provided in Ref. 1. These are based solely upon takeoff gross weight, and give remarkably good correlations to most existing aircraft.

Fuselage fineness ratio is the ratio between the fuselage length and its maximum diameter. If the fuselage cross section is not a circle, an equivalent diameter is calculated from the cross-sectional area. Numerous design books such as the classic Hoerner Fluid Dynamic Drag (Ref. 8) indicate an optimum fineness ratio of around three. This assumes a fixed diameter; in other words,

Table 6.3 Fuselage length vs W_0 (lb or {kg})

Length = aW_0^C (ft or {m})	a	C
Sailplane—unpowered	0.86 {0.383}	0.48
Sailplane—powered	0.71 {0.316}	0.48
Homebuilt—metal/wood	3.68 {1.35}	0.23
Homebuilt—composite	3.50 {1.28}	0.23
General aviation—single engine	4.37 {1.6}	0.23
General aviation—twin engine	0.86 {0.366}	0.42
Agricultural aircraft	4.04 {1.48}	0.23
Twin turboprop	0.37 {0.169}	0.51
Flying boat	1.05 {0.439}	0.40
Jet trainer	0.79 {0.333}	0.41
Jet fighter	0.93 {0.389}	0.39
Military cargo/bomber	0.23 {0.104}	0.50
Jet transport	0.67 {0.287}	0.43

"what length minimizes total drag given a certain maximum fuselage diameter?" This would be of concern in a design where there is a specific layout requirement that forces a large maximum cross-section area, such as side-by-side seating for two people.

A fineness ratio of three might not provide enough tail moment arm (see the following), and so a tail boom can be added, with a smooth fairing from the front part of the fuselage. This creates the streamlined "tadpole" shape characteristic of many sailplanes and several recent small airplanes.

For a larger aircraft where tight packaging is desired and there is not a requirement for a large cross-section area, a more important question would be "what fineness ratio minimizes drag for a given total volume enclosed?" A recent analytical optimization study (Ref. 141) found that if volume is held constant then the optimum fineness ratio for subsonic aircraft is somewhere between 6 and 8. Interestingly enough, this matches the fineness ratios of most successful airships.

Supersonic drag is typically minimized by a fineness ratio of about 14, but that is very design dependent.

Whatever fuselage fineness ratio is thought to be optimal, when making the actual design layout the various "real-world" constraints such as cockpit and payload shape must take priority. For most design efforts the realities of packaging the internal components will ultimately establish the fuselage length and diameter—but it is good to know the optimal fineness ratio as a layout goal.

Wing

The actual wing size can now be determined simply as the takeoff gross weight divided by the takeoff wing loading. Remember that this is the reference area of the theoretical, trapezoidal wing, and includes the area extending into the aircraft centerline.

Tail Volume Coefficient

For the initial layout, a historical approach is used for the estimation of tail size. The effectiveness of a tail in generating a moment about the center of gravity is proportional to the force (i.e., lift) produced by the tail and to the tail moment arm.

The primary purpose of a tail is to counter the moments produced by the wing. Thus, it would be expected that the tail size would be in some way related to the wing size. In fact, there is a directly proportional relationship between the two, as can be determined by examining the moment equations presented in Chapter 16. Therefore, the tail area divided by the wing area should show some consistent relationship for different aircraft, if the effects of tail moment arm could be accounted for.

The force due to tail lift is proportional to the tail area. Thus, the tail effectiveness is proportional to the tail area times the tail moment arm. This product has units of volume, which leads to the "tail volume coefficient" method for initial estimation of tail size.

Rendering this parameter nondimensional requires dividing by some quantity with units of length. For a vertical tail, the wing yawing moments which must be countered are most directly related to the wing span b_W. This leads to the "vertical tail volume coefficient," as defined by Eq. (6.26). For a horizontal tail or canard, the pitching moments which must be countered are most directly related to the wing mean chord ($\overline{C}_W$). This leads to the "horizontal tail volume coefficient," as shown by Eq. (6.27).

$$c_{VT} = \frac{L_{VT} S_{VT}}{b_W S_W} \tag{6.26}$$

$$c_{HT} = \frac{L_{HT} S_{HT}}{\overline{C}_W S_W} \tag{6.27}$$

Note that the moment arm (L) is commonly approximated as the distance from the tail quarter-chord (i.e., 25% of the mean chord length measured back from the leading edge of the mean chord) to the wing quarter-chord.

The definition of tail moment arm is shown in Fig. 6.2, along with the definitions of tail area. Observe that the horizontal tail area is commonly measured to the aircraft centerline, while a canard's area is commonly considered to include only the exposed area. If twin vertical tails are used, the vertical tail area is the sum of the two.

Table 6.4 provides typical values for volume coefficients for different classes of aircraft. These values (conservative averages based upon data in Refs. 1 and 11), are used in Eqs. (6.28) or (6.29) to calculate tail area.

(Incidentally, Ref. 11 compiles a tremendous amount of aircraft data and is highly recommended for every designer's library.)

$$S_{VT} = c_{VT} b_W S_W / L_{VT} \tag{6.28}$$

$$S_{HT} = c_{HT} \overline{C}_W S_W / L_{HT} \tag{6.29}$$

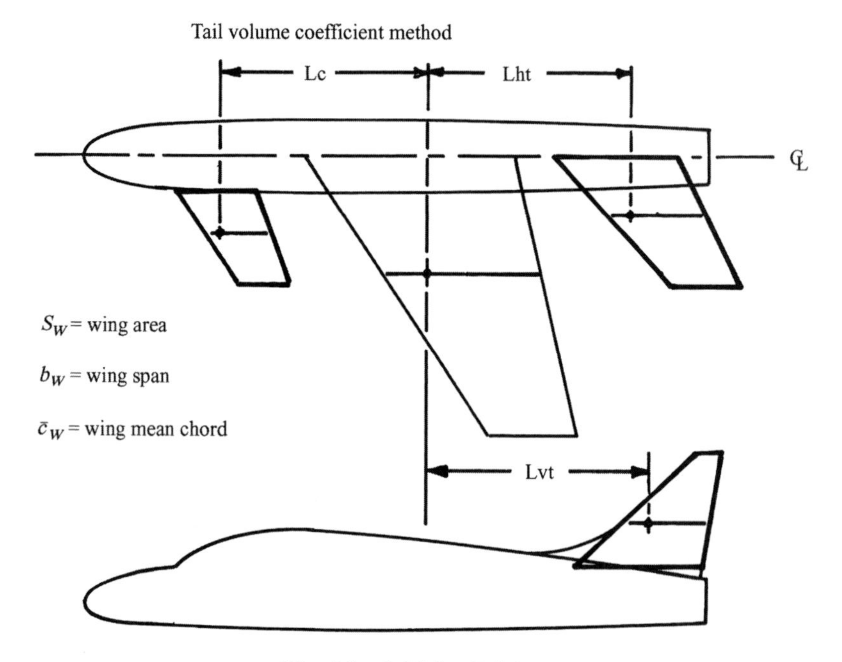

Fig. 6.2 Initial tail sizing.

To calculate tail size, the moment arm must be estimated. This can be approximated at this stage of design by a percent of the fuselage length as previously estimated.

For an aircraft with a front-mounted propeller engine, the tail arm is about 60% of the fuselage length. For an aircraft with the engines on the wings, the

Table 6.4 Tail volume coefficient

	Typical values	
	Horizontal c_{HT}	Vertical c_{VT}
Sailplane	0.50	0.02
Homebuilt	0.50	0.04
General aviation—single engine	0.70	0.04
General aviation—twin engine	0.80	0.07
Agricultural	0.50	0.04
Twin turboprop	0.90	0.08
Flying boat	0.70	0.06
Jet trainer	0.70	0.06
Jet fighter	0.40	0.07
Military cargo/bomber	1.00	0.08
Jet transport	1.00	0.09

tail arm is about 50–55% of the fuselage length. For aft-mounted engines the tail arm is about 45–50% of the fuselage length. A sailplane has a tail moment arm of about 65% of the fuselage length.

For an all-moving tail, the volume coefficient can be reduced by about 10–15%. For a "T-tail," the vertical-tail volume coefficient can be reduced by approximately 5% due to the end-plate effect, and the horizontal tail volume coefficient can be reduced by about 5% due to the clean air seen by the horizontal. Similarly, the horizontal tail volume coefficient for an "H-tail" can be reduced by about 5%.

For an aircraft which uses a "V-tail," the required horizontal and vertical tail sizes should be estimated as before. Then the V surfaces should be sized to provide the same total surface area (Ref. 3) as required for conventional tails. The tail dihedral angle should be set to the arctangent of the square root of the ratio between the required vertical and horizontal tail areas. This should be near 45 deg.

The horizontal tail volume coefficient for an aircraft with a control-type canard is approximately 0.1, based upon the relatively few aircraft of this type that have flown. For canard aircraft there is a much wider variation in the tail moment arm. Typically, the canard aircraft will have a moment arm of about 30–50% of the fuselage length.

For a lifting canard aircraft, the volume coefficient method isn't applicable. Instead, an area split must be selected by the designer. The required total wing area is then allocated accordingly. Typically, the area split allocates about 25% to the canard and 75% to the wing, although there can be wide variation. A 50-50 split produces a tandem-wing aircraft.

For an airplane with a computerized "active" flight control system, the statistically estimated tail areas may be reduced by approximately 10% provided that trim, engine-out, and nosewheel liftoff requirements can be met. These are discussed in Chapter 16.

6.6 Control-Surface Sizing

The primary control surfaces are the ailerons (roll), elevator (pitch), and rudder (yaw). Final sizing of these surfaces is based upon dynamic analysis of control effectiveness, including structural bending and control-system effects. For initial design, the following guidelines are offered.

The required aileron area can be estimated from Fig. 6.3, an updated version of a figure from Ref. 12. In span, the ailerons typically extend from about 50% to about 90% of the span. In some aircraft, the ailerons extend all the way out to the wing tips. This extra 10% provides little control effectiveness due to the vortex flow at the wing tips, but can provide a location for an aileron mass balance (see the following).

Wing flaps occupy the part of the wing span inboard of the ailerons. If a large maximum lift coefficient is required, the flap span should be as large as possible. One way of accomplishing this is through the use of spoilers rather than ailerons. Spoilers are plates located forward of the flaps on the top of the wing, typically aft of the maximum thickness point. Spoilers are deflected upward into the slipstream to reduce the wing's lift. Deploying the spoiler on one wing will cause a large rolling moment.

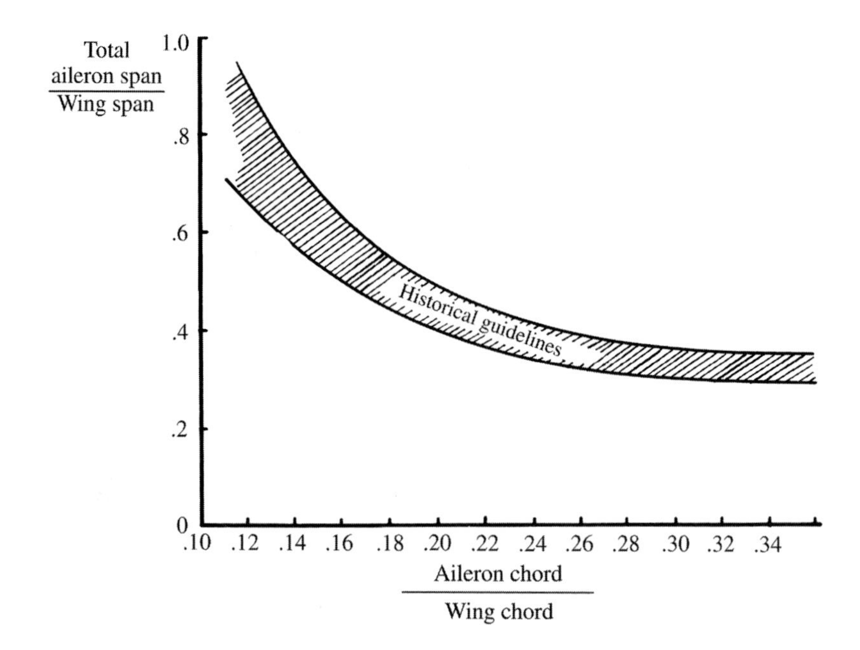

Fig. 6.3 Aileron guidelines.

Spoilers are commonly used on jet transports to augment roll control at low speed, and can also be used to reduce lift and add drag during the landing rollout. However, because spoilers have very nonlinear response characteristics, they are difficult to implement for roll control when using a manual flight control system.

High-speed aircraft can experience a phenomenon known as "aileron reversal" in which the air loads placed upon a deflected aileron are so great that the wing itself is twisted. At some speed, the wing may twist so much that the rolling moment produced by the twist will exceed the rolling moment produced by the aileron, causing the aircraft to roll the wrong way.

To avoid this, many transport jets use an auxiliary, inboard aileron for high-speed roll control. Spoilers can also be used for this purpose. Several military fighters rely upon "rolling tails" (horizontal tails capable of being deflected nonsymmetrically) to achieve the same result.

Elevators and rudders generally begin at the side of the fuselage and extend to the tip of the tail or to about 90% of the tail span. High-speed aircraft sometimes use rudders of large chord which only extend to about 50% of the span. This avoids a rudder effectiveness problem similar to aileron reversal. Guidelines for preliminary control surface sizing are offered in Table 6.5.

Control surfaces are usually tapered in chord by the same ratio as the wing or tail surface so that the control surface maintains a constant percent chord (Fig. 6.4). This allows spars to be straight-tapered rather than curved. Ailerons and flaps are typically about 15–25% of the wing chord. Rudders and elevators are typically about 25–50% of the tail chord.

Table 6.5 Control surface sizing guidelines

Aircraft	Elevator C_e/C	Rudder C_r/C
Fighter/attack	0.30[a]	0.30
Jet transport	0.25[b]	0.32
Jet trainer	0.35	0.35
Biz jet	0.32[b]	0.30
GA single	0.45	0.40
GA twin	0.36	0.46
Sailplane	0.43	0.40

[a]Supersonic usually all-moving only.
[b]Often all-moving plus elevator.

Control-surface "flutter," a rapid oscillation of the surface caused by the airloads, can tear off the control surface or even the whole wing. Flutter tendencies are minimized by using mass balancing and aerodynamic balancing. Flutter is discussed in more detail in Chapter 8.

Mass balancing refers to the addition of weight forward of the control-surface hingeline to counterbalance the weight of the control surface aft of the hingeline. This greatly reduces flutter tendencies. To minimize the weight penalty, the balance weight should be located as far forward as possible. Some aircraft mount the balance weight on a boom flush to the wing tip. Others bury the mass balance within the wing, mounted on a boom attached to the control surface.

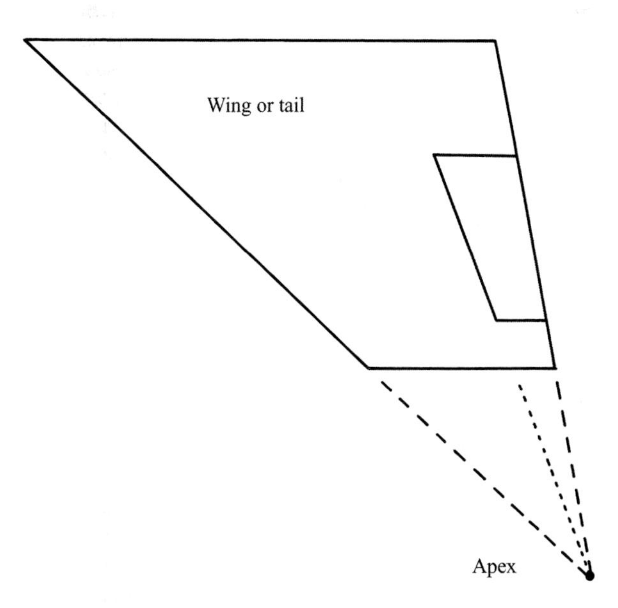

Fig. 6.4 Constant-percent chord control surface.

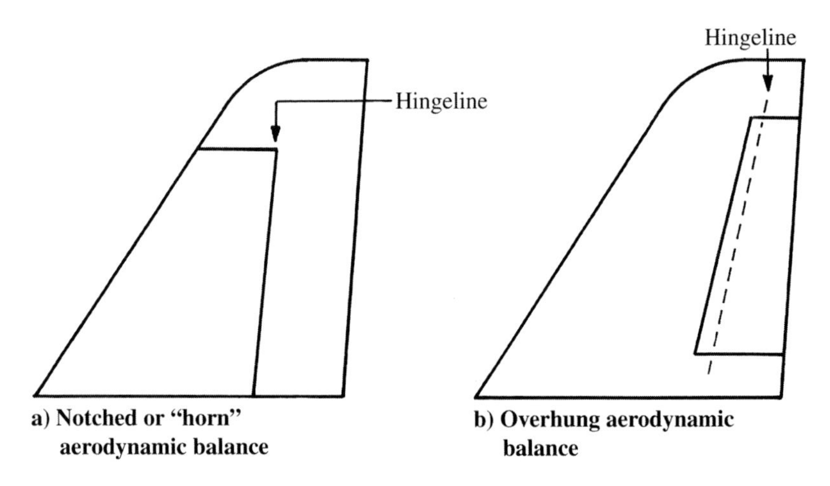

Fig. 6.5 Aerodynamic balance.

An aerodynamic balance is a portion of the control surface in front of the hinge line. This lessens the control force required to deflect the surface, and helps to reduce flutter tendencies.

The aerodynamic balance can be a notched part of the control surface (Fig. 6.5a), an overhung portion of the control surface (Fig. 6.5b), or a combination of the two. The notched balance is not suitable for ailerons or for any surface in high-speed flight. The hinge axis should be no farther aft than about 20% of the average chord of the control surface.

An old naval architects' approximation for balanced rudders can be used for a first layout of the hingeline of a balanced control surface, as follows: Break the control surface into spanwise strips. For a movable surface trailing a fixed surface, assume the center of pressure is at 0.33 of the movable chord length. For a movable surface in the freestream, as in the top of the rudder in Fig. 6.5, assume that the center of pressure is at 0.20 of the chord length. Add up the centers of pressure, weighted by the areas, to find an overall center of pressure and make sure that the hingeline is well ahead of it. Then, don't trust the result—use a more sophisticated analysis method as soon as possible.

The horizontal tail for a manually controlled aircraft is usually configured such that the elevator will have a hinge line perpendicular to the aircraft centerline. This permits connecting the left- and right-hand elevator surfaces with a torque tube, which reduces elevator flutter tendencies.

Some aircraft have no separate elevator. Instead, the entire horizontal tail is mounted on a spindle to provide variable tail incidence. This provides outstanding "elevator" effectiveness but is somewhat heavy. Some general-aviation aircraft use such an all-moving tail, but it is most common for supersonic aircraft, where it can be used to trim the rearward shift in aerodynamic center that occurs at supersonic speeds.

A few aircraft such as the F-23, SR-71, and North American F-107 have used all-moving vertical tails to increase control authority.

7
Configuration Layout and Loft

7.1 Introduction

The process of aircraft conceptual design includes numerous statistical estimations, analytical predictions, and numerical optimizations. However, the end product of aircraft design is a drawing. While the analytical tasks are vitally important, the designer must remember that these tasks serve only to influence the drawing, for it is the drawing alone that ultimately will be used to fabricate the aircraft.

All of the analysis efforts to date were performed to guide the designer in the layout of the initial drawing. Once that is completed, a detailed analysis can be conducted to resize the aircraft and determine its actual performance. This is discussed in Chapters 12–19.

This detailed analysis is time-consuming and costly, so it is essential that the initial drawing be credible. Otherwise, substantial effort will be wasted upon analyzing an unrealistic aircraft.

This chapter and Chapters 8–11 discuss the key concepts required to develop a credible initial drawing of a conceptual aircraft design. These concepts include the development of a smooth, producible, and aerodynamically acceptable external geometry, the installation of the internal features such as the crew station, payload, landing gear, and fuel system, and the integration of the propulsion system.

Real-world considerations that must be met by the design include the correct relationship between the aerodynamic center and the center of gravity, the proper amount of pilot outside visibility, and sufficient internal access for production and maintenance.

7.2 End Products of Configuration Layout

The outputs of the configuration layout task will be design drawings of several types as well as the geometric information required for further analysis.

The design layout process generally begins with a number of conceptual sketches. Figure 7.1 illustrates an actual, unretouched sketch from a fighter conceptual design study (Ref. 13). As can be seen, these sketches are crude and quickly done, but depict the major ideas which the designer intends to incorporate into the actual design layout.

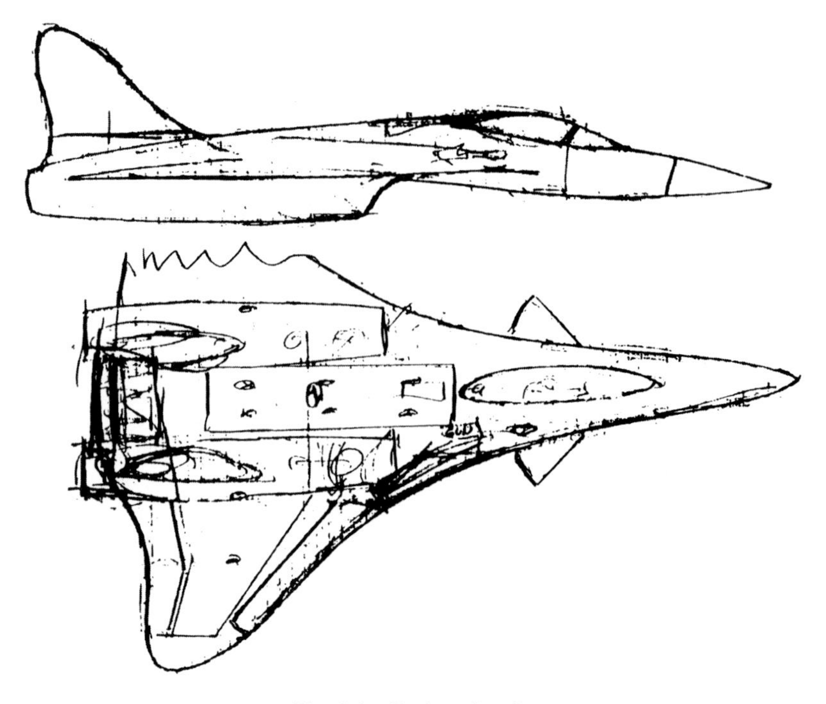

Fig. 7.1 Design sketch.

A good sketch will show the overall aerodynamic concept and indicate the locations of the major internal components. These should include the landing gear, crew station, payload or passenger compartment, propulsion system, fuel tanks, and any unique internal components such as a large radar. Conceptual sketches are not usually shown to anybody after the actual layout is developed, but may be used among the design engineers to discuss novel ideas before they begin the layout.

The actual design layout is developed using the techniques to be discussed in the following chapters.

Figure 7.2 is the initial design developed from the sketch shown as Fig. 7.1. A computer-aided conceptual design system was used to develop a three-dimensional geometric model of the aircraft concept (Ref. 14). The design techniques are similar whether a computer or a drafting board is used for the initial design.

Figure 7.3 shows a drafting table design layout, Rockwell's entry in the competition to build the X-29 Forward Sweep Demonstrator. This drawing typifies initial design layouts developed by major airframe companies during design studies.

A design layout such as those shown in Figs. 7.2 and 7.3 represents the primary input into the analysis and optimization tasks discussed in Chapters 12–19. Three other inputs must be prepared by the designer: the wetted-area plot (Fig. 7.4), volume distribution plot (Fig. 7.5), and fuel-volume plots for

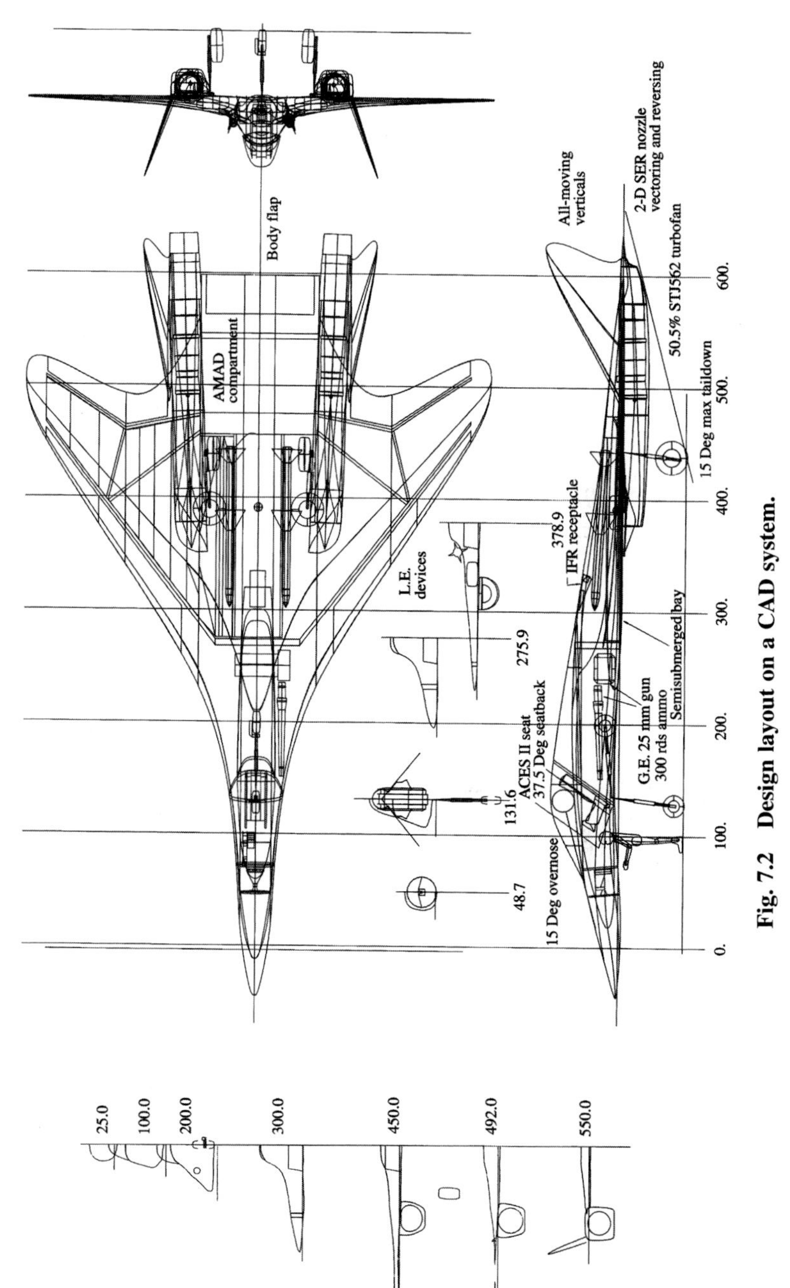

Fig. 7.2 Design layout on a CAD system.

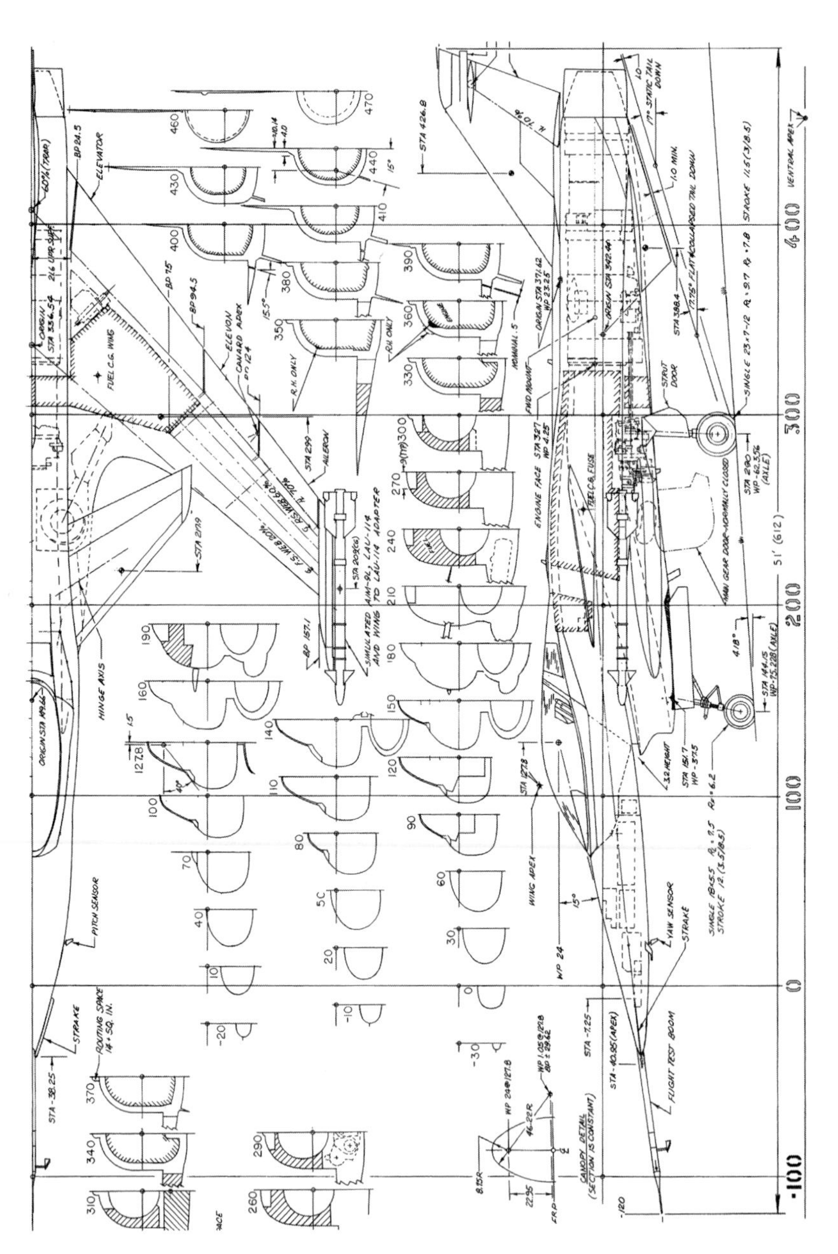

Fig. 7.3 FSW design layout.

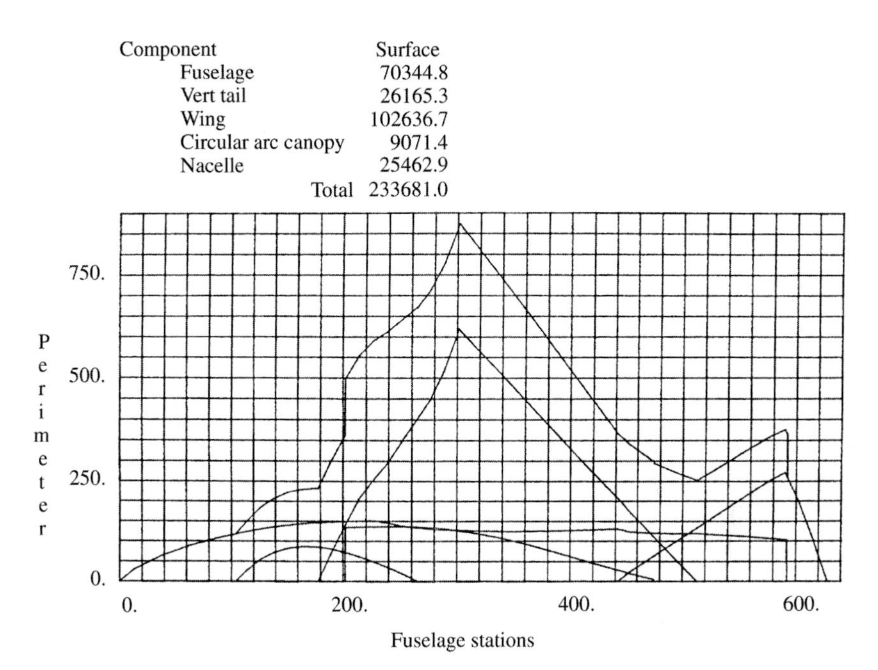

Component	Surface
Fuselage	70344.8
Vert tail	26165.3
Wing	102636.7
Circular arc canopy	9071.4
Nacelle	25462.9
Total	233681.0

Fig. 7.4 Wetted area plot.

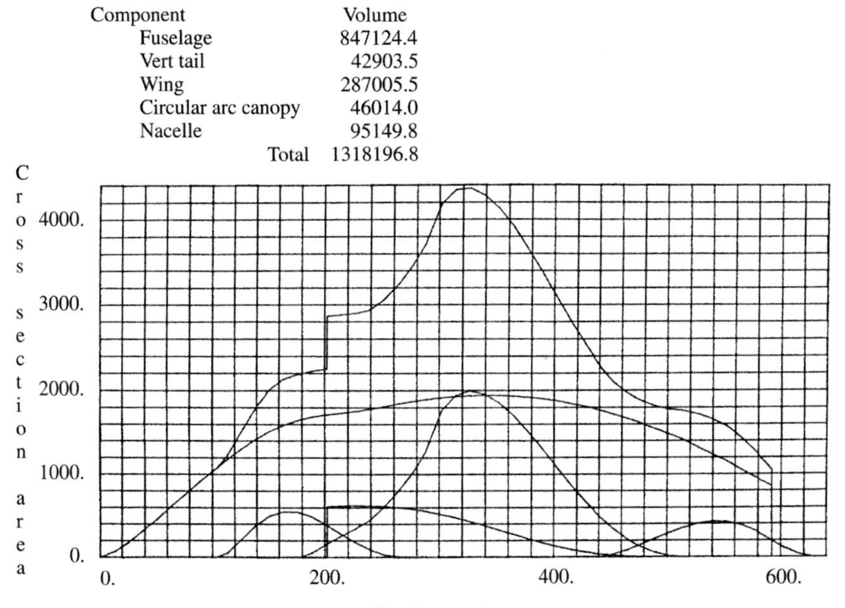

Component	Volume
Fuselage	847124.4
Vert tail	42903.5
Wing	287005.5
Circular arc canopy	46014.0
Nacelle	95149.8
Total	1318196.8

Fig. 7.5 Volume distribution plot.

the fuel tanks. Preparation of the wetted-area and volume plots is discussed later in this chapter; the fuel-volume determination is discussed in Chapter 10.

Once the design has been analyzed, optimized, and redrawn for a number of iterations of the conceptual design process, a more detailed drawing can be prepared. Called the "inboard profile" drawing, this depicts in much greater detail the internal arrangement of the subsystems. Figure 7.6 illustrates the inboard profile prepared for the design of Fig. 7.3. A companion drawing, not shown, would depict the internal arrangement at 20–50 cross-sectional locations.

The inboard profile is far more detailed than the initial layout. For example, while the initial layout may merely indicate an avionics bay based upon a statistical estimate of the required avionics volume, the inboard profile drawing will depict the actual location of every piece of avionics (i.e., "black boxes") as well as the required wire bundles and cooling ducts.

The inboard profile is generally a team project, and takes many weeks. During the preparation of the inboard profile, it is not uncommon to find that the initial layout must be changed to provide enough room for everything. As this can result in weeks of lost effort, it is imperative that the initial layout be as well thought out as possible.

Figure 7.7 shows a side-view inboard profile prepared in 1942 for an early variant of the P-51. This detailed drawing shows virtually every internal system, including control bellcranks, radio boxes, and fuel lines. Preparation of such a detailed drawing goes beyond the scope of this book, but aspiring designers should be aware of them.

At about the same time that the inboard profile drawing is being prepared, a "lines control" drawing may be prepared that refines and details the external geometry definition provided on the initial layout. Again, such a detailed drawing goes beyond the scope of this book. Also, most major companies now use computer-aided design and lofting systems that do not require a lines control drawing, and the inboard profile development can be done in three dimensions with automatic checking for component clearances.

After the inboard profile drawing has been prepared, an "inboard isometric" drawing (Fig. 7.8) may be prepared. It will usually be prepared by the art group for the purpose of illustration only, and be used in briefings and proposals. Such a drawing is frequently prepared and published by aviation magazines for existing aircraft. (In fact, the magazine illustrations are usually better than those prepared by the aircraft companies!)

7.3 Conic Lofting

"Lofting" is the process of defining the external geometry of the aircraft. "Production lofting," the most detailed form of lofting, provides an exact, mathematical definition of the entire aircraft including such minor details as the intake and exhaust ducts for the air conditioning.

A production-loft definition is expected to be accurate to within a few hundredths of an inch (or less) over the entire aircraft. This allows the different parts of the aircraft to be designed and fabricated at different plant sites yet fit together perfectly during final assembly. Most aircraft companies now use

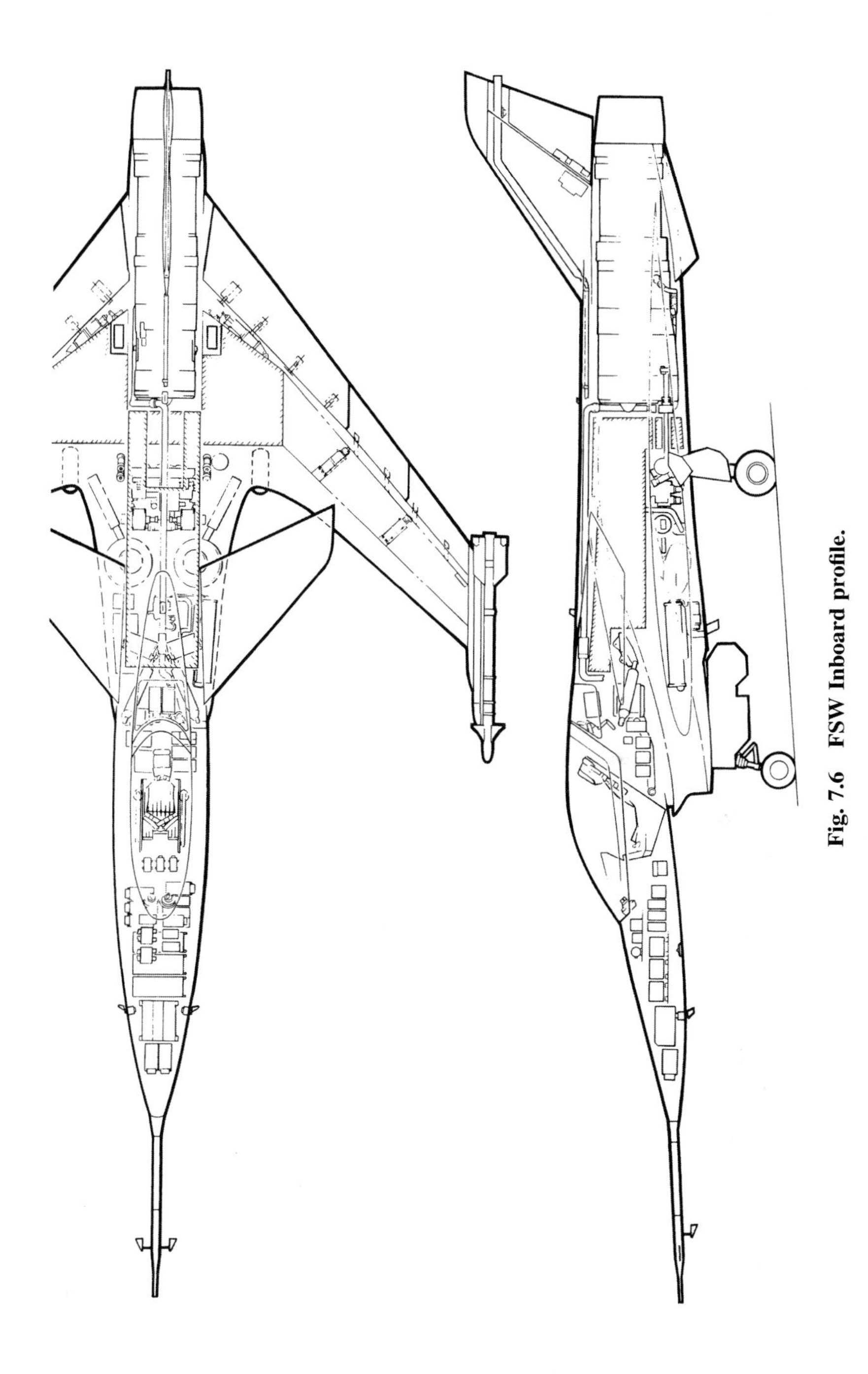

Fig. 7.6 FSW Inboard profile.

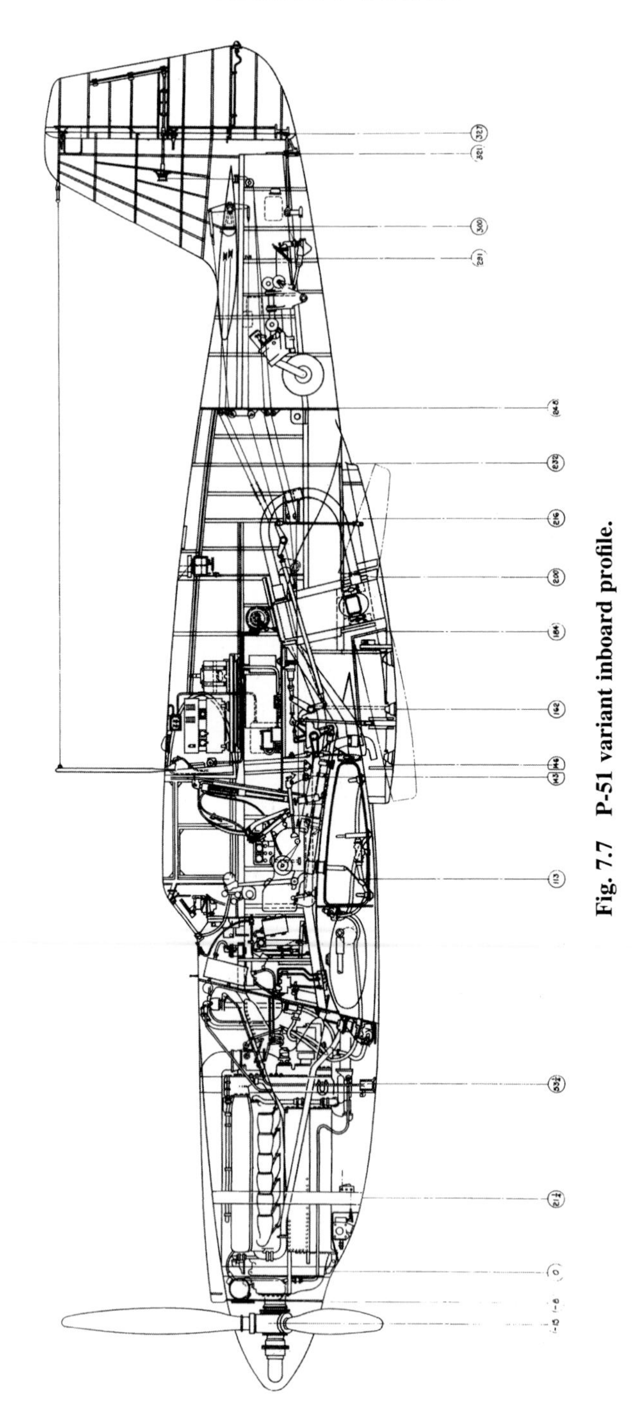

Fig. 7.7 P-51 variant inboard profile.

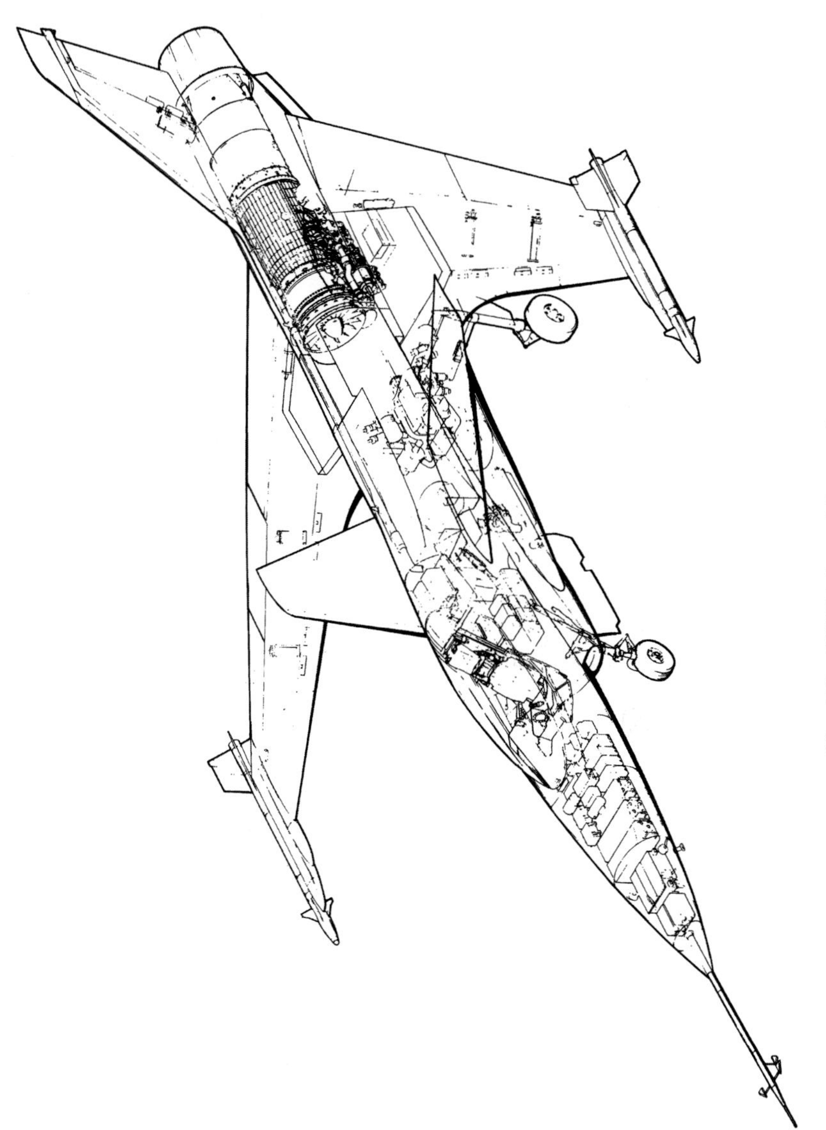

Fig. 7.8 FSW inboard isometric.

computer-aided design and loft systems that incorporate methods discussed in Ref. 80. These systems are so accurate that different parts of the aircraft can be designed and built in different locations, yet will fit together perfectly.

For an initial layout it is not necessary to go into as much detail. However, the overall lofting of the fuselage, wing, tails, and nacelles must be defined sufficiently to show that these major components will properly enclose the required internal components and fuel tanks while providing a smooth aerodynamic contour.

Lofting gets its name from shipbuilding. The definition of the hull shape was done in the loft over the shipyard, using enormous drawings. To provide a smooth longitudinal contour, points taken from the desired cross sections were connected longitudinally on the drawing by flexible "splines," long, thin wood or plastic rulers held down at certain points by lead "ducks" (pointed weights—see Fig. 7.9).

This technique was used for early aircraft lofting, but suffers from two disadvantages. First, it requires a lot of trial and error to achieve a smooth surface both in cross section and longitudinally.

Second, and perhaps more important, this method does not provide a unique mathematical definition of the surface. To create a new cross section requires a tremendous amount of drafting effort, especially for a canted cross section (i.e., a cross-sectional cut at some angle other than perpendicular to the centerline of the aircraft). In addition to the time involved, this method is prone to mismatch errors.

A new method of lofting was used for the first time on the P-51 Mustang (Ref. 15). This method, now considered traditional, is based upon a mathematical curve form known as the "conic."

The two great advantages of the conic method are the wide variety of curves that it can represent and the ease with which it can be constructed on the drafting table.

While many other forms of lofting are in use, conic lofting has been the most widely used. Also, an understanding of conic lofting provides the necessary foundation to learn the other forms of lofting, including computer-aided lofting.

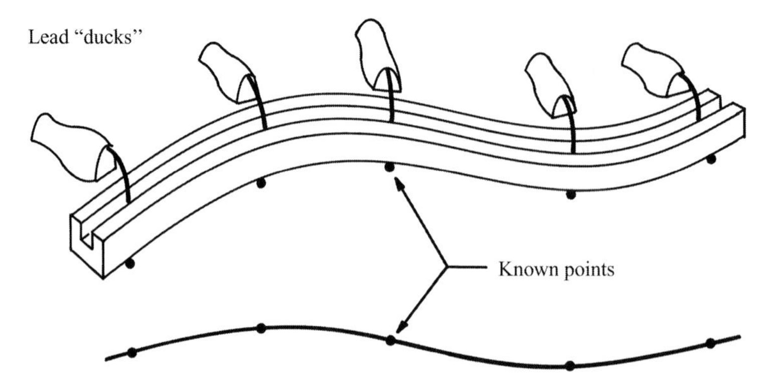

Fig. 7.9 Spline lofting.

A conic is a second-degree curve whose family includes the circle, ellipse, parabola, and hyperbola. The generalized form of the conic is given in Eq. (7.1). The conic is best visualized as a slanted cut through a right circular cone (Fig. 7.10). A number of specialized conic equations are provided in Ref. 80.

$$C_1X^2 + C_2XY + C_3Y^2 + C_4X + C_5Y + C_6 = 0 \tag{7.1}$$

The shape of the conic depends upon the angle of the cut through the cone. If the cut is flat (i.e., perpendicular to the axis of the cone), then the resulting curve will be a circle; if somewhat slanted, an ellipse; if exactly parallel to the opposite side, a parabola. A greater cut angle yields a hyperbola.

A conic curve is constructed from the desired start and end points ("A" and "B"), and the desired tangent angles at those points. These tangent angles intersect at point "C." The shape of the conic between the points A and B is defined by some shoulder point "S." (The points labeled "E" in Fig. 7.10 are a special type of shoulder point, discussed later.) Figure 7.11 illustrates the rapid graphical layout of a conic curve.

The first illustration in Fig. 7.11 shows the given points A, B, C, and S. In the second illustration, lines have been drawn from A and B, passing through S.

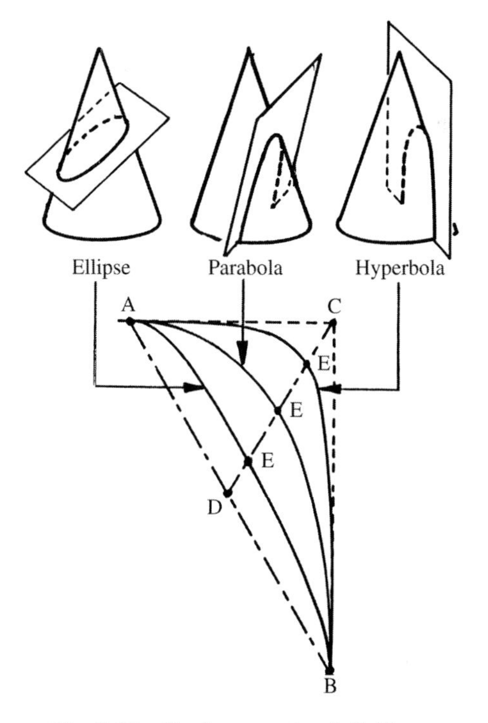

Fig. 7.10 Conic geometry definition.

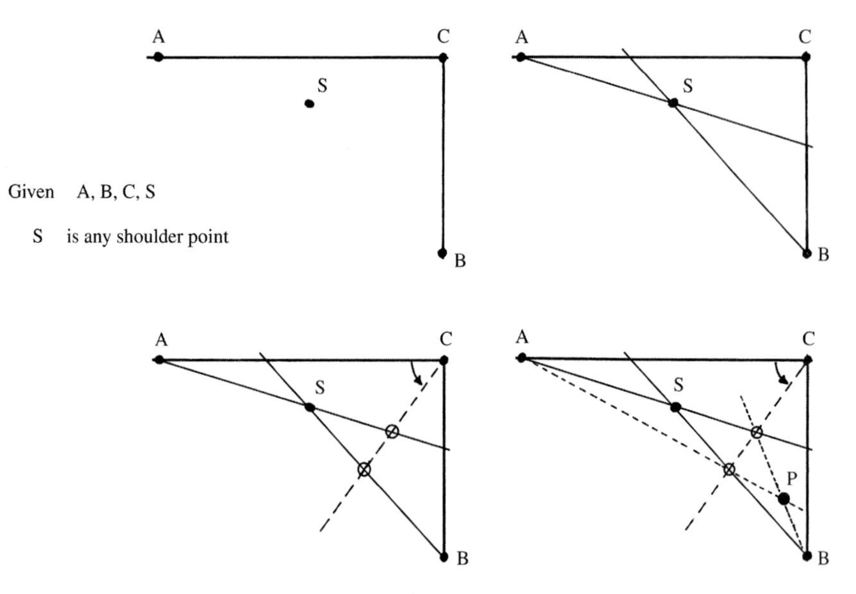

Fig. 7.11 Conic layout.

The remaining illustrations show the generation of one point on the conic. In the third illustration a line is drawn from point *C* at an arbitrary angle. Note the points where this line interesects the *A-S* and *B-S* lines.

Lines are now drawn from *A* and *B* through the points found in the last step. The intersection of these lines is a point "*P*," which is on the desired conic curve.

To generate additional points, the last two steps are repeated. Another line is drawn from point *C* at another arbitrary angle, and then the lines from *A* and *B* are drawn and their intersection is found. When enough points have been generated, a French curve is used to draw the conic.

While this procedure seems complicated at first, with a little practice a good designer can construct an accurate conic in less than a minute. Figure 7.12 illustrates a conic curve generated in this manner. Note that it is not necessary to completely draw the various lines, as it is only their intersections which are of interest.

7.4 Conic Fuselage Development

Longitudinal Control Lines

To create a smoothly lofted fuselage using conics, it is necessary only to ensure that the points *A, B, C,* and *S* in each of the various cross sections can be connected longitudinally by a smooth line. Figure 7.13 shows the upper half of a simple fuselage, in which the *A, B, C,* and *S* points in three cross sections are connected by smooth longitudinal lines. These are called "longitudinal control lines" because they control the shapes of the conic cross sections.

Figure 7.14 shows the side and top views of these longitudinal control lines. Since the cross sections are tangent to horizontal at the top of the fuselage, the

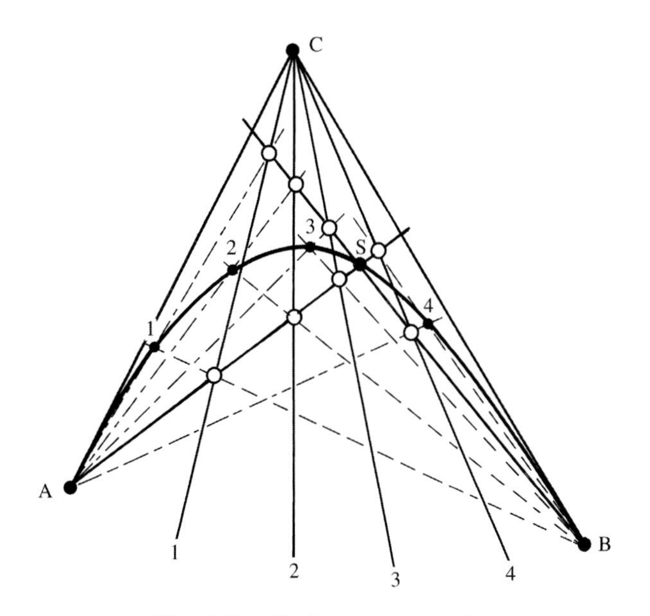

Fig. 7.12 Conic layout example.

A and *C* lines are identical in side view. Similarily, the cross sections are tangent to vertical at the side of the fuselage, so the *B* and *C* lines are identical in top view. This is common, but not required.

In Fig. 7.14, the longitudinal control lines are used to create a new cross section, in between the second and third cross sections previously defined.

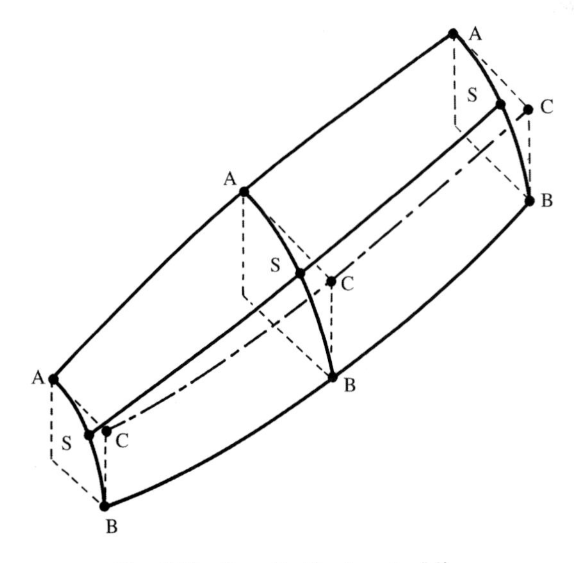

Fig. 7.13 Longitudinal control lines.

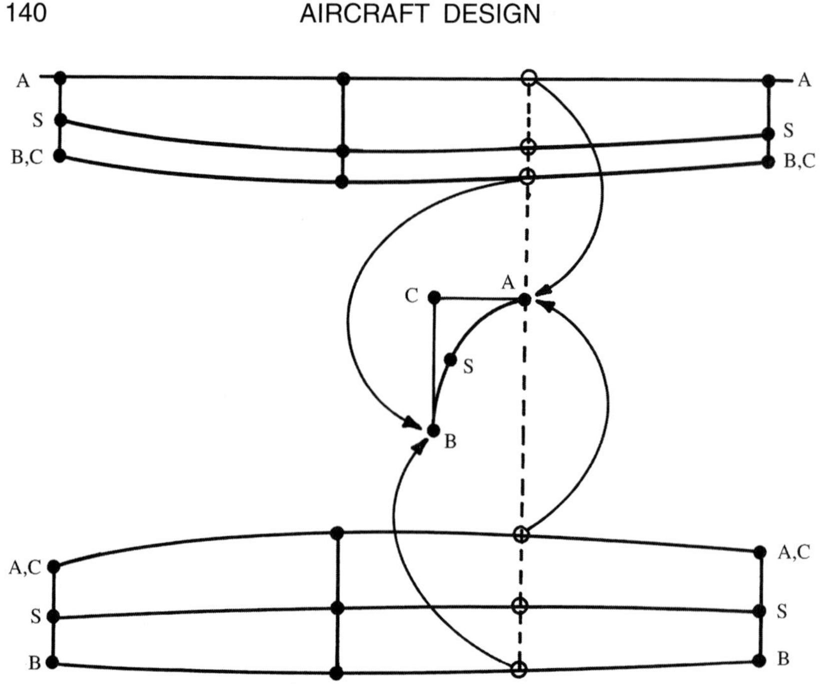

Fig. 7.14 Cross-section development from longitudinal control lines.

This new cross section is created by measuring, from the longitudinal control lines, the positions of the *A, B, C,* and *S* points at the desired location of the new cross section.

As is shown for point *A*, each point is defined by two measurements, one from side view and one from top view. From these points the new cross section can be drawn using the conic layout procedure illustrated in Fig. 7.11.

The original cross sections that are used to develop the longitudinal control lines are called the "control cross sections" or "control stations." These cross sections are drawn to enclose the various internal components, such as the cockpit or engine.

Control stations can also be drawn to match some required shape. For example, the last cross section of a single-engined jet fighter with a conventional round nozzle would have to be a circle of the diameter of the nozzle.

Typically, some 5–10 control stations will be required to develop a fuselage that meets all geometric requirements. The remaining cross sections of the fuselage can then be drawn from the longitudinal control lines developed from these control stations.

Fuselage Lofting Example

Figure 7.15 illustrates a common application of conic lofting to define a fighter fuselage for an initial layout. Five control stations are required for this example. Station 0 is the nose, which is a single point. All the longitudinal control lines must originate there.

Station 120 is established for this example by the requirements for the cockpit (Chapter 9). This station is approximately circular in shape, and is defined using

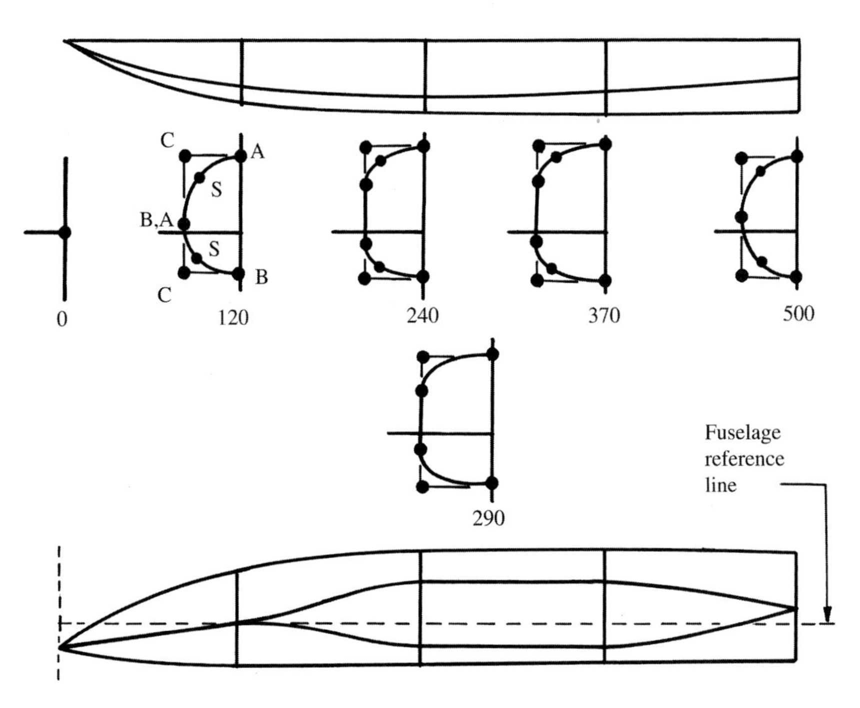

Fig. 7.15 Typical fuselage lofting.

two conics (upper and lower). Each conic has its own *A*, *B*, *C*, and *S* points. Note that the *B* (end) point of the upper conic is identical to the *A* (start) point of the lower conic.

Station 240 has a flat side to provide for a side-mounted inlet as can be seen on the F-4, the MiG-23, the SAAB Gripen, and many other aircraft. At this station, the end points of the upper and lower conics are moved apart vertically, with the area between them defined as a straight line. Note in side view that the longitudinal control lines separate smoothly, not suddenly. This is to ensure a smooth longitudinal contour.

Station 370 is similar to station 240, with a relatively square cross-sectional shape. This could allow room for the landing gear or perhaps to attach a low wing to the side of the fuselage, without a drag-producing acute angle.

Station 500 is a circular cross section, to allow for a connection with a round exhaust nozzle. The longitudinal control lines come back together in a smooth fashion, as shown.

These five control stations are then used to create the longitudinal control lines. From those lines, additional cross sections can be created as desired. Section 290 was created in such a fashion, by measuring the conic control points from the longitudinal control lines and then drawing the conics as previously described.

Figure 7.15 shows only the fuselage lofting. The canopy, inlet duct, and inlet duct fairing would be lofted in a similar fashion, using longitudinal control lines through a few control stations.

Conic Shape Parameter

One problem arises with this method of initial lofting. The locations of the shoulder points (S) can be difficult to control, creating conics either too square (shoulder point too close to point C) or too flat (shoulder point too far away from point C). An alternate technique using conics involves a parameter which directly controls the shoulder point's distance from the point C.

The points labeled E in Fig. 7.10 are conic shoulder points which happen to lie upon the line D-C. "D" is the point exactly midway between A and B. Such a shoulder point E determines the "conic shape parameter (ρ)," as defined in the following equation:

$$\rho = |\overline{DE}|/|\overline{DC}| \tag{7.2}$$

where

$$|\overline{AD}| = |\overline{BD}| \tag{7.3}$$

Referring to Fig. 7.10, the shoulder points labeled E are based upon the ρ values required to obtain the ellipse, parabola, or hyperbola forms of the conic. These are given below, along with the ρ value that defines a circle (a special form of the ellipse):

$$\begin{aligned} \text{Hyperbola:} \quad & \rho > 0.5 \\ \text{Parabola:} \quad & \rho = 0.5 \\ \text{Ellipse:} \quad & \rho < 0.5 \\ \text{Circle:} \quad & \rho = 0.4142 \ \text{ and } \ |\overline{AC}| = |\overline{BC}| \end{aligned} \tag{7.4}$$

The conic shape parameter allows the designer to specify the conic curve's distance from the point C. A conic with a large ρ value (approaching 1.0) will be nearly square, with the shoulder point almost touching the point C. A conic with a small ρ value (approaching 0.0) will nearly resemble the straight line from A-B. The parameter ρ can be used to control the longitudinal fairing of a fuselage more easily.

Figure 7.16 shows the use of the conic shape parameter (ρ) to lay out a conic. Points A, B, and C are known, but the shoulder point S is not known. However, the value of ρ is given.

In the illustration on the right side of Fig. 7.16, the line A-B has been drawn and bisected to find the point D. The shoulder point S is found by measuring along line D-C, starting at D, by a distance equal to ρ times the total length of line D-C. Once the shoulder point is found, the conic can be drawn as illustrated in Fig. 7.11.

By using this approach, a fuselage can be lofted without the use of a longitudinal control line to control the location of the shoulder points. If ρ is specified to be some constant value for all of the cross sections, then the designer need only control the conic endpoints and tangent intersection points. To permit the fuselage ends to be circular in shape, the value of ρ would be fixed at 0.4142.

Greater flexibility can be attained by allowing ρ to vary longitudinally. For example, the fuselage of Fig. 7.15 requires a ρ value of 0.4142 at both ends to

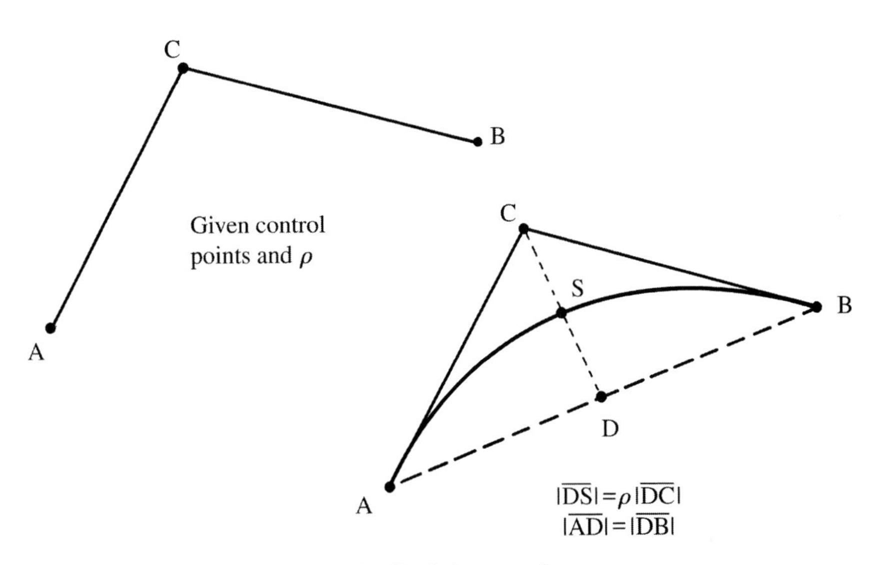

Fig. 7.16 Conic layout using ρ.

allow a circular shape, but the values of ρ at the middle of the fuselage are higher, perhaps around 0.7.

An "auxiliary control line" can be used to control the value of ρ graphically, as shown in Fig. 7.17. Note the auxiliary control line for ρ at the bottom. If the value of ρ varies smoothly from nose to tail, and the conic endpoints and tangent intersection point are controlled with smooth longitudinal lines, then the resulting fuselage surface will be smooth.

In Fig. 7.17 the upper conic has a constant ρ value of 0.4142, while the lower conic has a ρ value varying from 0.4142 at the nose and tail to about 0.6 at the middle of the fuselage. This has the effect of "squaring" the lower fuselage to provide more room for the landing gear.

Figure 7.18 shows the use of ρ to develop the cross sections labeled *A* and *B*. Observe the development of the upper and lower conics by the method shown previously in Fig. 7.16, and the use of different ρ values for the upper and lower conics.

Thus far, no mention has been made of the method for developing the longitudinal control lines and auxiliary control lines. During production lofting, these control lines would be defined mathematically, using conics or some form of polynomial.

For initial layouts, sufficient accuracy can be obtained graphically through the use of the flexible splines discussed earlier. Points are taken from the control cross sections and plotted in side and top view, then connected longitudinally using a spline to draft a smooth line. In fact, a designer with a "good eye" can obtain sufficient smoothness using a French curve if spline and ducks are not available.

Figure 7.19 shows an illustrative example of the conic-developed loft lines for an exotically shaped aircraft, the supersonic SAAB J-35 Draken (Dragon). In this isometric view you can see the longitudinal control scheme for fuselage, nacelle, canopy, and inlet duct, and can also see the lines definition for wing and tail. Such

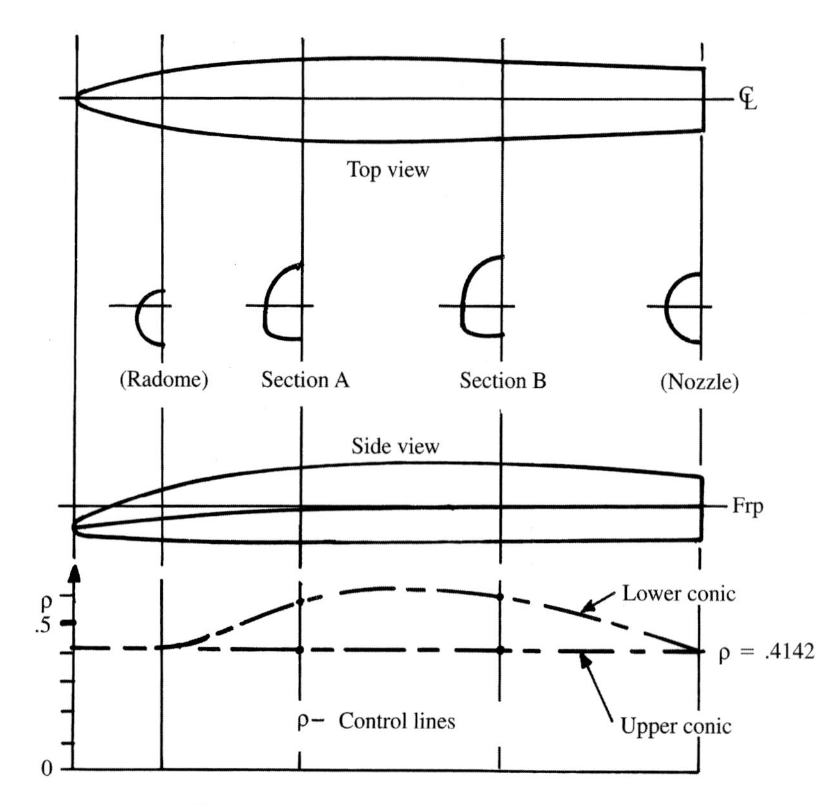

Fig. 7.17 Conic fuselage development using ρ.

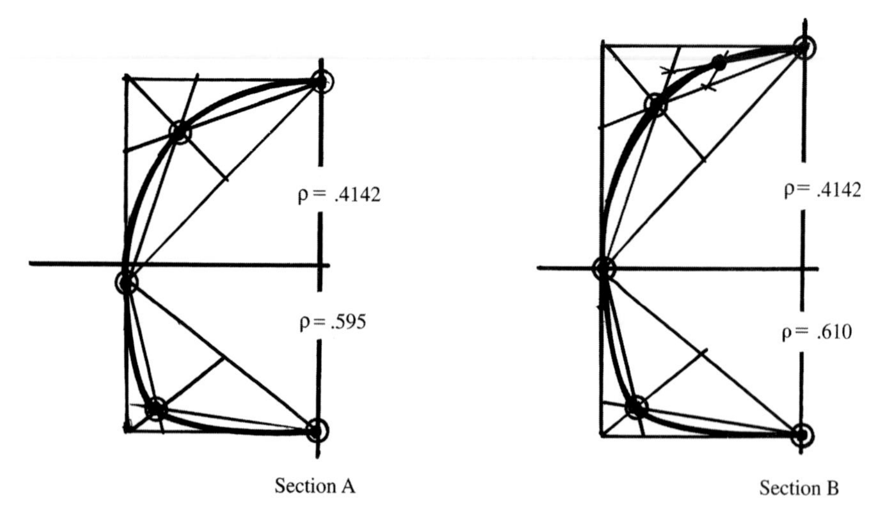

Fig. 7.18 Cross-section development using ρ.

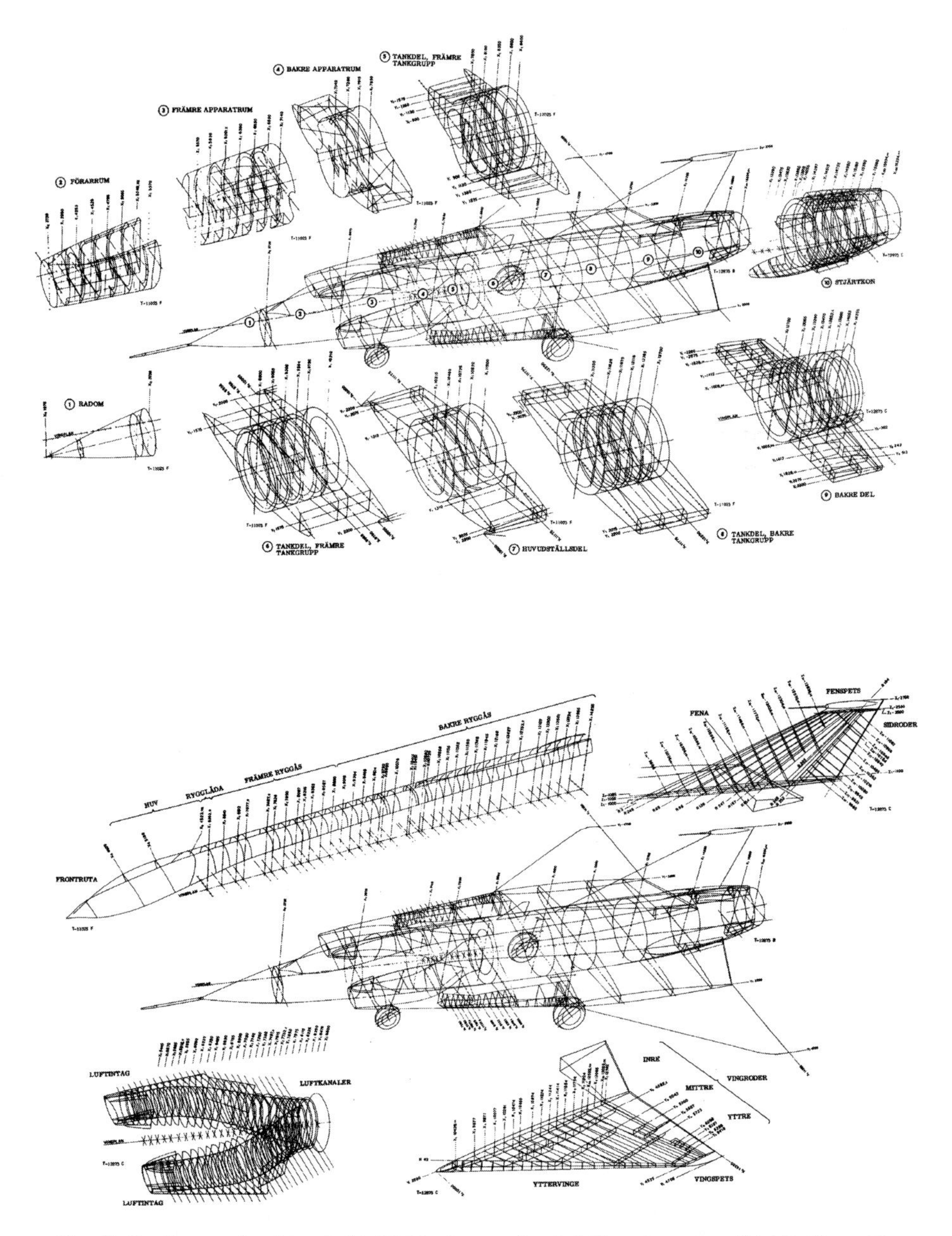

Fig. 7.19 Isometric view of SAAB Draken major loft lines (courtesy SAAB Aircraft).

a detailed loft definition is not normally done until sometime in preliminary design. But, a good designer will consider the overall loft definition even from the earliest conceptual design layout.

7.5 Flat-Wrap Fuselage Lofting

An important cost driver for aircraft fabrication is the amount of compound-curvature used in lofting the aircraft. Compound-curvature implies the existence of surface curvature in all directions for some point on the surface.

For example, a ball is entirely composed of compound-curvature surfaces. A flat sheet has no curvature, compound or otherwise. A cylinder is curved, but only in one direction, so it does not have any compound curvature. Instead, a cylinder or any other surface with curvature in only one direction is said to be "flat-wrapped."

If a surface is flat-wrapped, it can be constructed by "wrapping" a flat sheet around its cross sections. For aircraft fabrication, this allows the skins to be cut from flat sheets and bent to the desired skin contours.

This is far cheaper than the construction technique for a surface with compound curvature. Compound curvature requires that the skins be shaped by a stretching or stamping operation, which entails expensive tools and extra fabrication steps.

A recent example of the benefits of flat wrap occurred during the design and fabrication of the X-31 Enhanced Fighter Maneuver demonstrator. Rockwell (North American Aircraft) wisely included manufacturing personnel in the program early enough that they were able to point out a problem: the compound curves of the aft fuselage would require hot die forming, and since the material of choice for the area around the engine was titanium, the die itself would cost on the order of $400,000 (1999 dollars) and be the pacing item in the fabrication schedule. By changing the last 30 in. {76 cm} of the aft fuselage to a flat-wrap shape, titanium sheet could be bent to shape with no forming required.

Aircraft applications of flat-wrap lofting must be defined in the initial loft definition used for the conceptual layout. There are several ways of lofting a surface so that it is flat-wrapped. The simplest technique uses a constant cross section. For example, a commercial airliner usually has the identical circular-cross-sectional shape over most of its length. In fact, any cross-section shape will produce a flat-wrap surface if it is held constant in the longitudinal direction.

Often an identical cross-sectional shape will not be desired, yet a flat-wrap lofting may be attained. If the same cross-sectional shape is maintained but linearly scaled in size, a flat-wrap contour is produced. For example, a cone is a flat-wrap surface produced by linearly scaling a circular cross section.

Many aircraft have a tailcone that, although not circular in cross section, is linearly scaled to produce a flat-wrap surface. This can be accomplished with conics by maintaining identical tangent angles and ρ value, using straight longitudinal control lines, and maintaining the lengths AC and BC in constant proportion.

Sometimes it is necessary to vary the shape of the cross sections other than by scaling. Flat wrap cannot be exactly maintained in such cases using conics. A more sophisticated technique (beyond the scope of this book) must be used.

However, flat wrap can be closely approximated in most such cases on two conditions. First, the longitudinal control lines must be straight. This includes the line controlling the shoulder point (S). If the conic shape parameter (ρ) is

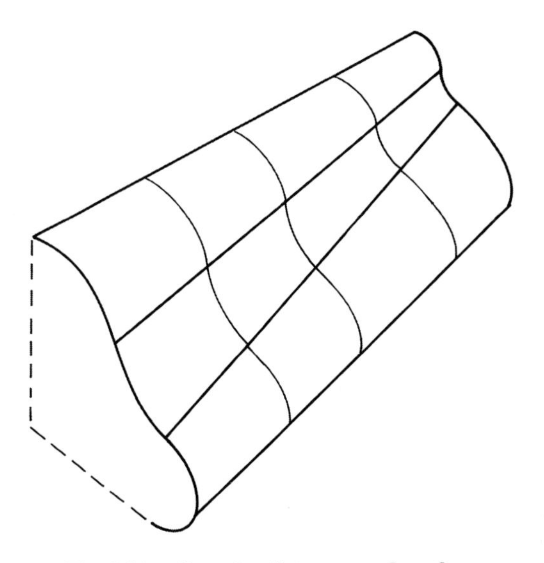

Fig. 7.20 Complex flat-wrapped surface.

used instead of a shoulder-point control line, then the ρ value must be either constant or linearly varied.

Second, the tangent angles of the conics must not change longitudinally. If the tangent angles are all either horizontal or vertical, as in Figs. 7.15 and 7.17, this condition can easily be met.

Figure 7.20 shows such a complex flat-wrapped surface. The fuselage is defined by five conics plus a straight-line, flat underside. The "bump" on top could represent the back of the canopy, and grows smaller toward the rear of the fuselage. While the conics change shape and size, their endpoints hold the same tangent angles.

It is important to realize that the use of flat-wrap lofting for a fuselage represents a compromise. While flat-wrap surfaces are easier and cheaper to fabricate, they are less desirable from an aerodynamic viewpoint. For example, a smoothly contoured teardrop shape will have less drag than a flat-wrap cylinder with a nosecone and tailcone.

7.6 Circle-to-Square Adapter

A common problem in lofting is the "circle-to-square adapter." For example, the inlet duct of many supersonic jet aircraft is approximately square at the air inlet, yet must attain a circular shape at the engine front-face. Modern, two-dimensional nozzles also require a circle-to-square adapter.

Flat-wrap can be attained for a circle-to-square adapter by constructing the adapter of interlocking, *V*-shaped segments, each of which is itself flat-wrapped (Fig. 7.21).

The flat sides of the square section taper to points that just touch the circular section. Similarly, the cone-shaped sides of the circular section taper to points that touch the corners of the square section. Note the "rounded-off square"

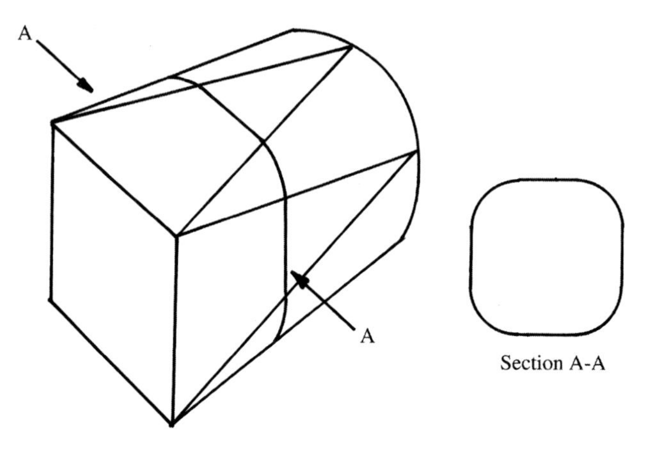

Fig. 7.21 Circle-to-square adapter.

shape of the intermediate sections. The connecting surfaces must be straight longitudinally for a flat-wrap surface to be maintained.

7.7 Fuselage Loft Verification

The use of smooth longitudinal control lines defining conic cross sections assures a smooth fuselage; but sometimes it is necessary to deviate from this type of definition. For example, if a part of the fuselage is to be flat-wrapped, it may be difficult to smoothly connect the sraight control lines for the flat-wrap part of the fuselage with the curved control lines for the rest of the fuselage.

Also, it may be desirable to have two different flat-wrap parts of the fuselage that are directly connected, resulting in an unavoidable break in the smoothness of the longitudinal control lines. In such cases the designer should evaluate the resulting contours to ensure that any breaks in the longitudinal smoothness are not too extreme.

Sometimes a designer will be asked to evaluate a design created by someone else. This is common practice at government design offices. In addition to the analytical evaluations for performance and range, the designer should evaluate the design layout to ensure that the cross sections shown are in fact smooth longitudinally.

These smoothness evaluations are performed using a technique borrowed from shipbuilding. Hull contours are evaluated for smoothness by laying out the “waterlines.” If a ship is floating in the water, the line around the hull where the surface of the water intersects the hull is a waterline. For good ship performance, this waterline should be smooth in the longitudinal direction.

If the hull is raised partly out of the water some arbitrary distance, a new waterline is formed. Hull designers check for hull smoothness by laying out a large number of these waterlines, each separated in height by some arbitrary distance. If all the waterlines have smooth contours, then the hull is smooth.

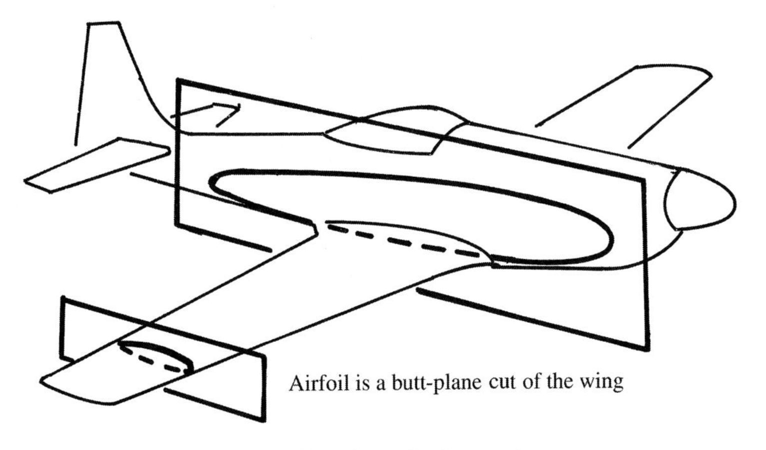

Fig. 7.22 Buttock-plane cut.

Such horizontal waterline cuts can be used for evaluation of the smoothness of an aircraft fuselage; but it is more common to use vertically oriented cuts known as "buttock-plane cuts" (Fig. 7.22).

Buttock-plane ("butt-plane") cuts form the intersection of the aircraft with vertical planes defined by their distance from the aircraft centerline. For example, "butt-plane 30" is the contour created by intersecting a vertical plane with the fuselage at a distance of 30 in. from the centerline.

Note in Fig. 7.22 that the butt-plane cuts are oriented such that the airfoil is a butt-plane cut of the wing. It is for this reason that butt-plane cuts are more commonly used for aircraft than waterlines.

Figure 7.23 illustrates the development of butt-plane cuts. Vertical lines are drawn on each cross section, indicating the locations of the arbitrarily selected butt-planes. The points where these vertical lines intersect the cross sections are transferred to the side-view drawing and connected longitudinally. If the fuselage surface is smooth, then these longitudinal lines for the different butt-planes will all be smooth.

Buttock-plane cuts can also be used to generate new cross sections. Once the butt-plane cuts are developed as in Fig. 7.23, a new cross section can be developed by transferring the vertical locations of the butt-plane cuts to the cross section desired, and then drawing a smooth cross-sectional contour using those points.

Sometimes this method is easier than developing the longitudinal control lines for conic fuselage lofting. This is most likely when the surface is highly irregular, such as the forebody of a blended wing-body aircraft like the B-1B.

7.8 Wing/Tail Layout and Loft

Reference Wing/Tail Layout

Chapter 4 described the selection of the basic geometric parameters for the wing and tails. These parameters include the aspect ratio (A), taper ratio (λ), sweep, dihedral, and thickness. Also, the selection of an appropriate airfoil was

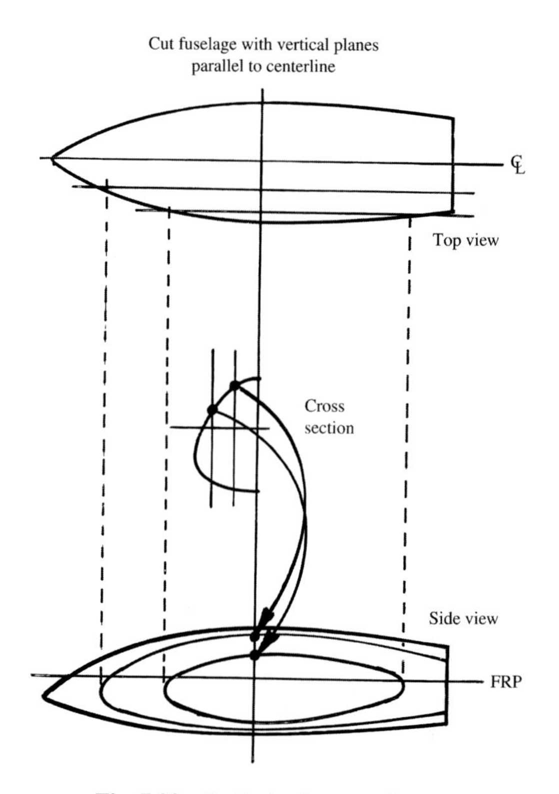

Fig. 7.23 Buttock-plane cut layout.

considered. In Chapter 6, the actual sizes for the wing, tails, and fuselage were defined, based upon an initial estimate for the takeoff gross weight.

From these parameters, the geometric dimensions necessary for layout of the reference (trapezoidal) wing or tail can be obtained, as shown in Fig. 7.24 and defined by the following equations:

$$b = \sqrt{AS} \tag{7.5}$$

$$C_{\text{root}} = \frac{2S}{b(1+\lambda)} \tag{7.6}$$

$$C_{\text{tip}} = \lambda C_{\text{root}} \tag{7.7}$$

$$\bar{C} = \left(\frac{2}{3}\right) C_{\text{root}} \frac{1+\lambda+\lambda^2}{1+\lambda} \tag{7.8}$$

$$\bar{Y} = \left(\frac{b}{6}\right)\left(\frac{1+2\lambda}{1+\lambda}\right) \tag{7.9}$$

For a vertical tail, $\bar{Y}$ is twice the value calculated in Eq. (7.9).

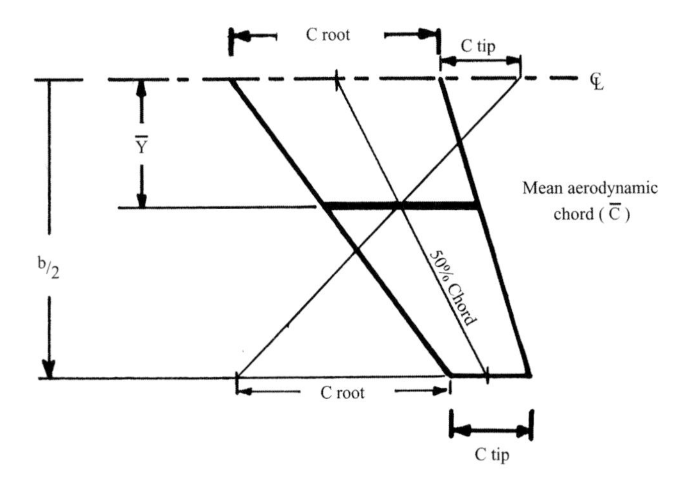

Fig. 7.24 Reference (trapezoidal) wing/tail.

Figure 7.24 also shows a quick graphical method of determining the spanwise (Y) location of the mean aerodynamic chord (MAC or $\bar{C}$), which is mathematically obtained by Eq. (7.9). The location of the mean chord is obtained graphically as the intersection of the 50%-chord line and a line drawn from a point located at the tip chord length behind the root chord to a point at the root chord length ahead of the tip chord.

Wing Location with Respect to the Fuselage

The location and length of the MAC is important because the wing is located on the aircraft so that some selected percent of the MAC is aligned with the aircraft center of gravity. This provides a first estimate of the wing position to attain the required stability characteristics.

For a stable aircraft with an aft tail, the wing should be initially located such that the aircraft center of gravity is at about 30% of the mean aerodynamic chord. When the effects of the fuselage and tail are considered, this will cause the center of gravity to be at about 25% of the total subsonic aerodynamic center of the aircraft.

For an unstable aircraft with an aft tail, the location of the wing depends upon the selected level of instability, but will usually be such that the center of gravity is at about 40% of the mean aerodynamic chord.

For a canard aircraft, such rules of thumb are far less reliable due to the canard downwash and its influence upon the wing. For a control-type canard with a computerized flight control system (i.e., unstable aircraft), the wing can be initially placed such that the aircraft center of gravity is at about 15–20% of the wing's mean aerodynamic chord.

For a lifting-type canard, the mean aerodynamic chords of the wing and canard should both be determined, and a point at about the 15% MAC for each should be identified (20–25% for an unstable aircraft). Then the combined MAC location can be determined as the average of these percentage MAC locations for the wing and canard, weighted by their respective areas. Note that this is a very crude estimate!

Chapter 15 provides a quick method of approximating the aircraft center of gravity once the locations of the major internal components are known.

After the initial layout is completed and analyzed using the methods of Chapters 12–19, the wing will probably be moved and the tails resized to meet all required stability and control characteristics. Hopefully the initial estimates will be close enough so that major changes will not be needed.

Wing/Tail Lofting

The reference (trapezoidal) wing and tails are positioned with respect to the fuselage using the methods just discussed. During the layout process, the actual, exposed wing and tails will be drawn.

The reference wing is defined to the aircraft centerline, and is based upon the projected area (i.e., dihedral does not affect the top view of the reference wing). The actual, exposed wing begins at the side of the fuselage and includes the effect of the dihedral upon the true-view area. The dihedral angle increases the actual wing area equivalent to dividing by the cosine of the dihedral angle.

Also, the actual wing planform may not be trapezoidal. Figure 7.25 illustrates several of the many nontrapezoidal wing variations. A typical rounded wing tip is shown in Fig. 7.25a. This and other wing-tip shapes have already been discussed. The straightened-out trailing edge shown in Fig. 7.25b increases the flap chord and provides increased wing thickness for the landing gear.

Figure 7.25c illustrates a "leading-edge extension (LEX)," which increases lift for combat maneuvering (see Chapter 12). A highly blended wing/body is shown in Fig. 7.25d, in which the actual wing looks very little like the reference wing. This type of wing is used to minimize the transonic and supersonic shocks.

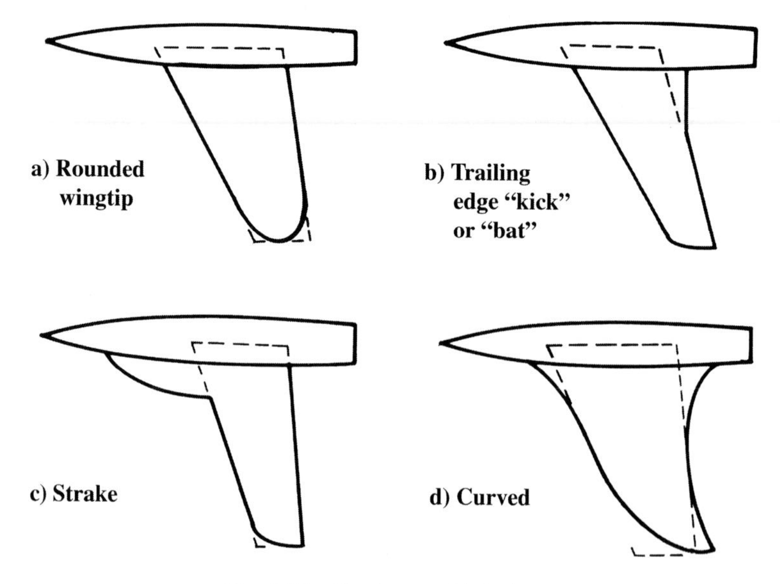

Fig. 7.25 Nontrapezoidal wings.

Once the designer has settled upon the actual wing and tail planforms, their surfaces must be lofted to provide accurate cross sections. These are required to verify that there is sufficient room for the fuel tanks, landing gear, spars, and other internal components. During production design, this lofting would be done using conics or some other mathematical surface definition.

For initial design, simpler methods of wing and tail lofting can be used. These rely upon the assumption that the airfoil coordinates themselves are smoothly lofted. This is an excellent assumption, as otherwise the airfoil performance would be poor.

If the wing or tail uses the identical airfoil section and thickness ratio at all span stations, and is without twist, the airfoils can be drawn simply by scaling the airfoil coordinates to fit the chord lengths of the selected spanwise locations.

It is customary to lightly draw the airfoils on the top view of the wing, superimposing them on their chordline (Fig. 7.26). This layout procedure simplifies the generation of cross sections, as will be discussed later. For initial design purposes the airfoils can be quickly drawn using only a few scaled coordinate points for the top and bottom surfaces.

If twist is incorporated, the incidence at each span station must be determined and the chord line rotated accordingly before the airfoil is drawn. Since the chord length is defined in top view, the chord length at each spanwise station must be increased equivalent to dividing by the cosine of the appropriate incidence angle.

Instead of calculating the twist, an auxiliary twist control line may be constructed behind the wing. The airfoil incidence at each span station can then be read from the control line (Fig. 7.27).

A wing with a complicated aerodynamic design may have the twist, camber, and thickness all varying from root to tip. These spanwise variations can be lofted by using a separate auxiliary control line for each, as shown in Fig. 7.28. The airfoil coordinate points must be calculated by separating the airfoil into its camber line and thickness distribution, scaling them as indicated by the auxiliary control lines, and recombining them. Such a complicated wing design is not normally accomplished until much later in the design process.

For a wing such as shown in Fig. 7.28, the complex curvatures of the wing surface may present difficulties. A spar running from root to tip may very well

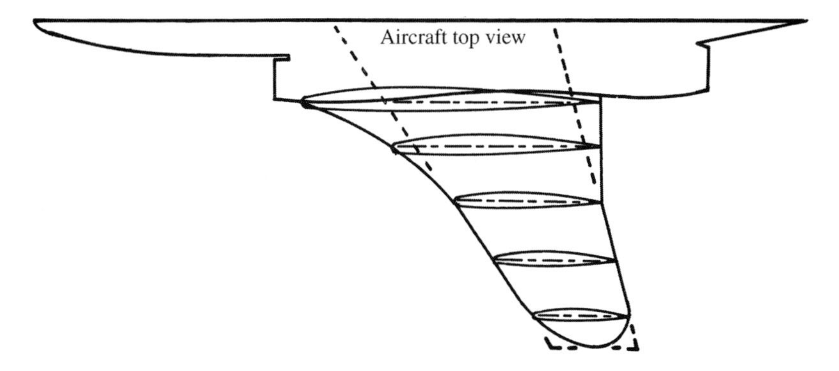

Fig. 7.26 Airfoil layout on wing planform.

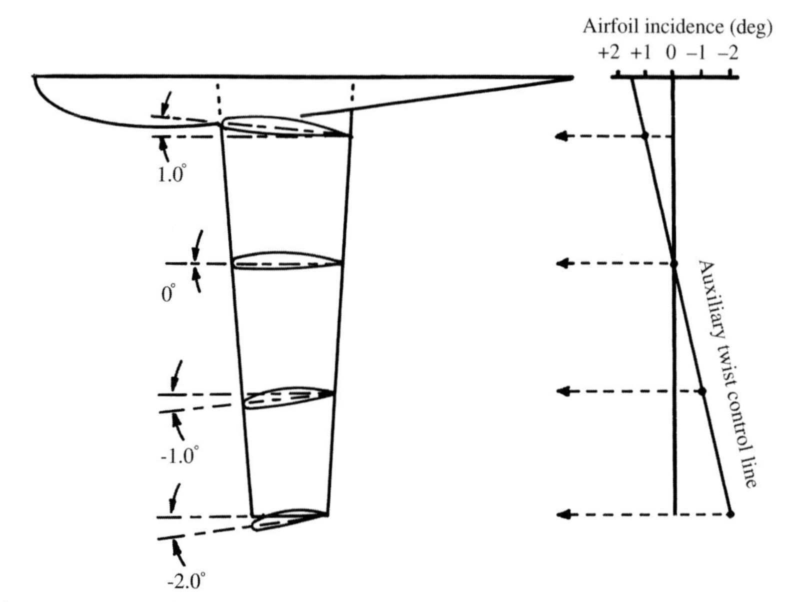

Fig. 7.27 Airfoil layout with twist.

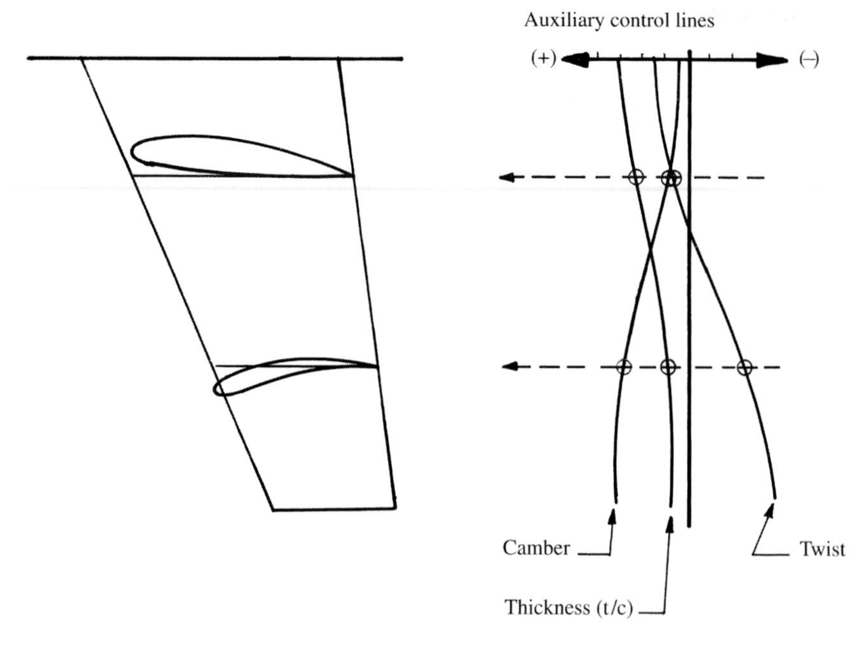

Fig. 7.28 Wing airfoil layout—nonlinear variations.

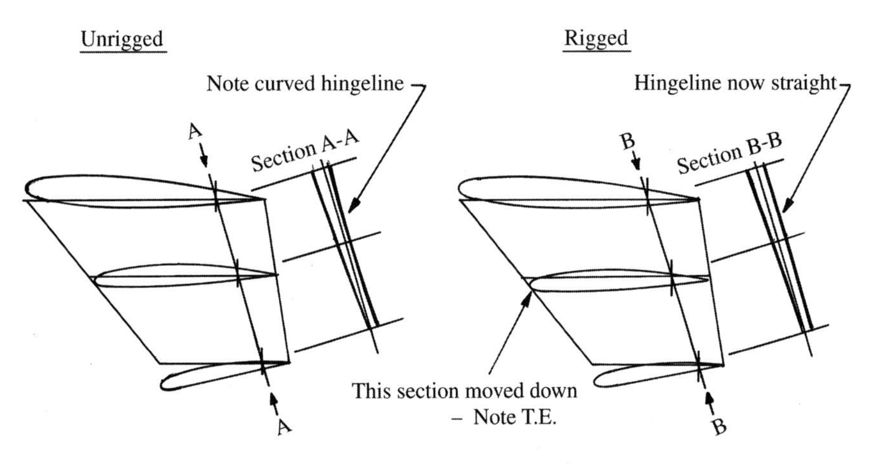

Fig. 7.29 Wing airfoil rigging.

be so curved that it is structurally undesirable. Even worse, the hinge lines for the ailerons and flaps may not lie in a straight line. As curved hinge lines are impossible, the ailerons and flaps may have to be broken into a number of surfaces unless the wing surface can be modified to straighten the hingeline.

This is done by "wing rigging" (not to be confused with the rigging of a biplane wing)—the process of vertically shifting the airfoil sections until some desired spanwise line is straight.

Figure 7.29 illustrates a complex wing in which the aileron hingeline, Section A-A, is curved. On the right side of the figure is the same wing with the midspan airfoil moved downward a few inches. This provides a straight hingeline shown as Section B-B.

Airfoil Linear Interpolation

Most wings are initially defined by a root airfoil and a tip airfoil, which may be different, and their incidence angles or relative twist. Frequently the tip airfoil will be selected for gentle stall characteristics while the root airfoil is selected for best performance. The resulting wing has good overall performance, with good stall characteristics because the tip will stall after the root. The airfoils between the root and tip can be quickly developed by one of two methods.

Linear interpolation, the easiest method, is depicted in Fig. 7.30. Here the new airfoils are created as "weighted averages" of the root and tip airfoils. Linearly interpolated airfoils have section properties that are approximately the interpolation of the section properties of the root and tip airfoils.

(Some modern laminar airfoils will not provide interpolated section characteristics. Instead, the interpolated airfoils must each be separately analyzed.)

The intermediate airfoils are linearly interpolated by a five-step process. The root and tip airfoils are drawn (step 1). A constant percent-chord line is drawn connecting the root and tip airfoil, and vertical lines are drawn from the intersection of that line with the chordlines (step 2). The airfoil points found at those vertical lines are "swung down" to the chord line, using an arc centered at the intersection of the

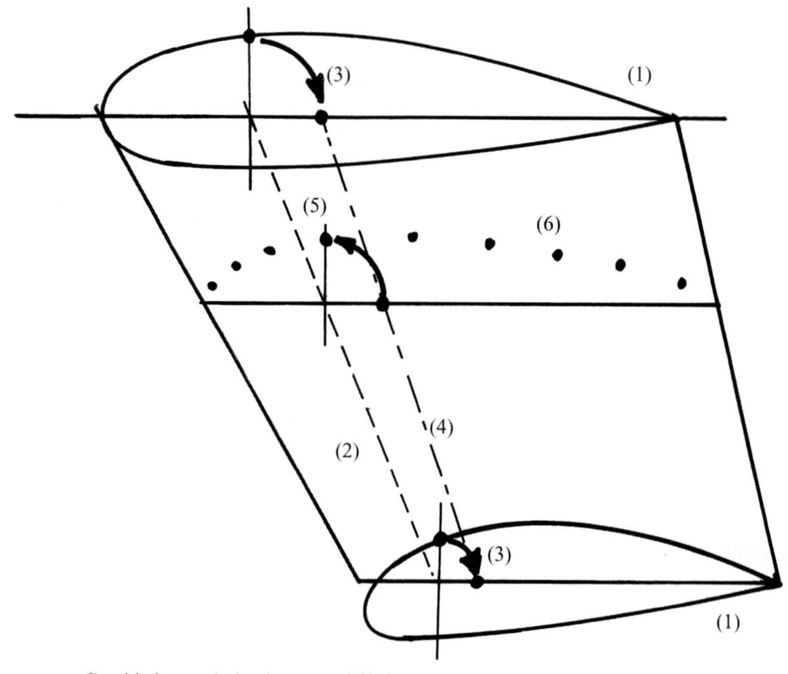

Graphic interpolation between differing root and tip airfoils
1–Superimpose root and tip airfoils on planform
2–Draw line at some constant percent of chord
3–Swing airfoil point down onto chord reference line
4–Connect root and tip points from 3
5–Swing point up to new airfoil location
6–Repeat for other percent chord lines

Fig. 7.30 Wing airfoil layout—linear interpolation.

chord line and the vertical line (step 3). These "swung down" points for the root and tip airfoils are then connected by a straight line (step 4).

At the desired location of an interpolated airfoil, a chord line is drawn. The intersection of that chord line with the line drawn in step 4 defines the chordwise location of a point on the interpolated airfoil. In step 5 this point is "swung up" to its thickness location by an arc centered at the intersection of the chord line and the spanwise percent-chord line from step 2.

This process is repeated for as many points as are needed to draw the new airfoil. Then the process is repeated to draw other airfoils. While it seems complicated, a wing can be developed using this method in about 15 min by an experienced designer. (However, a computer does this instantly!)

Airfoil Flat-Wrap Interpolation

The linear-interpolation method doesn't necessarily provide a flat-wrap surface. In laying out a fuselage for flat-wrap, it was necessary to hold the same tangent angle for the conics in the different cross sections. The same is true for wings.

To provide a flat-wrap wing, it is necessary to interpolate between airfoil coordinates with the same slope (i.e., tangent angle). The linear interpolation method connects points based upon their percent of chord. If the wing is twisted or the airfoils are dissimilar, the surface slopes may be different for airfoil points that are at the same percent of chord. This requires a modification to the method just described.

Figure 7.31 illustrates this modification. The only difference is in step 2. Previously a spanwise line was drawn connecting constant percent chord locations on the chordline. To obtain a flat-wrap surface this spanwise line must be drawn connecting locations on the chordline that have the same surface slope. Note in the figure how the tip chord has the indicated slope at a more-aft percent location of the chord than does the root chord.

A number of composite homebuilts are being fabricated by a method long used for model airplanes. A large block of foam (usually polystyrene) is cut directly to the desired wing shape using a hot-wire cutter which is guided by

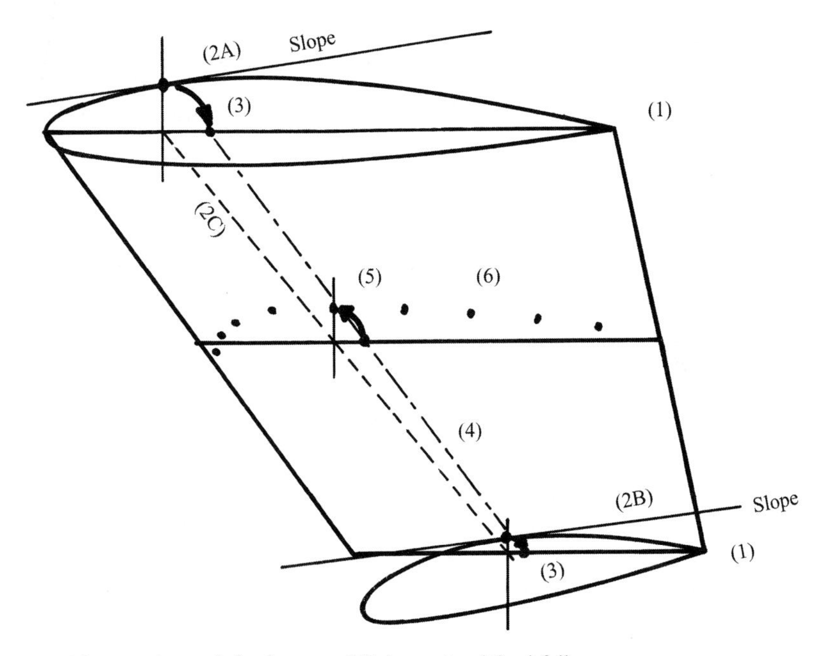

Flat-wrap interpolation between differing root and tip airfoils
1 – Superimpose root and tip airfoils on planform
2A – For a point on the root airfoil, find the slope
2B – Find the point on the tip airfoil with the same slope
2C – Connect the percent chord points from (2A) and (2B)
3 – At root and tip, swing points down onto chord reference line
4 – Connect the points from (3)
5 – Swing point up to new airfoil location
6 – Repeat for other points

Fig. 7.31 Wing airfoil layout—flat-wrap.

root and tip airfoil templates attached to the foam block. The templates have tic-marks that are numbered. The wire is guided around the templates by two home-builders, one of whom calls out the numbers of the tic-marks.

If the tic-marks are at constant percent-chord locations, and the wing has dissimilar airfoils or appreciable twist, this method will produce a linearly interpolated instead of flat-wrap surface. If the wing is to be covered by fiber-glass, this will pose no problem as the fiberglass cloth will easily conform to the slight amount of compound-curvature present.

However, if the wing is to be covered by sheet metal or plywood, the linearly interpolated foam surface will be depressed relative to the flat-wrapped skin. This could reduce the strength of the skin bonding. It is conceivable that such a wing could fail in flight for this simple reason. Who said lofting is not important?

Wing/Tail Cross-Section Layout

Wing lofting during initial design permits verification that the fuel and other internal components will fit within the wing. This requires the development of wing and tail cross sections oriented perpendicular to the aircraft centerline.

Such cross sections can be easily developed once the airfoils are drawn onto the top view of the wing. Figure 7.32 illustrates the development of one such cross section.

To develop a wing (or tail) cross section, vertical lines are drawn on the cross section at the spanwise location of the airfoils shown on the wing top view. Also, the wing reference plane is shown at the appropriate wing dihedral angle. Then

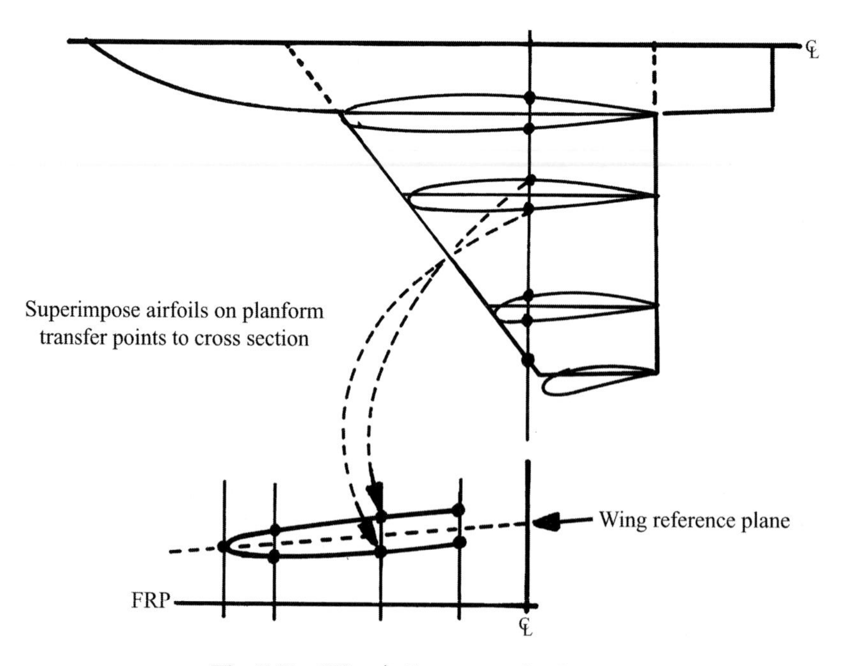

Fig. 7.32 Wing/tail cross-section layout.

the airfoil upper and lower points are measured relative to the plane of the wing, and drawn accordingly on the cross section. The cross-section shape can then be drawn using French curves.

The same procedure can be used to develop section cuts at angles other than perpendicular to the aircraft centerline. The sections of Fig. 7.29 labeled A-A and B-B were developed in this manner.

Wing Fillets

For improved aerodynamic efficiency, the wing-fuselage connection of many aircraft is smoothly blended using a "wing fillet" (Fig. 7.33). A wing fillet is generally defined by a circular arc of varying radius, tangent to both the wing and fuselage. Typically a wing fillet has a radius of about 10% of the root-chord length.

The fillet circular arc is perpendicular to the wing surface, so the arc is in a purely vertical plane only at the maximum thickness point of the wing. At the leading edge, the arc is in a horizontal plane.

The fillet arc radius may be constant, or may be varied using an auxiliary radius control line, as shown in Fig. 7.33. Note that the starting radius must be equal to the fillet radius shown in the wing top view. Also, the fillet radius is usually increasing towards the rear of the aircraft, to minimize airflow separation.

Some aircraft have a fillet only on the rear part of the wing. In this case the fillet starts, with zero radius, at the wing's maximum thickness point.

For initial layout purposes the fillet is frequently "eyeballed." Only a few of the 10 or 15 aircraft cross sections developed for an initial layout will show the wing fillet, so a fillet radius that "looks good" can be used.

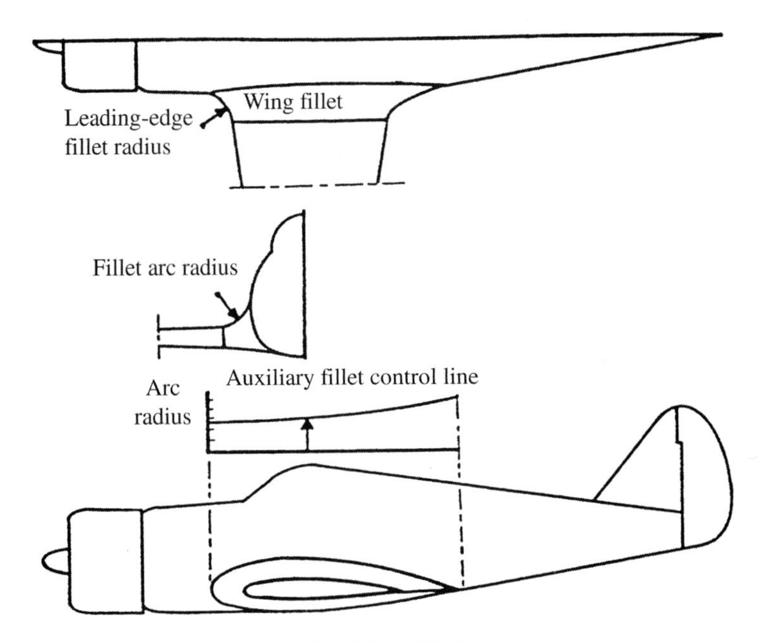

Fig. 7.33 Wing fillet layout.

Winglet Design

Winglets were presented in Chapter 4 as devices to reduce induced drag, especially for a wing with a fairly high span loading. They have been used on many aircraft in the last 20 years, and are especially beneficial when an existing design is being recertified to a higher takeoff weight. This increases span loading, which increases induced drag, unless the wing span is also increased. That is difficult and expensive. Instead, winglets can be added to avoid an induced drag increase without the need to extend the wing span.

Fundamentally, the winglet works by producing a side force (inward-pointing "lift") that has a slight forward component because of the rotation of the vortex over the top of the wing tip. If there is no side force, then there is no winglet effect. Thus, the winglet must be wing-like (hence the name), with both camber and angle of attack to the local flow. The winglet also acts as an endplate, allowing the wing to generate more lift near the tip.

There are many types, shapes, sizes, and geometries of winglets. Almost every year, a new variation on the basic theme is proposed. Figure 7.34 illustrates what some call the "classic" winglet as defined by R. Whitcomb, the original developer. The upper winglet should begin at the place where the wing-tip airfoil has its maximum thickness, and it should be swept about the same as the wing. It should be at least as tall as the tip chord of the wing, and even taller is better because the drag reduction is roughly proportional to the winglet height. The camber of the winglet should be greater than that of the wing to ensure sufficient side force, and it should have a 4-deg leading-edge-out incidence angle. Typically, the winglet t/c is about 8%.

The bottom winglet panel, seen on the original winglet concept and on a number of aircraft, does not really contribute all that much to the drag reduction. Since it sticks below the wing, it threatens to scrape on the ground if the aircraft is rolled too much, so it is not included on many winglet-equipped aircraft. If included, it should be twisted with its root at a 7-deg incidence and its tip at 11 deg.

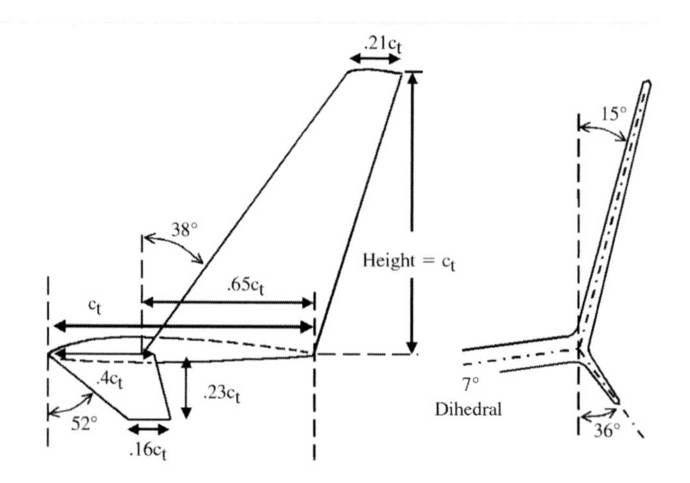

Fig. 7.34 Winglet design guidelines (after NASA N76-26163, R. Whitcomb).

While not on the original winglet concept, it appears that a substantial drag reduction can be obtained by smoothly curving the wing tip upward to the winglet (when seen from the front) rather than having the winglet be a separate piece attached to the top of the wing.

One danger with the winglet is that it is adding mass behind the elastic axis of the wing. Flutter tendencies must always be considered, and a detailed aeroelastic analysis should be performed to determine if structure stiffening will be required. Since that will add weight, it may reduce the benefit of the winglet.

7.9. Wetted Area Determination

Aircraft wetted area (S_{wet}), the total exposed surface area, can be visualized as the area of the external parts of the aircraft that would get wet if it were dipped into water. The wetted area must be calculated for drag estimation, as it is the major contributor to friction drag.

The wing and tail wetted areas can be approximated from their planforms, as shown in Fig. 7.35. The wetted area is estimated by multiplying the true-view exposed planform area ($S_{exposed}$) times a factor based upon the wing or tail thinkness ratio.

If a wing or tail were paper-thin, the wetted area would be exactly twice the true planform area (i.e., top and bottom). The effect of finite thickness is to increase the wetted area, as approximated by Eqs. (7.10) or (7.11). Note that the true exposed planform area is the projected (top-view) area divided by the cosine of the dihedral angle.

If $t/c < 0.05$

$$S_{wet} = 2.003\, S_{exposed} \tag{7.10}$$

If $t/c > 0.05$

$$S_{wet} = S_{exposed}\, [1.977 + 0.52\,(t/c)] \tag{7.11}$$

The exposed area shown in Fig. 7.35 can be measured from the drawing in several ways. A professional designer will have access to a "planimeter," a

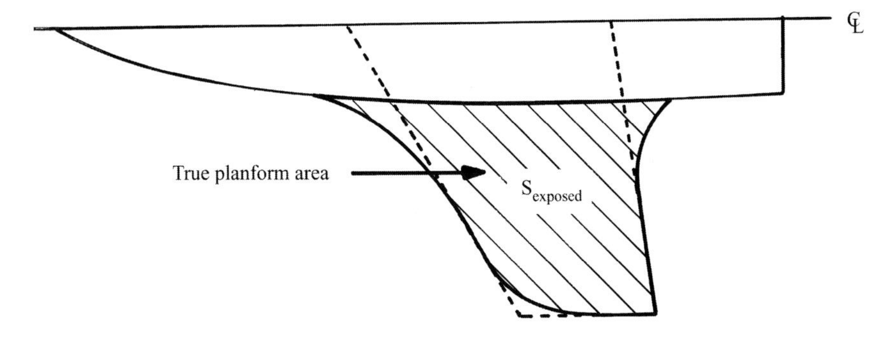

Fig. 7.35 Wing/tail wetted area estimate.

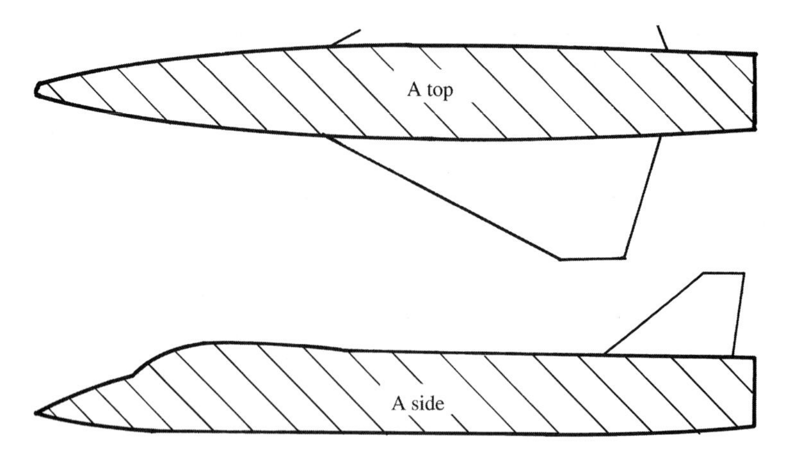

Fig. 7.36 Quick fuselage wetted area estimate.

mechanical device for measuring areas. Use of the planimeter is a dying art as the computer replaces the drafting board. Alternatively, the area can be measured by tracing onto graph paper and "counting squares."

The wetted area of the fuselage can be initially estimated using just the side and top views of the aircraft by the method shown in Fig. 7.36. The side- and top-view projected areas of the fuselage are measured from the drawing, and the values are averaged.

For a long, thin body circular in cross section, this average projected area times π will yield the surface wetted area. If the body is rectangular in cross section, the wetted area will be four times the average projected area. For typical aircraft, Eq. (7.12) provides a reasonable approximation.

$$S_{\text{wet}} \cong 3.4\left(\frac{A_{\text{top}} + A_{\text{side}}}{2}\right) \tag{7.12}$$

A more accurate estimation of wetted area can be obtained by graphical integration using a number of fuselage cross sections. If the perimeters of the cross sections are measured and plotted vs longitudinal location, using the same units on the graph, then the integrated area under the resulting curve gives the wetted area (Fig. 7.37).

Perimeters can be measured using a professional's "map-measure," or approximated using a piece of scrap paper. Simply follow around the perimeter of the cross section making tic marks on the paper, and then measure the total length using a ruler.

Note that the cross-sectional perimeter measurements should not include the portions where components join, such as at the wing-fuselage intersection. These areas are not "wetted."

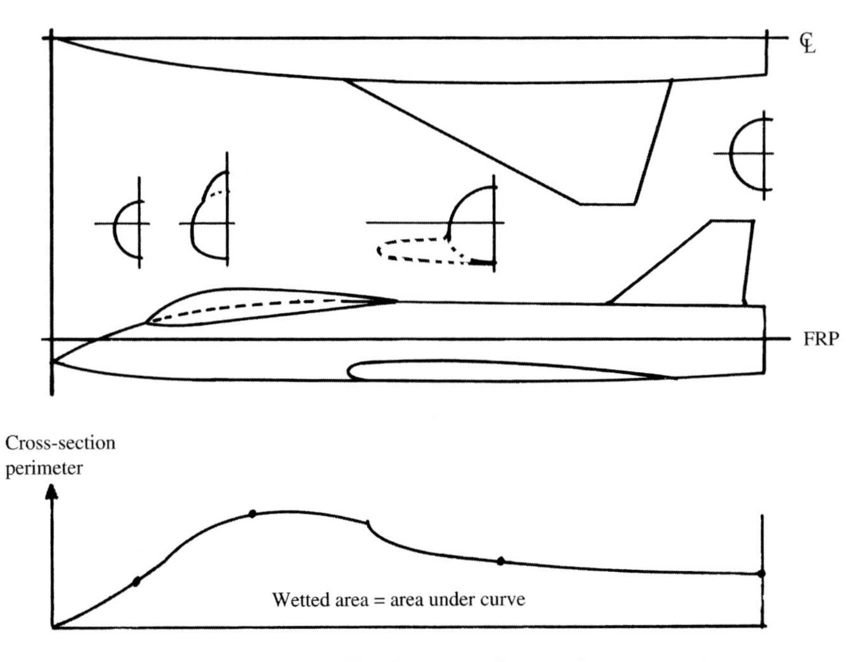

Fig. 7.37 Fuselage wetted area plot.

7.10 Volume Determination

The aircraft internal volume can be used as a measure of the reasonableness of a new design, by comparing the volume to existing aircraft of similar weight and type. This is frequently done by customer engineering groups, using statistical data bases that correlate internal volume with takeoff gross weight for different classes of aircraft. An aircraft with a less-than-typical internal volume will probably be tightly packed, which makes for poor maintainability.

Aircraft internal volume can be estimated in a similar fashion to the wetted-area estimation. A crude estimate of the fuselage internal volume can be made using Eq. (7.13), which uses the side and top view projected areas as used in Eq. (7.12). "L" in Eq. (7.13) is the fuselage length.

$$\text{Vol} \cong 3.4 \frac{(A_{\text{top}})(A_{\text{side}})}{4L} \tag{7.13}$$

A more accurate estimate of internal volume can be found by a graphical integration process much like that used for wetted area determination. The cross-section areas of a number of cross sections are measured and plotted vs longitudinal location, using consistent units (typically inches horizontally and square inches vertically on the graph). The area under the resulting curve is the volume, as shown in Fig. 7.38.

To obtain reasonable accuracy, cross sections should be plotted and measured anywhere that the cross-sectional area changes substantially. This typically

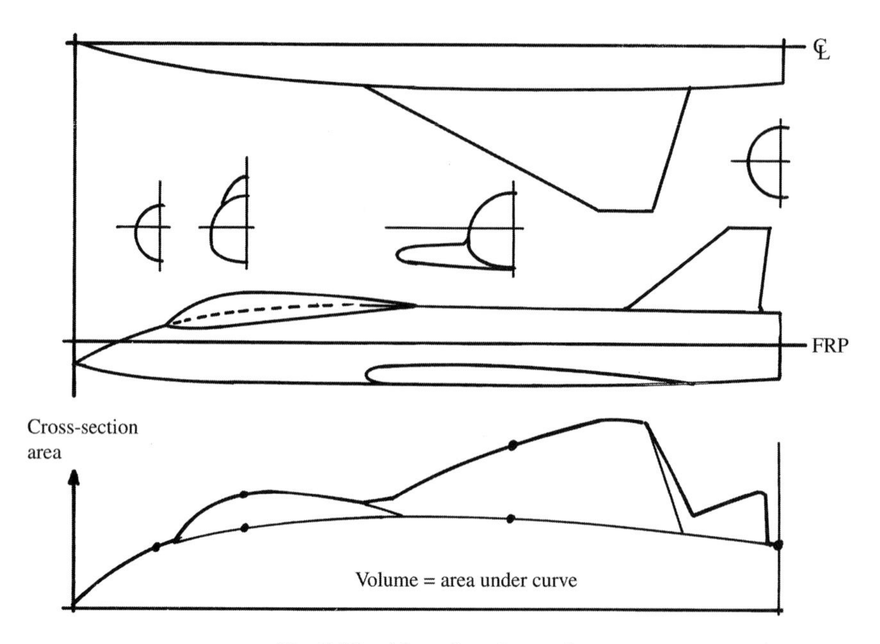

Fig. 7.38 Aircraft volume plot.

includes the start of an inlet duct, the start and end of a canopy, and where a wing or tail begins and ends.

Another use of the "volume distribution plot" is to predict and minimize supersonic wave drag and transonic drag rise. This will be discussed in Chapter 12.

7.11 Use of Computer-Aided Design (CAD) in Conceptual Design

Today, the previous discussion of drafting table techniques and tricks sounds almost quaint. Everyone, from students to grizzled industry veterans, uses a CAD system of some sort for most design work. For the most part, this is desirable, and the old notion that an engineering student simply must take a drafting class to be employable is obsolete.

Modern CAD systems are amazingly powerful and offer excellent graphical user interfaces, accurate surface definitions, realistic photo-like rendering capabilities, and sophisticated data management systems, even on a personal computer. Design capabilities allow creation of every imaginable type of geometry, and various CAD systems have specific geometry creation tools to simplify development of certain design components and features. The best modern CAD systems have virtually automated design of certain parts, such as hydraulic tubing and access doors. In one CAD program the hydraulic system designer can simply indicate, in three dimensions, the desired path of a hydraulic line and the system will create the tubing at the proper diameter, construct bends with diameters that can be fabricated without cracking, and include the proper fittings, couplings, and brackets—all automatically. The future will see more and more such automation of the design of common parts and systems.

Furthermore, through the industry usage of modern CAD systems the entire aircraft is being designed digitally, allowing the use of virtual rather than actual mock-ups. This saves time and money and does a better job of identifying and fixing component interference problems and potential difficulties in fabrication and maintenance of the aircraft. The digital product definition also improves prototype fabrication and aircraft production. Transference of the design data to computer-aided manufacturing (CAM) becomes almost trivially easy, and the resulting parts fit together perfectly. Altogether, the integrated use of CAD and CAM has been, in this author's opinion, the single greatest improvement in cost and quality that the aircraft industry has ever seen.

However, there can be problems with too great of a willingness to "let the CAD system do it." First of all, with a CAD system there is a tendency to let the computer lead you in the "easy" direction. If it is easy to retract the landing gear directly inward with your CAD system, you may do so even if a better design would result from having it retract inward and forward at a difficult-to-construct oblique angle. If you can easily calculate the volume of a square fuel tank, but don't know how to get the volume of a complicated tank wrapped around the inlet duct, guess which one you are likely to design!

Another problem is the actual calculation of the volumes, wetted areas, and other dimensions critical to your analysis of your design. Sometimes a CAD system may confidently display an incorrect answer! For example, we may model the wing as a collection of airfoils connected by a mathematical surface, and may readily calculate the wetted area of the wing itself. However, where that wing intersects the fuselage we must cut away the surface of the wing where it penetrates the fuselage, and cut away the fuselage where the wing covers it. It is possible in many CAD systems to forget to account for, or double-account for, the wing root airfoil "wetted area" that must be removed from the fuselage and not included with the wing! This potential problem is minimized if true "solid models" are (properly!) employed. Other examples include the inlet front and the back end of a fuselage or nacelle with a jet engine, or the front of a propeller nacelle, where the exhaust or intake areas must not be included. Even a solid model could accidentally give the wrong answer in this case, failing to understand that the "hole" isn't there!

For this reason it is STRONGLY recommended that all CAD users start by doing a trivially simple "aircraft design" consisting of a tube-plus-cone fuselage and a simple wing, where the correct wetted areas and volumes can be easily calculated by hand and compared with the answer from the CAD system.

Yet another problem for students is that the aircraft design course can easily become the "learn how to use a certain CAD system" course. There is not enough time in a semester course to really learn how to do conceptual design, and ANY time spent learning which button produces which geometry is time NOT spent learning the philosophy, methods, and techniques of aircraft conceptual design.

In industry, a real but subtle problem is that, with a CAD system, everybody's designs look good whether they are or are not! When everybody was using a drafting table, you could usually tell from drafting technique that a design was done by a beginner and therefore whether the design needed to be reviewed extra carefully. Today, it "takes one to know one"—you must be a pretty good designer yourself to know if a design you are looking at was done properly!

As was briefly mentioned in Chapter 2, CAD tools that are to be used during conceptual design should be tailored toward the fluid environment and the unique tasks of aircraft conceptual design. Quite simply, what is done during conceptual design, the things that are critical, and the tasks that are boring and repetitive (and therefore ideal for computerization) are different from those in other, later phases of aircraft design.

A perfect example is the wing trapezoidal geometry, i.e., the wing area, aspect ratio, taper ratio, and sweep. During detail design, it is out of the question to change the wing trapezoidal geometry, no matter how much the design of, say, a certain wing rib would be improved as a result. During conceptual design, those parameters are constantly being changed, almost every week in the early stages. Conceptual designers need capabilities to change these instantly and to have the computer automatically revise to the new geometry the wing's nontrapezoidal shaping and also revise the geometries of any parts made from the wing, such as wing fuel tanks, flaps, ailerons, spars, ribs, and possibly even wing carry-through structure and landing gear attachments. With most CAD systems, these must be completely recreated after the wing trapezoidal geometry is revised.

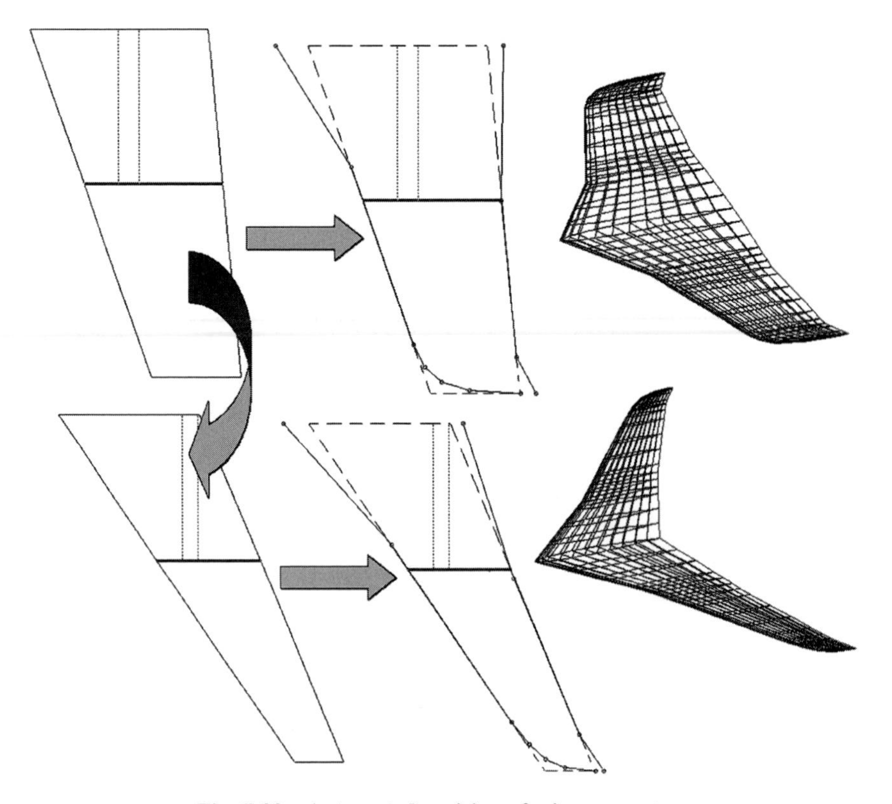

Fig. 7.39 Automated revision of wing geometry.

Some CAD systems do have new modules that can be used to accomplish this automatically, but they require a very time-consuming creation of flow-down logic done by the designer prior to "automatically" making the change. A good conceptual CAD system would have that capability built into the system so that all that the designer would have to do is to enter the revised geometric parameter (such as aspect ratio).

Figure 7.39 shows an example of such an automatic revision of the nontrapezoidal geometry from changes to the geometric trapezoidal parameters, done with the RDS-Professional program, described in Ref. 95. At the upper left is trapezoidal wing geometry. To its right is the wing created from it, with a swept-back tip, leading-edge strake, and trailing-edge kick. Below is the revised trapezoidal geometry after the aspect ratio, taper ratio, and sweep are changed in response to some optimization. To its right is the resulting wing geometry including the same swept-back tip, leading-edge strake, and trailing-edge kick.

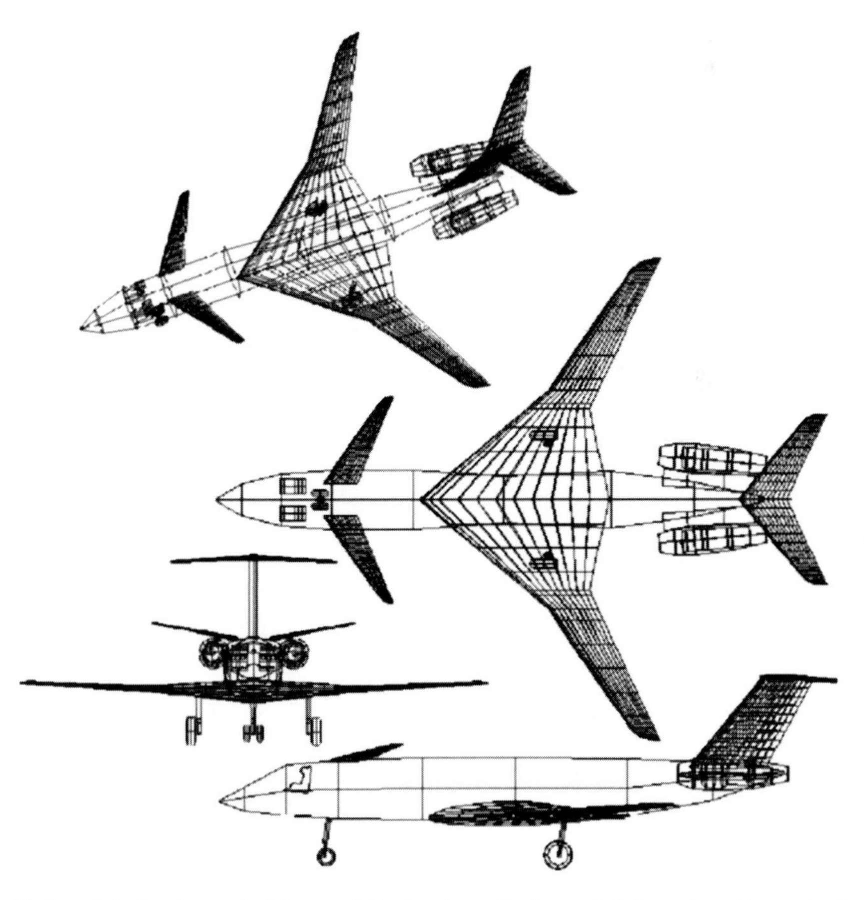

Notional design layout: Advanced Technology Business Jet (courtesy Conceptual Research Corp.).

C-17 Globemaster (NASA photo by Jim Ross).

8
Special Considerations in Configuration Layout

8.1 Introduction

The previous chapter discussed the mechanics of configuration layout. Later chapters will focus on the required provisions for specific internal components, such as the crew station and landing gear. This chapter discusses a number of important intangible considerations that the designer should consider when making the initial layout. These include aerodynamics, structures, detectability, vulnerability, producibility, and maintainability. All of these are numerically analyzed in later stages of the design process. During configuration layout, the designer must consider their impact in a qualitative sense.

8.2 Aerodynamic Considerations

Design Arrangement

The overall arrangement and smoothness of the fuselage can have a major effect upon aerodynamic efficiency. A poorly designed aircraft can have excessive flow separation, transonic drag rise, and supersonic wave drag. Also, a poor wing-fuselage arrangement can cause lift losses or disruption of the desired elliptical lift distribution, and can even cause bad handling qualities including spin tendencies.

Aerodynamic analysis will be discussed in Chapter 12 and a variety of first-order estimation methods will be presented. During concept layout, the designer must consider the requirements for aerodynamics based upon experience and a "good eye."

Minimization of wetted area is the most powerful aerodynamic consideration for virtually all aircraft. Wetted area directly affects the friction drag. Fuselage wetted area is minimized by tight internal packaging and a low fineness ratio (i.e., a short, fat fuselage). However, excessively tight packaging should be avoided for maintainability considerations. Also, a short, fat fuselage will have high supersonic wave drag.

Another major driver for good aerodynamic design during fuselage layout is the development of smooth longitudinal contours. These can be provided by the use of smooth longitudinal control lines. Generally, longitudinal breaks in contour should follow a radius at least equal to the fuselage diameter at that point.

To prevent separation of the airflow, the aft-fuselage deviation from the free-stream direction should not exceed 10–12 deg (Fig. 8.1). However, the air inflow induced by a pusher-propeller will prevent separation despite contour angles of up to 30 deg or more.

A lower-surface upsweep of about 25 deg can be tolerated for a rear-loading transport aircraft provided that the fuselage lower corners are fairly sharp. This causes a vortex-flow pattern that reduces the drag penalty. In general, aft-fuselage upsweep should be minimized as much as possible, especially for high-speed aircraft.

The importance of well-designed wing fillets has already been discussed. Fillets are especially important for low-wing, high-speed aircraft such as jet transports.

"Base area" is any unfaired, rearward-facing blunt surface. Base area causes extremely high drag due to the low pressure experienced by the rearward-facing surface (see Chapter 12).

However, a base area between or very near the jet exhausts may be "filled-in" by the pressure field of the exhaust, partially alleviating the drag penalty. The T-38 has such a base area between its nozzles. A base area fill-in effect is difficult to predict.

The aerodynamic interaction between different components should be visualized in designing the aircraft. For example, a canard should not be located such that its wake might enter the engine inlets at any possible angle of attack. Wake ingestion can stall or even destroy a jet engine.

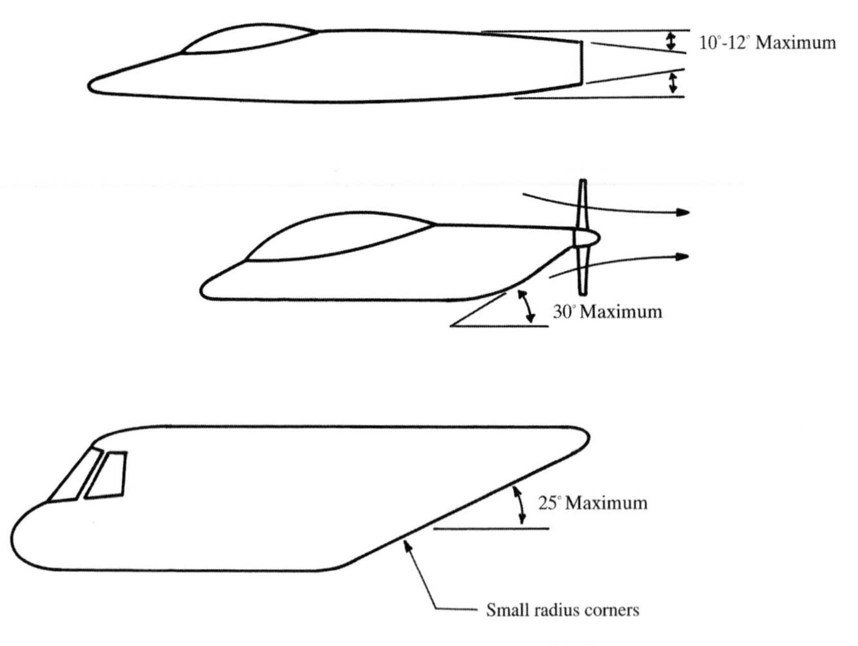

Fig. 8.1 Longitudinal contour guidelines.

If an aircraft's forebody has sharp lower corners, a separated vortex can be expected at high angle of attack. This could also be ingested by the inlets, with bad results. Also, such a vortex could unpredictably affect the wing or tail surfaces.

Isobar Tailoring

In Chapter 4 the importance of wing sweep for delaying the formation of shocks over the wing was discussed. It was explained that the shocks are formed over the top of the wing due to the increased velocity causing the air to go supersonic. It was also explained that theoretically this could be proven to depend not on the actual velocity of the air over the top of the wing but by the velocity perpendicular to the leading edge. Sweeping the wing causes this velocity to appear to be reduced, so shock formation is delayed.

Actually, the wing sweep theory is based not just on leading-edge sweep, but on the sweep of the wing pressure "isobars." Isobars are lines connecting regions with the same pressure. This is illustrated in Fig. 8.2. At the upper left there is an airfoil with its pressure contours shown, and four pressures are depicted with dots. To the right is a top view of part of the wing with those same four dots shown and lines (isobars) connecting those dots with other points on the top of the wing having the same pressure.

The complete wing shown in Fig. 8.2 illustrates two common problems with "real" wings. At the root, the isobars from the left and right sides of the wing

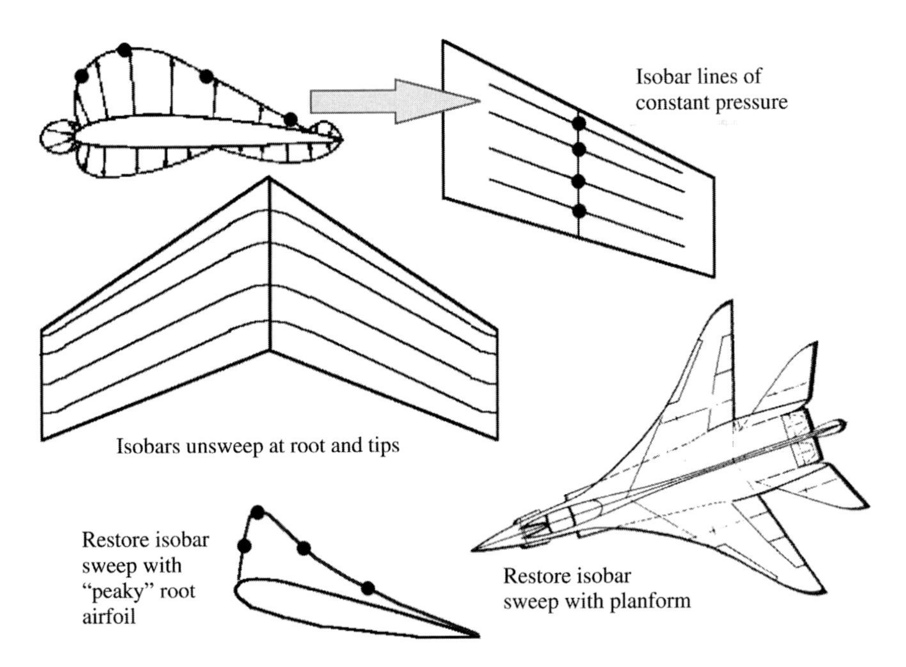

Fig. 8.2 Isobar tailoring for shock suppression.

cannot meet in a "V." Instead, they are joined by a rounded-off corner. As a result, this swept wing has no sweep at all right at the root, at least according to the wing sweep theory. This causes shocks near the wing root, and is a very real problem. Something similar happens at the wing tips, as shown.

To solve the isobar unsweep problem at the root, two aerodynamic strategies can be employed. One is to exaggerate the wing sweep near the root, blending the wing in a smooth fashion into the forebody of the fuselage. This is seen on the B-1B and was featured on the North American Rockwell F-X (F-15 Proposal) design, shown at the right of Fig. 8.2.

Another approach, commonly used on large airliners, is to "pull" the wing root isobars forward by using a strange airfoil shape at the root that is specially designed to have its pressure peak very near its nose. Such an airfoil tends to have a large nose radius, a fairly flat top, and, oddly enough, negative camber. This negative camber tends to create negative lift, so it must be placed at a high nose-up twist angle to maintain a good spanwise lift distribution.

Such design tricks are beyond the scope of early conceptual design but can be approximated on the initial design layout based on similar aircraft. At a later date, an airfoil/wing optimization code will be run to define the best airfoil geometries.

Supersonic Area Rule

For supersonic aircraft, the greatest aerodynamic impact upon the configuration layout results from the desire to minimize supersonic wave drag, a pressure drag due to the formation of shocks. This is analytically related to the longitudinal change in the aircraft's total cross-sectional area. In fact, wave drag is calculated using the second derivative (i.e., curvature) of the volume-distribution plot as shown in Fig. 7.38.

Thus, a "good" volume distribution from a wave-drag viewpoint has the required total internal volume distributed longitudinally in a fashion that minimizes curvature in the volume-distribution plot. Several mathematical solutions to this problem have been found for simple bodies-of-revolution, with the Sears–Haack body (Fig. 8.3; see Ref. 16) having the lowest wave drag.

If an aircraft could be designed with a volume plot shaped like the Sears–Haack volume distribution it would have the minimum wave drag at Mach 1.0 for a given length and total internal volume. (What happens at higher Mach numbers is discussed in Chapter 12, but for initial layout purposes the minimization of wave drag at Mach 1.0 is a suitable goal in most cases.)

However, it is usually impossible to exactly or even approximately match the Sears–Haack shape for a real aircraft. Fortunately, major drag reductions can be obtained simply by smoothing the volume distribution shape.

As shown in Fig. 8.4, the main contributors to the cross-sectional area are the wing and the fuselage. A typical fuselage with a trapezoidal wing will have an irregularly shaped volume distribution with the maximum cross-sectional area located near the center of the wing. By "squeezing" the fuselage at that point, the volume-distribution shape can be smoothed and the maximum cross-sectional area reduced.

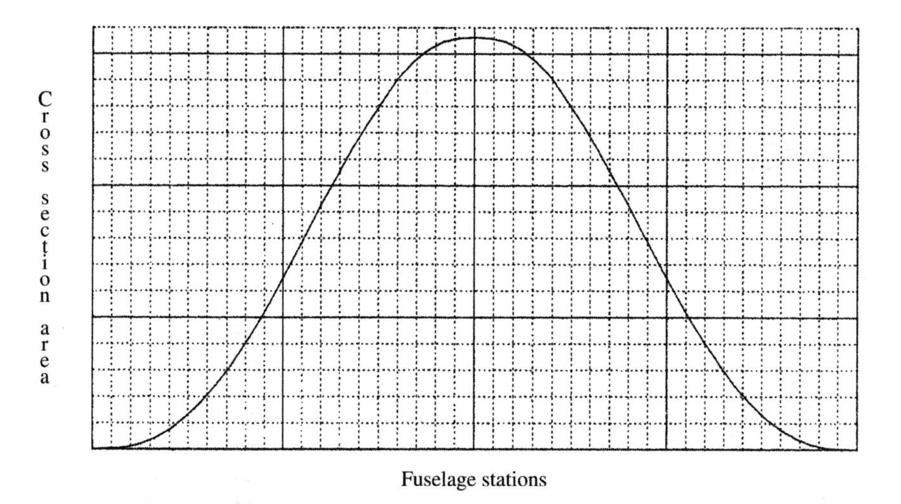

Fig. 8.3 Sears–Haack volume distribution.

This design technique, developed by R. Whitcomb of the NACA (Ref. 17), is referred to as "area-ruling" or "coke-bottling" and can reduce the wave drag by as much as 50%. Note that the volume removed at the center of the fuselage must be provided elsewhere, either by lengthening the fuselage or by increasing its cross-sectional area in other places.

While area-ruling was developed for minimization of supersonic drag, there is reason to believe that even low-speed aircraft can benefit from it to some extent. The airflow over the wing tends to separate toward the trailing edge. If an aircraft is designed such that the fuselage is increasing in cross-sectional area toward the

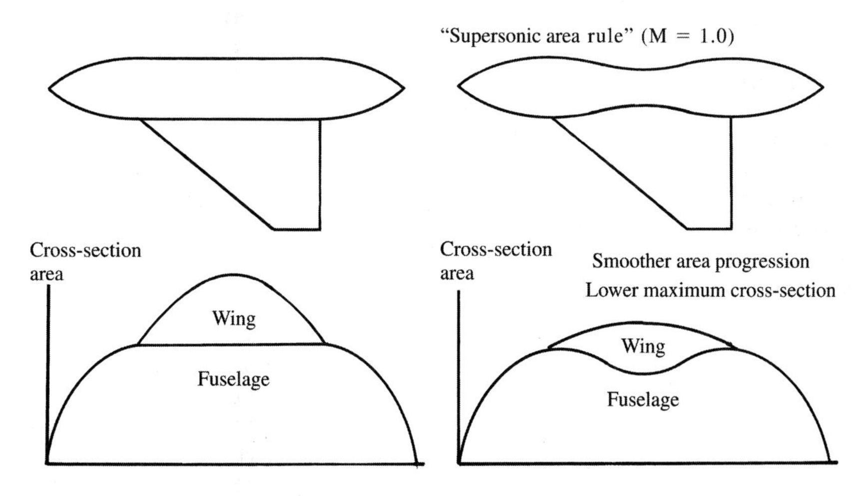

Fig. 8.4 Design for low wave drag.

wing trailing edge, this may "push" air onto the wing, thus reducing the tendency to separate. The Wittman Tailwind, which is remarkably efficient, uses this approach.

Compression Lift

A very successful yet almost forgotten aerodynamic concept can be used to improve lift-to-drag ratio at supersonic speeds. Compression lift was apparently conceived by two researchers at NACA Langley in 1954, and used by R. Child and G. Owl of North American Aviation to configure a huge supersonic bomber that literally rode its own shock wave, the B-70.

Any body shape will create shock waves at supersonic speeds, forming at the nose and at anyplace else where the cross-section area is increasing. These shocks trail back at approximately the Mach angle [arcsine $(1/M)$], as shown in Fig. 8.5. In the B-70, the inlet duct was faired back into a wide nacelle, with a steadily widening cross-sectional area until a maximum was reached (Fig. 8.6). Engines and payload were carried in this nacelle, which created a strong shock on either side with greatly increased static pressures behind the shocks. By placing the wing above these shocks, the increased pressure beneath the wing provided free lift—roughly 30% of the total lift required!

The B-70 also used fold-down wing tips. As can be seen, these reflected the shocks from the nacelle creating even more shocks under the wing—more free

Fig. 8.5 Supersonic shocks.

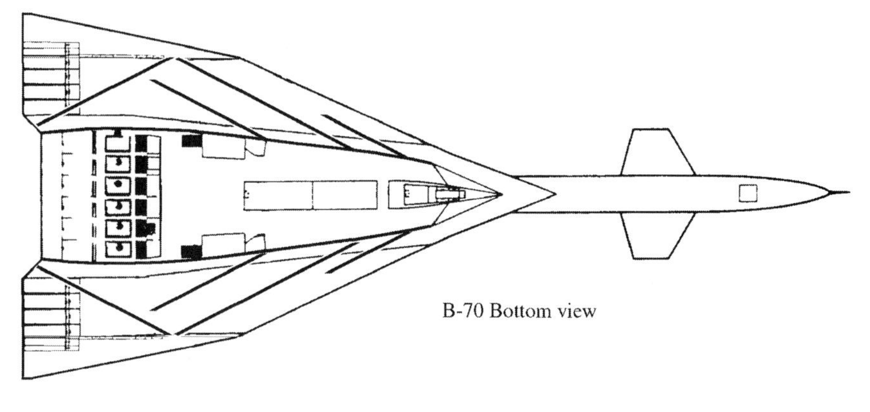

Fig. 8.6 Compression lift.

lift! Furthermore, they solved the two big stability problems inherent in supersonic flight. First, the aerodynamic center moves considerably to the rear requiring some way to move it forward at supersonic speeds, or to move the center of gravity to the rear. Folding down the wing tips does the former. Also, at supersonic speeds the effectiveness of a vertical tail usually reduces. Folding down the wing tips helps this problem, too.

In fact, the B-70 is an excellent example of synergistic design. The design features are all working together, and many of the design components do more than one task and offer more than one benefit.

The NAAF-X proposal shown in Fig. 8.2 was configured for compression lift as well as isobar sweep.

Design "Fixes"

Real airplanes have many "things" on them that are not seen on a conceptual design layout. Some of these things are equipment such as antennas and lights, and some of these things are minor design details such as fuel drains and cooling vents that are not normally considered during conceptual design. Some of these things, though, are fixes to aerodynamic problems discovered later in design development or flight test. In conceptual design we think we have no such aerodynamic problems, and if we did, we would revise the overall arrangement to avoid them. Later on, it is too difficult to change the overall geometry, and so if unexpected problems are found, they must be fixed in some other way.

Aerodynamic problems are most often attributable to two phenomena: separation of the flow or formation of an unwanted, "bad" vortex. Typical devices to fix aerodynamic problems are shown in Fig. 8.7. These mostly work by creating and controlling "good" vortices.

Flow separation over a wing or fuselage often occurs because the air near the aircraft has been slowed down too much by viscous effects and no longer has much energy. When this low-energy air is asked to turn a corner, it simply can't, and separates instead. To fix this, a number of small plates are bent into

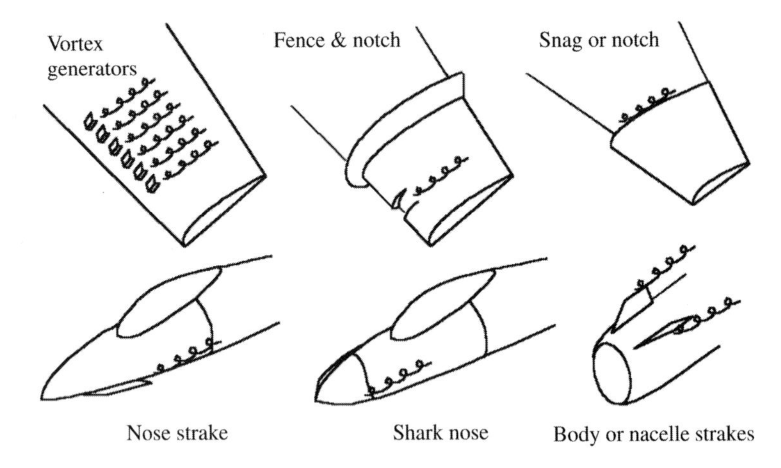

Fig. 8.7 Aerodynamic fixes.

an "L" shape and attached just before the region of separation, set at an angle to the flow. These create small vortices that stir up the air near the surface, bringing high-energy air into the boundary layer. This action permits the flow to follow a much-greater turn. Such "vortex generators" are commonly found on the tops of wings and near the back of a long fuselage, but can be found almost anywhere on airplanes except right at the nose!

The best locations for vortex generators to fix some particular problem are found by trial and error, both in the wind tunnel and in flight test. Strangely enough, the vortex generators cause almost no increase in parasitic drag, even on a flat plate. They are so small that they are mostly in the boundary layer, and their own effect on drag is negligible whereas, if they prevent separation, they can greatly reduce the total drag of the aircraft.

At high angle of attack, the flow experiences a disastrous form of separation called wing stall. Properly placed vortex generators can delay this, and are commonly found on wings for this purpose, but still don't allow the wing to reach its maximum lift.

Wing stall tends to start at the wing root and spread outward. By placing a "fence" just outboard of where the stall has been found to begin, the stall can be prevented from spreading outward until such a high angle of attack is reached that the outboard part of the wing stalls on its own.

A fence can also be used to cure a problem common in highly swept wings. The sweep of the wing tends to push the air outward, especially in the boundary layer where the air is low in energy. It is not uncommon for the boundary-layer air from the root of the wing to travel outward, all of the way to the tip of the wing. This increases boundary-layer thickness, and that tends to cause flow separation and wing stall. A fence can physically prevent that occurrence and can improve stall characteristics.

One can create a "virtual fence" by placing a notch or snag at the location just outboard of where the stall begins. These form a vortex that, like a fence, acts to separate the stalled from the unstalled flow and stop the stall from spreading.

The leading edge outboard of the wing notch can be cambered downward to further reduce the outboard wing panel's tendency to stall. Properly done, this can also greatly reduce spin tendencies and promote spin recovery, and is highly recommended for general aviation and training aircraft.

Nose strakes, or the similar, sharp-sided "shark nose," are used to force vortices to form simultaneously on both sides of the forebody at higher angles of attack. With a rounded forebody, at some high angle of attack such vortices will form but the vortex on one side may form sooner than on the other. Having a vortex on only one side of the forebody creates a strong suction force that can pull the nose to one side, causing a spin. Sharp edges on the nose fix this.

Finally, large strakes or fins can be strategically placed to form vortices and do something good. For example, the vertical tails of the F-18 were having structural fatigue problems resulting from an unexpected tendency of the vortices from the wing strakes to hit the vertical tails. To fix this, small upright strakes were added to the top of the aircraft to create vortices that divert the wing strake vortices. As can be imagined, they were not on the conceptual design layouts!

Many airliners have similar strakes on the engine nacelles. These can be used to improve flow over the wing flaps, or to fix a flow problem at the horizontal tail, or both. The DC-10, perhaps the first to use such nacelle strakes, needed them because the nacelle and pylon were causing the flow to separate resulting in a premature stall. The nacelle strakes fixed the separation and increased maximum lift.

The growth versions of the DC-9 had flow problems at the vertical tail, leading to directional stability reduction at moderate sideslip. Strakes below the cockpit were found to cure this problem, even though they are located about 100 ft {30 m} ahead of the tail.

Another type of vortex-generating strake called a "vortilon" is placed just below the wing leading edge and is aligned with the flight direction (it looks like a miniature engine pylon that lost its engine!). At high angle of attack, the local flow at the leading edge is diverted outward toward the wing tip so that the vortilon finds itself at an angle to the local flow and produces a vortex. This vortex wraps over the top of the wing and energizes the boundary layer while acting like a stall fence.

8.3 Structural Considerations

Load Paths

In most larger companies, the configuration designer is not ultimately responsible for the structural arrangement of the aircraft. That is the responsibility of the structural design group. However, a good configuration designer will consider the structural impacts of the general arrangement of the aircraft, and will in fact have at least an initial idea as to a workable structural arrangement.

The primary concern in the development of a good structural arrangement is the provision of efficient "load paths"—the structural elements by which opposing forces are connected. The primary forces to be resolved are the lift of the wing and the opposing weight of the major parts of the aircraft, such as the engines and payload. The size and weight of the structural members will be minimized by locating these opposing forces near to each other.

Carried to the extreme, this leads to the Flying Wing concept. In a flying wing the lift and weight forces can be located at virtually the same place. In the ideal case, the weight of the aircraft would be distributed along the span of the wing exactly as the lift is distributed (Fig. 8.8). This is referred to as "spanloading" and eliminates the need for a heavy wing structure to carry the weight of the fuselage to the opposing lift force exerted by the wing. The structure can then be sized by lesser requirements such as the landing-gear loads.

While ideal span-loading is rarely possible, the span-loading concept can be applied to more-conventional aircraft by spreading some of the heavy items such as engines out along the wing. This will yield noticeable weight savings, but must be balanced against the possible drag increase, especially if it requires a larger vertical tail to handle an engine-out situation.

If the opposing lift and weight forces cannot be located at the same place, then some structural path will be required to carry the load. The weight of structural members can be reduced by providing the shortest, straightest load path possible.

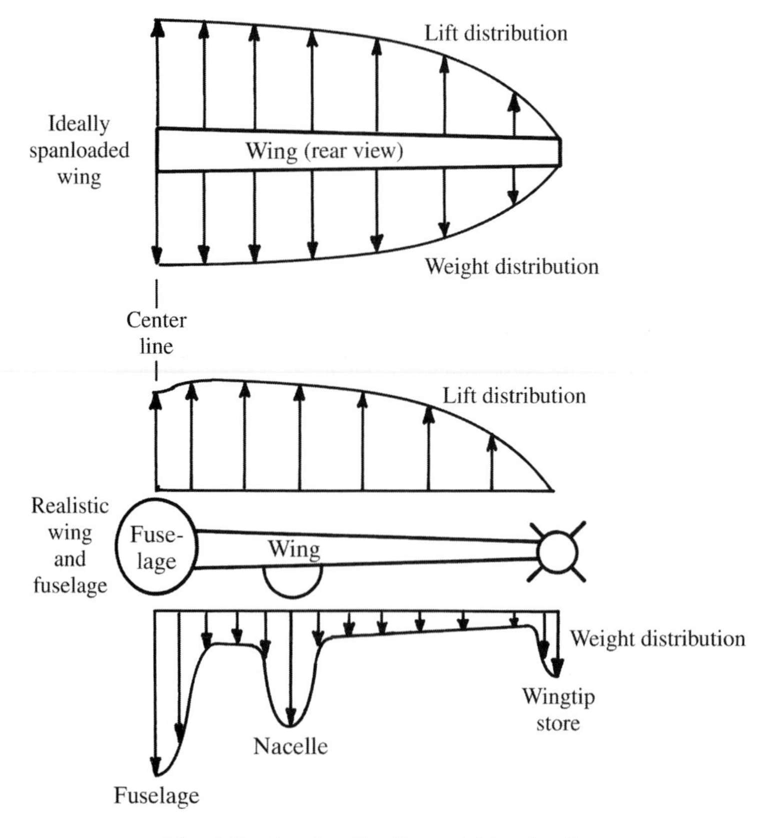

Fig. 8.8 Spanloading for weight reduction.

Figure 8.9 illustrates a structural arrangement for a small fighter. The major fuselage loads are carried to the wing by "longerons," which are typically I- or H-shaped extrusions running fore and aft and attached to the skin. Longerons are heavy, and their weight should be minimized by designing the aircraft so that they are as straight as possible.

For example, the lower longerons in Fig. 8.9 are high enough that they pass over the wing-carrythrough box. Had the longerons been placed lower, they would have required a kink to pass over the box.

On the other hand, the purpose of the longeron is to prevent fuselage bending. This implies that the lightest longeron structure occurs when the upper and lower longerons are as far apart vertically as possible. In Fig. 8.10 the longerons are farther apart, but this requires a kink to pass over the box. Only a trade study can ultimately determine which approach is lighter for any particular aircraft.

In some designs similar to Fig. 8.9, the lower longerons are placed near the bottom of the aircraft. A kink over the wing box is avoided by passing the longeron under or through the wing box. This minimizes weight but complicates both fabrication and repair of the aircraft.

For aircraft such as transports, which have fewer cutouts and concentrated loads than a fighter, the fuselage will be constructed with a large number of "stringers," which are distributed around the circumference of the fuselage (Fig. 8.11). Weight is minimized when the stringers are all straight and uninterrupted.

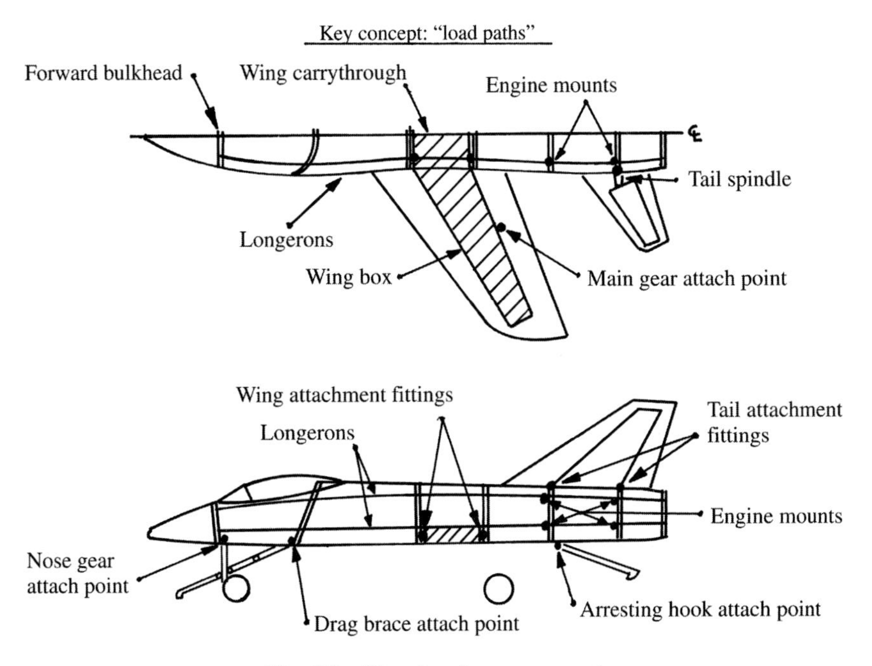

Fig. 8.9 Structural arrangement.

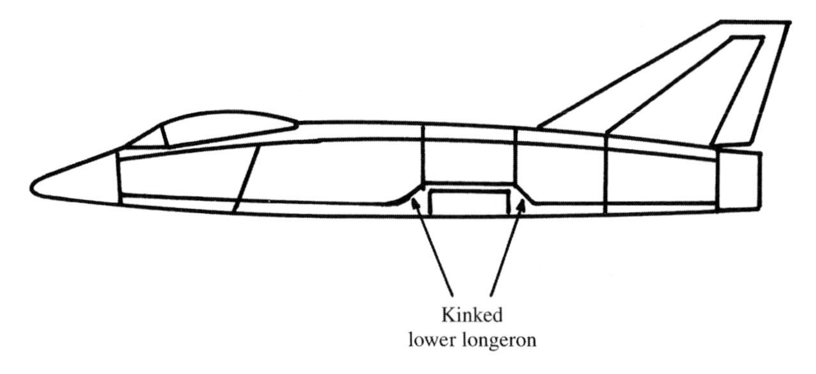

Fig. 8.10 Kinked lower longeron.

Another major structural element used to carry fuselage bending loads is the "keelson." This is like the keel on a boat, and it is a large beam placed at the bottom of the fuselage as shown in Fig. 8.11. A keelson is frequently used to carry the fuselage bending loads through the portion of the lower fuselage which is cut up by the wheel wells.

As the wing provides the lift force, load-path distances can be reduced by locating the heavy weight items as near to the wing as possible. Similarly, weight can be reduced by locating structural cutouts away from the wing.

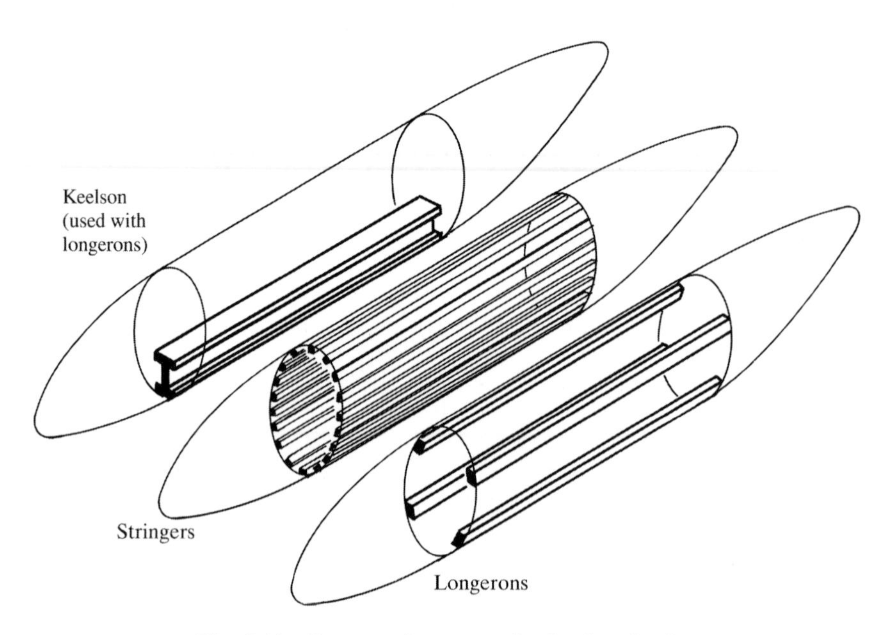

Fig. 8.11 Structural concepts for fuselage loads.

Required structural cutouts include the cockpit area and a variety of doors (passenger, weapons bay, landing gear, engine access, etc.).

An especially poor arrangement (seen on some older fighter aircraft) has the main landing gear retracting into the wing-box area, which requires a large cutout where the loads are the greatest.

When possible, structural cutouts should be avoided altogether. For example, a jet engine that is buried in the fuselage requires a cutout for the inlet, a cutout for the exhaust, and in most cases another cutout for removal of the engine. The resulting weight penalty compared to a podded engine must be balanced against the reduced drag of a buried engine installation.

Figure 8.9 illustrates another important concept in structural arrangement. Large concentrated loads such as the wing and landing gear attachments must be carried by a strong, heavy structural member such as a major fuselage bulkhead. The number of such heavy bulkheads can be minimized by arranging the aircraft so that the bulkheads each carry a number of concentrated loads, rather than requiring a separate bulkhead for each concentrated load.

In Fig. 8.9 the two bulkheads in the aft fuselage carry the loads for the engines, tails, and arresting hook. Had the tails and engine been located without this in mind, the structural designer would have had to provide four or five heavy bulkheads rather than the two shown.

Carrythrough Structure

The lift force on the wing produces a tremendous bending moment where the wing attaches to the fuselage. The means by which this bending moment is carried across the fuselage is a key parameter in the structural arrangement and will greatly influence both the structural weight and the aerodynamic drag of the aircraft.

Figure 8.12 illustrates the four major types of wing carrythrough structure. The "box carrythrough" is virtually standard for high-speed transports and general-aviation aircraft. The box carrythrough simply continues the wing box through the fuselage. The fuselage itself is not subjected to any of the bending moment of the wing, which minimizes fuselage weight.

However, the box carrythrough occupies a substantial amount of fuselage volume, and tends to add cross-sectional area at the worst possible place for wave drag, as already discussed. Also, the box carrythrough interferes with the longeron load-paths.

The "ring-frame" approach relies upon large, heavy bulkheads to carry the bending moment through the fuselage. The wing panels are attached to fittings on the side of these fuselage bulkheads. While this approach is usually heavier from a structural viewpoint, the resulting drag reduction at high speeds has led to the use of this approach for most modern fighters.

The "bending beam" carrythrough can be viewed as a compromise between these two approaches. Like the ring-frame approach, the wing panels are attached to the side of the fuselage to carry the lift forces. However, the bending moment is carried through the fuselage by one of several beams that connect the two wing panels. This approach has less of a fuselage volume increase than does the

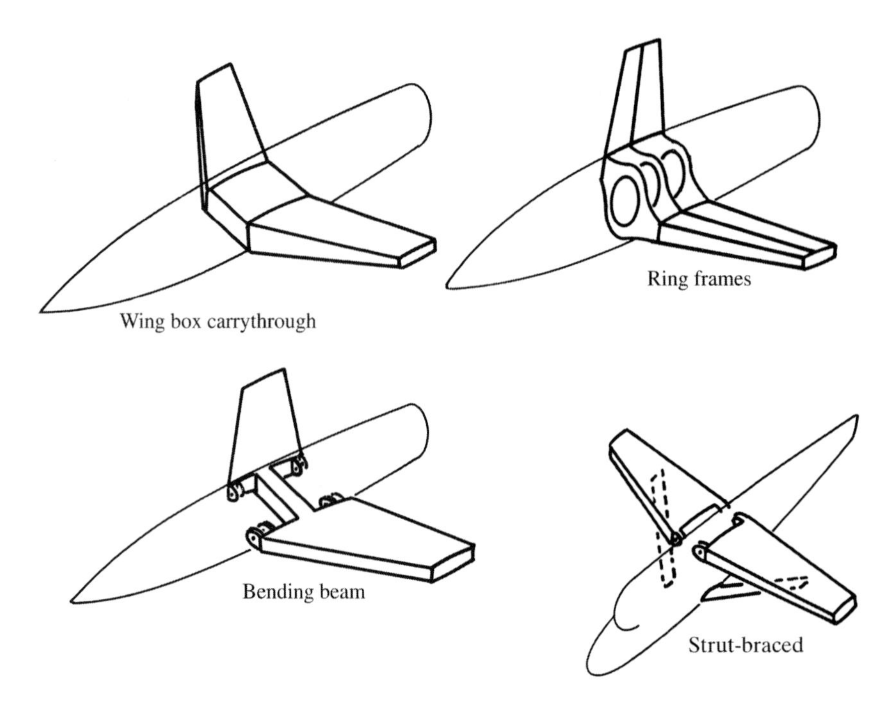

Fig. 8.12 Wing carrythrough structure.

box-carrythrough approach. The bending-beam carrythrough is common in sailplanes and is also seen on a number of advanced composite general aviation designs. Frequently there is a separate bending beam for each wing half, which simplifies manufacture.

Many light aircraft and slower transport aircraft use an external strut to carry the bending moments. While this approach is probably the lightest of all, it obviously has a substantial drag penalty at higher speeds.

Aircraft wings usually have the front spar at about 20–30% of the chord back from the leading edge. The rear spar is usually at about the 60–75% chord location. Additional spars may be located between the front and rear spars forming a "multispar" structure. Multispar structure is typical for large or high-speed aircraft.

If the wing skin over the spars is an integral part of the wing structure, a "wing box" is formed which in most cases provides the minimum weight.

Aircraft with the landing gear in the wing will usually have the gear located aft of the wing box, with a single trailing-edge spar behind the gear to carry the flap loads, as shown in Fig. 8.13.

Ribs carry the loads from the control surfaces, store stations, and landing gear to the spars and skins. A multispar wing box will have comparatively few ribs, located only where major loads occur.

Another form of wing structure, the "multirib" or "stringer panel" box, has only two spars, plus a large number of spanwise stringers attached to the wing skins. Numerous ribs are used to maintain the shape of the box under bending.

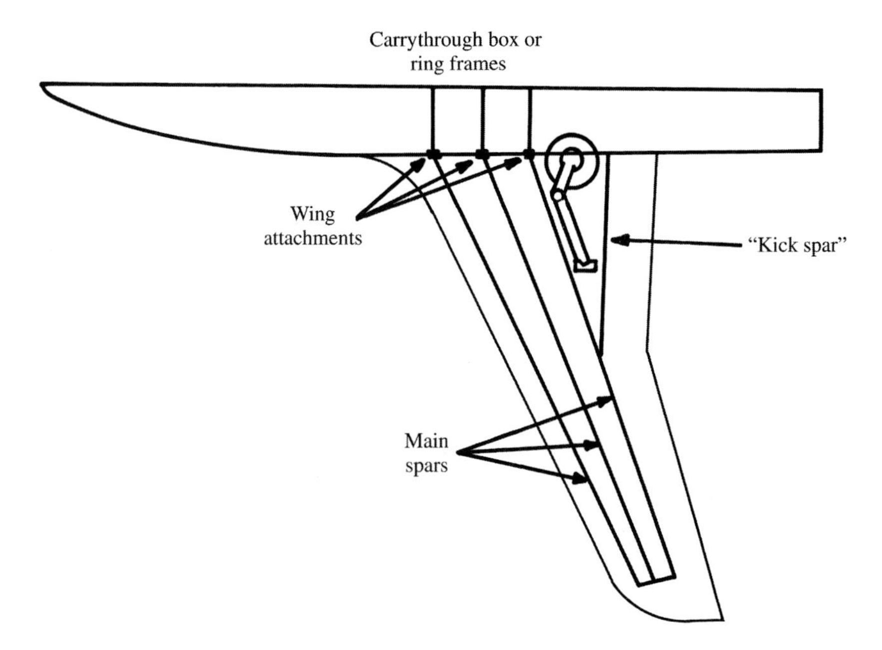

Fig. 8.13 Typical wing box structure.

Variable sweep and folding capability add considerably to the wing structural weight. On the other hand, use of a delta wing will reduce the structural weight. These are further discussed in Chapter 15.

Clearances and Allowances

First-order structural sizing will be discussed in Chapter 14. For initial layout purposes the designer must guess at the amount of clearance required for structure around the internal components. A good designer with a "calibrated eyeball" can prevent a lot of lost effort, for the aircraft may require substantial redesign if later structural analysis determines that more room is required for the structural members.

A large airliner will typically require about 4 in. {10 cm} of clearance from the inner wall of the passenger compartment to the outer skin ("moldline"). The structure of a conventional fighter fuselage will typically require about 2 in. {5 cm} of offset from the moldline for internal components. For a small general-aviation aircraft, 1 in. clearance or less may be acceptable.

The type of internal component will affect the required clearance. A jet engine contained within an aluminum or composite fuselage will require perhaps an additional inch of clearance to allow for a heat shield. The heat shield may be constructed of titanium, steel, or a heat-proof matting. On the other hand, an "integral" fuel tank in which the existing structure is simply sealed and filled with fuel will require no clearance other than the thickness of the skin.

There is no easy formula for the estimation of structural clearance. The designer must use judgment acquired through experience. The best way to gain this judgment other than actual design experience is by looking at existing designs.

Flutter

Flutter is an unfortunate dynamic interaction between the aerodynamics and the structure of an aircraft. It occurs when some structural deflection of the aircraft such as wing bending causes an aerodynamic load that tends to amplify the deflection during each oscillation until structural failure is reached. There are many possible flutter modes. An aileron with its center of mass well behind its hinge line will tend to lag when accelerated upwards by oscillating wing bending. This lagging is similar to a flap deflection, increasing the wing lift and amplifying the wing bending. On the way back down, the aileron lags upward, driving the wing down even further.

Similar flutter modes occur in elevators and rudders that have center of masses behind their hinge lines. Early Learjets were crashing because water was freezing inside the elevators behind the hinge line, causing flutter. This was difficult to uncover because, of course, the ice melted by the time the accident investigators got to the scene. Even a trim tab or servo tab may cause flutter if it has its center of mass behind its hinge line.

The solution to this control surface flutter is obvious: don't allow the center of mass to be behind the hinge line! Instead, add mass balancing in the form of weight ahead of the hinge line, and ruthlessly avoid weight behind it. A control surface is said to be statically balanced if its chordwise center of gravity is on its hinge line. Many World War II–vintage planes had fabric-covered control surfaces to keep the center of gravity forward to avoid flutter.

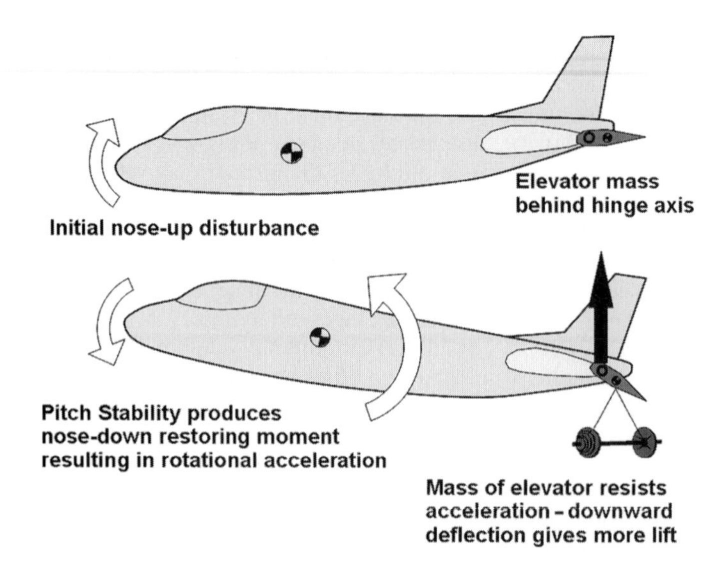

Fig. 8.14 Elevator lag pitching flutter.

Complete balancing of a control surface requires the product of inertia about the hinge axis to be zero. This leads to placing balance weights near the tips of control surfaces to reduce the product of inertia. Dynamic balance is obtained when a control surface moves with its wing or tail without any tendency towards relative rotation between the two, so they act as if they were welded together.

Control surface flutter is more likely if there is play (looseness) in the control linkages or play in the trim tab linkage. For this reason, stiff pushrod linkages are preferred over wire cables. Also, pilots should always inspect control linkages before flight.

The shaping of the control surfaces has an effect on flutter. They should never be convex, bulging out into the airflow, because it sets up unstable flow at the trailing edge. Instead, they should be flat-sided or concave. It is desirable to have a beveled trailing edge, and a control surface that is "fattened" at the hinge line will tend to reattach the flow, improving flutter characteristics.

It is bad for flutter if the natural frequency of the vibration of the aileron about its hinge is nearly the same as the wing natural bending frequency (analysis of this requires complicated calculations or shake-testing). It is also bad if the tip of the aileron is in the wing-tip vortex. For this reason, ailerons should not extend all the way to the wing tip. Another potential source of flutter problems is an excessive amount of aileron aerodynamic balance. If a deflecting aileron produces almost no restoring moments, flutter can result.

There is less of a tendency for a fuselage torsional flutter if the rudder is halfway below the fuselage rather than solely above it, mounted on the vertical tail. To increase their relative torsional stiffness, a rigid torque tube should connect left- and right-side elevators.

Another type of flutter that has nothing to do with control surface problems is called wing flexure-torsion binary flutter. In it, a torsional vibration or oscillation of the wing sets up aerodynamic forces in phase with an up-and-down wing bending flexural motion. The wing is bending up and down and twisting at the same time (in-phase oscillation) such that it has a positive angle of attack when the wing is going up and a negative angle of attack when it is going down. The resulting change in lift amplifies the up and down bending, possibly leading to flutter and divergence (wing breaks). This can be avoided by increasing the wing's torsional rigidity and by keeping the wing's chordwise center of gravity at or in front of the wing's structural elastic axis. In other words, avoid any weight behind roughly the middle of the wing, and try to give the wing a strong and rigid box structure.

Yet another type of flutter is a problem peculiar to high speed aircraft. Aerodynamic forces on structural panels can set up an in-and-out oil-can-like flutter, with the potential to rip the panel right off the aircraft. This is avoided by making sure that the panels do not have too great an unsupported length, or by using honeycomb panels or some other stiffened skin. This panel flutter is not typically addressed in conceptual design.

8.4 Radar Detectability

Ever since the dawn of military aviation, attempts have been made to reduce the detectability of aircraft. During World War I, the only "sensor" in use was the human eyeball. Camouflage paint in mottled patterns was used on both sides to reduce the chance of detection.

Radar (acronym for radio detection and ranging), the primary sensor used against aircraft today, consists of a transmitter antenna that broadcasts a directed beam of electromagnetic radio waves and a receiver antenna which picks up the faint radio waves that bounce off objects "illuminated" by the radio beam. Usually the transmitter and receiver antennas are collocated ("monostatic radar"), although some systems have them in different locations ("bistatic radar").

Detectability to radar has been a concern since radar was first used in World War II. "Chaff" was an early counter-radar "stealth" technology. Chaff, also called "Window," consists of bits of metal foil or metallized fibers dropped by an aircraft to create many radar echos that hide its actual echo return. Chaff is still useful against less-sophisticated radars.

Chaff obscures the actual location of the aircraft, but does not allow the aircraft to pass unnoticed. To avoid detection, the aircraft must return such a low amount of the transmitted radio beam that the receiver antenna cannot distinguish between it and the background radio static.

Radar stealth treatments go back further than many people realize. In World War II, German U-boats had RAM-covered periscopes to avoid detection. The first jet flying wing in the world, the Horten IX (which flew in 1945), used a charcoal-and-glue RAM and configuration shaping to reduce its signature. By 1960 the U.S. Air Force had flown a T-33 covered entirely with RAM and a B-47 with its inlets covered by screens and its exhausts obscured. U-2s were flown with RAM and other more-exotic RCS-reduction techniques. The North American Aviation Hound Dog air-launched cruise missile (operationally carried by the B-52) incorporated RAM, and NAA's F-X proposal for the F-15 had inlet ducts treated with RAM, in 1969.

By the 1970s most U.S. military aircraft companies had the ability to design for stealth in terms of overall configuration shaping. This included sloping the fuselage sidewalls; hiding inlets and engine front and rear faces; sweeping the edges of the wing, tail, and other edges; and similar fundamental stealth techniques. However, that is only part of the capability required to be a qualified "stealth house." Other key capabilities include the ability to analytically estimate signature, the technologies for stealth treatments of surfaces, edges, and details such as access doors and running lights, and the technology for "stealthy" integration of avionics, including radomes. Two companies were clear leaders in these areas and as a result, were the most successful in development of actual stealth aircraft. Lockheed gained such expertise through its development of spy planes (U-2 and SR-71), while Northrop apparently made a corporate decision in the 1960s that stealth was a critical emerging technology and invested accordingly (Ref. 97).

Development of radar stealth technology capable of making an aircraft operationally undetectable was accelerated by the DARPA Project Harvey begun around 1970 and named after the invisible rabbit in the play of the same name. This classified program led to the Have Blue flight test demonstrator, awarded to Lockheed over Northrop, the only other serious competitor. Have Blue led in turn to the operational F-117, which proved the operational worth of stealth.

The fundamental mathematical relationships governing radar cross section and other electromagnetic phenomena are Maxwell's equations, defined over 100 years ago. These, like the Navier–Stokes equations for aerodynamics (see Chapter 12), are complete governing equations and, if solved, would tell us everything we want to know! However, like the Navier–Stokes equations, they are currently impossible to solve exactly in their complete form for any complicated geometry, and so we solve simplified versions of the equations to attempt to predict RCS. These simplified versions of Maxwell's equations can only consider a limited version of the physical phenomena that cause radar energy to return.

The extent to which an object returns electromagnetic energy is the object's radar cross section (RCS). RCS is usually measured in square meters or in decibel square meters, with "zero dBsm" equal to 10 to the zero power, or 1 sqm. "Twenty dBsm" equals 10 to the second power, or 100 sqm. Because radar signal strength is an inverse function of the fourth power of the distance to the target, it takes a very substantial reduction in RCS to obtain a meaningful operational benefit.

Actually, the RCS of an aircraft is not a single number. The RCS is different for each "look-angle" (i.e., direction from the threat radar). When graphed in polar coordinates, RCS appears as shown in Fig. 8.15. These are actual data for the B-70 supersonic bomber. As can be seen, RCS varies widely from different directions, by almost four orders of magnitude for this design.

We use the expression "spikes" to describe directions from which the RCS of an aircraft is very high. These are typically perpendicular to the leading and trailing edges of the wing, perpendicular to the flat side of the aircraft unless it is properly shaped and treated, and directly off the nose and tail due to the inlets, nozzles, radome, and other features. For the B-70, huge spikes are evident to the sides perpendicular to the big flat sides of the nacelle. However, with a cruising speed of Mach 3.0 at almost 80,000 ft {24,300 m}, it would be difficult

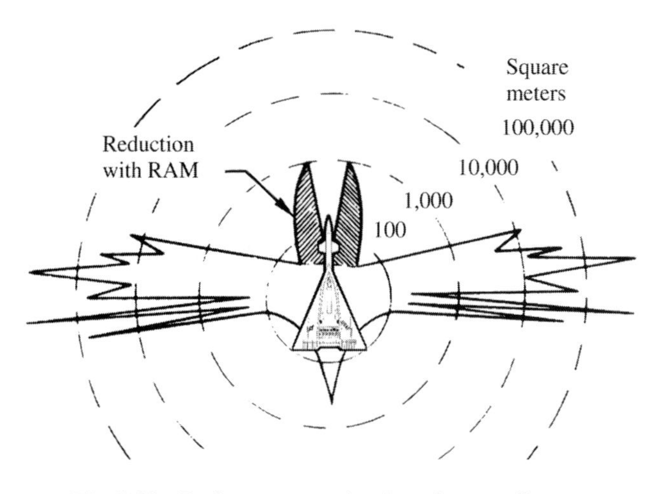

Fig. 8.15 Radar cross section in polar coordinates.

to intercept a B-70 that was already flying past. Unfortunately, there are also substantial spikes just off the nose, perpendicular to the leading edges of the canards, which have fairly low sweep. These spikes would have warned defenders that a B-70 was coming. As shown, treatment with RAM was under serious consideration for operational B-70s. This would have reduced the nose-on signature by several orders of magnitude.

Actual signature levels are, for obvious reasons, highly classified numbers. Reference 101 gives the signature of the B-52 as 100 sqm, or 20 dBsm, and the stealth-treated B-1B as 1 sqm, or 0 dBsm. The Lockheed A-12, similar to the SR-71 and highly treated for stealth using the technology of the early 1960s, is quoted in Ref. 101 as having an RCS of 0.014 sqm or −18 dBsm. Nonstealth fighters typically have nose-on signatures on the order of 10 sqm, or 10 dBsm. The stealthy MiG 1.42 fighter technology demonstrator is quoted by MiG as having an RCS of 0.1 sqm, or −10 dBsm. Reference 100 suggests that "where stealth is a primary design objective, RCS will probably be in the region of 0.01 to 0.1 square meter" (−20 to −10 dBsm).

RCS varies depending upon the frequency and polarization of the threat radar (see Ref. 21). The following comments relate to typical threat radars seen by military aircraft.

There are many electromagnetic phenomena that contribute to the RCS of an aircraft. These require different design approaches for RCS reduction and can produce conflicting design requirements. Figure 8.16 illustrates the major RCS contributors for a typical, untreated fighter aircraft.

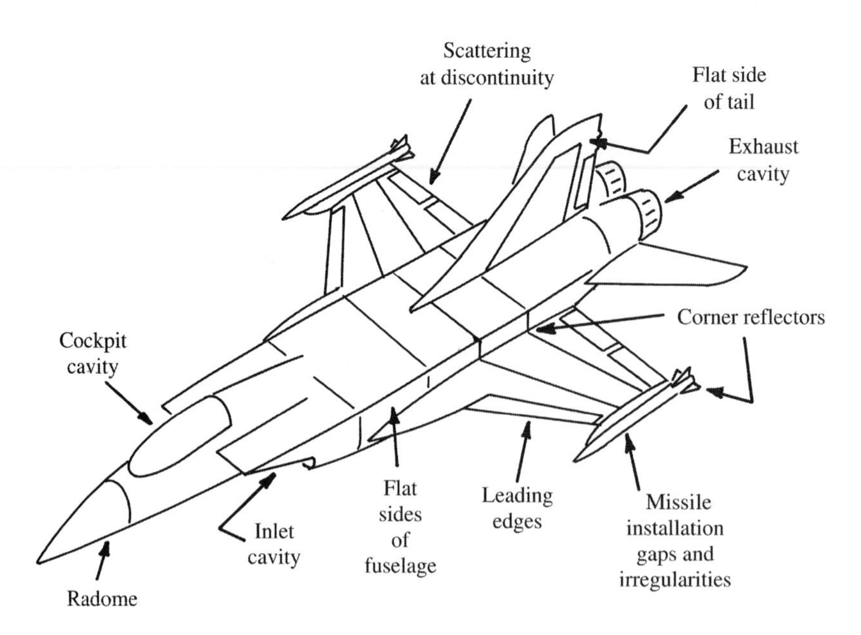

Fig. 8.16 Major RCS contributors.

One of the largest contributions to airframe RCS occurs any time a relatively flat surface of the aircraft is perpendicular to the incoming radar beam. Imagine shining a flashlight at a shiny aircraft in a dark hanger. Any spots where the beam is reflected directly back at you will have an enormous RCS contribution.

Typically this "specular return" occurs on the flat sides of the aircraft fuselage and along an upright vertical tail (when the radar is abeam the aircraft). To prevent these RCS "spikes," the designer may slope the fuselage sides, angle the vertical tails, and so on, so that there are no flat surfaces presented toward the radar (Fig. 8.17).

Note that this RCS reduction approach assumes that the designer knows where the threat radar will be located relative to the aircraft. This information is usually provided by the operations-analysis department or by the customer as a design driver. Also, this assumes a monostatic radar.

Another area of the aircraft that can present a perpendicular bounce for the radar is the round leading edge of the wing and tail surfaces. If the aircraft is primarily designed for low detectability by a nose-on threat radar, the wings and tails can be highly swept to reduce their contribution to RCS. Note that this and many other approaches to reducing the RCS will produce a penalty in aerodynamic efficiency.

Aircraft cavities such as inlet front faces and engine exhausts create a radar return perpendicular to the plane of the opening. All around the opening there will be small perpendicular bounces. When the threat radar is at a direction

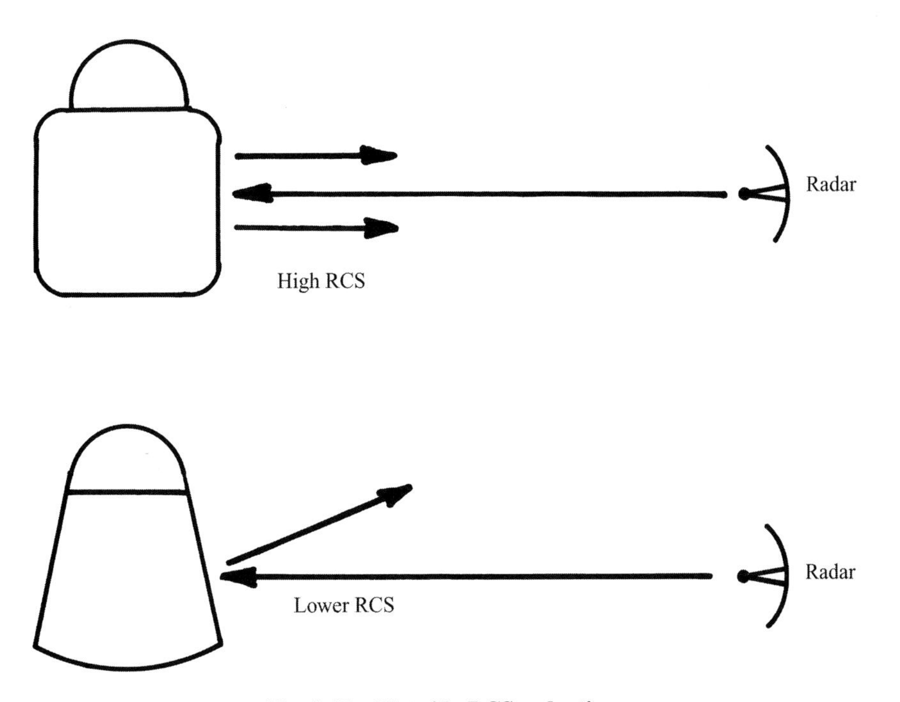

Fig. 8.17 Flat side RCS reduction.

perpendicular to the opening, those small bounces will be "in phase" and so will sum to a single large return. This is avoided by sweeping the plane of the opening well away from the expected directions of threat radars, as can be seen on the F-22, B-1B, F/A-18E, and other designs. To further reduce this RCS contribution, the inlet lips are often treated with radar absorbers.

It is also important to avoid any "corner reflectors," i.e., intersecting surfaces that form approximately a right angle, as shown in Fig. 8.16 at the wing-fuselage junction.

Another contributor to airframe RCS occurs due to the electromagnetic currents that build up on the skin when illuminated by a radar. These currents flow across the skin until they hit a discontinuity such as at a sharp trailing edge, a wing tip, a control surface, or a crack around a removable panel or door. At a discontinuity, the currents "scatter," or radiate electromagnetic energy, some of which is transmitted back to the radar (Fig. 8.18).

This effect is much lower in intensity than the specular return, but is still sufficient for detection. The effect is strongest when the discontinuity is straight and perpendicular to the radar beam. Thus, the discontinuities such as at the wing and tail trailing edges are usually swept to minimize the detectability from the front. Carried to the extreme, this leads to diamond- or sawtooth-shaped edges on every door, access plate, and other discontinuity on the aircraft, as seen on the B-2 and F-117.

This scattering of surface currents actually represents three different types of radar returns. If the surface is being directly illuminated by the radar, a surface discontinuity causes diffraction, which is the same phenomenon that causes a rainbow. Diffraction will occur not just at a physical edge, such as a wing trailing edge, but any place that the surface has a sharp corner. There is even an apparent

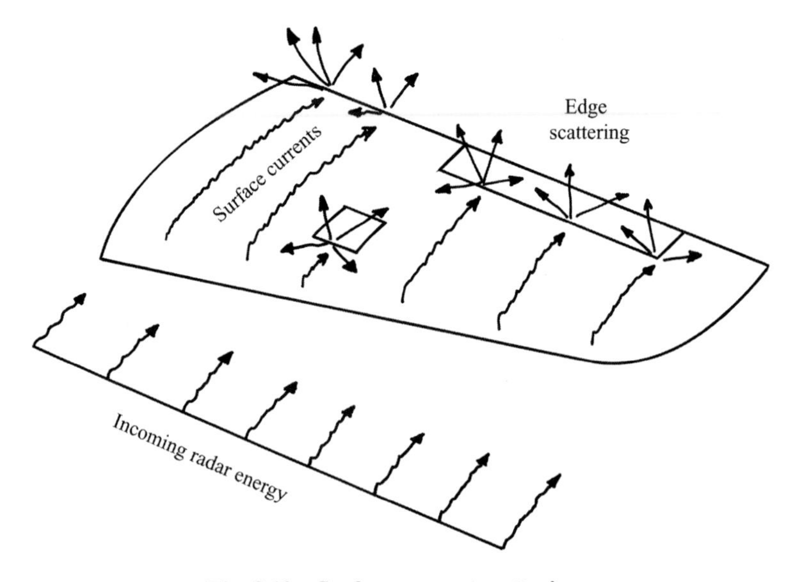

Fig. 8.18 Surface current scatterings.

shadow edge anyplace where the surface is no longer illuminated by the radar energy, such as at the transition from front to back of a wing airfoil when illuminated from the front.

Two other radar returns, from traveling waves and creeping waves, are caused by the flow of electromagnetic energy from the front to the back, nonilluminated side of the body. Traveling waves occur when a sharp discontinuity is reached and the energy (which cannot be destroyed) travels back to the front where it reradiates. Traveling waves and edge diffraction both call for the avoidance of trailing edges perpendicular to the threat radar.

Creeping waves occur when the backside of the body is smoothly curved, and so the energy "creeps" all the way around the body, slightly radiating as it goes around. Radar absorbers, as discussed next, are useful for suppressing creeping waves. References 98 and 99 are suggested for discussions of RCS theory.

First-generation stealth designs such as the Lockheed F-117 and the never-constructed North American Rockwell "Surprise Fighter" relied upon faceted shaping in which the aircraft shape is constructed of interlocking flat triangles and trapezoids. This has advantages in ease of construction and signature analysis, but offers a large number of sharp edges to create diffraction returns, and so is no longer in favor (Ref. 92).

Current stealth design begins with the aiming of the RCS spikes. For good stealth design, we "aim them where the bad guys ain't," based on requirements as to what signatures are required from what directions. This starts with a decision as to what directions pose a severe threat, and what directions pose a lesser threat. Typically, we assume that the forward direction poses a severe threat, because we are flying toward someone that we plan to attack. Toward the rear we have a severe threat because we have just attacked them and are running away, and they are really angry! Around the sides of the aircraft at similar altitudes we assume a lesser, but still significant threat, based on the supposition that opponents at our altitude can most readily attack us. Directly above and below us are not as likely to pose a threat. The design requirements of a stealth aircraft will include desired levels of RCS from different directions (azimuth and elevation), such as, "no greater than (classified) decibel square meters within (classified) degrees off the nose at plus-or-minus (classified) degrees elevation."

Given the definition of the directions of likely threat, we can select where to aim the spikes by appropriately aligning the spike-producing features such as wing leading and trailing edges. With current technology, all aircraft will have at least four spikes, namely the perpendicular bounces and edge diffractions from the leading and trailing edges of the wing. From the directions of those spikes, a threat radar will be able to see the aircraft anyway, so we align any additional spikes in the same direction rather than allowing a spike at another direction. To paraphrase an old song, "one big spike is better than two little spikes," when it comes to stealth design.

For example, the edge diffractions off the wing trailing edge when seen from the front will create a spike perpendicular to that trailing edge. By setting the wing trailing edge to the same angle as a wing leading edge, the spikes align. If we align the left trailing edge with the right leading edge, we get a diamond planform as seen to the left of Fig. 8.19. Aligning the left trailing edge with

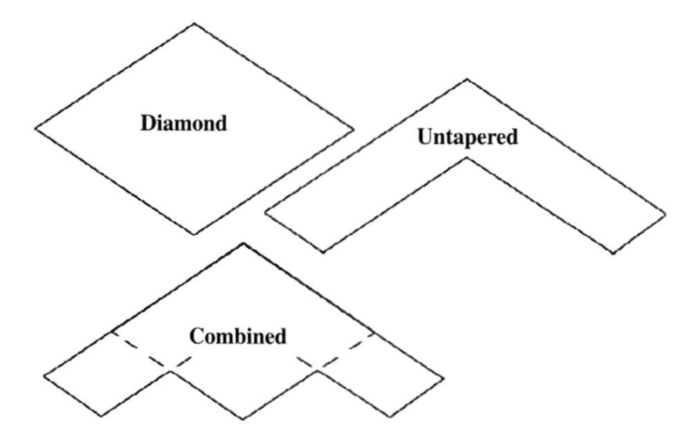

Fig. 8.19 "Aiming the spikes" for RCS reduction.

the left leading edge generates an untapered planform (highly swept to avoid spikes to the front and rear).

As was discussed in Chapter 4, an untapered wing (taper ratio of one) or a wing with a taper ratio of zero as in the diamond wing are the worst possible wings from an aerodynamic standpoint. The untapered wing has excessive lift outboard, especially if swept, while the diamond wing has insufficient lift outboard to form an elliptical lift distribution as desired for minimum drag due to lift. However, if we combine these aerodynamically bad planforms as shown at the bottom of Fig. 8.19 and carefully twist and camber the resulting planform we can obtain a fairly good aerodynamic efficiency. This illustration is quite similar to the original B-2 configuration, but it was later revised to the current configuration to better balance the design.

Smaller, but nontrivial spikes also arise from the edges of an access door, landing gear door, or weapons bay door. Where possible, we design such doors rotated roughly 45 deg so that the edges align with the existing spikes from the wing leading and trailing edges, creating the characteristic diamond shape. If this is not feasible, we put sawtooth edges on the doors to avoid strong spikes forward and to the rear.

This design approach leads to an aircraft planform composed entirely of straight, highly swept lines, much like the first-generation stealth designs. However, the desire to eliminate the edge diffractions caused by the facets of first-generation stealth now produces designs in which cross-sectional shapes are smooth, not sharp-edged. The steep angles on the fuselage sides as shown in Fig. 8.17 are employed to prevent broadside perpendicular bounce returns, but these angled sides flow smoothly over the top and bottom of the fuselage. Such shaping can be seen on the B-2, F-22, F-23, and F-35 Joint Strike Fighter (JSF), and is apparent in this notional fighter design developed for pre-JSF requirements trade studies at RAND Corporation (Ref. 29, see Fig. 8.20).

RCS can also be reduced simply by eliminating parts of the aircraft. A horizontal tail that does not exist cannot contribute to the radar return! Modern computerized flight controls combined with the use of vectored-thrust engines

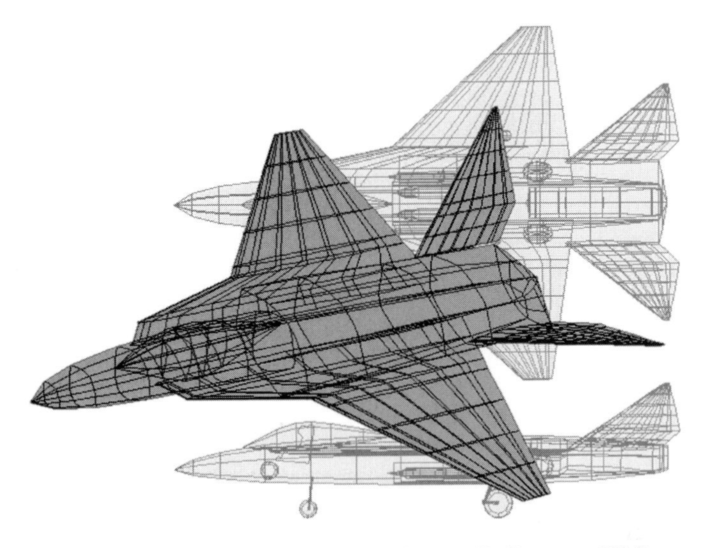

Fig. 8.20 Next-generation attack fighter (D. Raymer 1994).

can solve many of the difficulties of the tailless configuration. This author expects that eventually, fighters will be designed with neither vertical nor horizontal tails (no canards, either) to minimize signature, with vectored nozzles and forebody vortex control used to control the aircraft.

Similarly, RCS can be reduced if the nacelles can be eliminated through the use of buried engines, or better yet, by eliminating the entire fuselage through the use of the flying-wing concept. This approach is used in the Northrop B-2.

In addition to reshaping the aircraft, detectability can be reduced through the use of skin materials that absorb radar energy. Such materials, called radar absorbing materials (RAM), are typically composites such as fiberglass embedded with carbon or ferrite particles.

These particles are heated by the radar electromagnetic waves, thus absorbing some of the energy. This will reduce (not eliminate!) the radar return due to perpendicular bounce; it can also reduce the surface currents and thus reduce the RCS due to scattering at sharp edges. The thickness of the radar absorbing material should be about one-fourth of the wavelength of the threat radar. RAM can be applied parasitically, to the outside of the structure as attached non-load-bearing panels or even as a paint. RAM can also be built into the aircraft's structural material, which is then called radar absorbing structure, or RAS. A typical RAS is a honeycomb panel with Kevlar–epoxy skins that are transparent to radar, an inner skin of graphite–epoxy that is reflective of radar, and a Nomex honeycomb core in between that includes radar absorbers in increasing density from the outside to the inside so as to gradually trap the radar energy.

Each time the radar energy bounces off RAM, it loses more energy (attenuates), so the geometry should be designed to force multiple bounces. This is especially suitable in an inlet duct, where a long and curved duct will cause the radar energy to bounce of the sidewalls many times. Eventually, most of the radar energy will be absorbed.

As there are many types of RAM and similar treatments, no quick estimate for the weight impact of their use can be provided here. However, one can probably assume that such use will reduce or eliminate any weight savings otherwise assumed for the use of composite materials.

For most existing aircraft, the airframe is not the largest contributor to RCS, especially nose-on. A conventional radome, covering the aircraft's own radar, is transparent to radar for obvious reasons. Therefore, it is also transparent to the threat radar, allowing the threat radar's beam to bounce off the forward bulkhead and electronic equipment within the radome.

Even worse, the aircraft's own radar antenna, when illuminated by a threat radar, can produce a radar magnification effect much like a cat's eye. These effects can be reduced with a "bandpass" radome, which is transparent to only one radar frequency (that of the aircraft's radar).

Other huge contributors to the RCS for a conventional aircraft are the inlet and exhaust cavities. Radar energy gets into these cavities, bounces off the engine parts, and sprays back out the cavity towards the threat radar. Also, these cavities represent additional surface discontinuities.

The best solution for reducing these RCS contributions is to hide them from the expected threat locations. For example, inlets can be hidden from ground-based radars by locating them on top of the aircraft (Fig. 8.21). Exhausts can be hidden through the use of two-dimensional nozzles.

The F-117 used a mesh screen at the front of the inlet duct to keep out the radar energy. For this to work, the mesh must be smaller than the radar's wavelength, leading to a loss of inlet pressure recovery that in turn reduces thrust and

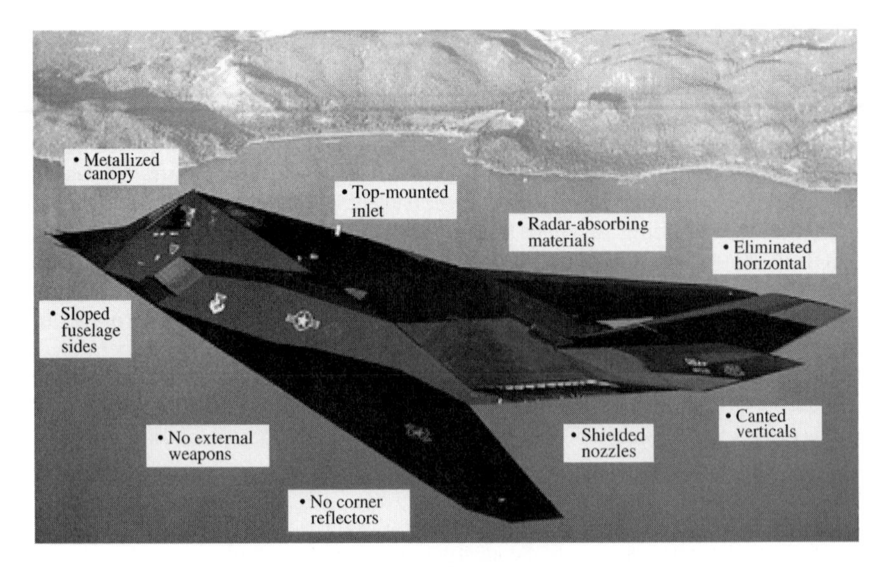

Fig. 8.21 Detectability reduction approaches.

increases fuel consumption. Also, icing becomes a concern. More recent stealth designs allow the radar energy into the inlet duct, but use RAM inside the duct to absorb it as described above. Also, if the radar energy is allowed inside, some provision must be made for hiding any direct view of the engine front face from the outside. This can be done by extreme snaking of the duct, or by putting curved vanes or an onion-shaped bulb in front of the engine. If such devices are put into the duct, care must be taken that the mean flowpath does not decrease in cross-sectional area, and provisions for anti-icing may be required.

Cockpits provide a radar return for a similar reason. The radar energy enters the cockpit, bounces around off the equipment inside, and then reradiates back outside. One solution for this is to thinly coat the canopy with some conductive metal such as gold, causing the canopy to reflect the radar energy away.

Finally, the aircraft's weapons can have a major impact on RCS. Missiles and bombs have fins that form natural corner reflectors. The carriage and release mechanisms have numerous corner reflectors, cavities, and surface discontinuities. Gun ports present yet another kind of cavity. The only real solution for these problems is to put all the weapons inside, behind closed doors. However, the weight, volume, and complexity penalties of this approach must be carefully considered.

Electronic countermeasures (ECM)—devices to trick the threat radar—usually consist of some sort of radar receiver that picks up the threat radar emissions, and some sort of transmit antenna to send a deceiving signal back to the threat radar. The many techniques for tricking radar (and ECM) go beyond the scope of this book. However, designers should be aware that there is a tradeoff between the aircraft's RCS level and the required amount of ECM.

8.5 Infrared Detectability

Infrared (IR) detectability also concerns the aircraft designer. Many short-range air-to-air and ground-to-air missiles rely upon IR seekers. Modern IR sensors are sensitive enough to detect not only the radiation emitted by the engine exhaust and hot parts, but also that emitted by the whole aircraft skin due to aerodynamic heating at transonic and supersonic speeds. Also, sensors can detect the solar IR radiation that reflects off the skin and cockpit transparencies (windows).

Of several approaches for reduction of IR detectability, one of the most potent reduces engine exhaust temperatures through the use of a high-bypass-ratio engine. This reduces both exhaust and hot-part temperatures. However, depending upon such an engine for IR reduction may result in selecting one that is less than optimal for aircraft sizing, which increases aircraft weight and cost.

Emissions from the exposed engine hot-parts (primarily the inside of the nozzle) can be reduced by cooling them with air bled off the engine compressor. This will also increase fuel consumption slightly. Another approach hides the nozzles from the expected location of the threat IR sensor. For example, the H-tails of the A-10 hide the nozzles from some angles. Unfortunately, the worst-case threat location is from the rear, and it is difficult to shield the nozzles from that direction!

Plume emissions are reduced by quickly mixing the exhaust with the outside air. As mentioned, a high-bypass engine is the best way of accomplishing this. Mixing can also be enhanced by the use of a wide, thin nozzle rather than a circular one. Another technique is to angle the exhaust upward or downward relative to the freestream. This will have an obvious thrust penalty, however.

Sun glint in the IR frequencies can be somewhat reduced by the use of special paints that have low IR reflectivity. Cockpit transparencies (which can't be painted!) can be shaped with all flat sides to prevent continuous tracking by an IR sensor.

Emissions due to aerodynamic heat are best controlled by slowing the aircraft down.

IR missiles can sometimes be tricked by throwing out a flare that burns to produce approximately the same IR frequencies as the aircraft. However, modern IR seekers are getting better at identifying which hot source is the actual aircraft.

IR fundamentals are more thoroughly discussed in Ref. 18.

8.6 Visual Detectability

The human eyeball is still a potent aircraft-detection sensor. On a clear day, an aircraft or its contrail may be spotted visually before detection by the onboard radar of a typical fighter. Also, fighter aircraft usually have radar only in front, which leaves the eyeball as the primary detector for spotting threat aircraft that are abeam or above.

Visual detection depends upon the size of aircraft and its color and intensity contrast with the background. In simulated combat, pilots of the small F-5 can frequently spot the much-larger F-15s well before the F-5s are seen. However, aircraft size is determined by the mission requirements and cannot be arbitrarily reduced.

Background contrast is reduced primarily with camouflage paints, using colors and surface textures that cause the aircraft to reflect light at an intensity and color equal to that of the background. This requires assumptions as to the appropriate background as well as the lighting conditions.

Frequently aircraft will have a lighter paint on the bottom, because the background for look-up angles is the sky. Current camouflage paint schemes are dirty blue-grey for sky backgrounds and dull, mottled grey-greens and grey-browns for ground backgrounds.

Different parts of the aircraft can contrast against each other, which increases detectability. To counter this, paint colors can be varied to lighten the dark areas, such as where one part of the aircraft casts a shadow on another. Also, small lights can sometimes be used to fill in a shadow spot.

Canopy glint is also a problem for visual detection. The use of flat transparencies can be applied as previously discussed, but will tend to detract from the pilot's outside viewing.

At night, aircraft are visually detected mostly by engine and exhaust glow and by glint off the transparencies. These can be reduced by techniques previously discussed for IR and glint suppression.

There are also psychological aspects to visual detection. If the aircraft does not look like an aircraft, the human mind may ignore it. The irregular mottled patterns used for camouflage paints exploit this tendency.

In air-to-air combat, seconds are precious. If a pilot is confused as to the opponent's orientation, the opponent may obtain favorable positioning. To this end, some aircraft have even had fake canopies painted on the underside. Forward-swept and oblique wings may also provide momentary disorientation.

8.7 Aural Signature

Aural signature (noise) is important for civilian as well as military aircraft. Commercial airports frequently have antinoise ordinances that restrict some aircraft. Aircraft noise is largely caused by airflow shear layers, primarily due to the engine exhaust.

A small-diameter, high-velocity jet exhaust produces the greatest noise, while a large-diameter propeller with a low tip-speed produces the least noise. A turbofan falls somewhere in between. Blade shaping and internal duct shaping can somewhat reduce noise.

Piston exhaust stacks are also a source of noise. This noise can be controlled with mufflers, and by aiming the exhaust stacks away from the ground and possibly over the wings.

Within the aircraft, noise is primarily caused by the engines. Well-designed engine mounts, mufflers, and insulation materials can be used to reduce the noise. Internal noise will be created if the exhaust from a piston engine impinges upon any part of the aircraft, especially the cabin. A new technology, "active sound suppression," uses a microphone to detect noise in the cabin then employs a speaker to send a noise signal 180 deg out of phase, cancelling the cabin noise. While not perfect, this system works well on aircraft such as the SAAB 2000.

Wing-mounted propellers can have a tremendous effect on internal noise. All propellers should have a minimum clearance to the fuselage of about 1 ft {30 cm}, and should perferably have a minimum clearance of about one-half of the propeller radius.

However, the greater the propeller clearance, the larger the vertical tail must be to counter the engine-out yaw.

Jet engines mounted on the aft fuselage (DC-9, B727, etc.) should be located as far away from the fuselage as structurally permitted to reduce cabin noise.

8.8 Vulnerability Considerations

Vulnerability concerns the ability of the aircraft to sustain battle damage, continue flying, and return to base. An aircraft can be "killed" in many ways. A single bullet through a nonredundant elevator actuator is as bad as a big missile up the tailpipe!

"Vulnerable area" is a key concept. This refers to the product of the projected area (square feet or meters) of the aircraft components, times the probability that each component will, if struck, cause the aircraft to be lost. Vulnerable area is different for each threat direction. Typical components with a high aircraft kill probability (near 1.0) are the crew compartment, engine (if single-engined), fuel tanks (unless self-sealing), and weapons. Figure 8.22 shows a typical vulnerable area calculation.

When assessing the vulnerability of an aircraft, the first step is to determine the ways in which it can be "killed." Referred to as a "failure modes and effects analysis" (FMEA), or "damage modes and effects analysis" (DMEA), this step will typically be performed during the later stages of conceptual design. The FMEA considers both the ways in which battle damage can affect individual aircraft components, and the ways in which damage to each component will affect the other components.

During initial configuration layout, the designer should strive to avoid certain features known to cause vulnerability problems. Fire is the greatest danger to a battle-damaged aircraft. Not only is the fuel highly flammable, but so is the hydraulic fluid. The second Have Blue stealth demonstrator crashed due to a crack in a weld in a hydraulic line, which sprayed fuel on the engine.

Also, combat aircraft carry gun ammo, bombs, and missiles. An aircraft may survive a burst of cannon shells only to explode from a fire in the ammo box.

If at all possible, fuel should not be located over or around the engines and inlet ducts. While tanks can be made self-sealing to a small puncture, a large hole will allow fuel to ignite on the hot engine. The pylon-mounted engines on

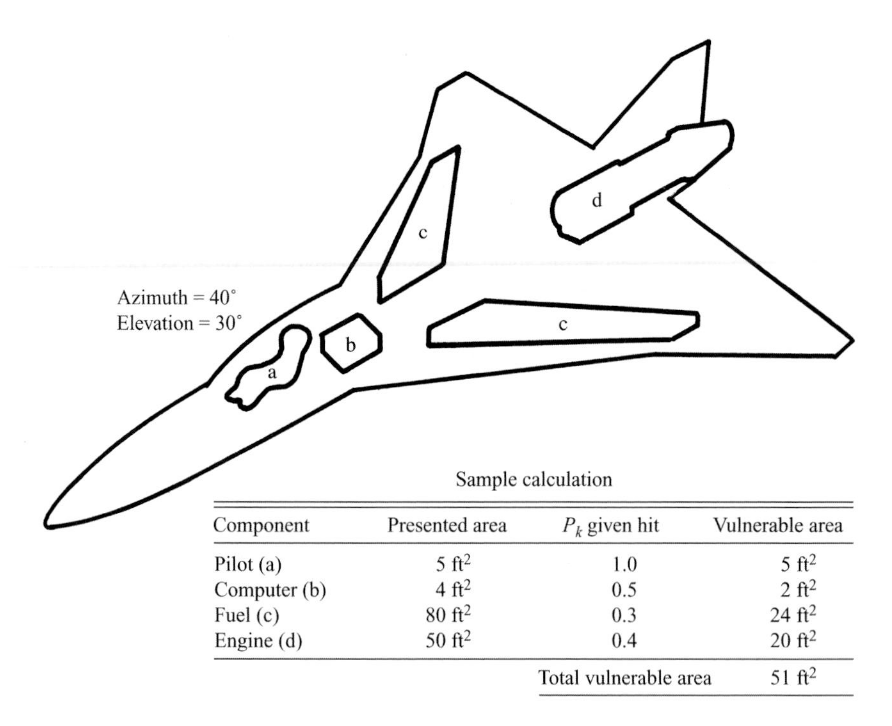

Sample calculation

Component	Presented area	P_k given hit	Vulnerable area
Pilot (a)	5 ft^2	1.0	5 ft^2
Computer (b)	4 ft^2	0.5	2 ft^2
Fuel (c)	80 ft^2	0.3	24 ft^2
Engine (d)	50 ft^2	0.4	20 ft^2
		Total vulnerable area	51 ft^2

Fig. 8.22 Vulnerable area calculation.

the A-10 insure that leaking fuel cannot ignite on the engines. Similarily, hydraulic lines and reservoirs should be located away from the engines.

Firewalls should be used to prevent the spread of flames beyond a burning engine bay. Engine bays, fuel bays, and weapon bays should have a fire-suppression system.

When an engine is struck, turbine and compressor blades can fly off at high speeds. Avoid placing critical components such as hydraulic lines or weapons anyplace where they could be damaged by an exploding engine. Also, a twin-engine aircraft should have enough separation between engines to prevent damage to the good engine. If twin engines are together in the fuselage, a combined firewall and containment shield should separate them. This requires at least 1 ft {30 cm} of clearance between engines.

Propeller blades can fly off either from battle damage or during a wheels-up landing. Critical components, especially the crew and passenger compartments, shouldn't be placed within a 5-deg arc of the propeller disk.

Avoid placing guns, bombs, or fuel near the crew compartment. Fuel should not be placed in the fuselage of a passenger plane.

Redundancy of critical components can be used to allow the survival of the aircraft when a critical component is hit or fails for any other reason. Typical components that could be redundant include the hydraulic system, electrical system, flight control system, and fuel system. Note that while redundancy improves the survivability and reliability, it worsens the maintenance requirements because there are more components to fail.

While normally considered a topic for military aircraft, the concepts for reducing vulnerability also apply to civil aircraft. FMEA should be conducted to minimize the possibility that a failure or damage in one system can cause the aircraft to crash. Also, cargo and equipment bays could have fire detection and possibly suppression equipment.

For more information on vulnerability, Ref. 18 is again suggested.

8.9 Crashworthiness Considerations

Airplanes crash. Careful design can reduce the probability of injury in a moderate crash. Several suggestions have already been mentioned, including positioning the propellers so that the blades will not strike anyone if they fly off during a crash. Also mentioned was the desire to avoid placing fuel tanks in the fuselage of a passenger airplane (although fuel in the wing box carrythrough structure is usually acceptable).

To protect the crew and passengers in the event of a crash, the aircraft should be designed to act like a shock absorber. A shock absorber works by deflecting in a controlled fashion, spreading the load from a sudden impact over a specified distance (the "stroke") and over time (see Chapter 11). The aircraft's structure can be designed to work the same way, crushing in a controlled fashion over distance and time. Helicopters are routinely designed in this way, with extensive analysis and test of the deflections of the structure during a crash.

For aircraft, one can see the benefits of collapsing structure very starkly when studying accidents of low-wing general-aviation aircraft. It is tragically common that the back-seat passengers will survive a crash, while the pilot and front seat passenger, who are sitting on the hard, noncollapsing wing box, will not survive. There is some concern that composite structures, which tend to be very stiff and do not deflect so readily during a crash, may be less survivable in accidents.

Figure 8.23 shows several other design suggestions that were learned the hard way. A normal, vertical firewall in a propeller aircraft has a sharp lower corner which tends to dig into the ground, stopping the aircraft dangerously fast. Sloping the lower part of the firewall back as shown will prevent digging in, therefore reducing the deceleration.

For a large passenger aircraft, the floor should not be supported by braces from the lower part of the fuselage. As shown, these braces may push upward through the floor in the even of a crash, unless special collapsing braces are used.

Common sense will avoid many crashworthiness problems. For example, things will break loose and fly forward during a crash. Therefore, do not put heavy items behind and/or above people. This sounds obvious, but there are some aircraft with the engine in a pod above and behind the cockpit.

There are also some military jets with large fuel tanks directly behind the cockpit, offering the opportunity to be bathed in jet fuel during a crash. However, the pilot would probably try to eject rather than ride out a crash bad enough to rupture the fuel tanks.

One should also consider secondary damage. For example, landing gear and engine nacelles will frequently be ripped away during a crash. If possible, they should be located so that they do not rip open fuel tanks in the process.

Some form of protection should be provided in the not-unlikely event that the aircraft flips over during a crash. This is lacking in several small homebuilt designs.

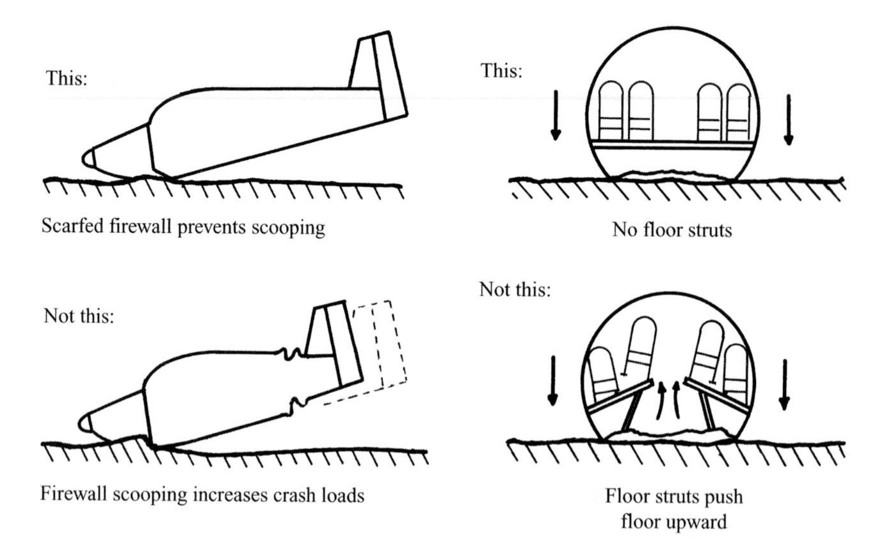

Fig. 8.23 Crashworthiness design.

8.10 Producibility Considerations

It is often said that aircraft are bought "by the pound." While it is true that aircraft cost is most directly related to weight, there is also a strong cost impact due to the materials selected, the fabrication processes and tooling required (forging, stamping, molding, etc.), and the assembly manhours.

The configuration designer does not usually determine the materials used or exactly how the aircraft will be fabricated. However, the ease of producing the aircraft can be greatly facilitated by the overall design layout.

One impact the configuration designer has upon producibility is the extent to which flat-wrap structure is incorporated. This has a major impact upon the tooling costs and fabrication manhours, as discussed in the last chapter.

Part commonality can also reduce production costs. If possible, the left and right main landing gear should be identical (left–right common). It may be desirable to use uncambered horizontal tails to allow left–right commonality even if a slight aerodynamic penalty results. In some cases the wing airfoil can be slightly reshaped to allow left-right common ailerons.

Forgings are the most expensive type of structure in common usage, and are also usually the longest-lead-time items for production tooling. Forgings may be required whenever a high load passes through a small area. Forgings are used for landing-gear struts, wing-sweep pivots, and all-moving tail pivots (trunnions). The designer should avoid, if possible, such highly loaded structure.

Installation of internal components and routing of hydraulic lines, electrical wiring, and cooling ducts comprise another major production cost due to the large amount of manual labor required. To ease installation of components and routing, avoid the tight internal packaging so desirable for reduced wetted area and wave drag. When evaluating proposed designs, government design boards will compare the overall aircraft density (weight divided by volume) with historical data for similar aircraft to ensure packaging realism.

Routing can be simplified through provision of a clearly defined "routing tunnel." This can be internal or, as shown in Fig. 8.24, an external and nonstructural fairing that typically runs along the spine or belly of the aircraft. However, if all routing is concentrated in one area the aircraft vulnerability will be drastically worsened.

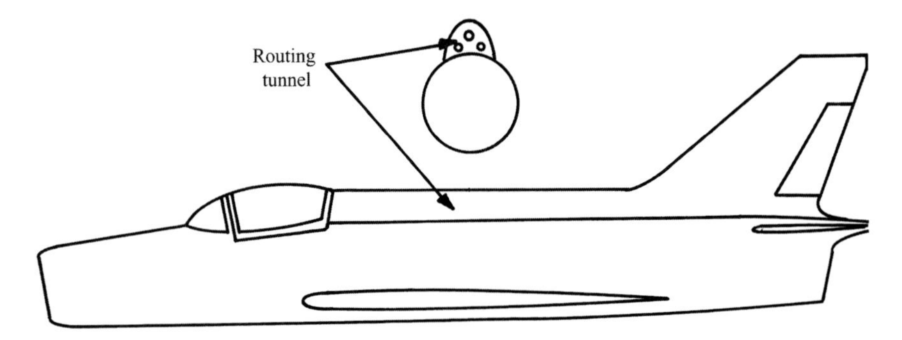

Fig. 8.24 External routing tunnel.

Routing can be reduced by careful placement of the internal components. For example, the avionics and the crew station will both require cooling air ("environmental control"). If the avionics, crew station, and environmental control system (ECS) can be located near to each other, the routing distances will be minimized.

Sometimes clever design can reduce routing. The Rutan Defiant, a "push-pull" twin-engined design, uses completely separate electrical systems for the front and rear engines, including separate batteries. This requires an extra battery, but a trade study determined that the extra battery weighs less than the otherwise-required electrical cable, and eliminates the front-to-rear routing requirement.

Another factor for producibility concerns manufacturing breaks. Aircraft are built in subassemblies as shown in Fig. 8.25. Typically, a large aircraft will be built up from a cockpit, an aft-fuselage, and a number of midfuselage subassemblies. A small aircraft may be built from only two or three subassemblies.

It is important that the designer consider where the subassembly breaks will occur, and avoid placing components across the convenient break locations. Figure 8.26 shows a typical fighter with a fuselage production break located just aft of the cockpit. This is very common because the cockpit pressure vessel should not be broken for fabrication.

In the upper design, the nose wheel well is divided by the production break, which prevents fully assembling the nose-wheel linkages before the two subassemblies are connected. The lower illustration shows a better arrangement.

Design for producibility requires experience that no book can provide. A good understanding of structural design and fabrication and the basic principles of

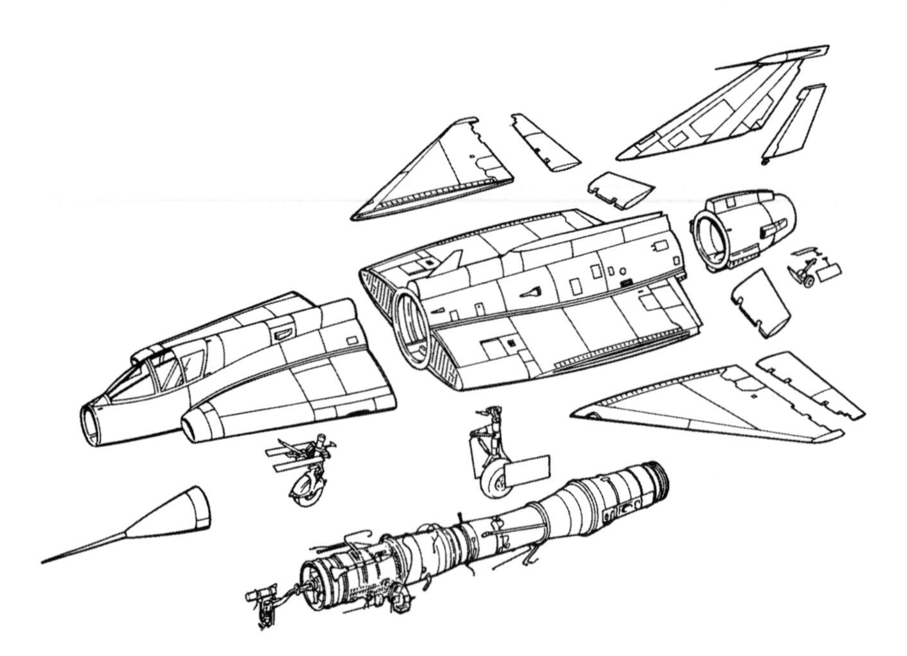

Fig. 8.25 Production subassemblies of SAAB Draken (courtesy SAAB Aircraft).

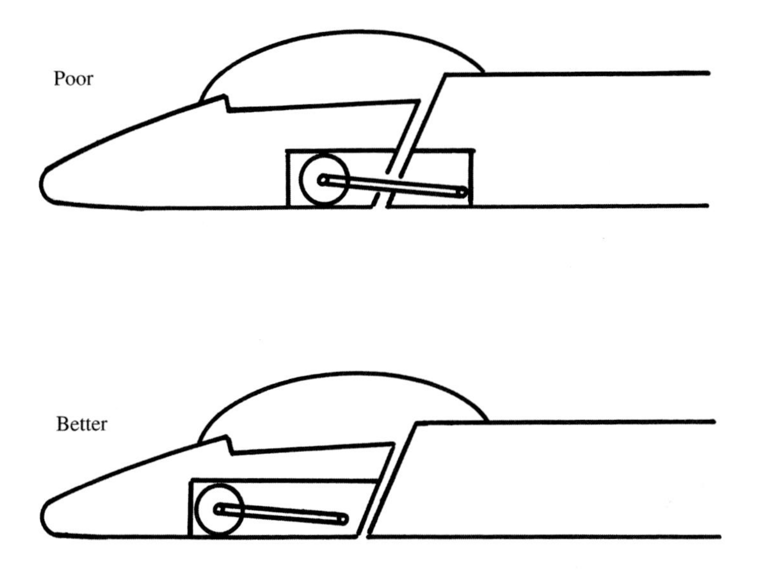

Fig. 8.26 Production breaks.

operation for the major subsystems provides the background for developing producible designs. The following material provides a brief introduction to aircraft fabrication.

While there have been tremendous advances in aircraft production in recent years, much of the modern factory would be recognizable to a manufacturing engineer from the Wright Brothers' days. Aircraft production, then and now, involves the application of the mechanical arts of machining, forming, finishing, joining, assembly, and testing.

Machining involves the removal of a carefully controlled amount of material from a part, typically by the application of a cutting tool via relative motion between the part and the tool. The cutting tool is generally based upon the inclined wedge, and acts to peel away a thin shaving of the part (a drill bit can be seen as a set of inclined wedges positioned radially around an axis). The relative motion between tool and part can be rotational, as with the drill, lathe, and mill, or it can be translational, as with the broach and planer.

Forming refers to the numerous ways in which materials, especially metals, are changed in shape other than by machining. Forming includes casting, forging, extruding, stamping, punching, bending, and drawing. In casting, the metal is brought up to its melting temperature then poured into a mold. Forging involves forcing nonmolten metal into a mold through pressure or impact. Extrusion is the process of forcing metal to flow out a hole with the desired cross-sectional area, creating shaped bar stock. Stamping and punching are used to cut out shapes and holes in sheet metal. Bending is self-explanatory, and drawing is the process of forcing sheet metal into a form creating cup-like geometries.

Finishing encompasses a number of processes applied to formed and/or machined parts. Some finish processes include further material removal to create a smoother surface, such as deburring, lapping, and finish grinding. Other finish processes, such as painting, anodization, and plating, involve application of a surface coating.

Composite fabrication is sufficiently unlike metal fabrication that it deserves special mention. In thermoset composite production, a liquid or pliable semisolid plastic material undergoes a chemical change into a new, solid material, usually accompanied by the application of heat and/or pressure. For aircraft applications the plastic "matrix" material is reinforced by a fiber, typically of graphite material. Thermoset composite manufacture is unique in that the material itself is produced at the same time and place as the part. A second class of composites, the thermoplastics, involves a plastic matrix that is heated in a mold until it deforms readily, assuming the shape of the mold. Composite fabrication is further described in Chapter 14.

Joining is simply the attachment of parts together, by processes including brazing, soldering, welding, bonding, riveting, and bolting. All these processes historically have a high manual-labor content, and all are being automated to various extents in modern factories. For example, modern car factories have long lines of robotic spot-welders attaching body panels. Automatic riveting machines, applicable for simple geometries such as rivets in a row down a wing spar, can be found in the modern aircraft factory.

Assembly is the process of combining parts and subassemblies into the final product. Assembly usually involves joining operations such as riveting or bolting, but is distinguished from joining by the greater level of completeness of the subassemblies. For example, when you attach a wing skin to the wing ribs it is "joining," but when you attach the wing to the fuselage, it is "assembly."

Testing is a key part of the manufacturing process. In traditional factories, testing was generally done by random selection of finished product and was frequently of a destructive nature. While helping to keep average quality up, such random destructive testing did not insure that any given part was acceptable because the only parts known by testing to be acceptable were destroyed in the process!

Today's factories are tending toward nondestructive testing techniques such as magnaflux, ultrasonic, and nuclear magnetic resonance, and are also applying advanced statistical techniques to better select samples to test and to determine the corrective action required.

CAD/CAM, or computer-aided design/computer-aided manufacture, is a generic term for the many different ways in which computers are being used in design and manufacture. Typically CAD/CAM refers specifically to the use of computers for component design, and the use of the resulting CAD data base as the input for the programming of numerically-controlled machinery and robots (as described below). The benefits of CAD/CAM are well-established and include improved design quality, reduction in design time and/or increase in the number of design iterations possible, earlier discovery of errors, integration of design, analysis, and manufacturing engineering, and facilitation of training.

Automation refers to almost any use of computerized equipment during manufacture. However, the generic term "automation" is most frequently applied to tasks such as riveting, parts retrieval, and process control (such as autoclave cycling), whereas the more specific terms "numerical control" and "robotics" are used as described next.

Numerical control (NC) programming refers to the creation of digital instructions that command a computer-controlled machine tool such as a mill or lathe. This area is probably one of the highest leverage in terms of reducing cost and improving quality. While machine tools themselves have experienced little fundamental change in this century (this author knows of a company making high-tech wind turbines on a 100-year-old lathe!), the application of numerical control replacing the skilled but bored machinist has had a tremendous effect on productivity and quality.

The most sophisticated subset of automation is robotics, in which a computer-controlled machine performs tasks involving highly complex motions which previously might have been performed by a human. Note that it is the ability to physically manipulate objects in response to programming which distinguishes the robot from other forms of automation or mechanism. Robotics examples include part pickup and positioning, painting, composite-ply laydown, material handling, simple assembly, and welding, and are usually limited to "semi-skilled" jobs, at least to date.

A key robotics technology for composites is in the labor-intensive tape lay-up process. Programmable robot arms with tape dispenser end effectors are widely used to place the prepreg. Computer-controlled filament dispensers are being used to wind approximately round bodies such as tanks, and even entire fuselages. Also, autoclave cycle control is widely automated.

Rapid prototyping of parts without tools is being performed using a new technique known as stereolithography (SLA), which can produce plastic prototype parts in a day or less. SLA works by mathematically slicing CAD designs into thin cross sections, which are traced one at a time by an ultraviolet laser beam on a vat of photosensitive chemicals that solidify as they are irradiated. After each layer is completed, an "elevator" holding the part moves down slightly and the next layer is solidified on top of it. While to date only certain types of relatively fragile plastics may be used by SLA devices, the plastic prototypes can then be used to create molds for strong epoxy or aluminum parts. Other technologies are being developed to create 3-D parts directly from a CAD data base, but so far none can create a strong piece other than by machining or by casting using SLA to make a mold.

8.11 Maintainability Considerations

Maintainability means simply the ease with which the aircraft can be fixed. "Reliability and Maintainability" (R&M) are frequently bundled together and measured in "Maintenance Manhours Per Flighthour" (MMH/FH). MMH/FHs range from less than one for a small private aircraft to well over a hundred for a sophisticated supersonic bomber or interceptor.

Reliability is usually out of the hands of the conceptual designer. Reliability depends largely upon the detail design of the avionics, engines, and other subsystems. The configuration designer can only negatively impact reliability by placing delicate components, such as avionics, too near to vibration and heat sources such as the engines.

Anybody who has attempted to repair a car will already know what the major driver is for maintainability. Getting at the internal components frequently takes longer than fixing them! Accessibility depends upon the packaging density, number and location of doors, and number of components that must be removed to get at the broken component.

For large aircraft, just getting to the access doors can be a major undertaking. Many airliners have the APU (auxiliary power unit; see Chapter 11) installed in the tail, 20 ft {6 m} off the ground! This is acceptable for airliners because they are serviced at major airports where work platforms can be rolled into position. This may pose a problem for military aircraft that are expected to operate away from main bases.

Figure 8.27 shows the actual servicing diagram for the B-70 supersonic bomber, which is so large that a tall man can barely touch its bottom. Notice the extra access panels near the engines and near the cockpit (for avionics servicing). For all its size, though, an engine on the B-70 could be changed in 25 min—still a good time today!

Packaging density has already been discussed. The number and location of doors on modern fighters have greatly improved over prior-generation designs. Frequently the ratio between the total area of the access doors and the total

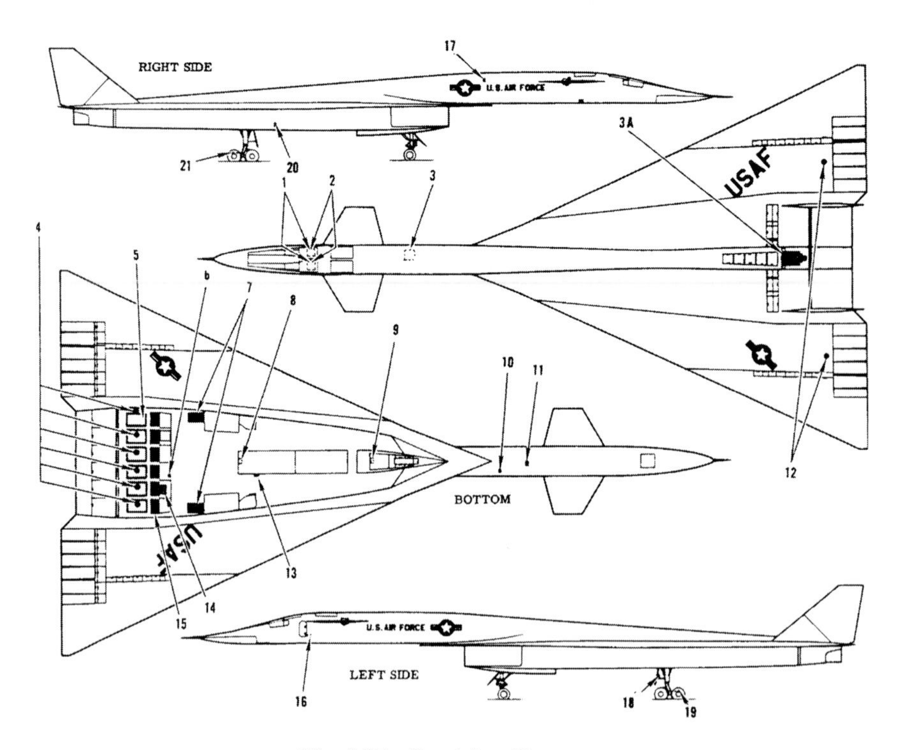

Fig. 8.27 Servicing diagram.

wetted area of the aircraft's fuselage is used as a measure of merit, with modern fighters approaching a value of one-half.

A structural weight penalty must be paid for such access. This leads to the temptation to use "structural doors" that carry skin loads via heavy hinges and latches. These are always more difficult to open than non-loadbearing doors because the airframe's deflection from its own weight will bind the latches and hinges. In extreme cases, the aircraft must be supported on jacks or a cradle to open these structural doors.

As a general rule, the best access should be provided to the components that break the most often or require the most routine maintenance. Engine access doors should definitely be provided that allow most of the engine to be exposed. Also, large doors should be provided for the avionics compartment, hydraulic pumps, actuators, electrical generators, environmental control system, auxiliary power unit, and gun bay.

The worst feature an aircraft can have for maintainability is a requirement for major structural disassembly to access or remove a component. For example, the V/STOL AV-8B Harrier requires that the entire wing be removed before removing the engine. Several aircraft require removal of a part of the longeron to remove the wing.

Similarly, the designer should avoid placing internal components such that one must be removed to get to another. In the F-4 Phantom, an ejection seat must be removed to get to the radio (a high-break-rate item). It is not uncommon for the ejection seat to be damaged during this process. "One-deep" design will avoid such problems.

B-70 with wingtips drooped for supersonic flight (U.S. Air Force photo).

9
Crew Station, Passengers, and Payload

9.1 Introduction

At the conceptual design level it is not necessary to go into the details of crew-station design, such as the actual design and location of controls and instruments, or the details of passenger and payload provisions. However, the basic geometry of the crew station and payload/passenger compartment must be considered so that the subsequent detailed cockpit design and payload integration efforts will not require revision of the overall aircraft.

This chapter presents dimensions and "rule-of-thumb" design guidance for conceptual layout of aircraft crew stations, passenger compartments, payload compartments, and weapons installations. Information for more detailed design efforts is contained in the various civilian and military specifications and in sub-system vendors' design data packages.

9.2 Crew Station

The crew station will affect the conceptual design primarily in the vision requirements. Requirements for unobstructed outside vision for the pilot can determine both the location of the cockpit and the fuselage shape in the vicinity of the cockpit.

For example, the pilot must be able to see the runway while on final approach, so the nose of the aircraft must slope away from the pilot's eye at some specified angle. While this may produce greater drag than a more streamlined nose, the need for safety overrides drag considerations. Similarly, the need for overside vison may prevent locating the cockpit directly above the wing.

When laying out an aircraft's cockpit, it is first necessary to decide what range of pilot sizes to accommodate. For most military aircraft, the design requirements include accommodation of the 5th to the 95th percentile of male pilots (i.e., a pilot height range of 65.2–73.1 in. {1.66–1.86 m}). Due to the expense of designing aircraft that will accommodate smaller or larger pilots, the services exclude such people from pilot training.

Women are now entering the military flying profession in substantial numbers. Future military aircraft will require the accommodation of approximately the 20th percentile female (roughly 5 ft {1.5 m} tall) and larger. This may affect the detailed layout of cockpit controls and displays, but should have little impact upon conceptual cockpit layout as described next.

General-aviation cockpits are designed to whatever range of pilot sizes the marketing department feels is needed for customer appeal, but typically are comfortable only for those under about 72 in. {1.83 m}. Commercial-airliner cockpits are designed to accommodate pilot sizes similar to those of military aircraft.

Figure 9.1 shows a typical pilot figure useful for conceptual design layout. This 95th percentile pilot, based upon dimensions from Ref. 22, includes allowances for boots and a helmet. A cockpit designed for this size of pilot will usually provide sufficient cockpit space for adjustable seats and controls to accommodate down to the 5th percentile of pilots.

Designers sometimes copy such a figure onto cardboard in a standard design scale such as 20-to-1, cut out the pieces, and connect them with pins to produce a movable manikin. This is placed on the drawing, positioned as desired, and traced onto the layout. A computer-aided aircraft design system can incorporate a built-in pilot manikin (see Ref. 14).

Dimensions for a typical cockpit sized to fit the 95th-percentile pilot are shown in Fig. 9.2. The two key reference points for cockpit layout are shown. The seat reference point, where the seat pan meets the back, is the reference for the floor height and the leg-room requirement. The pilot's eye point is used for defining the overnose angle, transparency grazing angle, and pilot's head clearance (10-in. {25 cm} radius).

This cockpit layout uses a typical 13-deg seatback angle, but seatback angles of 30 deg are in use (F-16), and angles of up to 70 deg have been considered for advanced fighter studies. This entails a substantial penalty in outside vision for

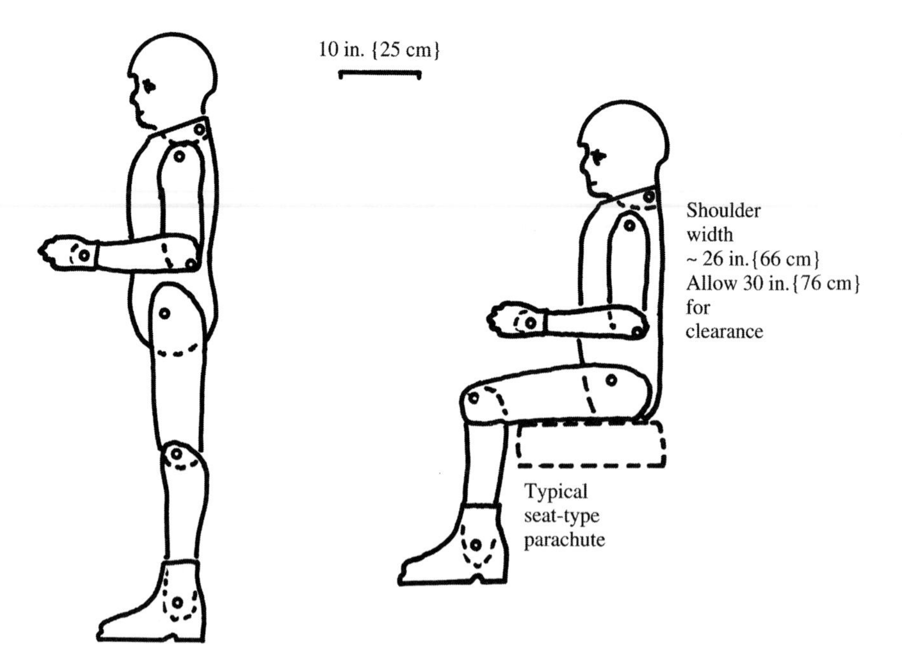

Fig. 9.1 Average 95th percentile pilot.

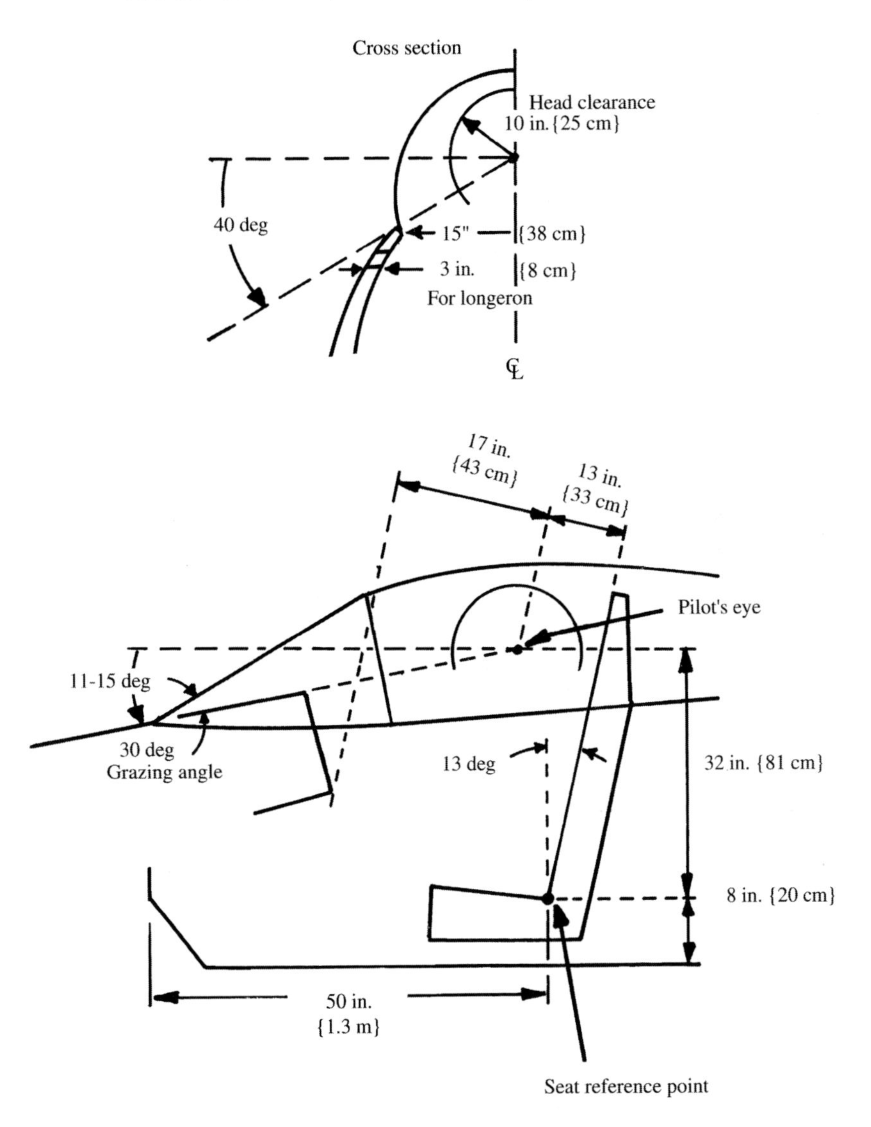

Fig. 9.2 Typical fighter cockpit.

the pilot, but can improve his ability to withstand high-*g* turns and also can reduce drag because of a reduction in the cockpit height.

When designing a reclined-seat cockpit, rotate both the seat and the pilot's eye point about the seat reference point, and then use the new position of the pilot's eye to check overnose vision.

Overnose vision is critical for safety especially during landing, and is also important for air-to-air combat. Military specifications typically require 17-deg overnose vision for transports and bombers, and 11–15 deg for fighter and attack aircraft. Military trainer aircraft in which the instructor pilot sits behind the student require 5-deg vision from the back seat over the top of the front seat.

Various military specifications and design handbooks provide detailed requirements for the layout of the cockpit of fighters, transports, bombers, and other military aircraft.

General-aviation aircraft land in a fairly level attitude, and so have overnose vision angles of only about 5–10 deg. Many of the older designs have such a small overnose vision angle that the pilot loses sight of the runway from the time of flare until the aircraft is on the ground and the nose is lowered.

Civilian transports frequently have a much greater overnose vision angle, such as the Lockheed L-1011 with an overnose vision angle of 21 deg. Civilian overnose vision angles must be calculated for each aircraft based upon the ability of the pilot to see and react to the approach lights at decision height (100 ft {30.5 m}) during minimum weather conditions (1200-ft {366 m} runway visual range). The higher the approach speed, the greater the overnose vision angle must be.

Reference 23 details a graphical technique for determining the required overnose angle, but it can only be applied after the initial aircraft layout is complete and the exact location of the pilot's eye and the main landing gear is known. For initial layout, Eq. (9.1) is a close approximation, based upon the aircraft angle of attack during approach and the approach speed.

$$\begin{aligned}\alpha_{\text{overnose}} &\cong \alpha_{\text{approach}} + 0.07\ V_{\text{approach}}\ (V \text{ in knots})\\ &= \alpha_{\text{approach}} + 0.04\ V_{\text{approach}}\ (V \text{ in km/hr})\end{aligned} \tag{9.1}$$

Figure 9.2 shows an over-the-side vision requirement of 40 deg, measured from the pilot's eye location on centerline. This is typical for fighters and attack aircraft. For bombers and transports, it is desirable that the pilot be able to look down at a 35-deg angle without head movement, and at a 70-deg angle when the pilot's head is pressed against the cockpit glass. This would also be reasonable for general-aviation aircraft, but many general-aviation aircraft have a low wing blocking the downward view.

The vision angle looking upward is also important. Transport and bomber aircraft should have unobstructed vision forwards and upwards to at least 20 deg above the horizon. Fighters should have completely unobstructed vision above and all the way to the tail of the aircraft. Any canopy structure should be no more than 2 in. {5 cm} wide to avoid blocking vision.

The transparency grazing angle shown in Fig. 9.2 is the smallest angle between the pilot's line of vision and the cockpit windscreen. If this angle becomes too small, the transparency of the glass or plexiglass will become substantially reduced, and under adverse lighting conditions the pilot may only see a reflection of the top of the instrument panel instead of whatever is in front of the aircraft! For this reason, a minimum grazing angle of 30 deg is recommended.

The cockpit of a transport aircraft must contain anywhere from two to four crew members as well as provisions for radios, instruments, and stowage of map cases and overnight bags. Reference 23 suggests an overall length of about 150 in. {3.8 m} for a four-crewmember cockpit, 130 in. {3.3 m} for three crewmembers, and 100 in. {2.5 m} for a two-crewmember cockpit.

The cockpit dimensions shown in Fig. 9.2 will provide enough room for most military ejection seats. An ejection seat is required for safe escape when flying at a speed that gives a dynamic pressure above about 230 psf {11 kN/m^2} (equal to 260 knots {481 km/hr} at sea level).

At speeds approaching Mach 1 at sea level (dynamic pressure above 1200 {58 kN/m^2}), even an ejection seat is unsafe and an encapsulated seat or separable crew capsule must be used. These are heavy and complex. A separable crew capsule is seen on the FB-111 and the prototype B-1A. The latter, including seats for four crew members, instruments, and some avionics, weighed about 9000 lb {4,082 kg}.

9.3 Passenger Compartment

The actual cabin arrangement for a commercial aircraft is determined more by marketing than by regulations. Figure 9.3 defines the dimensions of interest. "Pitch" of the seats is defined as the distance from the back of one seat to the back of the next. Pitch includes fore and aft seat length as well as leg room. "Headroom" is the height from the floor to the roof over the seats. For many smaller aircraft the sidewall of the fuselage cuts off a portion of the outer seat's headroom, as shown. In such a case it is important to ensure that the outer passenger has a 10-in. {25-cm} clearance radius about the eye position.

Table 9.1 provides typical dimensions and data for passenger compartments with first-class, economy, or high-density seating. This information (based upon Refs. 23, 24, and others) can be used to lay out a cabin floor plan.

Sad to say, today the typical design values presented in this table are rarely used in practice. Recent measurements of actual seats indicate that the airlines

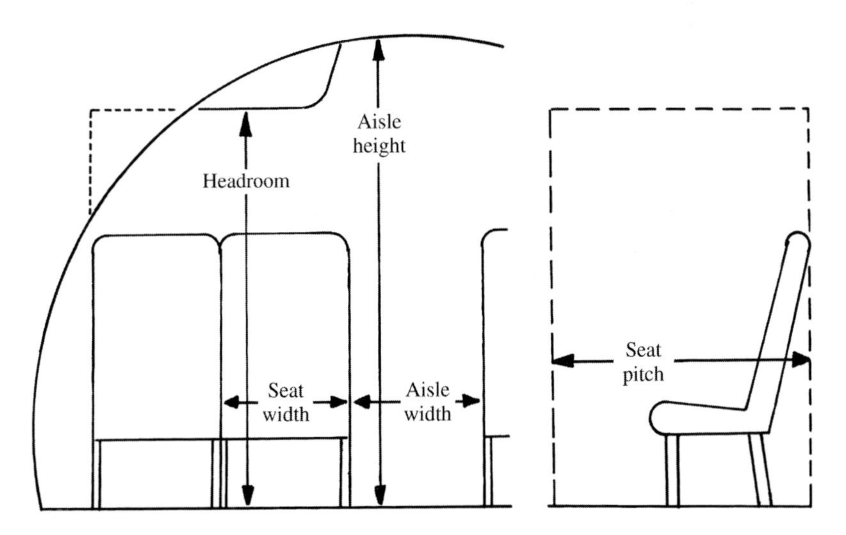

Fig. 9.3 Commercial passenger allowances.

Table 9.1 Typical passenger compartment data

	First class	Economy	High density/ small aircraft
Seat pitch (in. or cm)	38–40 {97–102}	34–36 {86–91}	30–32 {76–81}
Seat width (in. or cm)	20–28 {51–71}	17–22 {43–56}	16–18 {41–46}
Headroom (in. or cm)	>65 {165}	>65 {165}	—
Aisle width (in. or cm)	20–28 {51–71}	18–20 {46–51}	≥12 {30}
Aisle height (in. or cm)	>76 {193}	>76 {193}	>60 {152}
Passengers per cabin staff (international-domestic)	16–20	31–36	≤50
Passengers per lavatory (40″ × 40″) {1 m × 1 m}	10–20	40–60	40–60
Galley volume per passenger (ft^3 or m^3 per passenger)	5–8 {0.14–0.23}	1–2 {0.03–0.06}	0–1 {0–0.03}

are using roughly 31 in. pitch and 17 in. width {79 × 43 cm} for economy seats on commercial jets. Such cramped quarters, in years past, were only inflicted upon passengers flying short commuter flights. However, it is probably good to design the aircraft to the larger dimensions in the table—so the airlines can cram in more rows after they have bought the plane.

There should be no more than three seats accessed from one aisle, so an aircraft with more than six seats abreast will require two aisles. Also, doors and entry aisles are required for approximately every 10–20 rows of seats. These usually include closet space, and occupy 40–60 in. {1–1.5 m} of cabin length each.

Passengers can be assumed to weigh an average of 180 lb {82 kg} (dressed and with carry-on bags), and to bring about 40–60 lb {18–27 kg} of checked luggage. A current trend toward more carry-on luggage and less checked luggage has been overflowing the current aircrafts' capacity for overhead stowage of bags.

The cabin cross section and cargo bay dimensions (see the following) are used to determine the internal diameter of the fuselage. The fuselage external diameter is then determined by estimating the required structural thickness. This ranges from 1 in. {2.5 cm} for a small business or utility transport to about 4 in. {10 cm} for a jumbo jet.

9.4 Cargo Provisions

Cargo must be carried in a secure fashion to prevent shifting while in flight. Larger civilian transports use standard cargo containers that are pre-loaded with cargo and luggage and then placed into the belly of the aircraft. During conceptual design, it is best to attempt to use an existing container rather than requiring purchase of a large inventory of new containers.

Two of the more widely used cargo containers are shown in Fig. 9.4. Of the smaller transports, the Boeing 727 is the most widely used, and the 727 container shown is available at virtually every commercial airport.

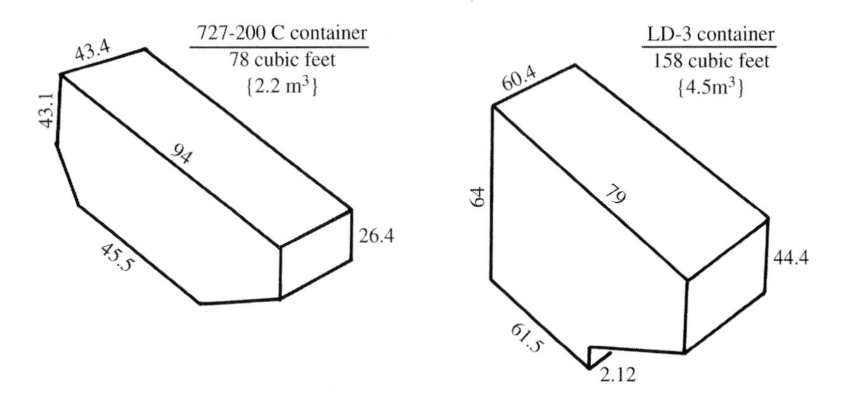

Fig. 9.4 Cargo containers.

The "Lower Deck" LD-3 container is used by all of the widebody transports. The B-747 carries 30 LD-3s plus 1000 ft^3 {28.3 m^3} of bulk cargo volume (noncontainered). The L-1011 carries 16 LD-3s plus 700 ft^3 {19.8 m^3} of bulk cargo volume, and the DC-10 and Airbus A-300 each carry 14 LD-3s plus 805 {22.8} and 565 ft^3 {16 m^3}, respectively, of bulk cargo volume.

To accommodate these containers, the belly cargo compartments require doors measuring approximately 70 in. {1.8 m} on a side. As was discussed in the section on wing vertical placement, low-wing transports usually have two belly cargo compartments, one forward of the wing box and one aft.

The cargo volume per passenger of a civilian transport ranges from about 8.6–15.6 ft^3 {0.24–0.44 m^3} per passenger (Ref. 24). The smaller number represents a small short-haul jet (DC-9). The larger number represents a transcontinental jet (B-747). The DC-10, L-1011, Airbus, and B-767 all have about 11 ft^3 {0.31 m^3} per passenger. Note that these volumes provide room for paid cargo as well as passenger luggage.

Smaller transports do not use cargo containers, but instead rely upon hand-loading of the cargo compartment. For such aircraft a cargo provision of 6–8 ft^3 {0.17–0.23 m^3} per passenger is reasonable.

Military transports use flat pallets to preload cargo. Cargo is placed upon these pallets, tied down, and covered with a tarp. The most common pallet measures 88 by 108 in. {2.2 by 2.7 m}.

Military transports must have their cargo compartment floor approximately 4–5 ft {1.4 m} off the ground to allow direct loading and unloading of cargo from a truck bed at air bases without cargo-handling facilities. However, the military does use some commercial aircraft for cargo transport and has pallet loaders capable of raising to a floor height of 13 ft {4 m} at the major Military Airlift Command bases.

The cross section of the cargo compartment is extremely important for a military transport aircraft. The C-5, largest of the U.S. military transports, and the newer C-17 are sized to carry so-called outsized cargo, which includes M-60 tanks, helicopters, and large trucks. The C-5 cargo bay is 19 ft wide, 13.5 ft high, and 121 ft long {5.8 × 4.1 × 36.9 m}. It can carry a payload of 263,000 lb {119,295 kg}.

The C-130 is used for troop and supply delivery to the front lines, and cannot carry outsized cargo. Its cargo bay measures 10′3″ wide, 9′2″ high, and 41′5″ long {3.1 × 2.8 × 12.7 m}.

9.5 Weapons Carriage

Carriage of weapons is the purpose of most military aircraft. Traditional weapons include guns, bombs, and missiles. Lasers and other exotic technologies may someday become feasible as airborne weapons, but will not be discussed here.

The weapons are a substantial portion of the aircraft's total weight. This requires that the weapons be located near the aircraft's center of gravity. Otherwise the aircraft would pitch up or down when the weapons are released.

Missiles differ from bombs primarily in that missiles are powered. Today, virtually all missiles are also guided in some fashion. Many bombs are "dumb," or unguided, and are placed upon a target by some bombsight mechanism or computer that releases them at the proper position and velocity so that they free-fall to the desired target. "Smartbombs," have some guidance mechanism, typically homing on a laser spot or guiding to a GPS (Global Positioning System) coordinate.

Missiles are launched from the aircraft in one of two ways. Most of the smaller missiles such as the AIM-9 are rail-launched. A rail-launcher is mounted to the aircraft, usually at the wing tip or on a pylon under the wing. Attached to the missile are several mounting lugs, which slide onto the rail as shown on Fig. 9.5. For launch, the missile motor powers the missile down the rail and free of the aircraft.

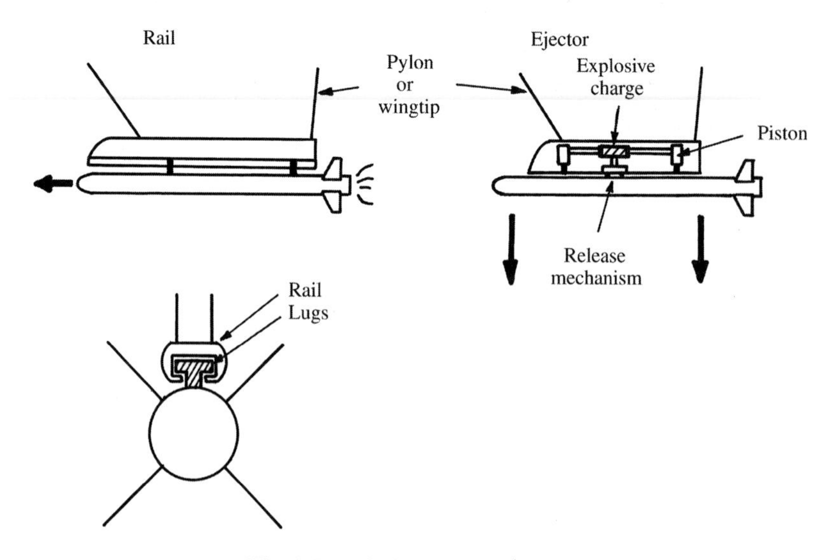

Fig. 9.5 Missile carriage/launch.

Ejection-launch is used mainly for larger missiles. The missile is attached to the aircraft through hooks which are capable of quick-release, powered by an explosive charge. This explosive charge also powers two pistons that shove the missile away from the aircraft at an extremely high acceleration. The missile motor is lit after it clears the aircraft by some specified distance.

Bombs can also be ejected, or can simply be released and allowed to fall free of the aircraft.

There are four options for weapons carriage. Each has pros and cons, depending upon the application. External carriage is the lightest and simplest, and offers the most flexibility for carrying alternate weapon stores.

While most fighter aircraft are designed to an air-to-air role, the ability to perform an additional air-to-ground role is often imposed. To avoid penalizing the aircraft's performance when "clean" (i.e., set up for dogfighting), most fighter aircraft have "hardpoints" under the wing and fuselage to which weapon pylons can be attached, as shown in Fig. 9.6. These are used to carry additional external weapons, and are removed for maximum dogfighting performance.

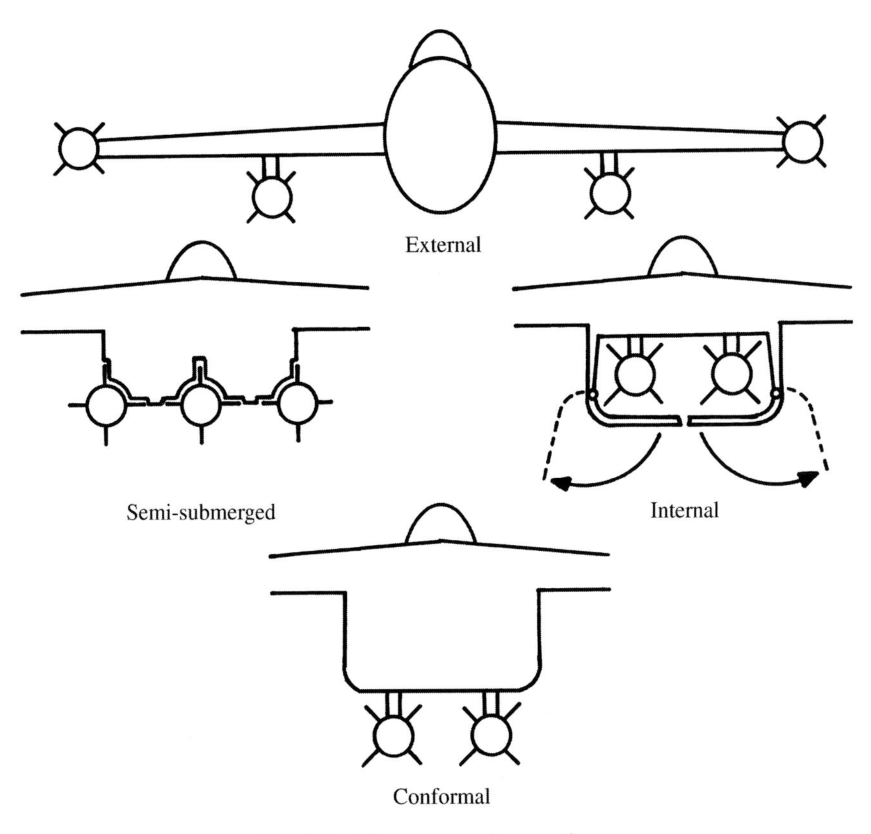

Fig. 9.6 Weapon carriage options.

Most fighter aircraft can also carry external fuel tanks on the weapons pylons. These can be dropped when entering a dogfight, but are not dropped during long overwater ferry flights. Standard external fuel tanks include 150 and 600 gallon sizes {568 and 2271 liters}.

Externally carried weapons have extremely high drag. At near-sonic speeds, a load of external bombs can have more drag than the entire rest of the aircraft. Supersonic flight is virtually impossible with pylon-mounted external weapons, due to drag and buffeting. (Wing-tip-mounted missiles are small and have fairly low drag.)

To avoid these problems, semisubmerged or conformally carried weapons may be used. Conformal weapons mount flush to the bottom of the wing or fuselage. Semisubmerged weapons are half-submerged in an indentation on the aircraft. This is seen on the F-4 for air-to-air missiles.

Semisubmerged carriage offers a substantial reduction in drag, but reduces flexibility for carrying different weapons. Also, the indentations produce a structural weight penalty on the airplane. Conformal carriage does not intrude into the aircraft structure, but has slightly higher drag than the semisubmerged carriage.

The lowest-drag option for weapons carriage is internal. An internal weapons bay has been a standard feature of bombers for over 50 years, but has been seen on only a few fighters and fighter-bombers, such as the F-106, FB-111, and the F-22. This is partly because of the weight penalty imposed by an internal weapons bay and its required doors, but is also because of the prevalent desire to maximize dogfighting performance at the expense of alternate mission performance. However, only an internal weapons bay can completely eliminate the weapons' contribution to radar cross section, so the internal weapons bay may become common for fighters as well as bombers.

During conceptual layout, there are several aspects of weapons carriage that must be considered once the type of carriage is selected. Foremost is the need to remember the loading crew. They will be handling large, heavy, and extremely dangerous missiles and bombs. They may be working at night, in a snowstorm, on a rolling carrier deck, and under attack. Missiles must be physically attached to the mounting hooks or slid down the rail, then secured by a locking mechanism. Electrical connections must be made to the guidance mechanism, and the safety wire must be removed from the fusing mechanism. For an ejector-type launcher, the explosive charge must be inserted. All of this cannot be done if the designer, to reduce drag, has provided only a few inches of clearance around the missile. The loading crew absolutely must have sufficient room in which to work.

Clearance around the missiles and bombs is also important for safety. To ensure that the weapons never strike the ground, the designer should provide at least a 3-in. {8-cm} clearance to the ground in all aircraft attitudes. This includes the worst-case bad landing in which one tire and shock-strut are completely flat, the aircraft is at its maximum tail-down attitude (usually 15 deg or more), and the aircraft is in a 5-deg roll. The minimum clearance should be doubled if the airplane is to operate from rough runways.

If weapons are mounted near each other, there should be a clearance on the order of 3 in. {8 cm} between them. There should also be a foot or more {30 cm} clearance between weapons and a propeller disk.

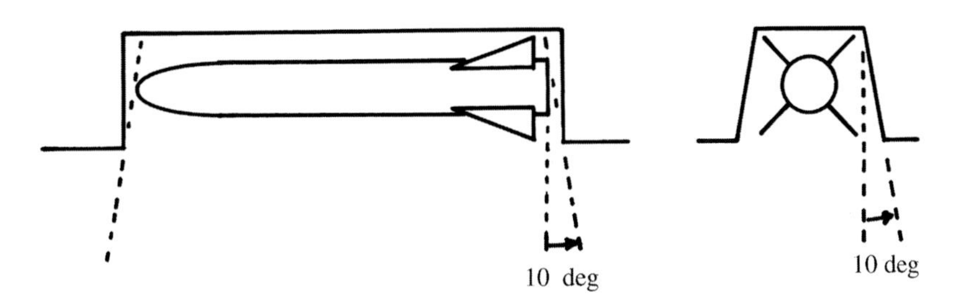

Fig. 9.7 Weapon release clearance.

The path taken by missiles or bombs when launched must be considered. For rail-launched missiles, there should be at least a 10-deg cone of clearance between any part of the aircraft and the launch direction of the missile. Also, the designer must consider the effects of the missile exhaust blast on the aircraft's structure.

For an ejector-launched or free-fall released weapon, there should be a fall line clearance of 10 deg off the vertical down from any part of the missile to any part of the aircraft or other weapons as shown in Fig. 9.7.

A special type of internal weapons carriage is the rotary weapons bay, as shown in Fig. 9.8. This allows launching all of the weapons through a single, smaller door. At supersonic speeds it can be difficult or impossible to launch weapons out of a bay due to buffeting and airloads that tend to push the weapon back into the bay. A single smaller door reduces these tendencies. Also, the rotary launcher simplifies installation of multiple weapons into a single bay. In fact, it is possible to design a rotary launcher that can be pre-loaded with weapons and loaded full into the aircraft.

9.6 Gun Installation

The gun has been the primary weapon of the air-to-air fighter since the first World War I scout pilot took a shot at an opposing scout pilot with a handgun. For a time during the 1950s it was felt that the then-new air-to-air missiles would replace the gun, and in fact several fighters such as the F-4 and F-104 were originally designed without guns. History proved that missiles cannot be solely relied upon, and all new fighters are being designed with guns.

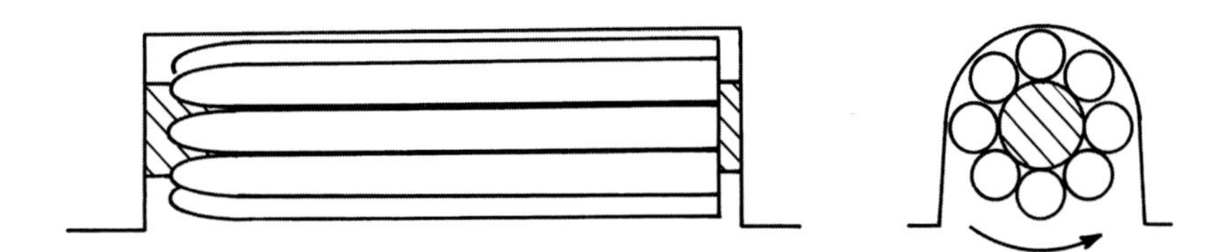

Fig. 9.8 Rotary weapons bay.

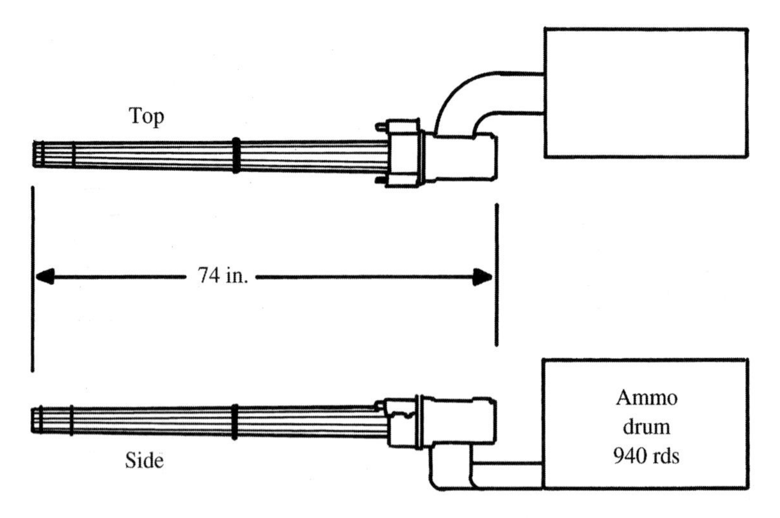

Fig. 9.9 M61 Vulcan gun.

The standard U.S. air-to-air gun today is the M61A1 Vulcan six-barrel Gatling gun, shown in Fig. 9.9. This is used in the F-15, F-16, F-18, and others. Note the ammunition container. This must be located near the aft end of the gun. Rounds of ammo are fed out of the container (drum) through feed chutes and into the gun. Ammo is loaded into the drum by attaching an ammo loading cart to the feed chute shown. The door to this loading chute must be accessible from the ground.

An air-to-air gun such as the M61A1 can produce a recoil force on the order of 2 tons {18 kN}. A large antitank gun such as the GAU-8 used in the A-10 can produce a recoil force five times greater. To avoid a sudden yawing motion from firing, guns should be located as near as possible to the centerline of the aircraft. On the A-10, the nose landing gear is offset to one side to allow the gun to be exactly on the centerline. This extreme is not necessary for the smaller air-to-air guns.

When a gun is fired, it produces a bright flash and a large cloud of smoke. The gun muzzle should be located so that these do not obscure the pilot's vision. Also, being very noisy, a gun should be located away from the cockpit.

The cloud of smoke produced by a gun can easily stall a jet engine if sucked into the inlet. This should also be considered when locating a gun. It is possible to capture the smoke produced when firing a gun, avoiding both pilot visibility and engine intake problems, by having the gun fire through a collection chamber. This adds weight and volume to the aircraft.

10
Propulsion and Fuel System Integration

10.1 Introduction

This section treats the integration and layout of the propulsion system into the overall vehicle design, not the calculation of installed propulsion performance. Propulsion analysis methods are covered in Chapter 13.

To develop the propulsion system layout, it is necessary to know the actual dimensions and installation requirements of the engine as well as its supporting equipment such as inlet ducts, nozzles, or propellers. Also, the fuel system including the fuel tanks must be defined.

10.2 Propulsion Selection

Figure 10.1 illustrates the major options for aircraft propulsion. All aircraft engines operate by compressing outside air, mixing it with fuel, burning the mixture, and extracting energy from the resulting high-pressure hot gases. In a piston-prop, these steps are done intermittently in the cylinders via the reciprocating pistons. In a turbine engine, these steps are done continuously, but in three distinct parts of the engine.

The piston-prop was the first form of aircraft propulsion. The continuing evolution of the piston engine, producing better power-to-weight, lower fuel consumption, less drag, more thrust, and greater reliability, was a major driver in the advancement of the aircraft. By the dawn of the jet era, a 5500-hp {4100 kW} piston-prop engine was in development. Today piston-props are mainly limited to light airplanes and some agricultural aircraft.

Piston-prop engines have two advantages. They are cheap, and they have the lowest fuel consumption. However, they are heavy and produce a lot of noise and vibration. Also, the propeller by its very nature produces less and less thrust as velocity increases.

The turbine engine consists of a compressor, a burner, and a turbine. These separately perform the three functions of the reciprocating piston in a piston engine.

The compressor takes the air delivered by the inlet system and compresses it to many times atmospheric pressure. This compressed air passes to the burner, where fuel is injected and mixed with the air and the resulting mixture ignited.

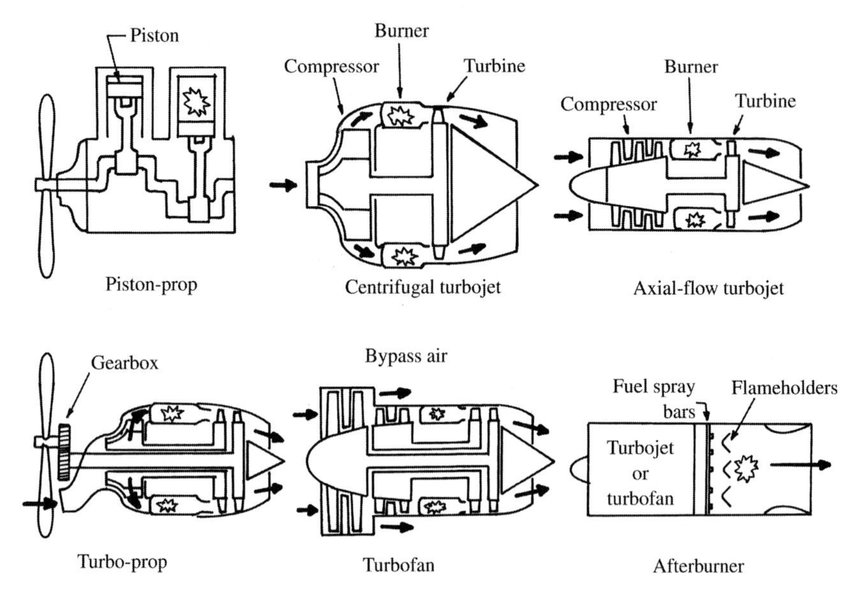

Fig. 10.1 Propulsion system options.

The hot gases could be immediately expelled out of the rear to provide thrust, but are first passed through a turbine to extract enough mechanical power to drive the compressor. It is interesting to note that one early jet engine used a separate piston engine to drive the compressor.

There are two types of compressors. The centrifugal compressor relies upon centrifugal force to "fling" the air into an increasingly narrow channel, which raises the pressure. In contrast, an axial compressor relies upon blade aerodynamics to force the air into an increasingly narrow channel. An axial compressor typically has about 6–10 stages, each of which consists of a rotor (i.e., rotating) disk of blades and a stator (i.e., stationary) disk of blades. The rotors tend to swirl the air, so the stators are used to remove the swirl.

The axial compressor, relying upon blade aerodynamics, is intolerant to distortions in the incoming air such as swirl or pressure variations. These distortions can stall the blades, causing a loss of compression and a possible engine flame-out.

The centrifugal compressor is much more forgiving of inlet distortion, but causes the engine to have a substantially higher frontal area, which increases aircraft drag. Also, a centrifugal compressor cannot provide as great a pressure increase (pressure ratio) as an axial compressor. Several smaller turbine engines use a centrifugal compressor behind an axial compressor to attempt to get the best of both types.

The turboprop and turbofan engines both use a turbine to extract mechanical power from the exhaust gases. This mechanical power is used to accelerate a larger mass of outside air, which increases efficiency at lower speeds.

For the turboprop engine, the outside air is accelerated by a conventional propeller. The "prop-fan" or "unducted fan" is essentially a turboprop with an advanced aerodynamics propeller capable of near-sonic speeds.

For the turbofan engine, the air is accelerated with a ducted fan of one or several stages. This accelerated air is then split, with part remaining in the engine for further compression and burning, and the remainder being "bypassed" around the engine to exit unburned.

The bypass ratio is the mass-flow ratio of the bypassed air to the air that goes into the core of the engine. Bypass ratio ranges from as high as 6 to as low as 0.25 (the so-called leaky turbojet).

The ideal turbine engine would inject enough fuel to completely combust all of the compressed air, producing maximum thrust for a given engine size. Unfortunately, this stoichiometric air/fuel mixture ratio of about 15 to 1 produces temperatures far greater than the capabilities of known materials, and would therefore burn up the turbine blades.

To lower the temperature seen by the turbine blades, excess air is used. Currently engines are limited to a turbine temperature of about 2000–2500°F, which requires an air/fuel mixture ratio of about 60 to 1. Thus, only about a quarter of the captured and compressed air is actually used for combustion. The exhaust is 75% unused hot air.

If fuel is injected into this largely uncombusted hot air, it will mix and burn. This will raise the thrust as much as a factor of two, and is known as "afterburning." Unfortunately, afterburning is inefficient in terms of fuel usage. The fuel flow required to produce a pound of thrust in afterburner is approximately double that used to produce a pound of thrust during normal engine operations.

Due to the high temperatures produced, afterburning must be done downstream of the turbine. Also, it is usually necessary to divert part of the compressor air to cool the walls of the afterburner and nozzles. Addition of an afterburner will approximately double the length of a turbojet or turbofan engine.

If the aircraft is traveling fast enough, the inlet duct alone will compress the air enough to burn if fuel is added. This is the principle of a "ramjet." Ramjets must be traveling at above Mach 3 to become competitive with a turbojet in terms of efficiency.

A "scramjet" is a ramjet that can operate with supersonic internal flow and combustion. Scramjets are largely unproven as of this writing, and are probably suitable only for operation above Mach 5 or 6. Ramjets and scramjets require some other form of propulsion for takeoff and acceleration to the high Mach numbers they require for operation.

The selection of the type of propulsion system—piston-prop, turboprop, turbofan, turbojet, ramjet—will usually be obvious from the design requirements. Aircraft maximum speed limits the choices, as shown in Fig. 10.2. In most cases there is no reason to select a propulsion system other than the lowest on the chart for the design Mach number. Fuel-consumption trends have been shown in Fig. 3.3.

The choice between a piston-prop and a turboprop can depend upon several additional factors. The turboprop uses more fuel than a piston prop of the same horsepower, but is substantially lighter and more reliable. Also, turboprops are usually quieter. For these reasons turbine engines have largely replaced piston

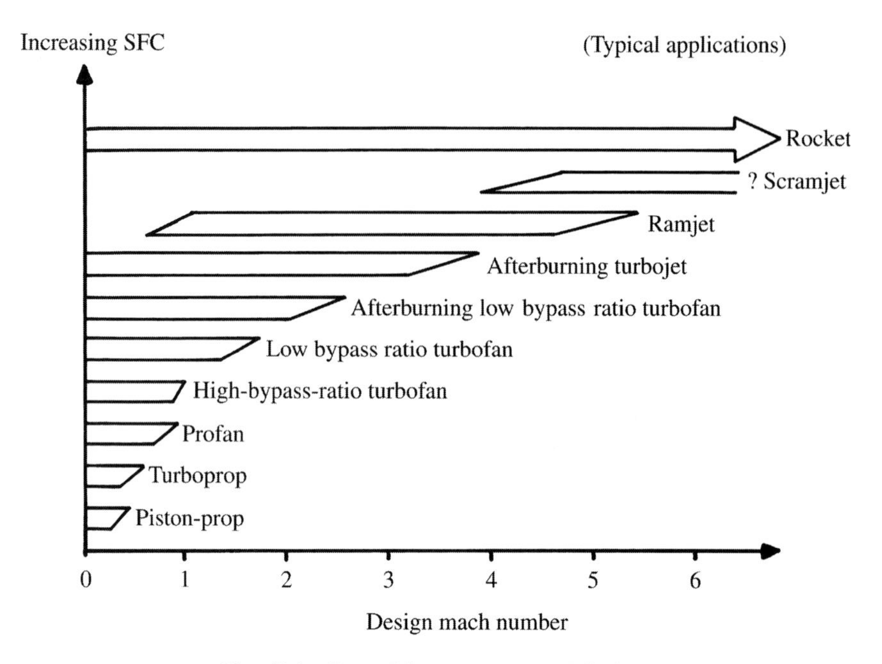

Fig. 10.2 Propulsion system speed limits.

engines for most helicopters, business twins, and short-range commuter airplanes regardless of design speed. However, piston-props are substantially cheaper and will likely remain the only choice for light aircraft for a long time.

10.3 Jet-Engine Integration

Integration of a jet engine into an aircraft conceptual design is very complicated. There are many calculations that must be made prior to the design layout, especially of the required thrust level (to pick or scale the engine) and the size of the inlet duct. The design layout must depict the engine properly with reasonable allowances for clearance for cooling air flowing around the engine, and for access to and removal of the engine. Engine controls and fuel lines must be considered, and engine-driven accessories must be depicted if there is any question about their fitting into the design.

There must be strong aircraft structure at the locations of the engine motor mounts. These can be found on the engine company's installation drawing. For commercial engines these are typically on the top, one toward the front, and one toward the back. For military engines there are typically one on the top toward the front, and one on each side somewhere in the middle of the engine, or vice-versa.

Figure 10.3 depicts a jet engine installation including inlet ducts, a remotely mounted nozzle (to better balance this particular design), control lines, fuel lines and fuel system components, and various engine-driven accessories

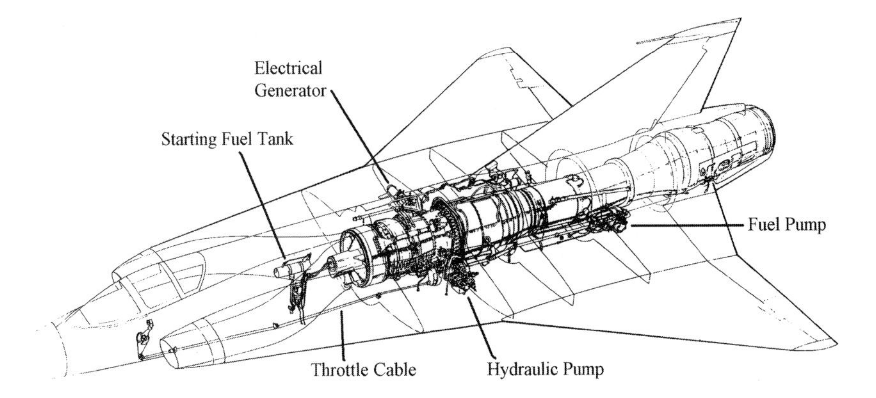

Fig. 10.3 RM6 engine installation of SAAB Draken (courtesy SAAB Aircraft).

such as hydraulic pumps and electrical generators. Note the clearance around the engine for cooling airflow, and the use of ringframe wing carrythrough structure.

Engine Dimensions

If the aircraft is designed using an existing, off-the-shelf engine, the dimensions are obtained from the manufacturer. If a "rubber" engine is being used, the dimensions for the engine must be obtained by scaling from some nominal engine size by whatever scale factor is required to provide the desired thrust. The nominal engine can be obtained by several methods.

In the major aircraft companies, designers can obtain estimated data for hypothetical rubber engines from the engine companies. These data are presented for a nominal engine size, and precise scaling laws are provided. Appendix E provides data for several hypothetical advanced engines.

Better yet, engine companies sometimes provide a "parametric deck," a computer program that will provide performance and dimensional data for an arbitrary advanced-technology engine based upon inputs such as bypass ratio, overall pressure ratio, and turbine-inlet temperature. This kind of program, which provides great flexibility for early trade studies, goes beyond the scope of this book.

Another method for defining a nominal engine assumes that the new engine will be a scaled version of an existing one, perhaps with some performance improvement due to the use of newer technologies. For example, in designing a new fighter one could start with the dimensions and performance charts of the P&W F-100 turbofan, which powers the F-15 and F-16.

To approximate the improvements due to advanced technologies, one could assume, say, a 10 or 20% reduction in fuel consumption and a similar reduction in weight. This would reflect the better materials, higher operating temperatures, and more efficient compressors and turbines that could be built today.

Figure 10.4 illustrates the dimensions that must be scaled from the nominal engine. The scale factor "SF" is the ratio between the required thrust and the

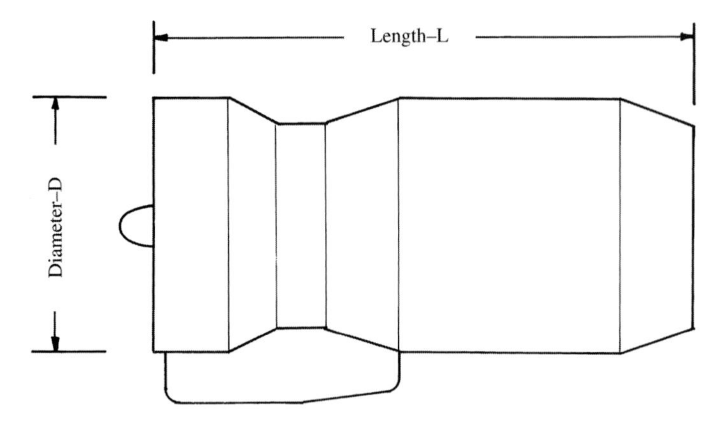

Scale factor − SF = T req. / T actual

Fig. 10.4 Engine scaling.

actual thrust of the nominal engine. Equations (10.1–10.3) show how length, diameter, and weight vary with the scale factor for the typical jet engine.

$$L = L_{\text{actual}}(\text{SF})^{0.4} \tag{10.1}$$

$$D = D_{\text{actual}}(\text{SF})^{0.5} \tag{10.2}$$

$$W = W_{\text{actual}}(\text{SF})^{1.1} \tag{10.3}$$

Although statistically derived, these equations make intuitive sense. Thrust is roughly proportional to the mass flow of air used by the engine, which is related to the cross-sectional area of the engine. Since area is proportional to the square of the diameter, it follows that the diameter should be proportional to the square root of the thrust scale-factor.

Note the engine-accessories package beneath the engine. The accessories include fuel pumps, oil pumps, power-takeoff gearboxes, and engine control boxes. The location and size of the accessory package vary widely for different types of engines. In the absence of a drawing, the accessory package can be assumed to extend below the engine to a radius of about 20–40% greater than the engine radius. On some engines these accessories have been located in the compressor spinner or other places.

If a parametric deck is unavailable, and no existing engines come close enough to the desired characteristics to be rubberized and updated as just described, then a parametric statistical approach can be used to define the nominal engine.

Equations (10.4–10.15) define two first-order statistical jet-engine models based upon data from Ref. 1. One model is for subsonic nonafterburning engines such as found on commercial transports, and covers a bypass-ratio range from zero to about six. The other model is for afterburning engines for supersonic fighters and bombers ($M < 2.5$), and includes bypass ratios from zero to just under one.

Nonafterburning engines:

$$W = 0.084T^{1.1}e^{(-0.045\ \text{BPR})}\ (\text{lb}) = 14.7T^{1.1}e^{(-0.045\ \text{BPR})}\ (\text{kg}) \tag{10.4}$$

$$L = 0.185T^{0.4}M^{0.2}\ (\text{ft}) = 0.49T^{0.4}M^{0.2}\ (\text{m}) \tag{10.5}$$

$$D = 0.033T^{0.5}e^{(0.04\ \text{BPR})}\ (\text{ft}) = 0.15T^{0.5}e^{(0.04\ \text{BPR})}\ (\text{m}) \tag{10.6}$$

$$\text{SFC}_{\max T} = 0.67e^{(-0.12\ \text{BPR})}\ (1/\text{hr}) = 19e^{(-0.12\ \text{BPR})}\ (\text{mg/Ns}) \tag{10.7}$$

$$T_{\text{cruise}} = 0.60T^{0.9}e^{(0.02\ \text{BPR})}\ (\text{lb}) = 0.35T^{0.9}e^{(0.02\ \text{BPR})}\ (\text{kN}) \tag{10.8}$$

$$\text{SFC}_{\text{cruise}} = 0.88e^{(-0.05\ \text{BPR})}\ (1/\text{hr}) = 25e^{(-0.05\ \text{BPR})}\ (\text{mg/Ns}) \tag{10.9}$$

Afterburning engines:

$$W = 0.063T^{1.1}M^{0.25}e^{(-0.81\ \text{BPR})}\ (\text{lb}) = 11.1T^{1.1}M^{0.25}e^{(-0.81\ \text{BPR})}\ (\text{kg}) \tag{10.10}$$

$$L = 0.255T^{0.4}M^{0.2}\ (\text{ft}) = 0.68T^{0.4}M^{0.2}\ (\text{m}) \tag{10.11}$$

$$D = 0.024T^{0.5}e^{(0.04\ \text{BPR})}\ (\text{ft}) = 0.11T^{0.5}e^{(0.04\ \text{BPR})}\ (\text{m}) \tag{10.12}$$

$$\text{SFC}_{\max T} = 2.1e^{(-0.12\ \text{BPR})}\ (1/\text{hr}) = 60e^{(-0.12\ \text{BPR})}\ (\text{mg/Ns}) \tag{10.13}$$

$$T_{\text{cruise}} = 2.4T^{0.74}e^{(0.023\ \text{BPR})}\ (\text{lb}) = 0.59T^{0.74}e^{(0.023\ \text{BPR})}\ (\text{kN}) \tag{10.14}$$

$$\text{SFC}_{\text{cruise}} = 1.04e^{(-0.186\ \text{BPR})}\ (1/\text{hr}) = 30e^{(-0.186\ \text{BPR})}\ (\text{mg/Ns}) \tag{10.15}$$

where

W = weight (lb or kg)
T = takeoff thrust (lb or kN)
BPR = bypass ratio
M = max Mach number

and cruise is at approximately 36,000 ft {11,000 m} and $0.9M$.

These equations represent a very unsophisticated model for initial estimation of engine dimensions. They should not be applied beyond the given bypass-ratio and speed ranges. Also, these equations represent today's state of the art. Improvement factors should be applied to approximate future engines. For a next-generation engine this author recommends, as a crude approximation, a 20% reduction in SFC, weight, and length for a given maximum thrust.

Reference 46 is recommended for the theory and practice of jet-engine design.

Inlet Geometry

Turbojet and turbofan engines are incapable of efficient operation unless the air entering them is slowed to a speed of about Mach 0.4–0.5. This is to keep the tip speed of the compressor blades below sonic speed relative to the incoming air. Slowing down the incoming air is the primary purpose of an inlet system.

The installed performance of a jet engine greatly depends upon the air-inlet system. The type and geometry of the inlet and inlet duct will determine the pressure loss and distortion of the air supplied to the engine, which will affect the installed thrust and fuel consumption. Roughly speaking, a 1% reduction in inlet pressure recovery (total pressure delivered to the engine divided by free-stream total pressure) will reduce thrust by about 1.3%.

Also, the inlet's external geometry including the cowl and boundary-layer diverter will greatly influence the aircraft drag. As discussed in Chapter 13, this drag due to propulsion is counted as a reduction in the installed thrust.

There are four basic types of inlets, as shown in Fig. 10.5. The NACA flush inlet was used by several early jet aircraft but is rarely seen today for aircraft propulsion systems because of its poor pressure recovery (i.e., large losses). At the subsonic speeds for which the NACA inlet is suitable, a pitot-type inlet will have virtually 100% pressure recovery vs about 90% for a well-designed NACA inlet. However, the NACA inlet tends to reduce aircraft wetted area and weight if the engine is in the fuselage.

The NACA inlet is regularly used for applications in which pressure recovery is less important, such as the intakes for cooling air or for turbine-powered auxiliary power units. The BD-5J, a jet version of the BD-5 home-built, used the NACA inlet, probably to minimize the redesign effort.

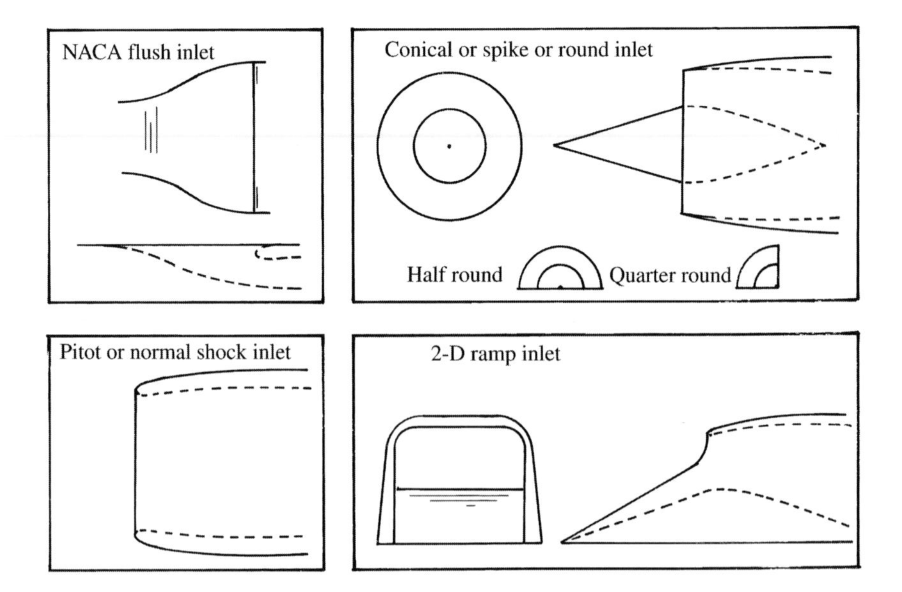

Fig. 10.5 Inlet types.

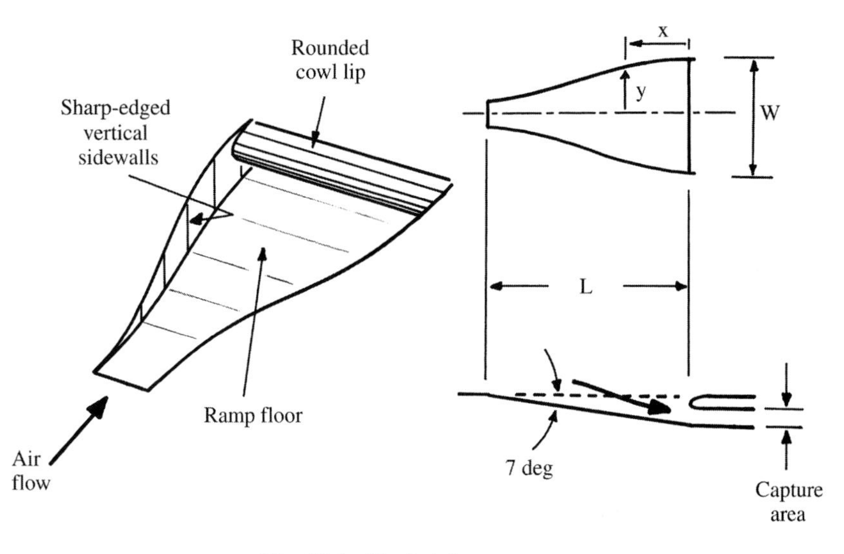

Fig. 10.6 Flush inlet geometry.

Figure 10.6 and Table 10.1 provide dimensions for laying out a good NACA flush inlet. This inlet will provide as high as 92% pressure recovery when operating at a mass flow ratio of 0.5 (i.e., air mass flow through inlet is 0.5 times the mass flow through the same cross-sectional area in the freestream).

The pitot inlet is simply a forward-facing hole. It works very well subsonically and fairly well at low supersonic speeds. It is also called a "normal shock inlet" when used for supersonic flight ("normal" meaning perpendicular in this case). Figure 10.7 gives design guidance for pitot inlets.

The cowl lip radius has a major influence upon engine performance and aircraft drag. A large lip radius tends to minimize distortion, especially at high

Table 10.1 Flush inlet wall geometry

x/L	$\frac{y}{W/2}$
1.0	0.083
0.9	0.160
0.8	0.236
0.7	0.313
0.6	0.389
0.5	0.466
0.4	0.614
0.3	0.766
0.2	0.916
0.1	0.996
0.0	1.000

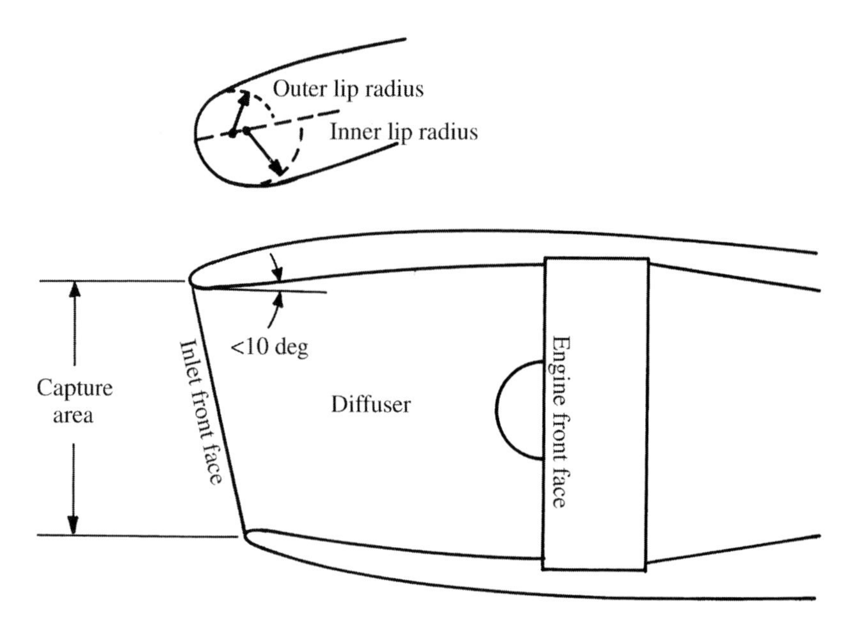

Fig. 10.7 Pitot (normal shock) inlet layout.

angles of attack and sideslip. Also, a large lip radius will readily accommodate the additional air required for takeoff thrust, when the ram air effect is small. However, a large lip radius will produce shock-separated flow on the outside of the inlet as the speed of sound is approached, and that greatly increases the drag.

For supersonic jets, the cowl lip should be nearly sharp. Typically the lip radius will be about 3–5% of the inlet front face radius. For subsonic jets, the lip radius ranges from 6–10% of the inlet radius.

To minimize distortion the lip radius on a subsonic inlet is frequently greater on the inside than the outside, with perhaps an 8% inner radius and a 4% outer radius. Also, a number of aircraft have a lip radius on the lower part of the inlet up to 50% greater than that on the upper lip. This reduces the effects of angle of attack during takeoff and landing.

Note that the inlet front face may not be perpendicular to the engine axis. The desired front-face orientation depends upon the location of the inlet and the aircraft's angle-of-attack range. Normally the inlet should be about perpendicular to the local flow direction during cruise. If the aircraft is to operate at large angles of attack, it may be desirable to compromise between these angles and the angle at cruise.

The remaining inlet types shown in Fig. 10.5 are for supersonic aircraft, and offer improvements over the performance of the normal shock inlet at higher supersonic speeds. The conical inlet (also called a spike, round, or axisymmetric inlet) exploits the shock patterns created by supersonic flow over a cone. Similarly, the two-dimensional ramp inlet (also called a "*D*-inlet") uses the flow over a wedge.

The spike inlet is typically lighter and has slightly better pressure recovery (1.5%), but has higher cowl drag and involves much more complicated mechanisms to produce variable geometry. The ramp inlet tends to be used more for speeds up to about Mach 2, while the spike inlet tends to be used above that speed.

Any inlet must slow the air to about half the speed of sound before it reaches the engine. The final transition from supersonic to subsonic speed always occurs through a normal shock. The pressure recovery through a shock depends upon the strength of the shock, which is related to the speed reduction through the shock.

In other words, a normal shock used to slow air from Mach 2 down to subsonic speeds will have a far worse pressure recovery than a normal shock used to slow the air from Mach 1.1 to subsonic speeds (72% vs 99.9%). For this reason a normal-shock inlet is rarely used for prolonged operation above Mach 1.4.

An oblique shock, however, does not reduce the air speed all the way to subsonic. The speed reduction and pressure recovery through an oblique shock depends upon the angle of the wedge or cone used to establish the shock. For example, a 10-deg wedge in Mach 2 flow creates an oblique shock at 39 deg that reduces the flow speed to Mach 1.66 (see NACA TR 1135). This gives a pressure loss of only 1.4% (i.e., pressure recovery of 98.6%).

If the Mach 1.66 air downstream of this oblique shock is then run into a normal shock inlet, it will slow to Mach 0.65, with a pressure recovery of 87.2%. The total pressure recovery from Mach 2 to subsonic speed is 98.6 times 87.2, or 86%. Thus, use of an oblique shock before the normal shock has improved pressure recovery for this example Mach 2 inlet from 72% to 86%. (Note that this is far from optimal. A well-designed Mach 2 inlet with one oblique shock will approach a 95% pressure recovery.)

This illustrates the principle of the external-compression inlet shown in Fig. 10.8. The preceding example is a two-shock system, one external and one normal. The greater the number of oblique shocks employed, the better the pressure recovery.

The theoretical optimal is the isentropic ramp inlet, which corresponds to infinitely many oblique shocks and produces a pressure recovery of 100% (ignoring friction losses). The pure isentropic ramp inlet works properly at only its design Mach number, and is seen only rarely except on "one-speed" drones such as the Lockheed D-21, which uses an isentropic cone optimized to its cruise speed. However, isentropic ramps are frequently used in combination with flat wedge ramps, such as on the Concorde SST.

Figure 10.9 illustrates a typical three-shock external-compression inlet. This illustration could be a side view of a 2-D inlet or a section view through a spike inlet. Note that the second ramp has a variable angle, and can collapse to open a larger duct opening for subsonic flight.

Some form of boundary-layer bleed is required on the ramp to prevent shock-induced separation on the ramp. The bled air is usually dumped overboard out a rearward-facing hole above the inlet duct.

Not shown are suck-in (or blow-in) and bypass doors in the diffuser section that may be required to provide extra air to the engine for takeoff or get rid of excess air during high-speed operation.

For initial layout, the overall length of the external portion of the inlet can be estimated by assuming an initial ramp angle (10–20 deg) and determining the shock angle for the design Mach number using standard shock charts such as in NACA TR 1135. The cowl lip should be placed just aft of the shock.

The throat area should be about 70–80% of the engine front-face area.

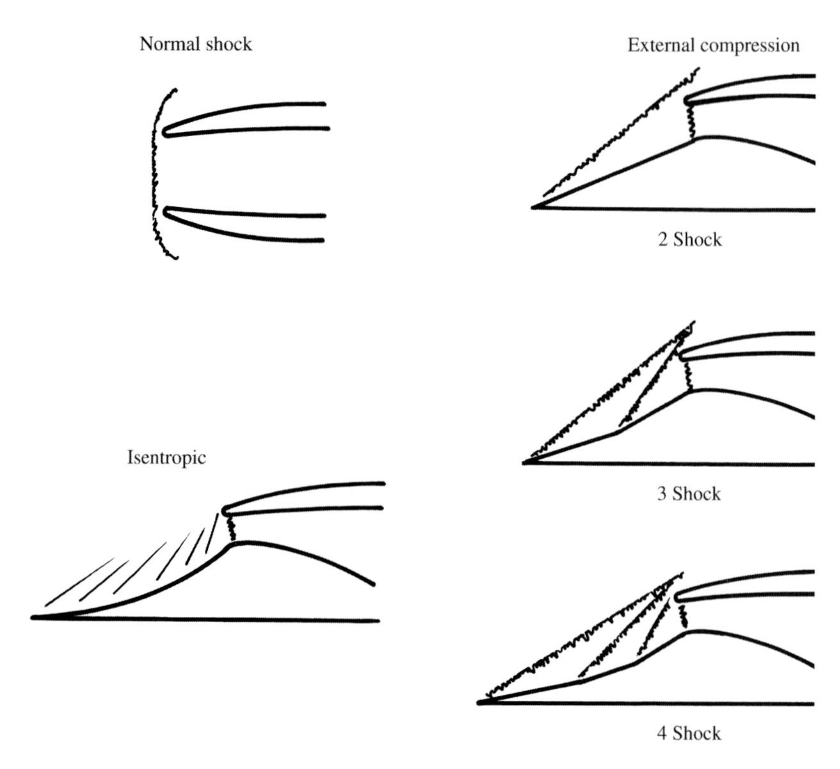

Fig. 10.8 Supersonic inlets—external shocks.

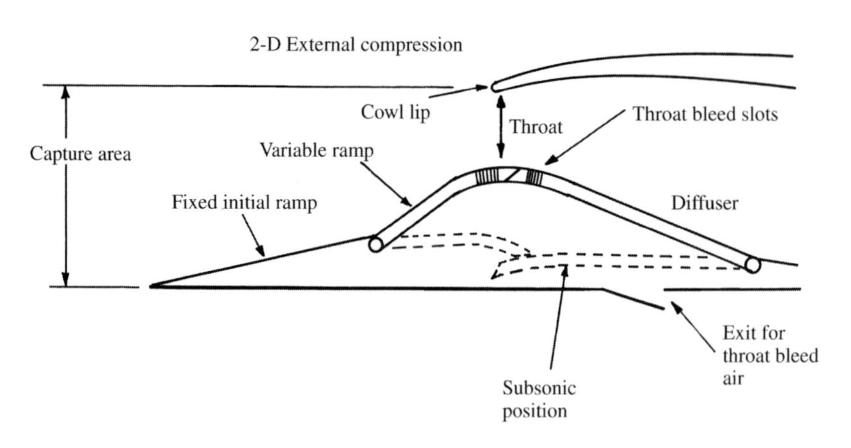

Fig. 10.9 Variable inlet geometry.

The speed limitation on external compression inlets is due to the flow turning angle introduced by the shocks. A wedge turns the flow parallel to the wedge angle, while a cone turns the flow to an angle slightly less than the cone angle.

At speeds approaching Mach 3, the required oblique shocks to obtain good pressure recovery will introduce a total flow turning of about 40 deg. This air must be turned by the outside cowl lip back to the freestream direction, which may not be possible without either separation or an excessively large lip radius that will increase aircraft drag.

One form of inlet system introduces no outside flow turning: the internal compression inlet, as shown in Fig. 10.10. In this inlet a pair of inward-facing ramps produce oblique shocks that cross upstream of the final normal shock.

This form of shock system can be very efficient when operating properly at its design Mach number. However, this inlet must be "started." If it is simply placed into supersonic flow, a normal shock will form across its front. To start the inlet and produce the efficient shock structure shown in Fig. 10.10, it is necessary to "suck" the normal shock down to the throat by opening doors downstream. Once formed, the desired shock structure is unstable. Any deviation in flow condition, such as temperature, pressure, or angle of attack can cause an

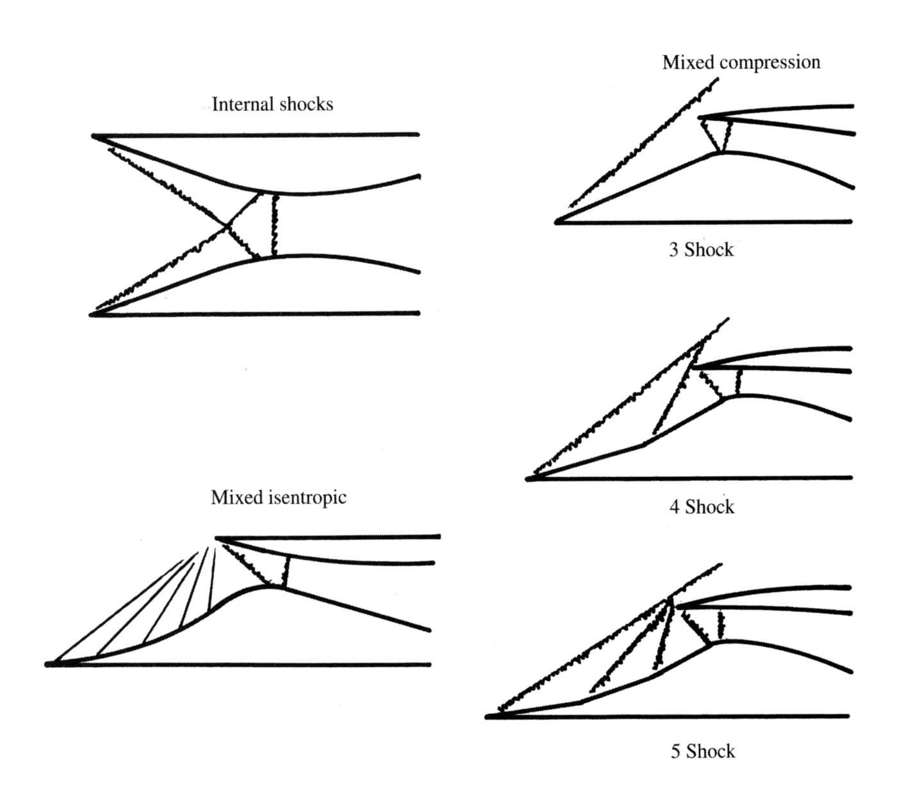

Fig. 10.10 Supersonic inlets—internal and mixed.

"unstart" in which the normal shock pops out of the duct. This can stall the engine.

The "mixed compression inlet" as shown in Fig. 10.10 uses both external and internal compression to provide high efficiency over a wide Mach number range, with an acceptable amount of external flow turning. Typically one or more external oblique shocks will feed a single internal oblique shock, followed by a final normal shock.

Such an inlet has been used for most aircraft designed to fly above Mach 2.5, including the B-70, which has a 2-D inlet (Fig. 10.11) and the SR-71, which has an axisymmetric inlet. Unstart remains a problem for this type of inlet. Automatically opening doors are used to control unstart.

Mixed-compression inlets are complex, and can be defined only by detailed propulsion analysis beyond the scope of this book. Reference 25 is recommended. The rules of thumb just provided for the dimensions of external-compression inlets give a reasonable first approximation for mixed-compression inlets.

The "diffuser" is the interior portion of an inlet where the subsonic flow is further slowed down to the speed required by the engine. Thus, a diffuser is increasing in cross-sectional area from front to back.

The required length of a diffuser depends upon the application. For a subsonic aircraft such as a commercial transport, the diffuser should be as short as possible without exceeding an internal angle of about 10 deg. Typically, this produces a pitot inlet with a length about equal to its front-face diameter.

For a supersonic application, the theoretical diffuser length for maximum efficiency is about eight times the diameter. Lengths longer than eight times the diameter are permissible but have internal friction losses as well as an additional weight penalty.

A supersonic diffuser shorter than about four times the diameter may produce some internal flow separation, but the weight savings can exceed the engine performance penalty. Diffusers as short as two times the diameter have been used with axisymmetric spike inlets.

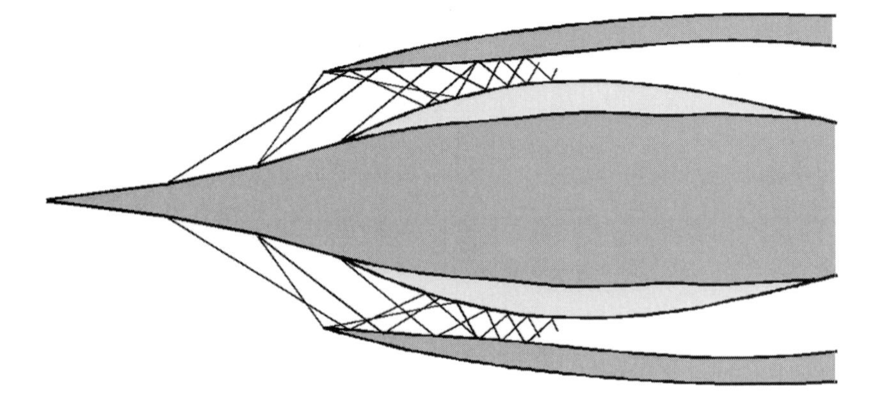

Fig. 10.11 B-70 inlet shock system.

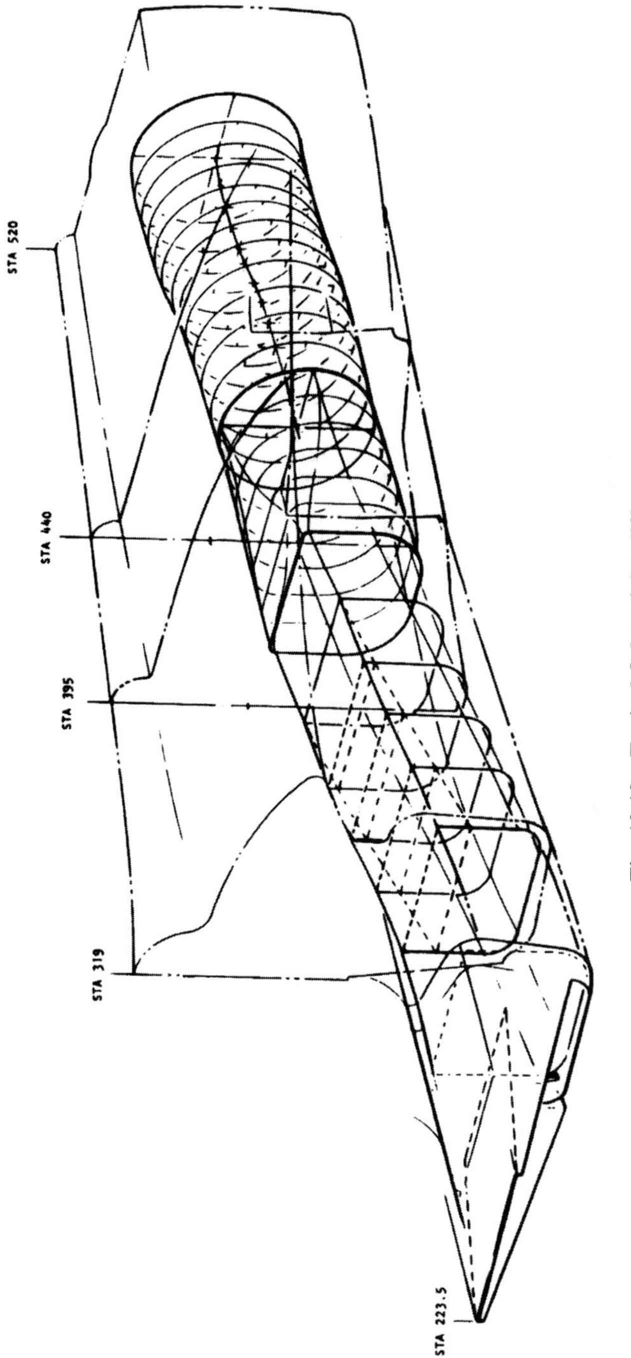

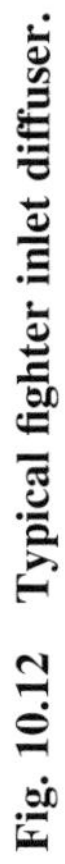
Fig. 10.12 Typical fighter inlet diffuser.

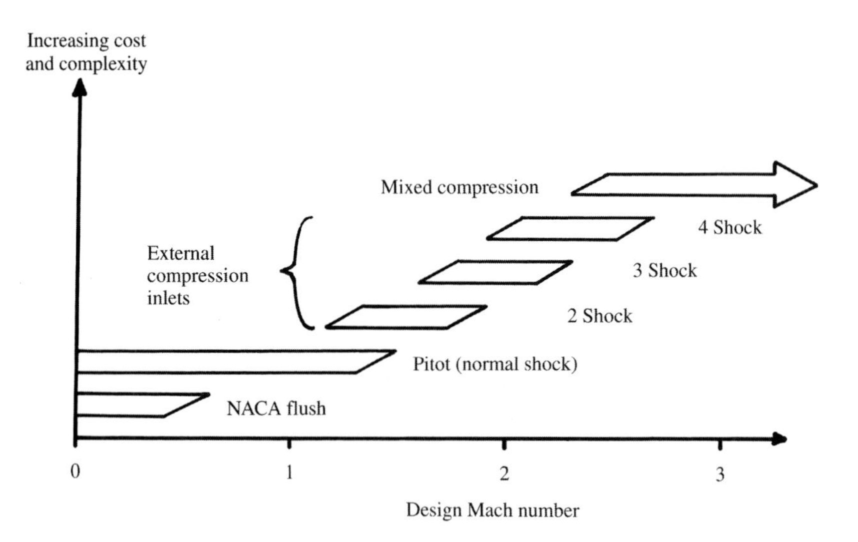

Fig. 10.13 Inlet applicability.

For a long diffuser it is important to verify that the cross-sectional area of the flowpath is smoothly increasing from the inlet front face back to the engine. This verification is done with a volume-distribution plot of the inlet duct, constructed in the same fashion as the aircraft volume plot shown in Fig. 7.38. An example of a smooth, long fighter diffuser is shown in Fig. 10.12, from the North American F-X proposal.

To reduce distortion, some aircraft use a diffuser oversized about 5% that "pinches" the flow down to the engine front-face diameter in a very short distance just before the engine.

Figure 10.13 summarizes the selection criteria for different inlets, based upon design Mach number. Note that these are approximate criteria, and may be overruled by special considerations. Estimated pressure recoveries of these inlets are provided in Chapter 13.

Inlet Location

The inlet location can have almost as great an effect on engine performance as the inlet geometry. If the inlet is located where it can ingest a vortex off the fuselage or a separated wake from a wing, the resulting inlet-flow distortion can stall the engine. The F-111 had tremendous problems with its inlets, which were tucked up under the intersection of the wing and fuselage. The A-10 required a fixed slot on the inboard wing leading edge to cure a wake-ingestion problem.

Figure 10.14 illustrates the various options for inlet location for buried engine installations. The nose location offers the inlet a completely clean airflow, and was used in most early fighters including the F-86 and MiG 21 as a way of ensuring that the fuselage would not cause distortion problems. However,

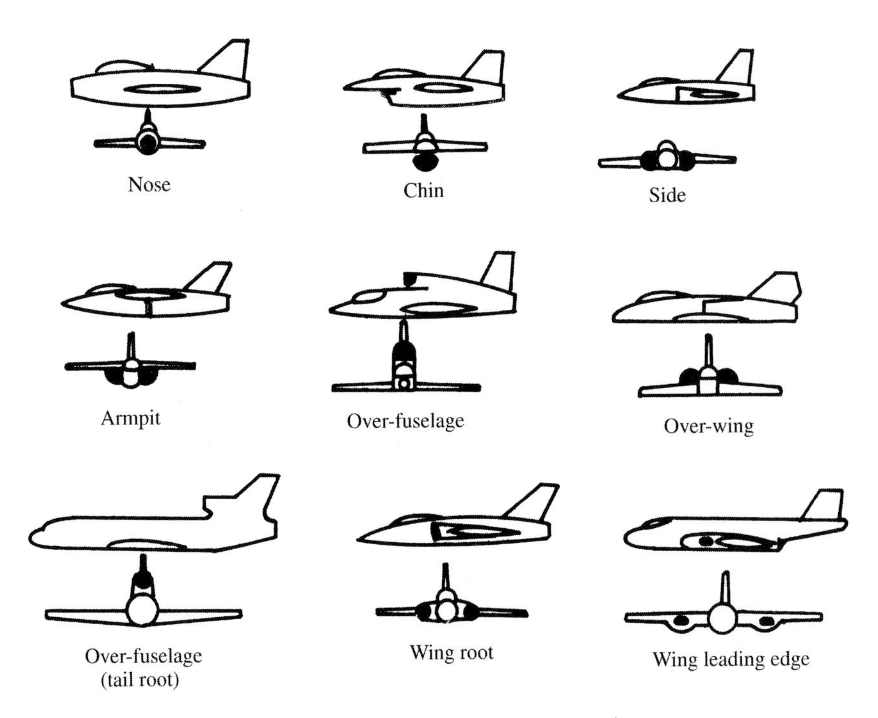

Fig. 10.14 Inlet locations—buried engines.

the nose inlet requires a very long internal duct, which is heavy, has high losses, and occupies much of the fuselage volume.

The chin inlet as seen on the F-16 has most of the advantages of the nose inlet but a shorter duct length. The chin inlet is especially good at high angle of attack because the fuselage forebody helps to turn the flow into it.

The location of the nose landing gear is a problem. It cannot be placed forward of the inlet because it would block and distort the flow, and also the nose wheel would tend to throw water and rocks into the inlet. Instead, it is usually placed immediately behind the inlet, which requires that the cowl be deep enough to hold the retracted gear, which can increase cowl drag. Also, the cowl must be strong enough to carry the nose-gear loads.

If two engines are used, twin inlets can be placed in the chin position with the nose wheel located between them. This was used on the North American Rockwell proposal for the F-15, and is seen on the Sukhoi Su-27.

Another problem with the chin inlet is foreign-object ingestion by suction. As a rule of thumb, all inlets should be located a height above the runway equal to at least 80% of the inlet's height if using a low-bypass-ratio engine, and at least 50% of the inlet's height for a high-bypass-ratio engine.

Side-mounted inlets are now virtually standard for aircraft with twin engines in the fuselage. Side inlets provide short ducts and relatively clean air. Side-mounted inlets can have problems at high angles of attack due to the vortex shed off the lower corner of the forebody. This is especially severe if the forward fuselage has a fairly square shape.

If side-mounted inlets are used with a single engine, a split duct must be used. Split ducts are prone to a pressure instability that can stall the engine. To minimize this risk, it is best to keep the two halves of the duct separate all the way to the engine front face, although several aircraft have flown with the duct halves rejoined well forward of the engine.

A side inlet at the intersection of the fuselage and a high wing is called an "armpit" inlet. It is risky! The combined boundary layers of the forebody and wing can produce a boundary layer in the wing-fuselage corner that is too thick to remove. (Boundary-layer removal is discussed later.) This type of inlet is especially prone to distortion at angle of attack and sideslip. In many cases however, the armpit inlet does offer a very short internal duct.

An over-fuselage inlet is much like an inverted chin inlet, and has a short duct length but without the problems of nose-wheel location. This was used on the unusual F-107. The upper-fuselage inlet is poor at high angle of attack because the forebody blanks the airflow, although careful forebody design can create vortices tailored to guide the flow into the inlet. Also, many pilots fear that they may be sucked down the inlet if forced to bail out manually.

Placed over the wing and near the fuselage, an inlet encounters problems similar to those of an inverted-armpit inlet. It also suffers at angle of attack.

An inlet above the aft fuselage for a buried engine is used on the L-1011 and B-727, with the inlet located at the root of the vertical tail. This arrangement allows the engine exhaust to be placed at the rear of the fuselage, which tends to reduce fuselage separation and drag. The buried engine with a tail inlet must use an "*S*-duct." This requires careful design to avoid internal separation. Also, the inlet should be well above the fuselage to avoid ingesting the thick boundary layer.

Inlets set into the wing leading edge can reduce the total aircraft wetted area by eliminating the need for a separate inlet cowl. However, these inlets can disturb the flow over the wing and increase its weight. The wing-root position may also ingest disturbed air off the fuselage.

A podded engine has higher wetted area than a buried engine, but offers substantial advantages that have made it standard for commercial and business jets. Podded engines place the inlet away from the fuselage, providing undisturbed air with a very short inlet duct. Podded engines produce less noise in the cabin because the engine and exhaust are away from the fuselage. Podded engines are usually easier to get to for maintenance. Most are mounted on pylons, but they can also be mounted conformal to the wing or fuselage. Various options are shown in Fig. 10.15.

The wing-mounted podded engine is the most commonly used engine installation for jet transports. The engines are accessible from the ground and well away from the cabin. The weight of the engines out along the wing provides a "span-loading" effect, which helps reduce wing weight. The jet exhaust can be directed downward by flaps which greatly increases lift for short takeoff.

On the negative side, the presence of pods and pylons can disturb the airflow on the wing, increasing drag and reducing lift. To minimize this, the pylons should not extend above and around the wing leading edge, as was seen on one early jet transport.

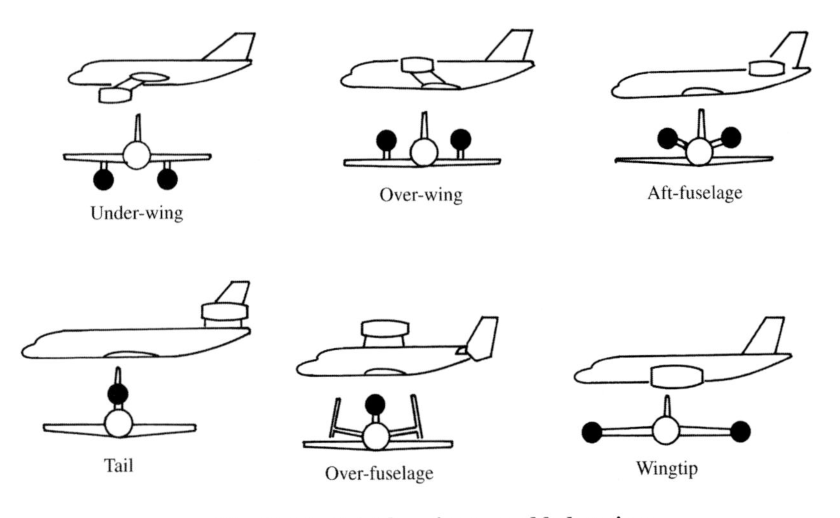

Fig. 10.15 Inlet locations—podded engines.

On the basis of years of wind-tunnel study, design charts for pylon-mounted engines have been prepared that minimize the interference effects of the nacelle pod on the wing. As a classical rule-of-thumb, the inlet for a wing-mounted podded engine should be located approximately two inlet diameters forward and one inlet diameter below the wing leading edge. However, modern computational fluid dynamic (CFD) methods now allow designing a wing-mounted nacelle much closer to the wing, or even conformal to the wing, without incurring substantial drag increase due to interference. This will be further discussed in Chapter 12.

The wing-mounted nacelle should be angled nose down by about 2–4 deg, and canted nose inward about 2 deg to align it to the local flow under the wing.

To reduce foreign-object ingestion by suction, the inlet of a high-bypass engine should be located about half a diameter above the ground. This requirement increases the required landing-gear height of the under-wing arrangement.

The over-wing podded nacelle reduces the landing-gear height and reduces noise on the ground, but is difficult to get to for maintenance. The inlets can be forward of the wing to minimize distortion, or above it. If an over-wing nacelle is conformal to the wing, the exhaust can be directed over the top of flaps, which, through Coanda effect, turn the flow downward for increased lift.

The other standard engine installation for jet transports is the aft-fuselage mount, usually with a T-tail. This eliminates the wing-interference effects of wing-mounted engines, and allows a short landing gear. However, it increases the cabin noise at the rear of the aircraft.

Also, aft-mounting of the engines tends to move the center of gravity aft, which requires shifting the entire fuselage forward relative to the wing. This shortens the tail moment arm and increases the amount of fuselage forward of the wing, and that necessitates a larger vertical and horizontal tail.

To align the aft nacelle with the local flow, a nose-up pitch of 2–4 deg and a nose outward cant of 2 deg are recommended.

The Illyushin Il-76 uses four aft-podded engines in two twin-engine pods. The B-727 and Hawker–Siddeley Trident combine aft-fuselage podded engines with a buried engine using an inlet over the tail.

The DC-10 combines wing-mounted engine pods with a tail-mounted podded engine. This is similar to the tail-mounted inlet for a buried engine like the L-1011, but eliminates the need for an *S*-duct. However, this arrangement increases the tail weight and doesn't have the fuselage drag-reduction effect. All told, the two installations are probably equivalent.

The supersonic Tupolev Tu-22 (Blinder) uses twin engines, pod-mounted on the tail, but this arrangement has not been seen on later Soviet supersonic designs.

The over-fuselage podded engine has been used only rarely, such as to add a jet engine to the turboprop Rockwell OV-10. Access and cabin noise are undesirable for this installation.

The wing-tip-mounted engine has an obvious engine-out controllability problem. It was used on the Soviet supersonic Myasishchev M-52 (Bounder), which also had under-wing engine pods.

Capture-Area Calculation

Figure 10.16 provides a quick method of estimating the required inlet capture area. This statistical method is based upon the design Mach number and the engine mass flow.

To determine the required capture area, the engine's mass flow is multiplied by the value from Fig. 10.16. If mass flow is not known, it may be estimated as 26 times the square of the engine front-face diameter in feet {127 times meters squared}. If engine front face diameter is not known, it may be estimated as 80% of maximum diameter.

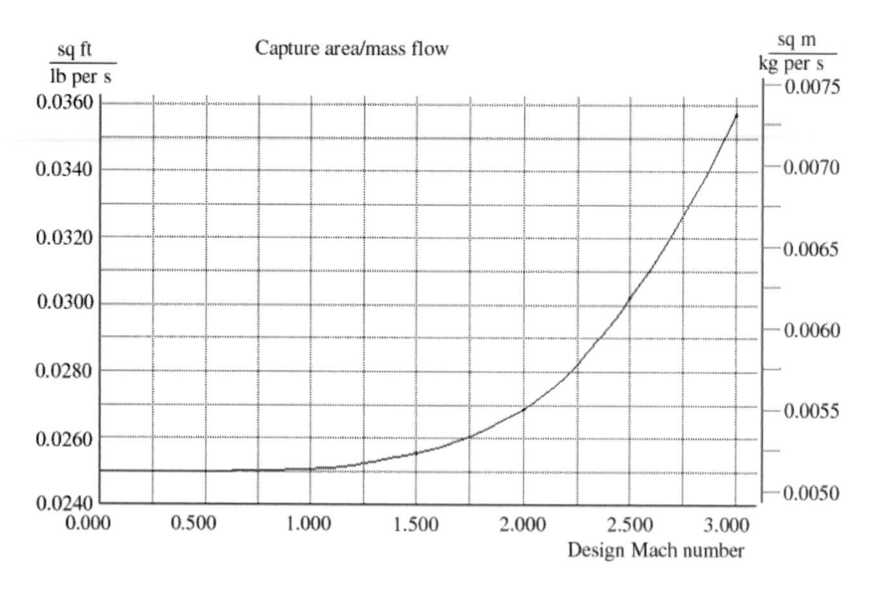

Fig. 10.16 Preliminary capture area sizing.

This capture-area estimation is adequate for initial layout and rough analysis, but not accurate enough for a good configuration layout. A better estimate of capture area should be made during configuration layout based upon the actual mass flow of the engine, as described next.

In a jet propulsion system, the engine is the boss. It takes the amount of air it wants, not what the inlet wants to give it. If the inlet is providing more air than the engine wants, the inlet must spill the excess out the front. If the inlet is not providing what the engine needs, it will attempt to suck in the extra air required.

The inlet capture area must be sized to provide sufficient air to the engine at all aircraft speeds. For many aircraft the capture area must also provide "secondary air" for cooling and environmental control, and also provide for the air bled off the inlet ramps to prevent boundary-layer buildup.

Figure 10.17 defines the capture area for a subsonic inlet. A typical subsonic jet inlet is sized for cruise at about Mach 0.8–0.9, and the inlet must slow the air to about Mach 0.4 for most engines. Because this is subsonic flow, the inlet does not need to do all the work itself. As shown in Fig. 10.17, the expansion associated with slowing the flow from its freestream velocity at "infinity" takes place about half within and half outside the inlet duct.

The area at the inlet front face is both the capture area and the throat area. It can be calculated from the following isentropic compressible flow relationship:

$$\frac{A_{\text{throat}}}{A_{\text{engine}}} = \frac{(A/A^*)_{\text{throat}}}{(A/A^*)_{\text{engine}}} \tag{10.16}$$

$$\frac{A}{A^*} = \frac{1}{M}\left(\frac{1+0.2M^2}{1.2}\right)^3 \tag{10.17}$$

where A^* is the area of the same flow at sonic speed.

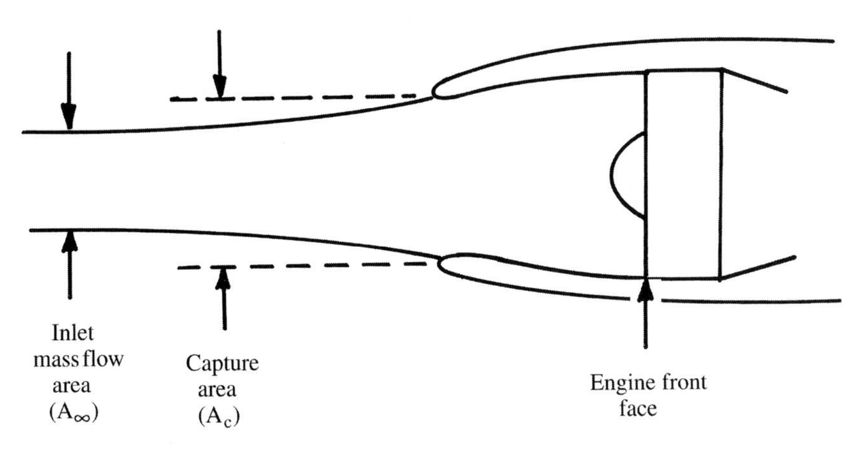

Fig. 10.17 Subsonic inlet capture area.

For a typical inlet designed to a cruise speed of Mach 0.8, the inlet must slow the air from about Mach 0.6 down to Mach 0.4. The air is slowed from Mach 0.8–0.6 outside the inlet.

Equations (10.16) and (10.17) give the ratio between throat area and engine front-face area as 1.188/1.59, or 0.75 (for this example). Taking the square root gives a diameter ratio of about 0.88, which is reasonable. Note that a subsonic inlet generally does not require bleed air, since secondary air is obtained from separate, small NACA flush inlets in most subsonic aircraft.

Equations (10.16) and (10.17) may be used to determine the capture area for a supersonic pitot inlet with negligible bleed or secondary airflow by finding the Mach number behind the normal shock from NACA TR 1135. For other supersonic inlets, the required capture area must be determined by considering the airflow requirements for the engine, bleed, and secondary airflows.

In sizing a supersonic inlet, a variety of flight conditions must be considered to find the largest required capture area. Typically this will be at the aircraft maximum Mach number, but may also occur during takeoff or subsonic cruise. If the maximum required capture area occurs during takeoff, consider using auxiliary suck-in (or blow-in) doors during takeoff. This allows the inlet to be sized to another, lesser capture area requirement.

To avoid the confusion of differing velocities and densities, the airflow is defined by mass flow. Mass flow is related to flow conditions by Eq. (10.18).

$$\dot{m} = \rho V A \tag{10.18}$$

A word of caution—users of British units (fps) sometimes multiply mass flow times "g" (32.2 ft/s^2) to obtain mass flow in pounds-mass per ft^2, rather than the more-correct slugs per ft^2.

Figure 10.18 defines the capture area geometry for a supersonic ramp or cone inlet. This inlet is shown at the design case, which is known as "shock-on-cowl." At this Mach number and ramp angle, the initial oblique shock is almost touching the cowl lip. If the auxiliary doors are shut and the shock is on cowl, the geometric capture area is providing exactly the right amount of air for the engine, bleed, and secondary flow.

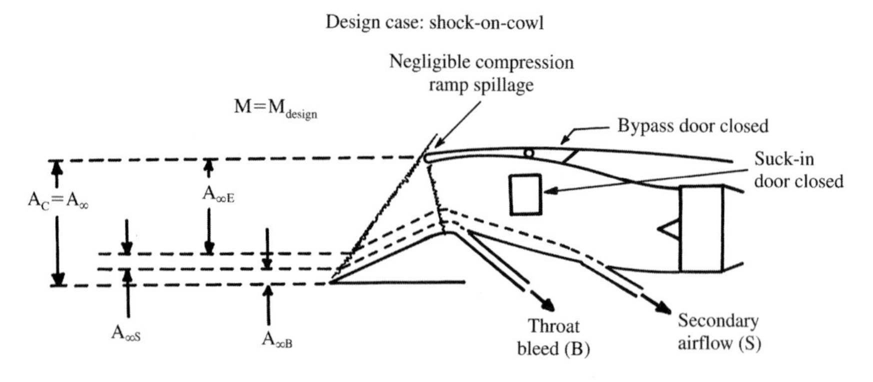

Fig. 10.18 Supersonic inlet capture area—on design.

Usually the inlet ramp geometry provides shock-on-cowl at about Mach 0.1–0.2 above the aircraft's maximum speed, giving a safety margin for speed overshoot and engine mass-flow fluctuations.

If the total mass flow required by the engine, bleed, and secondary flow is known, then Eq. (10.18) can be solved for the required cross-sectional area upstream of the inlet (at "infinity") using the freestream values for density and velocity. This calculated area is identical to the capture area in the design case (shock-on-cowl) since all of the air in the capture area is going into the inlet.

The required engine mass flow is provided by the engine manufacturer, and is a function of the Mach number, altitude, and throttle setting (percent power). Usually the manufacturer's data should be increased by 3% to allow for manufacturing tolerances.

The secondary airflow requirements are accurately determined by an evaluation of the aircraft's subsystems such as environmental control. For initial capture-area estimation, Table 10.2 (from Ref. 26) provides secondary airflow as a fraction of engine mass flow.

Inlet boundary layer bleed should also be determined analytically, but can be approximated using Fig. 10.19, taken from Ref. 27. This estimates the required extra capture area for bleed as a percent of the capture area required for the engine and secondary airflow.

The capture area is therefore determined as in Eq. (10.19), using Table 10.2 and Fig. 10.19.

$$A_{\text{capture}} = \left[\frac{\dot{m}_e(1 + \dot{m}_s/\dot{m}_e)}{g\rho_\infty V_\infty}\right]\left(1 + \frac{A_B}{A_C}\right) \tag{10.19}$$

Figure 10.18 shows the inlet operating at its design condition, shock-on-cowl, where the geometric capture area equals the freestream area of the air actually taken into the inlet and used. If the freestream Mach number is reduced, the oblique shock angle drops, which moves the oblique shock in front of the cowl, as shown in Fig. 10.20a.

Table 10.2 Secondary airflow (typical) (Ref. 26)

System	$\dot{m}_s/\dot{m}_e$
Engine	
Nacelle cooling	0–0.04
Oil cooling	0–0.01
Ejector nozzle air	0.04–0.20
Hydraulic system cooling	0–0.01
Environmental control system cooling air (if taken from inlet)	0.02–0.05
Typical totals	
Fighter	0.20
Transport	0.03

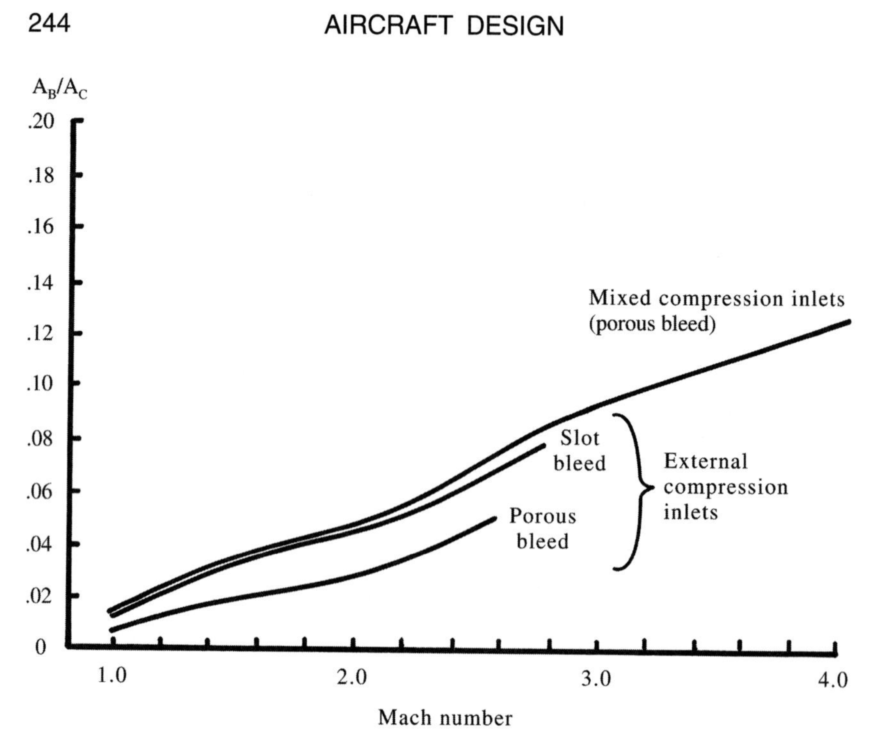

Fig. 10.19 Typical boundary layer bleed area.

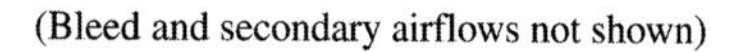

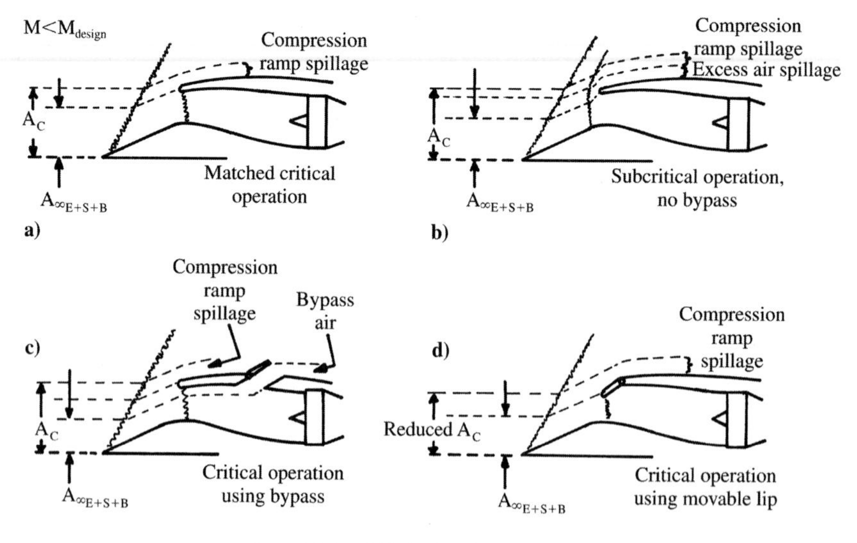

Fig. 10.20 Off-design inlet operation.

Since the airflow is parallel to the ramp, it can be seen that the freestream cross-sectional area of the air that actually goes into the inlet has been reduced. Part of the air defined by the geometric capture area is now spilled after being compressed. This represents wasted work and increased drag compared to the case of shock-on-cowl.

If the mass-flow demand exactly equals the mass flow shown going into the inlet in Fig. 10.20a (i.e., capture area less compression-ramp spillage), then the engine and inlet duct are still "matched" and the normal shock will be at the cowl lip, as shown in Fig. 10.20a.

However, the engine demand is usually reduced at a slower speed. The excess air is simply rejected by the inlet, as shown in Fig. 10.20b. (Remember, the engine is the boss!) This pushes the normal shock forward of the inlet and creates a much larger spillage drag than for the matched condition.

Two approaches to move the normal shock back to the cowl lip are shown in Figs. 10.20c and 10.20d. By opening a bypass door in the diffusor section, the excess air can be taken into the inlet and thrown away before reaching the engine. While an inlet bypass will create some additional drag, the total is reduced compared with the case in Fig. 10.20b.

(Do not confuse inlet bypass air with the engine bypass air. Inlet bypass air is dumped out of the inlet before it reaches the engine, and is therefore not a contributor to thrust. Engine bypass air is exited after being accelerated by the compressor, and does contribute to thrust.)

Another approach for returning the normal shock to the inlet lip is to move the cowl lip down, reducing the capture area as shown in Fig. 10.20d. This is complex and heavy to mechanize, and virtually impossible for an axisymmetric inlet.

It would also be possible to translate the ramp or spike fore and aft to maintain shock-on-cowl at different Mach numbers. However, spike translation is used on the SR-71, not for this, but to change the throat area.

The ratio between the airflow actually going into the inlet and the total possible airflow (i.e., the airflow of the capture area) is called the "capture area ratio," or "inlet mass flow ratio." The total mass flow actually going into the inlet is the mass flow required for the engine plus secondary airflow plus bleed airflow plus inlet bypass air, if any.

Capture-area ratio is calculated by determining the required mass flow and dividing by the mass flow through the capture area far upstream [Eq. (10.20)]. Note that capture-area ratio is generally critical for conditions in which the inlet bypass doors are closed (no bypass mass flow).

$$\frac{A_\infty}{A_C} = \frac{\dot{m}_e + \dot{m}_s + \dot{m}_{\mathrm{BL}} + \dot{m}_{\mathrm{bypass}}}{g\rho_\infty V_\infty A_C} \tag{10.20}$$

The capture-area ratio in subsonic flow can be greater than, equal to, or less than 1. In supersonic flow, it can only be equal to or less than 1.

Boundary-Layer Diverter

Any object moving through air will build up a boundary layer on its surface. In the last section, boundary-layer bleed was included in the capture-area calculation. This boundary-layer bleed was used to remove the low-energy boundary-layer air from the compression ramps, to prevent shock-induced separation.

The aircraft's forebody builds up its own boundary layer. If this low-energy, turbulent air is allowed to enter the engine, it can reduce engine performance subsonically and prevent proper inlet operation supersonically. Unless the aircraft's inlets are very near the nose (within two to four inlet diameters), some form of boundary-layer removal should be used just in front of the inlet.

The four major varieties of boundary-layer diverter are shown in Fig. 10.21. The step diverter is suitable only for subsonic aircraft, and relies upon the boundary layer itself for operation. The boundary layer consists of low-energy air, compared to the air outside of the boundary layer.

The step diverter works by forcing the boundary-layer air to either climb the step, pushing aside high-energy air outside the boundary layer, or to follow the step, pushing aside other boundary-layer air which is of lower energy. If the step diverter is properly shaped, the latter option prevails.

The step diverter should have an airfoil-like shape that is faired smoothly to the nacelle. The diverter should extend about one inlet diameter forward of the inlet, and should have a depth equal to roughly 2–4% of the forebody length ahead of the inlet.

The boundary-layer bypass duct (simply a separate inlet duct) admits the boundary-layer air and ducts it to an aft-facing hole. The internal duct shape should expand roughly 30% from intake to exit to compensate for the internal friction losses.

The suction form of boundary-layer diverter is similar. The boundary-layer air is removed by suction through holes or slots just forward of the inlet and ducted to an aft-facing hole. This type of diverter does not benefit from the ram impact of the boundary-layer air, and therefore does not work as well.

The channel diverter (Fig. 10.22) is the most common boundary-layer diverter for supersonic aircraft. It provides the best performance and the least weight in most cases. The inlet front face is located some distance away from the fuselage,

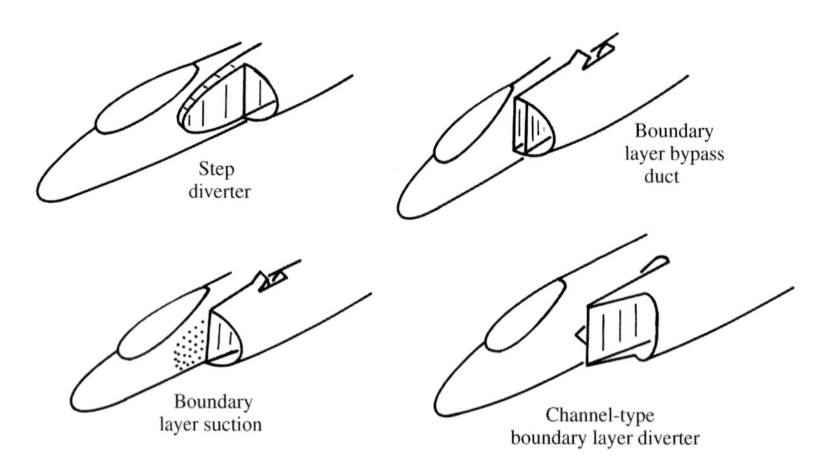

Fig. 10.21 Boundary-layer removal.

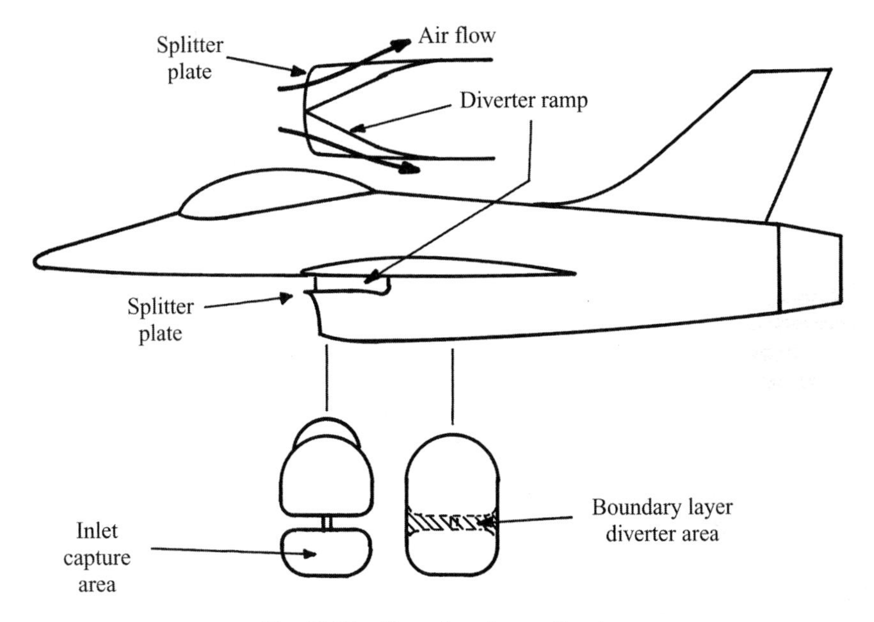

Fig. 10.22 Boundary-layer diverter.

with a "splitter plate" to ensure that the boundary-layer air does not get into the inlet. The boundary-layer air is caught between the splitter plate and the fuselage, and pushed out of the resulting channel by the diverter ramps. The diverter ramps should have an angle of no more than about 30 deg.

The required depth of a boundary-layer diverter depends on the depth of the boundary layer itself. This cannot be simply calculated. The classic boundary-layer equations assume a flat plate, which is unlike a fuselage forebody. The 3-D effects of a real forebody tend to reduce boundary-layer buildup compared to a flat plate.

A very good rule of thumb for the required thickness of a boundary-layer diverter is that it should be between 1 and 3% of the fuselage length in front of the inlet, with the larger number for fighters that go to high angle of attack.

As will be discussed in Chapter 12, the drag of a boundary-layer diverter depends upon its frontal area. During conceptual layout, the fuselage and inlet should be designed to minimize this area, shown shaded in Fig. 10.22.

Nozzle Integration

The fundamental problem in jet engine nozzle design is the mismatch in desired exit areas at different speeds, altitudes, and thrust settings. The engine can be viewed as a producer of high-pressure subsonic gases. The nozzle accelerates those gases to the desired exit speed, which is controlled by the exit area.

The nozzle must converge to accelerate the exhaust gases to a high subsonic exit speed. If the desired exit speed is supersonic, a converging-diverging nozzle is required.

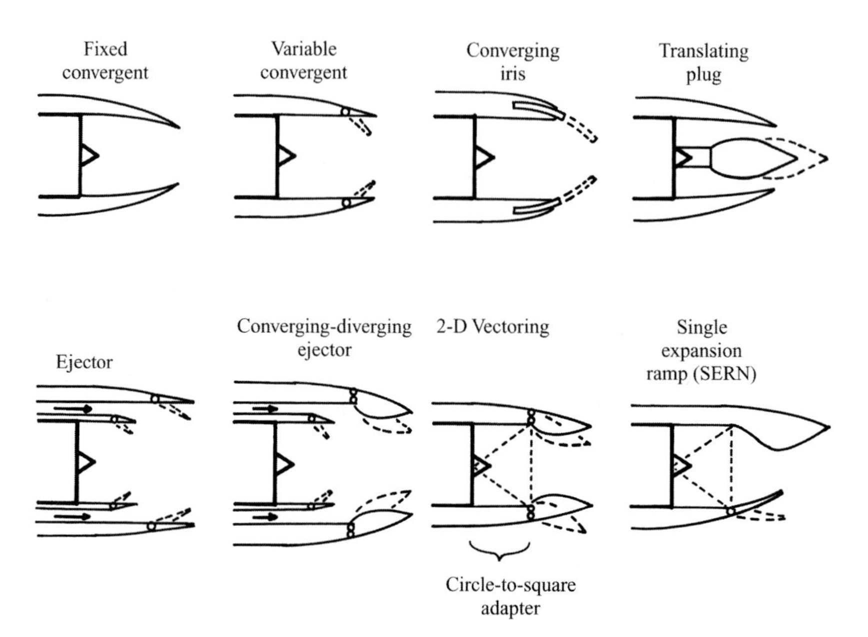

Fig. 10.23 Types of nozzles.

The exit area to obtain a desired exhaust velocity depends upon the engine mass flow (i.e., percent power). This is especially a problem with afterburning engines in which the desired exit area for supersonic afterburning operation can be three times the desired area for subsonic, part-thrust operation.

Typical nozzles are shown in Fig. 10.23. In the past, the nozzle of a jet engine was considered an integral part of the engine, to be installed on the aircraft without question or change. This is still the case for subsonic commercial aircraft, but is changing for supersonic military aircraft due to the emergence of 2-D and other advanced nozzles.

The fixed convergent nozzle is almost universally used for subsonic commercial turbojet and turbofan engines. The nozzle exit area is selected for cruise efficiency, resulting in a loss of theoretical performance at lower speeds. However, the gain in simplicity and weight reduction of the fixed nozzle more than makes up for the performance loss in most subsonic applications.

For an aircraft that occasionally flies at high-subsonic to low-supersonic speeds, a variable-area convergent nozzle allows a better match between low-speed, part-thrust operation and the maximum speed and thrust conditions. The nozzle shown has a fixed outer surface, which causes a "base" area when the nozzle inside is in the closed position.

Such a nozzle was used on many early transonic fighters, but is not typically used today. Instead, the convergent-iris nozzle is used to vary the area of a convergent nozzle without introducing a base area.

Another means to vary the exit area of a convergent nozzle is the translating plug. This was used on the engine for the Me-262, the first jet to be employed in combat in substantial numbers. The plug slides aft to decrease exit area.

The ejector nozzle takes engine bypass air that has been used to cool the afterburner and ejects it into the exhaust air, thus cooling the nozzle as well. The variable-geometry convergent-divergent ejector nozzle is used in supersonic jet aircraft. It allows varying the nozzle exit area for maximum engine performance throughout the flight envelope. The most advanced versions can also independently vary the throat area.

If an existing engine is used in the design, or if a hypothetical engine data package has been obtained from an engine company, the nozzle areas will be provided for the design flight regime. If not, the nozzle areas must be estimated because they have a substantial effect upon the calculated aircraft wave drag and boattail drag.

For initial design layout, a reasonable approximation can be made based upon the estimated capture area. For a subsonic convergent nozzle or a convergent-divergent nozzle in the closed position, the required exit area is approximately 0.5–0.7 times the capture area. For maximum supersonic afterburning operation, the required exit area is about 1.2–1.6 times the capture area.

As mentioned, nozzle arrangement can have a substantial effect on boattail drag. This is the drag caused by the separation on the outside of the nozzle and aft fuselage. To reduce boattail drag to acceptable levels, the closure angles on the aft fuselage should be kept below 15 deg, and the angles outside of the nozzle should be kept below 20 deg in the nozzle-closed position.

Jet engines mounted next to each other produce an interference effect that reduces net thrust. To minimize this, the nozzles should be separated by about one to two times their maximum exit diameter. The area between them should taper down like the back of an airfoil, terminating just before the nozzles. However, this arrangement increases weight and wetted area so many fighters have twin engines mounted right next to each other despite the increased interference.

Engine Cooling Provisions

A critical problem in the design integration of a jet engine is the heat put out by an operating engine and the need for engine bay cooling. Many aircraft such as the F-22 have their aft fuselage built mostly of titanium because the temperatures around the engines are too high for aluminum or most composite materials. Even the B-70, which was fabricated largely of high-temperature stainless steel, needed an elaborate system of cooling around the (six!) engines, as shown in Fig. 10.24. At the top of the figure is the operation at low speeds, when inlet duct bypass air, inlet boundary layer bleed air, and additional air taken in by ground cooling doors are all used for cooling. As can be seen, a cooling shroud surrounds each engine to prevent excess heat from getting to the aircraft structure.

At the bottom of Fig. 10.24, the normal operational mode up to Mach 3.0 is shown. Cooling air is taken from the inlet, just upstream of the engine, and used along with air taken from the boundary layer bleed. At all speeds, part of the cooling air is ejected through the engine nozzle, and part is allowed to exit to the rear around the engines.

During conceptual design, some allowance for engine cooling should be made based on similar aircraft. Do not "shrink-wrap" the aircraft's outside skin around the engine. You must provide room for cooling and possibly an engine shroud,

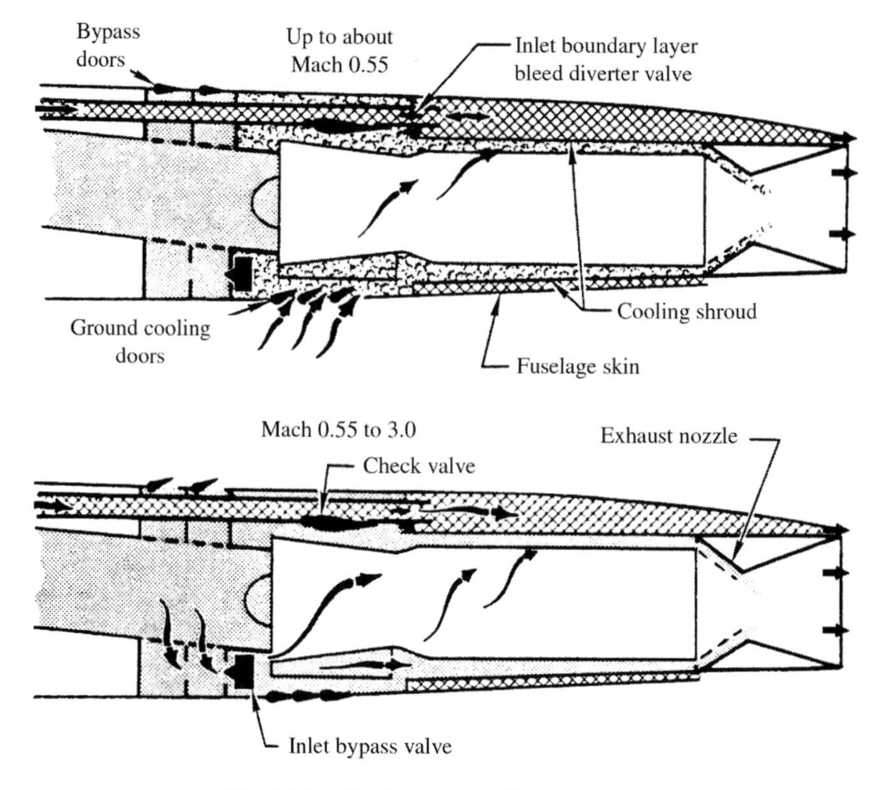

Fig. 10.24 B-70 engine cooling provisions.

along with clearance for the engine and structural depth for the fuselage or nacelle around the engine.

10.4 Propeller-Engine Integration

Propeller Sizing

The actual details of the propeller design, such as the blade shape and twist, are not required to lay out a propeller-engine aircraft. But the diameter of the propeller, the dimensions of the engine, and the required inlets and exhausts must be determined.

Generally speaking, the larger the propeller diameter, the more efficient the propeller will be. The old rule of thumb was "keep it as long as possible, as long as possible." The limitation on length is the propeller tip speed, which should be kept below sonic speed.

The tip of a propeller follows a helical path through the air. Tip speed is the vector sum of the rotational speed [Eq. (10.21)] and the aircraft's forward speed as defined in Eq. (10.22):

$$(V_{\text{tip}})_{\text{static}} = \pi n D \tag{10.21}$$

where

n = rotational rate obtained from engine data
D = diameter

$$(V_{tip})_{helical} = \sqrt{V_{tip}^2 + V^2} \tag{10.22}$$

Note: watch the units! Rotation rate is normally given as revolutions per minute (rpm) and must be converted to revolutions per second by dividing by 60.

At sea level the helical tip speed of a metal propeller should not exceed 950 fps {290 m/s}. A wooden propeller, which must be thicker, should be kept below 850 fps {260 m/s}. If noise is of concern, the upper limit for metal or wood should be about 700 fps {213 m/s} during takeoff.

Equation (10.23) provides an estimate of the propeller diameter as a function of horsepower or kilowatts. The propeller diameters obtained from these equations should be compared to the maximum diameters obtained from tip-speed considerations, and the smaller of the two values used for initial layout.

$$D = K_p \sqrt[4]{Power} \tag{10.23}$$

where

As forward velocity increases, the angle of attack seen by the blades of a fixed-pitch propeller will decrease. This limits the thrust obtained at higher speeds. If the fixed pitch is increased, the blades will tend to stall at low speeds, which reduces low-speed thrust. A fixed-pitch propeller is called a "cruise prop" or "climb prop" depending upon the flight regime the designer has decided to emphasize.

A variable-pitch propeller can be used to improve thrust across a broad speed range. A controllable-pitch propeller has its pitch directly controlled by the pilot through a lever alongside the throttle. A constant-speed propeller is automatically controlled in pitch to maintain the engine at its optimal rpm.

Most aircraft propellers have a "spinner," a cone- or bullet-shaped fairing at the hub. The inner part of the propeller contributes very little to the thrust. A spinner pushes the air out to where the propeller is more efficient. Also, a spinner streamlines the nacelle. Ideally, the spinner should cover the propeller out to about 25% of the radius, although most spinners are not that large.

To further streamline the nacelle, some aircraft designers use a "prop extension," a short shaft which locates the propeller 2–4 in. {5–10 cm} farther forward (or aft) of the engine. If the propeller is located much farther away

	British units	Metric units
No. blades	K_p	K_p
2	1.7	0.56
3	1.6	0.52
4+	1.5	0.49
Power units	hp	kW
Diameter units	ft	m

Method modified from Ref. 28, with special thanks to D. Gerren for aircraft data collection to update these equations.

from the engine, a complicated drive shaft with a separate bearing support for the propeller must be used. This type of installation was used in the P-39, which had a piston engine behind the cockpit and a drive shaft to the forward-mounted propeller. Similarly, the BD-5 had a drive shaft to a rear-mounted pusher propeller.

Propeller Location

A matrix of possible propeller locations is shown in Fig. 10.25. A tractor installation has the propeller in front of its attachment point (usually the motor). A pusher location has the propeller behind the attachment point.

The Wright Flyer was a pusher. However, the tractor location has been standard for most of the history of aviation. The conventional tractor location puts the heavy engine up front, which tends to shorten the forebody, allowing a smaller tail area and improved stability. The tractor location also provides a ready source of cooling air, and places the propeller in undisturbed air.

The pusher location has been used on a number of recent designs because of several advantages. Most important, it can reduce aircraft skin friction drag because the pusher location allows the aircraft to fly in undisturbed air. With a tractor propeller the aircraft flies in the turbulence from the propeller wake.

The fuselage-mounted pusher propeller can allow a reduction in aircraft wetted area by shortening the fuselage. The inflow caused by the propeller allows a much steeper fuselage closure angle without flow separation than

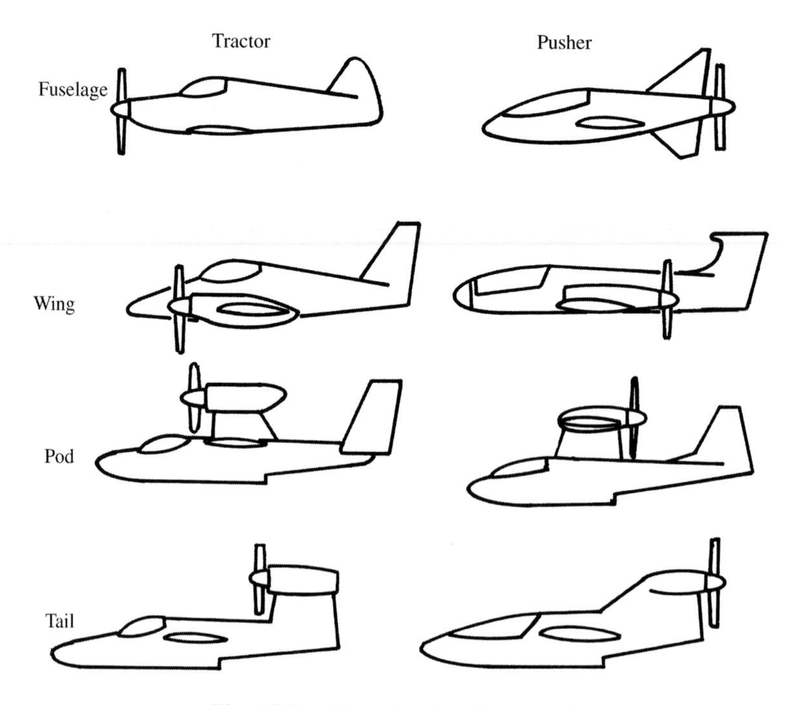

Fig. 10.25 Propeller location matrix.

otherwise possible. The canard–pusher combination is especially favorable because the canard requires a shorter tail arm than the aft tail.

The pusher propeller reduces cabin noise because the engine exhaust is pointed away from the cabin, and because the windscreen is not buffetted by propwash. Also, the pusher arrangement usually improves the pilot's outside vision, and danger from fire, smoke, and CO_2 is reduced.

However, the pusher configuration suffers several disadvantages. First, the propeller has reduced efficiency because it is forced to work with disturbed airflow off the fuselage, wing, and tails. Also, with weight to the rear the tails need to be larger.

The pusher propeller may require longer landing gear because the aft location causes the propeller to dip closer to the runway as the nose is lifted for takeoff. The propeller should have at least 9 in. {23 cm} of clearance in all attitudes.

The pusher propeller is also more likely to be damaged by rocks thrown up by the wheels. A pusher location for a turboprop propeller can create problems due to the engine exhaust impinging upon the propeller.

The Cessna Skymaster and Rutan Defiant use a combination of pusher and tractor engines on the fuselage, to eliminate engine-out yawing.

Wing mounting of the engines is normally used for multi-engine designs. Wing mounting of engines reduces wing structural weight through a span-loading effect, and reduces fuselage drag by removing the fuselage from the propeller wake.

Wing mounting of engines introduces engine-out controllability problems that force an increase in the size of the rudder and vertical tail. Also, care must be taken to ensure that the crew compartment is not located within plus or minus 5 deg of the propeller disk, in case a blade is thrown through the fuselage.

Most twin-engine aircraft are of low-wing design. For these, the location of the engine and propeller on the wing requires a longer landing gear. Frequently the propeller will be raised above the plane of the wing to reduce landing-gear height. This causes additional interference between the wing and propeller.

The wing-mounted pusher arrangement has been seen on the Beech Starship and B-36. This arrangement tends to lengthen the forebody and require a very long landing gear.

Also, the propeller is half in the wake from under the wing and half in the wake from over the wing. The pressure differences between these two wakes can cause the propeller to lose efficiency and produce vibrations. This is minimized by locating the propeller as far as possible behind the wing.

Upper fuselage pods and tail-mounted pods tend to be used only for seaplane and amphibian designs, which need a huge clearance between the water and the propeller (minimum of 18 in. {46 cm}, preferably one propeller diameter). The high thrust line can cause undesirable control characteristics in which application of power for an emergency go-around produces a nose-down pitching moment.

Engine-Size Estimation

The required power has previously been calculated. The dimensions of an engine producing this power must now be determined. In propeller aircraft design it is far more common to size the aircraft to a known, fixed-size engine as opposed to the rubber-engine aircraft sizing more common early in jet-aircraft design.

In fact, most propeller-aircraft designs are based around some production engine, probably because very few new piston or turboprop engines have been designed and certified. Most piston engines in production were designed three decades ago. The high cost of developing and certifying a new engine, and the relatively small market, prevent new engines from appearing.

However, rubber-engine trade studies can point to the optimal existing engine. Also, the use of rubber-engine trade studies for comparison of alternate technologies (such as composite vs aluminum structure) can prevent a bias in the results due to the use of a fixed engine size.

If a production engine is to be used, dimensional and installation data can be obtained from the manufacturer. If a rubber-engine is to be used, an existing engine can be scaled using the scaling equations defined in Table 10.3. Alternatively, the statistical models defined in Table 10.4 can be used to define a nominal engine. The equations in Tables 10.3 and 10.4 were developed by the author from data taken from Ref. 1.

Tables 10.3 and 10.4 include equations for four different types of propeller powerplant. The horizontally opposed piston engine sees most use today. In-line and radial engines were common up to the 1950s, but are rare today in the Western countries. In former Soviet-block countries large radial engines are still in production for agricultural, utility, and aerobatic aircraft. The radial arrangement provides better piston cooling for a high-horsepower piston engine.

While the general-aviation piston-prop aircraft will continue to rely on the old, reliable horizontally opposed engine for many years to come, there are some exciting new developments underway. Teledyne-Continental is developing an all-new 200 hp {150 kW} engine with some novel features. It is two-stroke rather than four. Two-stroke engines, typical for small motorcycles, usually offer more power but at the expense of noise and fuel consumption. However, this new engine is being designed to be the quietest production piston aircraft engine ever, and will attain fuel consumption of 0.36 lb/hr-hp {0.06 mg/W-s} lower than current four-stroke aircraft engines. And, it will burn cheaper jet fuel rather than gasoline!

Piston-Engine Installation

Piston engines have special installation requirements that can greatly affect the configuration layout. These are illustrated in Fig. 10.26.

Table 10.3 Scaling laws for piston and turboprop engines

$X_{scaled} = X_{actual} SF^{b}$; b from table values
$SF = \text{power}_{scaled}/\text{power}_{actual}$

	Piston engines			
X	Opposed	In-line	Radial	Turboprop
Weight	0.78	0.78	0.809	0.803
Length	0.424	4.24	0.310	3.730
Diameter	—[a]	—[a]	0.130	0.120

[a]Width and height vary insignificantly within ±50% horsepower.

Table 10.4 Piston and turboprop statistical models

	Piston engines							
	Opposed		In-line		Radial		Turboprop	
X	a	b	a	b	a	b	a	b
			British: X = $a(bhp)^b$ *(lb or ft)*					
Weight	5.47	0.780	5.22	0.780	4.90	0.809	1.67	0.803
Length	0.32	0.424	0.49	0.424	0.52	0.310	0.35	0.373
Diameter	Width 2.6–2.8 ft Height 1.8–2.1 ft		Width 1.4–1.6 ft Height 2–2.2 ft		1.7	0.130	0.8	0.120
Typical propeller rpm	2770		2770		2300			
Applicable bhp range	60–500		100–300		200–2000		400–5000	
			Metric: X = $a(power)^b$ *(kg or m)*					
Weight	3.12	0.780	2.98	0.780	2.82	0.809	0.96	0.803
Length	0.11	0.424	0.17	0.424	0.174	0.310	0.12	0.373
Diameter	Width 0.8–0.9 Height 0.6–0.7		Width 0.4–0.5 Height 0.6–0.7		0.54	0.130	0.25	0.120
Typical propeller rpm	2770		2770		2300			
Applicable power range, kW	45–370		75–225		150–1500		300–3728	

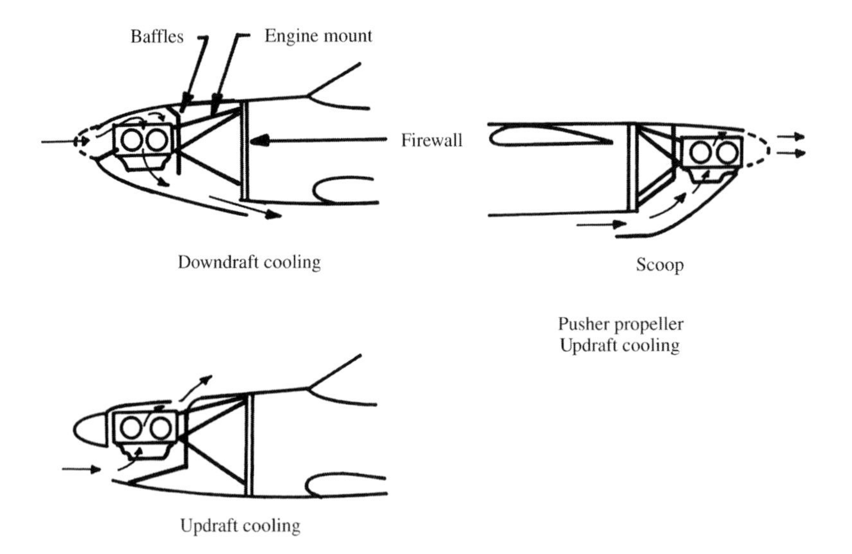

Fig. 10.26 Piston engine installation.

Cooling is a major concern. Up to 10% of the engine's horsepower can be wasted by the drag associated with taking in cooling air, passing it over the engine, and exiting it.

To minimize this cooling drag, the cooling-air mass flow should be kept as small as possible and used as efficiently as possible. Typical air-cooled engines need about 1 lb of cooling-air mass flow per second per 100 hp of the engine {~0.6 kgps per 100 kW power}. Optimization studies indicate that the best intake slows the air to 30% of the aircraft flight speed (climb speed in the worst case). This results in the following equation for piston engine cooling area sizing (Ref. 130):

$$\text{Cooling Intake Area: (fps)}\, A_{\text{cooling}} = \frac{\text{bhp}}{2.2 V_{\text{climb}}}\ (\text{ft}^2) \qquad (10.24)$$

$$\text{(mks)}\ A_{\text{cooling}} = \frac{P}{58 V_{\text{climb}}}\ (\text{m}^2) \qquad (10.25)$$

Power is in horsepower or kilowatts. V_{climb} is the climb speed in feet per second or meters per second. This is usually the critical condition for cooling.

Despite an old rule of thumb that says that the exit area should be 30% larger than the intake area, recent analysis has shown that an exit area slightly *smaller* than the intake is actually better. For preliminary layout this author suggests designing to a ratio $A_{\text{exit}}/A_{\text{inlet}}$ of 0.8 and providing adjustable cowl flaps that open to a ratio of 2 or more. Adjustable cowl flaps let us change the exit area in flight, which changes the cooling airflow. It is not necessary to vary the cooling intake area because the cooling airflow always adjusts to the exit area. If you really do not want a variable exit, try an exit that is 30% larger than the intake but make it easy to modify the exit area during the flight-test program.

For tractor engines, the cooling-air intake is usually located directly in front of the engine cylinders. The air is diverted over the top of the engine by "baffles," which are flat sheets of metal that direct the airflow within the engine compartment. The air then flows down through and around the cylinders into the area beneath the engine, and then exits through an aft-facing hole below the fuselage. This is referred to as "down-draft" cooling.

Down-draft cooling exits the air beneath the fuselage, which is a high-pressure area and therefore a poor place to exit air. "Up-draft" cooling flows the cooling air upward through the cylinders and exits it into low-pressure air above the fuselage, creating more efficient cooling flow due to a suction effect.

However, updraft cooling dumps hot air in front of the windscreen; this can heat up the cabin. An engine oil leak can coat the windscreen with black oil. Aircraft engines have the exhaust pipes below the cylinders, so updraft cooling causes the cooling air to be heated by the exhaust pipes before reaching the cylinders.

For pusher engines, cooling is much more difficult. On the ground a front-mounted propeller blows air into the cooling intakes. This is not the case for a pusher engine. Also, the cooling-air intakes for a pusher engine are at the rear of the fuselage where the boundary layer is thick and slow-moving. For these reasons virtually all piston-pushers use updraft cooling with a large scoop mounted below the fuselage. Also, internal fans are sometimes used to improve cooling on pusher configurations.

Figure 10.26 also shows the motor mount and firewall. The motor mount—usually fabricated from welded steel tubing—transfers the engine loads to the corners of the fuselage or the longerons. Typically the motor mount extends the engine forward of the firewall by about half the length of the engine. This extra space is used for location of the battery and nosewheel steering linkages.

The firewall is typically a 0.015-in. {0.4-mm steel sheet (stainless or galvanized) attached to the first structural bulkhead of the fuselage or nacelle. Its purpose is to prevent a fire in the engine compartment from damaging the aircraft structure or spreading into the rest of the aircraft.

The firewall should not be broken with cutouts (such as for a retractable nosewheel). All controls, hoses, and wires that pass through the firewall have to be sealed with fireproof fittings.

Piston-engine installation is covered in depth in Ref. 29.

10.5 Fuel System

An aircraft fuel system includes the fuel tanks, fuel lines, fuel pumps, vents, and fuel-management controls. Usually the tanks themselves are the only components that affect the overall aircraft layout, although the winglets on the round-the-world Rutan Voyager were added solely to raise the fuel vents above the wing tanks when the wing tips bent down to the runway on takeoff.

There are three types of fuel tanks: discrete, bladder, and integral. Discrete tanks are fuel containers that are separately fabricated and mounted in the aircraft by bolts or straps. Discrete tanks are normally used only for small general-aviation and homebuilt aircraft. Discrete tanks are usually shaped like the front of an airfoil and placed at the inboard wing leading edge, or are placed in the fuselage directly behind the engine and above the pilot's feet.

Bladder tanks are made by stuffing a shaped rubber bag into a cavity in the structure. The rubber bag is thick, causing the loss of about 10% of the available fuel volume. However, bladders are widely used because they can be made self-sealing. If a bullet passes through a self-sealing tank, the rubber will fill in the hole preventing a large fuel loss and fire hazard. This offers a major improvement in aircraft survivability as approximately a third of combat losses are attributed to hits in the fuel tanks.

Integral tanks are cavities within the airframe structure that are sealed to form a fuel tank. Ideally, an integral tank would be created simply by sealing existing structure such as wing boxes and cavities created between two fuselage bulkheads.

Despite years of research, integral tanks are still prone to leaks as witnessed upon the introduction of the B-1B into service. Because of the fire hazard in the event of a leak or battle damage, integral tanks should not be used near personnel compartments, inlet ducts, gun bays, or engines.

The fire hazard of an integral tank can be reduced by filling the tank with a porous foam material, but some fuel volume is lost. Approximately 2.5% of the fuel volume is displaced by the foam. In addition, another 2.5% of the volume is lost because the foam tends to absorb fuel. This increases the unusable fuel weight. Furthermore, the foam itself weighs roughly 1.3 lb per cubic ft {21 kg/m^3}.

The required volume of the fuel tanks is based upon the total required fuel, as calculated during the mission sizing. Densities for various fuels are provided in Table 10.5. The lower values represent hot-day densities. The higher densities are the 0°F values. The actual fuel volume required can be calculated using the fact that 7.5 gallon occupy 1 ft^3 (or 1000 liters equals 1 m^3).

If fuel tanks of simple geometry are used, the tank volume can be calculated directly. Wing-box fuel volume can be approximated by assuming a tapered box shape. For complex integral and bladder tanks, the tank volume is determined using a fuel-volume plot as shown in Fig. 10.27. This is constructed by measuring the cross-sectional area of the tanks at various fuselage locations, then plotting those cross-sectional areas on a volume plot similar to the aircraft volume plot previously discussed.

If a discrete tank is used, the actually available internal volume can be calculated by subtracting the wall thickness from the external dimensions. For integral and bladder tanks, the available tank volume must be reduced from the measured value to allow for wall thickness, internal structure, and bladder thickness.

A rule of thumb is to assume that 85% of the volume measured to the external skin surface is usable for integral wing tanks and 92% is usable for integral

Table 10.5 Fuel densities (lb/gal or kg/liter)

	Average actual density		
	0°F	100°F	Mil-spec density
Aviation gasoline	6.1 {0.73}	5.7 {0.68}	6.0 {0.72}
JP-4	6.7 {0.80}	6.4 {0.77}	6.5 {0.78}
JP-5	7.2 {0.86}	6.8 {0.82}	6.8 {0.82}
JP-8/JETA1	—	—	6.7 {0.80}

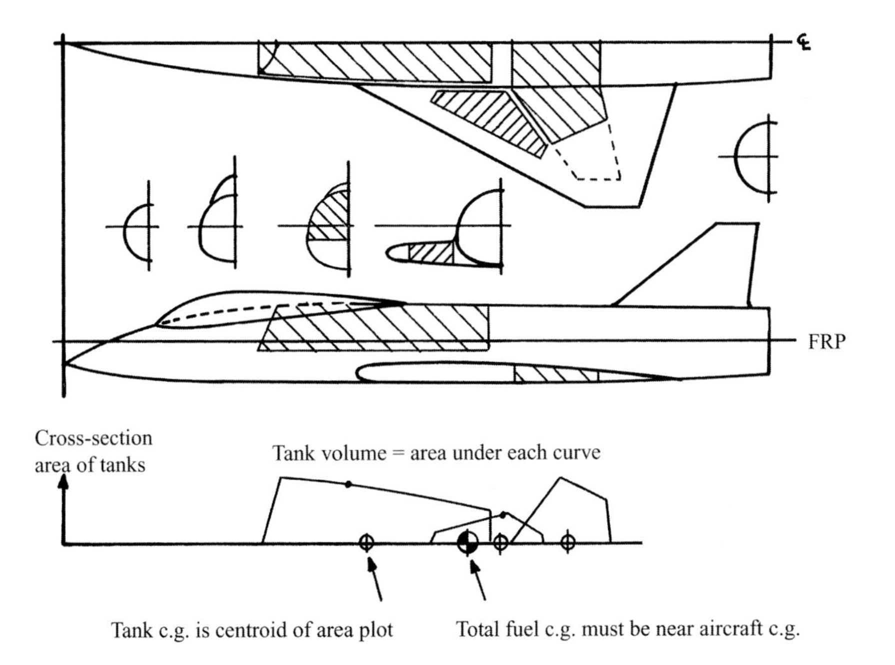

Fig. 10.27 Fuel tank volume plotting.

fuselage tanks. If bladder tanks are used, the values become 77% for wing tanks and 83% for fuselage tanks.

Note in Fig. 10.27 that the fuel volume plot allows the estimation of the center of gravity (c.g.) for each fuel tank, which is the centroid of the area plotted for the tank. The total fuel c.g. is simply the weighted average of the individual tank c.g., and should be close to the aircraft c.g.

Another consideration for fuel system design is the fact that, as shown in Table 10.5, the fuel expands substantially from cold to warm. Fuel is often pumped into the aircraft from below-ground, hence cold storage tanks. On a hot day it will expand substantially as it warms up. It may be desirable to provide perhaps 3–5% extra fuel volume to allow for this. On the F-18 and several airliners, there are "expansion" tanks in the vertical tail into which the fuel can flow as it warms and expands.

Fuel tanks can also be used to aerodynamically optimize the aircraft. As will be discussed in Chapter 12, the (usually) downward tail lift force required to trim the aircraft causes a "trim drag," which is greater when the c.g. is more toward the front of the aircraft. This is especially a problem in supersonic flight when the wing center of lift moves toward the rear, requiring even more tail-down trim load. To minimize this trim drag, aircraft such as the Concorde SST and the B-70 pump fuel toward the rear when cruising altitude and speed are reached. Many commercial aircraft have "trim tanks" in the horizontal tail for this purpose. Figure 10.28 shows the B-70 fuel system, which is most of the internal volume of the aircraft.

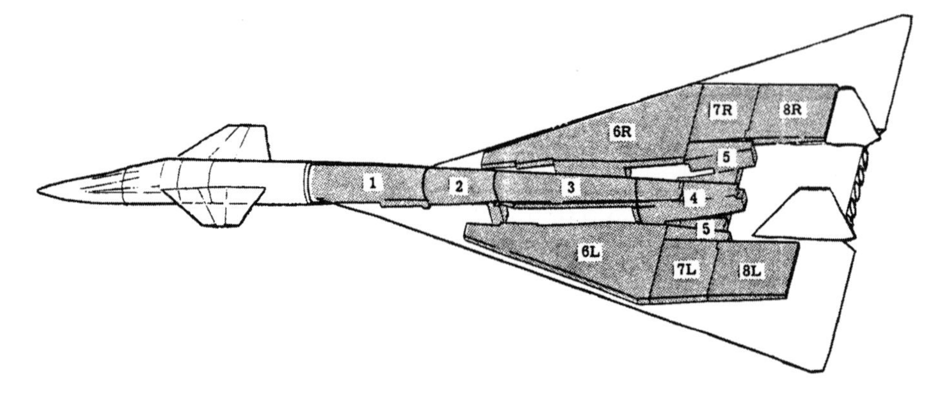

Fig. 10.28 B-70 fuel system.

Even subsonic commercial aircraft can benefit from keeping the c.g. as far to the rear as stability and safety will allows. The MD-11, a-growth version of the DC-10, has a fuel tank in the horizontal tail into which fuel is pumped during cruise to keep the c.g. at the aft-most limit.

One final aspect of fuel system design, for military aircraft, is the provision of in-flight refueling capability. There are two options. The U.S. Air Force uses a "boom" system, while the rest of the world uses a "probe-and-drogue" system. In the boom system, special tanker aircraft are equipped with a fueling boom positioned at the bottom rear of the tanker aircraft. This is "flown" by a boom operator and extended into a receptacle on the top of the aircraft needing fuel (which merely holds position under the aircraft). The refueling receptacle must be mounted somewhere fairly near the centerline of the aircraft, toward the front, but should not be directly in front of the pilot due to the fuel that is always spilled during disconnect. Boom operation can be seen in the opening sequence of the classic movie, *Dr. Strangelove.*

In the probe-and-drogue system, the tanker aircraft extends a drogue (parachute-like device) with a parachute-like "basket," with a plug-in receptacle in the middle. The receiving aircraft has a probe, basically a pipe extending forward, which must be flown into the basket. The probe can be fixed to the outside of the aircraft, but this will add a large drag penalty. Instead, most aircraft have a retracting probe. The probe must be easily visible to the pilot to facilitate "hitting" the basket, and is usually located on the right side of the aircraft just forward of the canopy.

Boom systems allow higher fuel flow rates and are more forgiving of pilot error and fatigue, but require a fleet of dedicated and expensive tanker aircraft. Also, probe-and-drogue systems can be installed in pods that look like external fuel tanks and can be bolted onto different aircraft. This even allows "buddy" tanking, where, say, two F-18s take off and fly halfway to the target where one aircraft gives the other most of the remaining fuel then flies home, allowing the other aircraft to strike at a much greater range than otherwise possible.

11
Landing Gear And Subsystems

11.1 Introduction

Of all the many internal components that must be defined in an aircraft configuration design layout, the landing gear will usually cause the most trouble. The tires and shock strut must be just the right size, and if the plane gets heavier than planned, they must get larger. The wheels must be placed in the correct down locations for takeoff and landing or, believe it or not, the airplane may crash. If retractable, the landing gear must somehow fold into the aircraft without chopping up the structure, obliterating the fuel tanks, or bulging out into the slipstream.

This chapter covers landing gear design, as typified by Fig. 11.1, as well as installation of other major aircraft subsystems including hydraulics, electrical, and pneumatic. Also, definition and installation of aircraft avionics is described.

11.2 Landing Gear Arrangements

The common options for landing gear arrangement are shown in Fig. 11.2. The single main gear is used for many sailplanes because of its simplicity. The wheel can be forward of the center of gravity (c.g.), as shown here, or can be aft of the c.g. with a skid under the cockpit.

"Bicycle" gear has two main wheels, fore and aft of the c.g., with small "outrigger" wheels on the wings to prevent the aircraft from tipping sideways. The bicycle landing gear has the aft wheel so far behind the c.g. that the aircraft must takeoff and land in a flat attitude, which limits this type of gear to aircraft with high lift at low angles of attack (i.e., high-aspect-ratio wings with large camber and/or flaps). Bicycle gear has been used mainly on aircraft with narrow fuselage and wide wing span such as the B-47.

The "taildragger" landing gear has two main wheels forward of the c.g. and an auxiliary wheel at the tail. Taildragger gear is also called conventional landing gear, because it was the most widely used arrangement during the first 40 years of aviation. Taildragger gear provides more propeller clearance, has less drag and weight, and allows the wing to generate more lift for rough-field operation than does tricycle gear.

However, taildragger landing gear is inherently unstable. If the aircraft starts to turn, the location of the c.g. behind the main gear causes the turn to get tighter until a "ground loop" is encountered, and the aircraft either drags a wing tip,

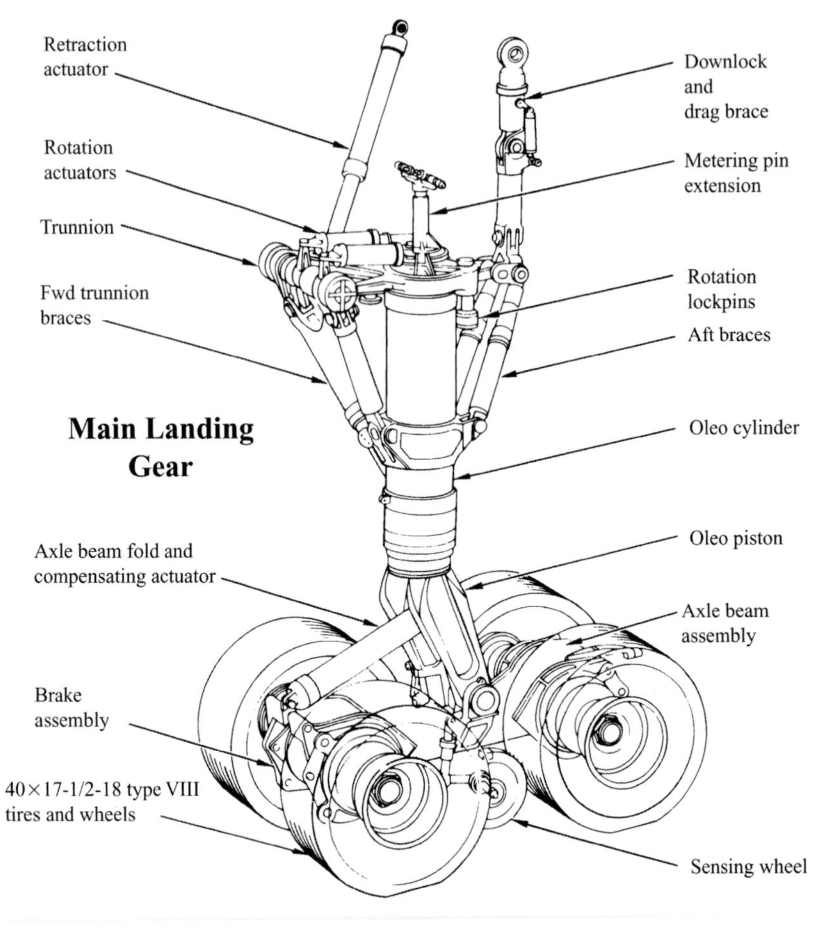

Fig. 11.1 Typical multiwheel main landing gear.

collapses the landing gear, or runs off the side of the runway. To prevent this, the pilot of a taildragger aircraft must align the aircraft almost perfectly with the runway at touchdown, and "dance" on the rudder pedals until the aircraft stops.

The most commonly used arrangement today is the "tricycle" gear, with two main wheels aft of the c.g. and an auxiliary wheel forward of the c.g. With a tricycle landing gear, the c.g. is ahead of the main wheels so the aircraft is stable on the ground and can be landed at a fairly large "crab" angle (i.e., nose not aligned with the runway). Also, tricycle landing gear improves forward visibility on the ground and permits a flat cabin floor for passenger and cargo loading.

Quadricycle gear is much like bicycle gear but with wheels at the sides of the fuselage. Quadricycle gear also requires a flat takeoff and landing attitude. It is used on the B-52 and several cargo planes where it has the advantage of permitting a cargo floor very low to the ground.

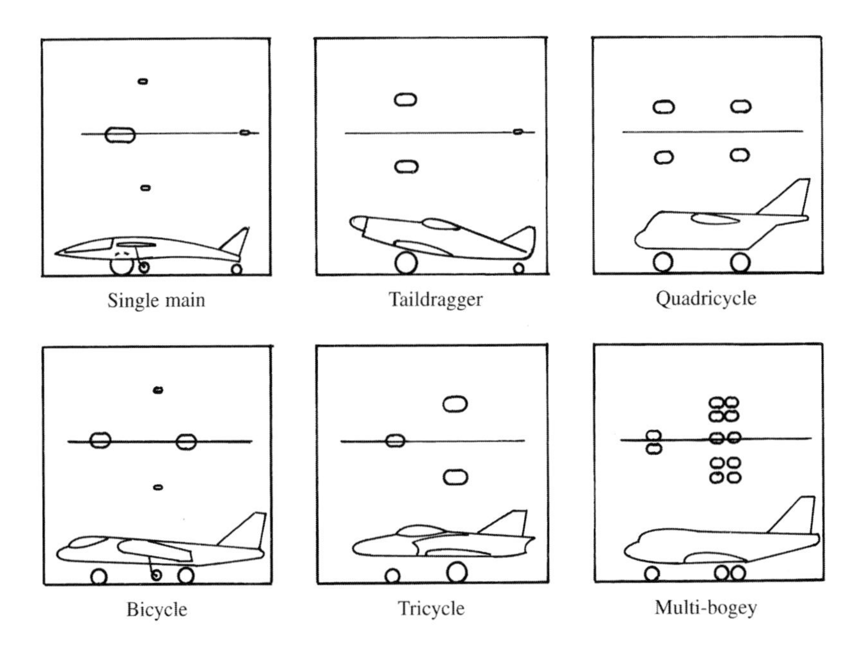

Fig. 11.2 Landing gear arrangements.

The gear arrangements just described are also seen with two, four, or more wheels in place of the single wheels shown in Fig. 11.2. As aircraft weights become larger, the required wheel size for a single wheel capable of holding the aircraft's weight becomes too large. Then multiple wheels are used to share the load between reasonably sized tires.

Also, it is very common to use twin nose-wheels to retain some control in the event of a nosewheel flat tire. Similarly, multiple main wheels (i.e., total of four or more) are desirable for safety. When multiple wheels are used in tandem, they are attached to a structural element called a "bogey," or "truck," or "axle beam" that is attached to the end of the shock-absorber strut (see Fig. 11.1).

Typically an aircraft weighing under about 50,000 lb {22,680 kg} will use a single main wheel per strut, although for safety in the event of a flat tire it is always better to use two wheels per strut. Between 50,000 and 150,000 lb {22,680–68,040 kg}, two wheels per strut are typical. Two wheels per strut are sometimes used for aircraft weighing up to about 250,000 lb {113,400 kg}.

Between aircraft weights of about 200,000 and 400,000 lb {90,720–181,440 kg} the four-wheel bogey is usually employed; for aircraft over 400,000 lb {181,440 kg} four bogeys, each with four or six wheels, spread the total aircraft load across the runway pavement.

Except for light aircraft and a few fighters, most aircraft use twin nosewheels to retain control in the event of a flat nose tire. Carrier-based aircraft must use twin nosewheels at least 19 in. {483 cm} in diameter to straddle the

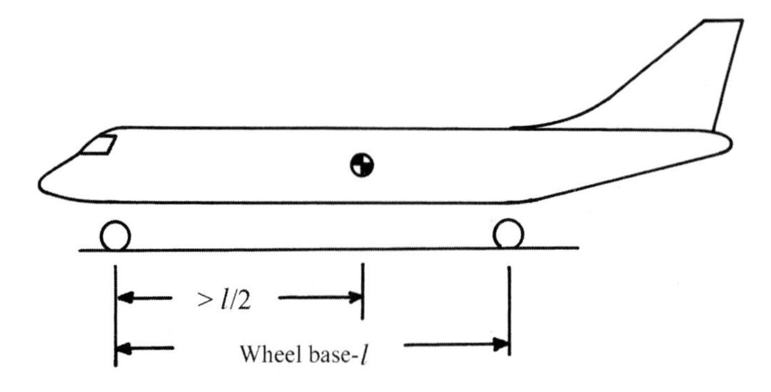

Fig. 11.3 Bicycle landing gear.

catapult-launching mechanism. The massive C-5 employs four nosewheels to spread the tire load, permitting operation off of relatively soft fields.

Guidelines for layout of a bicycle landing gear are shown in Fig. 11.3. The c.g. should be aft of the midpoint between the two wheels.

The requirements for taildragger gear are shown in Fig. 11.4. The tail-down angle should be about 10–15 deg with the gear in the static position (i.e., tires and shock absorbers compressed the amount seen when the aircraft is stationary on the ground at takeoff gross weight).

There must be 9-in. {23-cm} propeller ground clearance in the takeoff attitude.

The c.g. (most forward and most aft) should fall between 16–25 deg back from vertical measured from the main wheel location. If the c.g. is too far forward the aircraft will tend to nose over, and if it is too far back it will tend to groundloop.

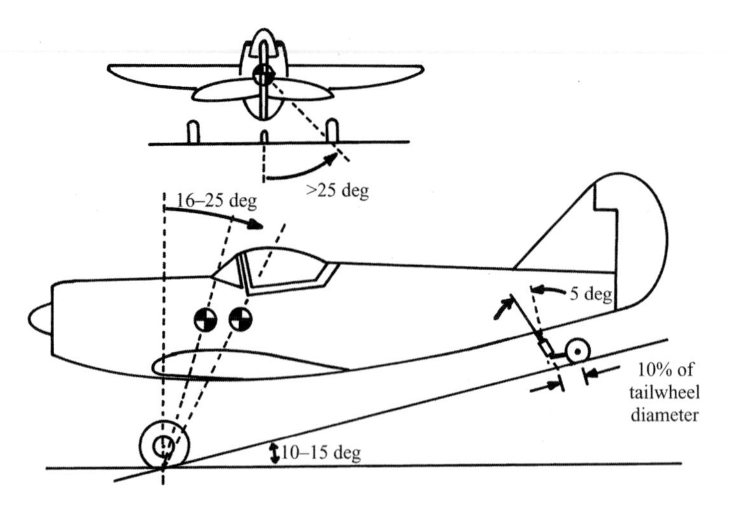

Fig. 11.4 Taildragger landing gear.

To prevent the aircraft from overturning the main wheels should be laterally separated beyond a 25-deg angle off the c.g., as measured from the rear in a tail-down attitude.

The layout of tricycle landing gear as shown in Fig. 11.5 is even more complex. The length of the landing gear must be set so that the tail does not hit the ground on landing. This is measured from the wheel in the static position assuming an aircraft angle of attack for landing that gives 90% of the maximum lift. This ranges from about 10–15 deg for most types of aircraft.

There must be 7-in. {18-cm} propeller ground clearance.

The "tipback angle" is the maximum aircraft nose-up attitude with the tail touching the ground and the strut fully extended. To prevent the aircraft from tipping back on its tail, the angle off the vertical from the main wheel position to the c.g. should be greater than the tipback angle or 15 deg, whichever is larger.

For carrier-based aircraft this angle frequently exceeds 25 deg, implying that the c.g. for carrier-based aircraft is well forward of the main wheels. This ensures that the rolling of the deck will not cause an aircraft to tip back on its tail.

However, a tipback angle much over 25 deg also makes it difficult to lift the nose for a runway takeoff, and can lead to catastrophic "porpoising" on takeoff. If the nosewheel is carrying over 20% of the aircraft's weight, the main gear is probably too far aft relative to the c.g.

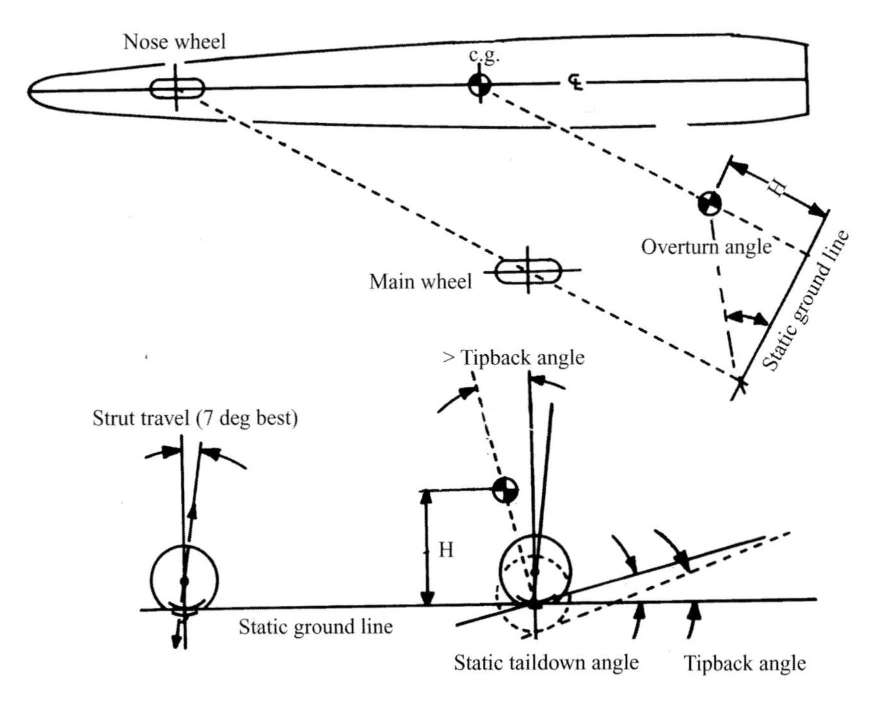

Fig. 11.5 Tricycle landing gear geometry.

On the other hand, if the nosewheel is carrying less than 5% of the aircraft's weight, there will not be enough nose-wheel traction to steer the aircraft. The optimum range for the percentage of the aircraft's weight that is carried by the nosewheel is about 8–15%, for the most-aft and most-forward c.g. positions.

The "overturn angle" is a measure of the aircraft's tendency to overturn when taxied around a sharp corner. This is measured as the angle from the c.g. to the main wheel, seen from the rear at a location where the main wheel is aligned with the nosewheel. For most aircraft this angle should be no greater than 63 deg (54 deg for carrier-based aircraft).

Figure 11.5 also shows the desired strut-travel angle as about 7 deg. This optimal angle allows the tire to move upward and backward when a large bump is encountered, thus tending to smooth out the ride. However, any strut-travel angle from purely vertical to about 10-deg aft of vertical is acceptable. Strut geometry in which the tire must move forward as it moves up is undesirable.

11.3 Tire Sizing

Strictly speaking, the "wheel" is the circular metal object upon which the rubber "tire" is mounted. The "brake" inside the wheel slows the aircraft by increasing the rolling friction. However, the term "wheel" is frequently used to mean the entire wheel/brake/tire assembly.

The tires are sized to carry the weight of the aircraft. Typically the main tires carry about 90% of the total aircraft weight. Nose tires carry only about 10% of the static load but experience higher dynamic loads during landing.

For early conceptual design, the engineer can copy the tire sizes of a similar design or use a statistical approach. Table 11.1 provides equations developed from data in Ref. 1 for rapidly estimating main tire sizes (assuming that the main tires carry about 90% of the aircraft weight).

Table 11.1 Statistical tire sizing

	Diameter		Width	
	A	B	A	B
British units: Main wheels diameter or width (in.) $= A\,W_W^B$				
General aviation	1.51	0.349	0.7150	0.312
Business twin	2.69	0.251	1.170	0.216
Transport/bomber	1.63	0.315	0.1043	0.480
Jet fighter/trainer	1.59	0.302	0.0980	0.467
Metric units: Main wheels diameter or width (cm) $= A\,W_W^B$				
General aviation	5.1	0.349	2.3	0.312
Business twin	8.3	0.251	3.5	0.216
Transport/bomber	5.3	0.315	0.39	0.480
Jet fighter/trainer	5.1	0.302	0.36	0.467

W_W = Weight on wheel.

These calculated values for diameter and width should be increased about 30% if the aircraft is to operate from rough unpaved runways.

Nose tires can be assumed to be about 60–100% the size of the main tires. The front tires of a bicycle or quadricycle-gear aircraft are usually the same size as the main tires. Taildragger aft tires are about a quarter to a third the size of the main tires.

For a finished design layout, the actual tires to be used must be selected from a manufacturer's catalog. This selection is usually based upon the smallest tire rated to carry the calculated static and dynamic loads.

Calculation of the static loads on the tires is illustrated in Fig. 11.6 and Eqs. (11.1–11.3). The additional dynamic load on the nose tires under a 10 fps braking deceleration is given in Eq. (11.4). Note that these loads are divided by the total number of main or nose tires to get the load per tire (wheel) "W_w," which is used for tire selection.

$$(\text{Max Static Load}) = W\frac{N_a}{B} \tag{11.1}$$

$$(\text{Max Static Load})_{\text{nose}} = W\frac{M_f}{B} \tag{11.2}$$

$$(\text{Min Static Load})_{\text{nose}} = W\frac{M_a}{B} \tag{11.3}$$

$$(\text{Dynamic Braking Load})_{\text{nose}} = \frac{10HW}{gB} \tag{11.4}$$

Equation (11.4) assumes a braking coefficient (μ) of 0.3, which is typical for hard runways. This results in a deceleration of 10 ft/s^2 {3 m/s^2}.

To ensure that the nose gear is not carrying too much or too little of the load, the parameter (M_a/B) should be greater than 0.05, and the parameter (M_f/B) should be less than 0.20 (0.08 and 0.15 preferred).

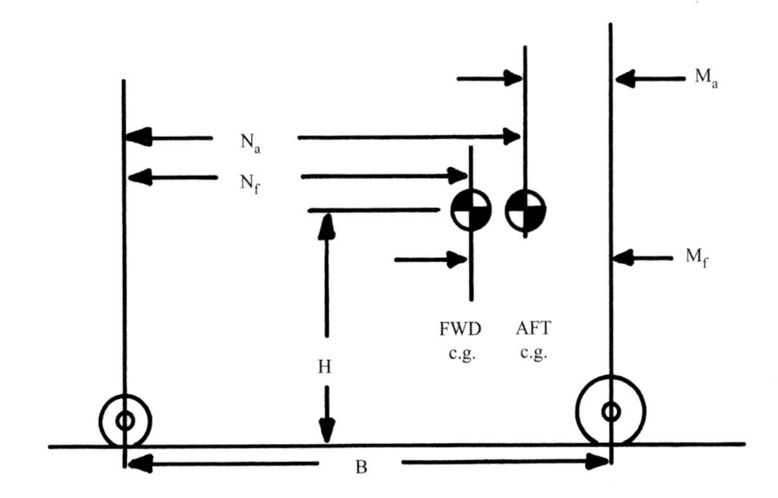

Fig. 11.6 Wheel load geometry.

If an airplane is to be operated under FAR 25 provisions, a 7% margin should be added to all calculated wheel loads. Also, it is common to add an additional 25% to the loads to allow for later growth of the aircraft design.

Table 11.2 summarizes design data for typical tires (Ref. 30), including maximum load ratings, inflation pressures at that load, maximum landing speed, tire width, diameter, and "rolling radius" (i.e., radius when under load, typically two-thirds of tire radius). For a complete listing of available tires, a "tire book" can be obtained from the tire manufacturers.

The data in Table 11.2 are organized by tire type. A Type III tire, used for most piston-engined aircraft, has a wide tread and low internal pressure. The identifying numbers for a Type III tire, such as 8.50–10, refer to the approximate tire width (8.2–8.7 in.) and wheel rim diameter (10 in.). The tire outside diameter must be obtained from a tire book such as Ref. 30.

Table 11.2 Tire data

Size	Speed, mph	Max load, lb	Infl, psi	Max width, in.	Max diam, in.	Rolling radius	Wheel diam	Number of plies
				Type III				
5.00-4	120	1,200	55	5.05	13.25	5.2	4.0	6
5.00-4	120	2,200	95	5.05	13.25	5.2	4.0	12
7.00-8	120	2,400	46	7.30	20.85	8.3	8.0	6
8.50-10	120	3,250	41	9.05	26.30	10.4	10.0	6
8.50-10	120	4,400	55	8.70	25.65	10.2	10.0	8
9.50-16	160	9,250	90	9.70	33.35	13.9	16.0	10
12.50-16	160	12,800	75	12.75	38.45	15.6	16.0	12
20.00-20	174kt	46,500	125	20.10	56.00	22.1	20.0	26
				Type VII				
16 × 4.4	210	1,100	55	4.45	16.00	6.9	8.0	4
18 × 4.4	174kt	2,100	100	4.45	17.90	7.9	10.0	6
18 × 4.4	217kt	4,350	225	4.45	17.90	7.9	10.0	12
24 × 5.5	174kt	11,500	355	5.75	24.15	10.6	14.0	16
30 × 7.7	230	16,500	270	7.85	29.40	12.7	16.0	18
36 × 11	217kt	26,000	235	11.50	35.10	14.7	16.0	24
40 × 14	174kt	33,500	200	14.00	39.80	16.5	16.0	28
46 × 16	225	48,000	245	16.00	45.25	19.0	20.0	32
50 × 18	225	41,770	155	17.50	49.50	20.4	20.0	26
				Three-Part Name				
18 × 4.25-10	210	2,300	100	4.70	18.25	7.9	10.0	6
21 × 7.25-10	210	5,150	135	7.20	21.25	9.0	10.0	10
28 × 9.00-12	156kt	16,650	235	8.85	27.60	11.6	12.0	22
37 × 14.0-14	225	25,000	160	14.0	37.0	15.1	14.0	24
47 × 18-18	195kt	43,700	175	17.9	46.9	19.2	18.0	30
52 × 20.5-23	235	63,700	195	20.5	52.0	21.3	23.0	30

Type VII tires, used by most jet aircraft, operate under higher internal pressures, which reduces their size. Also, the Type VII tires are designed for higher landing speeds. They are identified by their approximate external dimensions. For example, an 18 3 × 5.7 tire has an outside diameter of 17.25–17.8 in. and a width of 5.25–5.6 in. These actual numbers are obtained from a tire book.

For a while, the newest and highest-pressure tires were called Type VIII, but are now simply referred to as "new design" or "three-part name" tires. Designed for specific requirements, they are identified by outside diameter, width, and rim diameter. For example, a 36 × 10.00-18 tire has diameter ranging from 35.75–36.6 in., width from 9.75–10.3 in., and rim diameter of 18 in.

Tires are selected by finding the smallest tire that will carry the calculated maximum loads (W_w). For the nose tire the total dynamic load must be carried as well as the maximum static load.

However, a tire is permitted to carry more dynamic load than the rated static value found in Table 11.2 or a tire book. A Type III tire is permitted a dynamic load of 1.4 times the static value. A Type VII or a new design tire is allowed to carry 1.3 times the static value.

The total dynamic nosewheel load (static plus dynamic) should therefore be divided by 1.4 or 1.3, and then used to pick the nose tire size for dynamic load. Both static and dynamic loadings should be used to determine a minimum size of nosewheel tire. Then the larger of the two should be selected.

As a tire ages, it loses ability to withstand its own internal pressure. This causes it to swell in size by about 2 or 3% in diameter and 4% in width. This swelling should be allowed for in designing the wheel wells and retraction geometry, or the wheel pants for a fixed-gear aircraft.

A tire supports a load almost entirely by its internal pressure. The load-carrying ability of the sidewalls and tread can be ignored. The weight carried by the tire (W_w) is simply the inflation pressure (P) times the tire's contact area with the pavement (A_p, also called footprint area), as shown in Fig. 11.7 and defined in Eq. (11.5).

$$W_w = PA_p \tag{11.5}$$

$$A_p = 2.3\sqrt{wd}\left(\frac{d}{2} - R_r\right) \quad (\text{in.}^2 \text{ or cm}^2) \tag{11.6}$$

Equation (11.6), from Ref. 31, relates the pavement contact area (A_p) to the tire width (w), diameter (d), and rolling radius (R_r). [Note: do not confuse tire width (w) with weight on the wheel (W_w)].

Usually a tire is kept at about the same rolling radius as given in the tire book even when being used for a lower-than-maximum load. From Eq. (11.5), the internal pressure must therefore be proportionally reduced when a tire is operated at a lower than maximum load.

Operating a tire at a lower internal pressure will greatly improve tire life. Roughly speaking, operating a tire at half of its maximum rated load (hence pressure) will increase the number of landings obtained from the tire by a factor of six. However, this requires a larger tire causing greater drag, weight, and wheel-well size.

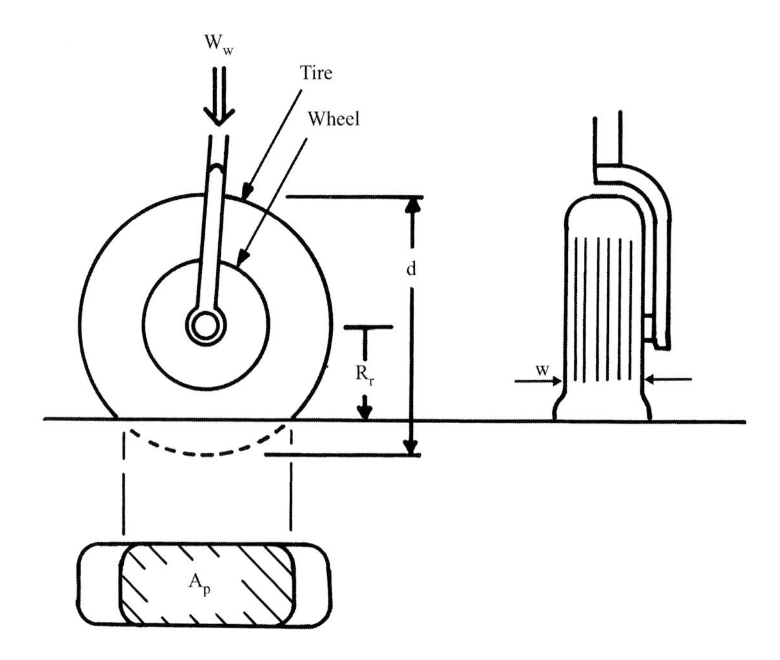

Fig. 11.7 Tire contact area.

Tire pressure should also be reduced (i.e., larger tires used) if the aircraft will operate from soft or rough runways. Actual determination of the tire size for a particular soft-field landing requirement is very complex (see Ref. 32). As a rough estimate, tires should be sized to keep internal pressures below the values in Table 11.3 (Ref. 33) for the desired application.

Sometimes the diameter of the tires is set by the braking requirements. Aircraft brakes are similar to automobile disk brakes and are usually placed inside the

Table 11.3 Recommended tire pressures

Surface	Maximum pressure	
	psi	{kPa}
Aircraft carrier	200+	1380+
Major military airfield	200	1380
Major civil airfield	120	828
Tarmac runway, good foundation	70–90	480–620
Tarmac runway, poor foundation	50–70	345–480
Temporary metal runway	50–70	345–480
Dry grass on hard soil	45–60	310–415
Wet grass on soft soil	30–45	210–310
Hard packed sand	40–60	275–415
Soft sand	25–35	170–240

wheels in all but the smallest aircraft. The wheel typically has a rim diameter of about half the total diameter of the tire mounted on it. Wheel rim diameters are provided in Table 11.2 or a tire book.

The brakes must absorb the kinetic energy of the aircraft at touchdown, less the energy absorbed by aerodynamic drag and thrust reversing. These can be ignored by assuming that the brakes are applied when the aircraft has slowed to stall speed. This yields Eq. (11.7), which must be divided by the number of wheels with brakes to get the kinetic energy that must be absorbed by each brake. Note that while Western design practice puts brakes only on the main wheels, several former Soviet block designs also put brakes on the nose wheels.

$$KE_{\text{braking}} = \frac{1}{2}\frac{W_{\text{landing}}}{g}V_{\text{stall}}^2 \tag{11.7}$$

The landing weight in Eq. (11.7) is not the same as the weight at the end of the design mission. To allow an emergency landing shortly after takeoff, the landing weight should be approximated as 80–100% of the takeoff weight.

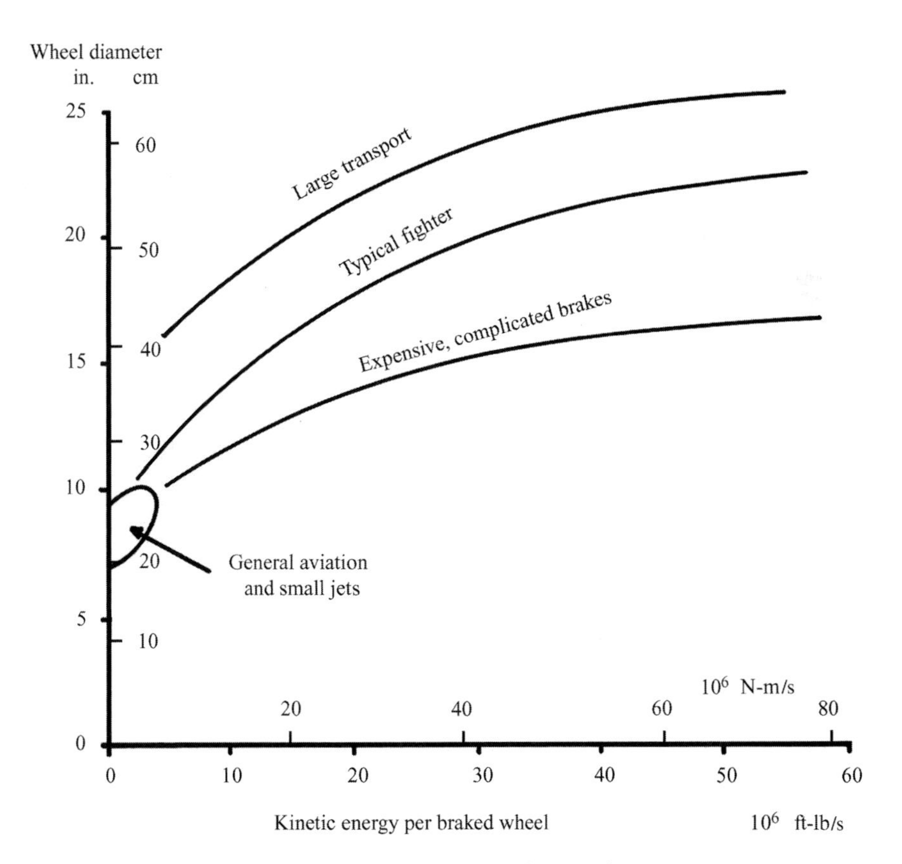

Fig. 11.8 Wheel diameter for braking.

A brake absorbs kinetic energy by turning it into heat. There is insufficient time during landing for much heat energy to be radiated to the air, so it must be absorbed in the mass of the brake material. The amount of heat a brake can tolerate depends upon its size.

Figure 11.8 provides a statistical estimate of the required wheel rim diameter to provide a brake that can absorb a given amount of kinetic energy. Note that if an aircraft is initially designed with inadequate brakes, additional braking ability can be obtained in the same-sized wheels but only at a much higher cost by using exotic materials and complex design.

If the wheel rim diameter as estimated from Fig. 11.8 is larger than the rim diameter of the selected tire, a larger tire must be used. Alternatively, a brake that protrudes laterally from the tire can be used but will require a larger wheel well.

11.4 Shock Absorbers

Shock-Absorber Types

The landing gear must absorb the shock of a bad landing and smooth out the ride when taxiing. The more common forms of shock absorber are shown in Fig. 11.9.

The tires themselves provide some shock-absorbing ability by deflecting when a bump is encountered. Sailplanes and a few homebuilt aircraft have been built with rigid axles, relying solely upon the tires for shock absorbing.

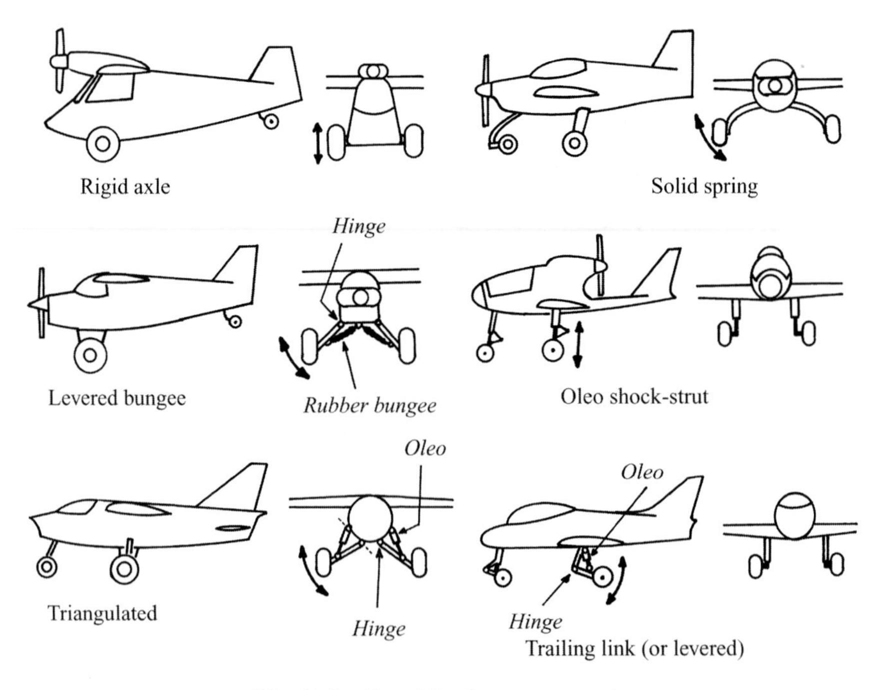

Fig. 11.9 Gear/shock arrangements.

Many World War I fighters used a rigid axle mounted with some vertical movement. The axle was attached to the aircraft with strong rubber chords ("bungees") that stretched as the axle moved upward. This is rarely seen today.

The solid spring gear is used in many general-aviation aircraft (especially Cessna products). The solid spring is as simple as possible, but is slightly heavier than other types of gear.

Note that the solid-spring gear deflects with some lateral motion instead of straight up and down. This lateral motion tends to scrub the tires sideways against the runway, wearing them out. The solid spring has no damping other than this scrubbing action. The aircraft thus tends to bounce a lot, much like a car with bad shock-absorbers.

The levered bungee-chord gear was very common in early light aircraft such as the Piper Cub. The gear leg is pivoted at the fuselage. Rubber bungee chords underneath the gear are stretched as the gear deflects upward and outward. In some aircraft a lever action compresses a rubber block providing the same effect in a less-draggy installation. This gear is light in weight but is high in drag. This gear also causes lateral scrubbing of the tires.

The oleopneumatic shock strut, or "oleo," is the most common type of shock-absorbing gear in use today (Fig. 11.10). The oleo concept was patented in 1915 as a recoil device for large cannons. The oleo combines a spring effect using compressed air with a damping effect using a piston which forces oil through a

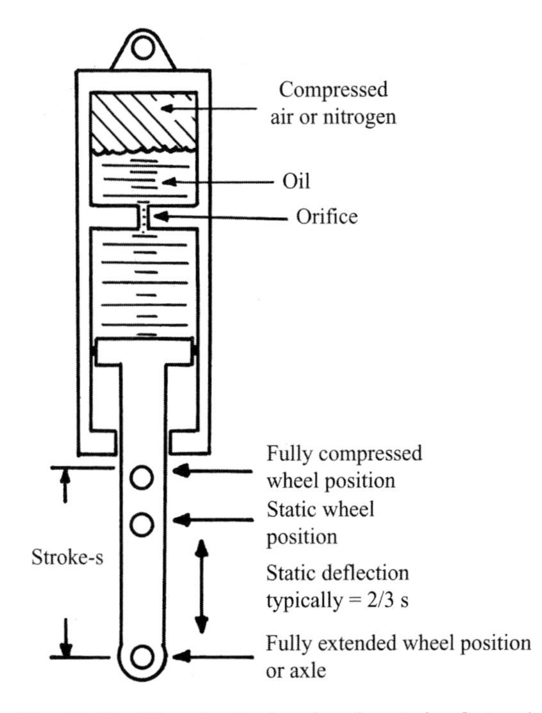

Fig. 11.10 Oleo shock absorber (most simple type).

small hole (orifice). For maximum efficiency, many oleos have a mechanism for varying the size of the orifice as the oleo compresses (metered orifice).

When used as a shock-strut, the oleo itself must provide the full required amount of wheel deflection, which can lengthen the total landing gear height. Also, the oleo strut must be strong enough to handle the lateral and braking loads of the wheels. To repair or replace the oleo strut, the entire wheel assembly must be removed because it is attached to the bottom of the strut.

The triangulated gear is similar to the levered bungee gear. When the triangulated gear is deflected, an oleopneumatic shock absorber is compressed. This provides a leveraged effect in which the oleo can be shorter than the required wheel travel. This is especially useful for carrier-based aircraft such as the A-7 that require large amounts of wheel travel to absorb the carrier-landing impact loads.

On a triangulated gear, the oleo can be replaced without removing the wheel assembly. The wheel lateral and braking loads are carried by the solid gear legs, which reduces the oleo weight. However, the complete triangulated gear is usually a little heavier than the oleo shock-strut gear. Also, there is a tire-scrubbing effect that shortens tire life.

The triangulated gear is sometimes seen on smaller aircraft using rubber blocks or springs in compression instead of an oleopneumatic shock absorber. The rubber blocks or springs can be inside the fuselage which streamlines the exposed part of the gear but requires the gear leg to support the aircraft's weight in a cantilevered fashion. This increases the gear weight.

The trailing-link, or levered, gear resembles the triangulated gear, but with the solid gear leg running aft rather than laterally. This gear is common for carrier-based aircraft such as the F-18 where it provides the large amounts of gear travel required for carrier landings. Typically the pivot point of the lower gear leg is slightly in front of the tire, less than one tire radius in front of the tire.

The levered gear allows the wheel to travel aft as it deflects. This is very desirable for operations on rough fields. If a large rock or other obstacle is encountered, the aft motion of the wheel gives it more time to ride over the obstacle. For this reason the levered gear was used on the North American Rockwell OV-10 and the Polish PZL-104 Wilga.

The triangulated gear and levered gear provide a mechanical lever effect that reduces the deflection of the shock-absorber oleo. However, this also increases the forces on the shock-absorber oleo, which increases its required diameter. The mechanical advantage of the triangulated or levered gear is determined from the actual dimensions of the gear layout, and is used to size the shock absorber.

Stroke Determination

The required deflection of the shock-absorbing system (the "stroke") depends upon the vertical velocity at touchdown, the shock-absorbing material, and the amount of wing lift still available after touchdown. As a rough rule-of-thumb, the stroke in inches approximately equals the vertical velocity at touchdown in (ft/s).

The vertical velocity (or "sink speed") at touchdown is established in various specifications for different types of aircraft. Most aircraft require 10 ft/s {3 m/s} vertical velocity capability. This is substantially above the 4–5 ft/s {1–1.5 m/s} that most passengers would consider a "bad" landing.

While most Air Force aircraft require only 10 ft/s {3 m/s} Air Force trainer aircraft require 13 ft/s {4 m/s}. Due to steeper descent angles, Ref. 32 suggests 15 ft/s {4.6 m/s} for short takeoff and landing (STOL) aircraft.

Carrier-based naval aircraft require 20 ft/s {6 m/s} or more vertical velocity, which is much like a controlled crash! This is the reason that carrier-based aircraft tend to use triangulated or levered gear, which provide longer strokes than shock-strut gear.

In most cases it may be assumed that the wing is still creating lift equal to the aircraft's weight during the time that the shock absorber is deflecting. The detailed shock-absorber calculations for FAR-23 aircraft must assume that only two-thirds of the aircraft's weight is supported by the wing during touchdown. However, this can be ignored for initial stroke calculations.

The vertical energy of the aircraft, which must be absorbed during the landing, is defined in Eq. (11.8). This kinetic energy is absorbed by the work of deflecting the shock absorber and tire.

$$KE_{\text{vertical}} = \left(\frac{1}{2}\right)\left(\frac{W_{\text{landing}}}{g}\right)V_{\text{vertical}}^2 \tag{11.8}$$

where W = total aircraft weight.

If the shock absorber were perfectly efficient, the energy absorbed by deflection would be simply the load times the deflection. Actual efficiencies of shock absorbers range from 0.5–0.9, as provided in Table 11.4. The actual energy absorbed by deflection is defined in Eq. (11.9).

$$KE_{\text{absorbed}} = \eta L S \tag{11.9}$$

where

η = shock-absorbing efficiency
L = average total load during deflection (not lift!)
S = stroke

Table 11.4 Shock absorber efficiency

Type	Efficiency η
Steel leaf spring	0.50
Steel coil spring	0.62
Air spring	0.45
Rubber block	0.60
Rubber bungee	0.58
Oleopneumatic	
–Fixed orifice	0.65–0.80
–Metered orifice	0.75–0.90
Tire	0.47

For tires it is assumed that the tire deflects only to its rolling radius, so the stroke (S_T) of a tire is equal to half the diameter minus the rolling radius.

Combining Eqs. (11.8) and (11.9) and assuming that the shock absorber and tire both deflect to absorb the vertical kinetic energy yields:

$$\left(\frac{1}{2}\right)\left(\frac{W_{\text{landing}}}{g}\right)V_{\text{vertical}}^2 = (\eta LS)_{\text{shock absorber}} + (\eta_T LS_T)_{\text{tire}} \tag{11.10}$$

Note in Eq. (11.10) that the number of shock absorbers does not enter into the equation. Remember that L is the average total load on the shock absorbers during deflection. The number of shock absorbers affects the diameter of the shock absorbers but not the required stroke.

The shock absorbers and tires act together to decelerate the aircraft from the landing vertical velocity to zero vertical velocity. The vertical deceleration rate is called the gear load factor (N_{gear}). Gear load factor is the average total load summed for all of the shock absorbers divided by the landing weight, and is assumed to be constant during touchdown. N_{gear} is defined in Eq. (11.11) and typically equals three:

$$N_{\text{gear}} = L/W_{\text{landing}} \tag{11.11}$$

The gear load factor determines how much load the gear passes to the airframe, which affects the airframe structural weight as well as crew and passenger comfort during the landing. Table 11.5 provides typical gear load factors permitted for various types of aircraft.

Substituting Eq. (11.11) into Eq. (11.10) yields Eq. (11.12) for shock-absorber stroke. Note that the equation for stroke does not include any terms containing the aircraft weight. For the same required landing vertical velocity and gear load-factor, an airliner and an ultralight would require the same stroke!

$$S = \frac{V_{\text{vertical}}^2}{2g\eta N_{\text{gear}}} - \frac{\eta_T}{\eta} S_T \tag{11.12}$$

The stroke calculated by Eq. (11.12) should be increased by about 1 in. {3 cm} as a safety margin. Also, a stroke of 8 in. {20 cm} is usually considered a minimum, and at least 10–12 in. {25–30 cm} is desirable for most aircraft.

Table 11.5 Gear load factors

Aircraft type	N_{gear}
Large bomber	2.0–3
Commercial	2.7–3
General aviation	3
Air Force fighter	3.0–4
Navy fighter	5.0–6

Nosewheel stroke is generally set equal to or slightly larger than mainwheel stroke to provide a smooth ride while taxiing.

Note that the stroke defined by Eq. (11.12) is a vertical distance. If a type of gear is used that produces some lateral motion of the wheel, the total distance the wheel moves must provide the required stroke in a vertical direction.

Equation (11.12) defines the total stroke required. If a levered or triangulated gear is used, the required stroke of the oleo, bungee, or rubber block is reduced and is determined by dividing the total stroke by the mechanical advantage. The actual load on the shock absorber is increased (multiplied) by the mechanical advantage.

Oleo Sizing

The actual dimensions of an oleo shock absorber or shock strut can now be estimated. The total oleo stroke is known. For most types of aircraft the static position is approximately 66% of the distance from the fully extended to the fully compressed position (see Fig. 11.10). For large transport aircraft, the static position is about 84% of stroke above the fully extended position. For a general-aviation aircraft the static position is typically about 60% of stroke above the extended position.

The total length of the oleo including the stroke distance and the fixed portion of the oleo will be approximately 2.5 times the stroke. For an aircraft with the desired gear attachment point close to the ground, this minimum oleo length may require going to a levered gear.

Oleo diameter is determined by the load carried by the oleo. The mainwheel oleo load is the static load found from Eq. (11.1) divided by the number of mainwheel oleos (usually two). The nosewheel oleo load is the sum of the static and dynamic loads [Eqs. (11.2) and (11.4)]. These loads must be increased by the mechanical advantage if levered or triangulated gear is used.

The oleo carries its load by the internal pressure of compressed air, applied across a piston. Typically an oleo has an internal pressure (P) of 1800 psi {12,415 kPa}. Internal diameter is determined from the relationship which states that force equals pressure times area. The external diameter is typically 30% greater than the piston diameter, so the external oleo diameter can be approximated by Eq. (11.13):

$$D_{\text{oleo}} = 1.3\sqrt{\frac{4L_{\text{oleo}}}{P\pi}} \tag{11.13}$$

where L_{oleo} = load on the oleo.

Solid-Spring Gear Sizing

Figure 11.11 illustrates the deflection geometry for a solid-spring gear leg. The total stroke as determined by Eq. (11.12) is the vertical component of the deflection of the gear leg. Note that the wheel is mounted so that it is vertical when the gear leg is deflected under the static load. This provides even tire wear.

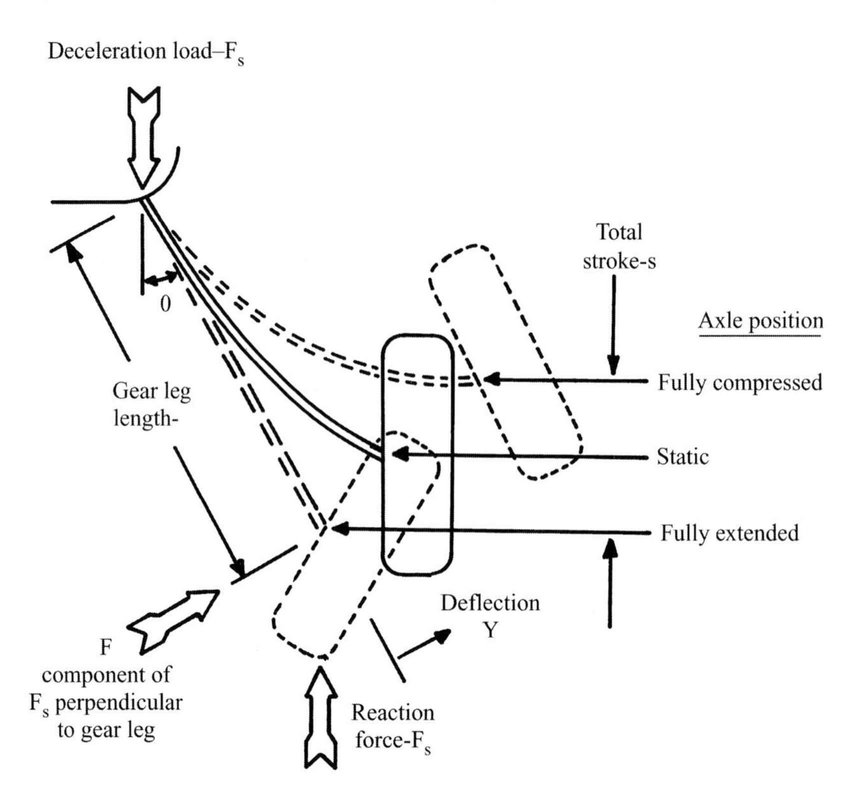

Fig. 11.11 Solid spring gear deflection.

If the gear leg is not excessively tapered, it may be approximated as a constant-cross-section bending beam using the average values of beam width (w) and thickness (t).

The load on the gear leg in the fully deflected position is the force required to produce the gear load factor, N_{gear}. Assuming there are two gear legs yields Eq. (11.14); the component of the load on the gear that is perpendicular to the gear leg is defined by Eq. (11.15):

$$F_s = WN_{gear}/2 \tag{11.14}$$

$$F = F_s(\sin\theta) \tag{11.15}$$

The deflection y perpendicular to the gear leg is related to the stroke by Eq. (11.16), and is calculated by the structural bending-beam equation [Eq. (11.17)—see Chapter 14]. Substituting Eqs. (11.14–11.16) into Eq. (11.17) yields the equation for the stroke S of a solid-spring gear leg:

$$S = y \sin\theta \tag{11.16}$$

$$y = Fl^3/3EI \tag{11.17}$$

$$S = F_s(\sin^2\theta)\frac{l^3}{3EI} \tag{11.18}$$

where

I = beam's moment of inertia
E = material modulus of elasticity

For a rectangular-cross-section gear leg, the moment of inertia is defined by Eq. (11.19):

$$I = \frac{wt^3}{12} \tag{11.19}$$

The static deflection is also determined using Eq. (11.18), but from the static wheel-load for the force (F_s).

The methods of Chapter 14 should be used to ensure that the stresses in the gear legs remain well below their material's maximum values.

11.5 Castoring-Wheel Geometry

For ground steering, a nosewheel or tailwheel must be capable of being castored (turned). The castoring can introduce static and dynamic stability problems causing "wheel shimmy," a rapid side-to-side motion of the wheel that can tear the landing gear off the airplane.

Prevention of shimmy is accomplished by selection of the rake angle and trail, as shown in Fig. 11.12. In some cases a frictional shimmy damper is also used to prevent shimmy. This can be a separate hydraulic plunger or simply a pivot with a lot of friction.

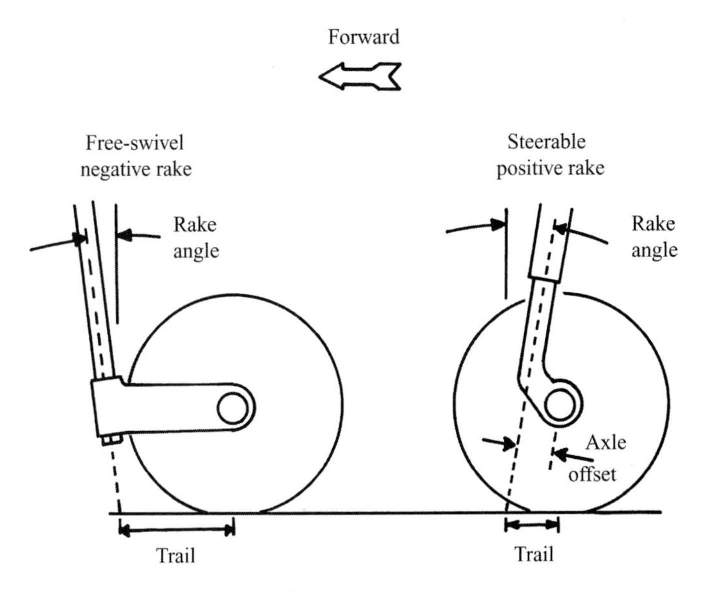

Fig. 11.12 Castoring wheel geometry.

If the castoring wheel is free to swivel, shimmy can be prevented by using a small negative angle of rake (4–6 deg), and trail equal to 0.2–1.2 times the tire radius (this is also typical for tailwheels). If the trail is less than the tire radius, a shimmy damper may be required.

If the nosewheel is free to swivel, the pilot must steer the aircraft on the ground using only the brakes. This increases brake wear and presents a great danger if one brake fails during takeoff or landing.

Note that tailwheels are always designed as if they are free to swivel. Steerable tailwheels are connected to the rudder pedals by soft springs that don't affect the wheel dynamics. Most larger tailwheel aircraft have provisions for locking the tailwheel during takeoff and landing.

For most tricycle-geared aircraft, a steering linkage is connected to the rudder pedals or a separate steering wheel, providing positive control of the turning angle. A key objective in the design of a steerable nosewheel is to reduce the required control forces while retaining dynamic stability.

This is done by minimizing the trail through use of a positive rake angle. Note that the weight of the aircraft tends to cause this gear to "flop over," i.e., the gear is statically unstable. This is prevented by the control linkages.

For a large aircraft with a steerable nosewheel, the rake angle should be about 7 deg positive, and the trail should be at least 16% of the tire radius. For smaller aircraft, rake angles up to 15 deg and trail of about 20% are used.

11.6 Gear-Retraction Geometry

At this point, the required sizes for the wheels, tires, and shock absorbers are known, along with the required down locations of the wheels. The one remaining task is to find a "home for the gear" in the retracted position.

A poor location for the retracted gear can ruin an otherwise good design concept! A bad choice for the retracted position can chop up the aircraft structure (increasing weight), reduce the internal fuel volume, or create additional aerodynamic drag.

Figure 11.13 shows the options for main-landing-gear retracted positions. Locating the gear in the wing, in the fuselage, or in the wing-fuselage junction produces the smallest aerodynamic penalty but tends to chop up the structure. Gear in the wing reduces the size of the wing box, which increases weight and may reduce fuel volume. Gear in the fuselage or wing-fuselage junction may interfere with the longerons. However, the aerodynamic benefits of these arrangements outweigh the drawbacks for higher-speed aircraft.

Virtually all civilian jet transports retract the gear into the wing-fuselage junction. Most low-wing fighters retract the gear into the wing or wing-fuselage junction, while mid- and high-wing fighters retract the gear into the fuselage.

While some slower aircraft retract the gear into the wing, fuselage, or wing-fuselage junction, many retract the gear into the nacelles or a separate gear pod. This reduces weight significantly because the wing and fuselage structure is uninterrupted.

The wing-podded arrangement is rarely seen in Western aircraft designs (A-10), but is used in designs in former Soviet block countries even for jet

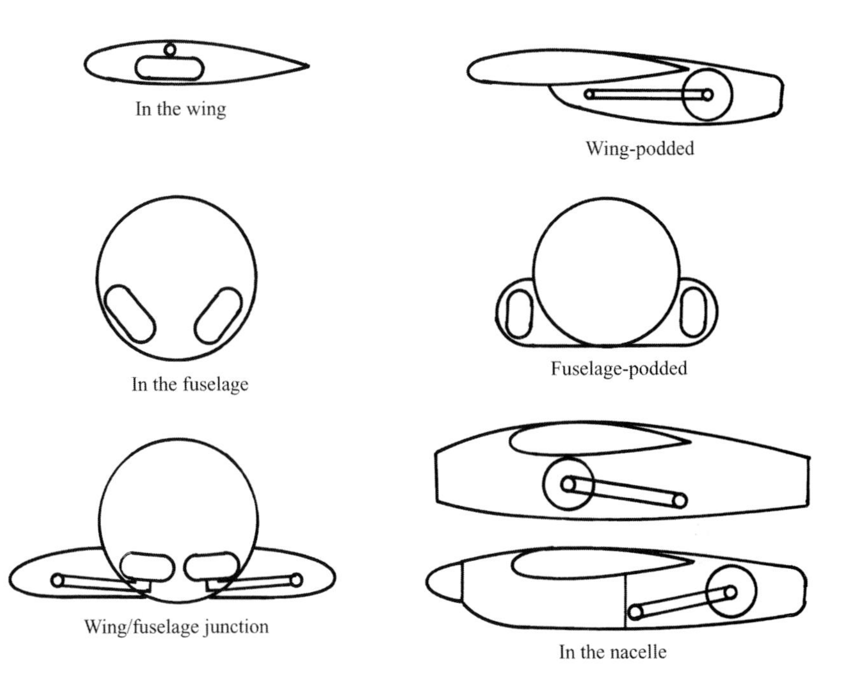

Fig. 11.13 "A home for the gear."

transports and bombers. The aerodynamic penalty is minimized by placing the pods at the trailing edge of the wing, where some "area-ruling" benefit is obtained.

The fuselage-podded arrangement is common for high-winged military transports where the fuselage must remain open for cargo. The drag penalty of the pods can be substantial.

Retraction of the gear into the nacelles behind the engine is typical for propeller-driven aircraft. For jet-engined aircraft, nacelle-mounted landing gear must go alongside the engine, which widens the nacelle, increasing the drag.

Most mechanisms for landing-gear retraction are based upon the four-bar linkage. This uses three members (the fourth bar being the aircraft structure) connected by pivots. The four-bar linkage provides a simple and lightweight gear because the loads pass through rigid members and simple pivots.

Several variations of four-bar linkage landing gear are shown in Fig. 11.14. The oldest form of four-bar linkage for landing-gear retraction is shown in front view (Fig. 11.14a), where the wheel is at the bottom of a vertical gear member attached to parallel arms, which in turn attach to the fuselage. The gear retracts by pivoting the arms upward and inward. This was widely used during the 1930s, and is seen today in modified form on the MiG-23.

Figure 11.14b shows the typical retraction arrangement for nosewheels. The diagonal arm is called a drag brace because it withstands the aerodynamic loads (as well as braking loads). The drag brace breaks at the middle for retraction.

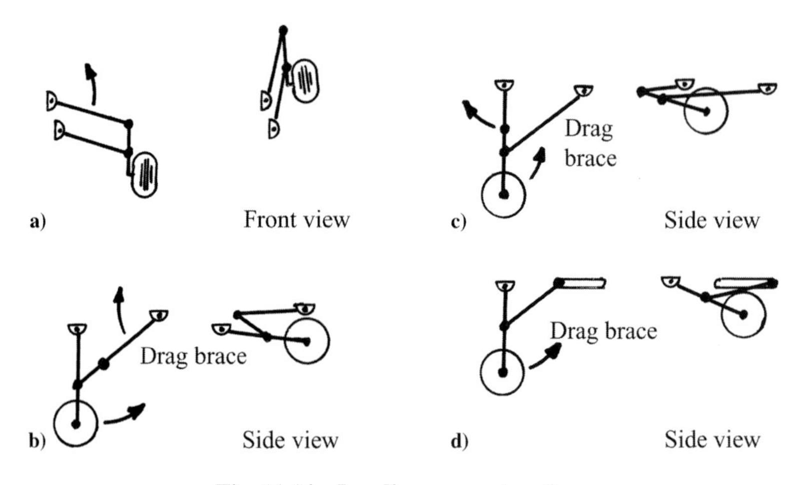

Fig. 11.14 Landing-gear retraction.

The drag brace may be behind the wheel with the gear retracting rearward or it may be in front of the wheel with the gear retracting forward. The latter is preferable because the air loads will blow the gear down in the event of a hydraulic failure.

In Fig. 11.4c the vertical gear member breaks for retraction instead of the drag brace. This has the advantage of reducing the length of the retracted gear, but is usually heavier. This gear was used on the DC-3 and several World War II bombers.

Figure 11.14d shows the use of a sliding pivot rather than a four-bar linkage. The sliding motion is frequently provided by a wormscrew mechanism that is rotated to retract the gear. This is usually heavier than a four-bar linkage because the entire length of the wormscrew must be strong enough for the landing-gear loads. However, this gear is very simple and compact.

The retraction concepts as shown in Fig. 11.14 are for nose- or mainwheels that retract in a fore or aft direction. However, the same basic concepts can be used for main wheels that retract inward or outward. These illustrations then become front views, and the tires are redrawn accordingly. The gear members labeled "drag brace" become "sway braces" because they provide lateral support for the gear in this arrangement.

There are dozens of additional geometries for gear retraction based upon the four-bar linkage and other concepts. For these, see Ref. 32 or 33.

The landing-gear leg is attached to the aircraft at the pivot point. This is determined as shown in Fig. 11.15. The pivot point can lie anywhere along the perpendicular bisector to the line connecting the up and down positions of the wheel.

Normally the gear strut is allowed to extend fully before retraction, as shown in Fig. 11.15, although it is possible to install a strut compressor that causes the gear to be retracted with the strut in the compressed position. This should be used only when internal space is absolutely unavailable for the fully extended strut.

It is also possible to provide a rotator mechanism or planing link that will change the angle between the gear leg and the wheel axis when the gear is

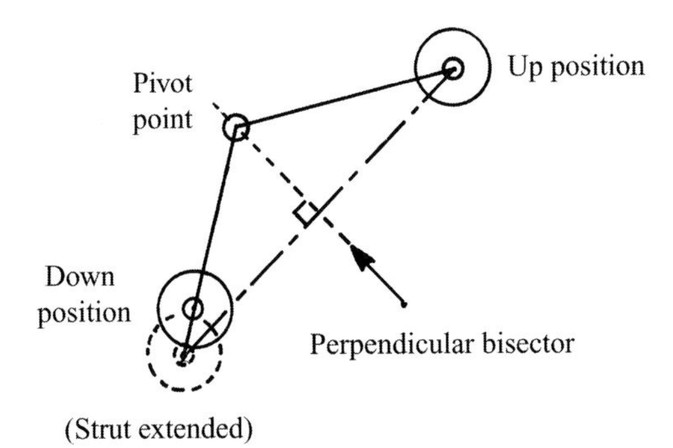

Fig. 11.15 Pivot point determination.

retracted. This is sometimes required to permit the wheel to lie flat inside the wheel well when retracted. This is fairly simple and is seen on many military aircraft such as the F-16. However, all such mechanisms should be avoided if possible due to the increased weight, complexity, and maintenance.

11.7 Seaplanes

Seaplanes were important during the early days of aviation because of the limited number of good airports. Early commercial over-water flights were made exclusively by seaplanes, for safety. Most early speed records were held by seaplanes (or floatplanes) because the use of water for takeoff allowed long takeoff runs and hence high wing-loadings.

Today the seaplane concept is largely restricted to sportplanes, bush planes, and search-and-rescue aircraft. However, the notion of a "sea-sitter" strategic bomber which loiters for weeks at an unknown ocean location is revived every few years.

The hull of a seaplane and the pontoon of a float plane are based on the planing-hull concept. The bottom is fairly flat, allowing the aircraft to skim (plane) on top of the water at high speeds. A step breaks the suction on the after body. A vertical discontinuity, as shown in Fig. 11.16, the step can be straight in planview, or it can have an elliptical shape in planview to reduce aerodynamic drag.

While a few small seaplanes have been built with flat bottoms, most use a *V*-shaped bottom to reduce the water-impact loads. The height of the *V* is called the deadrise, and the angle is the deadrise angle. Deadrise angle must be increased for higher landing speeds, and should roughly follow Eq. (11.20). Deadrise angle is increased toward the nose to about 30–40 deg to better cut through waves.

$$\alpha_{\text{deadrise}} \cong \frac{V}{2} - 10 \text{ deg} \tag{11.20}$$

where V = stall speed in miles per hour.

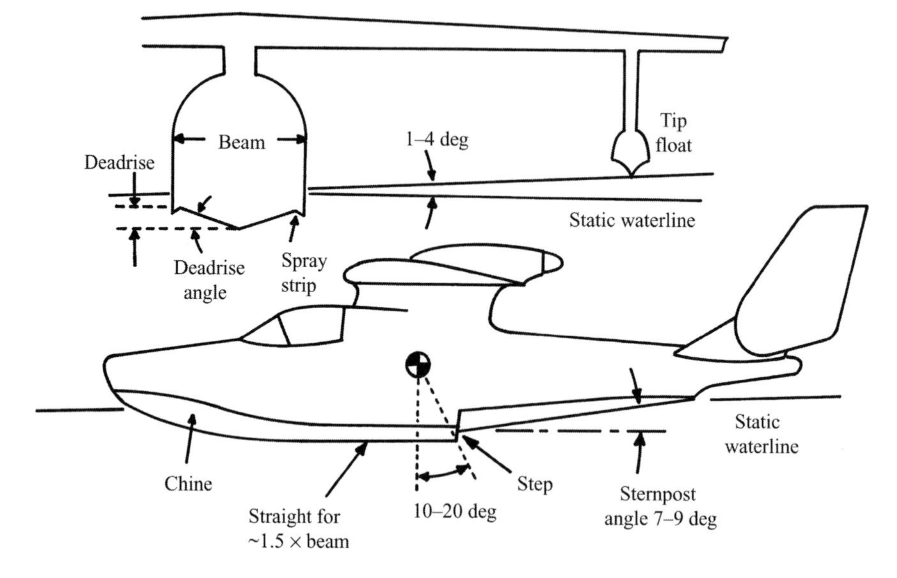

Fig. 11.16 Seaplane geometry.

To reduce water spray, spray strips can be attached to the edges of the bottom, as shown. These are angled about 30 deg below the horizon.

The ratio between the waterline length and "beam" (width) has a strong effect upon water resistance and landing impact. A wider hull has a lower water resistance due to its better planing ability but suffers a higher landing impact. Length-to-beam ratios vary from about six for a small seaplane to about 15 for a large one.

The step height should be about 5% of the beam. The step should be located on an angle about 10–20 deg behind the c.g. The bottom of the hull forward of the step should not be curved for a distance about equal to 1.5 times the beam. This is to reduce porpoising tendencies. Also, the hull bottom aft of the step (the sternpost) should angle upward about 8 deg.

For a true "flying boat" (i.e., seaplane with a boat-like fuselage), lateral stability on the water is usually provided by wing-mounted pontoons. These should be located such that they contact the water when the aircraft tips sideways about 1 deg.

Determination of the static waterline is done using a modification of the fuselage-volume plotting technique previously described. First, a static waterline is assumed and drawn on the configuration layout. Then, a volume plot is prepared from the cross-sectional areas of only those parts of the fuselage that are below the assumed waterline.

The area under the curve on the volume plot defines the fuselage submerged volume (i.e., below the waterline). This is multiplied by the density of water ($62.4\ \mathrm{lb/ft^3}$ or $1000\ \mathrm{kg/m^3}$) to determine the weight of aircraft that can be supported by that amount of displaced water.

The centroid of the area on the volume plot is the center of buoyancy, which should coincide with the c.g. If either the weight of aircraft supported or the center of buoyancy are incorrect, another waterline must be assumed and a new volume plot prepared.

Water-resistance drag is very difficult to estimate. It depends upon the mechanics of wave production, and can vary widely for similar hull shapes. Also, water-resistance drag varies with speed. A seaplane hull can have a maximum water resistance at "hump speed" equal to 20% or more of the aircraft's weight.

Seaplane design and analysis should be based upon some published set of test results for a known hull shape. These can be found in early NACA reports. The major U.S. facility for testing seaplane and ship hulls is the Naval Ship Research and Development Center (NSRDC) near Washington, D.C.

For a rough estimate of a seaplane's takeoff distance, it can be assumed that the water-resistance drag will average about 10–15% of the waterborne weight. This is analagous to a rolling-friction coefficient (μ) of 0.10–0.15 in the takeoff calculations provided in Chapter 17.

11.8 Subsystems

Aircraft subsystems include the hydraulic, electrical, pneumatic, and auxiliary/emergency power systems. Also, the avionics can be considered a subsystem (although, to the avionics engineers, the airframe is merely the "mobility subsystem" of their avionics package!).

In general, the subsystems do not have a major impact on the initial design layout. However, later in the design cycle the configuration designer will have to accommodate the needs of the various subsystems, so a brief introduction is provided next. No attempt is made to provide examples or rules of thumb because the subsystems hardware varies widely between different classes of aircraft. Reference 131 is highly recommended as an overview to subsystems. Reference 11 provides additional information on subsystems.

Hydraulics

A simplified hydraulic system is shown in Fig. 11.17. Hydraulic fluid, a light oil-like liquid, is pumped up to some specified pressure and stored in an accumulator (simply a holding tank).

When the valve is opened, the hydraulic fluid flows into the actuator where it presses against the piston, causing it to move and, in turn, moving the control

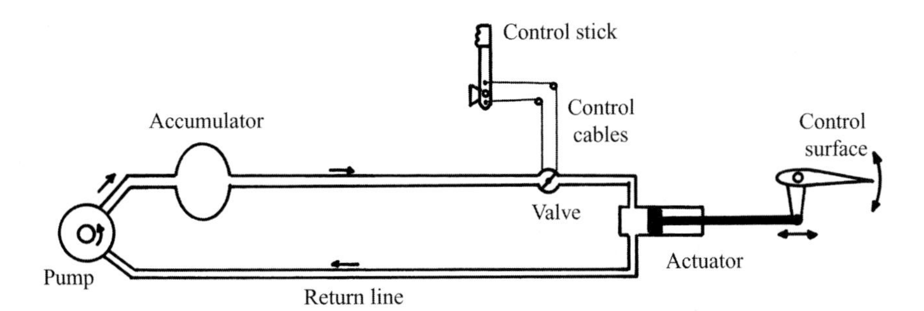

Fig. 11.17 Simplified hydraulic system.

surface. To move the control surface the other direction, an additional valve (not shown) admits hydraulic fluid to the back side of the piston. The hydraulic fluid returns to the pump by a return line.

To obtain rapid response, the valve must be very close to the actuator. The valve therefore cannot be in or near the cockpit, and instead is usually attached to the actuator.

In most current designs the pilot's control inputs are mechanically carried to the actuator by steel cables strung from the control wheel or rudder pedals to the valves on the actuators. In many new aircraft the pilot's inputs are carried electronically to electromechanical valves ("fly-by-wire"). In a "fly-by-light" aircraft, the pilot's control inputs are carried via fiber-optics.

Hydraulics are used for aircraft flight control as well as actuation of the flaps, landing gear, spoilers, speed brakes, and weapon bays. Flight-control hydraulic systems must also include some means of providing the proper control "feel" to the pilot. For example, the controls should become stiffer at higher speeds, and should become heavier in a tight, high-g turn. Such "feel" is provided by a combination of springs, bobweights, dashpots, and air bellows.

In most cases the hydraulic system will impact the aircraft conceptual design only in the provision of space for the hydraulic pumps, which are usually attached to the engines. These should be copied from a similar aircraft if better information is not available.

A new type of hydraulic actuation system has recently been flight tested on F-16 and F-18 research aircraft, and offers several advantages. The electrohydrostatic actuator (EHA) places an entire hydraulic system in miniature at each place an actuator is required. Rather than having a centralized hydraulic pump and accumulator, with high-pressure lines running all over the aircraft, each hydraulic actuator has its own small hydraulic pump (driven by an electric motor) and an accumulator. For an F-16-size aircraft, about 80 kW of electrical power are required. While this system probably weighs a little more than a next-generation conventional hydraulic system, it has the big advantage of allowing the entire unit to be replaced for maintenance, with only an electrical power and control signal connection to be disconnected and reattached. This offers large savings in maintenance time and cost.

Electrical System

An aircraft electrical system provides electrical power to the avionics, hydraulics, environmental-control, lighting, and other subsystems. The electrical system consists of batteries, generators, transformer-rectifiers (TRs), electrical controls, circuit breakers, and cables.

Aircraft generators usually produce alternating current (AC) and are located on or near the engines. TRs are used to convert the alternating current to direct current (DC). Aircraft batteries can be large and heavy if they are used as the only power source for starting.

The Eurofighter Typhoon is typical of modern fighter electrical system design. It has two engine-driven 30 kVA generators supplying 115/200 V, 400 Hz, three-phase AC power, and uses transformer rectifier units to produce 28V DC power. A battery supplies DC power to start the APU (auxiliary power unit; see the following), which then provides power to start the engines.

Typical of modern transport aircraft electrical system design, the Boeing 767 has two engine-driven 90 kVA generators supplying 115/200 V, 400 Hz, three-phase AC power, and has another 90 kVA generator attached to the APU for emergencies and ground power.

An advanced electrical technology that shows promise for replacing traditional hydraulic actuators is the use of electrical actuators. These are basically electric motors operating through gearboxes to directly drive the control surfaces, with no hydraulics at all. This is actually the same flight control actuation that has been used for decades by radio control flying models, but only now does it seem possible to build such a system for a real aircraft at a competitive weight and cost. Like the electro-hydrostatic actuator, this eliminates a centralized hydraulic pump and accumulator, with high-pressure lines running all over the aircraft, but probably weighs more than a next-generation conventional hydraulic system. It also has the advantage of allowing the entire unit to be replaced with only electrical power and control signal connections to be disconnected. Electrical actuators have been flight tested, and development is ongoing.

Electrically operated brakes are being tested on an F-16. These would eliminate brake hydraulic lines that are especially problem-prone because they must flex or rotate as the gear is retracted. The electric brakes operate like regular brakes, much like the disk brakes on a car, but are actuated by electric motors operating through ballscrews. They have roughly twice the response speed of hydraulic brakes. This could improve anti-skid capabilities. Weights should be comparable to a traditional hydraulic braking system.

Pneumatic/ECS System

The pneumatic system provides compressed air for pressurization, environmental control, anti-icing, and in some cases engine starting. Typically the pneumatic system uses pressurized air bled from the engine compressor.

This compressed air is cooled through a heat exchanger using outside air. This cooling air is taken from a flush inlet inside the inlet duct (i.e., inlet secondary airflow) or from a separate inlet usually located on the fuselage or at the front of the inlet boundary-layer diverter.

The cooled compressor air is then used for cockpit pressurization and avionics cooling. This is called the Environmental Control System (ECS). For anti-icing, the compressor bleed air goes uncooled through ducts to the wing leading edge, inlet cowls, and windshield.

Compressed air is sometimes used for starting other engines after one engine has been started by battery. Also, some military aircraft use a ground power cart that provides compressed air through a hose to start the engine.

Auxiliary/Emergency Power

Large or high-speed aircraft are completely dependent upon the hydraulic system for flight control. If the hydraulic pumps stop producing pressure for any reason, the aircraft will be uncontrollable. If the pumps are driven off the engines, an engine flame-out will cause an immediate loss of control.

For this reason, some form of emergency hydraulic power is required. Also, electrical power must be retained until the engines can be restarted. The three major forms of emergency power are the ram-air turbine (RAT), monopropellant emergency power unit (EPU), and jet-fuel EPU.

The ram-air turbine is a windmill extended into the slipstream. Alternatively, a small inlet duct can open to admit air into a turbine.

The monopropellant EPU uses a monopropellant fuel such as hydrazine to drive a turbine. The available monopropellants are all toxic and caustic, so monopropellant EPUs are undesirable for operational considerations.

However, they have the advantage of not requiring any inlet ducts and can be relied upon to provide immediate power regardless of aircraft altitude, velocity, or attitude. Monopropellant EPUs must be located such that a small fuel leak will not allow the caustic fuel to puddle in the aircraft structure, possibly dissolving it!

Jet-fuel EPUs are small jet engines that drive a turbine to produce emergency power. These may also be used to start the main engines (jet-fuel starter). While they do not require a separate and dangerous fuel, the jet-fuel EPUs require their own inlet duct.

Most commercial transports and an increasing number of military aircraft use a jet-fuel auxiliary power unit (APU). An APU is much like an EPU but is designed and installed to allow continuous operation.

Usually an APU is designed to provide ground power for air conditioning, cabin lighting, and engine starting. This frees the aircraft from any dependence upon ground power carts. The APU is also used for in-flight emergency power, and in some cases is run continuously in-flight for additional hydraulic and/or electrical power.

The APU is actually another jet engine, and its installation must receive attention in the earliest design layout. The APU requires its own inlet and exhaust ducts, and must be contained in a firewalled structure. APUs have fairly high maintenance requirements so access is important.

To avoid high levels of noise, the inlet and exhaust of an APU should be directed upward. For in-flight operation of the APU, the inlet should ideally be in a high-pressure area and the exhaust in a low-pressure area. Also, the inlet should not be located where the exhaust of the jet engines or APU can be ingested. The exhaust of an APU is hot and noisy, and should not impinge upon aircraft structure or ground personnel.

Transport aircraft usually have the APU in the tail, as shown in Fig. 11.18. This removes the APU from the vicinity of the passenger compartment to reduce noise. The APU firewall is of minimum size, and the APU is easily accessible from a workstand.

Military transports with the landing gear in fuselage-mounted pods can place the APU in the pod. This provides ground-level access to the APU, but requires increased firewall area.

Fighters usually have the APU in the fuselage, near the hydraulic pumps and generators. This requires a firewall that completely encloses the APU.

APU installation is discussed in detail in Ref. 34.

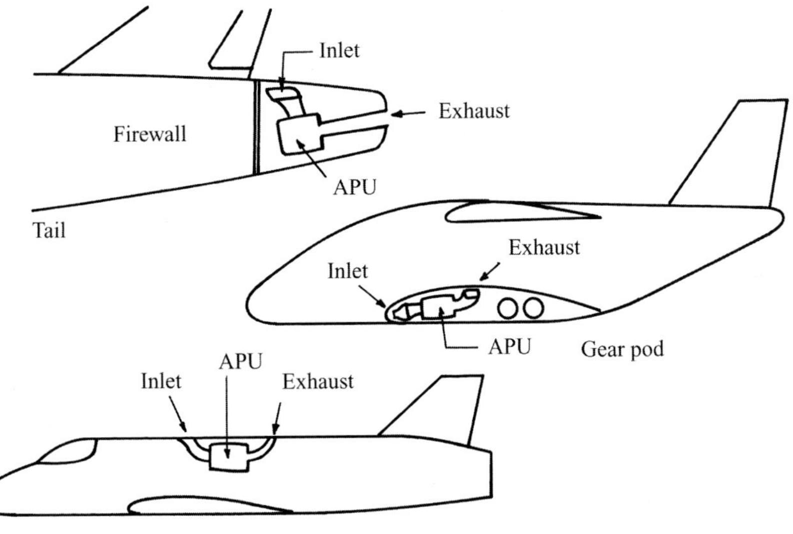

Fig. 11.18 APU installation.

Accessory Drives

Many high-performance aircraft have a special gearbox called an airframe mounted accessory drive, or AMAD. In earlier aircraft, the hydraulic pumps, electrical generators, starter motors, and other accessories turned by engine mechanical power would be directly mounted to the engine. These were typically below the engine but sometimes were in an oversized round fairing directly in front of the compressor. When the engine had to be removed, the accessories all had to be separately disconnected. The AMAD, as typically shown in Fig. 11.19, is simply a gearbox to which all engine-driven accessories are attached. The AMAD is itself attached to the engine by a spinning driveshaft, which is easily disconnected when the engine must be removed.

Avionics

Avionics, a contraction of "aviation electronics," includes radios, flight instruments, navigational aids, flight control computers, radar, infrared detectors, and other equipment. Whereas the limited avionics of early aircraft would simply be "bolted on" sometime before first flight, today's avionics are an integral part of the design process for most new aircraft. Furthermore, the cost of avionics has gone from being nearly insignificant to, for some military aircraft, approaching a third of total costs.

Avionics can be subdivided into various categories, which this author would characterize as Com/Nav, Mission Equipment, and Vehicle Management.

The earliest avionics were strictly for communication and navigation (Com/Nav), and included radios and various types of radio-based navigation

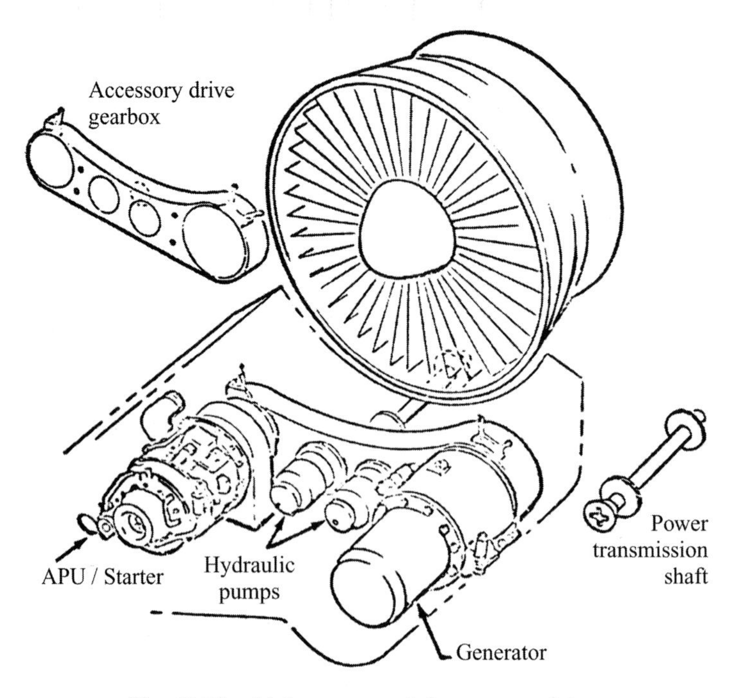

Fig. 11.19 Airframe mounted accessory drive.

aides including one of the simplest, a device that merely points the direction to a tuned-in radio station. The early avionics later included instruments and radio aids for in-weather flight and landings, and rudimentary autopilots capable of holding a heading and an altitude. After World War II, weather radar became available, along with radar transponders to assist ground radars in identifying aircraft under their control. The steady march of electronics progress has brought an incredible array of avionics gear to assist the pilot in basic communication, navigation, and related tasks, such as the GPS satellite navigation receiver and the modern autopilots that can fly the aircraft throughout its entire mission. Modern aircraft cockpit displays must also be considered avionics (actually, computer display screens) rather than the pressure gauges and electrical needles of before.

As electronics progressed, military aircraft began to rely on onboard electronics for the performance of the aircraft's mission. Early examples were the radar-carrying night fighters of World War II and the anti-submarine aircraft such as the Lockheed P-2V Neptune. Mission Equipment avionics now includes air-to-air and air-to-ground radar, electronic countermeasures, infrared seekers and sensors, infrared countermeasures, aircraft identification (friend or foe), gun and missile aiming, terrain-following autopilots, active electronic stealth techniques, and a host of other mission-specific systems. These mission equipment systems also require a lot of onboard computing power.

For the most part, Com/Nav and Mission Equipment avionics can still be thought of as "add-ons." The airplane would fly without them, and the pilots could always follow railroads as in Lindbergh's day. Vehicle Management avionics go a step beyond and are actually critical to the flight of the aircraft. A good example is the previously mentioned Fly-By-Wire technology wherein the aircraft is aerodynamically unstable but is artificially stabilized by a smart computer and some fast actuators. On the X-31, an air data sensor became iced up and couldn't tell the computer what was happening. A perfectly good aircraft crashed as a result of a failure in vehicle management avionics.

Today's military fighters and bombers, as well as the newer commercial transports, are not actually "flown" by their pilots (don't tell them—it hurts their feelings!). The pilots are only allowed to give suggestions to the flight control computer. If the pilot wants to bring the nose up, he or she may ask for a tail deflection to bring the nose up, but the computer may decide otherwise if the nose is already coming up, or is up too high already, or if bringing the nose up right now would aggravate some dynamic or structural problem. If the pilot's wish were directly granted, the plane would crash.

Vehicle management systems can go way beyond such active flight control computers. There are serious proposals to fly aircraft by commanding only altitude and turn rate, or by specifying map coordinates (waypoints) to which the aircraft would fly itself. If such capabilities are combined with automated takeoff and landing, then anybody could "fly" an airplane. In fact, the unmanned Global Hawk flies just this way—and in case of emergency a remote pilot on the ground cannot "grab the stick" because there is none!

Vehicle management systems can also damp out flutter tendencies, suppress structural oscillations, and even redistribute the spanwise lift of the wing to improve aerodynamic efficiency or to reduce structural bending and increase fatigue life. The Lockheed L-1011 was certified many years ago with a load alleviation system in which the ailerons are actively driven up and down to damp out wing flexing in gusts. The B-1B uses actively controlled structural mode control vanes (the mini-canards up near the nose) to damp out fuselage structural bending in response to gust excitations when flying at 200 ft altitude {61 m} at almost the speed of sound.

Needless to say, vehicle management systems require powerful and redundant computers.

A real problem facing the aerospace industry is the rapid increase in the required computer programming before a new aircraft can enter operation. In less than 20 years, the lines of code for a typical military aircraft have increased by several orders of magnitude. This has a large cost impact, both in development and in support costs, and also leads to the possibility of hidden "bugs" impacting mission or safety. When these computers crash, so do you!

For a simple general-aviation aircraft, design of the avionics system consists mostly of selecting the usual components from catalogs. FAA regulations specify what radios, navigation aids, and other equipment are required for various types of operation, and the customer/pilot community has a good idea of what other equipment they expect. Component weights, geometries, and power and cooling requirements can be obtained from the manufacturers.

For more sophisticated aircraft, the avionics design is a system integration problem of the same magnitude as the design of the aircraft itself. Industry aircraft designers usually work with avionics experts in the company, who deal with the various component manufacturers. For military aircraft, most of the mission equipment avionics will be new, and it will be very difficult to get any actual installation drawings for the avionics components because they haven't been designed yet! Component weights, geometries, and power and cooling requirements will have to be estimated by the avionics experts, but that will take six months to a year.

However, for initial layout it is only necessary to provide sufficient volume in the avionics bays. If the actual avionics components are not yet defined, the required volumes can be estimated from the estimated avionics system weight. The required avionics weight should be estimated by the avionics experts using statistical and analytical means based on the required functionality (radar range, power, etc....). These predictions of avionics weights and size must take into account the ongoing and relentless improvement in electronics technologies in the military, commercial, and consumer fields. There is a continuing trend toward miniaturization and integration, such as the active phased arrays that are currently revolutionizing the design of radars and allowing as much as a 30% reduction in weight in the next generation of radars and intercept receivers. Miniature active phased arrays are even allowing antennas to be mounted directly on the skin of the aircraft. This will produce "smart skins" in which various electronic devices including antennas and radar countermeasures are integrated into the aircraft's skin during fabrication. Similarly, equipment such as onboard computers and GPS receivers are getting smaller almost every year (or allowing far greater capability for roughly the same size).

That last comment leads to an interesting paradox. Avionics are shrinking with each passing year due to technology improvements, yet the total weight of avionics for a certain class of aircraft actually hasn't reduced much, if at all. This curious result is easily explained. The avionics are getting better, not smaller. In part this is driven by increased mission needs, but in part it is the result of human expectations. If the last fighter had 1,000 lb {454 kg} of avionics, nobody will be surprised if the new fighter has the same and the avionics designers will think of great new capabilities that they can provide for that weight! Therefore, a simple approximation suitable for aircraft designers is to estimate avionics from a historical fraction of aircraft weight ignoring any potential "weight reductions" from improved technology. Table 11.6 provides historical ratios between avionics weight and aircraft empty weight. Once weight is estimated by whatever means, volume can be estimated assuming that avionics has an average density of about 30–45 $\mathrm{lb/ft^3}$ {480–720 $\mathrm{kg/m^3}$}.

The location of the avionics bay(s) is very important. Avionics, like most electronics, are sensitive to vibration, shock, and heat. They should be located fairly near the crew station to shorten the wires running between avionics and crew. They must have adequate electrical power, and, very important, they require cooling air. Avionics must be accessible for maintenance.

With all this considered, most aircraft will have avionics bays just in front of the cockpit, or below it, or sometimes even in the cockpit for certain bombers and

Table 11.6 Avionics weights

	Typical values: $\frac{W_{\text{avionics}}}{W_{\text{empty}}}$
General aviation—single engine	0.01–0.03
Light twin	0.02–0.04
Turboprop transport	0.02–0.04
Business jet	0.04–0.05
Jet transport	0.01–0.02
Fighters	0.03–0.08
Bombers	0.06–0.08
Jet trainers	0.03–0.04

transports. Many fighter aircraft have, starting at the nose, a radome, the radar inside, the first fuselage bulkhead, an avionics bay, and the cockpit all in a line.

For general-aviation aircraft, most of the avionics are actually mounted in the instrument panel. In conceptual design you must allow enough room behind (forward of) the panel for the largest avionics box that will ever be installed, plus enough room for wires, cables, tubes, and access to the entire area, which often looks like spaghetti after all the wiring is installed.

The design layout must also account for the installation of radar, which will often drive the shape of the nose of the aircraft. Radar dish size and power requirements depend upon the desired detection range, RCS of the threat aircraft, and radar frequency. Again, the avionics experts should estimate this for you using the radar range equation, a consequence of Maxwell's equations. For initial design layout, radar size can be estimated from similar aircraft. Or, it can be assumed that a bomber will use a 40 in. {100 cm} radar, a large fighter will use a 35 in. {90 cm} radar, and a small fighter will use a roughly 22 in. {56 cm} radar. Transport aircraft radars are only for weather avoidance, and are very small relative to the size of the aircraft's nose.

Military aircraft often have other electronic sensors and equipment requiring apertures such as electronic countermeasures (ECM), infrared search and track (IRST), and IR jammers. These must be located so as to provide the required field of view, which can be checked with a vision plot as discussed in Chapter 9.

Harrier landing on a ship (U.S. Navy photo).

INTERMISSION: STEP-BY-STEP DEVELOPMENT OF A NEW DESIGN

Up to this point, the book has presented the things that you need to know and do to create a viable new aircraft configuration design. The rest of the book describes how you should analyze and optimize that design, and what you should learn from the initial design to be ready to make the next iteration, i.e., to redraw the aircraft a little bit better the next time. After perhaps 5–25 iterations, each one a full design layout with analysis, optimization, and study by an expanding team of experts, you may finally have a design good enough to believe it might be the "right" answer for the given requirements.

This brief intermission between the design layout and the design analysis/optimization portions of the book is offered to pull together for students a step-by-step framework for creating a new aircraft design. This section summarizes the previous chapters and gives a preview of the coming chapters. Two caveats must be noted.

First, the procedures presented next are only one designer's opinions and would be hotly debated by almost every other designer! Different designers have found different approaches that work best for themselves. Also, different types of aircraft tend to be designed differently. In design of military fighters, the actual arrangement of the configuration concept is crucial from the earliest notional trade studies because the concept approaches can differ so widely, and the concept arrangement has a huge impact on weight and drag. Therefore, we spend much of our time early on considering alternative layout approaches and tend to avoid early optimizations based on purely historical estimations. For commercial aircraft, the basic concept is likely to be a circular fuselage with conventional wings and tails, so historical weight and drag estimates can be more readily used for initial trade studies. In design of helicopters, the design and analysis of the rotor and rotorhead are so important that the actual shape of the fuselage is almost an afterthought!

Second, even for the same type of aircraft there are really no standard procedures for aircraft layout. Every aircraft demands its own procedure. The designer must develop the aircraft drawing while simultaneously considering a wide variety of requirements, design drivers, and good design practices. This process is impossible to describe or teach, and is only learned through practice and a high level of desire. But, as promised, the following describes how this one designer usually develops a new concept.

ONE: You must have design requirements to begin a design. As repeatedly stated in Chapters 1–3, the early stages of the conceptual design process should include study of the requirements themselves, leading to suggestions for revising the requirements to produce a better match between fundamental needs and the affordability of the resulting aircraft. But, you cannot draw the initial layout without a set of requirements, even if you know they will change later. These may come from the customer, or you may have to make up some reasonable and typical requirements.

Design requirements should include a description of the operational need, the likely customer, and how the aircraft will be operated. Specific required values should include payload and/or number of passengers, range, takeoff and landing distances, and certain flight speeds (maximum, cruise, and stall or approach). Required rate of climb under certain conditions (such as immediately after takeoff with an engine out) should be included or inferred by reference to FAR or Mil Specs. In addition to these fundamental design requirements, different classes of aircraft will have certain unique requirements. For airliners there are specific requirements such as takeoff noise, passenger compartment dimensions and doors, emergency descent rates, and payload handling. For military aircraft there will be in-flight performance specifications such as turn rate and acceleration, and other design requirements such as stealth and radar capabilities (which drive antenna size and hence forebody shaping).

Another part of the definition of design requirements is the selection of what level of technology to incorporate (Chapter 3). If the aircraft project is not going to enter detail design until some later date, we must pick a technology availability date (TAD), typically being the expected beginning of detail design. During the design and analysis of the aircraft, everybody will base their design approach and analysis on the technologies that are expected to be available by that date.

TWO: Before you proceed further, you should gather a lot of data. You need the geometries and weights of all the major components you expect to incorporate in the design, such as the APU, seats, payload containers, bombs, guns, galleys, and toilets. You should also identify some candidate engines and obtain geometric, weights, and performance data. You can use the statistical models of Chapter 10, or use and possibly scale an actual engine, or get data for a notional new engine from an engine company. You should also take the time to acquaint yourself with the most-current thinking as to aerodynamic and structural design, flight control systems, stealth techniques, and similar areas. This is an area where teamwork, including the use of an aircraft design integrated product team, can be of tremendous help. Other members of the team will have in-depth knowledge that can be brought to bear (but this author still believes that one person, not a team, conceives a given design!).

THREE: Design sketches of numerous configuration concepts should be done next. As has been repeatedly stated, do not begin a design project thinking that you know the right approach. Look at many possibilities, sketch up alternatives, and if possible, do several competing initial designs as described next, including analysis and optimization.

When you are going to begin a design layout of a particular candidate concept, the design sketch is used to firm up your overall vision of the aircraft you intend to design, and to rough out the aerodynamic configuration as well as to decide where you expect to put the flight crew, payload/passengers, landing gear,

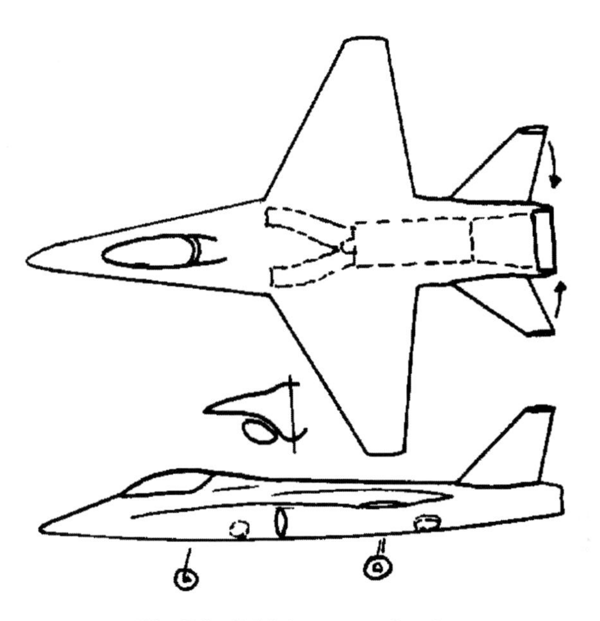

Fig. I.1 Initial concept sketch.

engine, fuel, and major subsystems. The sketch is also used to estimate the weights and drags for initial sizing calculations.

Fig. I.1 shows a typical sketch, for the DR-3 design example of Chapter 21. This and the DR-1 example illustrate the design methods and analysis/ optimization techniques of this book.

If you are designing an aircraft very much like prior aircraft (say, the next Airbus airliner or the next Cessna business jet), you may not need a concept sketch. You know roughly what the aircraft will look like without a sketch, and you can use historical data based on prior aircraft for initial estimates of weights and drags. But, this designer/author always does concept sketches. It's fun, and sometimes a better idea emerges to surprise you!

FOUR: From the design requirements, design sketch, and the selected engine, initial sizing calculations should be made to estimate design takeoff gross weight and fuel weight (Chapters 3 and 6), and the required T/W and W/S (Chapter 5). If you have a "canned" computer program available for rough initial sizing, by all means use it, but be very suspicious of the results until you have checked it with a few hand calculations!

Wing geometry must be initially selected (Chapter 4), along with tire sizes (Chapter 11) and fuel tank volumes (Chapter 10). For a jet you should estimate inlet capture area, or for a propeller aircraft, the propeller diameter (Chapter 10). The percent-chord locations of the wing front and back spar should be selected based on history and conversations with structures and aerodynamics experts, seeking a balance between room for structure and fuel, and room for large flaps and slats to maximize lift.

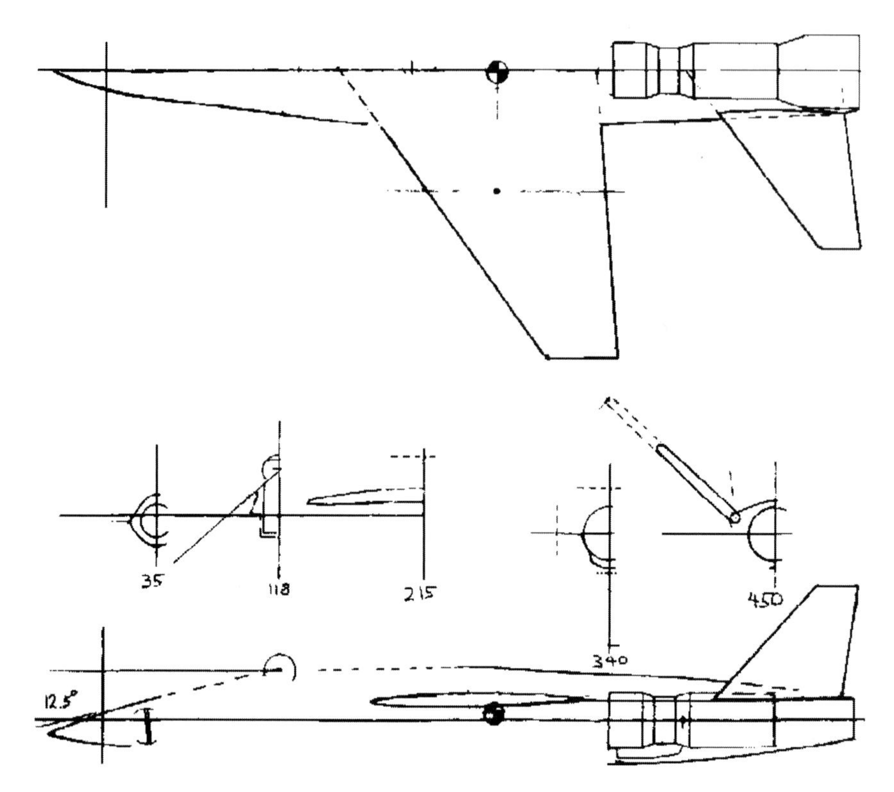

Fig. I.2 Beginning of design layout.

FIVE: Finally, you can begin the actual design layout. Start by calculating and laying out (or entering into the CAD system) the wing trapezoidal geometry, and place it on the drawing with respect to the desired c.g. location to get the desired stability (Chapter 7). Approximate where the fuselage begins and ends, and initially place the biggest components (probably engines, payload, and crew station). The fuselage length should be based upon prior aircraft or the statistical approach of Table 6.3, and should be modified as the drawing proceeds to contain the internal components. When fuselage length is approximately known, the tails can be sized using the tail volume coefficient method and added to the layout (Fig. I.2). As soon as possible, the landing gear should be located on the drawing.

SIX: Once the roughed-out layout is taking shape, the initial lofting approach is defined. The designer must select the number and locations of control stations, which are cross sections that define the shape (as opposed to the rest of the cross sections that are developed by fairing through the control stations). A minimum number of control stations should be used to avoid excessive "wiggling" of the resulting longitudinal contours. Control stations should be selected to ensure that all of the large internal components are properly

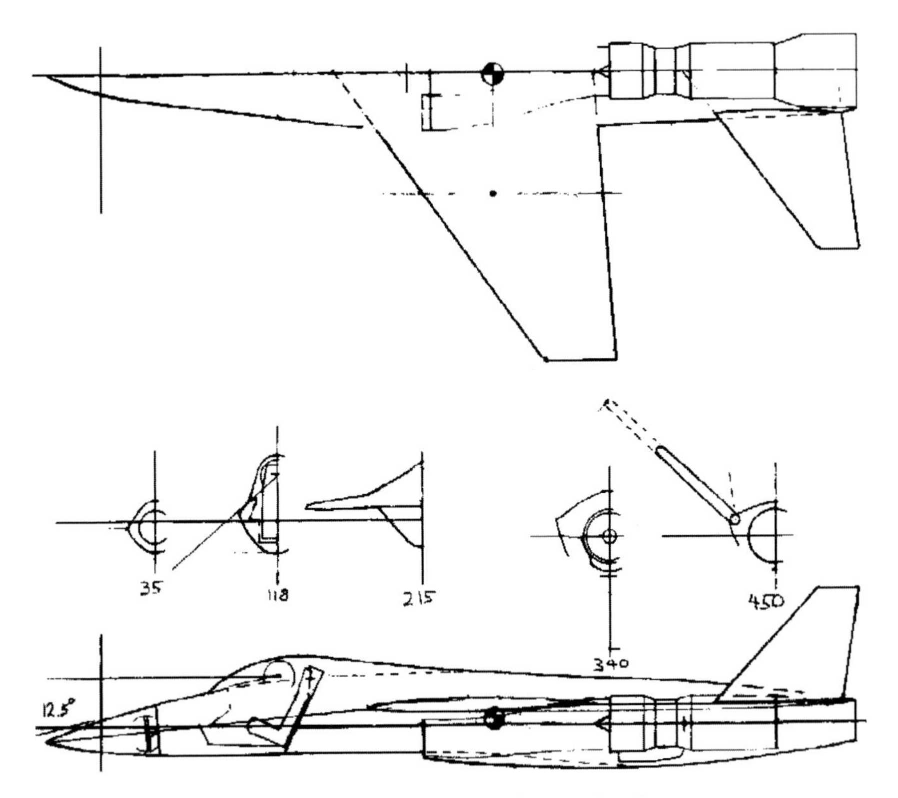

Fig. I.3 Design loft scheme and further development.

enclosed by the aircraft surface, so control stations are usually placed right at the location of the biggest components such as the engine and payload. Other control stations are placed to allow meeting geometric requirements such as overnose vision.

The designer must also decide on the number and type of curves that will define each cross section. This will determine the number of longitudinal control lines if a drafting table technique is used. On a CAD system, the number of curves in a cross section defines the number of "patches" that will be used for lofting the surface. Again, avoid a large number of defining curves to avoid wiggling.

Figure I.3 shows the design after key control stations are drawn and the loft schemes for fuselage, nacelle, inlet duct, and canopy are defined.

SEVEN: This step (Fig. I.4) involves detailing the landing gear and other systems, such as the avionics, payload, AMAD, and fuel system. Note that if the designer has not already planned for locating the landing gear before the loft scheme is defined, there is an excellent chance that it won't fit!

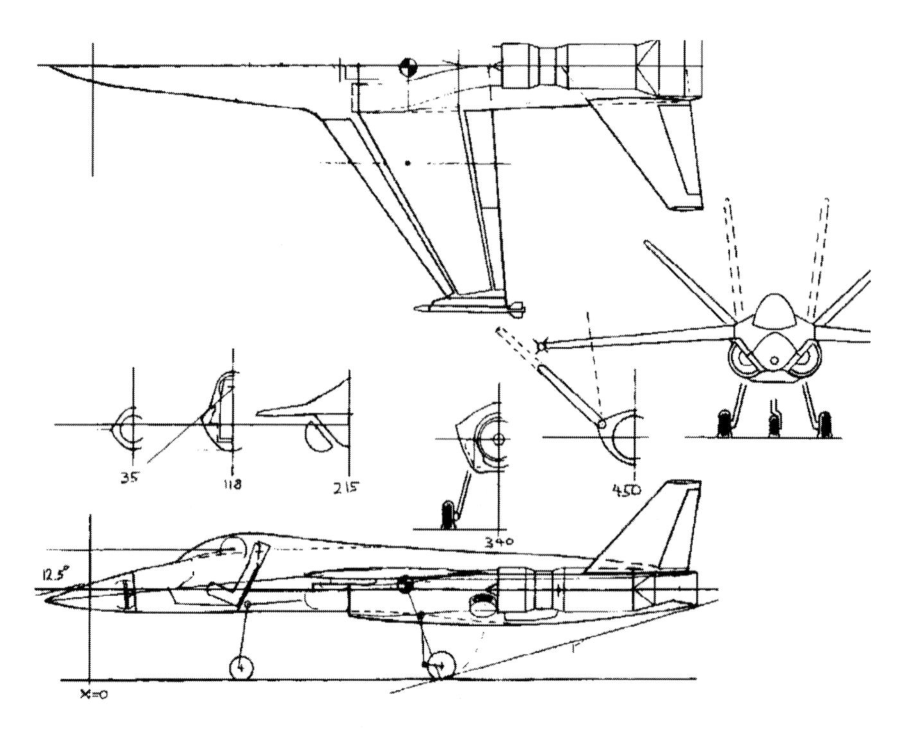

Fig. I.4 Landing gear and systems detailed.

EIGHT: Figure I.5 shows the final design layout, including fuel tanks, inlet details, and others. The wing and tail geometric parameters should be tabulated somewhere on the drawing (or attached to the CAD file), along with the estimated takeoff gross weight, fuel weight and volume, engine type and size (if not 100%), inlet capture area, propeller geometry, etc. This information will greatly aid those who later attempt to analyze the drawing.

NINE: The design layout must be *carefully* analyzed as to wetted areas, exposed planform areas, volume distribution, fuel volumes, and numerous lengths and other measurements needed for analysis. As mentioned before, be very wary of automatic CAD systems and always check the results for reasonableness using rough approximations such as those provided in Chapter 7.

TEN: The designer should now immediately begin preparations for redrawing the aircraft for the next design iteration. What didn't work out so well on the first drawing? Is the landing gear as simple as it could be? Did you have to use any design "tricks" to make something fit? Could the fuselage be made shorter, or could the wetted area be reduced some other way? Does the design have growth potential, or would a future fuselage stretch be impossible

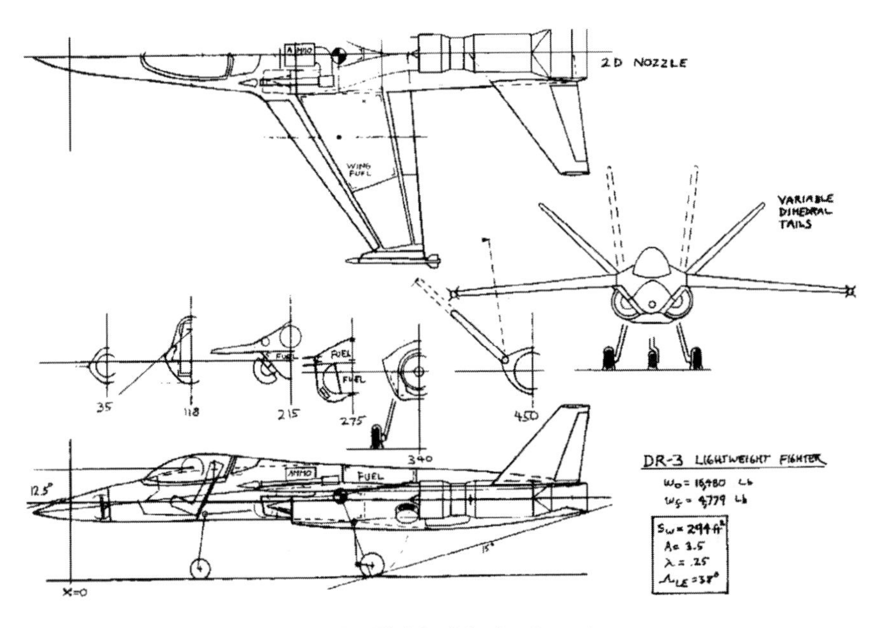

Fig. I.5 Finished design layout.

due to, say, tail-down ground angle? Do not "fall in love" with your design—there is always room for improvement.

The analysis process as discussed in the next chapters will result in a revised sizing calculation that will almost always tell you that the design you drew doesn't really work! Usually, the detailed calculations using actual numbers from your design indicate that the as-drawn aircraft cannot really meet the range or performance requirements. Also, optimization methods as discussed later will tell you that you should revise the T/W, W/S, and wing planform parameters.

So, get ready! Make notes on what to do next. Preparations for the next iteration should begin today! (Student design projects should always include a substantial write-up on what was learned and what will be done in the next iteration.)

Boeing 727 Tip Vortex Research Aircraft (NASA Photo).

12
Aerodynamics

12.1 Introduction

The previous chapters have presented methods for the design layout of a credible aircraft configuration. Initial sizing, wing geometry, engine installation, tail geometry, fuselage internal arrangement, and numerous other design topics have been discussed.

The initial sizing was based upon rough estimates of the aircraft's aerodynamics, weights, and propulsion characteristics. At that time we could not calculate the actual characteristics of the design because the aircraft had not been designed yet!

Now the aircraft design can be analyzed "as-drawn" to see if it actually meets the required mission range. If not, we will resize the aircraft until it does.

Also, a variety of trade studies can now be performed to determine the best combination of design parameters (T/W, W/S, aspect ratio, etc.) to meet the given mission and performance requirements at the minimum weight and cost.

Additional analysis on the as-drawn aircraft is also required at this time to ensure that stability and control requirements are met. In previous chapters, an approximate tail volume coefficient method was used for tail sizing. Now that the aircraft is drawn, we can analytically determine if the selected tail sizes are adequate.

These methods will be presented in Chapters 12–19. The overall objective of these chapters is the understanding of the sizing optimization and trade study process, not the presentation of particular theories and analysis methods. Presumably the student has spent several years studying aerodynamics, controls, structures, and propulsion.

The analysis techniques presented in these chapters illustrate the major parameters to be determined and provide realistic trends for trade studies. In many cases these are not the methods employed by the major aircraft companies, whose methods are highly computerized and cannot be presented in any single textbook. Also, each company uses many proprietary methods that are simply unavailable to students.

By using the simplified methods provided in this book, the student designer will experience the interaction of the major design variables but will not devote excessive time to the analytical tasks. This should leave more time for learning the basic principles of sizing optimization and trade studies, which are the same regardless of analytical techniques.

12.2 Aerodynamic Forces

Figure 12.1 shows the only two ways that the air mass and the airplane can act upon each other. As the aircraft moves forward, the air molecules slide over its skin. The molecules closest to the skin act as if they are stuck to it, moving with the aircraft (no-slip condition).

If the air molecules closest to the aircraft skin are moving with it, there must be slippage (or shear) between these molecules and the nonmoving molecules away from the aircraft. "Viscosity" is the honey-like tendency of air to resist shear deformation, which causes additional air near the aircraft skin to be dragged along with the aircraft. The force required to accelerate this boundary-layer air in the direction the aircraft is travelling produces skin-friction drag.

If the air molecules slide over each other (shear) in an orderly fashion, the flow is said to be "laminar." If the molecules shear in a disorderly fashion the flow is "turbulent." This produces a thicker boundary layer, indicating that more air molecules are dragged along with the aircraft, generating more skin-friction drag.

Airflow along a smooth plate becomes turbulent when the local Reynolds number reaches about one-half million, but can become turbulent at a lower Reynolds number if there is substantial skin roughness. Also, the curvature of the surface can either prevent or encourage the transition from laminar to turbulent flow.

As the aircraft moves forward, the air molecules are pushed aside. This causes the relative velocity of the air to vary about the aircraft. In some places, mostly toward the nose, the air is slowed down. In other places the air is speeded up relative to the freestream velocity.

According to Bernoulli's equation, the total pressure (static plus dynamic) along a subsonic streamline remains constant. If the local air velocity increases, the dynamic pressure has increased so the static pressure must decrease. Similarly, a reduction in local air velocity leads to an increase in static pressure.

Thus, the passage of the aircraft creates varying pressures around it, which push on the skin as shown in Fig. 12.1.

In fact, lift is created by forcing the air that travels over the top of the wing to travel faster than the air that passes under it. This is accomplished by the wing's

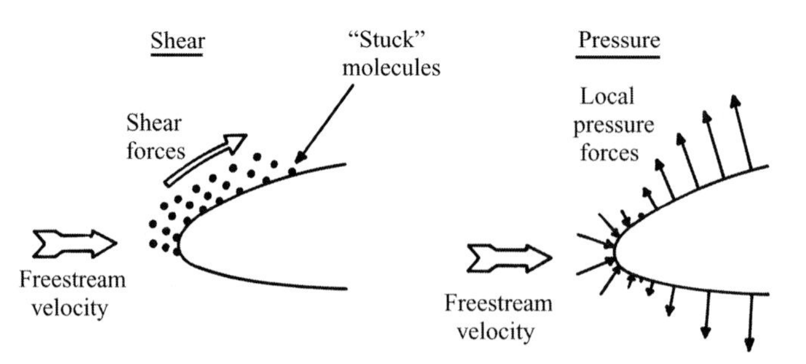

Fig. 12.1 Origin of aerodynamic forces.

angle of attack and/or wing camber. The resulting difference in air velocity creates a pressure differential between the upper and lower surfaces of the wing, which produces the lift that supports the aircraft.

If the aircraft is traveling near or above the speed of sound, additional pressure forces are produced by the shock waves around the aircraft. Shock waves result whenever supersonic flow is being slowed down.

All aerodynamic lift and drag forces result from the combination of shear and pressure forces. However, the dozens of classification schemes for aerodynamic forces can create considerable confusion because of overlapping terminology.

For example, the drag on a wing includes forces variously called airfoil profile drag, skin-friction drag, separation drag, parasite drag, camber drag, drag due to lift, wave drag, wave drag due to lift, interference drag—and so forth.

Figure 12.2 presents the various drag terminologies using a matrix that defines the drag type based upon the origin of the drag force (shear or pressure) and whether or not the drag is strongly related to the lift force being developed.

Drag forces not strongly related to lift are usually known as parasite drag or zero-lift drag. In subsonic cruising flight of a well-designed aircraft, the parasite drag consists mostly of skin-friction drag, which depends mostly upon the wetted area.

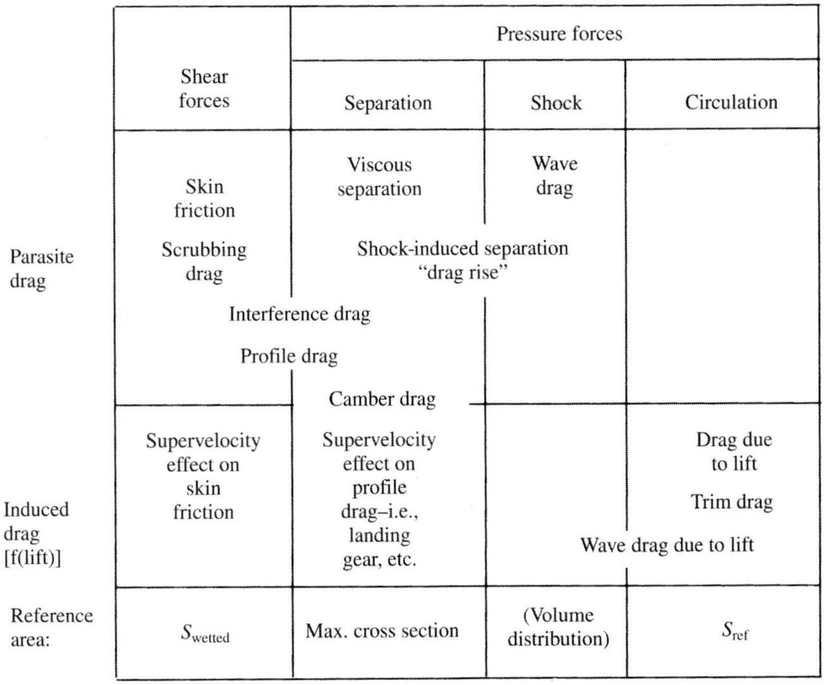

	Shear forces	Pressure forces		
		Separation	Shock	Circulation
Parasite drag	Skin friction Scrubbing drag	Viscous separation	Wave drag	
		Shock-induced separation "drag rise"		
	Interference drag Profile drag			
		Camber drag		
Induced drag [f(lift)]	Supervelocity effect on skin friction	Supervelocity effect on profile drag–i.e., landing gear, etc.		Drag due to lift Trim drag
			Wave drag due to lift	
Reference area:	S_{wetted}	Max. cross section	(Volume distribution)	S_{ref}

Fig. 12.2 Drag terminology matrix.

The skin-friction drag of a flat plate of the same wetted area as the aircraft can be determined for various Reynolds numbers and skin roughnesses using equations provided next. However, the actual parasite drag will be somewhat larger than this value, as will be shown later.

"Scrubbing drag" is an increase in the skin-friction drag due to the propwash or jet exhaust impinging upon the aircraft skin. This produces a higher effective air velocity and assures turbulent flow, both of which increase drag. It is for this reason that pusher-propellers are desirable, and that few modern jets have conformal nacelles in which the exhaust rubs along the aft fuselage.

There are three separate origins of the drag-producing pressure forces. The first, viscous separation, was the source of considerable difficulty during the early theoretical development of aerodynamics.

If the theoretical pressure forces in a perfect fluid are integrated over a streamlined body without flow separation, it is found that the pressures around the body that yield a drag force in the flight direction are exactly matched by the pressures around the body which yield a forward force. Thus, if skin friction is ignored the net drag is zero.

This was known to be false, and was called d'Alembert's paradox. The paradox was finally resolved by Prandtl who determined that the boundary layer, which is produced by viscosity, causes the flow to separate somewhere on the back half of the body. This prevents the full attainment of the forward-acting force, leaving a net drag force due to viscous separation. (See Ref. 35 for a more detailed discussion.)

Viscous separation drag, also called "form drag," depends upon the location of the separation point on the body. If the flow separates nearer to the front of the body, the drag is much higher than if it separates more towards the rear.

The location of the separation point depends largely upon the curvature of the body. Also, the separation point is affected by the amount of energy in the flow. Turbulent air has more energy than laminar air, so a turbulent boundary layer actually tends to delay separation.

If a body is small and flying at low speed, the Reynolds number will be so low that the flow will remain laminar resulting in separated flow. For this reason a small body may actually have a lower total drag when its skin is rough. This produces turbulent flow, which will remain attached longer than would laminar flow. The dimples on a golf ball are an example of this.

For a very long body such as the fuselage of an airliner, the turbulent boundary layer will become so thick that the air near the skin loses most of its energy. This causes separation near the tail of the aircraft, resulting in high "boattail drag."

To prevent this, small vanes perpendicular to the skin and angled to the airflow are placed just upstream of the separation point. These vanes produce vortices off their ends, which mix the boundary layer with higher-energy air from outside the boundary layer. This delays separation and reduces boattail drag. Such vortex generators are also used on wing and tail surfaces (see Chapter 8).

Viscous separation is largely responsible for the drag of irregular bodies such as landing gear and boundary-layer diverters. It also produces base drag, the pressure drag created by a "cutoff" aft fuselage.

The subsonic drag of a streamlined, nonlifting body consists solely of skin friction and viscous separation drag and is frequently called the profile drag. Profile drag is usually referenced to the maximum cross-sectional area of the body.

Note that the terms "profile drag" and "form drag" are frequently intermixed, although strictly speaking the profile drag is the sum of the form drag and the skin-friction drag. Also note that the term "profile drag" is sometimes used for the zero-lift drag of an airfoil.

Interference drag is the increase in the drag of the various aircraft components due to the change in the airflow caused by other components. For example, the fuselage generally causes an increase in the wing's drag by encouraging airflow separation at the wing root.

Interference drag usually results from an increase in viscous separation, although the skin-friction drag can also be increased if one component causes the airflow over another component to become turbulent or to increase in velocity.

"Wave drag" is the drag caused by the formation of shocks at supersonic and high subsonic speeds. At high subsonic speeds, the shocks form first on the upper surface of the wings because the airflow is accelerated as it passes over the wing.

Drag forces that are a strong function of lift are known as induced drag or drag due to lift. The induced drag is caused by the circulation about the airfoil that, for a three-dimensional wing, produces vortices in the airflow behind the wing. The energy required to produce these vortices is extracted from the wing as a drag force, and is proportional to the square of the lift.

Another way of looking at induced drag is that the higher-pressure air under the wing escapes around the wing tip to the wing upper surface, reducing the lift and causing the outer part of the wing to fly in an effective downwash. In other words, the wing is always flying uphill! This rotates the lift force vector toward the rear, so that a component of the lift is now in the drag direction.

To counter the pitching moment of the wing, the tail surfaces produce a lift force generally in the downward direction. The induced drag of the tail is called "trim drag." Trim drag also includes the additional lift required of the wing to counter any download produced by the tail.

When aircraft total drag vs lift is presented, the drag can be calculated with some fixed elevator deflection or it can be calculated using the varying elevator deflections required to trim the aircraft at each lift coefficient. This "trimmed" drag provides the correct data for use in performance calculations.

In supersonic flight there is a component of wave drag that changes as the lift changes. The creation of lift results from changes in the pressure around the aircraft. Wave drag is a pressure drag due to shock formation, and any changes in the pressures around the aircraft will change the location and strength of the shocks around it resulting in "wave drag due to lift." This drag is fairly small and is usually ignored in early conceptual design.

Two-dimensional (2-D) airfoil drag, or profile drag, is a combination of skin-friction drag and viscous separation drag. There is no drag due to lift for the 2-D airfoil because the lift force is perpendicular to the freestream direction. However, the profile drag increases as the angle of attack is increased, leading to some confusion.

This increase in 2-D airfoil drag is due to an increase in viscous separation caused by a greater pressure drop on the upper surface of the airfoil as the angle of attack is increased. This increase in profile drag with increasing angle of attack is not technically caused by the generation of lift, but does vary as the lift is varied.

Most preliminary drag-estimation methods do not actually use the airfoil profile drag data to determine total wing drag. Instead, the drag for an idealized wing with no camber or twist is determined, and then a separate camber drag is estimated. Often the camber drag term is included statistically in the drag-due-to-lift calculation even though it is not technically caused by the generation of lift!

Changing the lift on the wing changes the velocities above and below it. This change in local airflow velocity causes a small change in skin-friction drag. Sometimes called a "supervelocity" effect, this is minor and is usually ignored.

12.3 Aerodynamic Coefficients

Lift and drag forces are usually treated as nondimensional coefficients as defined in Eqs. (12.1) and (12.2). The wing reference area, S_{ref} or simply S, is the full trapezoidal area extending to the aircraft centerline. The dynamic pressure of the freestream air is called "q," as defined in Eq. (12.3),

$$L = qSC_L \tag{12.1}$$

$$D = qSC_D \tag{12.2}$$

where

$$q = \frac{1}{2}\rho V^2 \tag{12.3}$$

By definition, the lift force is perpendicular to the flight direction while the drag is parallel to the flight direction. Remember that the 2-D airfoil characteristics are denoted by lowercase subscripts (i.e., C_ℓ,) whereas the 3-D wing characteristics are denoted by uppercase subscripts (i.e., C_L).

Drag is normally spoken of as so many "counts" of drag, meaning the four digits to the right of the decimal place. For example, 38 counts of drag mean a drag coefficient of 0.0038.

Figure 12.3 illustrates the drag polar, which is the standard presentation format for aerodynamic data used in performance calculations. The drag polar is simply a plot of the coefficient of lift vs the coefficient of drag.

Note that the angle of attack (α) is indicated here by tic marks along the polar curve. This is not standard practice, but is useful for understanding the relationship between lift, drag, and angle of attack.

$$\text{Uncambered:} \qquad C_D = C_{D_0} + KC_L^2 \tag{12.4}$$

$$\text{Cambered:} \qquad C_D = C_{D_{\min}} + K(C_L - C_{L_{\text{min drag}}})^2 \tag{12.5}$$

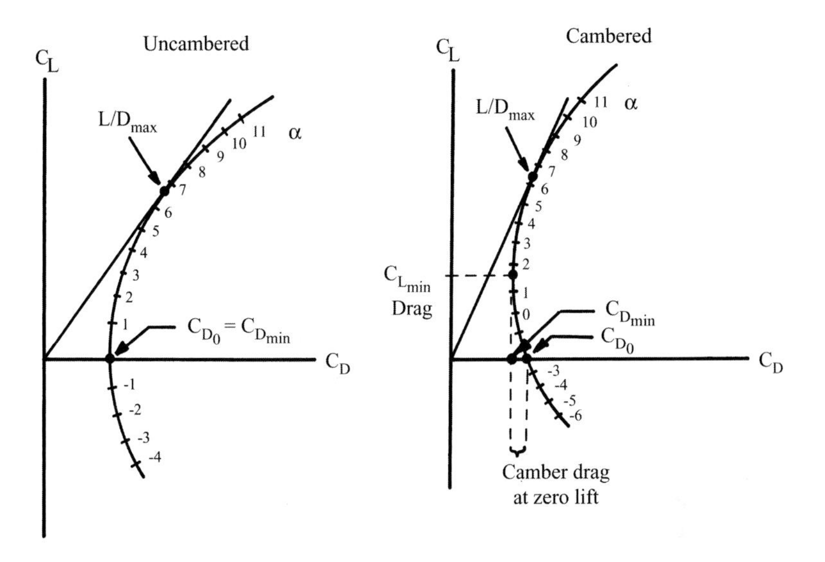

Fig. 12.3 Drag polar.

For an uncambered wing, the minimum drag (C_{D_0}) occurs when the lift is zero. The drag polar has an approximately parabolic shape, as defined by Eq. (12.4). The value of K will be discussed later.

For a cambered wing, the minimum drag ($C_{D_{\min}}$) occurs at some positive lift ($C_{L_{\min\,drag}}$). The drag polar also has a parabolic shape, but is offset vertically as defined by Eq. (12.5). For wings of moderate camber this offset is usually small, which implies that C_{D_0} approximately equals $C_{D_{\min}}$ and that Eq. (12.4) may be used.

The point at which a line from the origin is just tangent to the drag polar curve is the point of maximum lift-to-drag ratio. Note that this is not the point of minimum drag!

12.4 Lift

Figure 12.4 shows typical wing lift curves. The uncambered wing has no lift at zero angle of attack, while the cambered wing has a positive lift at zero angle of attack. A negative angle of attack is required to obtain zero lift with a cambered wing.

An old rule of thumb is that the negative angle of attack for zero lift in degrees equals the airfoil's percent camber (the maximum vertical displacement of the camber line divided by the chord).

Maximum lift is obtained at the stall angle of attack, beyond which the lift rapidly reduces. When a wing is stalled, most of the flow over the top has separated.

The slope of the lift curve is essentially linear except near the stall angle, allowing the lift coefficient below stall to be calculated simply as the lift-curve slope times the angle of attack (relative to the zero-lift angle). At the stall, the

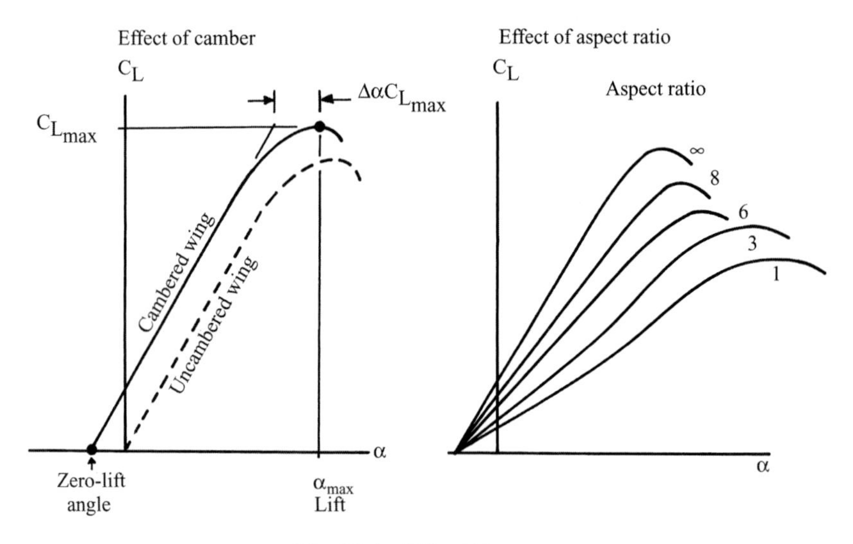

Fig. 12.4 Wing lift curve.

lift curve has become nonlinear such that the angle for maximum lift is greater than the linear value by an amount shown as $\Delta\alpha$ at $C_{L_{max}}$ in the figure.

Figure 12.4 also shows the effect of aspect ratio on lift. For an infinite-aspect-ratio wing (the 2-D airfoil case) the theoretical low-speed lift-curve slope is two times π (per radian).

Actual airfoils have lift-curve slopes between about 90 and 100% of the theoretical value. This percentage of the theoretical value is sometimes called the airfoil efficiency (η).

Reduction of aspect ratio reduces the lift-curve slope, as shown. At very low aspect ratios, the ability of the air to escape around the wing tips tends to prevent stalling even at very high angles of attack. Also note that the lift curve becomes nonlinear for very low aspect ratios.

Increasing the wing sweep has an effect similar to reducing the aspect ratio. A highly swept wing has a lift-curve slope much like the aspect-ratio three curve shown.

The effect of Mach number on the lift-curve slope is shown in Fig. 12.5. The 2-D airfoil lines represent upper boundaries for the no-sweep, infinite aspect-ratio wing. Real wings fall below these curves as shown.

Also, real wings follow a transition curve in the transonic regime between the upward-trending subsonic curve and the downward-trending supersonic curve. Note that a fat and unswept wing loses lift in the transonic regime whereas a thinner, swept wing does not.

The lift-curve slope is needed during conceptual design for three reasons. First, it is used to properly set the wing incidence angle. This can be especially important for a transport aircraft, in which the floor must be level during cruise. Also, the wing incidence angle influences the required fuselage angle of attack during takeoff and landing, which affects the aft-fuselage upsweep and/or landing gear length.

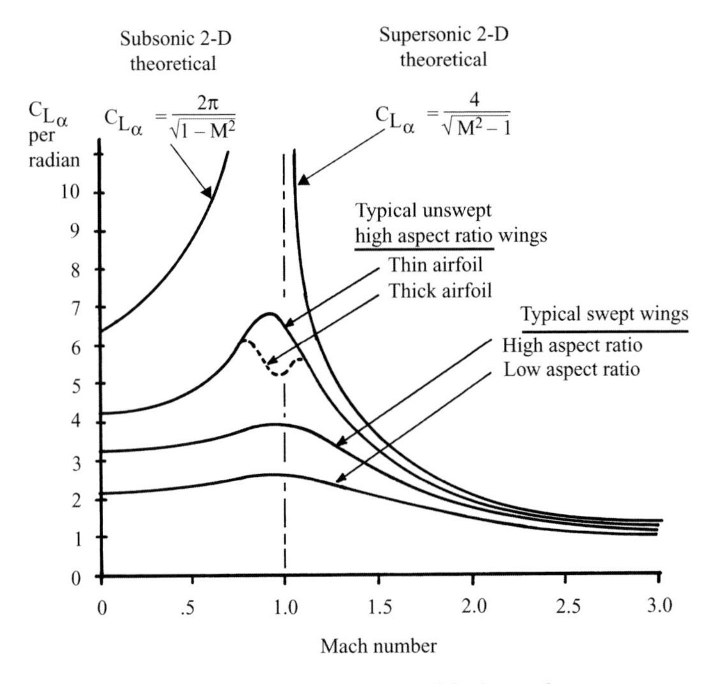

Fig. 12.5 Lift-curve slope vs Mach number.

Secondly, the methodology for calculating drag due to lift for high-performance aircraft uses the slope of the lift curve, as will be seen.

The third use of the lift-curve slope in conceptual design is for longitudinal-stability analysis, as discussed in Chapter 16.

Subsonic Lift-Curve Slope

Equation (12.6) is a semi-empirical formula from Ref. 36 for the complete wing lift-curve slope (per radian). This is accurate up to the drag-divergent Mach number, and reasonably accurate almost to Mach 1 for a swept wing.

$$C_{L\alpha} = \frac{2\pi A}{2 + \sqrt{4 + \dfrac{A^2\beta^2}{\eta^2}\left(1 + \dfrac{\tan^2 \Lambda_{\max t}}{\beta^2}\right)}}\left(\frac{S_{\text{exposed}}}{S_{\text{ref}}}\right)(F) \tag{12.6}$$

where

$$\beta^2 = 1 - M^2 \tag{12.7}$$

$$\eta = \frac{C_{\ell_\alpha}}{2\pi/\beta} \tag{12.8}$$

$\Lambda_{\max t}$ is the sweep of the wing at the chord location where the airfoil is thickest.

If the airfoil lift-curve slope as a function of Mach number is not known, the airfoil efficiency η can be approximated as about 0.95. (In several textbooks this term is dropped by assuming that η equals 1.0 at all Mach numbers.)

$S_{exposed}$ is the exposed wing planform, i.e., the wing reference area less the part of the wing covered by the fuselage. F is the fuselage lift factor [Eq. (12.9)] that accounts for the fact that the fuselage of diameter d creates some lift due to the "spill-over" of lift from the wing.

$$F = 1.07(1 + d/b)^2 \tag{12.9}$$

Sometimes the product $(S_{exposed}/S_{ref})F$ is greater than one, implying that the fuselage produces more lift than the portion of the wing it covers. This is unlikely and should probably be suppressed by setting this product to a value slightly less than 1.0, say 0.98.

The wing aspect ratio A is the geometric aspect ratio of the complete reference planform. The effective aspect ratio will be increased by wing endplates or winglets.

$$\text{Endplate:} \qquad A_{effective} = A(1 + 1.9h/b) \tag{12.10}$$

where h = endplate height.

$$\text{Winglet:} \qquad A_{effective} \cong 1.2A \tag{12.11}$$

These effective aspect ratios should also be used in the following induced drag calculations. Note that Eq. (12.11) for winglets is a crude approximation based upon limited data for wings of moderate aspect ratio.

The actual increase in effective aspect ratio due to the use of winglets is a function of velocity and lift coefficient and depends upon the selected airfoils and the relative location, geometry, and twist of the wing and winglet. Typically, a wing with higher aspect ratio will obtain less improvement by the use of winglets.

Supersonic Lift-Curve Slope

For a wing in purely supersonic flow, the lift-curve slope is ideally defined by Eq. (12.12), as shown in Fig. 12.5. A wing is considered to be in purely supersonic flow when the leading edge is "supersonic," i.e., when the Mach cone angle is greater than the leading-edge sweep [see Eq. (12.14)].

$$C_{L_\alpha} = 4/\beta \tag{12.12}$$

where

$$\beta = \sqrt{M^2 - 1} \tag{12.13}$$

when

$$M > 1/\cos \Lambda_{LE} \tag{12.14}$$

The actual lift-curve slope of a wing in supersonic flight is difficult to predict without use of a sophisticated computer program. The charts in Fig. 12.6 are probably the best approximate method available. They were defined in Ref. 37 and have been used in a number of textbooks.

These charts actually estimate the slope of the "normal force" coefficient (C_n), i.e., the lift-curve slope in a direction perpendicular to the surface of the wing. For low angles of attack, this is approximately equal to the lift-curve slope.

To use these charts, the wing aspect ratio, taper ratio, and leading-edge sweep are employed. The six charts each represent data for wings of a different taper ratio. If a chart for the actual taper ratio of a wing is not provided, interpolation must be used.

The term β [Eq. (12.13)] divided by the tangent of the leading-edge sweep is calculated and found on the horizontal axis of the chart. If this ratio is greater than 1.0, it is inverted and the right side of the chart must be used. Then the appropriate line is selected by calculating the wing aspect ratio times the tangent of the leading-edge sweep, and the vertical-axis value is read.

To obtain the approximate slope of the lift curve, this value is then divided by the tangent of the leading-edge sweep, if on the left side of the chart, or by β if on the right side of the chart.

As this value is referenced to the exposed planform of the wing, it must be multiplied by $(S_{\text{exposed}}/S_{\text{ref}})$ as in Eq. (12.6). Also, the value must be multiplied by F from Eq. (12.6) to account for the fuselage life effect.

Note that these charts give best results only for trapezoidal wings without kinks or strakes. For highly nontrapezoidal planforms, Ref. 37 contains additional estimation procedures. However, these charts are rarely used in industry where computerized "panel methods" are available. These are discussed later.

Transonic Lift-Curve Slope

In the transonic regime (roughly Mach 0.85–1.2 for a swept wing) there are no good initial-estimation methods for slope of the lift curve. It is suggested that the subsonic and supersonic values be plotted vs Mach number, and that a smooth curve be faired between the subsonic and supersonic values similar to the curves shown in Fig. 12.5.

Nonlinear Lift Effects

For a wing of very high sweep or very low aspect ratio (under two or three), the air escaping around the swept leading edge or wing tip will form a strong vortex that creates additional lift at a given angle of attack. This additional lift varies approximately by the square of the angle of attack. This nonlinear increase in the slope of the lift curve is difficult to estimate and can conservatively be ignored during early conceptual design.

(However, the increase in maximum lift due to vortex formation is very important. It will be discussed in the next section.)

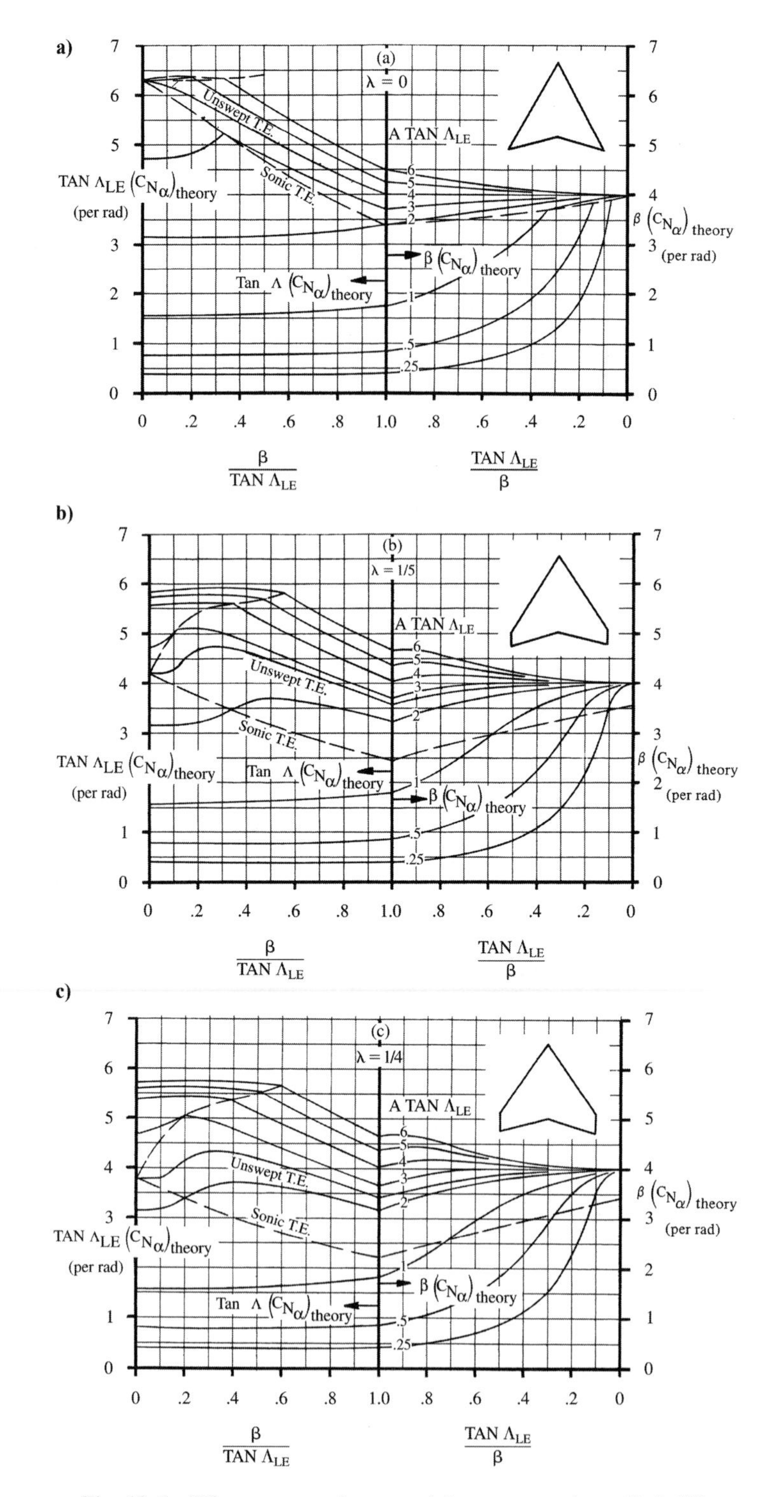

Fig. 12.6 Wing supersonic normal-force-curve slope (Ref. 37).

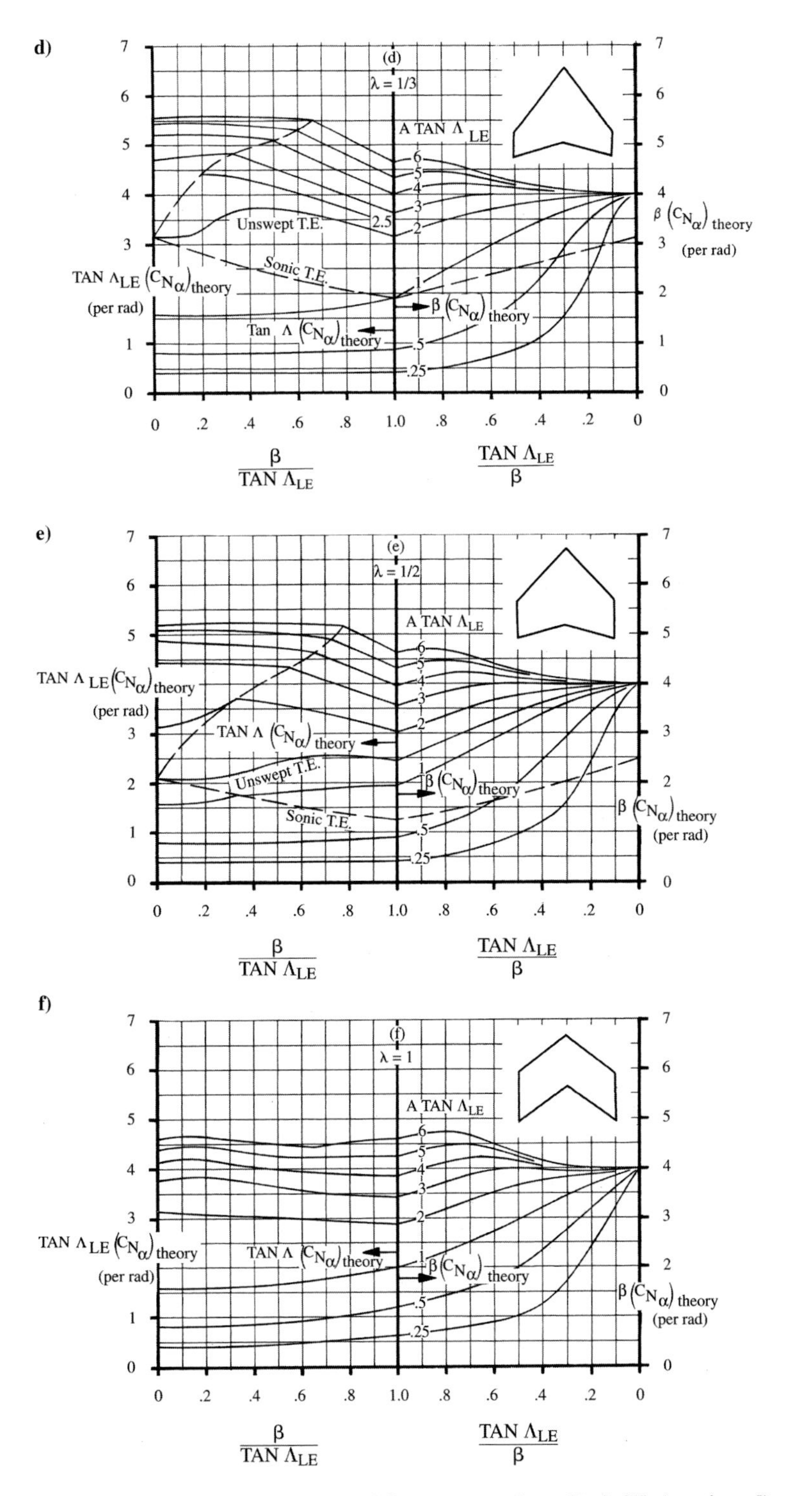

Fig. 12.6 Wing supersonic normal-force-curve slope (Ref. 37) (continued).

Maximum Lift (Clean)

The maximum lift coefficient of the wing will usually determine the wing area. This in turn will have a great influence upon the cruise drag. This strongly affects the aircraft takeoff weight to perform the design mission.

Thus, the maximum lift coefficient is critical in determining the aircraft weight; yet the estimation of maximum lift is probably the least reliable of all of the calculations used in aircraft conceptual design. Even refined wind-tunnel tests cannot predict maximum lift with great accuracy. Frequently an aircraft must be modified during flight test to achieve the estimated maximum lift.

For high-aspect-ratio wings with moderate sweep and a large airfoil leading-edge radius, the maximum lift depends mostly upon the airfoil characteristics. The maximum lift coefficient of the "clean" wing (i.e., without the use of flaps and other high-lift devices) will usually be about 90% of the airfoil's maximum lift as determined from the 2-D airfoil data at a similar Reynolds number (see typical data in Appendix D).

Sweeping the wing reduces the maximum lift, which can be found by multiplying the unswept maximum lift value by the cosine of the quarter-chord sweep [Eq. (12.15)]. This equation is reasonably valid for most subsonic aircraft of moderate sweep.

$$C_{L_{\max}} = 0.9 C_{\ell_{\max}} \cos \Lambda_{0.25c} \tag{12.15}$$

If a wing has a low aspect ratio or has substantial sweep and a relatively sharp leading edge, the maximum lift will be increased due to the formation of leading-edge vortices. This vortex formation is strongly affected by the shape of the upper surface of the leading edge.

Leading-edge shape could be defined by the airfoil nose radius. However, the nose radius alone does not take into account the effect of airfoil camber on the shape of the upper surface of the airfoil leading edge.

Instead, an arbitrary leading-edge sharpness parameter has been defined as the vertical separation between the points on the upper surface, which are 0.15% and 6% of the airfoil chord back from the leading edge (Fig. 12.7). The leading-edge

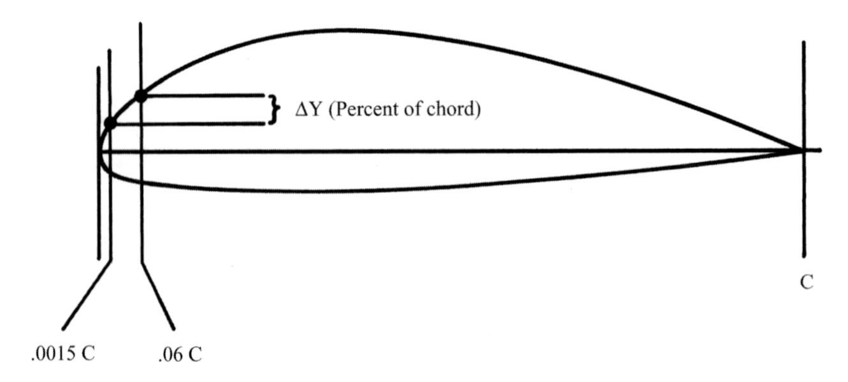

Fig. 12.7 Airfoil leading-edge sharpness parameter.

Table 12.1 Δy for common airfoils

Airfoil type	Δy
NACA 4 digit	26 t/c
NACA 5 digit	26 t/c
NACA 64 series	21.3 t/c
NACA 65 series	19.3 t/c
Biconvex	11.8 t/c

sharpness parameter (or Δy) as a function of thickness ratio for various airfoils is provided in Table 12.1.

The leading-edge sharpness parameter has been used in Ref. 37 to develop methods for the construction of the lift curve up to the stall, for low- or high-aspect-ratio wings. For high-aspect-ratio wings, Eq. (12.16) is used along with Figs. 12.8 and 12.9. The first term of Eq. (12.16) represents the maximum lift at Mach 0.2, and the second term represents the correction to a higher Mach number.

$$\text{High Aspect Ratio:} \qquad C_{L_{\max}} = C_{\ell_{\max}}\left(\frac{C_{L_{\max}}}{C_{\ell_{\max}}}\right) + \Delta C_{L_{\max}} \tag{12.16}$$

where $C_{\ell_{\max}}$ is the airfoil maximum lift coefficient at $M = 0.2$. This trapezoidal planform maximum lift can be adjusted for exposed planform and fuselage lift effects, as in Eq. (12.6).

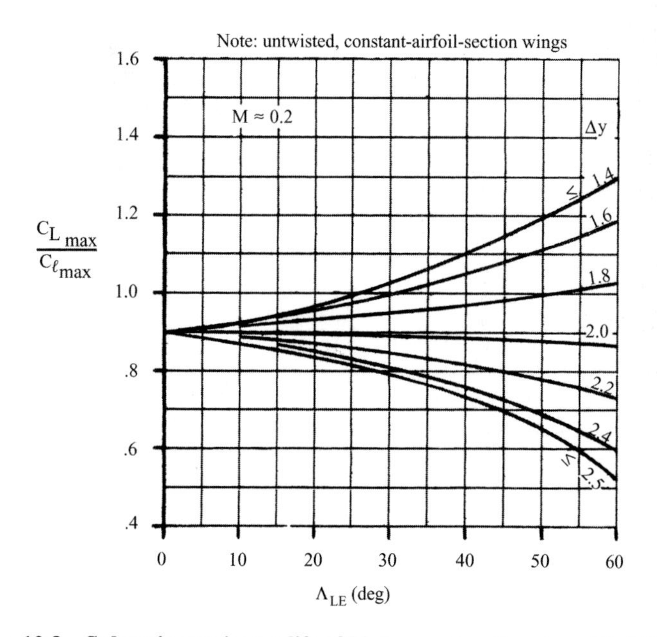

Fig. 12.8 Subsonic maximum lift of high-aspect-ratio wings (Ref. 37).

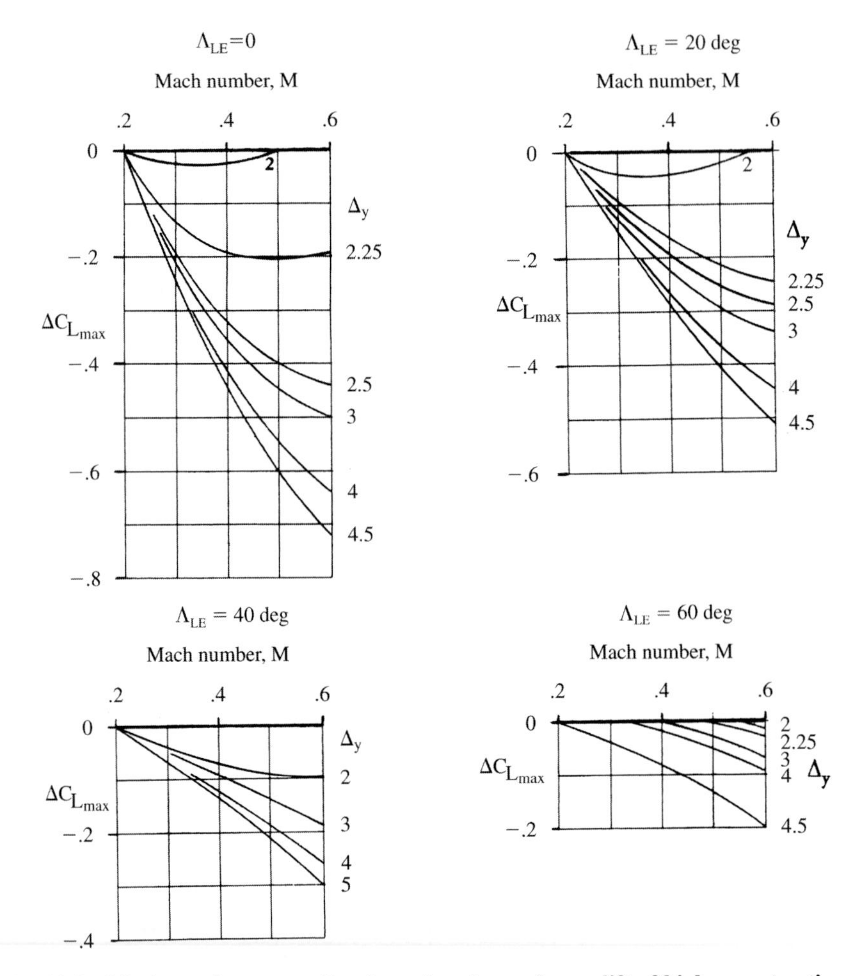

Fig. 12.9 Mach-number correction for subsonic maximum lift of high-aspect-ratio wings (Ref. 37).

The angle of attack for maximum lift is defined in Eq. (12.17) with the help of Fig. 12.10. Note that the first and second terms represent the angle of attack if the lift-curve slope were linear all the way up to stall. The second term may be approximated by the airfoil zero-lift angle, which is negative for a cambered airfoil. If the wing is twisted, the zero-lift angle is approximately the zero-lift angle at the mean chord location. The third term in Eq. (12.17) is a correction for the nonlinear effects of vortex flow.

$$\text{High Aspect Ratio:} \quad \alpha_{C_{L\max}} = \frac{C_{L\max}}{C_{L\alpha}} + \alpha_{0L} + \Delta\alpha_{C_{L\max}} \tag{12.17}$$

A different set of charts is used for a low-aspect-ratio wing, where vortex flow dominates the aerodynamics. For use of these charts, low aspect ratio is defined by Eq. (12.18), which uses the parameter C_1 from Fig. 12.11.

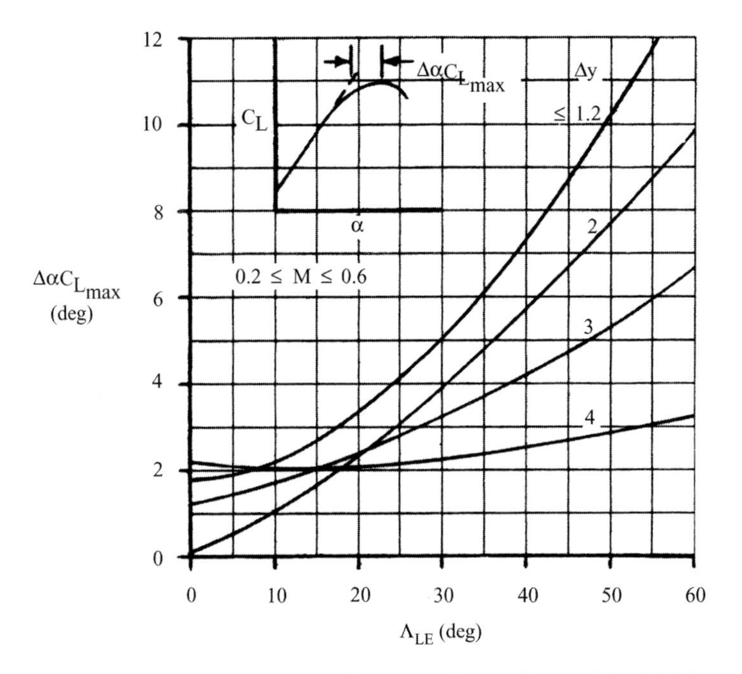

Fig. 12.10 Angle-of-attack increment for subsonic maximum lift of high-aspect-ratio wings (Ref. 37).

Maximum lift of a low-aspect-ratio wing is defined by Eq. (12.19) using Figs. 12.12 and 12.13. The angle of attack at maximum lift is defined by Eq. (12.20) using Figs. 12.14 and 12.15.

$$\text{Low Aspect Ratio if:} \qquad A \le \frac{3}{(C_1 + 1)(\cos \Lambda_{LE})} \tag{12.18}$$

$$\text{Low Aspect Ratio:} \qquad C_{L_{max}} = (C_{L_{max}})_{base} + \Delta C_{L_{max}} \tag{12.19}$$

$$\alpha_{C_{L_{max}}} = (\alpha_{C_{L_{max}}})_{base} + \Delta\alpha_{C_{L_{max}}} \tag{12.20}$$

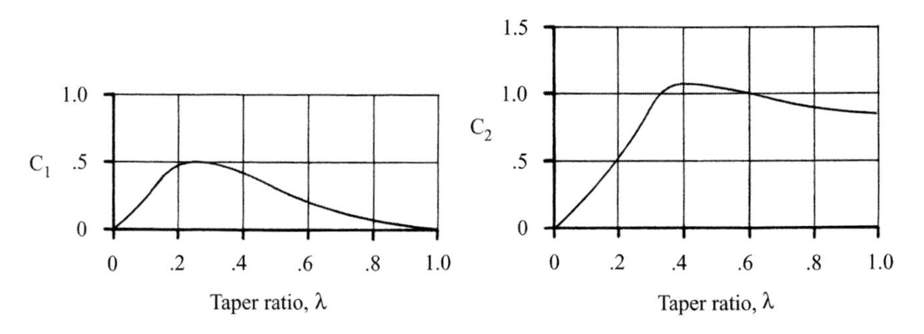

Fig. 12.11 Taper-ratio correction factors for low-aspect-ratio wings (Ref. 37).

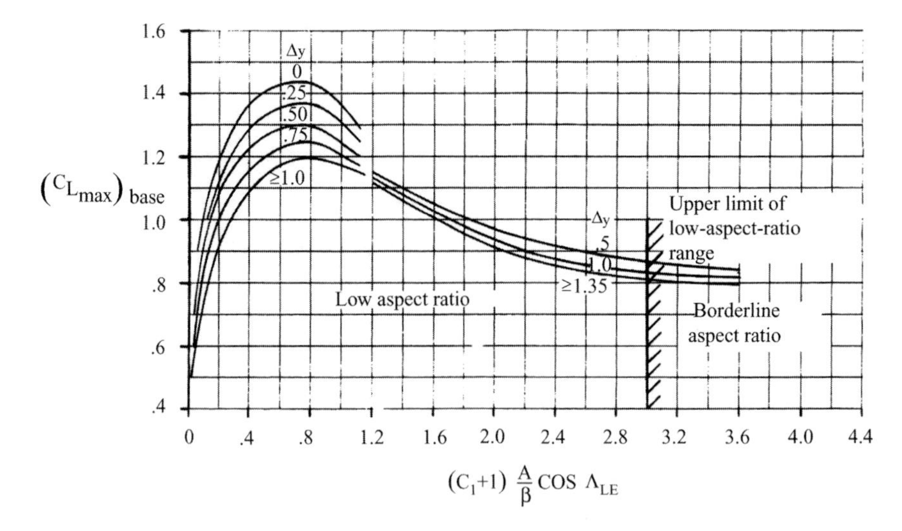

Fig. 12.12 Maximum subsonic lift of low-aspect-ratio wings (Ref. 37).

At transonic and supersonic speeds, the maximum lift a wing can achieve is usually limited by structural considerations rather than aerodynamics. Unless the aircraft is flying at a very high altitude, the available maximum lift at Mach 1 is usually enough to break the wings off!

Also, maximum lift is often limited by buffeting, controllability, or flexibility rather than by actual maximum lift. Wind-tunnel and flight-test data from similar designs are usually used to estimate the usable lift beyond the Mach limit of the preceding methods. Figure 12.16, developed by the author from various empirical sources, is a reasonable first approximation for normal designs. Determine the

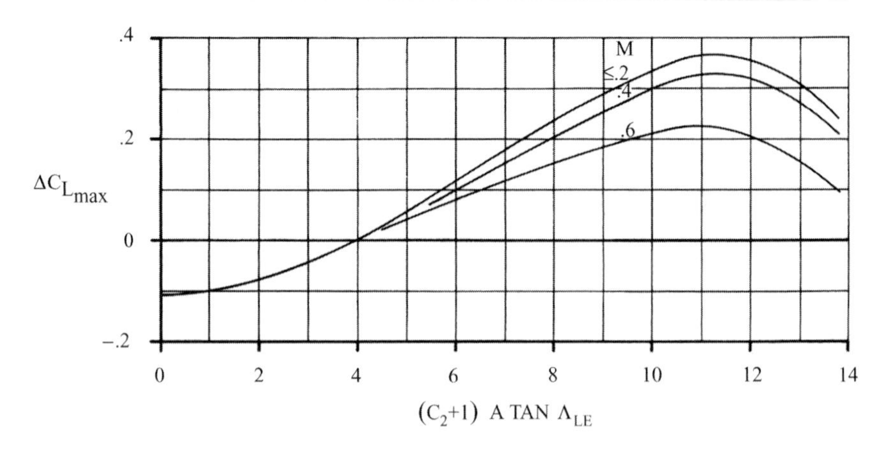

Fig. 12.13 Maximum-lift increment for low-aspect-ratio wings (Ref. 37).

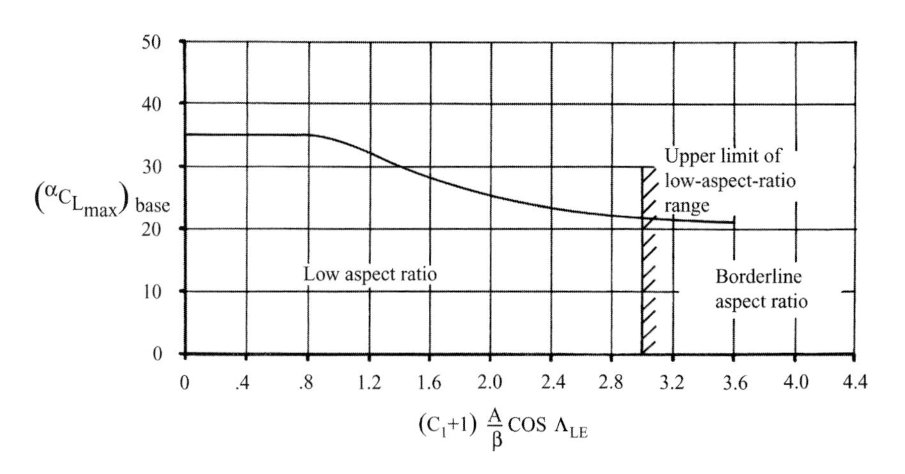

Fig. 12.14 Angle-of-attack for subsonic maximum lift of low-aspect-ratio wings (Ref. 37).

maximum lift coefficient at Mach 0.5 from the preceding methods, then multiply it by the factors from the curve.

Maximum Lift with High-Lift Devices

There is always a basic incompatibility in aircraft wing design. For cruise efficiency a wing should have little camber and should operate at a high wing-loading. For takeoff and landing a wing should have lots of lift, which means a lot of camber and a low wing-loading.

In the history of aviation almost every imaginable device for varying the wing camber and wing area has been attempted, including a wing with a telescoping

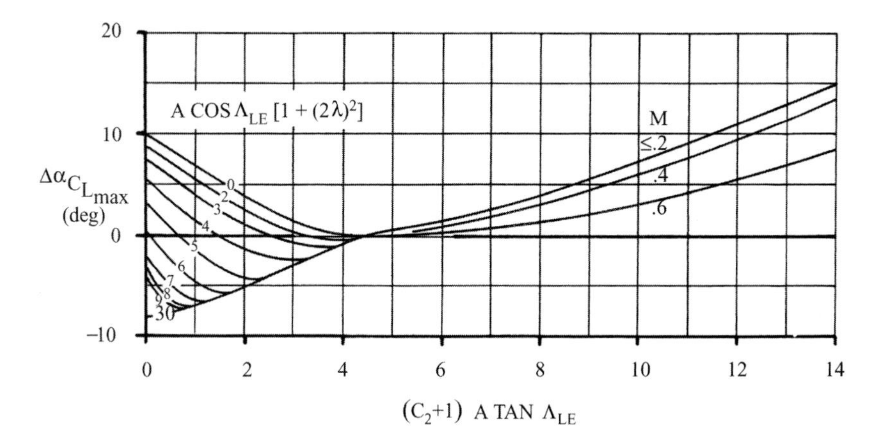

Fig. 12.15 Angle of attack increment for subsonic maximum lift of low-aspect-ratio wings (Ref. 37).

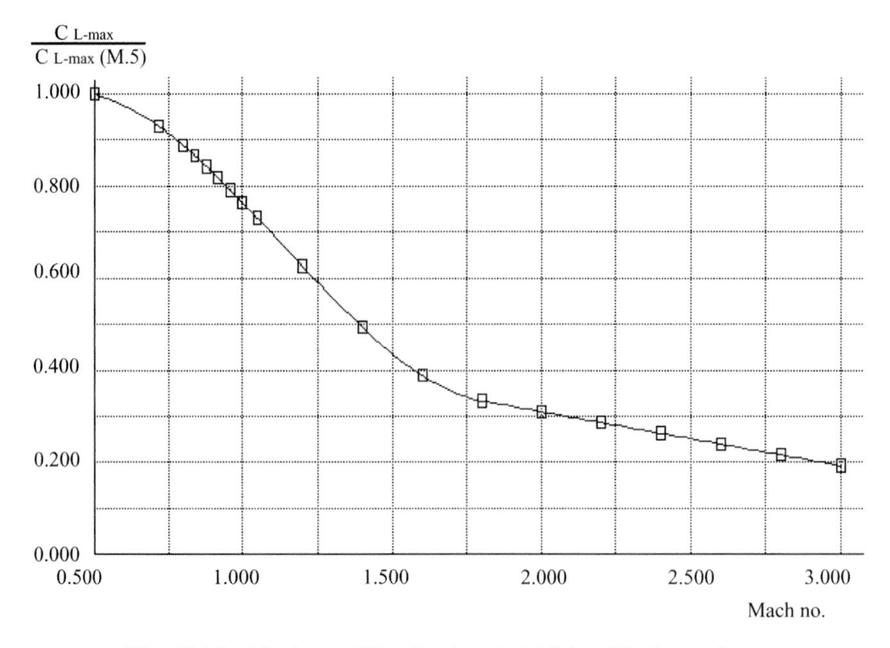

Fig. 12.16 Maximum lift adjustment at higher Mach numbers.

outer panel, a fabric membrane that unfurls behind the wing, a device that pivots out from the fuselage forming an extended flap, and even something called a "mutable" wing having variable span, camber, and sweep (Ref. 38).

Figure 12.17 illustrates the commonly used high lift flaps. The plain flap is simply a hinged portion of the airfoil, typically with a flap chord C_f of 30% of the airfoil chord. The plain flap increases lift by increasing camber. For a typical airfoil, the maximum lift occurs with a flap deflection angle of about 40–45 deg. Note that ailerons and other control surfaces are a form of plain flap.

The split flap is like the plain flap except that only the bottom surface of the airfoil is hinged. This produces virtually the same increase in lift as the plain flap. However, the split flap produces more drag and much less change in pitching moment, which may be useful in some designs. Split flaps are rarely used now but were common during World War II.

The slotted flap is a plain flap with a slot between the wing and the flap. This permits high-pressure air from beneath the wing to exit over the top of the flap, which tends to reduce separation. This increases lift and reduces drag.

The Fowler-type flap is like a slotted flap, but mechanized to slide rearward as it is deflected. This increases the wing area as well as the camber. Fowler flaps can be mechanized by a simple hinge located below the wing, or by some form of track arrangement contained within it.

To further improve the airflow over the Fowler flap, double- and even triple-slotted flaps are used on some airliners. These increase lift but at a considerable increase in cost and complexity.

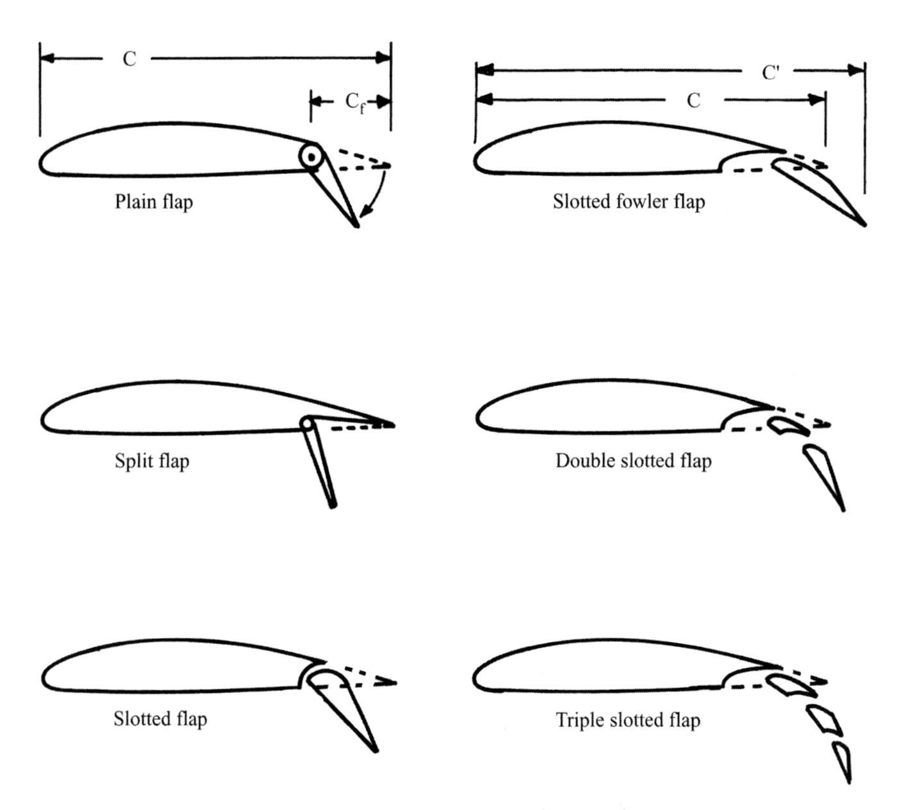

Fig. 12.17 Flap types.

Aft flaps do not increase the angle of stall. In fact, they tend to reduce the stall angle by increasing the pressure drop over the top of the airfoil, which promotes flow separation. To increase the stall angle, some form of leading-edge device is required, as shown in Fig. 12.18.

The leading-edge slot is simply a hole which permits high-pressure air from under the wing to blow over the top of the wing, delaying separation and stall. Usually such a slot is fixed, but may have closing doors to reduce drag at high speeds.

A leading-edge flap is a hinged portion of the leading edge that droops down to increase camber. This has the effect of increasing the curvature on the upper surface. The increase has been shown to be a major factor in determining maximum lift. Leading-edge flaps are usually used for improving the transonic maneuvering performance of high-speed fighters, which need a thin wing for supersonic flight.

A slotted leading-edge flap ("slat") provides increased camber, a slot, and an increase in wing area. Slats are the most widely used leading-edge device for both low-speed and transonic maneuvering. At transonic speeds, slats are also useful for reducing the buffetting tendency which may limit the usable lift. At Mach 0.9 the use of slats improved the usable lift of the F-4 by over 50%.

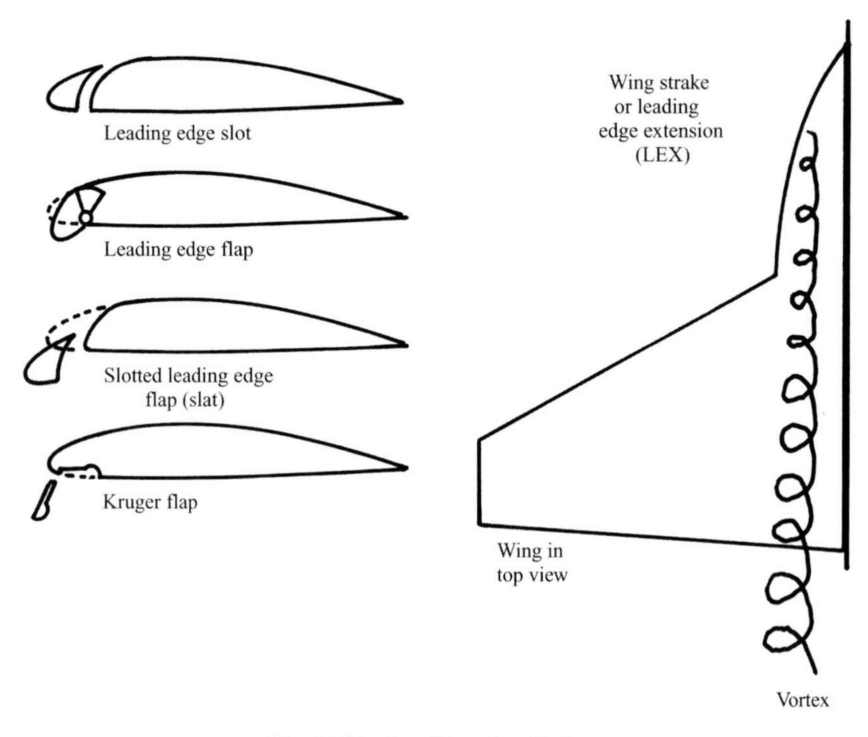

Fig. 12.18 Leading-edge devices.

The Kruger flap is used mostly by large airliners. It works as an air dam, forcing air up and over the top of the wing. Kruger flaps are lighter in weight than slats but produce higher drag at the lower angles of attack.

The wing strake, or "leading-edge extension" (LEX), is similar to the dorsal fin used on vertical tails. Like dorsal fins, the LEX at high angle of attack produces a vortex that delays separation and stall. Unfortunately, a LEX tends to promote pitch-up tendencies and so must be used with care.

Figure 12.19 illustrates the effects these high-lift devices have upon the lift curve of the wing. The nonextending flaps such as the plain, split, or slotted flaps act as an increase in camber, which moves the angle of zero lift to the left and increases the maximum lift. The slope of the lift curve remains unchanged, and the angle of stall is somewhat reduced.

An extending flap such as the Fowler type acts much like the other flaps as far as zero-lift angle and stall angle are concerned. However, the wing area is increased as the flap deflects, so the wing generates more lift at any given angle of attack compared to the nonextending flap.

Since the lift coefficient is referenced to the original wing area, not the extended wing area, the effective slope of the lift curve for an extending flap is increased by approximately the ratio of the total extended wing area to the original wing area.

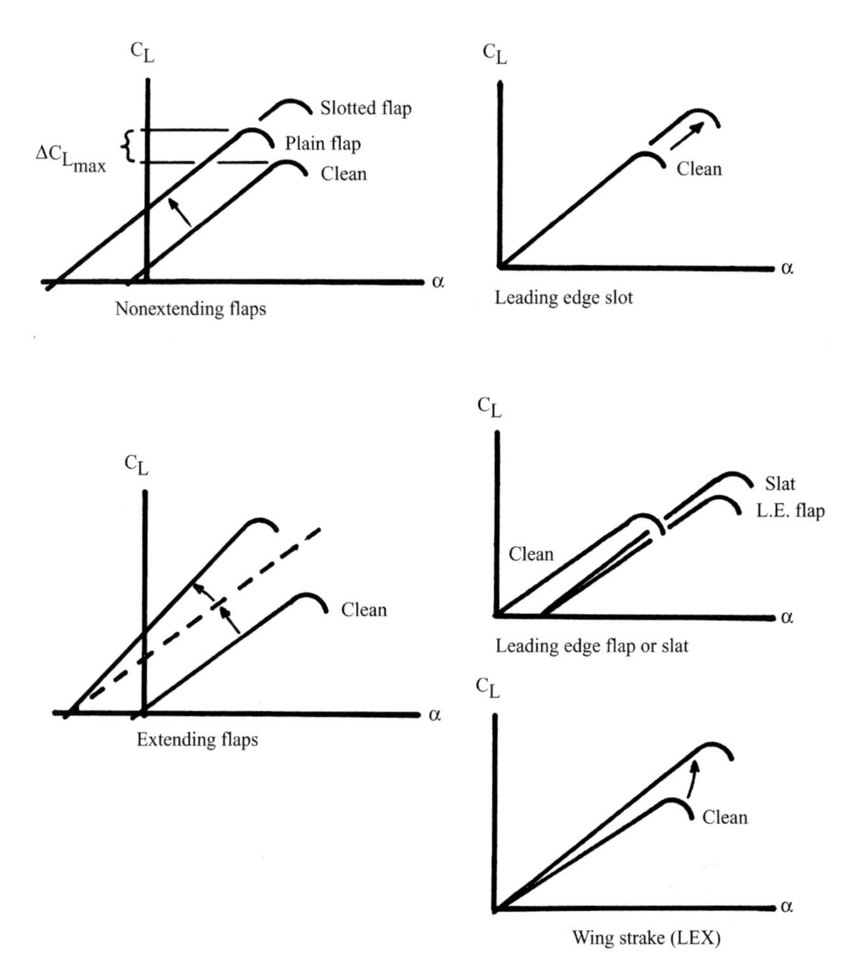

Fig. 12.19 Effects of high-lift devices.

Double- and triple-slotted flaps act much like single-slotted Fowler flaps, but the maximum lift is increased.

A leading-edge slot acts only to delay stall. A leading-edge flap or slat delays the stall, but also has the effect of reducing the lift at a given angle of attack (i.e., the lift curve moves to the right). This is because the droop in the leading edge acts as a reduction in the effective angle of attack as measured from the leading edge to the trailing edge. Note that a leading-edge slat, which increases wing area, also increases the slope of the lift curve much as does a Fowler flap.

Leading-edge devices alone do little to improve lift for takeoff and landing, because they are effective only at fairly high angles of attack. However, they are very useful when used in combination with trailing-edge flaps because they prevent premature airflow separation caused by the flaps.

The wing strake, or LEX, delays the stall at high angles of attack (over 20 deg). Also, the area of the LEX provides additional lift, thus increasing the slope of the lift curve. However, the LEX does little to increase lift at the angles of attack seen during takeoff and landing. The LEX does not delay the premature stall associated with trailing-edge flaps.

There are many complex methods for estimating the effects of high-lift devices, some of which are detailed in Ref. 37. Chapter 16 provides a method for estimating the lift increment due to a simple plain flap. For initial design, Eqs. (12.21) and (12.22) provide a reasonable estimate of the increase in maximum lift and the change in the zero-lift angle for various types of flaps and leading-edge devices when deployed at the optimum angle for high lift during landing.

$\Delta C_{\ell_{max}}$ values should be obtained from test data for the selected airfoil, or may be approximated from Table 12.2. For takeoff flap settings, lift increments of about 60–80% of these values should be used. The change in zero-lift angle for flaps in the 2-D case is approximately -15 deg at the landing setting, and -10 deg at the takeoff setting.

$$\Delta C_{L_{max}} = 0.9 \Delta C_{\ell_{max}} \left(\frac{S_{flapped}}{S_{ref}} \right) \cos \Lambda_{H.L.} \tag{12.21}$$

$$\Delta \alpha_{OL} = (\Delta \alpha_{OL})_{airfoil} \left(\frac{S_{flapped}}{S_{ref}} \right) \cos \Lambda_{H.L.} \tag{12.22}$$

In Eqs. (12.21) and (12.22), H.L. refers to the hinge line of the high-lift surface. $S_{flapped}$ is defined in Fig. 12.20. The lift increment for a leading-edge extension may be crudely estimated as 0.4 at high angles of attack.

Other methods for increasing the lift coefficient involve active flow control using either suction or blowing. Suction uses mechanical air pumps to suck the

Table 12.2 Approximate lift contributions of high-lift devices

High-lift device	$\Delta C_{\ell_{max}}$
Flaps	
Plain and split	0.9
Slotted	1.3
Fowler	1.3 c′/c
Double slotted	1.6 c′/c
Triple slotted	1.9 c′/c
Leading-edge devices	
Fixed slot	0.2
Leading edge flap	0.3
Kruger flap	0.3
Slat	0.4 c′/c

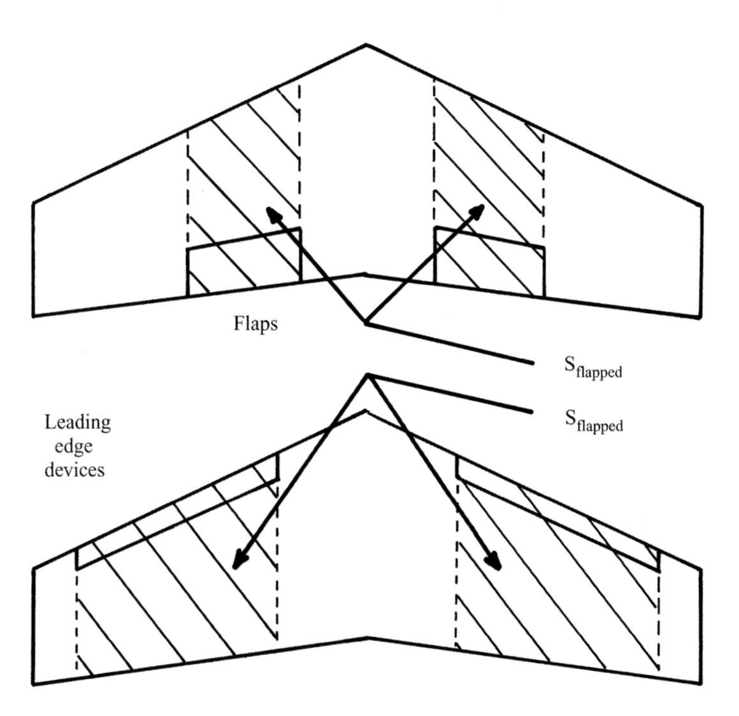

Fig. 12.20 "Flapped" wing area.

thickening boundary layer off the wing before it causes separation. This increases the stall angle of attack, and therefore increases maximum lift in a manner similar to leading-edge flaps.

Blowing uses compressor bleed air or compressed air provided by a mechanical air pump to prevent flow separation and increase the freestream-flow turning. Typically, the compressed air is exited through rearward-facing slots over the flaps or leading-edge flaps.

12.5 Parasite (Zero-Lift) Drag

Equivalent Skin-Friction Method

Two methods for the estimation of the parasite drag (C_{D_0}) are presented next. The first is based upon the fact that a well-designed aircraft in subsonic cruise will have parasite drag that is mostly skin-friction drag plus a small separation pressure drag. The latter is a fairly consistent percentage of the skin-friction drag for different classes of aircraft. This leads to the concept of an "equivalent skin friction coefficient" (C_{f_e}), which includes both skin-friction and separation drag.

C_{f_e} is multiplied by the aircraft's wetted area to obtain an initial estimate of parasite drag. This estimate [Eq. (12.23) and Table 12.3] is suitable for initial subsonic analysis and for checking the results of the more detailed method

Table 12.3 Equivalent skin friction coefficients

$C_{D_0} = C_{fe} \frac{S_{wet}}{S_{ref}}$	C_{f_e}-subsonic
Bomber and civil transport	0.0030
Military cargo (high upsweep fuselage)	0.0035
Air Force fighter	0.0035
Navy fighter	0.0040
Clean supersonic cruise aircraft	0.0025
Light aircraft–single engine	0.0055
Light aircraft–twin engine	0.0045
Prop seaplane	0.0065
Jet seaplane	0.0040

described in the next section.

$$C_{D_0} = C_{fe} \frac{S_{wet}}{S_{ref}} \tag{12.23}$$

Component Buildup Method

The component buildup method estimates the subsonic parasite drag of each component of the aircraft using a calculated flat-plate skin-friction drag coefficient (C_f) and a component "form factor" (FF) that estimates the pressure drag due to viscous separation. Then the interference effects on the component drag are estimated as a factor "Q" and the total component drag is determined as the product of the wetted area, C_f, FF, and Q.

(Note that the interference factor Q should not be confused with dynamic pressure q.)

Miscellaneous drags ($C_{D_{misc}}$) for special features of an aircraft such as flaps, unretracted landing gear, an upswept aft fuselage, and base area are then estimated and added to the total, along with estimated contributions for leakages and protuberances ($C_{D_{L\&P}}$). Subsonic parasite-drag buildup is shown in Eq. (12.24), where the subscript "c" indicates that those values are different for each component.

$$(C_{D_0})_{subsonic} = \frac{\Sigma(C_{f_c} FF_c Q_c S_{wet_c})}{S_{ref}} + C_{D_{misc}} + C_{D_{L\&P}} \tag{12.24}$$

For supersonic flight, the skin-friction contribution is simply the flat-plate skin friction coefficient times the wetted area. All supersonic pressure drag contributions (except base drag) are included in the wave-drag term, which is determined from the total aircraft volume distribution.

For transonic flight, a graphical interpolation between subsonic and supersonic values is used. Supersonic and transonic drag calculations are discussed later.

Flat-Plate Skin Friction Coefficient

The flat-plate skin friction coefficient C_f depends upon the Reynolds number, Mach number, and skin roughness. The most important factor affecting skin friction drag is the extent to which the aircraft has laminar flow over its surfaces.

At a local Reynolds number of one million, a surface with turbulent flow will have a friction drag coefficient as much as three times the drag coefficient of a surface with laminar flow. Laminar flow may be maintained if the local Reynolds number is below roughly half a million, and only if the skin is very smooth (molded composite or polished aluminum without rivets).

Most current aircraft have turbulent flow over virtually the entire wetted surface, although some laminar flow may be seen towards the front of the wings and tails. A typical current aircraft may have laminar flow over perhaps 10–20% of the wings and tails, and virtually no laminar flow over the fuselage.

A carefully designed modern composite aircraft such as the Piaggio GP180 can have laminar flow over as much as 50% of the wings and tails, and about 20–35% of the fuselage.

For the portion of the aircraft that has laminar flow, the flat-plate skin friction coefficient is expressed by Eq. (12.25). Note that laminar flow is unlikely at transonic or supersonic speeds, unless great attention is paid to shaping and surface smoothness.

$$\text{Laminar:} \qquad C_f = 1.328/\sqrt{R} \tag{12.25}$$

where R is the nondimensional Reynolds number defined as

$$R = \rho V \ell / \mu \tag{12.26}$$

The "ℓ" in Eq. (12.26) is the characteristic length. For a fuselage, ℓ is the total length. For a wing or tail, ℓ is the mean aerodynamic chord length.

For turbulent flow, which in most cases covers the whole aircraft, the flat-plate skin friction coefficient is determined by Eq. (12.27). Note that the second term in the denominator, the Mach number correction, goes to 1.0 for low-subsonic flight.

$$\text{Turbulent:} \quad C_f = \frac{0.455}{(\log_{10} R)^{2.58}(1 + 0.144M^2)^{0.65}} \tag{12.27}$$

Figure 12.21 depicts the flat-plate skin friction coefficient vs Reynolds number by these equations.

If the surface is relatively rough, the friction coefficient will be higher than indicated by Eq. (12.27). This is accounted for by the use of a "cutoff Reynolds number," which is determined from Eq. (12.28) or (12.29) using the characteristic length, ℓ (feet) and a skin-roughness value k based upon Table 12.4. The lower of the actual Reynolds number and the cutoff Reynolds number should be

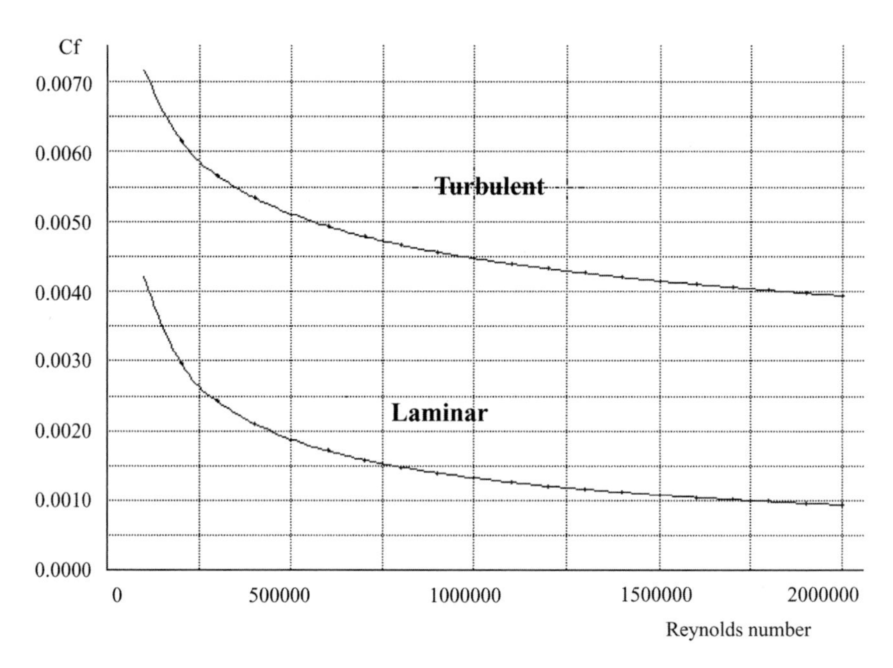

Fig. 12.21 Flat-plate skin friction coefficient vs Reynolds number.

used in Eq. (12.27).

$$\text{Subsonic:} \quad R_{\text{cutoff}} = 38.21(\ell/k)^{1.053} \tag{12.28}$$

$$\text{Transonic or Supersonic:} \quad R_{\text{cutoff}} = 44.62(\ell/k)^{1.053} M^{1.16} \tag{12.29}$$

Once laminar and turbulent flat-plate skin friction coefficients have been calculated, an average coefficient can be calculated as the weighted average of the two. This requires estimation of the percentage of laminar flow which can be attained. This estimation is a judgment call based on past experience as just discussed, and one must review the current literature to determine how much laminar flow can be attained with current state of the art.

Table 12.4 Skin roughness value k

Surface	k, ft	k, m
Camouflage paint on aluminum	3.33×10^{-5}	1.015×10^{-5}
Smooth paint	2.08×10^{-5}	0.634×10^{-5}
Production sheet metal	1.33×10^{-5}	0.405×10^{-5}
Polished sheet metal	0.50×10^{-5}	0.152×10^{-5}
Smooth molded composite	0.7×10^{-5}	0.052×10^{-5}

Component Form Factors

Form factors for subsonic-drag estimation are presented in Eqs. (12.30–12.32). These are considered valid up to the drag-divergent Mach number. In Eq. (12.30), the term "$(x/c)_m$" is the chordwise location of the airfoil maximum thickness point. For most low-speed airfoils, this is at about 0.3 of the chord. For high-speed airfoils this is at about 0.5 of the chord. Λ_m refers to the sweep of the maximum-thickness line.

Wing, Tail, Strut, and Pylon:

$$FF = \left[1 + \frac{0.6}{(x/c)_m}\left(\frac{t}{c}\right) + 100\left(\frac{t}{c}\right)^4\right]\left[1.34M^{0.18}(\cos\Lambda_m)^{0.28}\right] \tag{12.30}$$

Fuselage and Smooth Canopy:

$$FF = \left(1 + \frac{60}{f^3} + \frac{f}{400}\right) \tag{12.31}$$

Nacelle and Smooth External Store:

$$FF = 1 + (0.35/f) \tag{12.32}$$

where

$$f = \frac{\ell}{d} = \frac{\ell}{\sqrt{(4/\pi)A_{\max}}} \tag{12.33}$$

A tail surface with a hinged rudder or elevator will have a form factor about 10% higher than predicted by Eq. (12.30) due to the extra drag of the gap between the tail surface and its control surface.

Equation (12.31) is mainly used for estimation of the fuselage form factor, but can also be used for a blister or fairing such as a pod used for landing-gear stowage.

For a fuselage with a steep aft-fuselage closure angle in front of a pusher propeller, the separation drag will be lower than predicted using this form-factor equation.

A square-sided fuselage has a form factor about 30–40% higher than the value estimated with Eq. (12.31) due to additional separation caused by the corners. This can be somewhat reduced by rounding the corners. A flying-boat hull has a form factor about 50% higher, and a float has a form factor about three times the estimated value.

Equation (12.31) will predict the form factor for a smooth, one-piece fighter canopy such as seen on the F-16. For a typical two-piece canopy with a fixed but streamlined windscreen (i.e., F-15), the form factor calculated with Eq. (12.31) should be increased by about 40%. A canopy with a flat-sided windscreen has a form factor about three times the value estimated with Eq. (12.31).

The external boundary-layer diverter for an inlet mounted on the fuselage can have a large drag contribution. Equations (12.34) and (12.35) estimate the form factors to use for a double-wedge and single-wedge diverter, where the Reynolds number is determined using ℓ and the wetted area is defined as shown in Fig. 12.22. Remember to double the drag if there are two inlets.

$$\text{Double Wedge:} \quad FF = 1 + (d/\ell) \tag{12.34}$$

$$\text{Single Wedge:} \quad FF = 1 + (2d/\ell) \tag{12.35}$$

Component Interference Factors

Parasite drag is increased due to the mutual interference between components. For a nacelle or external store mounted directly on the fuselage or wing, the interference factor Q is about 1.5. If the nacelle or store is mounted less than about one diameter away, the Q factor is about 1.3. If it is mounted much beyond one diameter, the Q factor approaches 1.0. Wing-tip-mounted missiles have a Q factor of about 1.25.

For a high-wing, a midwing, or a well-filletted low wing, the interference will be negligible so the Q factor will be about 1.0. An unfilletted low wing can have a Q factor from about 1.1–1.4.

The fuselage has a negligible interference factor ($Q = 1.0$) in most cases. Also, $Q = 1.0$ for a boundary-layer diverter. For tail surfaces, interference ranges from about 3% ($Q = 1.03$) for a clean V-tail to about 8% for an H-tail. For a conventional tail, 4–5% may be assumed (Ref. 8).

Component parasite drags can now be determined using Eq. (12.24) and the skin friction coefficients, form factors, and interference factors.

Miscellaneous Drags

The drag of miscellaneous items can be determined separately using a variety of empirical graphs and equations, and then adding the results to the parasite drags just determined.

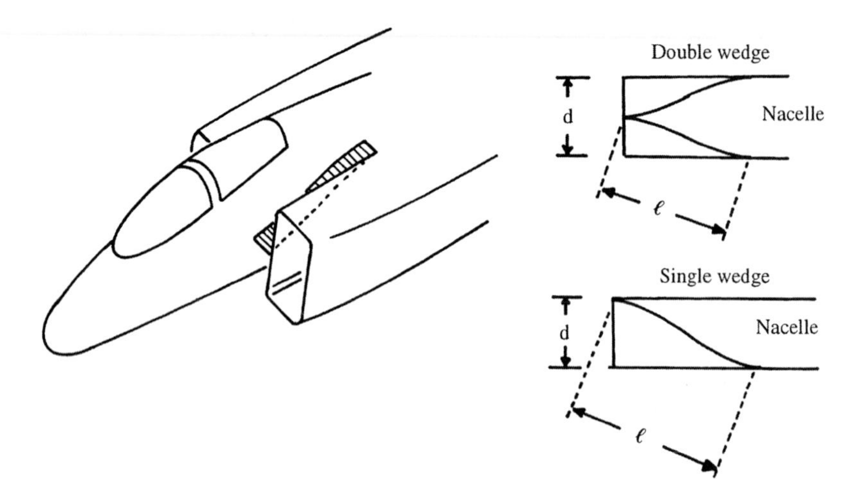

Fig. 12.22 Inlet boundary-layer diverter.

While the drag of smooth external stores can be estimated using Eq. (12.31), the majority of external stores is in fact not very smooth. Figures 12.23 and 12.24 provide drag estimates for external fuel tanks and weapons, presented as drag divided by dynamic pressure (D-over-q or D/q).

D/q has units of square feet (or meters) and so is sometimes called the "drag area." D/q divided by the wing reference area yields the miscellaneous parasite drag coefficient. Note that pylon and bomb-rack drag as estimated using Fig. 12.25 must be added to the store drag.

Most transport and cargo aircraft have a pronounced upsweep to the aft fuselage (Fig. 12.26). This increases the drag beyond the value calculated using Eq. (12.31). This extra drag is a complicated function of the fuselage cross-sectional shape and the aircraft angle of attack, but can be approximated using Eq. (12.36) where u is the upsweep angle (radians) of the fuselage centerline and A_{max} is the maximum cross-sectional area of the fuselage.

$$D/q_{\text{upsweep}} = 3.83u^{2.5}A_{\text{max}} \tag{12.36}$$

The landing-gear drag is best estimated by comparison to test data for a similar gear arrangement. Such data for a variety of aircraft are available in Refs. 7, 8, 28,

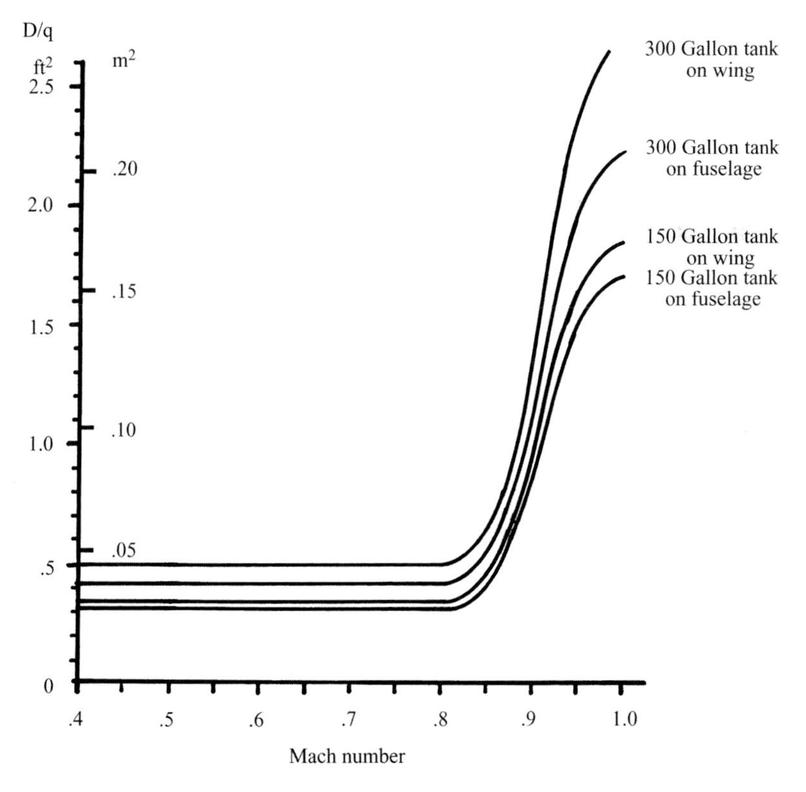

Fig. 12.23 External stores (fuel tanks) drag.

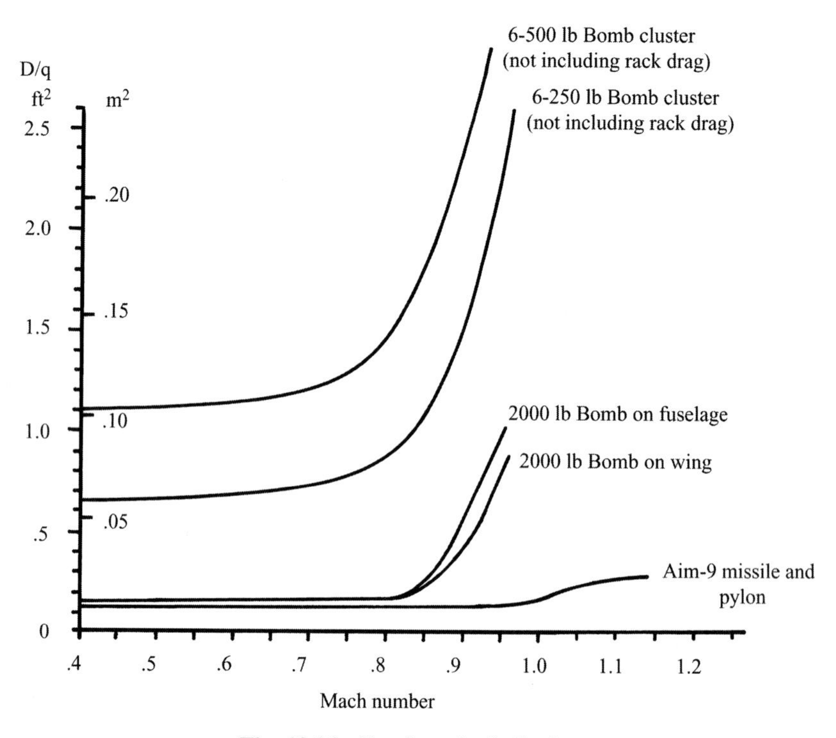

Fig. 12.24 Bomb and missile drag.

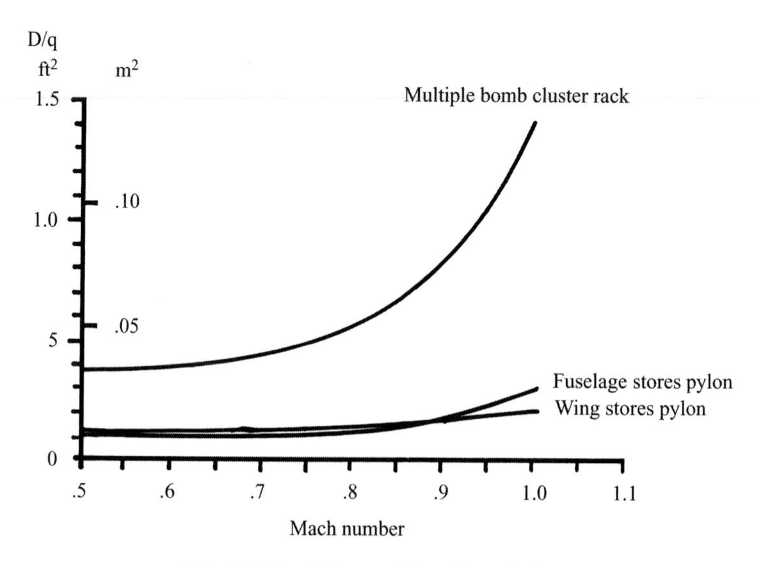

Fig. 12.25 Pylon and bomb rack drag.

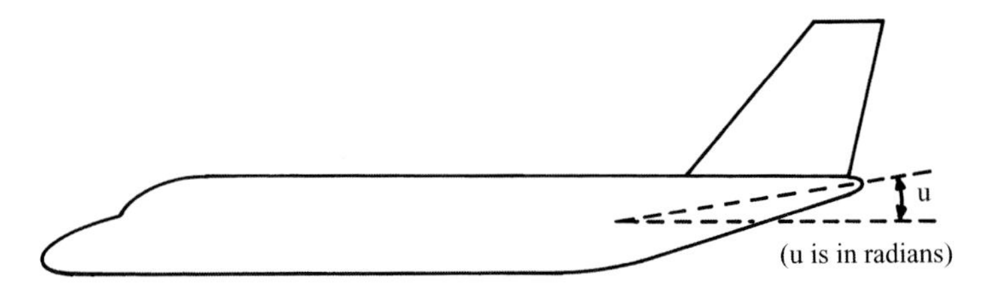

Fig. 12.26 Fuselage upsweep.

and others. If such data are not available, the gear drag can be estimated as the summation of the drags of the wheels, struts, and other gear components using the data in Table 12.5 (largely from Ref. 8).

These values times the frontal area of the indicated component yield D/q values, which must be divided by the wing reference area to obtain parasite-drag coefficients. To account for mutual interference, it is suggested that the sum of the gear component drags be multiplied by 1.2. Also, the total gear drag should be increased by about 7% for a retractable landing gear in which the gear wells are left open when the gear is down.

Note that landing-gear drag is actually a function of lift. The more lift the aircraft wing is producing, the greater the velocity of the airflow over the top of the wing and, conversely, the lesser the airflow velocity underneath the wing where the gear is located. Hence, at higher lift coefficients the gear drag is reduced. This can be ignored for initial analysis.

Strut, wire, and fitting data in Table 12.4 may also be used to estimate the extra drag for a braced wing or biplane. The optimal thickness ratio considering both aerodynamic and structural efficiency is about 0.19 for a strut in tension and about 0.23 for a strut in compression.

Flaps affect both the parasite and induced drag. The flap contribution to parasite drag is caused by the separated flow above the flap, and can be estimated using Eq. (12.37) for most types of flap. Note that this is referenced to wing area. Typically the flap deflection is about 60–70 deg for

Table 12.5 Landing gear component drags

	$\frac{D/q}{\text{Frontal area}}$
Regular wheel and tire	0.25
Second wheel and tire in tandem	0.15
Streamlined wheel and tire	0.18
Wheel and tire with fairing	0.13
Streamline strut ($1/6 < t/c < 1/3$)	0.05
Round strut or wire	0.30
Flat spring gear leg	1.40
Fork, bogey, irregular fitting	1.0–1.4

landing and about 20–40 deg for takeoff. Light aircraft usually take off with no flaps.

$$\Delta C_{D_{0_{\text{flap}}}} = F_{\text{flap}}(C_f/C)(S_{\text{flapped}}/S_{\text{ref}})(\delta_{\text{flap}} - 10) \tag{12.37}$$

where

δ_{flap} is in degrees

$F_{\text{flap}} = 0.0144$ for plain flaps $= 0.0074$ for slotted flaps

$C_f =$ chord length of flap (see Fig. 12.17)

Note that this is a very rough estimate, and that detailed calculations or wind-tunnel data should be obtained as soon as possible for accuracy. Also, the deflection of flaps changes the spanwise lift distribution so that a flap deflection actually increases the induced drag as well [see Eq. (12.62)].

Many aircraft have some form of speed brake. Typically these are plates that extend from the fuselage or wing. Fuselage-mounted speed brakes have a D/q of about 1.0 times the speed-brake frontal area, while wing-mounted speed brakes have a D/q of about 1.6 times their frontal area if mounted at about the 60% of chord location.

Speed brakes mounted on top of the wing will also disturb the airflow and spoil the lift, and so are called "spoilers." These further reduce landing distance by transferring more of the aircraft's weight to the landing gear which increases the braking action.

Base area produces a drag according to Eqs. (12.38) and (12.39) (Ref. 40). "A_{base}" includes any aft-facing flat surfaces as well as the projected aft-facing area for any portions of the aft fuselage that experience highly separated airflow. Roughly speaking, this should be expected any place where the aft fuselage angle to the freestream exceeds about 20 deg. As previously mentioned, a pusher propeller may prevent aft-fuselage separation despite an aft fuselage angle of 30 deg or more.

$$\text{Subsonic:} \quad (D/q)_{\text{base}} = [0.139 + 0.419(M - 0.161)^2]A_{\text{base}} \tag{12.38}$$

$$\text{Supersonic:} \quad (D/q)_{\text{base}} = [0.064 + 0.042(M - 3.84)^2]A_{\text{base}} \tag{12.39}$$

Fighter-type canopies have already been discussed. For transport and light-aircraft windshields that smoothly fair into the fuselage, an additional D/q of about 0.07 times the windshield frontal area is suggested. A sharp-edged, poorly faired windshield has an additional D/q of about 0.15 times its frontal area.

An open cockpit has a D/q of about 0.50 times the windshield frontal area. For an aircraft with an unenclosed cockpit, such as a hang-glider or ultralight a seated person has a D/q of about 6 ft^2 {0.56 m^2}. This reduces to a D/q of 1.2 ft^2 {0.11 m^2} in the prone position.

An arresting hook for carrier operation adds a D/q of about 0.15 ft^2 {0.014 m^2}. The smaller emergency arresting hook for Air Force aircraft adds a D/q of about 0.10 ft^2 {0.009 m^2} Machine-gun ports add a D/q of about 0.02 ft^2 {0.002 m^2} per gun. A cannon port such as for the *M*61 adds a D/q of about 0.2 ft^2 {0.019 m^2}.

Leakage and Protuberance Drag

Leaks and protuberances add drag that is difficult to predict by any method. Leakage drag is due to the tendency of an aircraft to "inhale" through holes and gaps in high-pressure zones, and "exhale" into the low-pressure zones. The momentum loss of the air "inhaled" contributes directly to drag, and the air "exhaled" tends to produce additional airflow separation.

Protuberances include antennas, lights, door edges, fuel vents, control surface external hinges, actuator fairings, and such manufacturing defects as protruding rivets and rough or misaligned skin panels. Typically these drag increments are estimated as a percent of the total parasite drag.

For a normal production aircraft, leaks and protuberance drags can be estimated as about 2–5% of the parasite drag for jet transports or bombers, 5–10% for propeller aircraft, and 10–15% for current-design fighters (5–10% for new-design fighters). If special care is taken during design and manufacturing, these drag increments can be reduced to near zero but at a considerable expense.

An aircraft with variable-sweep wings will have an additional protuberance drag of about 3% due to the gaps and steps of the wing pivot area.

Stopped-Propeller and Windmilling Engine Drags

The specifications for civilian and military aircraft require takeoff and climb capabilities following an engine failure. Not only does this reduce the available thrust, but the drag of the stopped propeller or windmilling engine must be considered.

Data on the drag of a stopped or windmilling propeller are normally obtained from the manufacturer. For a jet engine, detailed knowledge of the characteristics of the engine, inlet, and nozzle are required to estimate the drag from a stopped or windmilling engine. In the absence of such data, the following rough approximations can be used.

For a stopped propeller, Ref. 8 indicates that the subsonic drag coefficient will be about 0.1 based upon the total blade area if the propeller is feathered (turned so that the blades align with the airflow). If the propeller has fixed pitch and cannot be feathered, the drag coefficient is about 0.8.

To determine the total blade area, it is necessary to know or to estimate the propeller "solidity" (σ), the ratio between the total blade area and the propeller disk area. This can be shown to equal the number of blades divided by the blade aspect ratio and π.

For a typical blade aspect ratio of 8, the solidity will be 0.04 times the number of blades. A small piston-prop engine will generally use a two-bladed propeller. A fast piston-prop or a small turboprop will use a three-bladed propeller, while a large turboprop may use a four-bladed propeller.

Drag of a feathered propeller can be roughly estimated by Eq. (12.40). For an unfeathered, stopped propeller, the 0.1 term is replaced by 0.8.

$$(D/q)_{\text{feathered prop}} = 0.1\sigma A_{\text{propeller disk}} \tag{12.40}$$

For jet engines, Ref. 9 indicates that the subsonic drag coefficient of a windmilling turbojet engine will be about 0.3, referenced to the flow area at the engine's front face. Thus, the drag of a windmilling turbojet will be approximately

$$(D/q)_{\text{windmilling jet}} = 0.3A_{\text{engine front face}} \tag{12.41}$$

Supersonic Parasite Drag

The supersonic parasite drag is calculated in a similar fashion to the subsonic drag, with two exceptions. First, the supersonic skin friction drag does not include adjustments for form factors or interference effects (i.e., $FF = Q = 1.0$). Second, a new term, wave drag, is added. This accounts for the pressure drag due to shock formation. Supersonic parasite-drag buildup is defined in Eq. (12.42):

$$C_{D_{0\,\text{supersonic}}} = \frac{\Sigma(C_{f_c} S_{\text{wet}_c})}{S_{\text{ref}}} + C_{D_{\text{misc}}} + C_{D_{\text{L\,\&\,P}}} + C_{D_{\text{wave}}} \tag{12.42}$$

The supersonic turbulent skin friction coefficient was previously presented in Eq. (12.27), using the cutoff Reynolds number from Eq. (12.29).

Miscellaneous drag calculations for supersonic flight have already been presented, where appropriate. Many of the items that produce miscellaneous drag will not appear on a supersonic aircraft (floats, open cockpits, etc.!).

The drag due to leaks and protuberances in supersonic flight follows about the same percentages as just presented, applied to the skin friction drag only.

The wave drag in supersonic flight will often be greater than all the other drag put together. Wave drag is pressure drag due to shocks, and is a direct result of the way in which the aircraft's volume is distributed.

An ideal volume distribution is produced by the Sears–Haack body (Ref. 16), which was shown in Fig. 8.3. A Sears–Haack body, as defined by Eq. (12.43), has a wave drag as in Eq. (12.45). This is the minimum possible wave drag for any closed-end circular cross-section body of the same length and total volume.

$$\frac{r}{r_{\text{max}}} = \left[1 - \left(\frac{x}{\ell/2}\right)^2\right]^{0.75} \tag{12.43}$$

where

$r =$ the cross-section radius
$\ell =$ the longitudinal dimension

and

$$-\ell/2 \leq x \leq \ell/2 \tag{12.44}$$

$$(D/q)_{\text{wave}} = \frac{9\pi}{2}\left(\frac{A_{\max}}{\ell}\right)^2 \tag{12.45}$$

where $A_{\max}$ is the maximum cross-sectional area.

The linear area-rule theory says that the theoretical wave drag of an aircraft at Mach 1.0 is identical to the wave drag of a body of revolution with the same volume-distribution plot. In other words, the actual cross-sectional shape at a given longitudinal location has no effect on wave drag at Mach 1.0. All that matters is the cross-sectional area at each longitudinal location and the way that the cross-sectional area varies longitudinally.

This leads to the area-rule principle for minimizing wave drag. Wave drag at Mach 1.0 is minimized when the aircraft has a volume distribution identical to that of a Sears–Haack body. Drag is reduced when the volume distribution is changed to more resemble the Sears–Haack's, which has a minimal amount of longitudinal curvature.

As was discussed in Chapter 8, the wave drag at Mach 1.0 is directly related to the second derivative (i.e., curvature) of the longitudinal volume distribution. To minimize wave drag, the designer should try to arrange the configuration so that the volume distribution is smooth and bell-shaped.

For a typical aircraft, the wing tends to put a "bump" in the volume distribution. This bump can be reduced by pinching in the fuselage at the wing location, creating the characteristic "coke-bottle" area-ruled fuselage.

No realistic aircraft will have a volume distribution identical to that of a Sears–Haack body. However, a well-designed supersonic aircraft will have a theoretical wave drag at Mach 1.0 that is about twice the Sears–Haack value. Typical ratios of actual wave drag to the optimum Sears–Haack value will be used below as a first-order wave drag estimation method.

At Mach 1.0, shocks form at an angle of 90 deg to the freestream direction. At Mach numbers higher than 1.0, the shocks may form at an angle less than 90 deg. The "Mach angle" is the smallest angle at which a shock may form, representing a "zero-strength" shock. Mach angle is defined as arcsine $(1/M)$.

At Mach 1.0, the wave drag is based upon the aircraft's cross-sectional areas found by the intersection of the aircraft and an infinite plane set at an angle perpendicular (90 deg) to the freestream direction. At speeds higher than Mach 1.0, the wave drag still depends upon the volume distribution as before, but with one major exception.

At higher Mach numbers the volume distribution is based upon aircraft cross sections that are determined by intersecting the aircraft with "Mach planes," set at the appropriate Mach angle to the freestream direction.

A Mach plane may be rolled about the freestream direction to any roll angle. Figure 12.27 shows two roll angles. Note that the different Mach-plane roll angles produce entirely different volume-distribution plots. In the left illustration, the Mach-plane cut includes the fuselage and canopy plus a slice of the left wing.

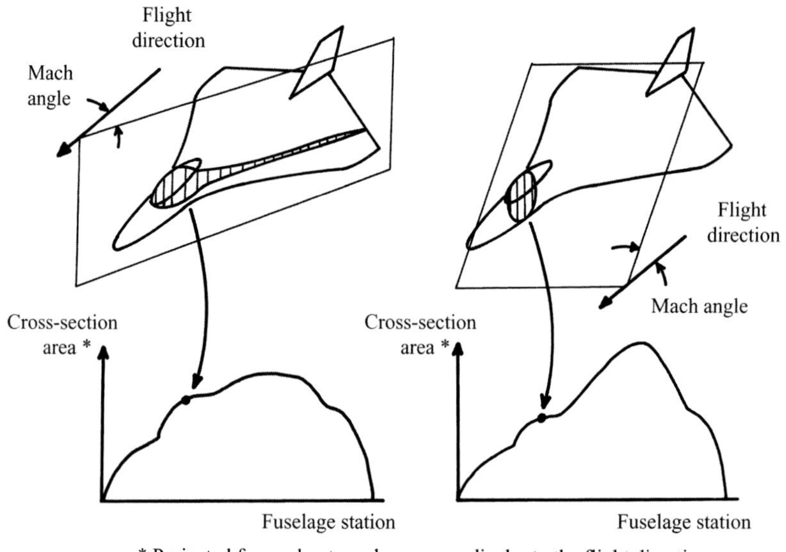

Fig. 12.27 Mach-plane cut volume distribution (two roll angles).

In the right illustration, only the fuselage and canopy are cut, producing a much smaller cross-sectional area at that location.

For each Mach-plane roll angle, a volume-distribution plot can be prepared by taking Mach-plane cuts at a number of longitudinal locations. According to linear wave drag theory (Ref. 41), the supersonic wave drag at Mach numbers greater than 1.0 is determined by averaging the wave drags of the Mach-plane-cut volume distributions for different roll angles.

This is the basis of the classic Harris wave drag code (Ref. 42). A simplified computer code suitable for university use is presented in Ref. 43.

The use of canted Mach-plane cuts to determine the volume distribution at Mach numbers greater than 1.0 requires a different approach to area-ruling. Pinching the fuselage at the wing location may smooth out the volume distribution for one Mach-plane roll angle, but may make the volume distribution even less smooth at another Mach-plane roll angle.

At higher Mach numbers it is very difficult to minimize total wave drag by "eyeball" area-ruling. Instead it is more profitable to smooth the entire configuration through wing-body blending, as seen on the B-1B and in the design concept of Fig. 7.3.

For preliminary wave drag analysis at $M \geq 1.2$, without use of a computer, a correlation to the Sears–Haack body wave drag is presented in Eq. (12.46), where the Sears–Haack D/q is from Eq. (12.45).

The maximum cross-sectional area (A_{max}) is determined from the aircraft volume-distribution plot. Inlet capture area should be subtracted from A_{max}. The length term ℓ is the aircraft length except that any portion of the aircraft with a constant cross-sectional area should be subtracted from the length.

If a design has the location of its maximum cross-section area far behind the midpoint of its fuselage, it should be assumed that the fuselage length is double the distance from nose to the location of maximum cross-section area. In such a case, though, there will probably be increased separation and hence larger base drag from this wedge-shaped design (Ref. 132).

$$(D/q)_{\text{wave}} = E_{\text{WD}}\left[1 - 0.386(M - 1.2)^{0.57}\left(1 - \frac{\pi\Lambda_{\text{LE-deg}}^{0.77}}{100}\right)\right](D/q)_{\text{Sears–Haack}} \tag{12.46}$$

E_{WD} is an empirical wave-drag efficiency factor and is the ratio between actual wave drag and the Sears–Haack value. For a perfect Sears–Haack body, $E_{\text{WD}} = 1.0$.

A very clean aircraft with a smooth volume distribution, such as a blended-delta-wing design, may have an E_{WD} as low as 1.2. A more typical supersonic fighter, bomber, or SST design has an E_{WD} of about 1.8–2.2. A poor supersonic design with a very bumpy volume distribution can have an E_{WD} of 2.5–3.0. The F-15, optimized for the dogfight instead of supersonic flight, has an E_{WD} of about 2.9 (Ref. 35).

Note, however, that this efficiency factor is less important in drag determination than the fineness ratio as represented by $(A_{\max}/\ell)$. This term is squared, which explains why area ruling that actually reduces $A_{\max}$ provides a far greater drag reduction than does merely smoothing the volume distribution without lowering $A_{\max}$.

The complicated middle term in Eq. (12.46) encased in square brackets represents the drop-off in wave drag coefficient as speed increases past Mach 1.2. This is probably due to the wing effects on the canted-cut volume distributions as described earlier. This author notes that this old empirical relationship seems overly optimistic and gets better results by replacing the 0.386 term with 0.2.

Transonic Parasite Drag

The transonic flow regime extends roughly from Mach 0.8–1.2. The increase in drag as an aircraft accelerates through the transonic regime, called the "drag rise," is due to the formation of shocks, and is in fact the transonic portion of wave drag.

The critical Mach number (M_{cr}) occurs when shocks first form on the aircraft. The drag divergent Mach number (M_{DD}) is the Mach number at which the formation of shocks begins to substantially affect the drag.*

The definition of what speed constitutes M_{DD} is arbitrary, and several definitions are in use. The Boeing definition is that M_{DD} is where the drag rise reaches 20 counts. M_{DD} (Boeing) is usually about 0.08 Mach above the critical Mach number. The Douglas definition, also used by the Air Force in Ref. 37, is that M_{DD} is the Mach number at which the rate of change in drag with Mach number ($\mathrm{d}C_{D_0}/\mathrm{d}M$) first reaches 0.10.

---

*Note that the terms M_{DD} and M_{cr} are often used interchangeably.

The Douglas M_{DD} is typically 0.06 Mach above the Boeing M_{DD}, and represents a drag rise of perhaps 80–100 counts. Jet transports usually cruise at about M_{DD} (Boeing), and have a maximum level speed of about M_{DD} (Douglas).

Shocks are formed on the top of the wing as a result of the increased airflow velocity, so M_{DD} reduces with an increased lift coefficient. For example, the Boeing 727 has an M_{DD} of about Mach 0.86 when the lift coefficient is only 0.1, but when the lift coefficient is increased to 0.3 the M_{DD} reduces to about Mach 0.82.

A preliminary estimate of wing M_{DD} (Boeing) is provided by Eq. (12.47) using Figs. 12.28 and 12.29. Figure 12.28 provides the wing drag divergence Mach number of an uncambered wing at zero lift. Figure 12.29 adjusts M_{DD} to the actual lift coefficient.

The last term in Eq. (12.47) is an adjustment for the wing design lift coefficient (i.e., camber and twist). Initially it can be assumed that the design lift coefficient is the same as the lift coefficient at cruise.

$$M_{DD} = M_{DD_{L=0}} LF_{DD} - 0.05 C_{L_{\text{design}}} \tag{12.47}$$

If the wing uses a supercritical airfoil, the actual thickness ratio should be multiplied by 0.6 before using these figures. This approximation is to account for the shock-delaying characteristics of the supercritical airfoil.

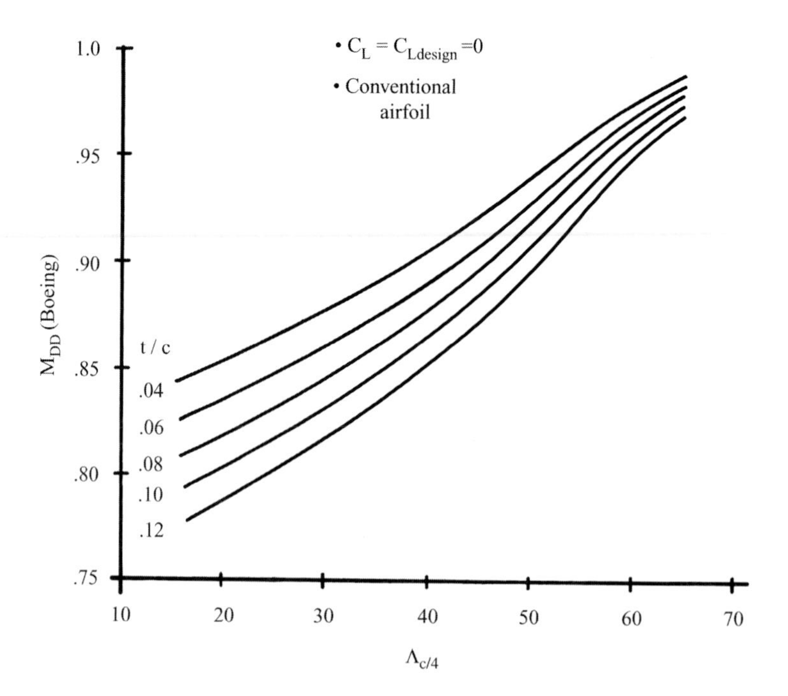

Fig. 12.28 Wing drag-divergence Mach number.

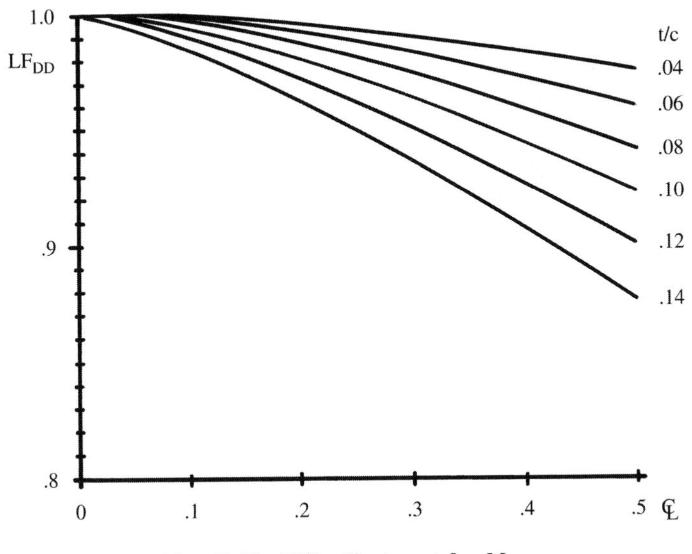

Fig. 12.29 Lift adjustment for M_{DD}.

M_{DD} changes with lift coefficient. Lift coefficient changes with weight and altitude, both of which may change during cruise. To be completely accurate, it is necessary to calculate M_{DD} for each point in the mission. For initial analysis, however, it is acceptable to use a single M_{DD} based upon a midmission weight and cruise altitude.

If the fuselage is relatively blunt, it will experience shock formation before the wing does. In this case, M_{DD} is set by the shape of the forebody. Body M_{DD} can be estimated using Fig. 12.30 (Ref. 44), where L_n is the length from the nose to the longitudinal location at which the fuselage cross section becomes essentially constant. The body diameter at that location is d. If the fuselage is noncircular, d is an equivalent diameter based upon the fuselage cross-sectional area. Determine both wing and fuselage M_{DD}, and use the lower value.

The linear-wave-drag analysis gives completely incorrect results in the transonic regime. This analysis is called "linear" because the higher-order, nonlinear terms have been dropped from the aerodynamic equations to permit computation.

Some of these dropped nonlinear terms account for any changes in the airflow longitudinal velocity. At high supersonic speeds these terms have little effect compared to the far greater aircraft velocity.

However, the drag rise at transonic speeds is largely caused by the increase in airflow velocity over the top of the wing. Thus, drag rise below Mach 1.0 is in fact caused by the terms that are dropped in the linear analysis! Transonic drag rise is therefore calculated to be zero by the linear-wave-drag methods. Only sophisticated nonlinear computational aerodynamic programs can give reasonable analytical results within the transonic regime.

Empirical methods for the calculation of the drag rise are presented in Ref. 37. These estimate the drag rise for the wing and fuselage separately, so the benefits

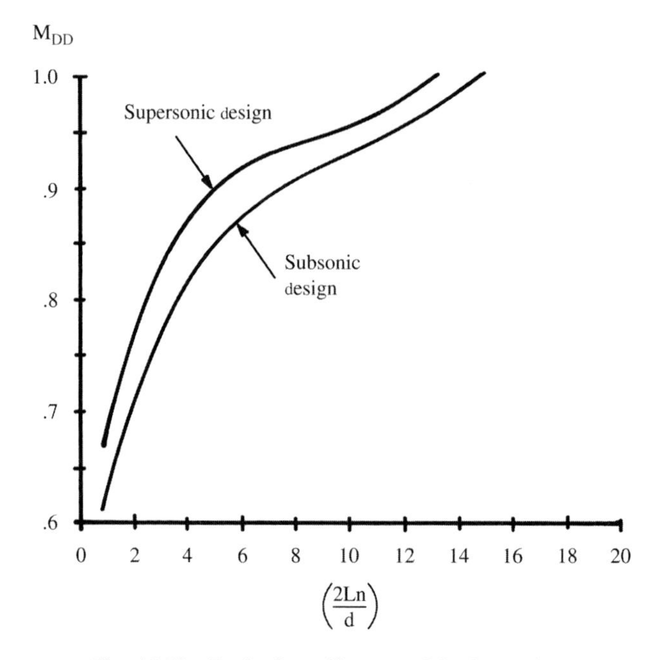

Fig. 12.30 Body drag-divergent Mach number.

of area ruling are ignored. These methods are very time-consuming and not very accurate, so an approximation technique is presented next.

For initial analysis the drag rise may be graphically estimated using a few rules of thumb, as shown in Fig. 12.31. The drag at and above Mach 1.2 (labeled A in the figure) is determined using Eq. (12.46) (divided by wing reference area). The drag at Mach 1.05 (labeled B) is typically equal to the drag at Mach 1.2.

The drag at Mach 1.0 (labeled C) is about half of the Mach 1.05 value. The drag rise at M_{DD} (just determined) is 0.002 by definition (labeled D). M_{cr}, the beginning of drag rise, is roughly 0.08 slower in Mach number than M_{DD} and is labeled E.

To complete the transonic-drag-rise curve from these points, draw a straight line through points B and C, extending almost to the horizontal axis. Then, draw a curve from M_{cr} through M_{DD} which fairs smoothly into the straight line as shown. If a smooth curve cannot be drawn, the M_{cr} point (E) should be moved until an approximately circular arc can be drawn. Finally, draw a smooth curve connecting B to A.

This crude technique may be used even for subsonic transport aircraft. The supersonic wave drag (point B) is determined from Eq. (12.46) although the aircraft will never fly at this speed. When calculating the Sears–Haack D/q for Eq. (12.46), remember to subtract from the aircraft length the portions of the aircraft where the cross-sectional area is constant. Also, data indicates that an E_{WD} of 4.0 will approximate a transport aircraft's drag rise.

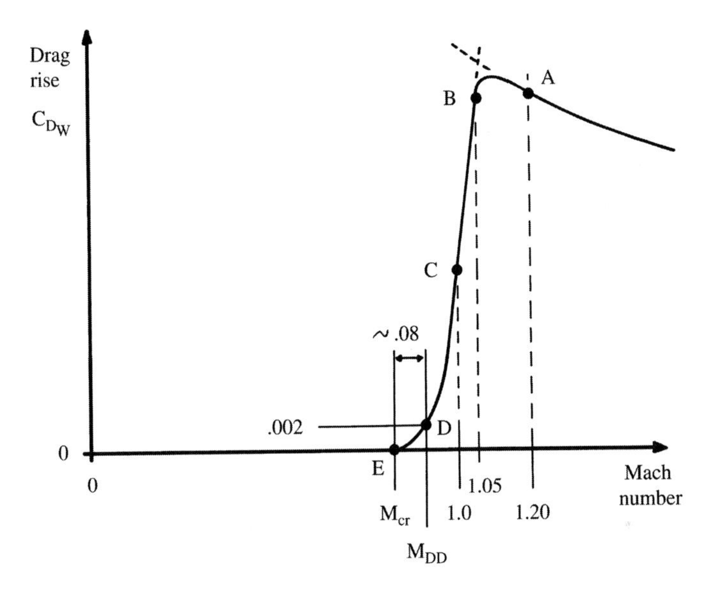

Fig. 12.31 Transonic drag rise estimation.

Complete Parasite-Drag Buildup

Figure 12.32 illustrates the complete buildup of parasite drag vs Mach number for subsonic, transonic, and supersonic flight. The subsonic drag consists of the skin friction drag, including form factor and interference, plus miscellaneous

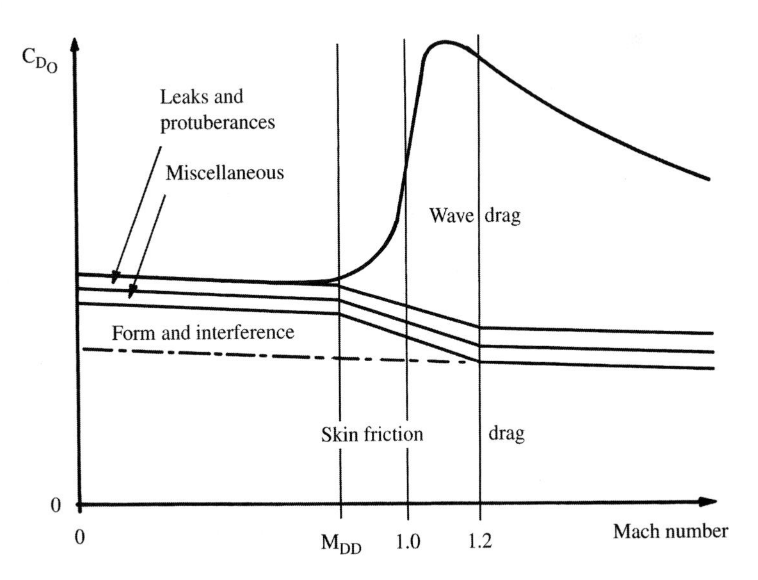

Fig. 12.32 Complete parasite drag vs Mach number.

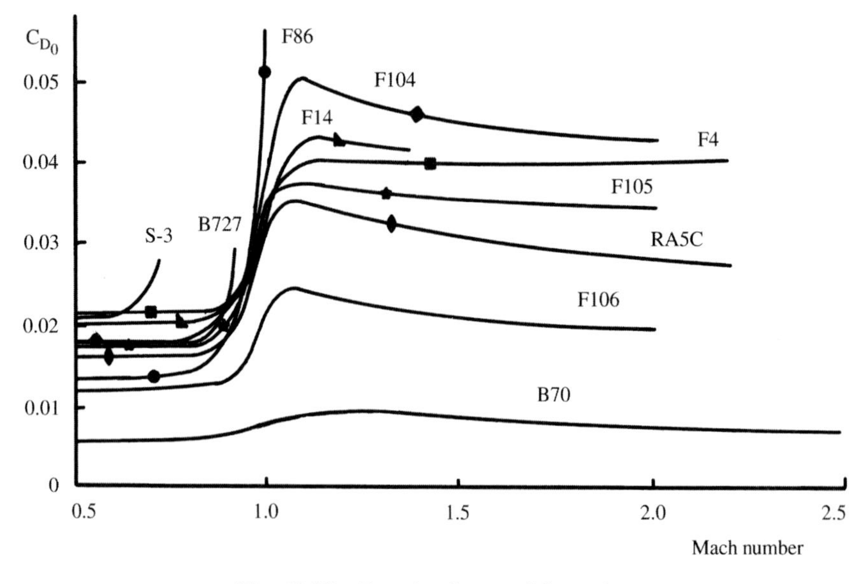

Fig. 12.33 Parasite drag and drag rise.

drag and leak and protuberance drag. The supersonic drag includes the flat-plate supersonic skin friction drag, miscellaneous drag, leak and protuberance drag, and wave drag.

In the transonic regime, the skin friction drag is estimated simply by drawing a straight line between the skin friction drag at M_{DD} (which includes form factor and interference) and the skin friction drag at $M1.2$ (which does not). This does not reflect any reduction in drag, merely a change in bookkeeping. The pressure drags represented by the form and interference terms at subsonic speeds are included in the wave-drag term at supersonic speeds.

In Fig. 12.33, the actual parasite drag and drag rise is shown for a number of aircraft.

12.6 Drag due to Lift (Induced Drag)

The induced-drag coefficient at moderate angles of attack is proportional to the square of the lift coefficient with a proportionality factor called the "drag-due-to-lift factor," or K [see Eq. (12.4)].

Two methods of estimating K will be presented. The first is the classical method based upon e, the Oswald span efficiency factor. Methods are presented for subsonic monoplanes and biplanes along with an empirical equation for supersonic speeds.

The second method for the estimation of K is based upon the concept of leading-edge suction and provides, for high-speed designs, a better estimate of

K, one that includes the effects of the change in viscous separation as lift coefficient is changed. This method also reflects the choice of the wing design lift coefficient on the drag due to lift at different lift coefficients.

Oswald Span Efficiency Method

According to classical wing theory, the induced-drag coefficient of a 3-D wing with an elliptical lift distribution equals the square of the lift coefficient divided by the product of aspect ratio and π. However, few wings actually have an elliptical lift distribution. Also, this doesn't take into account the wing separation drag.

The extra drag due to the nonelliptical lift distribution and the flow separation can be accounted for using e, the "Oswald[†] span efficiency factor." This effectively reduces the aspect ratio, producing the following equation for K:

$$K = \frac{1}{\pi A e} \tag{12.48}$$

The Oswald efficiency factor is typically between 0.7 and 0.85. Numerous estimation methods for e have been developed over the years, such as those by Glauert and Weissinger. These tend to produce results higher than the e values of real aircraft. More realistic estimation equations based upon actual aircraft (Ref. 45) are presented here:

$$\text{Straight-Wing Aircraft:} \quad e = 1.78(1 - 0.045A^{0.68}) - 0.64 \tag{12.49}$$

$$\text{Swept-Wing Aircraft:} \quad e = 4.61(1 - 0.045A^{0.68})(\cos \Lambda_{LE})^{0.15} - 3.1 \quad (\Lambda_{LE} > 30\,\text{deg}) \tag{12.50}$$

These equations should only be used with "normal" aspect ratios and sweeps and are not valid for high-aspect-ratio designs such as sailplanes. If the wing has end-plates or winglets, the effective aspect ratio from Eq. (12.10) or (12.11) should be used in Eq. (12.48).

Drag due to lift for a biplane was first analytically determined by Max Munk in 1922, based upon the calculation of an equivalent monoplane span providing the same wing area and the same drag.

Prandtl developed an interference factor (σ, shown in Fig. 12.34) that is used in Eq. (12.51) to determine a biplane span efficiency factor (Ref. 12). The biplane aspect ratio is the square of the longer span divided by the total area of both

[†]Presented in the doctoral dissertation of W. B. Oswald at California Institute of Technology.

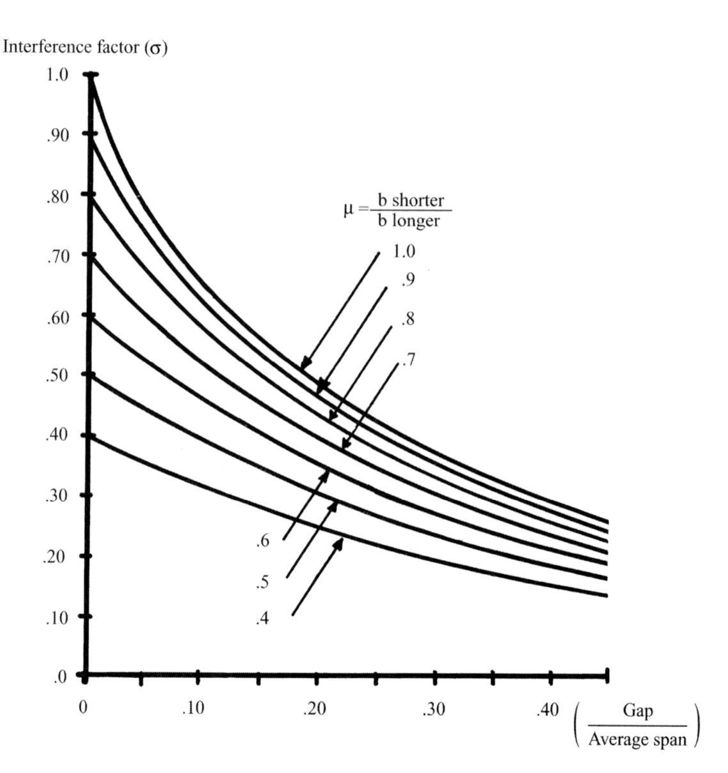

Fig. 12.34 Prandtl's biplane interference factor (Ref. 12).

wings.

$$\text{Biplane:} \quad e = \frac{\mu^2(1+r)^2}{\mu^2 + 2\sigma\mu r + r^2} \tag{12.51}$$

where

μ = shorter span/longer span
r = lift on shorter wing/lift on longer wing
(approximately = area of shorter wing/area of longer wing)

At supersonic speeds, the drag-due-to-lift factor K increases substantially. In terms of Oswald efficiency factor, e is reduced to approximately 0.3–0.5 at Mach 1.2. Equation (12.52) provides a quick estimate of K at supersonic speeds (Ref. 6), although the leading-edge suction method presented later is preferable.

$$\text{Supersonic Speeds:} \quad K = \frac{A(M^2 - 1)\cos\Lambda_{LE}}{(4A\sqrt{M^2-1}) - 2} \tag{12.52}$$

Leading-Edge-Suction Method

Drag at angle of attack is strongly affected by viscous separation. At high lift coefficients the drag polar breaks away from the parabolic shape represented by a fixed value of K in Eq. (12.4). The e method ignores this variation of K with lift coefficient. For a wing with a large leading-edge radius this is acceptable, but for most supersonic aircraft it gives a poor approximation.

A semi-empirical method for estimation of K allows for the variation of K with lift coefficient and Mach number; it is based upon the concept of "leading-edge suction." Figure 12.35 illustrates the concept. The thick airfoil on the left is at an angle of attack below that at which substantial separation occurs. The flow streamlines curve rapidly to follow the leading-edge radius over the top of the wing.

This rapid curvature creates a pressure drop on the upper part of the leading edge. The reduced pressure exerts a suction force on the leading edge in a forward direction. This "leading-edge suction" force S is shown at the bottom of the figure in a direction perpendicular to the normal force N.

If there is no viscous separation or induced downwash, the leading-edge suction force exactly balances the rearward component of the normal force and the airfoil experiences zero drag. This is the ideal 2-D case described by d'Alembert's paradox, and is called "100% leading-edge suction."

A 3-D wing is considered to have 100% leading-edge suction when the Oswald efficiency factor (e) exactly equals 1.0. When e equals 1.0, the induced-drag constant K exactly equals the inverse of the aspect ratio times π.

On the right side of Fig. 12.35 is a zero-thickness flat-plate airfoil. Even without the leading-edge separation, which will almost certainly occur, this airfoil must have higher drag because there is no forward-facing area for the leading-edge pressure forces to act against. All pressure forces for a zero-thickness flat plate must act in a direction perpendicular to the plate, shown as N. There is zero leading-edge suction, and the lift and induced drag are simply

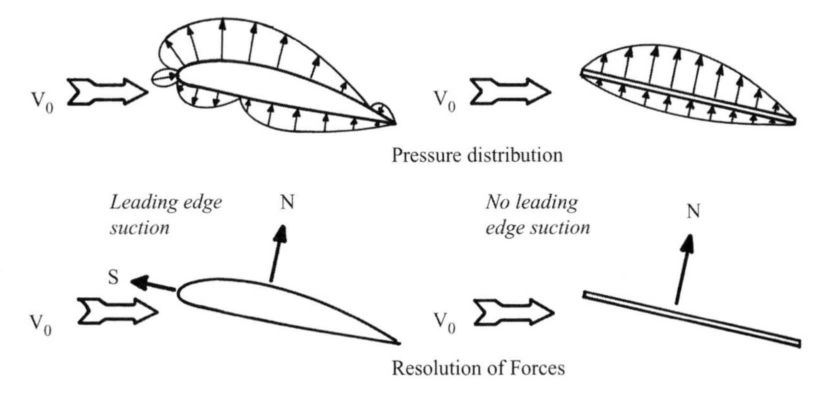

Fig. 12.35 Leading-edge suction definition.

N times the cosine or sine of the angle of attack [Eqs. (12.53) and (12.54)].

$$L = N \cos \alpha \tag{12.53}$$

$$D_i = N \sin \alpha = L \tan \alpha \tag{12.54}$$

or

$$C_{D_i} = C_L \tan \alpha \tag{12.55}$$

but (assuming α is small),

$$C_{D_i} = KC_L^2 \cong {}_\alpha C_L \tag{12.56}$$

so,

$$K = \frac{\alpha C_L}{C_L^2} = \frac{\alpha}{C_L} = \frac{1}{C_{L_\alpha}} \tag{12.57}$$

Thus, in the worst case of zero leading-edge suction, the drag-due-to-lift factor K is simply the inverse of the slope of the lift curve (in radians), as previously determined.

All real wings operate somewhere between 100 and 0% leading-edge suction. The percentage of leading-edge suction a wing attains is called S (not to be confused with the force S in Fig. 12.35).

During subsonic cruise, a wing with moderate sweep and a large leading-edge radius will have S equal to about 0.85–0.95 (85–95% leading-edge suction). The wing of a supersonic fighter in a high-g turn may have an S approaching zero.

The following method for calculating K for high-speed aircraft is based upon an empirical estimate of the actual percent of leading edge suction attainable by a wing, which is then applied to the calculated K values for 100 and 0% leading-edge suction. The actual K is calculated as a weighted average of the 100 and 0% K, as in Eq. (12.58):

$$K = SK_{100} + (1 - S)K_0 \tag{12.58}$$

The 0% K value is the inverse of the slope of the lift curve, as determined before. The 100% K value in subsonic flight is the inverse of the aspect ratio times π.

In transonic flight, starting at M_{dd}, the shock formation interferes with leading-edge suction. This increases the K value. When the leading edge becomes supersonic, the suction goes to zero, and so the K value equals the 0% K value.

This occurs at the speed at which the Mach angle (arcsine $1/M$) equals the leading-edge sweep. Above that speed the wing has zero leading-edge suction so the K value is always the inverse of the slope of the lift curve.

For initial analysis, the supersonic behavior of the 100% K line may be approximated by a smooth curve, as shown in Fig. 12.36. This shows the typical behavior of the 100 and 0% K values vs Mach number.

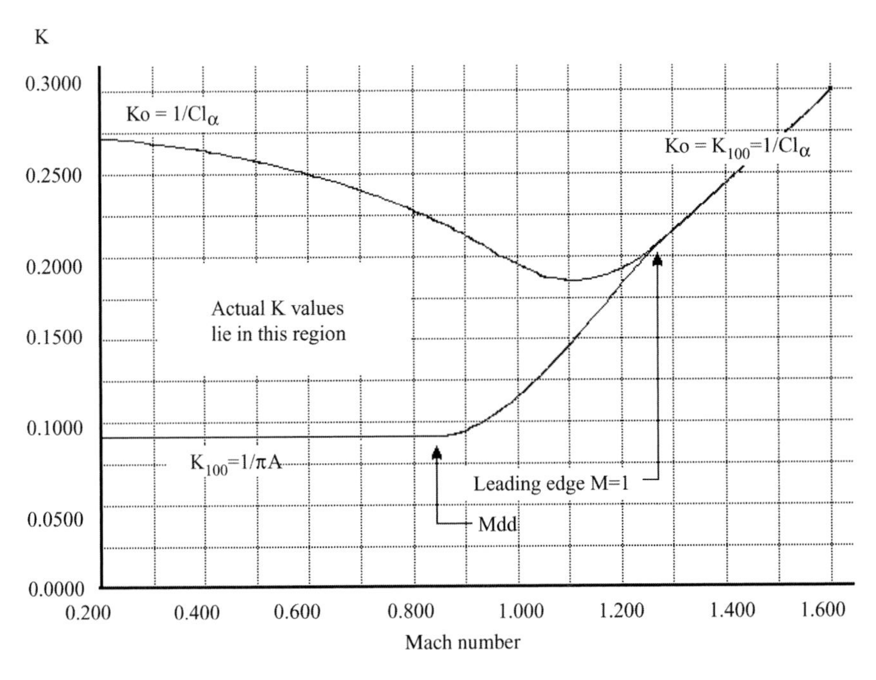

Fig. 12.36 0% and 100% *K* vs Mach number.

The only unknown remaining is the value of *S*, the percentage of leading-edge suction actually attained by the wing at the flight condition in question. *S* depends largely upon the leading-edge radius, and is also affected by the sweep and other geometric parameters.

S is also a strong function of the wing design lift coefficient and the actual lift coefficient. For any wing, the value of *S* is at a maximum when the wing is operating at the design lift coefficient. For most wings, *S* equals approximately 0.9 when operating at the design lift coefficient.

For a subsonic wing with large leading-edge radius and moderate sweep, the value of *S* will change very little with lift coefficient until the wing is near the stall angle of attack. For the thin, swept wings typical in supersonic aircraft, the value of *S* can change substantially with lift coefficient. A wing with an *S* of 0.9 at its design lift coefficient of 0.5 may have an *S* value less than 0.3 at a lift coefficient of 1.0.

Proper calculation of *S* for an actual wing is complex. An empirical approach may be used during conceptual design. Figure 12.37 provides a first-order estimate of the percent of leading-edge suction for a typical supersonic aircraft's wing, given the actual lift coefficient and the design lift coefficient (this determines which curve to use). Note that this chart assumes a well-designed wing, and at some later date the aerodynamics department must optimize the twist and camber to attain these values.

From Fig. 12.37 the leading-edge suction at various lift coefficients can be estimated. This allows adding curves to Fig. 12.36 that represent the estimated

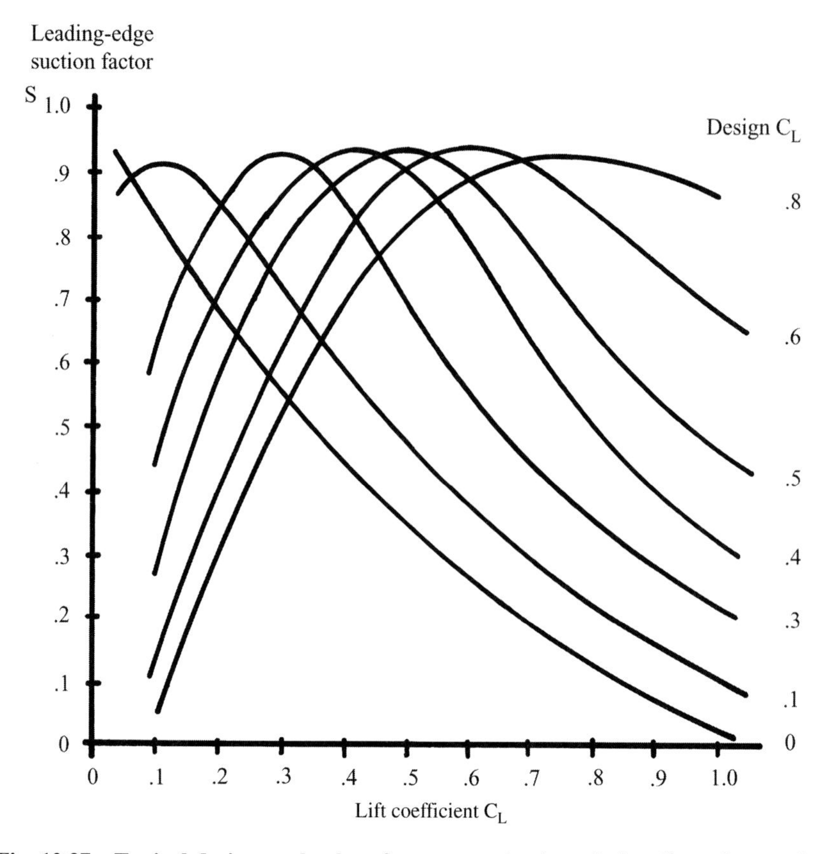

Fig. 12.37 Typical design goal values for supersonic aircraft. Leading-edge suction vs C_L.

K value for different lift coefficients as a function of Mach number, as in Fig. 12.38. These are then used for total drag estimation via Eq. (12.4).

As mentioned, a subsonic wing with large leading edge radius will see little change in S until near the stall. For such wings the left side of the suction curves of Fig. 12.37 should be replaced by a straight line at $S = 0.93$ (or greater if high aspect ratio; see the following).

For wings with high aspect ratios, the leading-edge suction schedule actually becomes a function of aspect ratio, with S increasing to values of 0.95–0.97 as higher aspect ratios are employed. The best way to use the leading-edge suction method for such a wing is to use suction test data from a similar geometry wing. If that is not available, a leading-edge suction schedule can be constructed by assuming that the Oswald span efficiency factor e at the design lift coefficient equals some specified value (typically $e = 0.8$), and solving for the equivalent S in Eq. (12.59). This S can be assumed to apply from zero lift up to about 0.1 C_L above the wing design C_L, after which it drops off to about 80% of the equivalent design S at the stall lift coefficient. While crude, this approximation correlates

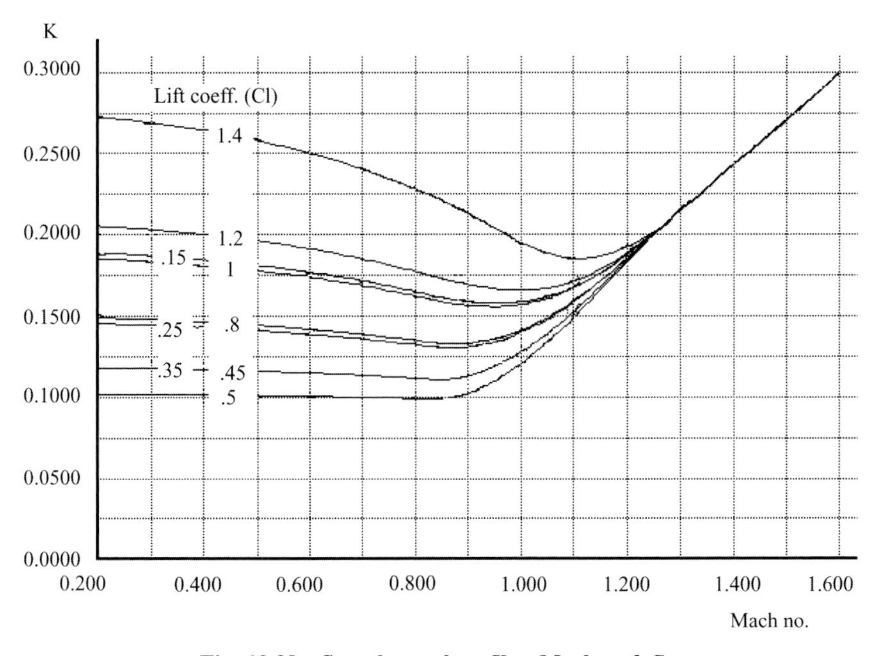

Fig. 12.38 Sample results—K vs Mach and C_L.

well with actual aircraft data and is more realistic than simply using Oswald's method with no adjustment for lift coefficient.

For the sake of comparison, Eqs. (12.59) and (12.60) relate S to e and ΔN (used in several other textbooks):

$$e = \frac{1}{(\pi A/C_{L\alpha})(1 - S) + S} \tag{12.59}$$

$$\Delta N = S\left(\frac{1}{C_{L\alpha}} - \frac{1}{\pi A}\right) \tag{12.60}$$

Trim Drag

The drag values used for performance calculations should include the trim drag. This additional drag is caused by the horizontal tail force required to balance (trim) the aircraft so that the total pitching moment about the aircraft c.g. will be zero for any given flight condition.

The tail usually trims the aircraft with a download that must be countered by additional lift from the wing. This produces an increase in the wing induced drag that must also be included in the trim drag, along with the drag due to lift of the tail itself. However, the tail is flying in the downwash off the wing so the direction of its downward lift is actually slightly forward. This reduces the trim drag.

Trim calculation is discussed in Chapter 16. The trim drag is determined using the preceding induced-drag methods once the tail lift force required for trim is known, taking into account the induced drag from the lift of the tail, the extra wing induced drag, and the parasitic drag of the deflected tail and/or elevator.

Ground Effect

When a wing is near the ground, say less than half the span away, the drag due to lift (K) can be substantially reduced. This is theoretically explained as a reduction in the induced downwash angle, but can be visualized as a trapping of a "cushion of air" under the wing. This effect is accounted for by multiplying K by the factor calculated in Eq. (12.61) (Ref. 70):

$$\frac{K_{\text{effective}}}{K} = \frac{33(h/b)^{1.5}}{1 + 33(h/b)^{1.5}} \tag{12.61}$$

where h is wing height above ground.

Flap Effect on Induced Drag

Previously the effect of flaps on parasitic drag was discussed, and a first-order estimation method was suggested [Eq. (12.37)]. The deflection of a flap has an additional effect that must be considered. When you deflect a flap, you expect an increase in lift, and that lift occurs in the part of the wing that has the flap (the "flapped" wing area as shown in Fig. 12.20). This extra lift in the vicinity of the flap affects the spanwise lift distribution and therefore, the drag due to lift. Our hard-won elliptical lift distribution is ruined, and we must adjust the drag due to lift upward.

This is best done with a computer program capable of predicting the effect of the flaps on lift distribution, and then converting that into drag due to lift. Alternatively, this can be estimated using wind tunnel data for a similar configuration. If neither is available, Refs. 8, 37, and 11 are recommended. As a first approximation, the following equation can be used based on the increase in lift due to the flap:

$$\Delta C_{D_i} = k_f^2 (\Delta C_{L_{\text{flap}}})^2 \cos \Lambda_{\bar{c}/4} \tag{12.62}$$

where $k_f = 0.14$ for full-span flaps and 0.28 for half-span flaps. This induced drag increment is added to the drag due to lift for the total lift using the clean wing drag due to lift factor.

12.7 Aerodynamic Codes and Computational Fluid Dynamics (CFD)

Industry Practice for Aerodynamic Estimation

The aerodynamic methods just presented do not reflect current industry practice. Aircraft companies rely upon linearized computer codes such as the Harris wave-drag code, the Sommer and Short skin friction code, and one of several panel codes such as USSAERO for induced effects. Newer panel codes such as PANAIR and QUADPAN are used to estimate the induced effects and the wave drag simultaneously and with better accuracy than the older codes.

These linearized computer codes can provide correct results only when the airflow around the aircraft is steady, unseparated, and does not contain any strong vortices. This is typically true only during cruising flight. Lift and drag at high angles of attack are estimated empirically using correlations to flight-test and wind-tunnel data for similar configurations.

The same is true for transonic lift and drag, where some of the very terms that are thrown away to linearize the equations are the longitudinal velocity-variation terms that produce the transonic shocks. Linearized wave-drag codes tend to overestimate the wave drag from Mach 1.0 to about Mach 1.2, and incorrectly predict zero drag rise below Mach 1.0. Empirical data are therefore used for the transonic regime.

Despite these problems, the standard industry practice of combining linearized computer codes with empirical data and corrections will produce good results in most cases. Actual flight-measured values of lift and drag are usually within about 2–10% of the estimates. Also, the estimates are the most accurate for the cruise portions of the flight where the most fuel is burned.

However, the fact that we can estimate a given design's lift and drag with reasonable accuracy does not guarantee that these methods will produce the best of all possible designs. Aerodynamic design has had to rely upon a trial-and-error process of design, analyze, test, and redesign.

Wind-tunnel testing offers a powerful tool for aircraft development. Unfortunately, the costs associated with detailed wind-tunnel testing tend to preclude an exhaustive evaluation of all possible designs. At a cost of several hundred thousand dollars per model, one is not likely to try something different just to see if it is better than the baseline design. Instead, the wind tunnel is largely used to verify that a given design is workable, and fix it if it is not.

It is sometimes difficult to identify the source of a problem during a wind-tunnel investigation because the wind tunnel "solves" all the flow equations simultaneously (i.e., viscous effects, vortex flow, induced effects, etc.). An unacceptable wiggle in the pitching-moment curve may be due to one of a number of causes, and the wind tunnel may not tell you which cause to fix!

Another problem with wind-tunnel testing is the Reynolds-number effect. Most wind tunnels cannot test at anything close to full-scale Reynolds numbers, resulting in substantial errors. Even worse, the optimal solution at a lower Reynolds number may not be the optimum at full-scale Reynolds numbers. Who would propose a full-scale test on an airfoil or complete aircraft design that is not the best one tested in the wind tunnel?

CFD Definitions

It is for these reasons that computational fluid dynamics (CFD) has rapidly become a key part of the aircraft design process. CFD is a catch-all phrase for a number of new computational techniques for aerodynamic analysis. It differs from prior aerodynamic codes by solving for the complete properties of the flowfield around the aircraft, rather than only on the surface of the aircraft.

CFD codes are based upon the Navier–Stokes (NS) equations, which were first derived in 1822. The NS equations completely describe the aerodynamics of a fluid (except for chemical-reaction effects at high temperatures).

NS includes equations based upon the existence of flow continuity, the conservation of momentum, and the conservation of energy. These are derived in many textbooks on theoretical and computational aerodynamics, and will not be repeated here.

The NS equations seem straightforward enough but cannot be analytically solved for any useful flow conditions. The author of Ref. 80 describes them as "some of the nastiest differential equations in theoretical physics."

The history of theoretical aerodynamics to date can largely be described as the quest for solvable simplifications of the NS equations. The classical lifting-line theory is one such simplification, as are the linearized wave-drag and panel codes, the Euler codes, and the various NS codes.

There is a complete hierarchy of aerodynamic codes depending upon how many flow phenomena are neglected from the full NS equations. While "direct numerical simulation" codes are beginning to solve the full NS for simplified geometrics and conditions, no currently practical codes for aircraft design solves the full NS equations, due to the difficulty in mathematically analyzing turbulence. Turbulence occurs at the molecular level, which would probably require gridding the flowfield with billions of molecule-sized grids.

The current so-called Navier–Stokes codes actually use a simplification in the handling of turbulence, which is the most difficult flow phenomena to analyze mathematically. Turbulence is handled with some type of separate statistically calibrated model apart from the NS solution.

The most sophisticated codes to date, the large eddy simulation codes, use a statistically based turbulence model for small-scale turbulence effects. Large eddy codes are capable of directly analyzing the larger turbulent eddies. The large eddy simulation is beyond the capabilities of current computers for a complex aircraft configuration, but has been used for simplified geometries.

The current state of the art for complex aircraft configurations, the Reynolds-averaged Navier–Stokes, has both large and small eddies (turbulence) modeled statistically. It is assumed the actual turbulence levels can be approximated by averaged levels in a grid, which simplifies the solution down to "only" solution of about 60 partial derivative equations! Reynolds-averaged codes can handle most of the complex flow phenomena that elude linearized codes, including vortex formation, separation, transonic effects, and unsteady effects.

Reynolds-averaged codes are being used on many projects to solve particular design problems where no other methods can give correct results. Unfortunately, Reynolds-averaged codes are still expensive to set up and run.

The NS workhorse for supersonic design analysis, the parabolized Navier–Stokes (PNS), drops the viscous terms in the streamwise direction, which ignores streamwise separation effects. However, with a good turbulence model the PNS codes give correct and illuminating results for many design problems.

If all viscosity effects are ignored and the flow is assumed to be steady, the Euler equations are derived from the NS equations. Euler codes are much cheaper to run than even PNS codes, and are widely used at this time. Actually, the invicid assumption is quite good outside of the boundary layer. The Euler codes can handle vortex formation, and with the addition of a separate boundary-layer code, can also realistically estimate viscous and separation effects.

The "potential flow" equations are further simplified from the Euler equations by dropping the rotational terms. This prevents the analysis of vortex flow, which is important at high angles of attack but is of lesser importance during cruise conditions. Potential flow codes can handle transonic shock formation and are very useful for transonic design compared to the linearized methods. The potential flow codes are not usually considered to be true "CFD," but are probably the most widely used aerodynamic codes that treat the entire flowfield rather than just the surface conditions. When a boundary-layer model is added, potential flow codes become even more useful for routine design analysis and optimization.

The "linearized" aerodynamic codes are based upon a further simplification to the potential flow equations by neglecting the higher-order terms. It is assumed that, since they involve small quantities multiplied by other small quantities, they must be very small and therefore negligible. At transonic speeds, however, these terms are not so small!

The linearized potential flow equations are the basis of the standard industry methods described at the beginning of this section. These include the Harris wave drag and the USSAERO and similar panel methods. With further simplifications, such classical methods as the lifting-line theory are derived.

To recap, only the large eddy, Reynolds-averaged, and PNS codes are considered to be true "Navier–Stokes" codes. However, the Euler, potential flow, and linearized aerodynamic codes are in fact successive simplifications of the NS equations. The choice of code for a given design problem depends upon the nature of the problem and the available budget (and not always in that order!).

Applications of CFD

CFD does not replace the wind tunnel. In fact, it really does not even reduce the number of wind-tunnel test hours, although it may cap the historical upward spiral in wind-tunnel hours (roughly one order of magnitude per decade).

CFD does permit you to design a better airplane by a truer understanding of the flowfield around it. Not only do the CFD codes determine the entire flowfield around the aircraft, but also, unlike the wind tunnel, the flowfield determination is done at the full-scale Reynolds number.

A perfect example of the use of CFD can be found at every major commercial airport in the country. The installation of the fuel-efficient CFM-56 engine on the Boeing 737 would not have been possible without the use of CFD, as described in Ref. 80.

The original Boeing 737 used the P&W JT8D, a low-bypass-ratio engine that was mounted in a wing-conformal nacelle. The nacelle barely cleared the ground, providing a minimum-weight landing gear.

When Boeing decided to develop an updated version of the 737, the CFM-56 engine was the logical choice as a modern fuel-efficient engine of the required thrust class. However, it has a diameter some 20% greater than the old engine. Furthermore, the CFM-56 exits its fan air up front like most modern turbofans. For this reason, a wing-conformal nacelle was not possible.

In an earlier chapter, the cited rough rule of thumb for podded jet nacelles said that the inlet should be about two inlet diameters forward of the wing and about one inlet diameter below it. A more-refined empirical method of locating a turbofan engine indicated that the geometry shown in Fig. 12.39a was the closest acceptable nacelle spacing. Clearly this posed a ground clearance problem!

The empirical rules for nacelle spacing were based upon years of trial and error in the wind tunnel. Closer spacings were found to increase cruise drag, although the wind-tunnel investigations had not clearly determined just exactly what this "interference" drag consisted of. Various suspects included increased skin friction due to supervelocity, increased separation, shock effects, and a change in the wing's spanwise lift distribution resulting in an increase in the induced drag.

Through the use of a nonlinear potential flow panel program (CFD state of the art in the 1970s), Boeing was able to determine that it was in fact the induced drag effect that was creating the "interference drag." This important piece of information had not been determined in 20 years of wind-tunnel testing!

With this information, Boeing was then able to contour a closely spaced podded nacelle to prevent any change in the wing's spanwise lift distribution. This was possible with CFD because the entire flowfield is solved, allowing the designers to study the streamlines and pressure fields resulting from various design changes. The designers sought to minimize the impact of the nacelle on the streamlines of the bare wing.

Figure 12.39b shows the result, namely, a nacelle of extremely close spacing to the wing that, nevertheless, has acceptable drag characteristics.

The formation and effects of vortices at high angles of attack represent another area of substantial concern to the designers of fighter aircraft. These vortices produce completely different lift, drag, and pitching-moment characteristics than those which would be predicted using the linear methods.

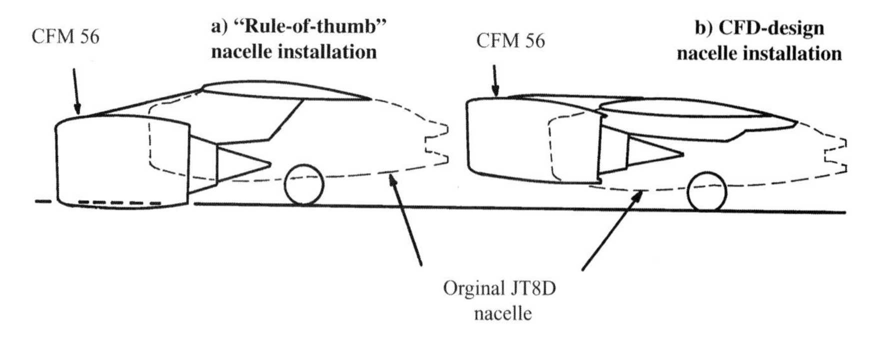

Fig. 12.39 CFD example: Boeing 737 nacelle (after R. Bengelink, AIAA Paper 88-2043).

Reference 81 details the CFD solution of a typical vortex-flow problem using a Lockheed Euler code called TEAM (Three-Dimensional Euler Aerodynamic Method). Figure 12.40 shows the close match between the calculated and the measured pressures over the double-delta configuration used in the study. The vortex region can be seen by the diagonal pressure contours in both calculated and measured plots.

Figure 12.41 illustrates the power of CFD to fully analyze the flowfield around the aircraft rather than just at the surface. This figure shows streamlines around the Ohio Airships Dynalifter, a hybrid airship with wings and a lifting body. CFD was used to predict drag and optimized shaping (Ref. 141).

Perhaps the most important application of CFD (and lesser aerodynamic methods) is the inverse problem: given a desired aerodynamic characteristic, what is the geometry to make it happen? This is the subject of vigorous research, and perhaps someday the computer really will be able to design the airplane"

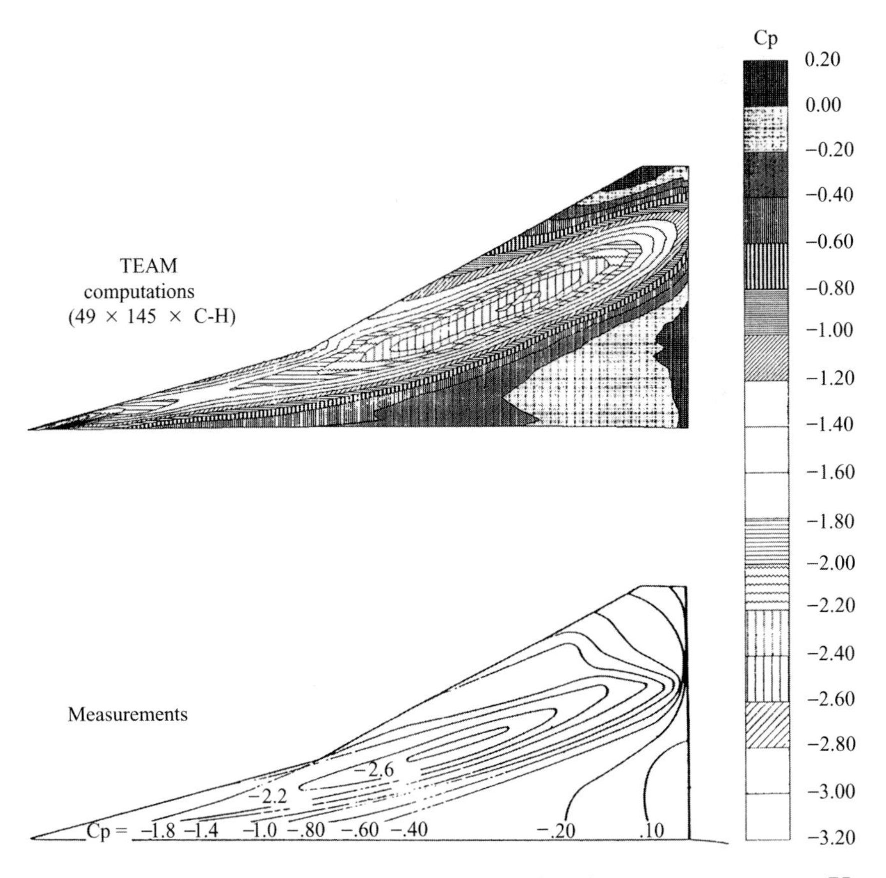

Fig. 12.40 Correlation of computed and measured surface pressure contours; 75-deg/62-deg double-delta wing body; $M_\infty = 0.3$; $\alpha = 20$ deg (Ref. 81; reprinted with permission).

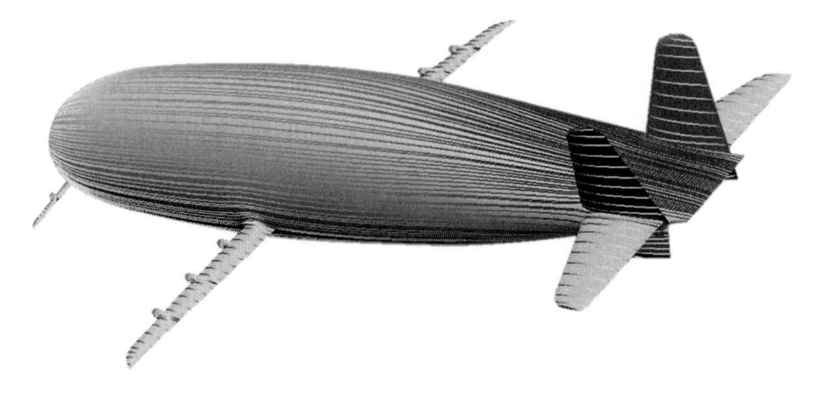

Fig. 12.41 CFD streamlines on dynamic lift airship (Ref. 133).

(at least, aerodynamically). Currently, the computer is pretty good at designing airfoils to obtain some specified pressure distribution in order to minimize shock formation, promote laminar flow, prevent viscous separation, and obtain a required amount of lift (although not all at the same time, usually!). Computational methods are also available to optimize the shape of simple fuselage-like bodies to minimize separation and promote laminar flow (but we still rely on wind tunnel and full scale testing to make sure the computer knows what it is talking about).

Emerging methods such as those described by Jameson et al. (Ref. 107) apply control theory techniques to optimize aerodynamic shapes using NS equations modeling viscous compressible flow. In these techniques, an initial aircraft geometry is modified iteratively using one cycle's result to specify an improved geometry for the next cycle. The geometric model can include a complete aircraft configuration, although only the wing airfoil shapes are being optimized. Results for a sample transport aircraft optimization show the elimination of wing shocks at Mach 0.83, and produce a 15-count drag reduction (8%).

More often, though, we use CFD to identify problems such as shocks, unwanted vortices, component interactions and flow separation, and then use designer intuition to revise the geometry in hopes of solving the problem. As the previous example showed, this works very well!

CFD Issues and Challenges

We have come a long way since 1879 when the annual proceedings of the British Royal Aeronautical Society could say, "Mathematics up to the present day has been quite useless to us in regard to flying" (quoted in Ref. 80). However, there are still many problems associated with the use of CFD for routinely solving aircraft design problems. Two problems are especially important: the influence of the turbulence model and the requirements for flow gridding.

The use of separate turbulence models for NS codes has been discussed. The results of the various NS codes are very sensitive to the turbulence model used, especially when separated flow is present.

CFD codes tend to produce reasonable-looking flowfields and pressures, but sometimes the integration of the calculated pressures yields lifts, drags, and moments which do not match experience. Reproduction of experimental data sometimes requires extensive "calibration" (i.e., fudging!) of the turbulence model. For this reason, CFD results are always somewhat suspect until the code has been checked against experimental data for a similar configuration.

The need to grid the entire flowfield around the aircraft presents another big problem for CFD users. "Gridding" refers to the breaking up of the space around the aircraft into numerous small blocks, or "cells," usually of roughly hexahedral shape. CFD methods calculate the flow properties within each cell, using various convergence schemes to equate the flow properties along the boundaries connecting the cells.

While gridding the space around a simple cylinder or a long wing can be easily automated, the gridding of the flowfield around a full aircraft cannot yet be fully automated and can take weeks. Figure 12.42 illustrates the complexity of the flowfield gridding. Note, for example, where the canopy meets the fuselage and where the cells must fan out in the empty region between wing and canard.

Gridding is especially important because the CFD results are highly sensitive to the shaping of the cells. You can actually get different answers for the same aircraft using two different gridding schemes. According to the author of

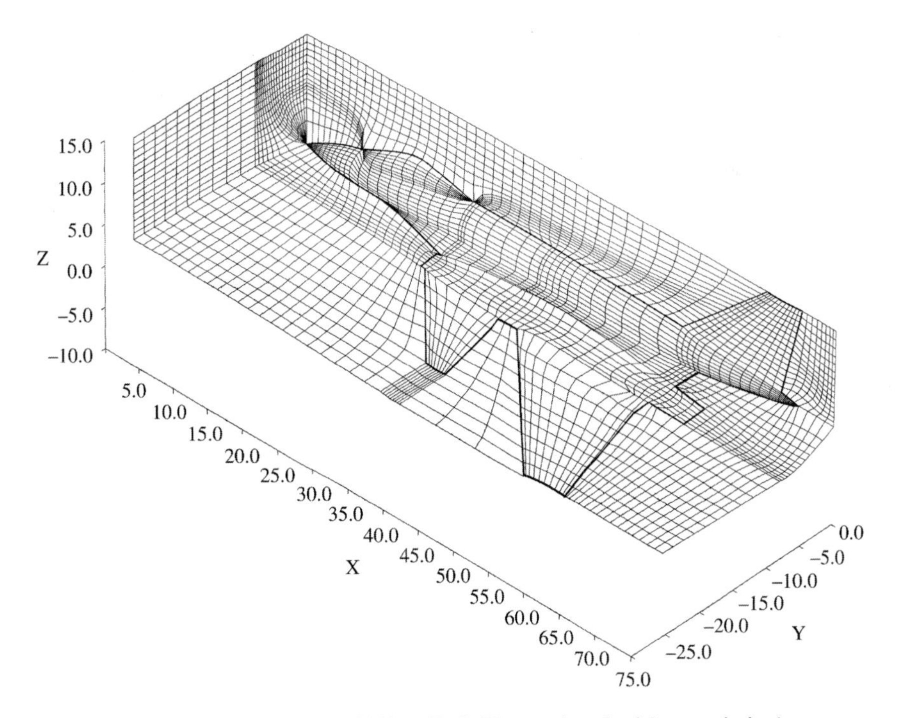

Fig. 12.42 Flowfield gridding (Ref. 82; reprinted with permission).

Ref. 82, "this sensitivity is more pronounced than that due to the type of mathematical model being used, e.g., NS vs Euler equations."

To address the gridding problem, researchers are investigating artificial intelligence (AI) approaches to gridding. Another approach is the computationally adaptive gridding in which the gridding scheme is automatically adjusted based upon the CFD results.

The most promising approach to developing automatic and instantaneous gridding for CFD is the "unstructured grid." The grid of Fig. 12.42 is highly structured. It begins at the outside boundaries as simple brick-like shapes, and transitions to the aircraft surface in a smooth and structured fashion. While this makes it easier for a human to create a reasonable grid, it is actually easier for an automated computer routine to create a grid that does not look like a brick wall with an aircraft-shaped hole in the center. Instead, the unstructured grid typically uses tetrahedral cells that connect at their vertices and can be placed as needed to complete a 3-D grid of the flowfield. Rather than a brick wall appearance, the unstructured grid looks like the work of a demented spider attempting to capture your aircraft! With the unstructured approach, the grids can often be generated in just a few days. (The goal of instantaneous gridding for complex aircraft configurations remains elusive.)

With unstructured grids it is relatively easy to automatically reduce the size of the cells where the flow properties are changing, such as around corners, to improve the accuracy of the CFD calculation. It is also easier to do this during execution (adaptive gridding), so that the beginning grid pattern need not be quite so perfect. Another advantage of the unstructured grids is that they lend themselves to parallel computer implementation, which provides reductions in execution time and computer costs.

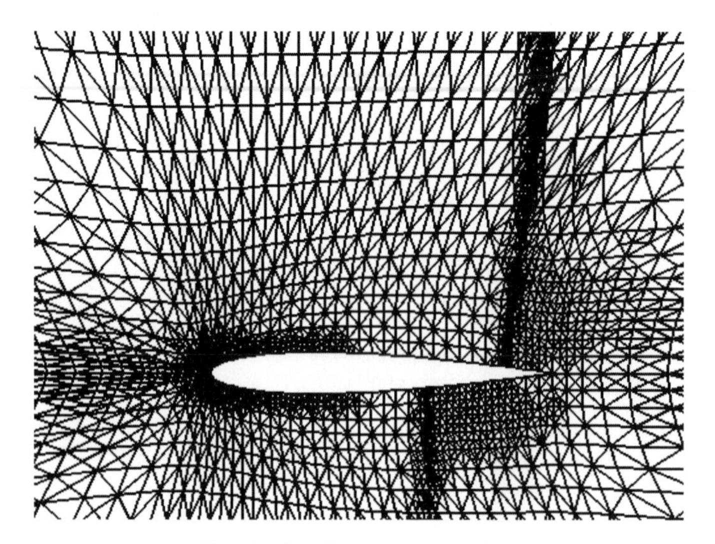

Fig. 12.43 Unstructured grid.

Problems of unstructured gridding include computational difficulties with viscous terms, tendencies to have regions of distorted cells, and computationally expensive flow solver routines (Ref. 102). But, unstructured grid CFD programs are gaining wider acceptance and offer great promise for the future. Figure 12.43 illustrates an unstructured grid for an airfoil. Note how during iteration the flow solver has clustered small grids where it found large changes in pressure, such as at the leading edge and through shocks. This increases computational accuracy (Ref. 118).

As with structural finite element methods, most working aerodynamic engineers will never be involved in writing CFD codes. The state of the art has advanced to the point where even big companies often use commercial or government codes such as USM3D and FLUENT rather than attempt to write and maintain their own proprietary codes. Instead, the working engineer must learn to use the available codes, understanding their applicability and limitations. Complete CFD analysis includes grid generation, flow definition (including initial conditions, properties, assumptions, and boundary conditions), calibration of the turbulence model, code execution, postprocessing, and evaluating and understanding the results. Finally, the CFD expert must provide the results to the aircraft designers in a manner that guides the improvement of the design.

CFD analysis of the Gee Bee Racer (courtesy Analytical Methods, Inc.).

Mig-29.

13 Propulsion

13.1 Introduction

All forms of aircraft propulsion develop thrust by pushing air (or hot gases) backward. In a simplified case the force obtained can be determined using Newton's equation ($F = ma$) by summing all the accelerations imparted to the air.

This is shown for fluid flow in Eq. (13.1), for the simplified example in Fig. 13.1. S is the cross-sectional area of the fluid acted upon by the propulsion system, V is the fluid velocity, and the zero subscript indicates the freestream condition. The propeller or jet engine is assumed to "magically" accelerate the air from velocity V_0 to V.

The rate of useful work done by the propulsion system, called the thrust power (P_t), equals the product of the thrust force and the aircraft velocity [Eq. (13.2)].

The change in kinetic energy (i.e., work) imparted to the fluid by the propulsion system is determined by the difference in fluid velocity. Taking the time derivative of work gives the power expended by this propulsion system, as shown in Eq. (13.3).

The propulsion efficiency (η_{PE}) is defined as the ratio of thrust power obtained to power expended, as shown in Eq. (13.4). Note that the efficiency is maximized when there is no change in fluid velocity. Unfortunately, at this condition Eq. (13.1) shows that the thrust is zero!

$$F = ma = \dot{m}\Delta V = (\rho V s)(V - V_0) = \rho s V(V - V_0) \tag{13.1}$$

$$P_t = F V_0 = \rho s V(V - V_0)V_0 \tag{13.2}$$

$$P_{t_{\text{expended}}} = \frac{\partial \Delta E}{\partial t} = \frac{1}{2}\dot{m}V^2 - \frac{1}{2}\dot{m}V_0^2 = \frac{1}{2}\rho V s(V^2 - V_0^2) = \frac{\rho s}{2}V(V^2 - V_0^2) \tag{13.3}$$

$$\eta_{PE} = \frac{P_t}{P_{t_{\text{expended}}}} = \frac{2}{V/V_0 + 1} \tag{13.4}$$

There is an unavoidable tradeoff between thrust and efficiency as determined by the ratio between exhaust and freestream fluid velocity. For maximum thrust, this ratio must be very high. For maximum efficiency, this ratio should be unity.

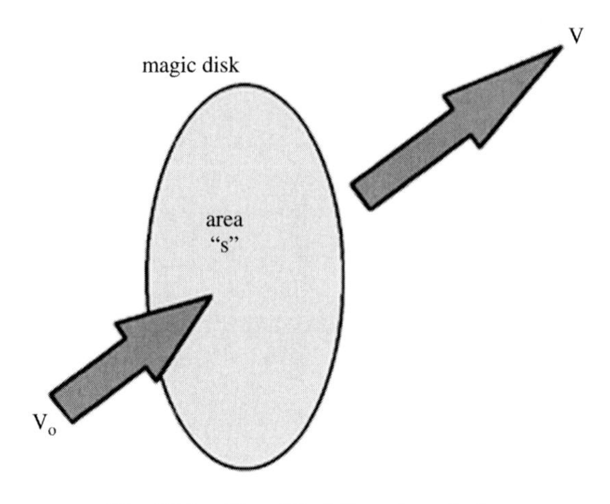

Fig. 13.1 Simplified thrust analysis.

If the exhaust velocity is reduced to little more than the freestream velocity for improved efficiency, the cross-sectional area of fluid affected by the propulsion system must approach infinity to maintain constant thrust. A typical turbojet will operate with the ratio of exhaust velocity to freestream velocity at well above 3.0, whereas a typical propeller aircraft will operate with this ratio at about 1.5.

The preceding analysis is too simplistic for actual thrust calculation. It falsely assumes that the fluid velocity is constant throughout the exhaust and that all of the accelerations experienced by the air mass occur at the propeller plane or within the jet engine.

Actually, the exhaust of a jet engine is usually at a higher pressure than the outside air, so the flow expands after leaving the nozzle. In other words, the air is still accelerating after the aircraft has passed.

For a propeller, the air-mass acceleration doesn't even occur at the propeller disk. Roughly half the air-mass acceleration occurs before reaching the propeller, and the other half occurs after passing the propeller.

Propulsion force estimation is also complicated by the fact that the propeller flowfield or jet intake and exhaust will influence the whole flowfield of the aircraft. It has already been mentioned that a pusher propeller will reduce the drag of a stubby aft fuselage by "sucking" air inward and preventing flow separation. Should this reduced drag be considered a part of the propulsive force because it is controlled with the throttle? What about the increased drag due to the propeller wake on a conventional airplane?

For a propeller aircraft, most of the propulsive force is exerted directly on the aircraft by the pull (or push) of the propeller itself through the propeller shaft. The propeller shaft is usually connected to the engine so the engine mounts actually pull (or push) the rest of the aircraft through the air.

For a jet aircraft the force exerted through the engine mounts may only be a third of the total propulsive force. Figure 13.2 shows the thrust contributors for a typical Mach 2.2 nacelle. The engine itself only contributes about 8% of the

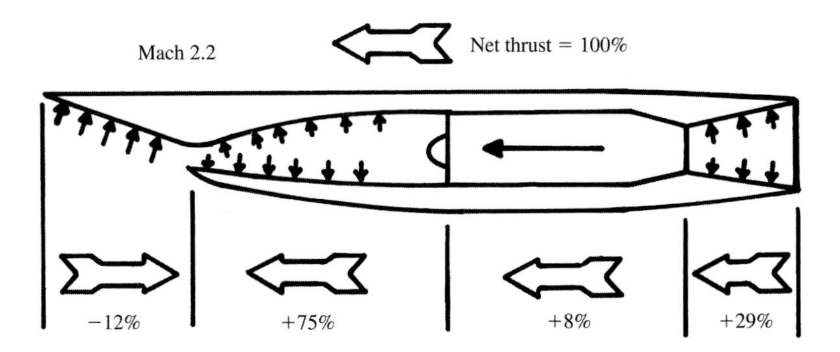

Fig. 13.2 Turbojet thrust contributors (North American Aviation A-5).

total. The nozzle, which generates thrust by expanding the high-pressure engine exhaust, contributes almost 30%.

The inlet system uses a system of shocks to slow the air to a subsonic speed. This creates a substantial drag. However, the subsonic expansion inside the inlet duct contributes a positive force that more than makes up for the external inlet system drag. In fact, the subsonic expansion inside the inlet duct is the largest single contributor of thrust! This illustrates the difficulty of calculating thrust by any simple model.

This chapter provides methods for estimating the net thrust provided by a propeller or jet engine as a part of the overall vehicle analysis and optimization. These methods are simplified to permit experiencing the whole design process within the time devoted to the typical college design course. The chapter will also introduce the vastly more complicated process of installed-thrust estimation used at major aerospace firms. Reference 46 provides a detailed treatment of jet-engine design and installation.

13.2 Jet-Engine Thrust Considerations

Before we begin the discussion of jet-engine installed thrust, a brief introduction to jet-engine cycle analysis and its effect on design is in order. As mentioned in Chapter 10, and shown in Fig. 13.3, a jet engine develops thrust by taking in air, compressing it (via the inlet duct and the compressor), mixing in fuel, burning the mixture, and expanding and accelerating the resulting high-pressure, high-temperature gases out the rear through a nozzle.

Fig. 13.3 Turbojet engine.

To provide power to drive the compressor, a turbine is placed in the exhaust stream which extracts mechanical power from the high-pressure gases. If greater thrust is required for a short period of time, an afterburner can be placed downstream of the turbine permitting the unburned air in the turbine exhaust to combust with additional fuel and thereby increase the exhaust velocity.

"Gross thrust" is produced as a result of the total momentum in the high-velocity exhaust stream. "Net thrust" is calculated as the gross thrust minus the "ram drag," which is the total momentum in the inlet stream. Note that the ram drag, which results from the deceleration of the air taken into the inlet, is included in the engine cycle analysis performed by the engine manufacturer to determine net uninstalled thrust.

Jet-engine cycle analysis, as detailed in Ref. 46 and other propulsion texts, is the straightforward application of the laws of thermodynamics to this Brayton engine cycle. In an "ideal" analysis, the efficiencies of components such as compressors and turbines are assumed to be 100%, i.e., no losses, and the resulting thrust for a given fuel flow, altitude, and Mach number is calculated. While optimistic, such ideal cycle analysis produces results in the "right ballpark" and illustrates the trends produced by varying such cycle parameters as overall pressure ratio, turbine inlet temperature, bypass ratio, and flight condition. These are discussed next.

One overriding factor in the determination of jet-engine performance is that the net thrust produced is roughly proportional to the air-mass flow (velocity × air density × airflow cross section) entering the engine. For a modern afterburning turbojet engine, roughly 100–130 lb of thrust (the "specific net thrust") is developed for each pound per second of air taken in by the engine {1–1.3 kN per kg/s}. For a turbofan engine, a specific net thrust of roughly 10–30 [0.1–0.3] can be obtained (sea level maximum static thrust).

An increase in air density such as at low altitude or low outside air temperature would therefore increase thrust by increasing mass flow. Hot day takeoffs from a high-elevation airport such as Denver pose problems because the reduction in air density causes a reduction in mass flow, and hence, thrust (compounded by the reduced wing lift in the thin air).

A simple first approximation of the effect of air density on jet engine thrust comes from mass-flow considerations. Thrust at other-than-sea-level standard conditions can be estimated as thrust at sea level for a given speed, multiplied by the ratio of air pressures and divided by the ratio of absolute air temperatures, relative to sea-level values. Pressures and temperatures for different altitudes can be found in Appendix B. As a rough but reasonable approximation for the standard atmosphere values, you can straight line from 100% thrust at sea level to 0 thrust at 55,000 ft {16,764 m} and get about the correct answer below 40,000 ft {13,716 m}.

Similarly, an increase in aircraft velocity also increases thrust due to ram effect increasing the mass flow. However, for a typical subsonic jet, the exhaust comes out the nozzle at a choked condition, and so the exit velocity equals the speed of sound regardless of aircraft velocity. As aircraft velocity approaches the speed of sound, the thrust is therefore reduced for the choked exit nozzle [see Eq. (13.1)]. When combined with the favorable ram effect, this results in a relatively constant thrust as velocity increases for the typical subsonic jet, dropping off as transonic speeds are reached.

For supersonic jet engines, a variable area, converging-diverging nozzle is typically employed which permits supersonic exhaust velocities. Therefore, the ram effect does cause the thrust to tend to increase with increasing velocity until at high Mach numbers where excessive total pressure losses occur in the inlet, resulting in thrust degradation. The Mach number at which inlet losses become excessive is determined by the number of shocks and the extent of variable geometry employed, as described in Chapter 10.

Thrust and propulsive efficiency are strongly affected by the engine's overall pressure ratio (OPR). OPR is the ratio of the pressures at the engine exhaust plane and inlet front face. This pressure ratio is a measure of the engine's ability to accelerate the exhaust, which produces thrust. OPRs usually range from about 15 to 1 to about 30 to 1.

Another key parameter which currently limits turbine engine performance is the turbine inlet temperature (TIT). As mentioned earlier, it would be desirable for maximum thrust and efficiency to combust at the stoichiometric air–fuel ratio of about 15 to 1. This produces temperatures far too high for current turbine materials, even using the best available cooling techniques. Instead, a "lean" mixture of about 60 to 1 (air to fuel) is used, with the extra air holding down the TIT to about 2000–2500°F {~1100–1400°C}. This results in less thrust and thermal efficiency, and so a key objective in propulsion technology development has always been the increase in allowable TIT. On average, it has been increased by 320°R {180 K} per decade since 1950.

To increase propulsive efficiency, the turbofan engine uses an oversized fan with some of the accelerated fan air "bypassed" around the engine, not being used for combustion (Fig. 13.4). This has the effect of allowing the engine to accelerate a larger cross-sectional area of air by a smaller change in velocity, which increases efficiency as determined by Eqs. (13.1) and (13.4). The bypass ratio was defined in Chapter 10 as the ratio of the mass flows of the bypassed air and the air that goes through the core of the engine to be used for combustion.

A higher bypass ratio, which enables the engine to accelerate a larger cross section of air, produces higher efficiency and hence greater thrust for a given expenditure of fuel. However, the fan alone cannot efficiently accelerate the air to transonic or supersonic exit speeds, and so this favorable effect works only

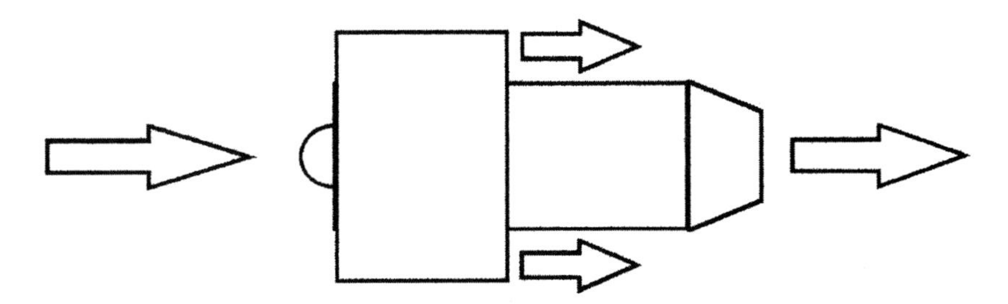

Fig. 13.4 Turbofan engine.

at lower speeds. Furthermore, the greater intake cross-section area of a high-bypass-ratio engine leads to a greater momentum ram drag, which increases roughly by the square of the airspeed. As was shown in Fig. 10.2, the high-bypass turbofan is best at subsonic speeds, giving way to the low-bypass ratio turbofan at the low supersonic speeds. At higher supersonic speeds, say over about Mach 2.2, the pure turbojet is superior.

13.3 Turbojet Installed Thrust

Chapter 10 described statistical methods for estimating installed thrust and specific fuel consumption for jet engines. These are suitable for initial sizing and performance estimation. For a more sophisticated analysis it is necessary to estimate analytically the installed performance of an existing engine or a proposed new engine from the uninstalled engine data.

Uninstalled engine data can be obtained from the engine manufacturer. Data for several conceptual engines are summarized in the appendices. For a new-design engine, the data must be estimated using cycle-analysis equations such as in Ref. 46. Usually the designers in an aircraft company will obtain the uninstalled-engine data for a proposed engine from the engine manufacturer, and then will correct the data for installation effects.

It is common early in design studies to approximate the performance of a new-design engine by a "fudge-factor" approach. An existing engine with approximately the same bypass ratio is selected, and its size, weight, and performance data are multiplied by factors based upon the expected improvements by applying advanced technologies.

For example, it might be assumed that an engine designed 10 years from now would have 25%-less specific fuel consumption, 30%-less length, and 30%-less weight compared with an existing engine. Such fudge-factors are based upon either historical trend analysis or an approximate cycle analysis for expected technology improvements.

It is assumed next that uninstalled-engine data are available, either from an engine manufacturer, a preliminary cycle analysis, or a fudge-factor approach based upon some given engine (such as those in the appendices).

13.4 Thrust-Drag Bookkeeping

Bookkeeping is not normally considered an engineering subject. However, the interactions between thrust and drag are so complex that only a bookkeeping-like approach can ensure that all forces have been counted once and only once. It is not at all uncommon to discover, halfway through an aircraft design project, that some minor drag item has been either included in both drag and thrust calculations or has been ignored by both departments under the assumption that it is being included by the other!

Each aircraft company develops its own system for thrust-drag bookkeeping. In fact, a different system may be developed for each design project. In most cases the guiding principle for determining whether a force is considered a part of the drag or the thrust is whether that force changes when the throttle setting is changed.

In an afterburning jet engine, for example, the nozzles open wide when the throttle is advanced to the afterburning position. This changes the aerodynamic drag on the outside of the nozzles, so the nozzle aerodynamic drag is counted as a reduction in the engine thrust in many thrust-drag bookkeeping systems.

In other thrust-drag bookkeeping systems, the nozzle drag is separated into two components: 1) the drag value at some fixed nozzle setting (usually full open), which is included in the aerodynamic drag calculation, and 2) the variation of drag as the nozzle setting is changed, which is included in the propulsion-installation calculations.

Either bookkeeping approach will give correct results providing that the Aerodynamics and Propulsion departments both understand it.

Thrust–drag bookkeeping becomes especially complex when sorting out the results of wind-tunnel testing. Different wind-tunnel models are used to test different thrust and drag items. The model used for determining basic aerodynamic and stability derivatives is usually unpowered, and a separate powered model is used to estimate propulsion effects. Lack of a mutually understood bookkeeping system by both the Aerodynamic and Propulsion departments will cause chaos.

The student should realize that the organization of this book assumes a thrust–drag bookkeeping system. Items presented in this chapter as reductions to thrust may be considered to be drag items in another bookkeeping system. Reference 35 contains a detailed review of the subject of thrust–drag bookkeeping.

13.5 Installed-Thrust Methodology

The actual available thrust used in performance calculations—called the "installed net propulsive force"—is the uninstalled thrust corrected for installation effects, minus the drag contributions that are assigned to the propulsion system by the selected thrust-drag bookkeeping system. This is depicted in Fig. 13.5.

The "manufacturer's uninstalled engine thrust" is obtained by cycle analysis and/or testing, or can be approximated using fudge-factors as already described. The manufacturer's engine data are based upon some assumed schedule of inlet pressure recovery vs Mach number. Inlet pressure recovery (P_1/P_0) is the total pressure at the engine front face (1) divided by the total pressure in the freestream (0).

For subsonic engines, it is frequently assumed that the inlet pressure recovery is perfect, i.e., $P_1/P_0 = 1.0$. For supersonic military aircraft a Mil-Spec formula is used. Inlet distortion, engine bleed, and engine power extraction are usually assumed by the manufacturer to be zero. Also, the engine data are based upon the manufacturer's nozzle design.

Note that the SFC values supplied with the engine are based upon this uninstalled-engine thrust, not the installed net propulsive force. When determining fuel usage, the SFC values must be ratioed up by the ratio between uninstalled-engine thrust and the net propulsive force (i.e., the thrust required for the desired performance).

The "installed-engine thrust" is the actual thrust generated by the engine when installed in the aircraft. This is obtained by correcting the thrust for the actual

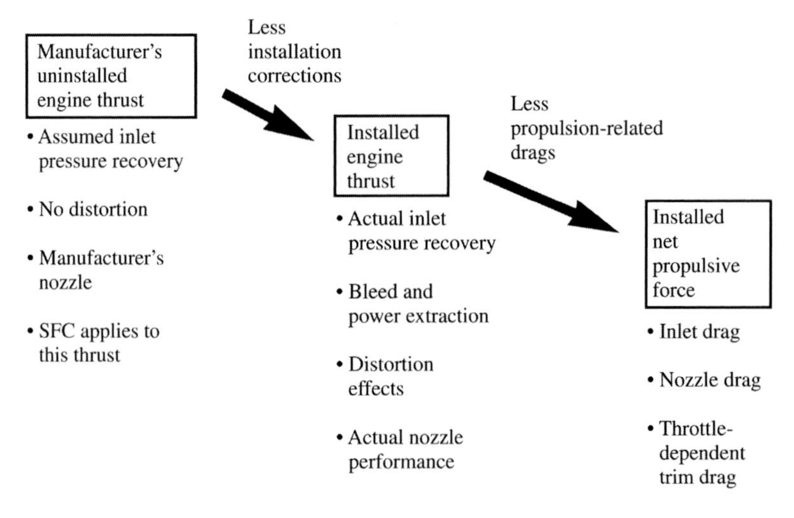

Fig. 13.5 Installed thrust methodology.

inlet pressure recovery and nozzle performance, and applying thrust losses to account for engine bleed and power extraction.

"Inlet distortion" refers to pressure and velocity variations in the air delivered to the engine. It primarily affects the allowable operating envelope of the engine.

The installed net propulsive force is the installed engine thrust minus the inlet, nozzle, and throttle-dependent trim drags. The steps depicted in Fig. 13.5 are detailed next.

Installed Engine Thrust Corrections

The manufacturer's uninstalled engine thrust is based upon an assumed inlet pressure-recovery. For a subsonic engine, it is typically assumed that the inlet has perfect recovery, i.e., 1.0. Supersonic military aircraft engines are usually defined using an inlet pressure-recovery of 1.0 at subsonic speeds and the inlet recovery of Eq. (13.5) (MIL-E-5008B) at supersonic speeds. Added to this is the pressure recovery loss due to internal flow in the inlet duct itself. Typically this adds 2–3% to the losses.

Figure 13.6 shows this reference inlet pressure-recovery plotted vs Mach number, compared to the recovery available for a normal-shock inlet and external compression inlets with one, two, and three ramps.

$$\left(\frac{P_1}{P_0}\right)_{\text{ref}} = 1 - 0.075(M_\infty - 1)^{1.35} \tag{13.5}$$

The external compression inlets of Fig. 13.6 are of movable ramp design with a perfectly optimized schedule of ramp angles as a function of Mach number. To determine the pressure recovery of a fixed or less-than-perfect inlet, the shock tables in NACA TR1135 can be used.

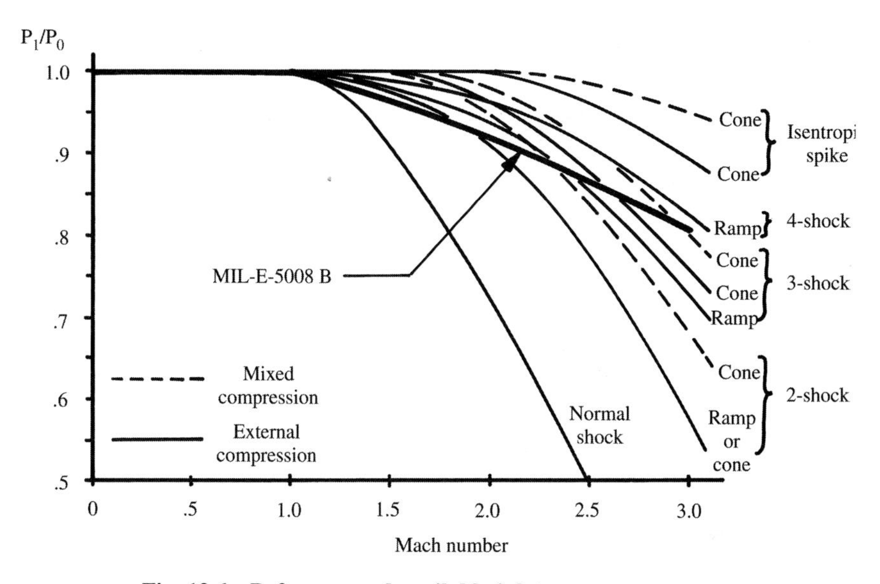

Fig. 13.6 Reference and available inlet pressure recovery.

The pressure losses inside the inlet duct must also be accounted for. These losses are determined by the length and diameter of the duct, the presence of bends in the duct, and the internal Mach number.

For initial evaluation of a typical inlet duct, an internal pressure recovery of 0.96 for a straight duct and 0.94 for an *S* duct may be used. The short duct of a subsonic podded nacelle will have a pressure recovery of 0.98 or better. More detailed estimation of inlet internal-pressure loss is based upon experimental data (see Ref. 27), and requires a separate evaluation at each Mach number.

Figure 13.7 provides the actual inlet pressure recoveries of some existing designs. This figure may be used for pressure-recovery estimation during early design studies.

Reducing inlet pressure recovery has a greater-than-proportional effect upon the engine thrust, as shown in Eq. (13.6). The inlet "ram recovery correction factor (C_{ram})" is provided by the manufacturer for various altitudes, Mach numbers, air temperatures, and thrust settings. Typically, C_{ram} ranges from 1.2–1.5. If the manufacturer's data are not available, C_{ram} may be approximated as 1.35 for subsonic flight and by Eq. (13.7) for supersonic flight.

$$\text{Percent thrust loss} = C_{ram}\left[\left(\frac{P_1}{P_0}\right)_{ref} - \left(\frac{P_1}{P_0}\right)_{actual}\right] \times [100] \tag{13.6}$$

$$\text{Supersonic:} \quad C_{ram} \cong 1.35 - 0.15(M_\infty - 1) \tag{13.7}$$

Actually, the inlet pressure recovery is a function of both Mach number and the inlet mass flow. At low speeds with a high throttle setting, the inlet "hole"

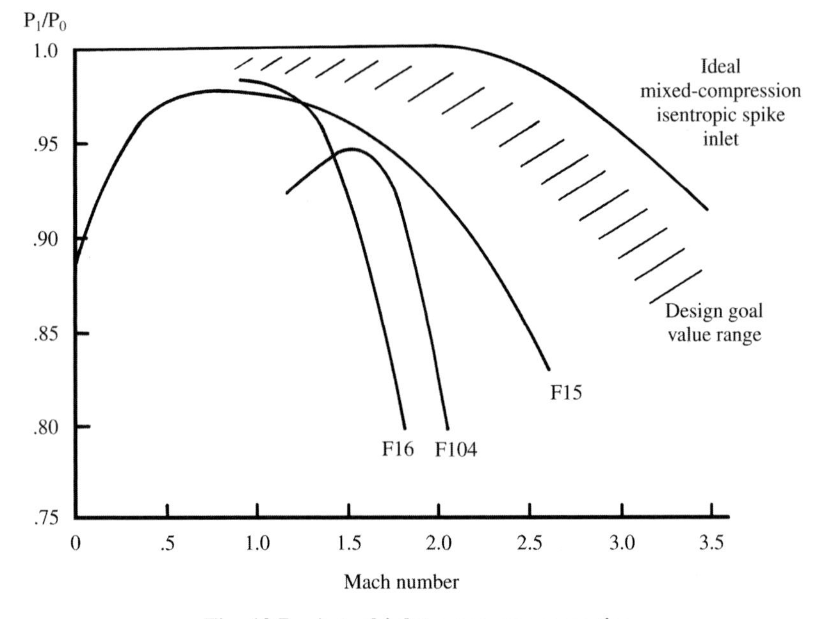

Fig. 13.7 Actual inlet pressure recoveries.

isn't big enough and the engine has to "suck" the air into the inlet duct. This causes a lower pressure recovery, as seen to the left of Fig. 13.7. (These represent full throttle pressure recoveries.) However, if the engine is demanding less airflow (lower throttle setting), the inlet can more readily meet the demand so the pressure recovery is higher.

For the static F-16 at maximum thrust (i.e., massflow), the pressure recovery drops to about 0.86. At half the maximum massflow, the static pressure recovery is over 0.96. At Mach 0.6, the pressure recovery difference from full to half massflow is only 2%, and it is less than that at higher Mach numbers.

This massflow variation in pressure recovery must be accounted for in detailed performance calculations, but can be neglected for conceptual design studies.

High-pressure air is bled from the engine compressor for cabin air, anti-icing, and other uses. This engine bleed air (not to be confused with inlet boundary-layer bleed and other forms of secondary airflow) exacts a thrust penalty that is also more than proportional to the percent of the total engine massflow extracted as bleed air.

Equation (13.8) illustrates this, where the "bleed correction factor (C_{bleed})" is provided by the manufacturer for various flight conditions. For initial analysis, C_{bleed} can be approximated as 2.0. The bleed massflow typically ranges from 1–5% of the engine massflow.

$$\text{Percent thrust loss} = C_{\text{bleed}}\left(\frac{\text{bleed massflow}}{\text{engine massflow}}\right) \times [100] \tag{13.8}$$

Installed engine thrust is also affected by horsepower extraction. Jet engines are equipped with rotating mechanical shafts turned by the turbines. The electrical generators, hydraulic pumps, and other such components connect to these shafts.

This extraction is typically less than 200 hp {150 kW} for a 30,000-lb-thrust {133 kN} engine, and usually has only a small effect upon installed thrust. Horsepower extraction is included in the cycle analysis used for detailed calculation of installed-engine thrust, but can be ignored for initial analysis.

As mentioned, moderate inlet distortion usually has little effect upon installed thrust, but can restrict the engine operating envelope. The effects of distortion are calculated later in the design process. For initial design, the guidelines previously suggested for location of inlets and for forebody shaping should avoid any later problems with inlet distortion.

Nozzle efficiency has a direct effect upon thrust. However, it is rare to use a nozzle other than that provided by the manufacturer. For cases in which a unique nozzle (such as a 2-D vectoring nozzle) is employed, the new nozzle can usually be designed to provide the same efficiency as the manufacturer's nozzle. (The drag effects of alternate nozzles are discussed later.)

For hot-day operation (if engine manufacturer's data are not available), thrust can be reduced about 0.42% per °R {0.75% per K}.

Installed Net Propulsive Force Corrections

The installed engine thrust is the actual thrust produced by the engine as installed in the aircraft. However, the engine creates three forms of drag that must be subtracted from the engine thrust to determine the thrust force actually available for propelling the aircraft. This propelling force, the installed net propulsive force, is the thrust value to be used for aircraft performance calculations.

Most of the engine-related drag is produced by the inlet as a result of a mismatch between the amount of air demanded by the engine and the amount of air that the inlet can supply at a given flight condition. When the inlet is providing exactly the amount of air the engine demands (mass flow ratio equals 1.0), the inlet drag is negligible.

The inlet must be sized to provide enough air at the worst-case condition, when the engine demands a lot of air. This sets the capture area. Most of the time the engine demands less air than an inlet with this capture area would like to provide (i.e., mass flow ratio is less than 1.0).

When the mass flow ratio is less than 1.0, the excess air must either be spilled before the air enters the inlet or bypassed around the engine via a duct that dumps it overboard (Fig. 13.8) or into an ejector-type engine nozzle.

The drag from air spilled before entering the inlet is called "spillage," or "additive" drag. Additive drag represents a loss in momentum of the air which is slowed and compressed by the external part of the inlet but not used by the engine. The additive drag is determined by calculating, for each flight Mach number and engine mass-flow ratio, the Mach numbers and pressures throughout the inlet and integrating the forces in the flight direction for the part of the air which is spilled.

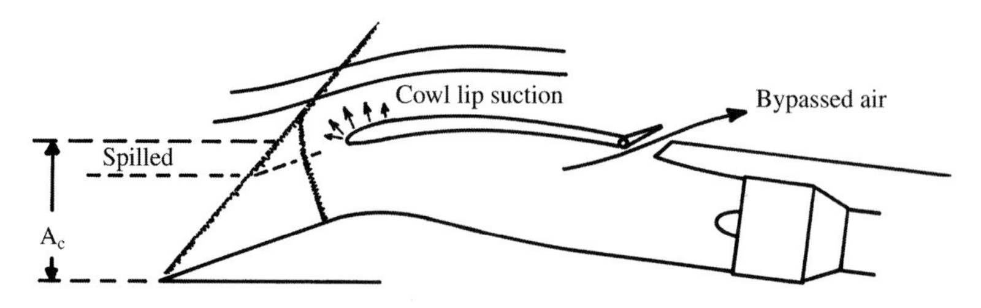

Fig. 13.8 Additive drag, cowl-lip suction, and bypass subcritical operation.

The spilled air will be turned back toward the freestream direction by the inlet cowl lip, producing a reduced pressure on the cowl. This provides a suction with a component in the forward direction, i.e., a thrust (as shown in Fig. 13.8). This cowl-lip suction reduces the additive drag by as much as 30–40% in the low-supersonic regime. For a subsonic jet with well-rounded cowl lips, this suction will virtually eliminate additive drag.

Even with cowl-lip suction, the additive drag under certain flight conditions could exceed 20% of the total aircraft drag. A penalty of this magnitude is never seen because the designers resort to inlet-air bypass whenever the additive drag is too great.

Allowing the excess air to enter the inlet and be dumped overboard or into an ejector nozzle will keep the inlet additive drag to a small value. The resulting bypass drag will be substantially less than the additive drag would have been. Bypass drag is calculated by summing the momentum loss experienced by the bypassed air.

Another form of inlet drag is the momentum loss associated with the inlet boundary-layer bleed. Air is bled through holes or slots on the inlet ramps and within the inlet to prevent shock-induced separation and to prevent the buildup of a thick turbulent boundary layer within the inlet duct. This air is dumped overboard out an aft-facing discharge exit, which is usually located a few feet behind the inlet.

(*Note:* Do not confuse inlet boundary-layer bleed with the inlet boundary-layer diverter. The diverter prevents the fuselage boundary-layer air from entering the inlet. Diverter drag has been accounted for in the aerodynamic chapter.)

Calculation of bleed, bypass, and additive drag including cowl-lip suction is a complicated procedure combining analytical and empirical methods. The textbook methods (see Refs. 10, 25, 26, and 27) are very time consuming and cannot account for the effects of the actual aircraft geometry, which may greatly affect both the inlet flowfield and the pressure loss through bleed and bypass ducts.

In a major aircraft company such calculations are made by propulsion specialists using complex computer programs. The results are included in the installed net propulsive force data that are provided to the sizing and performance analyst.

To permit rapid initial analysis and trade studies, Fig. 13.9 provides a "ballpark" estimate of inlet drag for a typical supersonic aircraft. This chart was prepared by the author using data from Ref. 47 and other sources, and

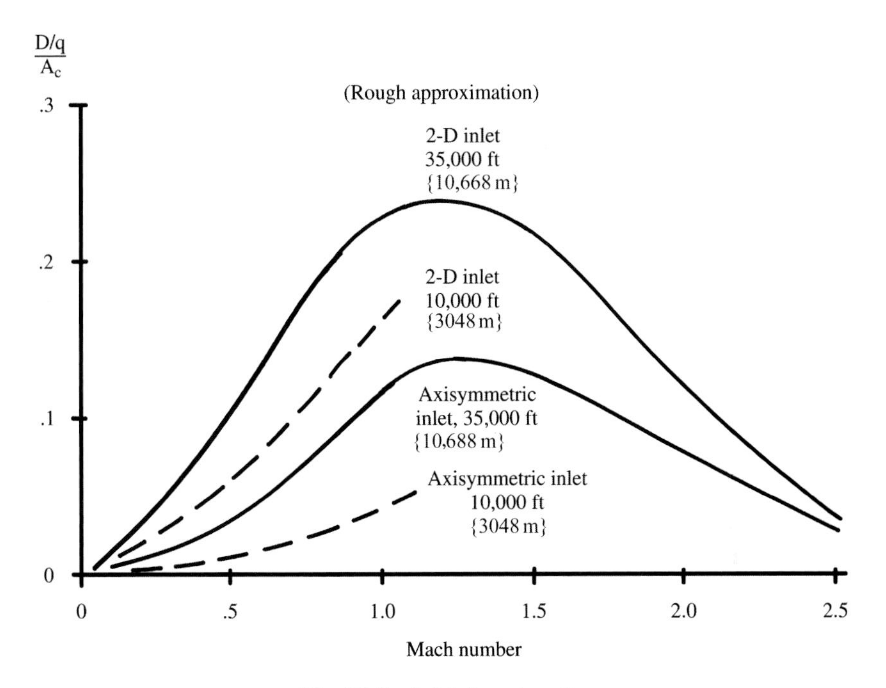

Fig. 13.9 Inlet drag trends.

should be used with great caution as it is merely typical data, not an estimate for any given inlet design.

This chart assumes that the engine is operating at a maximum dry or afterburning power setting, and that the inlet is operating at a corresponding mass-flow ratio. The chart does not reflect the increase in inlet drag experienced when the thrust setting is reduced (which reduces the mass-flow ratio). However, this chart should provide a reasonable approximation of inlet drag suitable for initial analysis and student design studies.

Nozzle drag varies with nozzle position as well as with the flight condition. To properly determine nozzle drag, the actual nozzle geometry as a function of throttle setting and flight condition must be known, and the drag calculated by taking into account the overall aircraft flowfield. As an initial approximation, the effect of nozzle position may be ignored and the nozzle drag estimated by the typical subsonic values shown in Table 13.1 (Ref. 10) for the nozzle types shown in Fig. 10.23.

The nozzle drag increases transonically and then drops off at supersonic speeds. For initial analysis the subsonic value may be assumed for all speeds. Note that these nozzle drags are referenced to the maximum cross-sectional area of the fuselage. For a subsonic, podded nacelle, the nozzle drag is negligible.

The remaining propulsion-system drag is the variation of trim drag with throttle setting. If the engine thrust axis is not through the center of gravity, any thrust change will cause a pitching moment. The trim force required to counter this

Table 13.1 Nozzle incremental drag (Ref. 10)

Nozzle type	Subsonic $\frac{D/q}{A_{\text{fuselage}}}$ [a]
Convergent	0.036–0.042
Convergent iris	0.001–0.020
Ejector	0.025–0.035
Variable ejector	0.010–0.020
Translating plug	0.015–0.020
2-D nozzle	0.005–0.015

[a]Referenced to fuselage maximum cross-section area.

moment is charged to the propulsion in most thrust-drag bookkeeping systems. For initial analysis this may be ignored unless the thrust line is substantially above or below the aircraft centerline.

Part Power Operation

Turbojet and turbofan engines do not like to operate at less than maximum thrust (or power) setting. When you throttle back, the reduction in thrust is more than proportional to the reduction in fuel flow, so the specific fuel consumption (SFC) increases. A noticeable increase in SFC typically begins when you throttle below about 90% power. For this reason, engine companies prepare "part-power tables" listing fuel consumption as a function of thrust at different altitudes and speeds. The installation analysis just described should also be applied to the part-power tables. This is a very laborious process!

The part-power effect on SFC can be approximated by a semi-empirical equation developed by Mattingly, coauthor of Ref. 46. This provides a realistic increase in SFC as thrust is reduced (Eq. 13.9).

$$\frac{c}{c_{\text{max dry}}} = \frac{0.1}{\left(\frac{T}{T_{\text{max dry}}}\right)} + \frac{0.24}{\left(\frac{T}{T_{\text{max dry}}}\right)^{0.8}} + 0.66\left(\frac{T}{T_{\text{max dry}}}\right)^{0.8} + 0.1M\left[\frac{1}{\left(\frac{T}{T_{\text{max dry}}}\right)} - \left(\frac{T}{T_{\text{max dry}}}\right)\right] \tag{13.9}$$

When a jet engine is throttled all the way back to "idle," neither the fuel flow nor the thrust actually go to zero. This residual thrust can be a real problem when you are trying to descend. If residual thrust divided by aircraft weight (T/W) is equal to the inverse of the lift-to-drag ratio ($1/L/D$), the aircraft cannot descend! Engine companies provide tables of idle thrust and fuel flow that should be used if available. If data are not available, a rough approximation is that idle SFC will be 1.5 times the max-dry SFC.

13.6 Piston-Engine Performance

The aircraft piston engine operates on the four-stroke Otto cycle used by automobiles. The thermodynamic theory of the Otto-cycle reciprocating engine is described in Refs. 28, 48, and 49. For design purposes the most important thing to know about the piston engine is that the power produced is directly proportional to the massflow of the air into the intake manifold. In fact, horsepower is approximately 620 times the air massflow (lb/s) {or power in kW = 1019 times massflow in kg/s}.

Mass flow into the engine is affected by the outside air density (altitude, temperature, and humidity) and intake manifold pressure. Equation (13.10) accounts for the air-density effect upon power, and is attributed to Gagg and Ferrar of the Wright Aeronautical Company (1934). This equation indicates that at an altitude of 20,000 ft {6100 m} a piston engine has less than half of its sea-level power.

$$\text{power} = \text{power}_{SL}\left(\frac{\rho}{\rho_0} - \frac{1-\rho/\rho_0}{7.55}\right) \tag{13.10}$$

The intake manifold is usually at atmospheric pressure. A forward-facing air-intake scoop can provide some small increase in manifold pressure at higher speeds. Large increases in manifold pressure require mechanical pumping via a "supercharger" or "turbosupercharger."

The supercharger is a centrifugal air compressor mechanically driven by a shaft from the engine. The amount of air compression available is proportional to engine rpm. The turbosupercharger, or "turbocharger," is driven by a turbine placed in the exhaust pipe. This recovers energy which would otherwise be wasted, and decouples the available amount of compression from the engine rpm.

Supercharging or turbocharging is usually used to maintain sea-level pressure in the intake manifold as the aircraft climbs. Typically the sea-level pressure can be maintained up to an altitude of about 15,000–20,000 ft {4500–6100 m}. Above this altitude the manifold pressure, and hence the power, drops. Figure 13.10 shows typical engine performance for nonsupercharged, supercharged, and turbocharged engines.

Supercharging or turbocharging may also be used to raise the intake manifold pressure above the sea-level value to provide additional power from a given engine. However, the increased internal pressures require a heavier engine for structural reasons.

Piston engine performance charts are provided by the manufacturer as a function of manifold pressure, altitude, and rpm.

Propeller Performance

A propeller is a rotating airfoil that generates thrust much as a wing generates lift. Like a wing, the propeller is designed to a particular flight condition. The propeller airfoil has a selected design lift coefficient (usually around 0.5), and the twist of the airfoil is selected to give the optimal airfoil angle of attack at the design condition.

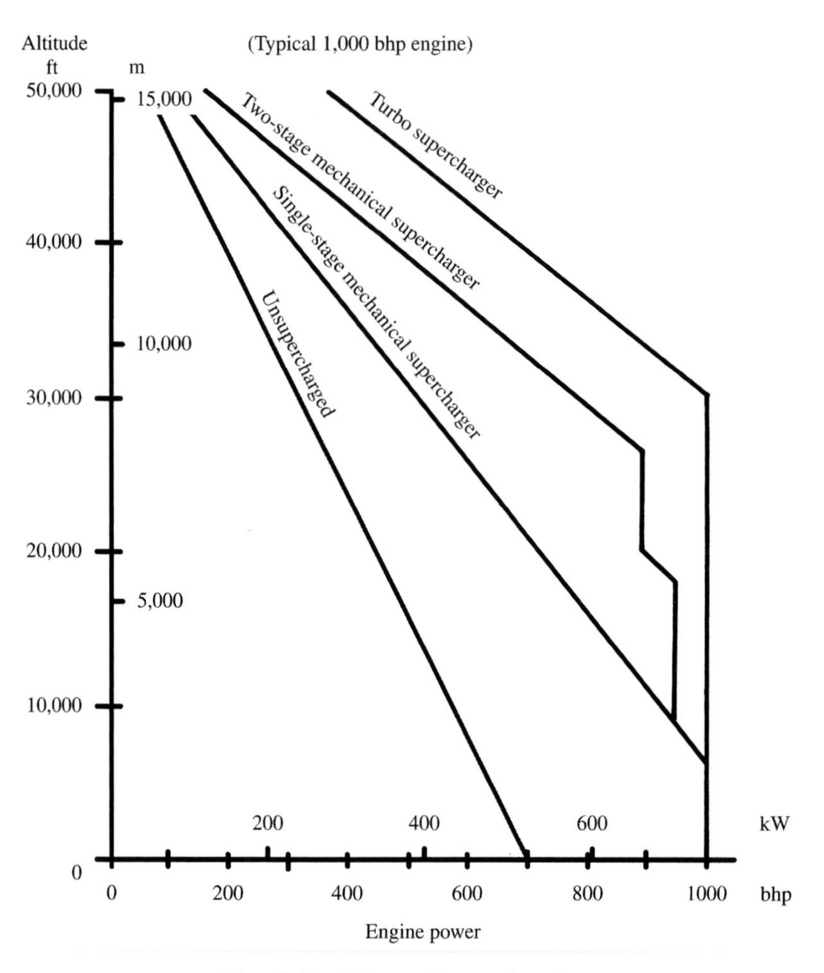

Fig. 13.10 Effects of supercharging.

Since the tangential velocities of the propeller airfoil sections increase with distance from the hub, the airfoils must be set at progressively reduced pitch-angles going from root to tip. The overall "pitch" of a propeller refers to the blade angle at 75% of the radius (70% in some books).

Propeller theory is well covered in many textbooks, such as Ref. 49. While theory is useful for propeller designers, the aircraft designers usually work with experimental propeller data provided by the propeller companies.

As for a wing, the properties of a propeller are expressed in coefficient form. Experimental data for design purposes are expressed using a variety of parameters and coefficients, as described next.

$$\text{Advance Ratio:} \quad J = V/nD \tag{13.11}$$

$$\text{Power Coefficient:} \quad c_P = \frac{P}{\rho n^3 D^5} = \frac{550\,\text{bhp}}{\rho n^3 D^5} \tag{13.12}$$

$$\text{Thrust Coefficient:} \quad c_T = T/\rho n^2 D^4 \tag{13.13}$$

$$\text{Speed-Power Coefficient:} \quad c_S = V\sqrt[5]{\rho/Pn^2} \tag{13.14}$$

$$\text{Activity Factor:} \quad AF_{\text{per blade}} = \frac{10^5}{D^5}\int_{0.15R}^{R} cr^3$$

$$= \frac{10^5 c_{\text{root}}}{16D}[0.25 - (1-\lambda)\,0.2] \tag{13.15}$$

$$\text{Propeller Efficiency:} \quad \eta_P = \frac{TV}{P} = \frac{TV}{550\,\text{bhp}} \tag{13.16}$$

$$\text{Thrust:} \quad T = P\eta_p/V = \frac{550\,\text{bhp}\,\eta_P}{V} \quad \text{(forward flight)} \tag{13.17}$$

or

$$T = \frac{c_T}{c_p}\frac{P}{nD} = \frac{c_T}{c_P}\frac{550\,\text{bhp}}{nD}\,\text{(static)} \tag{13.18}$$

where

T = thrust (lb or kN)
V = velocity (ft/s or m/s)
P = power (ft-lb/s or kW)
bhp = brake horsepower
n = rotation speed (rev/s)
D = propeller diameter (ft or m)
c = propeller airfoil chord (ft or m)

The advance ratio (equivalent to the wing angle of attack) is related to the distance the aircraft moves with one turn of the propeller. Advance ratio is sometimes called the "slip function" or "progression factor."

The power and thrust coefficients are nondimensional measures of those quantities, much like the wing lift coefficient. The speed–power coefficient is defined as the advance ratio raised to the fifth power divided by the power coefficient. The speed–power coefficient is nondimensional and does not involve the propeller diameter, which is useful for comparison between propellers of different sizes.

The activity factor is a measure of the effect of blade width and width distribution on the propeller and is a measure of the propeller's ability to absorb power. Activity factors range from about 90–200, with a typical light-aircraft activity factor being 100 and a typical large turboprop having an activity factor of 140. The final expression in Eq. (13.15) is the activity factor for a straight-tapered propeller blade of taper ratio λ.

Equation (13.16) relates the propeller efficiency, previously discussed in Chapters 3 and 10, to the advance ratio and the ratio of the thrust coefficient to the power coefficient. This ratio is used in Eq. (13.17) to determine the thrust at static conditions when the velocity is zero and the propeller-efficiency equation cannot be used for thrust determination.

Propeller data are available from the manufacturers as well as various NASA/NACA reports. These data are provided in many different formats using different combinations of the preceding parameters and coefficients. Whatever the format, Eq. (13.17) is ultimately used to determine the propeller thrust at a given flight condition.

Figures 13.11 and 13.12, propeller charts for static and forward flight (Ref. 50), have been chosen as typical of propellers used for modern light and business aircraft. These charts represent a three-bladed propeller with a design lift coefficient of 0.5 and an activity factor of 100.

For a two-bladed propeller, the forward-flight efficiencies are about 3% better than shown in Fig. 13.12, but the static thrust is about 5% less than shown in Fig. 13.11. The reverse trends are true for a four-bladed propeller. Also, a wooden propeller has an efficiency about 10% lower due to its greater thickness.

(If Ref. 50 is unavailable, Ref. 49 contains 43 pages of propeller charts taken from Ref. 50.)

If the propeller is of variable-pitch design, its pitch is adjusted to the optimum blade angle at each flight condition to produce a constant engine rpm regardless of the horsepower being produced.

The advance ratio and power coefficient are then independent variables and the propeller efficiency can be read in Fig. 13.12 for any combination of advance ratio and power coefficient that may occur in flight. Blade angle for the variable-pitch propeller can be read as a fallout parameter in Fig. 13.12.

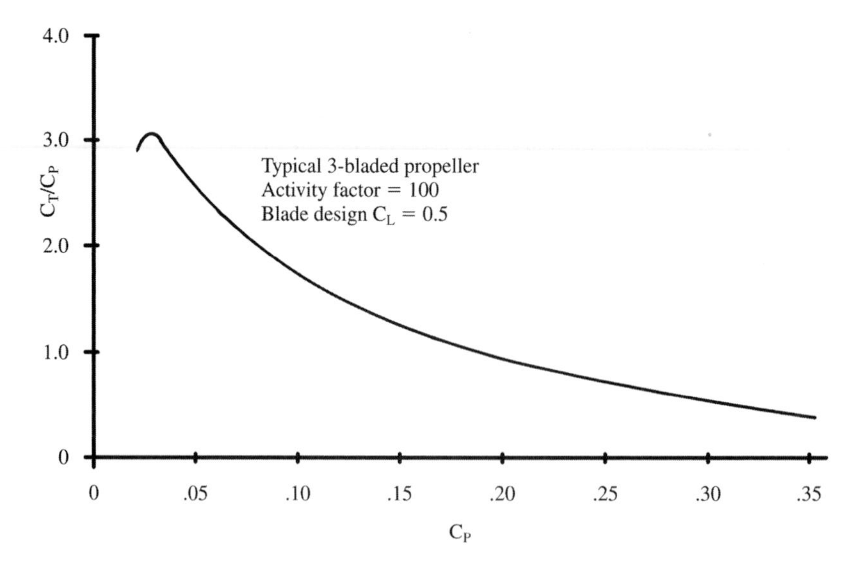

Fig. 13.11 Static propeller thrust (after Ref. 50).

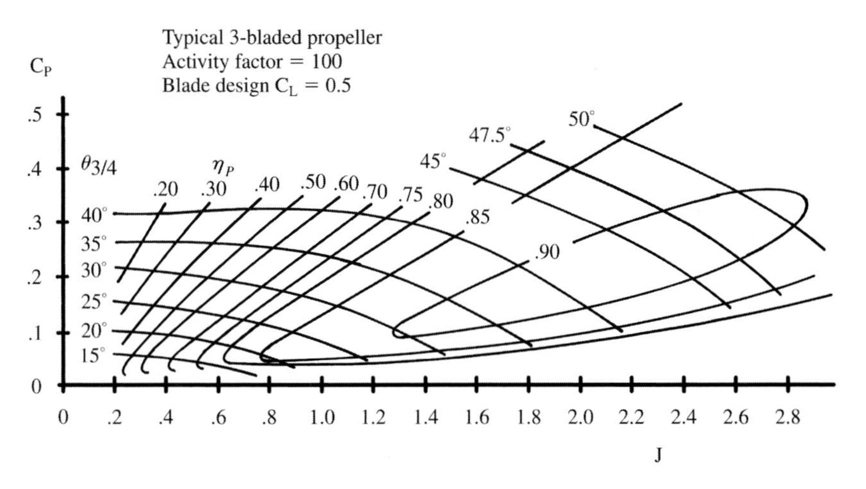

Fig. 13.12 Forward-flight thrust and efficiency (after Ref. 50).

Remember that the propeller thrust in forward flight is proportional to the inverse of the velocity, which would imply infinite thrust at zero velocity. Instead, the propeller produces the static-thrust value from Fig. 13.11 at zero velocity.

In the speed range from 0 to about 50 knots (such as during takeoff), the thrust varies in a fashion that can be represented by a smooth curve faired between the static-thrust value and the calculated forward-flight thrust.

If a fixed-pitch propeller is used, the blade angle cannot be varied in flight to maintain engine rpm at any flight condition. Since the rpm and therefore horsepower will vary with velocity, the efficiency and hence the thrust will be reduced at any speed other than the design speed.

Figure 13.12 could be used to determine the thrust from a fixed-pitch propeller by following the appropriate line for the selected blade angle. However, it is simpler to use the approximate method of Fig. 13.13 unless actual propeller data are available.

Figure 13.13 relates the fixed-pitch propeller efficiency at an off-design velocity and rpm to the on-design efficiency, which is attained by the propeller at some selected flight condition. The on-design efficiency is obtained from Fig. 13.12, which is also used to get the required blade angle for the design condition.

The static thrust of a fixed-pitch propeller will be less than is estimated using Fig. 13.11. A fixed-pitch propeller suffers at low speeds due to the high local angles of attack of the blades at low speeds and high rpms. As a rough approximation, it can be assumed that the static thrust is about 60% higher than the thrust at 100 knots.

These charts provide useful rough estimations of propeller performance, but actual charts for the selected propeller should be obtained from the manufacturer for any serious design effort.

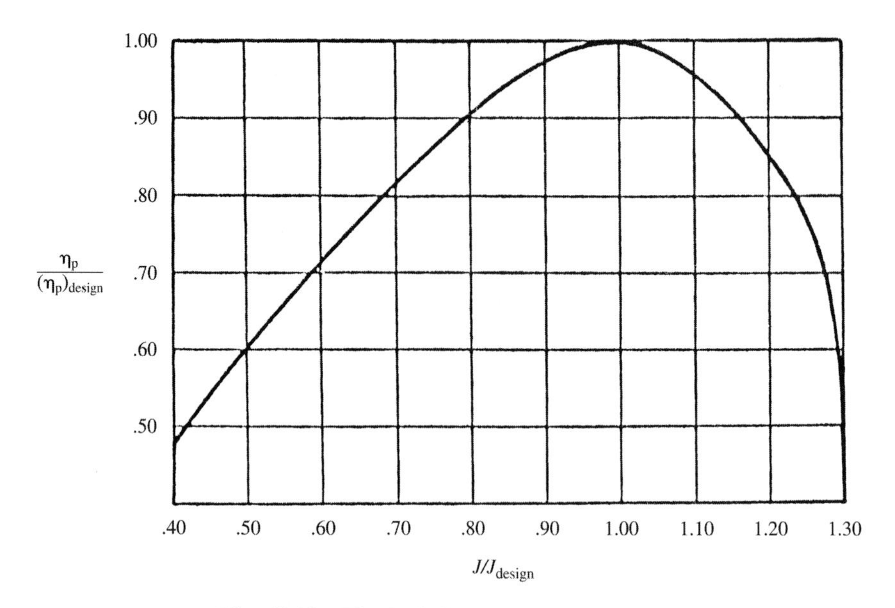

Fig. 13.13 Fixed-pitch propeller adjustment.

Piston-Prop Thrust Corrections

Propeller efficiency must be corrected for three important influences, namely, blockage, tip Mach, and scrubbing drag. Blockage refers to the effect of the nacelle immediately behind the propeller. It "blocks" the flow, causing it to slow down before it reaches the propeller. One way to correct for this is to adjust the advance ratio J prior to using a propeller efficiency chart such as Fig. 13.12. Equation (13.18), based on data from Ref. 23, is a reasonable first approximation for blockage and should be applied to J before J is used to find propeller efficiency:

$$J \text{ corrected} = J(1 - 0.329\, S_c/D^2) \tag{13.18}$$

where

S_c = maximum cross-section area of cowling immediately behind the propeller
D = propeller diameter

In Chapter 10 it was suggested that the propeller diameter be set so that the helical tip speed does not get too close to the speed of sound. However, for all but the slowest aircraft, tip Mach effects will reduce thrust at high speeds and rpm. Equation (13.19), based on data from Ref. 4, corrects the propeller efficiency after it is determined by Fig. 13.12 (or similar figure).

$$\eta_{p_{\text{corrected}}} = \eta_p - (M_{\text{tip}} - 0.89)\left(\frac{0.16}{0.48 - 3t/c}\right) \quad \text{for } M_{\text{tip}} > 0.89 \tag{13.19}$$

where

M_{tip} = tip Mach number = $\sqrt{V^2 + (\pi D)^2} \Big/ a$
a = speed of sound
t/c = propeller airfoil thickness-to-chord ratio

Scrubbing drag is the increase in aircraft drag resulting from the higher velocity and turbulence experienced by the parts of the aircraft within the propwash. This drag could be calculated by determining, for each flight condition, the increased dynamic pressure within the propwash and using that value for the component-drag calculation.

A simpler approach, called the SBAC (Society of British Aircraft Constructors) method, adjusts the propeller efficiency as in Eq. (13.20). The subscript "washed" refers to the parts of the aircraft which lie within the propwash. If the parasite-drag coefficient for the propwashed parts of the aircraft cannot be determined, 0.004 is a reasonable estimate.

$$\eta_{p_{\text{effective}}} = \eta_p\left[1 - \frac{1.558}{D^2}\frac{\rho}{\rho_0}\sum(C_{f_e}S_{\text{wet}})_{\text{washed}}\right] \tag{13.20}$$

where C_{f_e} is the equivalent skin-friction (parasite) drag coefficient, referenced to wetted area.

For a pusher-propeller configuration, the scrubbing drag is zero. However, the pusher propeller suffers a loss of efficiency due to the wake of the fuselage and wing. This loss is strongly affected by the actual aircraft configuration, and should equal about 2–5%.

Cooling drag represents the momentum loss of the air passed over the engine for cooling. This is highly dependent upon the detail design of the intake, baffles, and exit.

Miscellaneous engine drag includes the drag of the oil cooler, air intake, exhaust pipes, and other parts. Cooling and miscellaneous drags for a well-designed engine installation can be estimated by Eqs. (13.21) and (13.22) (Ref. 23). However, a typical light aircraft engine installation may experience cooling and miscellaneous drag levels 2–3 times the values estimated by these equations. Rather than use these equations, it is reasonable to assume that an expertly designed cooling system will produce a cooling drag equivalent to a 6% reduction in thrust, and a not-so-good system will produce an 8–10% reduction in thrust.

$$(D/q)_{\text{cooling}} = (4.9 \times 10^{-7})\frac{\text{bhp} \cdot T^2}{\sigma V} \quad (\text{ft}^2) \tag{13.21}$$

$$= 6 \times 10^{-8}\frac{PT^2}{\sigma V} \quad (\text{m}^2)$$

$$(D/q)_{\text{misc}} = (2 \times 10^{-4})\,\text{bhp} \quad (\text{ft}^2) \tag{13.22}$$

$$= 2.5 \times 10^{-5} P \quad (\text{m}^2)$$

where

T = air temperature (°R or K)
V = velocity (ft/s or m/s)
$\sigma = \rho/\rho_0$

13.7 Turboprop Performance

A turboprop is a jet engine that drives a propeller using a turbine in the exhaust. The jet exhaust retains some thrust capability, and can contribute as

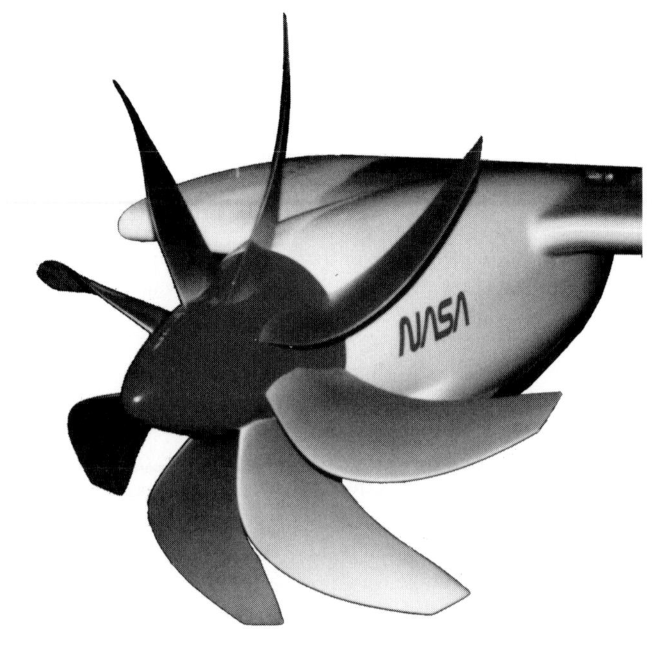

Fig. 13.14 Propfan.

much as 20% of the total thrust. For this reason the power rating of a turboprop engine includes the power equivalent of this residual thrust.

This power equivalent of residual thrust is arbitrarily calculated under static conditions as the residual thrust divided by 2.5. Under forward-flight conditions it is calculated using Eq. (13.17) assuming that the propeller efficiency $\eta_P = 0.80$. The total of the mechanical and thrust residual power, in horsepower, is called the equivalent shaft horsepower (ESHP).

Analysis of the turboprop is a hybrid between the jet and the piston-prop analysis. The engine is analyzed like a jet, including the inlet effects. The residual thrust is provided by the manufacturer as a horsepower equivalent. The propeller is analyzed as just described, including the scrubbing-drag term.

The conventional turboprop, like the piston-prop, is limited by tip Mach number to about Mach 0.7. The turboprop has higher efficiency than the piston-prop at Mach numbers greater than about 0.5 due to the residual jet thrust, but the conventional turboprop is no match for a turbofan engine at the higher subsonic speeds.

Recently, a new type of advanced propeller has been developed that offers good efficiencies up to about Mach 0.85. These are known as "propfans" or "unducted fans (UDF)" (Fig. 13.14). They are smaller in diameter than the regular propellers and feature numerous wide, thin, and swept blades. Test programs to date indicate that a well-designed propfan can retain propeller efficiencies of over 0.8 at speeds on the order of Mach 0.85.

Flying Boat (D. Raymer).

14
Structures and Loads

14.1 Introduction

In a large aircraft company, the conceptual designer may never do any structural analysis. The conceptual designer relies upon an experienced eye to ensure that sufficient space is provided for the required structural members. The only direct impact of structures during the initial stages of conceptual design is in the weights estimation. As will be shown in the next chapter, the statistical weights methods usually used in conceptual design do not require any actual structural analysis.

Designers at small aircraft companies and designers of homebuilt aircraft are more likely to perform an initial structural analysis as a part of the conceptual design process. This is especially true for a novel design concept such as the Rutan Voyager. To attain a design range of 26,000 miles {48,000 km}, the Voyager needed an empty-weight fraction of about 0.20 (!) and a wing aspect ratio over 30. Clearly, the knowledge that this was structurally possible was required before the design concept could be frozen.

Before the actual structural members can be sized and analyzed, the loads they will sustain must be determined. Aircraft loads estimation, a separate discipline of aerospace engineering, combines aerodynamics, structures, and weights.

In the past, the Loads Group was one of the larger in an aircraft company. Loads were estimated for each structural member of the aircraft using a combination of handbook techniques and wind-tunnel-data reduction.

Today's computer programs have mechanized much of the time-consuming work in loads estimation. Modern aerodynamic panel programs determine the airloads as an intermediate step toward determining aerodynamic coefficients. Also, modern wind tunnels employ computerized data reduction. These have reduced the workload so much that in some companies today there is no longer a separate loads group.

However, loads estimation remains a critical area because an error or faulty assumption will make the aircraft too heavy or will result in structural failure when the real loads are encountered in flight.

This chapter introduces the concepts of loads estimation and summarizes the subjects of aircraft materials and structural analysis. This material is presented from the viewpoint of the conceptual designer, and is not intended to serve as a general introduction to structures.

Furthermore, many of the methods presented are no longer in regular usage, having been supplanted by finite element methods, as discussed at the end of this chapter. The older methods are useful, however, for approximating the correct answer to ensure that the finite element method results are in the right "ballpark." Also, study of the classical methods is useful for learning the vocabulary of structural design.

14.2 Loads Categories

When one thinks of aircraft loads, the airloads due to high-g maneuvering come immediately to mind. While important, maneuvering loads are only a part of the total loads that must be withstood by the aircraft structure.

Table 14.1 lists the major load categories experienced by aircraft. Civil and military specifications [FAR Vol. III (23 and 25) and Mil-A-8860/8870] define specific loading conditions for these categories, as discussed later.

For each structural member of the aircraft, one of the loads listed in Table 14.1 will dominate. Figures 14.1 and 14.2 show typical critical loads for a fighter and a transport. Note that the lifting surfaces are almost always critical under the high-g maneuver conditions.

The largest load the aircraft is actually expected to encounter is called the limit, or applied, load. For the fighter of Fig. 14.1, the limit load on the wing occurs during an 8-g maneuver.

To provide a margin of safety, the aircraft structure is always designed to withstand a higher load than the limit load. The highest load the structure is designed to withstand without breaking is the "design," or "ultimate," load.

The "factor of safety" is the multiplier used on limit load to determine the design load. Since the 1930s the factor of safety has usually been 1.5. This

Table 14.1 Aircraft loads

Airloads	Landing	Other
–Maneuver	–Vertical load factor	–Towing
–Gust	–Spin-up	–Jacking
–Control deflection	–Spring-back	–Pressurization
–Component interaction	–Crabbed	–Bird strike
–Buffet	–One wheel	–Actuation
–Hailstones (3/4 in.)	–Arrested	–Crash
	–Braking	–Fuel pressure
Inertia loads	Takeoff	Powerplant
–Acceleration	–Catapult	–Thrust
–Rotation	–Aborted	–Torque
–Dynamic		–Gyroscopic
–Vibration	Taxi	–Vibration
–Flutter	–Bumps	–Duct pressure
	–Turning	–Hammershock
		–Prop/blade loss
		–Seizure

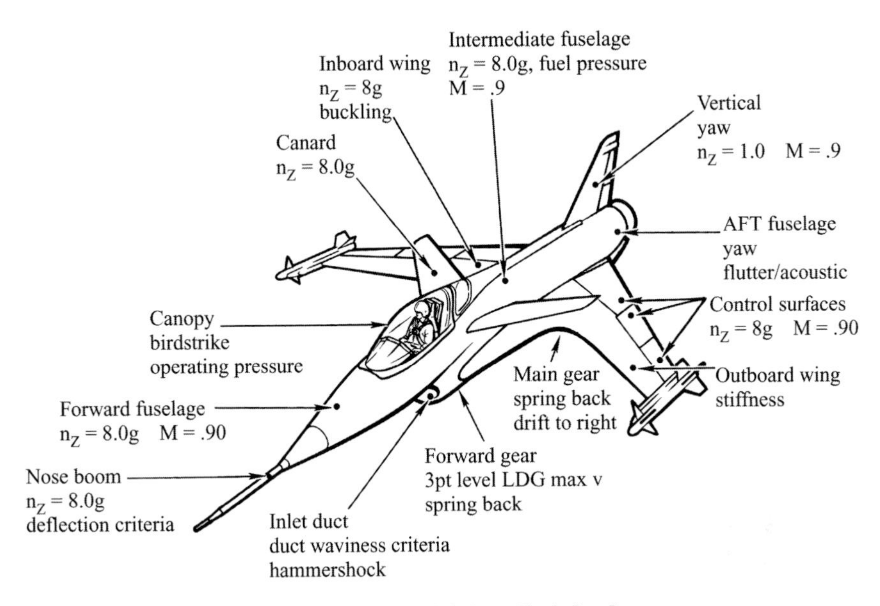

Fig. 14.1 Typical fighter limit loads.

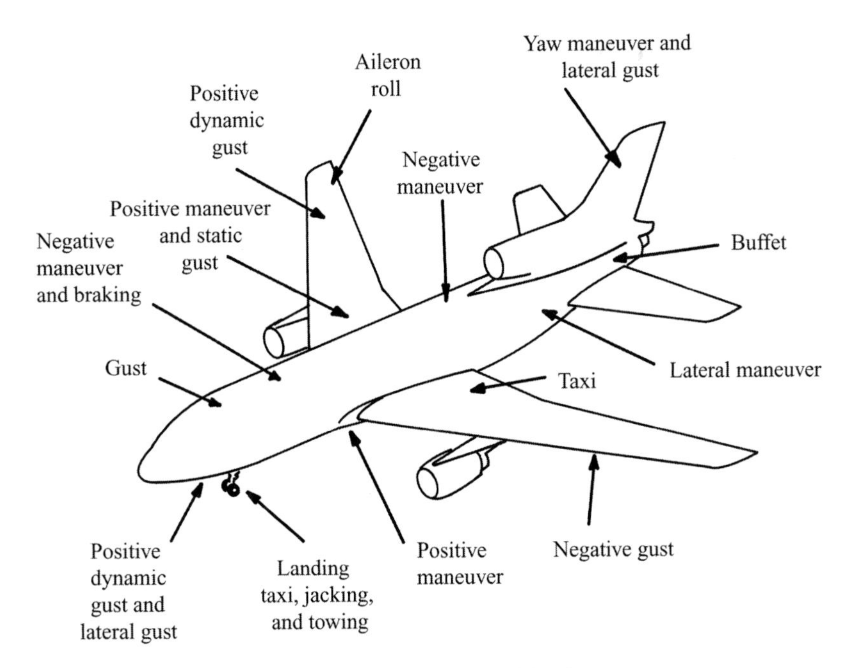

Fig. 14.2 L1011 critical loads.

was defined in an Air Corps specification based upon the ratio between the ultimate tensile load and yield load of 24ST aluminum alloy, and has proven to be suitable for other aircraft materials in most cases. For the fighter in Fig. 14.1, the design load for the wing structure would then be based upon a 12-g maneuver, above which the wing would break.

14.3 Air Loads

Maneuver Loads

The greatest air loads on an aircraft usually come from the generation of lift during high-g maneuvers. Even the fuselage is almost always structurally sized by the lift of the wing rather than by the air pressures produced directly on the fuselage.

Aircraft load factor (n) expresses the maneuvering of an aircraft as a multiple of the standard acceleration due to gravity ($g = 32.2\ \text{ft/s}^2 = 9.8\ \text{m/s}^2$). At lower speeds the highest load factor an aircraft may experience is limited by the maximum lift available.

At higher speeds the maximum load factor is limited to some arbitrary value based upon the expected use of the aircraft. The Wright Brothers designed their Flyer to a 5-g limit load. This remains a reasonable limit load factor for many types of aircraft. Table 14.2 lists typical limit load factors. Note that the required negative load factors are usually much less in magnitude than the positive values.

The V-n diagram depicts the aircraft limit load factor as a function of airspeed. The V-n diagram of Fig. 14.3 is typical for a general-aviation aircraft. Note that the maximum lift load factor equals 1.0 at level-flight stall speed, as would be expected. The aircraft can be stalled at a higher speed by trying to exceed the available load factor, such as in a steep turn.

The point labeled "high AOA" (angle of attack) is the slowest speed at which the maximum load factor can be reached without stalling. This part of the flight envelope is important because the load on the wing is approximately perpendicular to the flight direction, not the body-axis vertical direction.

At high angle of attack the load direction may actually be forward of the aircraft body-axis vertical direction, causing a forward load component on the wing structure (Fig. 14.4). During World War I, several aircraft had a problem

Table 14.2 Typical limit load factors

	n_{positive}	n_{negative}
General aviation—normal	2.5 to 3.8	−1 to −1.5
General aviation—utility	4.4	−1.8
General aviation—aerobatic	6	−3
Homebuilt	5	−2
Transport	3 to 4	−1 to −2
Strategic bomber	3	−1
Tactical bomber	4	−2
Fighter	6.5 to 9	−3 to −6

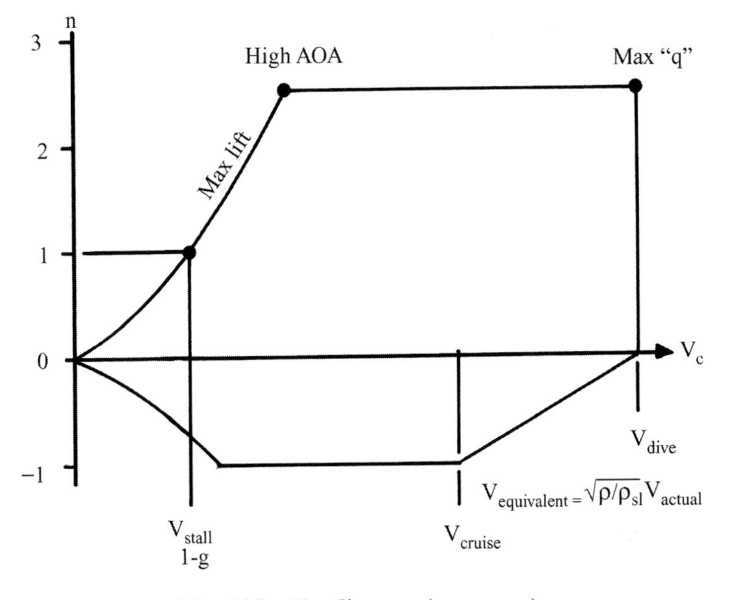

Fig. 14.3 *V-n* diagram (maneuver).

with the wings shedding forward due to this unexpected load. Velocity conversion methods are presented in Appendix C.

The aircraft maximum speed, or dive speed, at the right of the *V-n* diagram represents the maximum dynamic pressure q. The point representing maximum

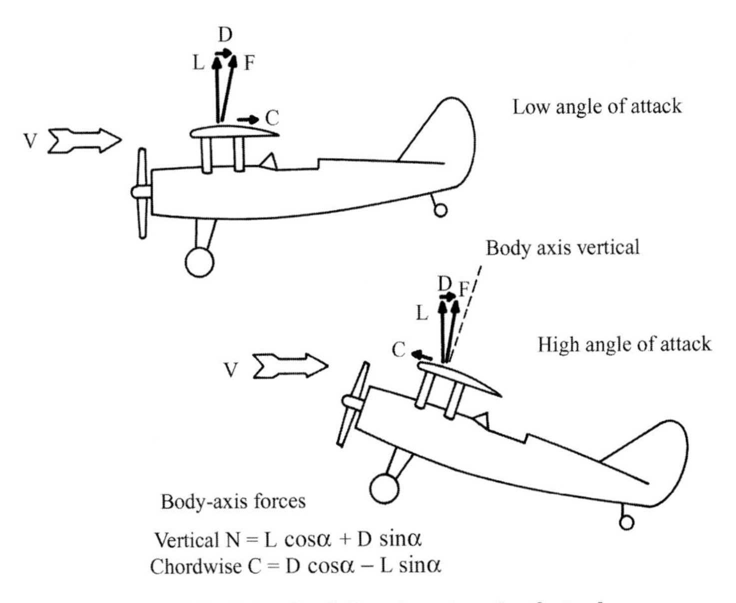

Fig. 14.4 Wing load direction at angle of attack.

q and maximum load factor is clearly important for structural sizing. At this condition the aircraft is at a fairly low angle of attack because of the high dynamic pressure, so the load is approximately vertical in the body axis.

For a subsonic aircraft, maximum or dive speed is typically 40–50% higher than the level-flight cruise speed. For a supersonic aircraft the maximum speed is typically about Mach 0.2 faster than maximum level-flight speed, although many fighters have enough thrust to accelerate past their maximum structural speed.

Note that aircraft speeds for loads calculation are in "equivalent" airspeeds V_e. An aircraft airspeed indicator uses a pitot probe to determine airspeed from the dynamic pressure, so the "airspeed" as measured by a pitot probe is based upon the dynamic pressure at the aircraft's velocity and altitude, and not the actual velocity. This dynamic pressure-based equivalent airspeed will be less than the actual airspeed at altitude due to the reduction in air density, as this expression describes:

$$V_e = \sqrt{\rho/\rho_{\text{SL}}}(V_{\text{actual}}) = \sqrt{\sigma}\,(V_{\text{actual}}) \tag{14.1}$$

For loads estimation, V_e is a convenient measure of velocity because it is constant with respect to dynamic pressure regardless of altitude. However, pilots must convert V_e to actual velocity to determine how fast they are really flying. Also, the dynamic pressure as measured by a pilot tube has a compressibility error at higher Mach numbers, so the "indicated" airspeed V_i as displayed to the pilot must be corrected for compressibility to produce the equivalent airspeed V_e, which can then be converted to actual airspeed.

Gust Loads

The loads experienced when the aircraft encounters a strong gust can exceed the manuever loads in some cases. For a transport aircraft flying near thunderstorms or encountering high-altitude "clear air turbulence," it is not unheard of to experience load factors due to gusts ranging from a negative 1.5 to a positive 3.5 g or more.

When an aircraft experiences a gust, the effect is an increase (or decrease) in angle of attack. Figure 14.5 illustrates the geometry for an upward gust of velocity U. The change in angle of attack, as shown in Eq. (14.2), is approximately U divided by V, the aircraft velocity. The change in aircraft lift is shown in Eq. (14.3) to be proportional to the gust velocity. The resulting change in load

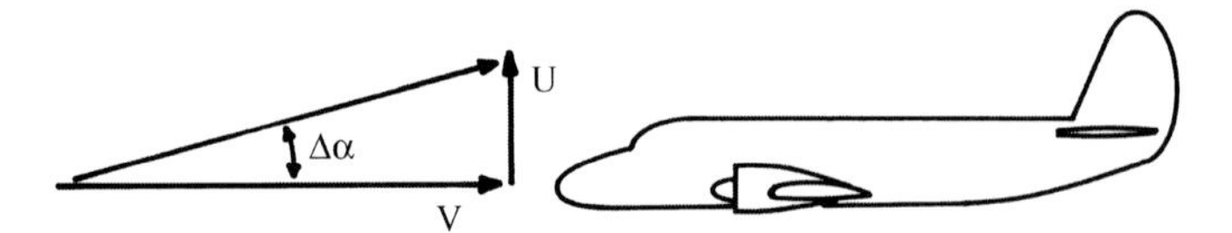

Fig. 14.5 Gust encounter.

factor is derived in Eq. (14.4).

$$\Delta\alpha = \tan^{-1}\frac{U}{V} \cong \frac{U}{V} \tag{14.2}$$

$$\Delta L = \tfrac{1}{2}\rho V^2 S(C_{L\alpha}\Delta\alpha) = \tfrac{1}{2}\rho VSC_{L\alpha}U \tag{14.3}$$

$$\Delta n = \frac{\Delta L}{W} = \frac{\rho UVC_{L\alpha}}{2W/S} \tag{14.4}$$

Figure 14.5 and Eq. (14.4) assume that the aircraft instantly encounters the gust and that it instantly affects the entire aircraft. These assumptions are unrealistic.

Gusts tend to follow a cosine-like intensity increase as the aircraft flies through, allowing it more time to react to the gust. This reduces the acceleration experienced by the aircraft by as much as 40%. To account for this effect, a statistical gust alleviation factor (K) has been devised and applied to measured gust data (U_{de}, discussed later). The gust velocity in Eq. (14.4) can be defined in the following terms (Ref. 51):

$$U = KU_{de} \tag{14.5}$$

where

$$\text{Subsonic:} \quad K = \frac{0.88\mu}{5.3+\mu} \tag{14.6}$$

$$\text{Supersonic:} \quad K = \frac{\mu^{1.03}}{6.95+\mu^{1.03}} \tag{14.7}$$

$$\text{Mass Ratio:} \quad \mu = \frac{2(W/S)}{\rho g\bar{c}C_{L\alpha}} \tag{14.8}$$

The mass ratio term accounts for the fact that a small, light plane encounters the gust more rapidly than a larger plane.

The design requirements for gust velocities are "derived" from flight-test data and are in "equivalent" airspeed (hence U_{de}). Actual accelerations experienced in flight have been applied to Eqs. (14.4–14.8) to determine what the vertical gust velocities must have been to produce those accelerations in the various flight-research aircraft employed.

For many years the standard vertical gust U_{de} has been 30 ft/s {9.1 m/s} (positive and negative). For most aircraft this produces roughly a 3-g positive load factor. This is still a suitable gust U_{de} for normal, utility, and aerobatic civil aircraft at speeds up to cruise speed. For higher speeds it may be assumed that U_{de} drops linearly to 15 ft/s {4.6 m/s} at maximum dive speed.

For transport and other classes of aircraft, a more detailed requirement of U_{de} is shown in Fig. 14.6 (data from Ref. 52). Note that the expected gusts are reduced at higher altitude. The maximum turbulence speed V_g may be specified in the design requirements or may be a fallout parameter.

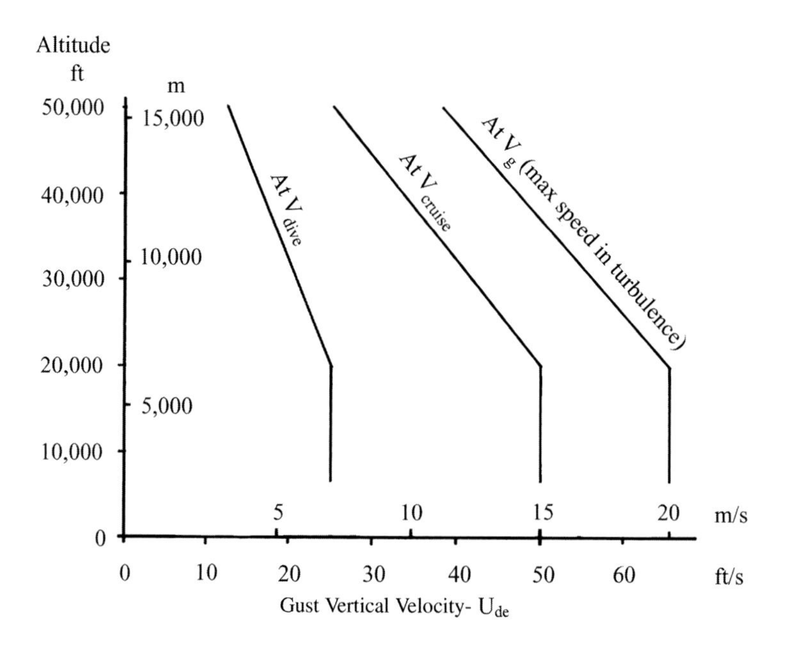

Fig. 14.6 Derived equivalent gust velocities (transport).

One interesting point concerning gusts is that, as shown in Eq. (14.4), the load factor due to a gust increases if the aircraft is lighter. This is counter to the natural assumption that an aircraft is more likely to have a structural failure if it is heavily loaded.

In fact, the change in lift due to a gust [Eq. (14.3)] is unaffected by aircraft weight, so the change in wing stress is the same in either case. However, if the aircraft is lighter the same lift increase will cause a greater vertical acceleration (load factor) so the rest of the aircraft will experience more stress.

Aeroelastic effects can also influence the load factor due to gusts. An aft-swept wing will bend up under load, which twists the wing and reduces the outboard angle of attack. This reduces total lift and also moves the spanwise lift distribution inboard, reducing the wing bending stress. An aft-swept wing will experience roughly 15% lower load factor due to a given gust than an unswept wing.

The gust load factors as calculated with Eqs. (14.4–14.8) and using the appropriate U_{de} (positive and negative) can then be plotted on a V-n diagram as shown in Fig. 14.7. It is assumed that the aircraft is in 1-g level flight when the gust is experienced. Few pilots will "pull g's" in severe turbulence conditions. The load factor between V_{dive}, V_{cruise}, and V_g is assumed to follow straight lines, as shown.

In Fig. 14.8, the V-n diagrams of Figs. 14.3 and 14.7 are combined to determine the most critical limit load-factor at each speed. Since the gust loads are greater than the assumed limit load, it may be desirable to raise the assumed limit load at all velocities, as shown by the dotted line. Remember that the structural design load factors will be 50% higher to provide a margin of safety.

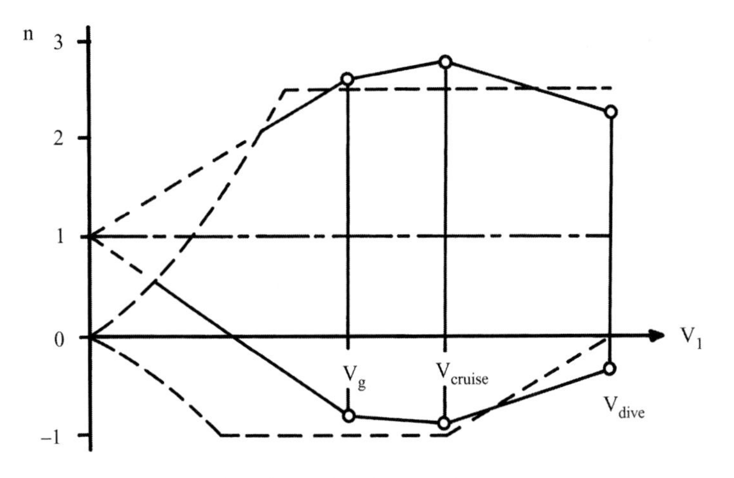

Fig. 14.7 *V-n* diagram (gust).

This method for estimation of gust loads is not as complete or accurate as the methods used at most large aircraft companies. The more accurate methods rely upon a power-spectral-density approach in which the gusts are included in an atmospheric transfer function and the actual aircraft dynamics are modeled. However, the methods just presented are useful for initial analysis and provide an introduction to the more detailed techniques (see Ref. 91).

Air Loads on Lifting Surfaces

Now that the *V-n* diagram is complete, the actual loads and load distributions on the lifting surfaces can be determined. In most cases this needs to be done only

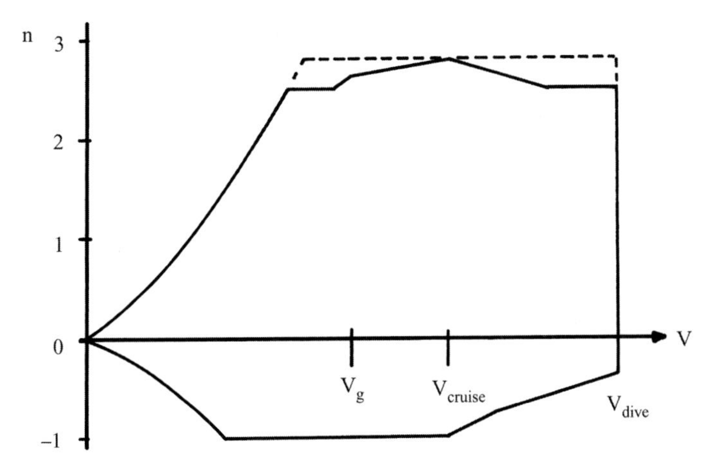

Fig. 14.8 Combined *V-n* diagram.

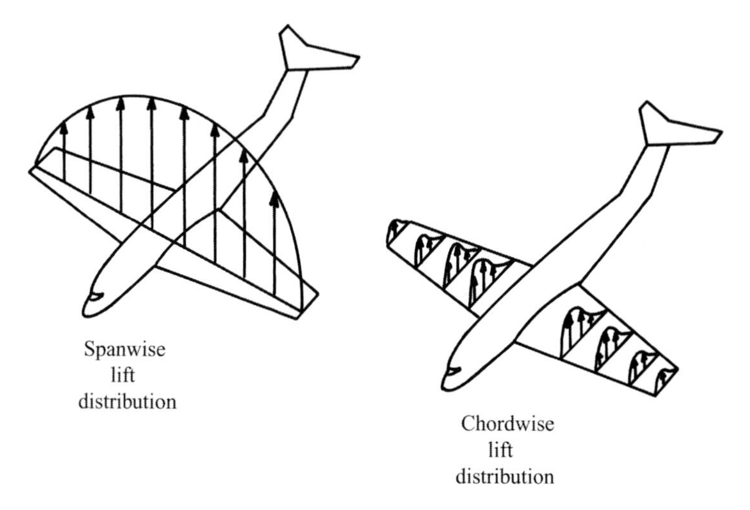

Fig. 14.9 Wing lift distribution.

at the high AOA and max q velocities (see Fig. 14.3) and any velocities where the gust load factor exceeds the assumed limit load factor.

The first step involves a stability-and-control calculation to determine the required lift on the horizontal tail to balance the wing pitching moment at the critical conditions. Note that the required tail lift will increase or decrease the required wing lift to attain the same load factor.

Complicated methods for estimating the lift on the trimmed tail and wing for a given load factor are presented in Chapter 16. These can be initially approximated by a simple summation of wing and tail moments about the aircraft center of gravity, ignoring the effects of downwash, thrust axis, etc.

Once the total lift on the wing and tail are known, the spanwise and chordwise load distributions can be determined (Fig. 14.9). Wind-tunnel and aerodynamic panel program data are used if available. For initial design and design of light aircraft, classical approximation methods give reasonably good results.

According to classical wing theory, the spanwise lift (or load) distribution is proportional to the circulation at each span station. A vortex lifting-line calculation will yield the spanwise lift distribution. For an elliptical planform wing, the lift and load distribution is of elliptical shape.

For a nonelliptical wing, a good semi-empirical method for spanwise load estimation is known as Schrenk's approximation (Ref. 53). This method assumes that the load distribution on an untwisted wing or tail has a shape that is the average of the actual planform shape and an elliptic shape of the same span and area (Fig. 14.10). The total area under the lift load curve must sum to the required total lift. Equations (14.9) and (14.11) describe the chord distributions of a trapezoidal and elliptical wing.

$$\text{Trapezoidal Chord:} \quad C(y) = C_r\left[1 - \frac{2y}{b}(1 - \lambda)\right] \tag{14.9}$$

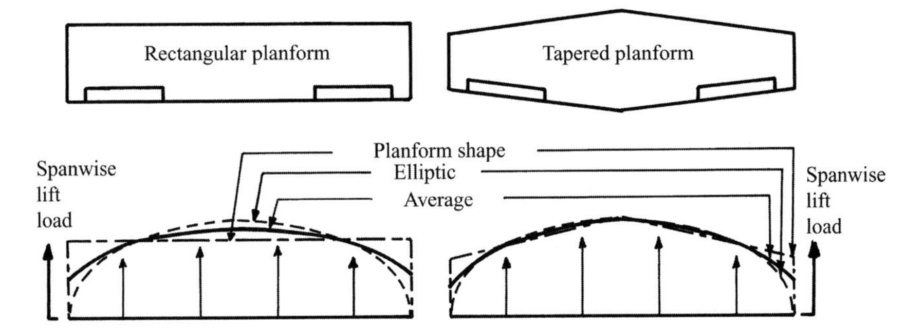

Fig. 14.10 Schrenk's approximation.

where

$$S = \frac{b}{2} C_r (1 + \lambda) \tag{14.10}$$

$$\text{Elliptical Chord:} \quad C(y) = \frac{4S}{\pi b} \sqrt{1 - \left(\frac{2y}{b}\right)^2} \tag{14.11}$$

Note in Fig. 14.10 that the load is assumed to continue to the centerline of the aircraft. This has proven to be a good assumption in subsonic flight. Also remember that if substantial dihedral is used, the perpendicular load on the wing is greater than the lift. Divide the lift by the cosine of the dihedral angle to get the perpendicular load.

If a wing has substantial geometric or aerodynamic twist, the effect upon spanwise lift-load distribution can be approximated by determining the load distribution when the wing is generating no net lift (the "basic load") and adding it to the "additional" load, which is determined as above for the net lift being produced (Ref. 54).

When a twisted wing has no net lift, part of the wing is generating an upload (usually the inboard section), and the rest of the wing is generating a download (usually the tips). The basic load can be approximated by ignoring the induced effects and basing the load at each spanwise station on the chord and section lift. The section lift is the section lift coefficient times the section's twist angle with respect to the wing angle of attack when no lift is being generated. This no-lift angle is approximately the angle of the mean aerodynamic chord, and must be found by trial and error.

Schrenk's approximation does not apply to highly swept planforms experiencing vortex flow. Vortex flow tends to greatly increase the loads at the wing tips. Loads for such a planform must be estimated using computers and wind tunnels.

The spanwise distribution of drag loads must also be considered, especially for fabric-covered aircraft in which drag loads are carried by internal "drag wires." Drag loads tend to be greatest near the wing tips and should be determined from wind-tunnel or aerodynamic panel program data.

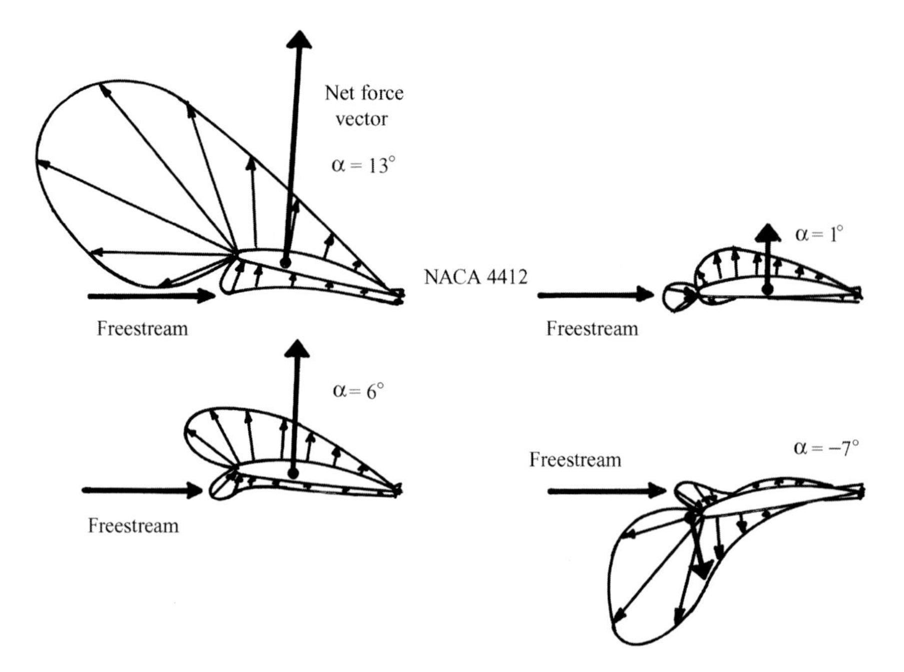

Fig. 14.11 Airfoil chordwise pressures.

As a first approximation, the spanwise distribution of drag loads can be roughly approximated as a constant 95% of the average drag loading from the root to 80% of the span, and 120% of the average loading from 80% of span to the wingtip.

The aerodynamic interaction of various aircraft components can produce additional loads. For example, the downwash from a canard will reduce the effective angle of attack of the inboard part of the wing. This moves the lift distribution of the wing outboard, producing greater wing bending stresses than expected.

A vortex from a leading-edge strake can cause vibrational stresses on any component of the aircraft it touches. The F-18 had a problem with vertical-tail fatigue for this reason. A similar problem can occur due to propeller propwash. These effects are difficult to predict, but must be considered during conceptual design.

Once the spanwise load distribution is known, the wing or tail bending stress can be determined as described in a later section. To determine torsional stresses, the airfoil moment coefficient is applied to spanwise strips and the total torsional moment is summed from tip to root. When wind-tunnel data are available, the torsional moments are summed from the chordwise pressure data.

Actual chordwise pressure distributions for a NACA 4412 airfoil at various angles of attack are shown in Fig. 14.11.

Airloads due to Control Deflection

Operation of the control surfaces produces airloads in several ways. The greatest impact is in the effect of the elevator on angle of attack and hence load factor.

The rudder's effect on yaw angle can also impose large loads. Deflection of control surfaces produces additional loads directly upon the wing or tail structure.

Maneuver speed, or pullup speed V_p, is the maximum speed at which the pilot can fully deflect the controls without damaging either the airframe or the controls themselves. For most aircraft the manuever speed is less than the maximum level cruise speed V_L.

Maneuver speed V_p is established in the design requirements or can be selected using an empirical relationship, Eq. (14.12). In this old but useful equation, aircraft weight W is in pounds, and so if using mks units, first multiply kg by 2.2. Velocities are in feet or meters per second. Stall speed V_s is with high lift devices deployed. The factor K_p is estimated in Eq. (14.13), but should not be allowed to fall below 0.5 or above 1.0. For general-aviation aircraft, K_p usually does not exceed 0.9.

$$V_p = V_s + K_p(V_L - V_s) \tag{14.12}$$

$$K_p = 0.15 + \frac{5400}{W + 3300} \tag{14.13}$$

At the selected maneuver speed, a control analysis using the methods of Chapter 16 determines the angle of attack or sideslip obtained by maximum control deflection. The airloads imposed upon the structure can then be determined.

Note that the instantaneous loads imposed by maximum aileron deflection while at maximum load factor (rolling pull-up) are frequently critical to the wing structure.

The maximum speed allowed with flaps down is also needed for estimation of the maximum loads on the flaps. Flap speed V_f will usually be twice the flaps-down stall speed.

Control deflection will typically provide a change in section lift coefficient of about 0.8–1.1 at 25-deg deflection. Estimation methods are provided in Chapter 16.

In the absence of better data, the change in airfoil moment coefficient can be estimated as (-0.01) times the control deflection in degrees. The additional load tends to be concentrated at the hingeline of the moving surface. Note that the deflection of a control surface increases the load on the fixed part of the airfoil as well as on the moving control surface.

For an aircraft with a manual flight-control system, the control loads may be limited by the strength of the pilot. For a stick-controlled aircraft, the pilot strength is limited to 167 lb {0.7 kN} for the elevator and to 67 lb {0.3 kN} for the ailerons. For a wheel-controlled aircraft, the pilot strength is limited to 200 lb {0.9 kN} for the elevator and to 53 (times the wheel diameter) in.-lb {0.1 times diameter N-m} for the ailerons. The rudder force is limited to 200 lb {0.9 kN}.

In addition to the maneuvering and control-surface loads just determined, the tail group of an aircraft is designed to withstand some arbitrarily determined loads at maneuver speed. These loads are based upon normal force coefficients (C_n) assuming that the spanwise load distribution is proportional to chord length. For the horizontal tail, the required C_n values are (-0.55) downward and (0.35) upward. For the vertical tail the required C_n value is (0.45).

14.4 Inertial Loads

Inertial loads reflect the resistance of mass to acceleration ($F = ma$). The various accelerations due to maneuver and gust, already described, establish the stresses for the aerodynamic surfaces.

Every object in the aircraft experiences a force equal to the object's weight times the aircraft load factor. This creates additional stresses throughout the aircraft, which must be determined. Note that the weight of the wing structure will produce torsional loads on the wing in addition to the aerodynamic torsional loads.

Inertial loads due to rotation must also be considered. For example, the tip tanks of a fighter rolling at a high rate will experience an outward centrifugal force. This force produces an outward load factor equal to the distance from the aircraft c.g. times the square of the rotation rate, divided by g.

A tangential acceleration force is produced throughout the aircraft by a rotational acceleration such as is caused by a gust, a sudden elevator deflection, or by nose-wheel impact. This force is equal to the distance from the aircraft c.g. times the angular acceleration, divided by g.

The loads produced by vibration and flutter are actually acceleration forces of a special nature. Calculation of these loads goes beyond this book. Proper design should avoid flutter and reduce vibrations to a negligible level.

14.5 Powerplant Loads

The engine mounts must obviously be able to withstand the thrust of the engine as well as its drag when stopped or windmilling. The mounts must also vertically support the weight of the engine times the design load factor. The engine mounts are usually designed to support a lateral load equal to one-third of the vertical design load. The mounts must withstand the gyroscopic loads caused by the rotating machinery (and propeller) at the maximum pitch and yaw rates.

For a propeller-powered aircraft, the engine mounts must withstand the torque of the engine times a safety factor based upon the number of cylinders. This reflects the greater jerkiness of an engine with few cylinders when one cylinder malfunctions. The torque moment load can be calculated from power and rotation rate.

For an engine with two cylinders, the safety factor is 4.0; with three cylinders, 3.0; and with four cylinders, 2.0. An engine with five or more cylinders requires a safety factor of 1.33. These safety factors are multiplied times the maximum torque in normal operation to obtain the design torque for the engine mounts.

For a jet engine, air loads within the inlet duct must be considered, as they will frequently bound a part of the flight envelope. At M3 at 65,000 ft {20,000 m}, the B-70 experienced inlet duct pressures of 4320 psf {207 kN/m^2}, which is 30 times the outside air pressure.

A pressure surge known as "hammershock" is especially severe. This is usually caused by pressure waves propagating forward from a compressor stall. "Duct buzz," caused by shock waves bouncing in and out of the duct in rapid oscillation, can overstress the structure and cause loss of thrust.

14.6 Landing-Gear Loads

The landing gear's main purpose is to reduce the landing loads to a level that can be withstood by the aircraft. Chapter 11 presented calculations for landing-gear stroke to yield an acceptable gear load-factor, as transmitted to the structure of the aircraft.

To analyze fully all the possible gear loads, a number of landing scenarios must be examined. These include a level landing, a tail-down landing, a one-wheel landing, and a crabbed landing. For certification the aircraft may be subjected to drop tests, in which an actual aircraft is dropped from a height of somewhere between 9.2–18.7 in. {23–48 cm}. The required drop distance typically will be 3.6 times the square root of the wing loading.

When the tires contact the ground, they are not rotating. During the brief fraction of a second it takes for them to spin up, they exert a large rearward force by friction with the runway. This spin-up force can be as much as half the vertical force due to landing.

When the tire is rotating at the correct speed, the rearward force is relieved and the gear strut "springs back" forward, overshooting the original position and producing a spring-back deflection load equal to or greater than the spin-up load.

Another landing-gear load, the braking load, can be estimated by assuming a braking coefficient of 0.8.

The load on the landing gear during retraction is usually based upon the airloads plus the assumption that the aircraft is in a 2-g turn. Other landing-gear loads such as taxiing and turning are usually of lesser importance, but must be considered during detail design of the landing gear and supporting structure.

14.7 Structures Fundamentals

Timoshenko's classic 1930 book *Strength of Materials* (Ref. 55) begins with this overall description of the action of structural members:

"We assume that a body consists of small particles, or molecules, between which forces are acting. These molecular forces resist the change in the form of the body which external forces tend to produce. If such external forces are applied to the body, its particles are displaced and the mutual displacements continue until equilibrium is established between the external and internal forces. It is said in such a case that the body is in a state of strain."

Thus, a structural member responds to a load by deforming in some fashion until the structure is pushing back with a force equal to the external load. The internal forces produced in response to the external load are called "stress," and the deformation of the structure is called "strain."

Figure 14.12 shows the three basic types of structural loading: tension, compression, and shear. The meanings of tension and compression should be clear from the illustration. Shear may be viewed as a combination of forces tending to cause the object to deform into two parts that slide with respect to each other. Scissors cut paper by application of shear. Figure 14.12 also shows the load on a rivet, a typical example of shear.

Figure 14.13 shows three other types of structural loading. These can be considered as variations and combinations of tension, compression, and shear. Bending due to a load at the end of a beam is a combination of tension and compression. The top part of the beam in Fig. 14.13 is in compression, while the bottom part is in tension.

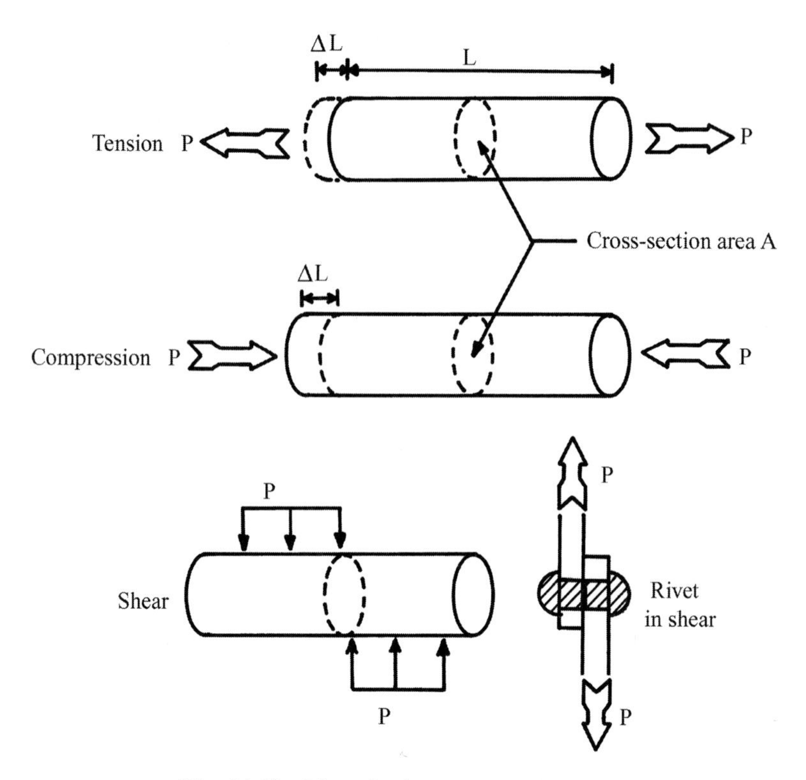

Fig. 14.12 Three basic structural loadings.

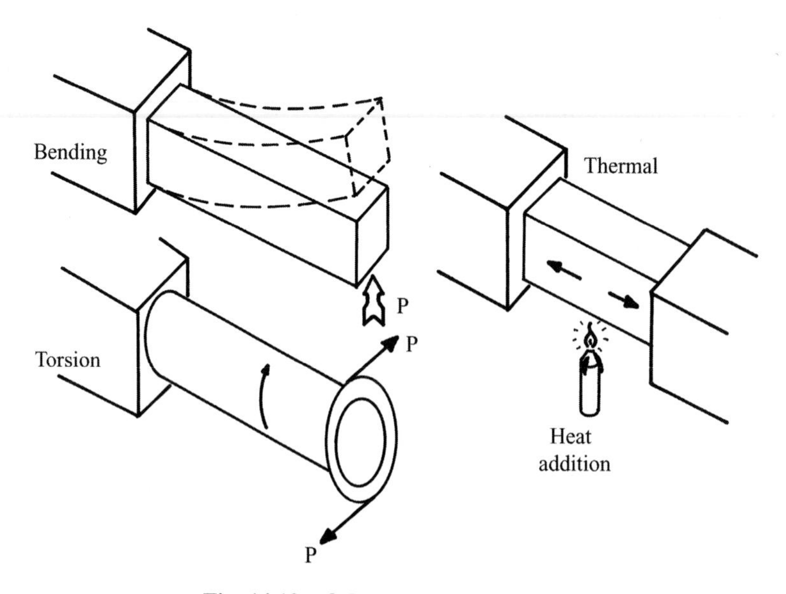

Fig. 14.13 Other structural loadings.

Torsion is due to a combination of forces producing a moment (torque) which tends to twist the object. Torsion produces tangential shear forces that resist the torque.

Thermal stresses are due to the expansion of materials with an increase in temperature. If a structural member is not free at one end, it will push against its supports as it is heated. This produces compression loads. Similarly, a severe reduction in material temperature will produce tension loads unless at least one end is free.

The unit stress (σ or F) is the stress force (P) per unit area [i.e., total stress divided by area—see Eq. (14.14)]. The unit strain (ϵ or e) is the deformation per unit length [i.e., total strain divided by length—see Eq. (14.15)].

$$\sigma = P/A \tag{14.14}$$

$$\epsilon = \Delta L/L \tag{14.15}$$

The relationship between stress (load) and strain (deformation) is critically important to the design of structure. Figure 14.14 illustrates a typical stress–strain diagram for an aluminum alloy. Over most of the stress range the strain is directly proportional to the stress (Hooke's law), with a constant of proportionality defined as Young's modulus, or the *modulus of elasticity* (E) [Eq. (14.16)].

$$E = \sigma/\epsilon \tag{14.16}$$

The highest stress level at which the strain is proportional to the stress is called the "proportional limit," and stresses less than this value are considered within the "elastic range." Within the elastic range a structure will return to its original shape when the load is removed.

At higher stress levels a permanent deformation (set) remains when the load is removed, as shown by the dotted line on Fig. 14.14. The "yield stress" is the stress level at which a substantial permanent set occurs.

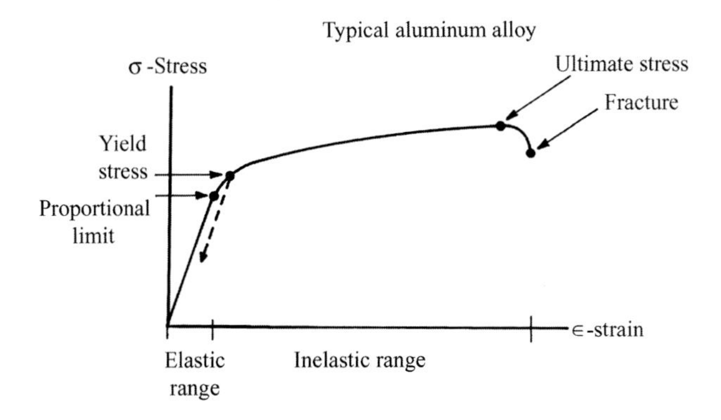

Fig. 14.14 Stress–strain diagram.

Yield stress is arbitrarily defined as a permanent set of 0.002 in. per inch {or meter per meter!} and is typically only slightly higher than the proportional limit. Above the yield stress is called the inelastic range.

Within the inelastic range, Hooke's law is no longer true and the modulus of elasticity can no longer be applied to Eq. (14.16) to determine the strain for a given stress. However, for some stress calculations it is useful to define an artificial modulus called the *tangent modulus* (E_t), which is the slope of the stress–strain curve at a given point in the inelastic range. This modulus cannot be applied to Eq. (14.16). The tangent modulus varies with stress and strain, and is plotted in material-property tables such as Ref. 61.

The ultimate stress is the highest stress level the material can withstand. Ultimate stress goes well past the elastic range. A material subjected to its ultimate stress will suffer a large and permanent set.

For aluminum alloys, ultimate stress is about 1.5 times the yield stress. If an aircraft is designed such that the application of a limit load factor causes some aluminum structural member to attain its yield stress, then the ultimate stress will not be reached until a load factor of 1.5 times the limit load factor is applied (i.e., at the design or ultimate load factor). However, when the aircraft exceeds its limit load factor some structural elements will be permanently deformed and must be repaired after the aircraft lands.

The specific strength of a material is defined as the ultimate stress divided by the material density. The specific stiffness is defined as the modulus of elasticity E divided by the material density. These parameters are useful for comparing the suitability of various materials for a given application.

Not all materials behave like the aluminum alloy of Fig. 14.14. Composites such as fiberglass and graphite–epoxy will fracture without warning at a stress just past the proportional limit (Ref. 56), as shown in Fig. 14.15. These materials do not have a "built in" 1.5 safety factor, so a safety factor must be assumed for design purposes.

Typically a safety factor for composites is assumed by designing to a stress level that provides a strain equal to two-thirds (i.e., $1/1.5$) of the strain at the ultimate stress level. If this stress level is higher than the proportional limit, then the proportional limit stress is used for designing to limit loads.

When a material elongates due to a tension load, the cross-sectional area decreases as shown in Fig. 14.16 (much exaggerated). Experimentation has shown that the ratio of lateral to axial strain is constant within the elastic range. This ratio (Poisson's ratio, μ or ν) is approximately 0.3 for steel and 0.33 for nonferrous materials such as aluminum.

The deformation due to shear, which was not shown in Fig. 14.12, is illustrated in Fig. 14.17. At the top is a bar subjected to a shear loading typical for a rivet, with a download and an upload separated by some very small distance. These loads are assumed to be provided by loads applied to two plates (not shown) that the bar or rivet connects.

The deformation of the bar is shown to the right. Shear introduces a kink within the material. The deformation is not a change in length, as with tension or compression, but instead is an angular deformation (shearing strain, or γ).

The upper-right illustration in Fig. 14.17 cannot be a complete free-body diagram because of the unbalanced moment of the two forces. Additional

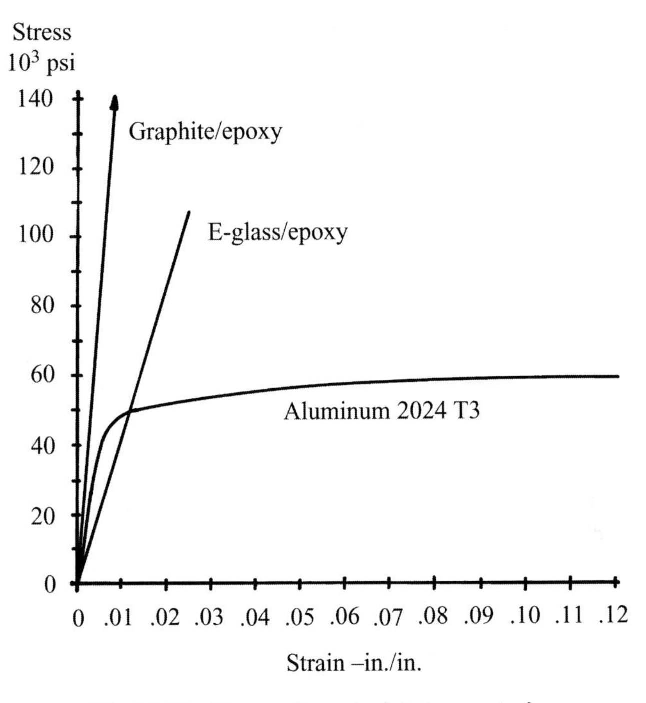

Fig. 14.15 Composite material stress–strain.

forces must exist to balance this moment. The lower right figure illustrates the total forces on a square element within the "kinked" portion of the bar. Again, the angle γ defines the shearing strain within the bar. The unit shear stress (τ) is defined in Eq. (14.17).

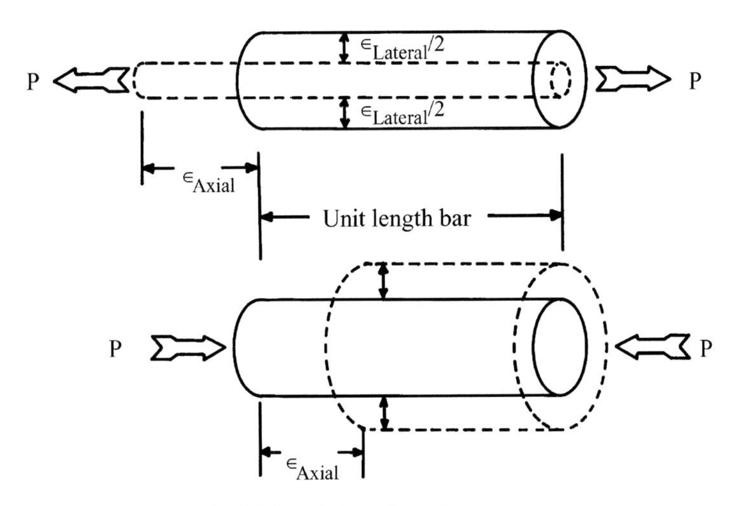

Fig. 14.16 Poisson's ratio.

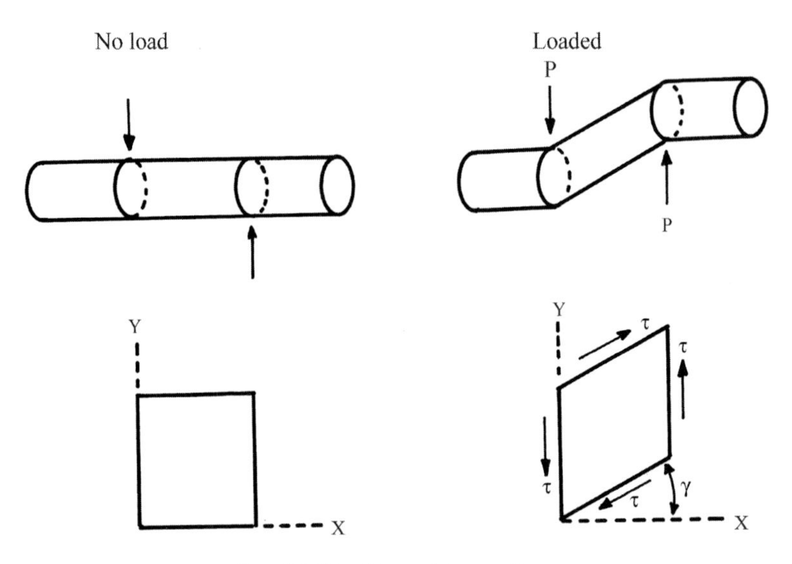

Fig. 14.17 Shear deformation.

These additional balancing forces, horizontal in the example in Fig. 14.17, are themselves shear forces that must be resisted by the material. For a riveted wing spar, the rivets that attach the shear web to the spar caps must be designed to resist these shear forces. Similarly, in a wood or composite wing box the glue that attaches the upper and lower covers must resist these shear forces.

Note in Fig. 14.17 that the transverse deformation (i.e., Y direction) due to the shear stress is equal to the longitudinal distance (X direction) from the point of no shear, times the shearing strain angle (γ) in radians, since γ is small.

As with tension or compression, there is a linear relationship between shear stress and shear strain provided that the shear force is below the proportional limit. The shear modulus, or *modulus of rigidity* (G), is defined in Eq. (14.18). Also, it can be shown that the shear modulus is related to the modulus of elasticity by Poisson's ratio (Ref. 55), as shown in Eq. (14.19).

$$\tau = P_{\text{shear}}/A \tag{14.17}$$

$$G = \tau/\gamma \tag{14.18}$$

$$G = \frac{E}{2(1+\mu)} \tag{14.19}$$

14.8 Material Selection

A number of properties are important to the selection of materials for an aircraft. The selection of the "best" material depends upon the application. Factors to be considered include yield and ultimate strength, stiffness, density, fracture toughness, fatigue crack resistance, creep, corrosion resistance, temperature limits, producibility, repairability, cost, and availability.

Strength, stiffness, and density have been discussed already. Fracture toughness measures the total energy per unit volume required to deflect the material to the fracture point, and is equivalent to the area under the stress–strain curve. A ductile material with a large amount of inelastic deformation prior to fracture will absorb more work energy in fracturing than a material with the same ultimate stress but with little inelastic deformation prior to fracture.

A material subjected to a repeated cyclic loading will eventually experience failure at a much lower stress than the ultimate stress. This "fatigue" effect is largely due to the formation and propagation of cracks, and is probably the single most common cause of aircraft material failure. There are many causes of fatigue, including gust loads, landing impact, and the vibrations of the engine and propeller.

Creep is the tendency of some materials to slowly and permanently deform under a low but sustained stress. For most aerospace materials, creep is a problem only at elevated temperatures. However, some titaniums, plastics, and composites will exhibit creep at room temperatures. Creep deformation data are presented in materials handbooks as a function of time, temperature, and stress loading.

Corrosion of aircraft materials has been a major problem since the early days of aviation. Aircraft materials are exposed to atmospheric moisture, salt-water spray, aircraft fuel, oils, hydraulic fluids, battery acid, engine exhaust products, missile plumes, gun gases, and even leaking toilets.

Furthermore, electrically dissimilar materials such as aluminum and graphite–epoxy composite will experience galvanic corrosion in which an

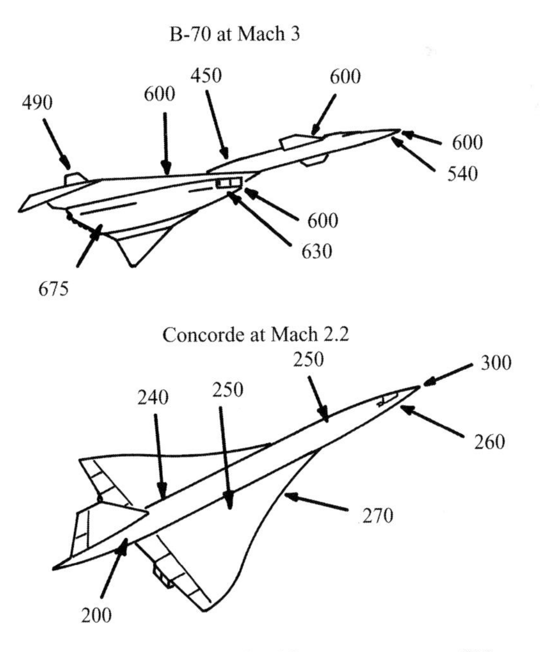

Fig. 14.18 Supersonic skin temperatures (°F).

electrical current is formed that deteriorates the more anodic material, converting it into ions or an oxide.

Corrosion of materials is greatly accelerated when the materials experience a sustained stress level. The corrosion products at the surface tend to form a protective coating that delays further corrosion. When the material is subjected to a tension stress, however, cracks in the protective coating are formed that accelerate the corrosion.

Once corrosion begins, it tends to follow cracks opened in the material by the stress. This "stress corrosion" can cause fracture at a stress level one-tenth the normal ultimate stress level. For this reason it is important to avoid manufacturing processes that leave residual tension stresses.

Operating temperature can play a major role in determining material suitability. Stainless steel or some other high-temperature material must be used as a firewall around the engine. For high-speed aircraft, aerodynamic heating may determine what materials may be used. Figure 14.18 shows typical skin temperatures at speeds of Mach 2.2 and 3.0.

The stagnation (total) temperature is the highest possible temperature due to aerodynamic heating [Eq. (14.20)]. Actual skin temperatures are difficult to

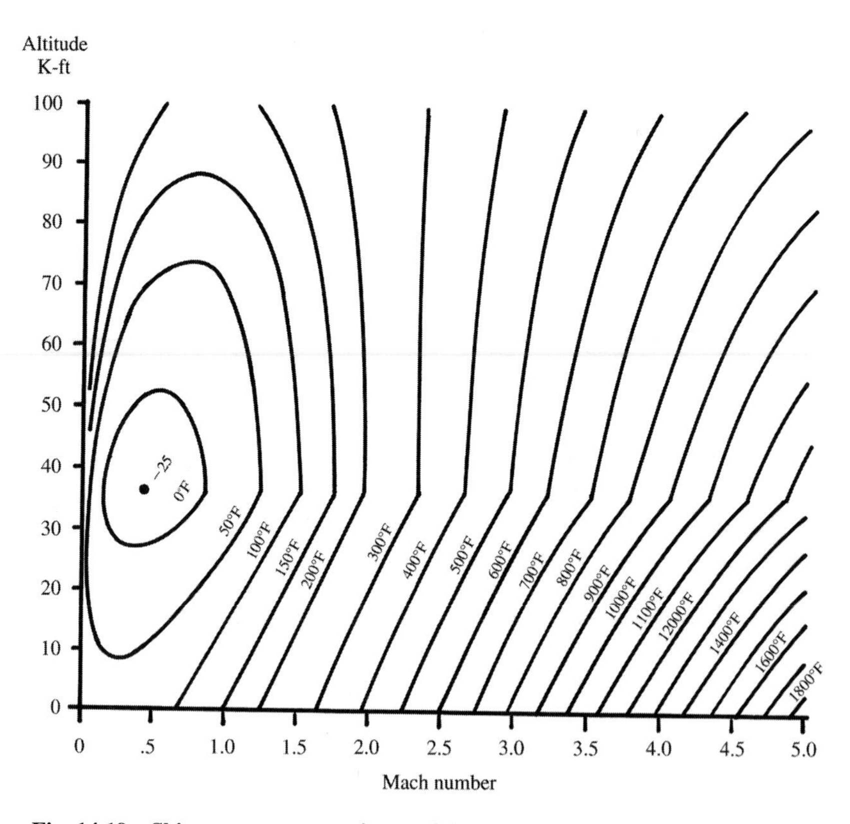

Fig. 14.19 Skin temperature estimate. Average values (°F), not leading edge.

calculate because they depend upon the airflow conditions, surface finish, and atmospheric conditions. Figure 14.19 provides a reasonable estimate of the expected skin temperatures over most of the airframe.

$$T_{\text{stagnation}} = T_{\text{ambient}}(1 + 0.2M^2) \quad (T \text{ in } ^\circ\text{R or K}) \tag{14.20}$$

Producibility and repairability are also important in material selection. As a rule, the better the material properties, the more difficult it is to work with.

For example, a major difficulty in the development of the SR-71 was in learning how to work with the selected titanium alloy. Similarly, composite materials offer a large reduction in weight, but pose problems both in fabrication and repair.

Cost is also important in material selection, both for raw material and fabrication. The better the material, the more it usually costs. Wood, mild steel, and standard aluminums are all relatively inexpensive. Titaniums and composites have high cost.

Another factor to consider is material availability. Titanium and some of the materials used to produce high-temperature alloys are obtained from unfriendly or unstable countries, and it is possible that the supply may someday be cut off. Also, aircraft-quality wood is in fairly short supply.

Figures 14.20–14.22 illustrate the materials selected for the Rockwell proposal for the X-29. These are typical of current fighter design practice. Note the stainless-steel heat shield and nozzle interface and the aluminum-honeycomb

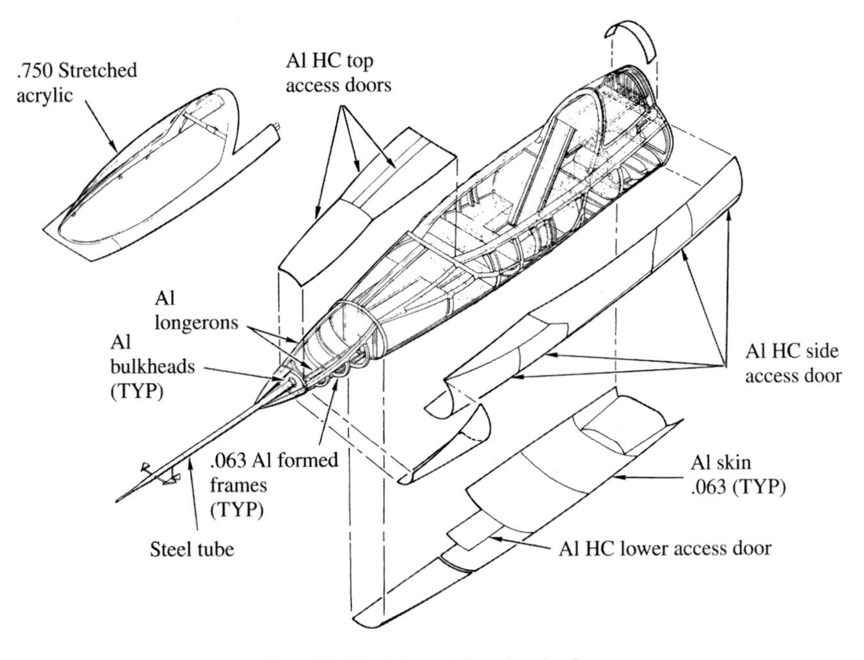

Fig. 14.20 Materials—forebody.

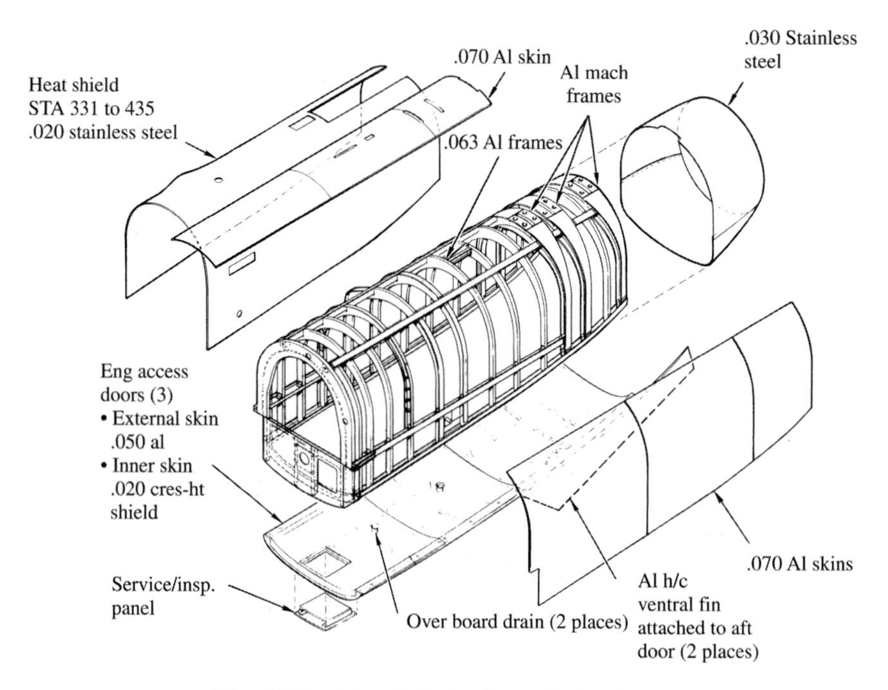

Fig. 14.21 Material selection—aft fuselage.

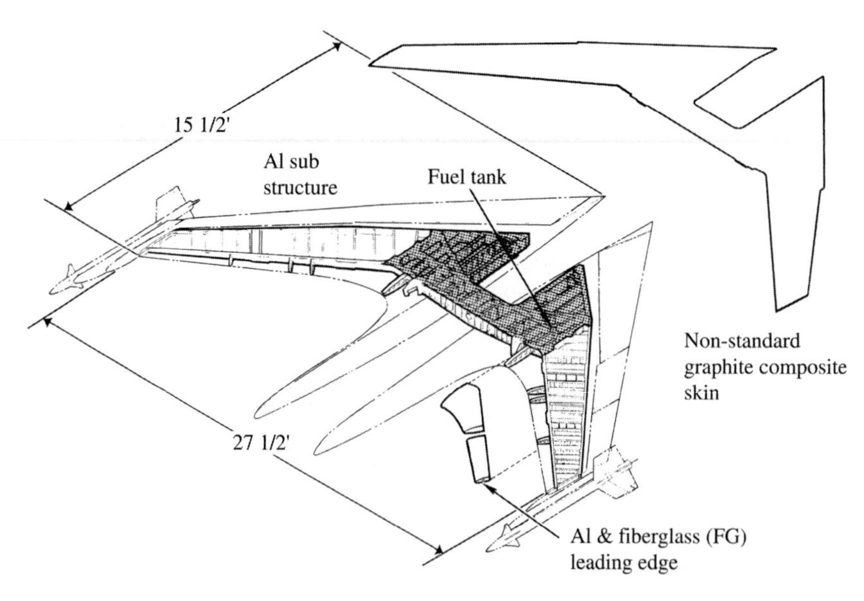

Fig. 14.22 Wing materials.

access doors. In a production fighter, the stretched acrylic windshield would be replaced by a bulletproof material.

14.9 Material Properties

This section covers various commonly used aircraft materials. Tables of representative material properties are at the back of this section.

Wood

The Wright Brothers selected spruce as the primary structural material for their aircraft, and it remained the material of choice for many years. Wood is rarely used today in production aircraft, but it is interesting to note that the Chinese have recently selected oak for the heat shield of a reentry vehicle!

Wood offers good strength-to-weight ratio and is easy to fabricate and repair. It is actually much like composite materials in that it has different properties in different directions. Wood makes a natural bending beam for wing spars because of the lengthwise fibers.

The wooden Hughes H-4 Hercules Flying Boat was built like a modern composite aircraft. Multiple thin plies of wood were placed in molds along with a resin glue and subjected to pressure during cure. Ply orientation was varied to give specific properties.

The disadvantages of wood are its sensitivity to moisture and its susceptibility to rot and insect damage. Wood must be regularly maintained and should not be left exposed to the elements. The Hughes H-4 looks virtually new today because it was kept in a climate-controlled hangar. Also, wood is produced by nature with poor quality control! Each piece of wood is unique so it requires craftsman-like skills to manufacture aircraft with wood.

Today wood is used largely in homebuilt and specialty, low-volume production aircraft. Wood has one additional advantage for homebuilders in that almost everyone knows how to saw, drill, and glue wood. However, the use of foam core and fiberglass–epoxy has largely replaced wood in home-built aircraft.

Aluminum

Aluminum remains by far the most widely used aircraft material. It has an excellent strength-to-weight ratio, is readily formed, is of moderate cost, and is resistant to chemical corrosion.

Aluminum is the most abundant metal in the Earth's crust, occurring mostly as silicates in clays. Discovered in 1827, it remained an expensive novelty until an electrical extraction method was developed in 1885. In 1856 aluminum cost $90 a pound. By 1935 the cost had dropped to 23 cents per pound. Inflation has raised this to several dollars per pound today depending upon its form.

Being relatively soft, pure aluminum is alloyed with other metals for aircraft use. The most common aluminum alloy is 2024 (or 24ST), sometimes called "duralumin." 2024 consists of 93.5% aluminum, 4.4% copper, 1.5% manganese, and 0.6% magnesium.

For high-strength applications, the 7075 alloy is widely used. 7075 is alloyed with zinc, magnesium, and copper. Since the corrosion resistance is lessened by alloying, aluminum sheet is frequently clad with a thin layer of pure aluminum. Newer alloys such as 7050 and 7010 have improved corrosion resistance and strength. An extensive discussion of aluminum alloys can be found in Ref. 105.

The strength and stiffness properties of aluminum are affected by the form (sheet, plate, bar, extrusion, or forging) and by heat treatment and tempering. In general, the stronger the aluminum, the more brittle it is.

While composite materials are considered the latest state of the art for lightweight aircraft structures, there are new aluminum alloys such as aluminum–lithium that offer nearly the same weight savings and can be formed by standard aluminum techniques. The Eurofighter Typhoon uses aluminum–lithium in the wing and tail leading edges. Aluminum will remain important in aircraft design for many years to come.

Steel

A major early advance in aircraft structures was the adoption of welded mild-steel tubing for the fuselage. Previously, aircraft such as the Sopwith Camel had fuselages of wire-braced wood construction that required constant maintenance. The steel-tube fuselage, used extensively by Fokker, greatly improved strength and required less maintenance.

Today steel is used for applications requiring high strength and fatigue resistance, such as wing attachment fittings. Also, steel is used wherever high temperatures are encountered such as for firewalls and engine mounts. The Mach 3 XB-70 (Fig. 14.18) was constructed largely of brazed steel honeycomb. This material proved strong at high temperatures but was extremely difficult to fabricate.

Steel is primarily an alloy of iron and carbon, with the carbon adding strength to the soft iron. As carbon content increases, strength and brittleness increase. Typical steel alloys have about 1% of carbon. Other materials such as chromium, molybdenum, nickel, and cobalt are alloyed with steel to provide various characteristics. The "stainless steel" alloys are commonly used where corrosion resistance is important.

The properties of steel are strongly influenced by heat treatment and tempering. The same alloy can have moderate strength and good ductility or can have much higher strength but at the expense of brittleness, depending upon the heat treatment and tempering employed.

Heat treatment begins by raising the temperature of the steel to about 1400–1600°F {760–870°C} at which point the carbon goes into solid solution with the iron. The rate at which the steel is then cooled defines the grain structure, which determines strength and ductility.

If the steel is slowly cooled by steadily reducing the temperature in the furnace (a process called annealing), a coarse grain structure is formed, and the steel is very ductile but weak. This is sometimes done before working with steel to make it easier to cut, drill, and bend.

If the heated steel is allowed to air-cool (to be "normalized"), it becomes much stronger but retains good ductility. Welded steel tubing structure is usually normalized after all welding is completed to return the steel around the welds to the original strength.

If quenched with water or oil, the steel becomes "martensitic" with a needle-like grain structure, great strength, and extreme brittleness.

To regain some ductility, the steel must be tempered by reheating it to about 1000°F {538°C} for an hour or more.

Standard heat-treatment and tempering processes are defined in material handbooks along with the resulting material properties.

Steel is very cheap, costing about one-sixth what aluminum does. Steel is also easy to fabricate.

Titanium

Titanium would seem to be the ideal aerospace material. It has a better strength-to-weight ratio and stiffness than aluminum, and is capable of temperatures almost as high as steel. Titanium is also corrosion-resistant.

However, titanium is difficult to form for these same reasons. Most titanium alloys must be formed at temperatures over 1000°F {538°C} and at very high forming stresses.

Also, titanium is seriously affected by any impurities that may be accidently introduced during forming. One of the worst impurity elements for "embrittling" titanium is hydrogen, followed by oxygen and nitrogen. After forming, titanium must be treated for embrittlement by chemical "pickling" or through heat treatment in a controlled environment.

Titanium is expensive, costing about five to ten times as much as aluminum. In the past it was more expensive to fabricate in titanium than aluminum, and "cost factors" of double or triple were applied to cost estimates. Today the technology has improved, and the cost of titanium fabrication is just slightly higher than aluminum fabrication.

To handle the aerodynamic heating of Mach 3 + flight, the structure of the SR-71 is about 93% titanium. The XB-70 uses a substantial amount of titanium in the forebody area. The midbody of the F-22 is largely titanium due to engine heating. Titanium is extensively used in jet-engine components, and is also used in lower-speed aircraft for such high-stress airframe components as landing gear beams and spindles for all-moving tails. Because it does not cause galvanic corrosion with graphite–epoxy, titanium is sometimes used as the substructure to graphite–epoxy skins.

Because of its material properties, titanium lends itself to a unique forming process called "SuperPlastic Forming/Diffusion Bonding." SPF/DB is a process where the titanium is placed in a press mold under extreme temperature and pressure such that it virtually "flows" to the shape of the mold. Furthermore, separate pieces of titanium are diffusion-bonded at the same time, forming a joint that is indistinguishable from the original metal. This process offers both cost reduction and the ability to form very complicated parts, all having the good material properties of titanium. The Euro-fighter Typhoon uses SPF/DB titanium for its canards rather than the originally intended composites, because of its better producibility.

Reference 57 gives a more detailed discussion of titanium and its alloys.

Magnesium

Magnesium has a good strength-to-weight ratio, tolerates high temperatures, and is easily formed, especially by casting, forging, and machining. It has been used for engine mounts, wheels, control hinges, brackets, stiffeners, fuel tanks, and even wings. However, magnesium is very prone to corrosion and must have a protective finish. Furthermore, it is flammable!

Mil Specs advise against the use of magnesium except to gain significant weight savings. Also, magnesium should not be used in areas that are difficult to inspect or where the protective finish would be eroded by rain (leading edges) or engine exhaust.

High-Temperature Nickel Alloys

Inconel, Rene 41, and Hastelloy are high-temperature nickel-based alloys suitable for hypersonic aircraft and reentry vehicles. Inconel was used extensively in the X-15, and Rene 41 was to have been used in the X-20 Dynasoar. Nickel alloy honeycomb sandwich is used for the stealth nozzles of the F-117. Hastelloy is used primarily in engine parts.

These alloys are substantially heavier than aluminum or titanium, and are difficult to form. For these reasons, the Space Shuttle uses an aluminum structure with heat-protective tiles. While a substantially lighter structure was obtained, the difficulties experienced with the tiles should be noted by the designers of the next-generation shuttle.

Composites

The greatest revolution in aircraft structures since the all-aluminum Northrop Alpha has been the ongoing adoption of composite materials for primary structure. In a typical aircraft part, the direct substitution of graphite–epoxy composite for aluminum yields a weight savings of 25%.

The F-22 and F/A-18 E/F are about 25% composites by structural weight. The AV-8B wing is almost entirely made of graphite–epoxy composite, and numerous military and commercial aircraft use composites for tails, flaps, and doors.

Composites consist of a reinforcing material suspended in a "matrix" material that stabilizes the reinforcing material and bonds it to adjacent reinforcing materials. Composite parts are usually molded, and may be cured at room conditions or at elevated temperature and pressure for greater strength and quality. Figure 14.23 shows the two major composite forms, filament-reinforced and whisker-reinforced.

In the whisker-reinforced composite, short strands of the reinforcing material are randomly located throughout the matrix. The most common example of this is chopped fiberglass, which is used for low-cost fabrication of boats and fast-food restaurant seats. Whisker reinforcing is sometimes used in advanced metal matrix composites such as boron-aluminum.

Most of the advanced composites used in aircraft structure are of the filament reinforced type because of outstanding strength-to-weight ratio. Also, filament composites may have their structural properties tailored to the expected loads in different directions.

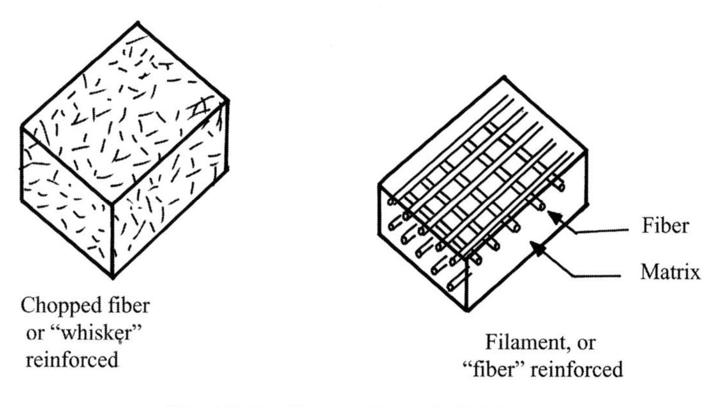

Fig. 14.23 Composite material types.

Metals and whisker-reinforced composites are isotropic, having the same material properties in all directions. Filament composites, like wood, are strongest in the direction the fibers are running. If a structural element such as a spar cap is to carry substantial load in only one direction, all the fibers can be oriented in that direction. This offers a tremendous weight savings.

Figure 14.24 shows four common arrangements for tailoring fiber orientation. In part (a), all fibers are aligned with the principle axis so the composite has maximum strength in that direction, and has little strength in other directions. Arrangement (b) offers strength in the vertical direction as well.

In (c), the fibers are at 45-deg angles with the principle axis. This provides strength in those two directions, and also provides good shear strength in the principle-axis direction. For this reason, this arrangement is commonly seen in a composite-wing-box shear web. Also, the 45-deg orientation is frequently used in structure that must resist torque.

Arrangement (d) combines (b) and (c), providing alternate layers (plies) of fibers at 0-, 45-, and 90-deg orientations. By varying the number of plies at these orientations the designer can obtain virtually any combination of tensile, compression, and shear strength in any desired directions.

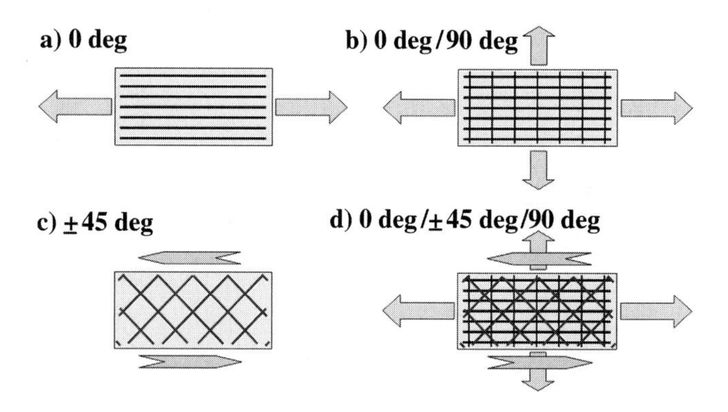

Fig. 14.24 Composite ply tailoring.

Another ply-orientation scheme uses plies that are 60 deg apart. Composites are sometimes designed with completely arbitrary ply directions to provide special characteristics.

Note that an odd number of plies is commonly used. This tends to reduce warpage, as has long been known by the makers of plywood.

The common forms of fiber used in composite production are shown in Fig. 14.25. The chopped form is simply sprayed or pressed into the mold. Unidirectional tape comes on large rolls and is placed in the mold by hand or by a robotic tape-laying machine. Tape is usually pre-impregnated ("prepreg") with the matrix material.

Fabrics may be bidirectional, with fibers running at 0 and 90 deg, or unidirectional, with the fibers running in one direction. (A few fibers run at 90 deg to bind the fabric together.) Fabrics may also be prepreg. Fabrics are sometimes called "broadgoods."

Prepreg tape and fabric is typically about 0.005–0.01 in. {0.01–0.03 cm} thick per ply.

In another form of composite, the individual filaments are wound around plugs to form shapes such as missile bodies and golf club shafts. This is called "filament-wound" construction.

There are a number of fiber and matrix materials used in composite aircraft structure. Fiberglass with an epoxy–resin matrix has been used for years for such nonstructural components as radomes and minor fairings. More recently, fiberglass–epoxy has been used by homebuilders.

While fiberglass–epoxy has good strength characteristics, its excessive flexibility (tensile E) prevents its use in highly loaded structure in commercial or

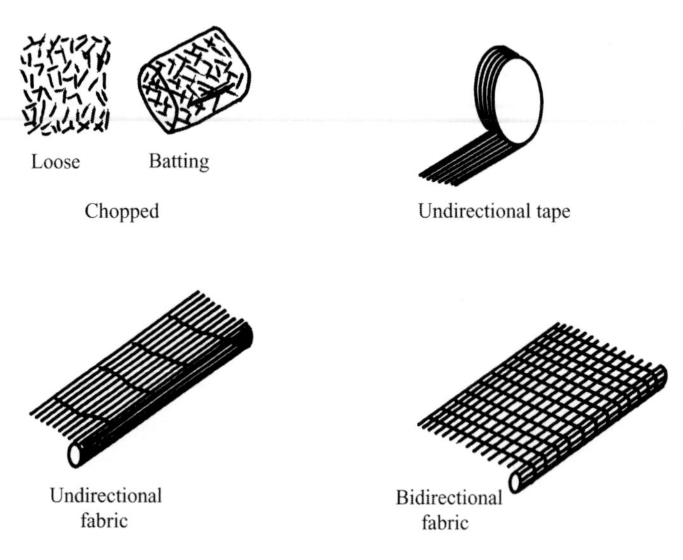

Fig. 14.25 Composite production forms.

military aircraft. However, it is cheap and easy to form, and is suitable for some applications.

The most commonly used advanced composite is graphite–epoxy, called "carbon-fiber composite" by the British who developed it. Graphite–epoxy composite has excellent strength-to-weight ratio and is not difficult to mold. It is substantially more expensive than aluminum at the present time (roughly 20 times), but unlike metals, little material is wasted in manufacturing operations such as milling and cutting from flat patterns.

Boron–epoxy was developed in the United States and initially used for complete part fabrication. An F-111 horizontal tail and F-4 rudder were built of boron–epoxy. However, boron–epoxy costs over four times as much as graphite–epoxy, so boron is used today largely to provide additional stiffness to graphite–epoxy parts, especially in compression.

Aramid, sold under the trade name Kevlar, is used with an epoxy matrix in lightly loaded applications. Aramid has a low compression strength, but exhibits much more gradual failure than other composites (i.e., less brittle). A graphite–aramid–epoxy hybrid composite offers more ductility than pure graphite–epoxy. It is used in the Boeing 757 for fairings and landing gear doors.

Composites using epoxy as the matrix are limited to maximum temperatures of about 350°F {177°C} and normally are not used in applications where temperatures will exceed 260°F {127°C}. For higher-temperature applications, several advanced matrix materials are in development. The polyimide resins show great promise. One polyimide, bismaleimide (BMI), shows good strength at 350°F {177°C}. A material called polymide shows good strength at up to 600°F {315°C} but is difficult to process.

The matrix materials just described are all "thermoset" resins, chemical mixtures that "cure," producing a change in the material's chemistry at the molecular level upon the application of heat. The thermoset process is not reversible. If the composite part is heated up again the thermosetting matrix does not revert to a liquid state.

In contrast, a "thermoplastic" matrix material does not undergo a chemical change when heated. It merely "runs," and can be heated up again and reformed. Much like the plastics used in model airplanes, thermoplastic materials can be readily formed with heat.

Thermoplastic materials under study for use as the matrix in aircraft structures include polyester, acrylic, polycarbonate, phenoxy, and polyethersulfone. Thermoplastic matrix materials can be used with the same fiber materials (graphite, boron, etc.) as the thermoset composites. Thermoplastics are especially good for higher temperature applications and where toughness is desired. The F-117s were retrofitted with graphite thermoplastic vertical tails, probably due to their proximity to the hot nozzles. Thermoset materials tend to be readily damaged, so thermoplastics are desirable for doors, access panels, and anywhere on the bottom where rocks may bounce up from the landing gear.

For higher-temperature, high-strength applications, "metal-matrix composites" are in development. These use metals such as aluminum or titanium as the matrix with boron, silicon carbide, or aramid as the fiber.

Composite materials offer impressive weight savings, but have problems too, one problem being a reluctance to accept concentrated loads. Joints and fittings must be used that smoothly spread the concentrated load out over the composite part. If a component such as a fuselage or wing has a large number of cutouts and doors, the fittings to spread out those concentrated loads may eliminate the weight savings. Wing attachment is another area where large and heavy metal fittings must be used to spread the load out into the composite skins. This is especially true where a composite wing is joined to a ring frame carrythrough structure. The Eurofighter (Typhoon) has about 70% of its structure built from graphite composite, and uses three large titanium root joints to attach each wing box to the fuselage carrythrough frames.

The strength of a composite is affected by moisture content, cure cycle, temperature exposure, ultraviolet exposure, and the exact ratio of fiber to matrix. These are difficult to control, and every composite part will probably have slightly different properties. Manufacturing voids are difficult to avoid or detect, and the scrapage rate for composite parts can be high (but is improving as composites are more widely used).

Composites in general are more likely to be damaged than aluminum. Unfortunately, mild damage to composites may occur internally after some impact, yet not show up by outside visual inspection. For this reason, composites must be designed to carry their full limit load after such nondetectable damage.

Furthermore, composites are difficult to repair because of the need to match strength and stiffness characteristics. A patch that is weak is obviously undesirable, but one that is overstrong can cause excessive deflection on adjoining areas, which can lead to fracture. Proper repair of an important composite part requires running a computer program to insure that the repaired part will match the original design specifications.

The properties of a composite material are not simply the algebraic sum of the properties of the individual ply layers. Actual material properties must be calculated using tensor calculus equations, such as are outlined in Ref. 58. Furthermore, extensive coupon testing is required to determine design allowables for the selected materials and ply orientation. Introductions to composites are provided in Refs. 59 and 83.

There is a designer's rule of thumb for composites called the Ten-Percent Rule (presented in Ref. 106), which gives a quick and reasonably good strength approximation for typical composites. This rule is valid for composites with plies oriented at 0 deg (i.e., the direction of the load), 90 deg, and $+/-45$ deg, and assumes that the 0 deg plies contribute their full strength while the other plies contribute only 10% of their full strength. In other words, simply add the number of plies times the strength per ply, but multiply all plies that are not running in the direction of the applied load by 0.10. Needless to say, this rough approximation is *only* for initial sizing purposes and should never be relied upon for a final design analysis!

Sandwich Construction

While not properly classed a "material," sandwich construction has special characteristics and is very important to aircraft design. A structural sandwich is composed of two face sheets bonded to and separated by a core (Fig. 14.26).

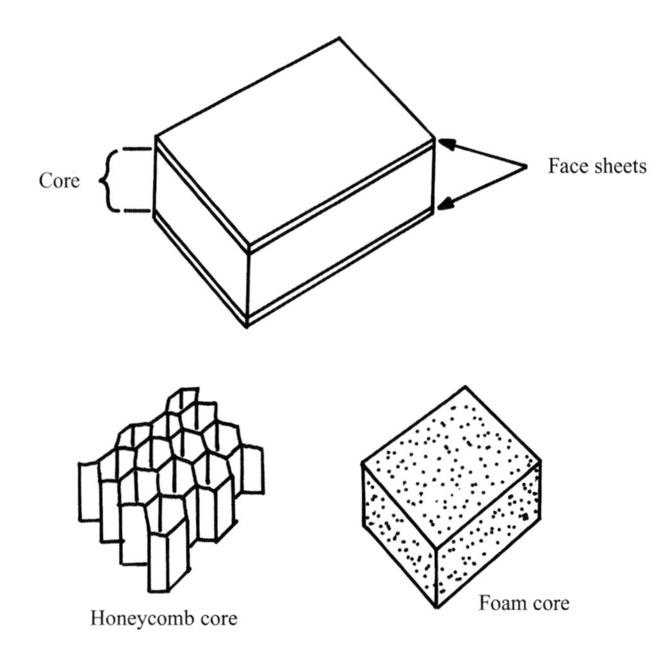

Fig. 14.26 Sandwich construction.

The face sheets can be of any material, but are typically aluminum, fiberglass–epoxy, or graphite–epoxy. The core is usually an aluminum or phenolic honeycomb material for commercial and military aircraft, but various types of rigid foam are used as the core in some cases. Many homebuilt aircraft today are constructed of foam-core sandwich with fiberglass composite skins. 70% of the B-70s airframe was stainless-steel honeycomb sandwich, typically 2 in. {5 cm} thick.

In a sandwich, the face sheets carry most of the tension and compression loads due to bending. The core carries most of the shear loads as well as the compression loads perpendicular to the skin. As with composites, joints and fittings are a problem with sandwich construction. Analysis of sandwich construction is discussed in Ref. 60.

Material-Property Tables

Tables 14.3–14.5 provide typical material properties for various metals, woods, and composites. Note that these are typical values only, and that actual material properties for use in detail design should be obtained from the producer or from a specification document such as Ref. 61.

For example, Ref. 61 contains 68 pages of design data on 2024 aluminum alone, covering many different forms, heat treatments, tempering, gauges, etc. The values for 2024 in Table 14.3 are merely typical, suitable for rough estimates and student design projects.

Table 14.3 Typical metal properties (room temperature)

Material	Density, lb/in.^{3}	Temp limit, °F	F_{tu} 10^3 psi	F_{ty} 10^3 psi	F_{cy} 10^3 psi	F_{su} 10^3 psi	E 10^6 psi	G 10^6 psi	Comments
Steel									
Aircraft steel (5 Cr-Mo-V)	0.281	1000	260	220	240	155	30	11	Heat treat to 1850°F
Low carbon steel (AISI 1025)	0.284	900	55	36	36	35	29	11	Shop use only today
Low alloy steel (D6AC-wrought)	0.283	1000	220	190	198	132	29	11	—
Chrom-moly steel (AISI 4130)									
sheet/plate/tubing	0.283	900	90	70	70	54	29	11	Widely used
wrought	0.283	900	180	163	173	108	29	11	—
Stainless steel (AM-350)	0.282	800	185	150	158	120	29	11	Good corrosion resistance
Stainless (PH 15-7 Mo-sheet/plate)	0.277	600	190	170	179	123	29	11	B-70 honeycomb material
Aluminum									
Aluminum-2017	0.101	250	55	32	32	33	10.4	3.95	—
Clad 2024 (24 st)-(sheet/plate)	0.100	250	61	45	37	37	10.7	4.0	Widely used, weldable
extrusions	0.100	250	70	52	49	34	10.8	4.1	
Clad 7178-T6 (78 st)-(sheet/plate)	0.102	250	80	71	71	48	10.3	3.9	High strength, not weldable,
extrusions	0.102	250	84	76	75	42	10.4	4.0	subject to stress corrosion
Clad 7075-T6-(sheet).	0.101	250	72	64	63	43	10.3	3.9	High strength, not weldable,
(forgings)	0.101	250	74	63	66	43	10.0	3.8	common in high-speed
(extrustions)	0.101	250	81	72	72	42	10.4	4.0	aircraft
Magnesium									
Magnesium HK 31A	0.0674	700	34	24	22	23	6.5	2.4	High-temperature, high
-HM 21A	0.0640	800	30	21	17	19	6.5	2.4	strength-to-weight, subject to corrosion
Titanium									Most-used titanium, including
Titanium-Ti-6A1-4V	0.160	750	160	145	154	100	16.0	6.2	B-70
-Ti-13V-11Cr-3A1	0.174	600–1000	170	160	162	105	15.5	—	SR-71 titanium
High-temperature nickel alloys									
Inconel X-750	0.300	1000–1500	155	100	100	101	31.0	11.0	X-15
Rene 41	0.298	1200–1800	168	127	135	107	31.6	12.1	X-20, very difficult to form
Hastelloy B	0.334	1400	100	45	—	—	30.8	—	Engine parts

Table 14.4 Wood properties (ANC-5)

		Parallel to grain				Perpendicular to grain	Parallel to grain	
	Density, lb/in.^{3}	F_{tu} 10^3 psi	F_{ty} 10^3 psi	F_{cu} 10^3 psi	F_{cy} 10^3 psi	F_{cu} 10^3 psi	F_s 10^3 psi	E 10^6 psi
Ash	0.024	14.8	8.9	7.0	5.3	2.3	1.4	1.46
Birch	0.026	15.5	9.5	7.3	5.5	1.6	1.3	1.78
African mahogany	0.019	10.8	7.9	5.7	4.3	1.4	1.0	1.28
Douglas fir	0.020	11.5	8.0	7.0	5.6	1.3	0.8	1.70
Western pine	0.016	9.3	6.0	5.3	4.2	0.8	0.6	1.31
Spruce	0.016	9.4	6.2	5.0	4.0	0.8	0.7	1.30

Table 14.5 Typical composite material properties (room temperature)

Material	Fiber orientation	Fiber, % volume	Density, lb/in.^{3}	Temp. limit, °F	F_{tu} (L) 10^3 psi	F_{tu} (T) 10^3 psi	F_{cu} (L) 10^3 psi	F_{cu} (T) 10^3 psi	F_{su} (LT) 10^3 psi	F_{isu} 10^3 psi	ϵ_{tu} (L) in./in.	ϵ_{tu} (T) in./in.	E_t (L) 10^6 psi	E_t (T) 10^6 psi	E_C (L) 10^6 psi	E_C (T) 10^6 psi	G (LT) 10^6 psi
High strength Graphite–epoxy	0	60	0.056	350	180.0	8.0	180.0	30.0	12	13	0.0087	0.0048	21.00	1.70	21.00	1.70	0.65
	±45	60	0.056	350	23.2	23.2	23.9	23.9	65.5	—	0.022	0.022	2.34	2.34	2.34	2.34	5.52
High-modulus Graphite–epoxy	0	60	0.056	350	110.0	4.0	100	20	9.0	10	0.0046	0.0025	25.00	1.70	25.00	1.70	0.65
	±45	60	0.058	350	16.9	16.9	18	18	43.2	—	0.012	0.012	2.38	2.38	2.38	2.38	6.46
Boron–epoxy	0	50	0.073	350	195	10.4	353	40	15.3	13	0.0065	0.004	30	2.7	30	2.7	0.70
Graphite–polyimide	0	—	—	—	204	4.85	111	18.5	8.5	—	—	0.0036	20	1.35	17.4	1.4	0.84
S-Fiberglass–epoxy	0	—	0.074	350	219	7.4	73.9	22.4	—	11	—	—	7.70	2.70	6.80	2.5	—
E-Fiberglss–epoxy	0	45	0.071	350	105	10.2	69	33	7.9	—	0.025	0.019	4.23	1.82	4.43	1.8	0.51
Aramid–epoxy	0	60	0.052	350	200	4.3	40	20	9	—	0.018	0.006	11	0.8	11	0.8	0.3

L = Longitudinal direction; T = transverse direction; F_{isu} = interlaminate shear stress (ultimate); t = tension; c = compression.

14.10 Structural-Analysis Fundamentals

The following sections will introduce the key equations for structural analysis of aircraft components. Derivations will not be presented as they are available in many references, such as 54, 55, and 60.

Properties of Sections

A number of geometric properties of cross sections are repeatedly used in structural calculations. Three of the most important—centroid, moment of inertia, and radius of gyration—are discussed next. Note that the cross sections of interest in tension and compression calculations are perpendicular to the stress, while in shear calculations they are in the plane of the shearing stress.

$$X_c = \frac{\Sigma x_i \; dA_i}{A} \tag{14.21}$$

$$Y_c = \frac{\Sigma y_i \; dA_i}{A} \tag{14.22}$$

The centroid of a cross section is the geometric center, or the point at which a flat cutout of the cross-section shape would balance. The coordinates of the centroid (X_c, Y_c) of an arbitrary shape (Fig. 14.27) are found from Eqs. (14.21) and (14.22). A symmetrical cross section always has its centroid on the axis of symmetry, and if a cross section is symmetric in two directions, the centroid is at the intersection of the two axes of symmetry.

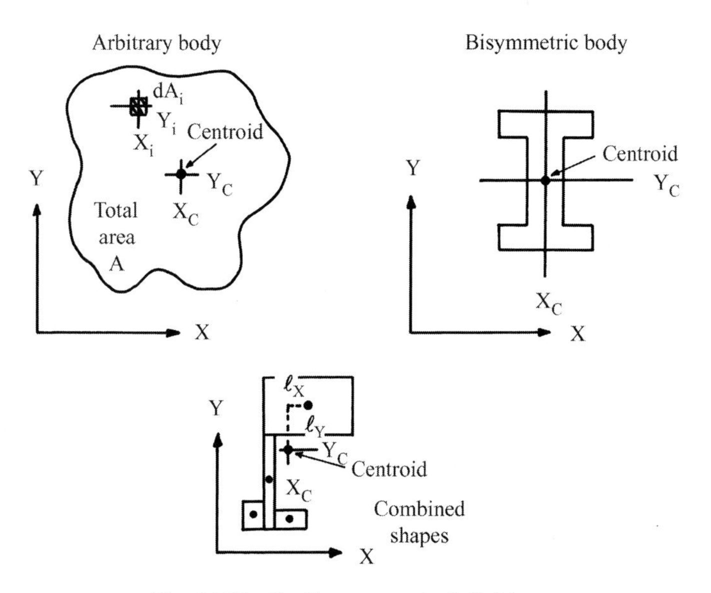

Fig. 14.27 Section property definitions.

A centroidal axis is any axis that passes through the centroid. An axis of symmetry is always a centroidal axis.

Centroids for simple shapes are provided in Table 14.6. The centroid of a complex shape built up from simple shapes can be determined using Eqs. (14.21) and (14.22) using the centroids and areas of the simple shapes.

The moment of inertia I is a difficult-to-define parameter that appears in bending and buckling equations. Moment of inertia can be viewed as the cross-section's resistance to rotation about some axis, assuming that the cross-sectional shape has unit mass. Moment of inertia is the sum of the elemental areas times the square of the distance to the selected axis [Eqs. (14.23) and (14.24)], and has units of length to the fourth power.

The polar moment of inertia (J or I_p) is the moment of inertia about an axis perpendicular to the cross section [Eq. (14.25)]; J is important in torsion calculations.

$$I_x = \Sigma y_i^2 \ \mathrm{d}A_i \tag{14.23}$$

$$I_y = \Sigma x_i^2 \ \mathrm{d}A_i \tag{14.24}$$

$$I_p = J = \Sigma r_i^2 \ \mathrm{d}A_i = I_x + I_y \tag{14.25}$$

Structural calculations usually require the moments of inertia about centroidal axes. Table 14.6 provides moments of inertia for simple shapes about their own centroidal axis. For a complex built-up shape, the combined centroid must be determined, then Eqs. (14.26) and (14.27) can be used to transfer the moments of inertia of the simple shapes to the combined centroidal axes. The "ℓ" terms are the x and y distances from the simple shapes' centroidal axes to the new axes (see Fig. 14.27, bottom).

Once the simple shapes' moments of inertia are transferred to the combined centroidal axes, the moments of inertia are added to determine the combined moment of inertia (I_x and I_y). The new J is determined from the new I_x and I_y using Eq. (14.25):

$$I_x = I_{x_c} + A\ell_y^2 \tag{14.26}$$

$$I_y = I_{y_c} + A\ell_x^2 \tag{14.27}$$

The radius of gyration ρ is the distance from the centroidal axis to a point at which the same moment of inertia would be obtained if all of the cross-sectional area were concentrated at that point. By Eq. (14.23), the moment of inertia is the total cross-sectional area times ρ squared; so ρ is obtained as follows:

$$\rho = \sqrt{I/A} \tag{14.28}$$

The main use of ρ is in column-buckling analysis. Also, the ρ values in Table 14.6 can be used to approximate I for the given shapes.

Table 14.6 Properties of simple sections

Illustrations	Area	Centroid		Moment of inertia		Radius of gyration	
		$\overline{X}$	$\overline{Y}$	I_x	I_y	ρ_x	ρ_y
SR-71 Blackbird (NASA photo by Jim Ross).	BH	$B/2$	$H/2$	$\frac{BH^3}{12}$	$\frac{HB^3}{12}$	$\frac{H}{\sqrt{12}}$	$\frac{B}{\sqrt{12}}$
SR-71 Blackbird (NASA photo by Jim Ross).	$BH\text{-}bh$	$B/2$	$H/2$	$\frac{BH^3 - bh^3}{12}$	$\frac{HB^3 - hb^3}{12}$	$\sqrt{\frac{BH^3 - bh^3}{12(BH - bh)}}$	$\sqrt{\frac{HB^3 - hb^3}{12(BH - bh)}}$
SR-71 Blackbird (NASA photo by Jim Ross).	πR^2	R	R	$\frac{\pi R^4}{4}$	$\frac{\pi R^4}{4}$	$R/2$	$R/2$
SR-71 Blackbird (NASA photo by Jim Ross).	$\pi(R^2 - r^2)$	R	R	$\frac{\pi(R^4 - r^4)}{4}$	$\frac{\pi(R^4 - r^4)}{4}$	$\frac{\sqrt{R^2 + r^2}}{2}$	$\frac{\sqrt{R^2 + r^2}}{2}$
SR-71 Blackbird (NASA photo by Jim Ross).	$\frac{BH}{2}$	O	$H/3$	$\frac{BH^3}{36}$	$\frac{B^3H}{48}$	$\frac{H}{\sqrt{18}}$	$\frac{B}{\sqrt{24}}$

Other cross-sectional properties such as the product of inertia and the principal axes will not be used in this overview of structures. See Refs. 54, 60, or other structures textbooks for more information about section properties.

Tension

Tension, the easiest stress to analyze, is simply the applied load divided by the cross-sectional area [Eq. (14.14), repeated next as Eq. (14.29)]. The shape of the cross section is unimportant in most cases.

The appropriate cross section is the smallest area in the loaded part. For example, if the part has rivet or bolt holes the smallest cross-sectional area will probably be where the holes are located, because the areas of the holes are not included for tensional calculations.

Usually the relevant cross section is perpendicular to the load. If a line of holes forms a natural "zipper" at an angle off the perpendicular, the part may fail there if the cross-sectional area along the zipper line is less than the smallest perpendicular cross section.

$$\sigma = P/A \tag{14.29}$$

Remember that the stress level at the limit load should be equal to or less than the yield stress or, for composite materials, the stress level corresponding to a strain equal to the ultimate strain capability of the material divided by the selected factor of safety (often 1.5, matching that used for metals).

Compression

The compression stress is also given by Eq. (14.29) (load divided by area). For the determination of the limit stress, this equation can only be applied to parts that are very short compared to cross-sectional dimensions (such as fittings) or to parts that are laterally constrained (such as spar caps and sandwich face sheets). Long unconstrained members in compression, called "columns" or "struts," are discussed next.

For short or laterally constrained parts in compression, the ultimate compressive strength is usually assumed to equal the tensile value. For ductile metals this is a conservative assumption as they never actually fail, but merely "squish" out and support the load by the increased area.

Rivet and bolt holes are included in the cross-sectional area calculation for compression because the rivets or bolts can carry compressive loads.

Columns in compression usually fail at a load well below that given by applying the ultimate stress to Eq. (14.29). Columns in compression fail either by "primary buckling" or by "local buckling."

An important parameter is the column's slenderness ratio: the column's effective length L_e divided by the cross-sectional radius of gyration [Eq. (14.30)]. The effective length of a column is determined by the end connections (pinned, fixed, or free) as shown in Fig. 14.28.

$$\text{Slenderness Ratio:} \quad \frac{L_e}{\rho} = \frac{L_e}{\sqrt{I/A}} \tag{14.30}$$

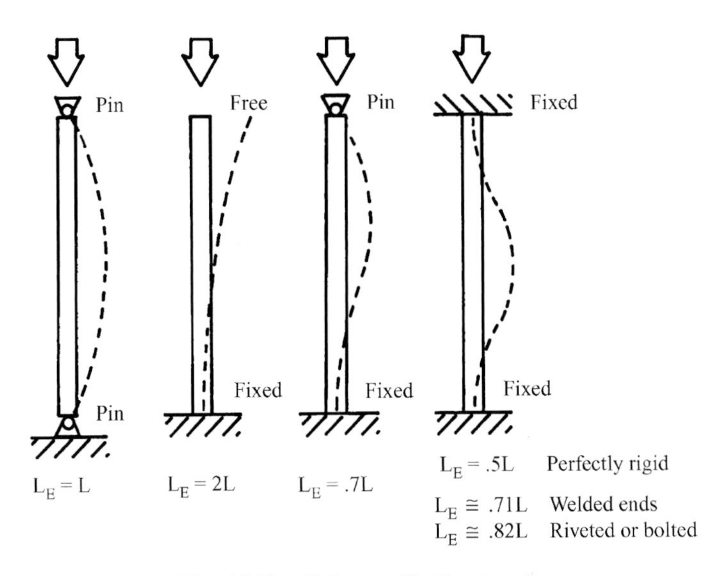

Fig. 14.28 Column effective length.

When you push down on an upright yardstick, the middle part bends outward in a direction perpendicular to the load. This bending action produces internal stresses much greater than the direct compression stress due to the applied load, and is called "primary column buckling." If the bending action after buckling involves stresses below the proportional limit, the column is said to experience "elastic buckling."

The highest compression load that will not cause this elastic column buckling—the so-called Euler load, or critical load P_c—will be determined from the Euler column equation [Eq. (14.31)]. The resulting compressive stress is found from Eq. (14.32).

Note in Eq. (14.31) that the total load a column can carry without buckling does not depend upon either the cross-sectional area or the ultimate compressive stress of the material! Only the column's effective length, its cross-sectional moment of inertia, and the material's modulus of elasticity affect the buckling load if the column is long.

$$P_c = \frac{\pi^2 EI}{L_e^2} \tag{14.31}$$

$$F_c = \frac{\pi^2 EI}{AL_e^2} = \frac{\pi^2 E}{(L/\rho)^2} \tag{14.32}$$

The buckling stresses of Eq. (14.32) are failure stresses and do not have any margin of safety. For design purposes the limit loads should be reduced, usually to two-thirds of these values.

A column with an open or highly irregular cross section may fail at a lower load due to cross-sectional twisting or deformation. Methods for analysis of such members can be found in Refs. 60 and 83.

Equation (14.31) implies that, as column length is reduced to zero, the Euler load goes to infinity. However, the compression stresses experienced due to bending in a buckled column are much greater than the applied load would directly produce. At some point, as column length is reduced the internal compressive stresses produced at the onset of buckling will exceed the proportional limit and the column will no longer be experiencing elastic buckling. This has the effect of reducing the buckling load compared to the Euler load.

The critical slenderness ratio defines the shortest length at which elastic buckling occurs. At a lower slenderness ratio, the stresses at buckling exceed the proportional limit. The column experiences inelastic buckling so the Euler equation cannot be used as shown. The critical slenderness ratio depends upon the material used. It is about 77 for 2024 aluminum, 51 for 7075 aluminum, 91.5 for 4130 steel, and 59–76 for alloy steel depending upon heat treatment. Most columns used in aircraft are below these critical slenderness values, so the elastic Euler equation cannot usually be used in aircraft column analysis.

The buckling load for inelastic buckling can be determined by Eq. (14.32), with one modification. The modulus of elasticity must be replaced by the tangent modulus, described previously. As the tangent modulus is a function of the stress, iteration is required to find the buckling load for a particular column. However, handbook graphs such as Fig. 14.29 are usually used for design (see Refs. 60 and 61).

As discussed at the beginning of this section, a very short "column" experiences pure compression without any danger of primary column buckling. This

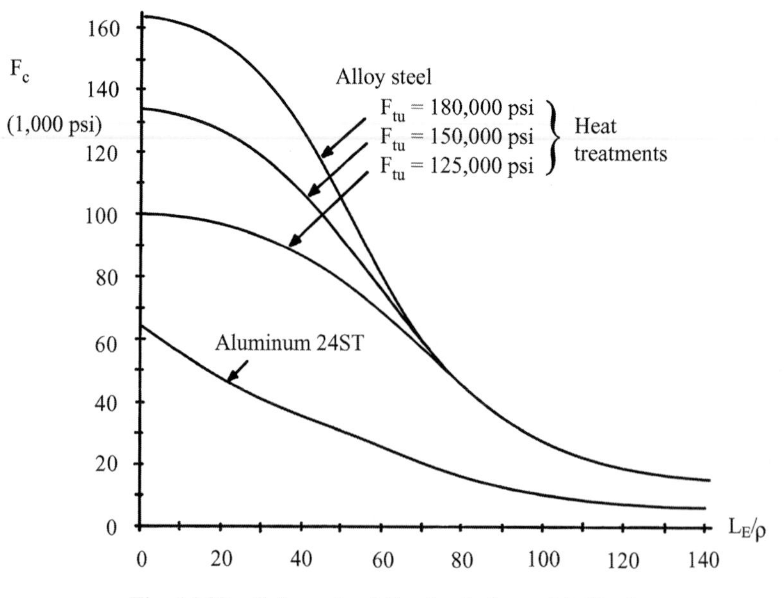

Fig. 14.29 Column buckling loads (round tubing).

is sometimes called "block compression." The compression yield value is used as the limit load, providing a cutoff value for the buckling load of a short column with either a solid cross section or with relatively thick walls (structural tubing). A column can usually be considered in block compression if the slenderness ratio is less than about 12.

When you step on an upright soda can, it fails in a form of local buckling called crippling, in which the walls of the cross section collapse without warning, and the load-carrying ability drops to virtually zero. This is typical for short columns with very thin walls. Methods for estimation of thin-wall crippling are found in Ref. 60. A rough estimate for the crippling stress of a thin-wall cylindrical tube is shown in Eq. (14.33), where t is the wall thickness and R is the radius.

$$F_{\text{crippling}} \cong 0.3(Et/R) \tag{14.33}$$

A flat sheet or panel under compression fails by buckling in a manner similar to a column. The buckling load [Eq. (14.34)] depends upon the length (a) in the load direction, the width (b), the thickness, and the manner in which the sides are constrained.

Clamped sides cannot rotate about their axis, and provide the greatest strength. Simply supported sides are equivalent to a pinned end on a column, and can rotate about their axis but cannot bend perpendicularly. A free side can rotate and bend perpendicularly and provides the least strength.

Figure 14.30 provides the buckling coefficient K for Eq. (14.34) based upon panel length-to-width ratio and end constraints. Most aircraft panels are clamped, but with some flexibility to rotate about the side axes. A K value between the clamped and simply supported values should be used in such a case.

$$F_{\text{buckling}} = KE(t/b)^2 \tag{14.34}$$

Truss Analysis

A truss is a structural arrangement in which the structural members (struts) carry only compression or tension loads ("columns" and "ties"). In the ideal truss, the struts are weightless and connected by frictionless pins. No loads are applied except at the pins, and no moments are applied anywhere. These ideal assumptions guarantee that the struts carry only compression or tension.

The strut loads calculated with these ideal assumptions are called primary truss loads. Additional loads such as those caused by the attachment of an aircraft component to the middle of a strut must be calculated separately and added to the primary load during analysis of each individual strut. The impact of rigid welded connections in a typical aircraft application is considered only in the definition of effective length in the column-buckling equation (see Fig. 14.28).

Truss structure was used extensively in welded steel-tube fuselages. Today the truss structure is largely used in piston-engine motor mounts, the ribs of large aircraft, and landing gear.

Figure 14.31 shows a typical truss structure, a light aircraft motor mount. For illustration purposes this will be analyzed as if it were a two-dimensional truss with only the three struts shown. Analysis of three-dimensional space structures will be discussed later.

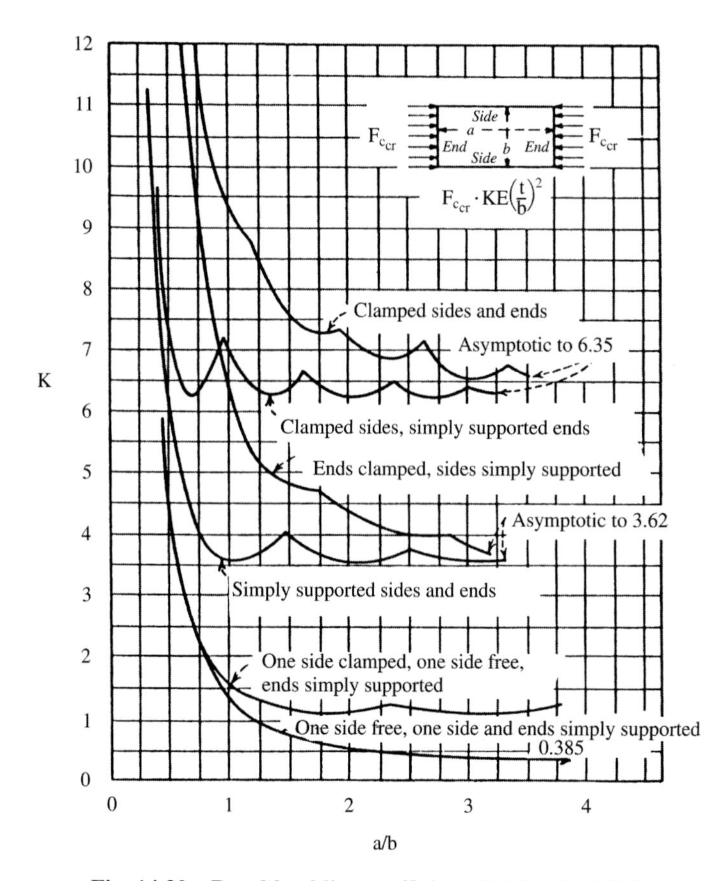

Fig. 14.30 Panel buckling coefficient (NACA TN3781).

The bottom of Fig. 14.31 shows an equivalent truss that includes the lines of force to the c.g. of the engine, and the vertical resisting forces due to the rigid attachment of the fuselage and engine to the truss. This equivalent truss can be solved by several methods.

The most general truss solution, the "method of joints," relies upon the fact that at each joint of the truss, the sums of the vertical and horizontal forces must each total zero.

To obtain a solution from the two equations (vertical and horizontal), the solution must begin at and always proceed to a joint with only two unknown struts. The method usually begins at a free joint with an applied external load, in this case at the engine load.

Figure 14.32 shows the forces at the joints. All the forces are shown as radiating outward from the joints so that a positive force is a tension and a negative force is a compression.

When summing forces at a joint, the positive or negative force is added to the sum if it is up (when summing vertical forces) or to the right (when summing horizontal forces), and subtracted if down or to the left. Confusion about the

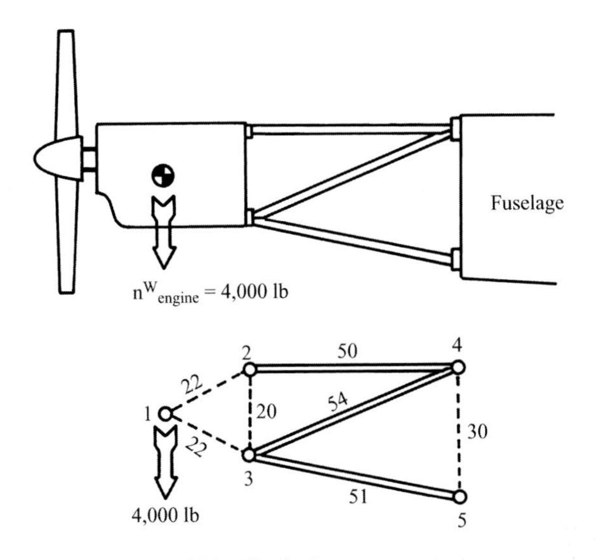

Fig. 14.31 Typical truss structure.

appropriate sign is the most common error in truss analysis. (The author did joint three wrong the first time!)

Joint one is at the engine's c.g. The unknown forces F_a and F_b must react to the engine load of 4000 lb. Solving the equations shown yields F_a of 4400 lb (tension) and F_b of -4400 lb (compression).

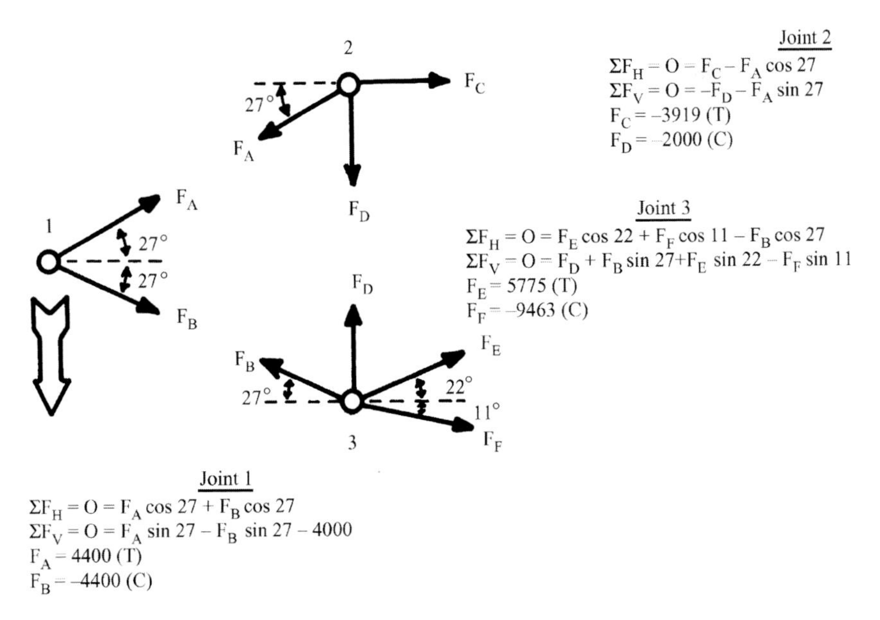

Fig. 14.32 Method of joints.

Selection of the next joint to analyze depends upon the number of unknown struts. At joint three, there are three unknown struts at this time, so we select joint two. Solving the equations yields F_c of 3919 lb (tension). F_d is found to be -2000 lb, a compression load on the engine due to the motor mount. If this load is in excess of what the engine can withstand, a vertical motor-mount strut should be welded between joints two and three.

At joint three there are now only two unknown strut loads. Solving the equations yields F_e of 5775 lb (tension) and F_f of -9463 lb (compression).

In some cases a quicker method can be employed to determine the forces in selected struts without having to solve the whole truss as in the method of joints. This quicker method is actually two methods, the "method of moments" for the upper and lower struts and the "method of shears" for the inner struts.

The top illustration of Fig. 14.33 shows the use of the method of moments to solve the force in the top strut of the motor mount. The whole structure is replaced by two rigid bodies connected by a pin, with rotation about the pin prevented by the unknown force in the strut under analysis. The moments about the pin are readily summed and solved for the unknown strut force, which is found to be 3919 lb.

A similar technique is shown in the middle illustration for the lower strut, which has a load of 9463 lb. Note that this technique, where applicable, allows direct solution for the desired unknown forces.

The lower illustration of Fig. 14.33 shows the use of the method of shears to solve for the inner strut. This method involves severing the structure along a plane that cuts only three members, the upper and lower strut and the inner strut under analysis.

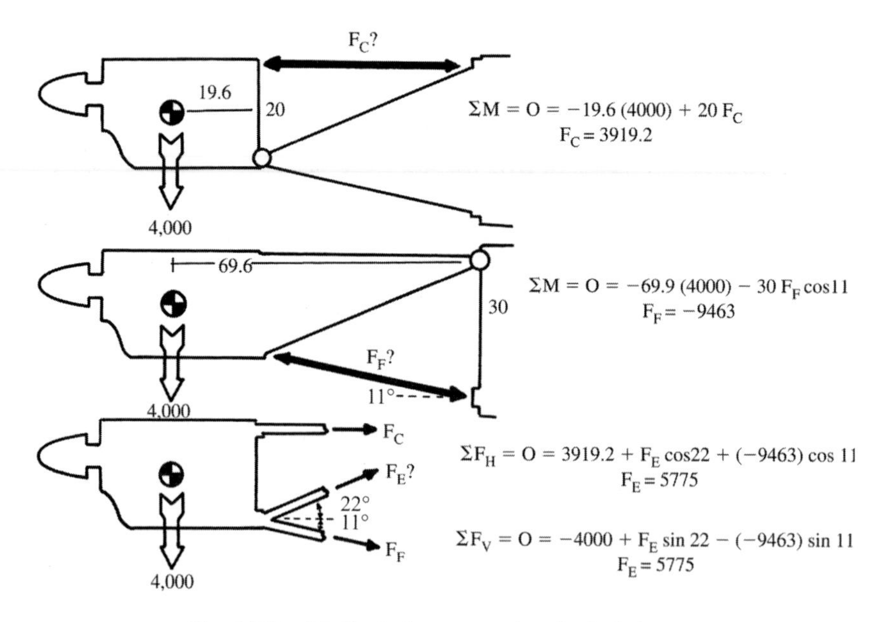

Fig. 14.33 Method of moments/method of shears.

The severed part of the structure is analyzed as a free body, summing either the vertical and horizontal forces, which must total zero. Note that by calculating the unknown strut force both ways (vertical and horizontal summation), a check of your result can be made. This example gives a result of 5775 lb.

These methods are only applicable if the truss structure is "statically determinate." In general, a truss is statically determinate if every strut can be cut by some plane that cuts only two other struts. This ensures that there is always a joint with only two unknown struts, permitting solution by the method of joints. For "indeterminate" trusses, more complicated methods based upon deflection analysis can be used (see Refs. 54 and 60), or a finite element structural analysis can be performed (see Sec. 14.11).

Once the loads in each member of the truss are known, the struts can be analyzed using the equations just presented for tension or compression. Use the appropriate effective length for welded, riveted, or bolted columns from Fig. 14.28. To provide an extra margin of safety, it is customary to assume that welded steel-tube motor mounts act as though the ends were pinned ($L_e = L$).

The 3-D trusses, or space structures, are solved similarly to the 2-D truss. Square cross-section 3-D trusses, such as a typical welded-tube fuselage, can sometimes be solved separately in side view and top view as 2-D structures. The resulting strut loads are then summed for the various members. This is permitted provided that the combined loads on all struts are within the elastic range.

For more complicated 3-D trusses, the method of joints can be applied using three equations and three unknown strut loads. This involves simultaneous solution of equations, e.g., with a simple computer iteration program. In some cases the moments about some selected point can be used to obtain the solution with less effort. Space structures are discussed in detail in Ref. 54.

Beam Shear and Bending

A common problem in aircraft design is the estimation of the shear and bending stresses in the wing spars or fuselage. This is a two-step process. First, the shear and bending moment distributions must be determined, and then the resulting stresses must be found.

Figure 14.34 shows a simple beam with a distributed vertical load. The beam is shown cut to depict internal forces. The right side of the beam being a free body, the sum of the vertical forces and the sum of the moments must equal zero.

If the severed part of the beam is to remain in vertical equilibrium, the externally applied vertical forces must be opposed by a vertical shear force within the cross section of the material, as shown. Thus, for any span station the shear force is simply the sum of the vertical loads outboard of that station, or the integral of a distributed load.

The moments produced by the vertical loads must be balanced by a moment at the cut cross section. This moment is equal to the summation of the discrete loads times their distance from the cut station, or the integral of a distributed load with respect to the distance from the cut.

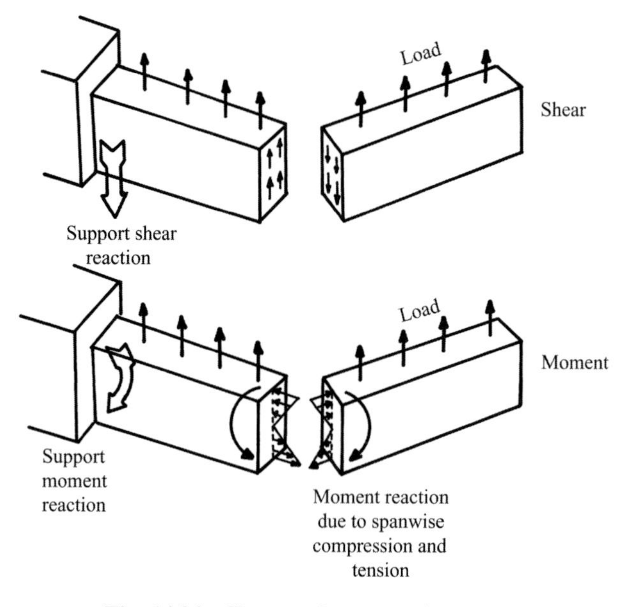

Fig. 14.34 Shear and moment in beams.

Figure 14.35 shows the typical loads on a wing. This shows the critical case of a rolling pull-up, with the additional lift load of full aileron deflection. The lift and wing-weight loads are distributed, while the nacelle weight is concentrated. Remember that wing and nacelle weights are multiplied by the aircraft load factor to determine the load on the wing.

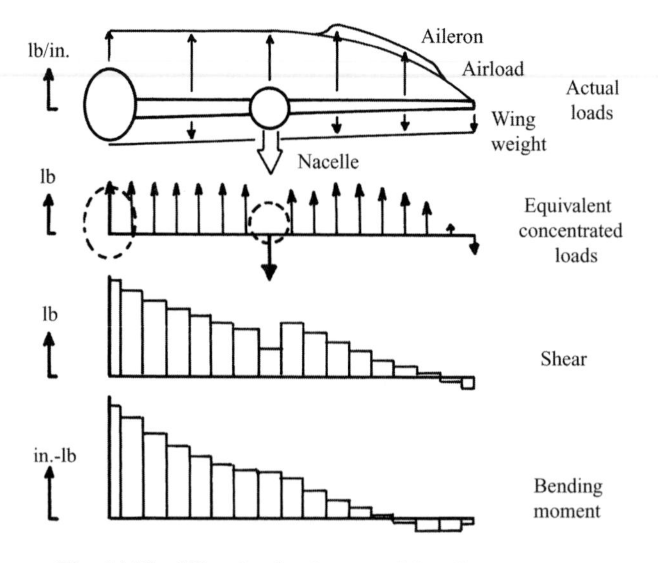

Fig. 14.35 Wing loads, shear, and bending moment.

The easiest way to calculate the shear and moment distribution along a wing is to replace the distributed loads (lift and wing weight) by concentrated loads. The lift distribution can be determined with Schrenk's approximation, just described. The wing weight will be determined in the next chapter, and can be assumed to be distributed proportional to the chord length.

Figure 14.36 shows the trapezoidal approximation for a distributed load, giving the total equivalent force and the spanwise location of that force. About 10 to 20 spanwise stations will provide an accurate enough approximation for initial design purposes.

Once the distributed loads are replaced by concentrated loads, determination of the shear and bending moment distributions is easy. The shear at each span station is the sum of the vertical loads outboard of that station. The shear is found by starting at the wing tip and working inward, adding the load at each station to the total of the outboard stations.

The bending moment can be found for each span station by multiplying the load at each outboard station times its distance from the span station. However, it is easier to graphically integrate by starting at the tip and working inward, adding to the total the area under the shear distribution at that station.

Referring back to Fig. 14.34, the bending moment at a cross-sectional cut is opposed by a combination of tension and compression forces in the spanwise direction. For a positive bending moment such as shown, the internal forces produce compression on the upper part of the beam and tension on the lower part. The vertical location in the beam at which there is no spanwise force due to bending is called the "neutral axis," and is at the centroid of the cross-sectional shape.

As long as the stresses remain within the elastic limit, the stresses vary linearly with vertical distance from the neutral axis regardless of the cross-sectional shape. These compression or tension stresses are found from Eq. (14.35) (for derivation, see Ref. 55), where M is the bending moment at the spanwise location and z is the vertical distance from the neutral axis. The maximum stresses due to bending are at the upper and lower surfaces.

$$\sigma_x = Mz/I_y \tag{14.35}$$

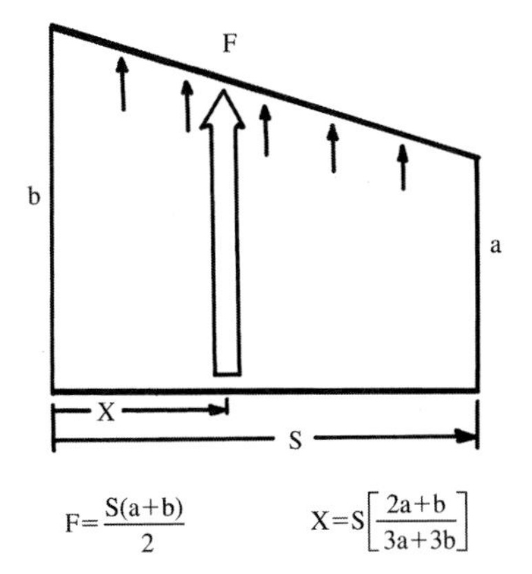

Fig. 14.36 Trapezoidal approximation for distributed loads.

The vertical shear stresses within a beam are not evenly distributed from top to bottom of the cross section, so the maximum shear stress within the material can not be calculated simply as the total shear divided by the cross-sectional area.

Referring back to Fig. 14.17, it should be remembered that the vertical shear stresses on an element are balanced by and equal to the horizontal shear stresses. One cannot exist without the other. Therefore, the vertical shear distribution must be related to the horizontal shears in the beam.

Figure 14.37 shows a beam in bending, with the vertical distribution of compression and tension stresses. The total horizontal force on any element is the horizontal stress at the element's vertical location times the elemental area. If this beam is split lengthwise as shown, the upper section has only leftward forces, so a shear force must be exerted along the cut.

This shear force must be the sum of the horizontal stresses times the elemental areas above the cut. This reaches a maximum at the neutral axis. At the upper and lower surfaces, this shear force is zero.

The bottom of Fig. 14.37 shows the resulting vertical distribution of shear forces, expressed as magnitude toward the right. (Do not be confused by this presentation; the shear forces are exerted in a vertical direction, but we show the magnitude to the right to illustrate the distribution of magnitude from top to bottom.)

$$\tau = \frac{V}{bI_y} \int_z^{h/2} z \, \mathrm{d}A \tag{14.36}$$

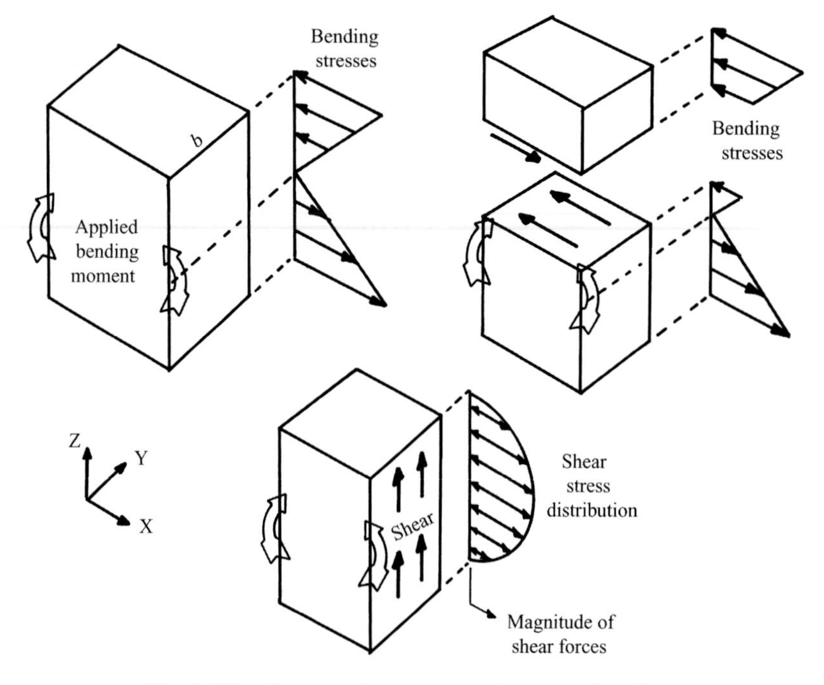

Fig. 14.37 Relationship between shear and bending.

Equation (14.36) describes this mathematically, where the integral term represents the area above the cut located at $z = z_1$. Note that the distribution of shear stresses depends upon the shape of the cross section. For a beam of rectangular cross section, the maximum shearing stress (at the neutral axis) is 1.5 times the averaged shearing stress (total shear divided by cross-sectional area). For a solid circular cross section, the maximum shearing stress is 1.33 times the averaged value.

Figure 14.38 shows a typical aircraft wing spar consisting of thick "spar caps" separated by a thin "shear web." The cross-sectional area of the shear web is insignificant compared to the area of the spar caps, so the caps absorb virtually all of the bending force (stress times area). The shear stress depends upon the cross-sectional area above the point of interest, and is therefore essentially constant within the thin shear web, as shown to the right.

In aircraft wing spar analysis, it is common to assume that the caps absorb all of the bending stresses and that the web (extended to the full depth of the spar) absorbs all of the shear. This is shown at the bottom of Fig. 14.38. It is also assumed that the shear is constant within the web and therefore the maximum shear stress equals the average shear stress (shear divided by web area).

The shear web will fail in buckling long before the material maximum shear stress is reached. Equation (14.37) defines the critical buckling shear stress for a shear web. The value of K is obtained from Fig. 14.39.

$$F_{\text{shear buckle}} = KE(t/b)^2 \tag{14.37}$$

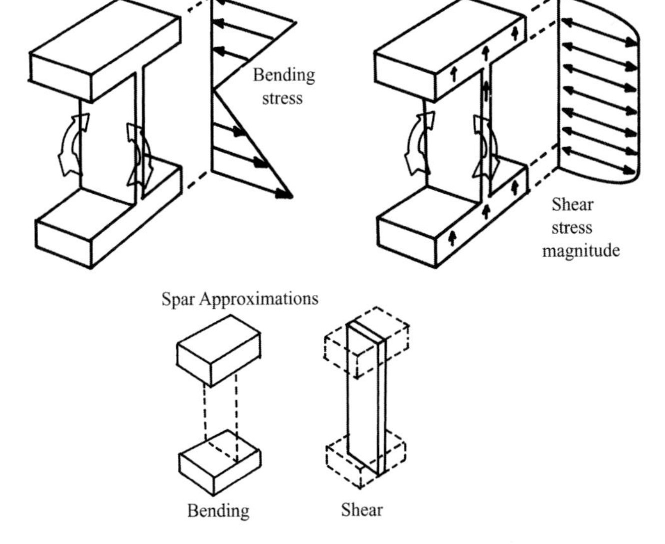

Fig. 14.38 Typical aircraft spar in bending and shear.

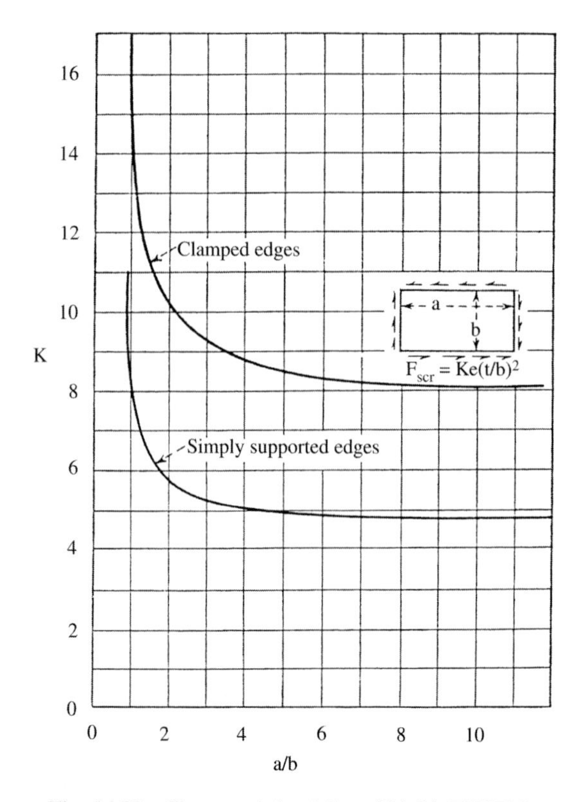

Fig. 14.39 Shear web buckling (NACA TN3781).

Braced-Wing Analysis

A wing braced with a strut will have the bending moments greatly reduced compared to a fully cantilevered wing. However, the analysis is more complex because of the spanwise compression loads exerted upon the wing by the strut. This can increase the bending moment by as much as a third compared to an analysis that ignores this compression effect.

Figure 14.40 shows a typical braced wing. The compression load P is the horizontal component of the force on the strut (S). The vertical component of S is found from summing the moments about the pin at the wing root, using the equivalent concentrated lift loads as discussed earlier.

The shear loads of the braced wing are analyzed as before, taking into account the large concentrated vertical load of the strut. The bending moment must be analyzed with special equations provided next.

The portion of the wing outboard of the strut is analyzed as before, and the bending moment at the strut location is determined (M_2). The root bending moment (M_1) is usually zero unless the hinge point is above or below the neutral axis, causing a bending moment due to the compression load P.

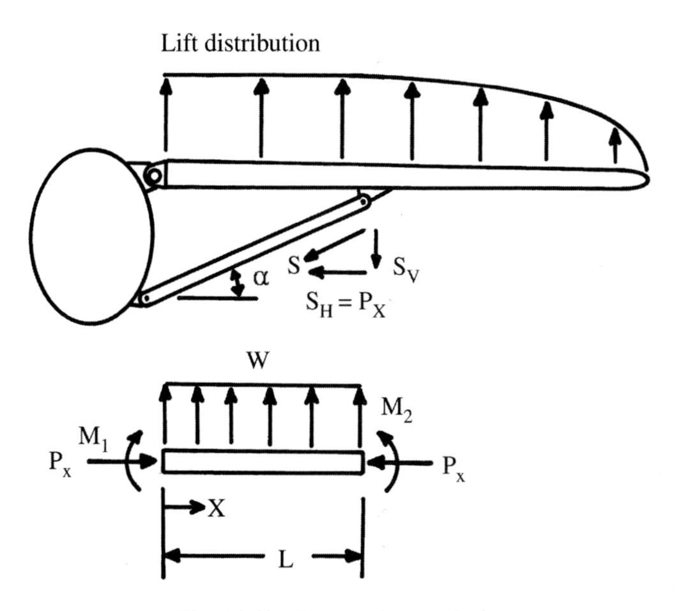

Fig. 14.40 Brace wing analysis.

The lift distribution on the portion of the wing inboard of the strut must be approximated by a uniform load distribution (w). This is usually a reasonable approximation inboard of the strut. The following equations describe bending-moment distribution, maximum bending moment, and spanwise location of the maximum bending moment (Ref. 60):

$$M(x) = C_1 \sin(x/j) + C_2 \cos(x/j) + wj^2 \tag{14.38}$$

$$M_{\max} = \frac{D_1}{\cos(x/j)} + wj^2 \tag{14.39}$$

$$\tan\left(\frac{x_m}{j}\right) = \frac{D_2 - D_1 \cos(L/j)}{D_1 \sin(L/j)} \tag{14.40}$$

where

$$j = \sqrt{EI/P} \tag{14.41}$$

$$C_1 = \frac{D_2 - D_1 \cos(L/j)}{\sin(L/j)} \tag{14.42}$$

$$C_2 = D_1 = M_1 - wj^2 \tag{14.43}$$

$$D_2 = M_2 - wj^2 \tag{14.44}$$

Torsion

Figure 14.41 shows a solid circular shaft in torsion. The applied torque T produces a twisting deformation ϕ that depends upon the length of the shaft. As shown at the right of the figure, the torque is resisted by shearing stresses that increase linearly with distance from the center—if the stresses remain within the elastic limit.

The shear stresses due to torsion are calculated with Eq. (14.45), and are at a maximum at the surface of the shaft ($r = R$). The angular deflection in radians is determined from Eq. (14.46). These equations also apply to circular tubing under torsion, using the appropriate value of Ip as provided previously.

$$\tau = Tr/I_p \tag{14.45}$$

$$\phi = TL/GI_p \tag{14.46}$$

For a noncircular member under torsion, the analysis is generally much more complex. Several special cases can be readily solved. A thin-walled, closed, cross-sectional member with constant wall thickness t, total cross-sectional area A, and cross-sectional perimeter s has shear stress and angular deflection as defined by Eqs. (14.47) and (14.48).

$$\tau = T/2At \tag{14.47}$$

$$\phi = \frac{TL}{G}\left(\frac{s}{4A^2t}\right) \tag{14.48}$$

Solid rectangular members may be analyzed with Eqs. (14.49) and (14.50) using the values from Table 14.7, where t is the thickness of the member and

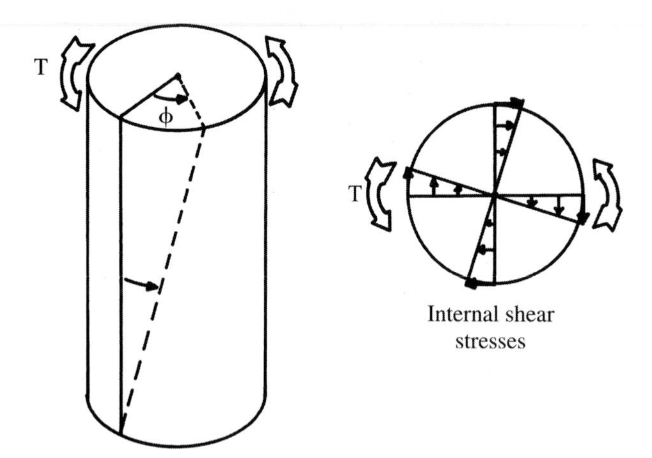

Fig. 14.41 Solid circular shaft in torsion.

Table 14.7 Torsion constants

b/t	1.00	1.50	1.75	2.00	2.50	3.00	4	6	8	10	∞
α	0.208	0.231	0.239	0.246	0.258	0.267	0.282	0.299	0.307	0.313	0.333
β	0.141	0.196	0.214	0.229	0.249	0.263	0.281	0.299	0.307	0.313	0.333

b is its width. These equations may also be applied to members bent up from flat sheet metal by "unwrapping" the member to find the total effective width.

$$\tau = \frac{T}{\alpha b t^2} \tag{14.49}$$

$$\phi = \frac{TL}{\beta b t^3 G} \tag{14.50}$$

Analysis of the torsional stresses in a complex shape such as a multicelled wing box goes beyond the scope of this book. See Ref. 60 for a discussion of such analysis.

14.11 Finite Element Structural Analysis

The structural-analysis methods just described, along with extensive handbooks and nomograms, have been used for many years for aircraft structural design. Today these methods are a dying art. Instead, virtually all major structural analysis is now performed using finite element computer programs. Even today's homebuilders have access to finite element programs using personal computers that are as powerful as the mainframe computers of the 1960s.

The *finite element method* (FEM) is based upon the concept of breaking the structure of the aircraft into numerous small "elements," much like the gridding of the air-mass for CFD. Equations describing the structural behavior of these finite elements are prepared using various approximations of the end-constraints and deflection shapes for the element.

The element equations are then linked together using matrix algebra so that the entire structure's response to a given external loading condition can be determined. The huge size of the matrices used for FEM analysis requires computers for solution of all but the most trivial cases.

Figure 14.42 illustrates the more commonly used finite elements. The aircraft structure must be modeled as a connected collection of one or more of these finite element shapes.

Selection of which element type to use is a matter of engineering judgment. Unfortunately, the selection of the element type can influence the results. Also, the selection of the size of the elements requires experience. As a general rule, the size of the elements should be reduced anywhere that the stress is expected to vary greatly. An example of this would be in the vicinity of a corner.

Figure 14.43 shows an FEM example in which the major structural members of a propfan research aircraft are modeled using the rectangular-plate finite

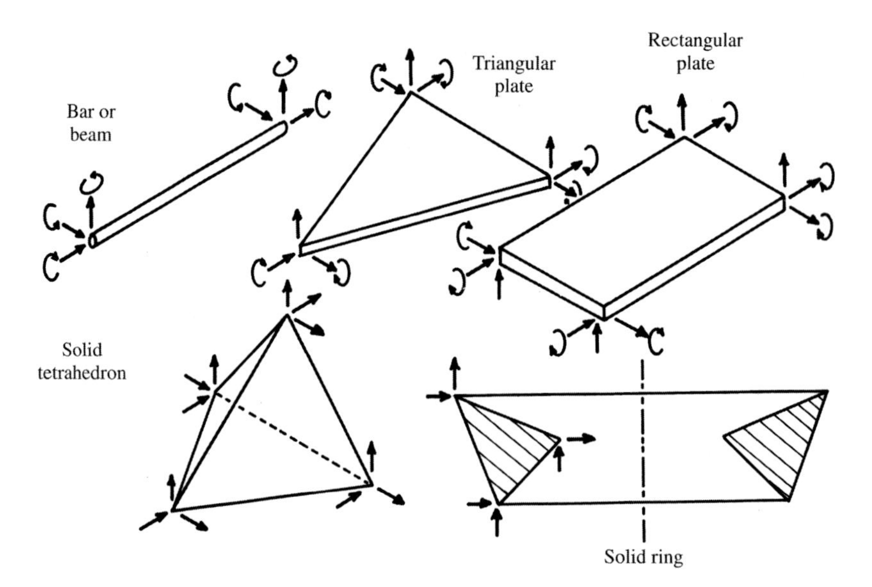

Fig. 14.42 Typical finite elements.

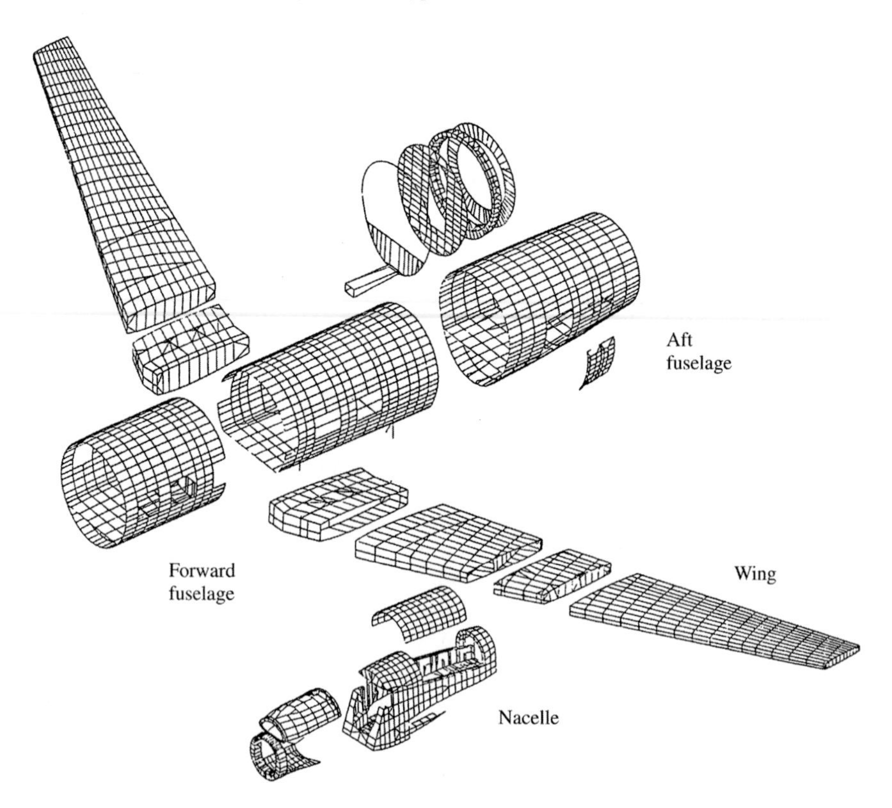

Fig. 14.43 Typical finite element model (courtesy Lockheed Martin).

element. As is the case for CFD gridding, the modeling of a complex structure for FEM analysis can be very time consuming.

Detailed derivations of the equations for the various finite element types shown in Fig. 14.42 are beyond the scope of this book (see Refs. 84 and 85). A simple example, the one-dimensional (1-D) bar, will be developed to illustrate the principles involved.

Figure 14.44 depicts a simple 1-D bar element with end-loadings P_1 and P_2, and end-deflections u_1 and u_2. For a static structural analysis, P_1 must equal the negative of P_2, although this is not true in a dynamic analysis. The cross-sectional area of the bar is shown as A. Note that while this example is a 1-D case, the deflected position is depicted slightly offset for clarity.

The strain ϵ is defined earlier in this chapter as the change in length divided by the original length L, as shown in Eq. (14.51). The stress σ is defined as the load divided by the cross-sectional area, and Young's modulus E is defined as the stress divided by the strain. This results in Eq. (14.52).

$$\epsilon = (u_1 - u_2)/L \tag{14.51}$$

$$E = \sigma/\epsilon = (P/A)/[(u_1 - u_2)/L] \tag{14.52}$$

or

$$P = \frac{EA}{L}(u_1 - u_2) \tag{14.53}$$

Applying a load P_1 yields Eq. (14.54). Similarly, applying a load P_2 results in Eq. (14.55). The change in signs of the deflections in Eq. (14.55) is due to the assumed directions of the two loads as drawn in the figure.

$$P_1 = \frac{EA}{L}(u_1 - u_2) \tag{14.54}$$

$$P_2 = \frac{EA}{L}(-u_1 + u_2) \tag{14.55}$$

Equations (14.54) and (14.55) can be combined into matrix form as shown in Eqs. (14.56) and (14.57). The k matrix is called the stiffness matrix because it

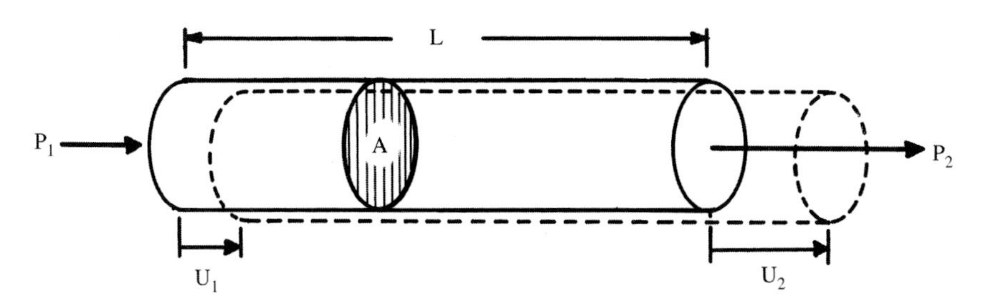

Fig. 14.44 Simple 1-D bar element.

relates the amount of deflection to the applied loads. The values within the k matrix are called stiffness coefficients.

The u matrix containing the deflection terms is called the "displacement vector." The P matrix is the force vector. (Letters other than P and u are frequently used for these terms, but for some reason k is almost always used for the stiffness matrix.)

$$\begin{Bmatrix} P_1 \\ P_2 \end{Bmatrix} = \begin{bmatrix} EA/L & -EA/L \\ -EA/L & EA/L \end{bmatrix} \begin{Bmatrix} u_1 \\ u_2 \end{Bmatrix} \tag{14.56}$$

$$\{P\} = [k]\{u\} \tag{14.57}$$

The values E, A, and L are known, so the stiffness matrix is known. By inverting the stiffness matrix, the deflections can be found for any loading condition.

This simple example could easily be solved by classical structure techniques. The power of FEM is in the assemblage of numerous finite elements.

Figure 14.45 shows a two-element assemblage using the 1-D bar element developed above. Two bars of different length and cross-sectional area are connected. The point where two (or more) finite elements are connected is called a "node" and is distinguished by the fact that at a node, the displacements of the connected finite elements are the same. Thus, u_2 represents both the displacement of the right end of the first element and the displacement of the left end of the second element.

From Eq. (14.56), the matrix equations for the left- and right-side elements can be written as Eqs. (14.58) and (14.59).

$$\begin{Bmatrix} P_1 \\ P_2 \end{Bmatrix} = \begin{bmatrix} EA_1/L_1 & -EA_1/L_1 \\ -EA_1/L_1 & EA_1/L_1 \end{bmatrix} \begin{Bmatrix} u_1 \\ u_2 \end{Bmatrix} \tag{14.58}$$

$$\begin{Bmatrix} P_2 \\ P_3 \end{Bmatrix} = \begin{bmatrix} EA_2/L_2 & -EA_2/L_2 \\ -EA_2/L_2 & EA_2/L_2 \end{bmatrix} \begin{Bmatrix} u_2 \\ u_3 \end{Bmatrix} \tag{14.59}$$

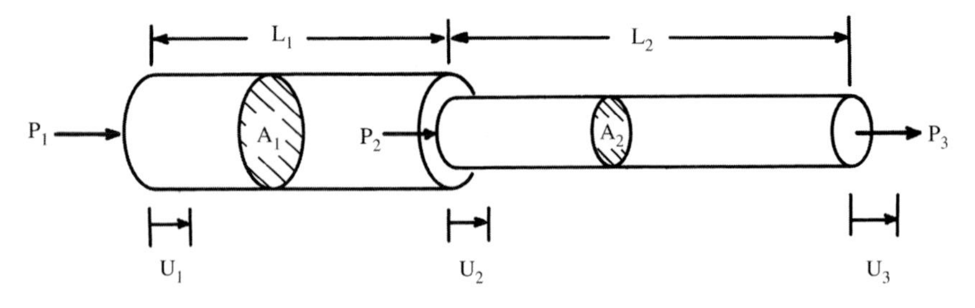

Fig. 14.45 One-dimensional bar FEM assembly.

Now the matrices can be assembled by merging the element matrices. This is shown in Eq. (14.60). Note that the "overlapping" terms at the node result from the nodal condition of identical deflection (u_2 in this case). These overlapping terms are added in forming the assembled matrix.

$$\begin{Bmatrix} P_1 \\ P_2 \\ P_3 \end{Bmatrix} = \begin{bmatrix} EA_1/L_1 & -EA_1/L_1 & 0 \\ -EA_1/L_1 & (EA_1/L_1 + EA_2/L_2) & -EA_2/L_2 \\ 0 & -EA_2/L_2 & EA_2/L_2 \end{bmatrix} \begin{Bmatrix} u_1 \\ u_2 \\ u_3 \end{Bmatrix} \tag{14.60}$$

This completes the FEM development for this example. The remaining work is strictly computation based upon the actual values of the variables in a given design problem. For example, Fig. 14.46 shows a two-bar structure in which the right side attaches to a wall, loads are as shown, and the dimensional and material values are as indicated. This produces the following:

$$\begin{Bmatrix} P_1 \\ P_2 \\ P_3 \end{Bmatrix} = \begin{bmatrix} (2.5 \times 10^7) & (-2.5 \times 10^7) & 0 \\ (-2.5 \times 10^7) & (3.4 \times 10^7) & (-9.2 \times 10^6) \\ 0 & (-9.2 \times 10^6) & (9.2 \times 10^6) \end{bmatrix} \begin{Bmatrix} u_1 \\ u_2 \\ u_3 \end{Bmatrix} \tag{14.61}$$

The 3×3 stiffness matrix in Eq. (14.61) can be inverted to find the deflections for any loading. This would first require determining the unknown wall-reaction load P_3.

Alternatively, we can simplify the FEM matrix solution by noting that the deflection at the wall u_3 is zero, so we can eliminate the third row and the third column from the martrix. This produces Eq. (14.62) with a 2×2 stiffness

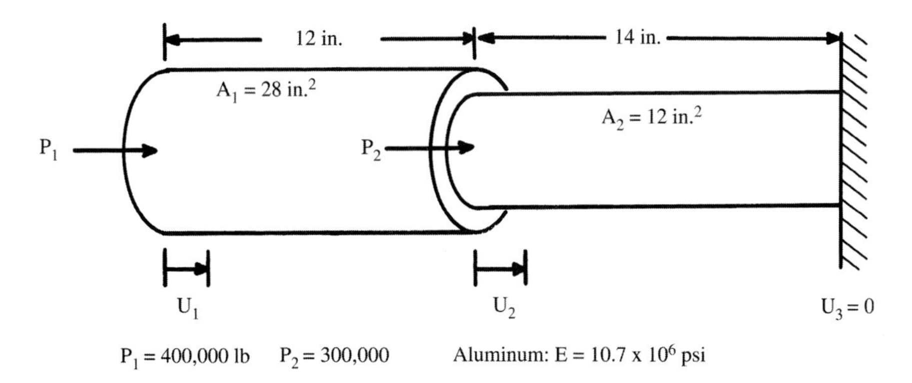

Fig. 14.46 FEM example.

matrix.

$$\begin{Bmatrix} P_1 \\ P_2 \end{Bmatrix} = \begin{bmatrix} (2.5 \times 10^7) & (-2.5 \times 10^7) \\ (-2.5 \times 10^7) & (3.4 \times 10^7) \end{bmatrix} \begin{Bmatrix} u_1 \\ u_2 \end{Bmatrix} \tag{14.62}$$

$$\begin{bmatrix} (1.5 \times 10^{-7}) & (1.1 \times 10^{-7}) \\ (1.1 \times 10^{-7}) & (1.1 \times 10^{-7}) \end{bmatrix} \begin{Bmatrix} P_1 \\ P_2 \end{Bmatrix} = \begin{Bmatrix} u_1 \\ u_2 \end{Bmatrix} \tag{14.63}$$

$$\begin{Bmatrix} 0.093 \\ 0.077 \end{Bmatrix} = \begin{Bmatrix} u_1 \\ u_2 \end{Bmatrix} \tag{14.64}$$

In Eq. (14.63) we have found the inverse of the reduced k matrix. By substituting the actual values of the loadings P, we determine the deflections as provided in Eq. (14.64). We can then use the deflections of the nodes to solve for the strain and stress, as follows:

$$\epsilon_1 = (0.093 - 0.077)/12 = 0.0013 \tag{14.65}$$

$$\epsilon_2 = (0.077 - 0)/14 = 0.0055 \tag{14.66}$$

$$\sigma_1 = 14{,}267 \text{ psi} \tag{14.67}$$

$$\sigma_2 = 58{,}850 \text{ psi} \tag{14.68}$$

This 1-D example does not illustrate the complications caused by 3-D geometry. For this simple example the deflections at the nodes produce identical changes in the length of the bars. Were the bars connected at some angle, the identical nodal deflections would produce different changes in bar lengths. Matrix direction-cosine terms must be used to keep track of these 3-D effects.

Most finite-element analyses use surface elements rather than simple bar elements. The triangle element shown in Fig. 14.42 is typical, and allows a complicated structure to be broken into numerous connected elements. These elements are assumed to be connected at the nodes (corners) where the deflections are identical.

Equations are prepared in matrix form describing how each element responds to loadings at its nodes. The element stiffness matrices are combined using appropriate direction cosine terms to account for 3-D geometry, and the combined matrix is inverted to solve for the deflections for a given loading.

For dynamic analysis, mass and damping terms are developed using matrix methods. These greatly increase the number of inputs required for the analysis.

Fortunately, working structural engineers do not need to develop their own FEM program every time they wish to analyze a structure. There are numerous "canned" FEM programs available, ranging from simple personal-computer ones to million-line programs.

The industry-standard FEM program is the NASTRAN (NASA Structural Analysis) program, developed years ago for NASA and continuously enhanced both by NASA and various private companies. NASTRAN handles virtually everything, but requires substantial experience to ensure that the results are meaningful. However, for complex structural analysis, some variant of NASTRAN will probably be in use for many years to come.

SR-71 Blackbird (NASA photo by Jim Ross).

World's largest aircraft—Antonov An-225 Mriya (Dream) (U.S. Air Force photo).

15
Weights

15.1 Introduction

The estimation of the weight of a conceptual aircraft is a critical part of the design process. The weights engineer interfaces with all other engineering groups, and serves as the "referee" during the design evolution.

Weights analysis per se does not form part of the aerospace engineering curriculum at most universities. It requires a broad background in aerospace structures, mechanical engineering, statistics, and other engineering disciplines.

There are many levels of weights analysis. Previous chapters have presented crude statistical techniques for estimating the empty weight for a given takeoff weight. These techniques estimate the empty weight directly and are only suitable for "first-pass" analysis.

More sophisticated weights methods estimate the weight of the various components of the aircraft and then sum for the total empty weight. In this chapter, two levels of component weights analysis will be presented.

The first is a crude component buildup based upon planform areas, wetted areas, and percents of gross weight. This technique is useful for initial balance calculations and can be used to check the results of the more detailed statistical methods.

The second uses detailed statistical equations for the various components. This technique is sufficiently detailed to provide a credible estimate of the weights of the major component groups. Those weights are usually reported in groupings as defined by MIL-STD-1374, or some similar groupings defined by company practice. MIL-STD-1374 goes into exhaustive detail (taxi lights, for example!), but at the conceptual level the weights are reported via a "Summary Group Weight Statement." A typical summary format appears as Table 15.1, where the empty weight groups are further classified into three major groupings (structure, propulsion, and equipment). This Group Weight Statement form is available as an Excel™ spreadsheet at the author's website, *www.aircraftdesign.com*.

The structures group consists of the load-carrying components of the aircraft. Note that it includes the inlet (air-induction-system) weight, as well as the nacelle (engine-section) weight including motor mounts and firewall provisions—despite their obvious relationship to the engine. The propulsion group contains only the engine-related equipment such as starters, exhaust, etc. The as-installed engine includes the propeller, if any.

Armament is broken down into fixed items, which are in the equipment groups, and expendable items, which are in the useful load. Sometimes a

Table 15.1 Group weight format

	Weight, lb	Loc., ft	Moment, ft-lb		Weight, lb	Loc., ft	Moment, ft-lb
Structures	**4,526**		**106,879**	**Equipment**	**4,067**		**80,646**
Wing	1,459.4	23.3	34,004	Flight controls	655.7	21.7	14,229
Horizontal tail	280.4	39.2	10,992	APU		0	0
Vertical tail		0	0	Instruments	122.8	10.0	1,228
Ventral tail		0	0	Hydraulics	171.7	21.7	3,726
Fuselage	1,574	21.7	34,156	Pneumatics		21.7	0
Main landing gear	631.5	23.8	15,030	Electrical	713.2	21.7	15,476
Nose landing gear	171.1	13.0	2,224	Avionics	989.8	10.0	9,898
Other landing gear		0	0	Armament		0	0
Engine mounts	39.1	33.0	1,290	Furnishings	217.6	6.2	1,3497
Firewall	58.8	33.0	1,940	Air conditioning	190.7	15.0	2,860.5
Engine section	21	33.0	693	Anti-icing			0
Air induction	291.1	22.5	6,550	Photographic			0
			0	Load and handling	5.3	15.0	79.5
			0	Misc. equipment and We	1,000	31.8	31,800
			0	Empty weight allowance	547	23.6	12,9237
Propulsion	**2,354**		**70,931**	**Total weight empty**	**11,495**	**23.6**	27,1379
Engine(s)—installed	1,517	33.0	50,061				
Accessory drive			0	**Useful load**	**4,985**		
Exhaust system			0	Crew	220	15.0	3,300
Engine cooling	172	33.0	5,676	Fuel—usable	3,836	22.3	85,551
Oil cooling	37.8	33.0	1,247	Fuel—trapped	39	22.3	864
Engine controls	20	33.0	660	Oil	50	33.0	1,650
Starter	39.5	15.7	620	Passengers			0
Fuel system/tanks	568	22.3	12,666	Cargo/payload	840	21.7	18,228
			0	Guns			0
			0	Ammunition	0	21.7	0
			0	Misc. useful load			
			0	**Takeoff gross weight**	**16,480**	**22.0**	362,744

judgment call is required. For example, a gun may be considered to be fixed equipment, or it may be viewed as readily removable and unimportant to flight and therefore a part of the useful load.

The takeoff gross weight—the sum of the empty weight and the useful load—reflects the weight at takeoff for the normal design mission. The flight design gross weight represents the aircraft weight at which the structure will withstand the design load factors. Usually this is the same as the takeoff weight, but some aircraft are designed assuming that maximum loads will not be permitted until the aircraft has taken off, climbed to altitude, and cruised some distance, thus burning off some fuel in the process. For military aircraft it is often assumed that flight design gross weight is takeoff weight but with only 50–60% of fuel remaining.

DCPR stands for "Defense Contractors Planning Report." The DCPR weight is important for cost estimation, and can be viewed as the weight of the parts of the aircraft that the manufacturer makes, as opposed to buys and installs. DCPR weight equals the empty weight less the weights of the wheels, brakes, tires, engines, starters, cooling fluids, fuel bladders, instruments, batteries, electrical power supplies/converters, avionics, armament, fire-control systems, air conditioning, and auxiliary power unit. DCPR weight is also referred to as AMPR weight (Aeronautical Manufacturers Planning Report), and as Airframe Unit Weight.

In a group weight statement, the distance to the weight datum (arbitrary reference point) is included, and the resulting moment is calculated. These are summed and divided by the total weight to determine the actual center-of-gravity (c.g.) location. The c.g. varies during flight as fuel is burned off and weapons expended.

To determine if the c.g. remains within the limits established by an aircraft stability and control analysis, a "c.g.-envelope" plot is prepared (Fig. 15.1).

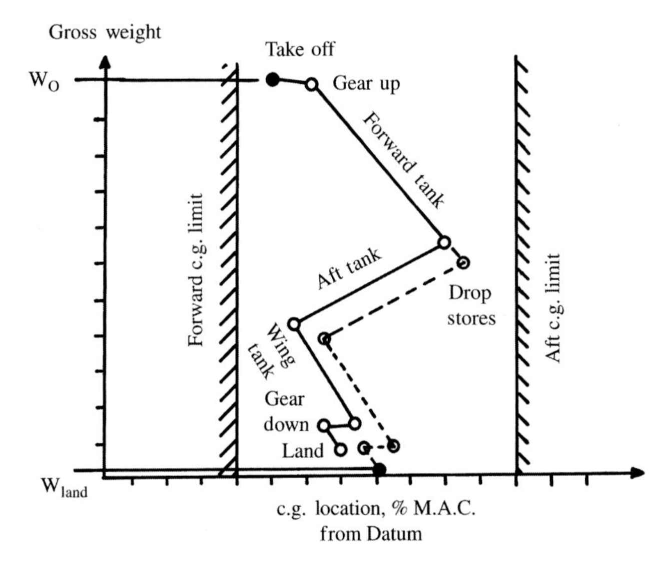

Fig. 15.1 Center-of-gravity envelope diagram.

The c.g. must remain within the specified limits as fuel is burned, and whether or not the weapons are expended. It is permissible to "sequence" the fuel tanks, selecting to burn fuel from different tanks at different times to keep the c.g. within limits. However, an automated fuel-management system must be used, and that imposes additional cost and complexity.

Note that the allowable limits on the c.g. vary with Mach number. At supersonic speeds the aerodynamic center moves rearward, so the forward-c.g. limit may have to move rearward to allow longitudinal trim at supersonic speeds. However, the aft-c.g. limit is often established by the size of the vertical tail, which loses effectiveness at supersonic speeds. This prevents moving the aft limit rearward at supersonic speeds, forcing a very narrow band of allowable limits.

15.2 Approximate Group Weights Method

Early in design it is desirable to do a rough c.g. estimate. Otherwise, substantial rework may be required after the c.g. is properly estimated. A rough c.g. estimate can be done with a crude statistical approach as provided in Table 15.2.

The wing and tail weights are determined from historical values for the weight per unit area of exposed planform area. The fuselage is similarly based upon its wetted area. The landing gear is estimated as a fraction of the takeoff gross weight. The installed engine weight is a multiple of the uninstalled engine weight. Finally, a catch-all weight for the remaining items of the empty weight is estimated as a fraction of the takeoff gross weight.

This technique also applies the approximate locations of the component c.g. as given in Table 15.2. The resulting c.g. estimate can then be compared to the desired c.g. location with respect to the wing aerodynamic center. Also, these approximate component weights can be used as a check of the more detailed statistical equations provided next.

15.3 Statistical Group Weights Method

A more refined estimate of the group weights applies statistical equations based upon sophisticated regression analysis. Development of these equations represents a major effort, and each company develops its own equations.

To acquire a statistical database for these equations, weights engineers must obtain group-weight statements and detailed aircraft drawings for as many current aircraft as possible. This sometimes requires weights engineers to trade group-weight statements much like baseball cards ("I'll trade you a T-45 for an F-16 and a C-5B"!)

The equations presented next typify those used in conceptual design by the major airframe companies, and cover fighter/attack, transport, and general-aviation aircraft. They have been taken from Refs. 62–64 and other sources. Definitions of the terms follow the equations. A critical term, W_{dg}, is the flight design gross weight. For military aircraft this is often less than the maximum takeoff weight. A common assumption is that only 50–60% of the fuel remains.

Table 15.2 Approximate empty weight buildup

Item	Fighters		Transports and bombers		General aviation (metal)		Multiplier	Approximate location
	lb/ft²	{kg/m²}	lb/ft²	{kg/m²}	lb/ft²	{kg/m²}		
Wing	9.0	{44}	10.0	{49}	2.5	{12}	$S_{\text{exposed planform}}$	40% MAC
Horizontal tail	4.0	{20}	5.5	{27}	2.0	{10}	$S_{\text{exposed planform}}$	40% MAC
Vertical tail	5.3	{26}	5.5	{27}	2.0	{10}	$S_{\text{exposed planform}}$	40% MAC
Fuselage	4.8	{23}	5.0	{24}	1.4	{7}	$S_{\text{wetted area}}$	40–50% length
Landing gear[a]	0.033	—	0.043	—	0.057	—	TOGW	—
	Navy: 0.045	—						
Installed engine	1.3	—	1.3	—	1.4	—	Engine weight	—
"All-else empty"	0.17	—	0.17	—	0.10	—	TOGW	40–50% length

[a]15% to nose gear; 85% to main gear; reduce gear weight by 0.014 W_0 if fixed gear.

It should be understood that there are no "right" answers in weights estimation until the first aircraft flies. However, these equations should provide a reasonable estimate of the group weights. Other, similar weights equations may be found in Refs. 10, 11, and 23. It's a good idea to calculate the weight of each component using several different equations and then select an average, reasonable result.

Table 15.3 Miscellaneous weights (approximate)

	Weight	
Component	lb	kg[a]
Missiles		
Harpoon (AGM-84 A)	1200	544
Phoenix (AIM-54 A)	1000	454
Sparrow (AIM-7)	500	227
Sidewinder (AIM-9)	200	91
Pylon and launcher	$0.12\ W_{missile}$	
M61 Gun		
Gun	250	113
940 rds ammunition	550	250
Commercial aircraft passenger (includes carry-on)	190	86
Seats		
Flight deck	60	27
Passenger	32	15
Troop	11	5
Instruments		
Altimeter, airspeed, accelerometer, rate of climb, clock, compass, turn & bank, Mach, tachometer, manifold pressure, etc.	1–2 each	0.5–1
Gyro horizon, directional gyro	4–6 each	2–3
Heads-up display	40	18
Lavatories		
Long-range aircraft	$1.11\ N_{pass}^{1.33}$	$0.5\ N_{pass}^{1.33}$
Short-range aircraft	$0.31\ N_{pass}^{1.33}$	$0.14\ N_{pass}^{1.33}$
Business/executive aircraft	$3.90\ N_{pass}^{1.33}$	$1.76\ N_{pass}^{1.33}$
Arresting gear		
Air Force-type	$0.002\ W_{dg}$	
Navy-type	$0.008\ W_{dg}$	
Catapult gear		
Navy carrier-based	$0.003\ W_{dg}$	
Folding wing		
Navy carrier-based	$0.06\ W_{wing}$	

[a]Mass equivalent of weight.

Needless to say, these equations are complicated, and it takes a lot of time to apply them successfully. Mistakes are easy, the most common being the use of limit load factor, where ultimate load factor N_z should be used. In the first design example of Chapter 21, this author used a pocket calculator for all of these calculations to "prove it could be done," and then made exactly this mistake. The RDS-Student computer program, available with this book, was created in part to help students with these calculations. (RDS is described at *www.aircraftdesign.com.*)

Reference 11 tabulates group weight statements for a number of aircraft. These can also be used to help select a reasonable weight estimate for the components by comparing the component weights as a fraction of the empty weight for a similar aircraft.

Table 15.3 tabulates various miscellaneous weights.

When the component weights are estimated using these or similar methods, they are tabulated in a format similar to that of Table 15.1 and are summed to determine the empty weight. Since the payload and crew weights are known, the fuel weight must be adjusted to yield the as-drawn takeoff weight that is the sum of the empty, payload, crew, and fuel weights. If the empty weight is higher than expected, there may be insufficient fuel to complete the design mission. This must be corrected by resizing and optimizing the aircraft as described in Chapter 19, *not* by simply increasing fuel weight for the as-drawn aircraft (which would invalidate the component weight predictions that were based on the as-drawn takeoff weight).

Fighter/Attack Weights (British Units, results in pounds)

$$W_{\text{wing}} = 0.0103\, K_{\text{dw}} K_{\text{vs}} (W_{\text{dg}} N_z)^{0.5} S_w^{0.622} A^{0.785} (t/c)_{\text{root}}^{-0.4} \times (1+\lambda)^{0.05} (\cos\Lambda)^{-1.0} S_{\text{csw}}^{0.04} \quad (15.1)$$

$$W_{\text{horizontal tail}} = 3.316 \left(1 + \frac{F_w}{B_h}\right)^{-2.0} \left(\frac{W_{\text{dg}} N_z}{1000}\right)^{0.260} S_{\text{ht}}^{0.806} \quad (15.2)$$

$$W_{\text{vertical tail}} = 0.452\, K_{\text{rht}} (1 + H_t/H_v)^{0.5} (W_{\text{dg}} N_z)^{0.488} S_{\text{vt}}^{0.718} M^{0.341} \times L_t^{-1.0} (1 + S_r/S_{\text{vt}})^{0.348} A_{\text{vt}}^{0.223} (1+\lambda)^{0.25} (\cos\Lambda_{\text{vt}})^{-0.323} \quad (15.3)$$

$$W_{\text{fuselage}} = 0.499\, K_{\text{dwf}} W_{\text{dg}}^{0.35} N_z^{0.25} L^{0.5} D^{0.849} W^{0.685} \quad (15.4)$$

$$W_{\text{main landing gear}} = K_{\text{cb}} K_{\text{tpg}} (W_l N_l)^{0.25} L_m^{0.973} \quad (15.5)$$

$$W_{\text{nose landing gear}} = (W_l N_l)^{0.290} L_n^{0.5} N_{\text{nw}}^{0.525} \quad (15.6)$$

$$W_{\text{engine mounts}} = 0.013\, N_{\text{en}}^{0.795} T^{0.579} N_z \quad (15.7)$$

$$W_{\text{firewall}} = 1.13\, S_{\text{fw}} \quad (15.8)$$

$$W_{\text{engine section}} = 0.01 W_{\text{en}}^{0.717} N_{\text{en}} N_z \tag{15.9}$$

$$W_{\text{air induction system}} = 13.29\, K_{\text{vg}} L_d^{0.643} K_d^{0.182}\, N_{\text{en}}^{1.498} (L_s/L_d)^{-0.373} D_e \tag{15.10}$$

where K_d and L_s are from Fig. 15.2.

$$W_{\text{tailpipe}} = 3.5 D_e L_{\text{tp}} N_{\text{en}} \tag{15.11}$$

$$W_{\text{engine cooling}} = 4.55 D_e L_{\text{sh}} N_{\text{en}} \tag{15.12}$$

$$W_{\text{oil cooling}} = 37.82 N_{\text{en}}^{1.023} \tag{15.13}$$

$$W_{\text{engine controls}} = 10.5 N_{\text{en}}^{1.008} L_{\text{ec}}^{0.222} \tag{15.14}$$

$$W_{\text{starter (pneumatic)}} = 0.025 T_e^{0.760} N_{\text{en}}^{0.72} \tag{15.15}$$

$$W_{\text{fuel system and tanks}} = 7.45 V_t^{0.47} \left(1 + \frac{V_i}{V_t}\right)^{-0.095} \times \left(1 + \frac{V_p}{V_t}\right) N_t^{0.066} N_{\text{en}}^{0.052} \left(\frac{T \cdot \text{SFC}}{1000}\right)^{0.249} \tag{15.16}$$

$$W_{\text{flight controls}} = 36.28 M^{0.003} S_{\text{cs}}^{0.489} N_s^{0.484} N_c^{0.127} \tag{15.17}$$

$$W_{\text{instruments}} = 8.0 + 36.37 N_{\text{en}}^{0.676} N_t^{0.237} + 26.4 (1 + N_{\text{ci}})^{1.356} \tag{15.18}$$

$$W_{\text{hydraulics}} = 37.23 K_{\text{vsh}} N_u^{0.664} \tag{15.19}$$

$$W_{\text{electrical}} = 172.2 K_{\text{mc}} R_{\text{kva}}^{0.152} N_c^{0.10} L_a^{0.10} N_{\text{gen}}^{0.091} \tag{15.20}$$

$$W_{\text{avionics}} = 2.117 W_{\text{uav}}^{0.933} \tag{15.21}$$

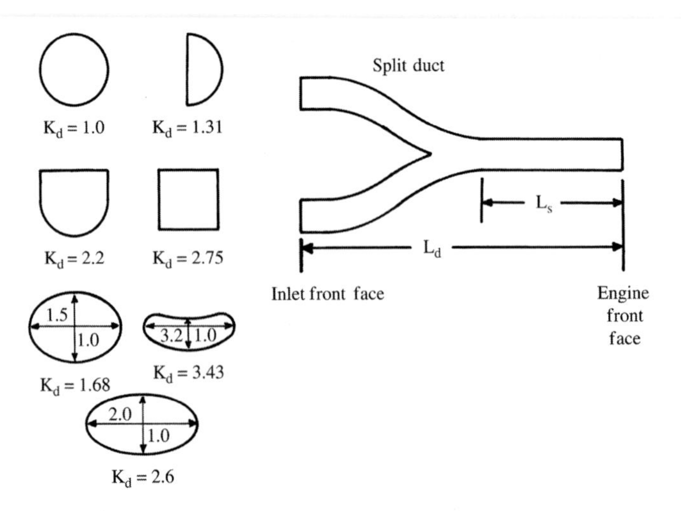

Fig. 15.2 Inlet duct geometry.

$$W_{\text{furnishings}} = 217.6N_c \text{ (includes seats)} \tag{15.22}$$

$$W_{\text{air conditioning and anti-ice}} = 201.6[(W_{\text{uav}} + 200N_c)/1000]^{0.735} \tag{15.23}$$

$$W_{\text{handling gear}} = 3.2 \times 10^{-4}\, W_{\text{dg}} \tag{15.24}$$

Cargo/Transport Weights (British Units, results in pounds)

$$W_{\text{wing}} = 0.0051(W_{\text{dg}}N_z)^{0.557} S_w^{0.649} A^{0.5} (t/c)_{\text{root}}^{-0.4} (1+\lambda)^{0.1} \times (\cos\Lambda)^{-1.0} S_{\text{csw}}^{0.1} \tag{15.25}$$

$$W_{\text{horizontal tail}} = 0.0379K_{\text{uht}}(1 + F_w/B_h)^{-0.25} W_{\text{dg}}^{0.639} N_z^{0.10} S_{\text{ht}}^{0.75} L_t^{-1.0} \times K_y^{0.704} (\cos\Lambda_{\text{ht}})^{-1.0} A_h^{0.166} (1 + S_e/S_{\text{ht}})^{0.1} \tag{15.26}$$

$$W_{\text{vertical tail}} = 0.0026(1 + H_t/H_v)^{0.225} W_{\text{dg}}^{0.556} N_z^{0.536} L_t^{-0.5} S_{\text{vt}}^{0.5} K_z^{0.875} \times (\cos\Lambda_{\text{vt}})^{-1} A_v^{0.35} (t/c)_{\text{root}}^{-0.5} \tag{15.27}$$

$$W_{\text{fuselage}} = 0.3280K_{\text{door}}K_{\text{Lg}}(W_{\text{dg}}N_z)^{0.5} L^{0.25} S_f^{0.302} \times (1 + K_{\text{ws}})^{0.04} (L/D)^{0.10} \tag{15.28}$$

$$W_{\text{main landing gear}} = 0.0106\, K_{\text{mp}} W_l^{0.888} N_l^{0.25} L_m^{0.4} N_{\text{mw}}^{0.321} N_{\text{mss}}^{-0.5} V_{\text{stall}}^{0.1} \tag{15.29}$$

$$W_{\text{nose landing gear}} = 0.032K_{\text{np}} W_l^{0.646}\, N_l^{0.2} L_n^{0.5}\, N_{\text{nw}}^{0.45} \tag{15.30}$$

$$W_{\text{nacelle group}} = 0.6724K_{\text{ng}} N_{\text{Lt}}^{0.10} N_w^{0.294} N_z^{0.119} W_{\text{ec}}^{0.611} \times N_{\text{en}}^{0.984} S_n^{0.224} \text{ (includes air induction)} \tag{15.31}$$

$$W_{\text{engine controls}} = 5.0N_{\text{en}} + 0.80L_{\text{ec}} \tag{15.32}$$

$$W_{\text{starter (pneumatic)}} = 49.19\left(\frac{N_{\text{en}} W_{\text{en}}}{1000}\right)^{0.541} \tag{15.33}$$

$$W_{\text{fuel system}} = 2.405V_t^{0.606} (1 + V_i/V_t)^{-1.0} (1 + V_p/V_t) N_t^{0.5} \tag{15.34}$$

$$W_{\text{flight controls}} = 145.9N_f^{0.554} (1 + N_m/N_f)^{-1.0} \times S_{\text{cs}}^{0.20} (I_y \times 10^{-6})^{0.07} \tag{15.35}$$

$$W_{\text{APU installed}} = 2.2W_{\text{APU uninstalled}} \tag{15.36}$$

$$W_{\text{instruments}} = 4.509K_r K_{\text{tp}} N_c^{0.541} N_{\text{en}} (L_f + B_w)^{0.5} \tag{15.37}$$

$$W_{\text{hydraulics}} = 0.2673N_f (L_f + B_w)^{0.937} \tag{15.38}$$

$$W_{\text{electrical}} = 7.291R_{\text{kva}}^{0.782} L_a^{0.346} N_{\text{gen}}^{0.10} \tag{15.39}$$

$$W_{\text{avionics}} = 1.73W_{\text{uav}}^{0.983} \tag{15.40}$$

$$W_{\text{furnishings}} = 0.0577 N_c^{0.1} W_c^{0.393} S_f^{0.75}$$

(does not include cargo handling gear or seats) (15.41)

$$W_{\text{air conditioning}} = 62.36 N_p^{0.25} (V_{\text{pr}}/1000)^{0.604} W_{\text{uav}}^{0.10} \qquad (15.42)$$

$$W_{\text{anti-ice}} = 0.002 W_{\text{dg}} \qquad (15.43)$$

$$W_{\text{handling gear}} = 3.0 \times 10^{-4} W_{\text{dg}} \qquad (15.44)$$

$$W_{\text{military cargo handling system}} = 2.4 \times (\text{cargo floor area, ft}^2) \qquad (15.45)$$

General-Aviation Weights (British Units, results in pounds)

$$W_{\text{wing}} = 0.036\, S_w^{0.758} W_{\text{fw}}^{0.0035} \left(\frac{A}{\cos^2 \Lambda}\right)^{0.6} q^{0.006} \lambda^{0.04} \left(\frac{100\, t/c}{\cos \Lambda}\right)^{-0.3}$$
$$\times (N_z W_{\text{dg}})^{0.49} \text{ (ignore second term if } W_{\text{fw}} = 0) \qquad (15.46)$$

$$W_{\text{horizontal tail}} = 0.016 (N_z W_{\text{dg}})^{0.414} q^{0.168} S_{\text{ht}}^{0.896} \left(\frac{100\, t/c}{\cos \Lambda}\right)^{-0.12}$$
$$\times \left(\frac{A}{\cos^2 \Lambda_{\text{ht}}}\right)^{0.043} \lambda_h^{-0.02} \qquad (15.47)$$

$$W_{\text{vertical tail}} = 0.073 \left(1 + 0.2 \frac{H_t}{H_v}\right) (N_z W_{\text{dg}})^{0.376}$$
$$\times q^{0.122} S_{\text{vt}}^{0.873} \left(\frac{100\, t/c}{\cos \Lambda_{\text{vt}}}\right)^{-0.49}$$
$$\times \left(\frac{A}{\cos^2 \Lambda_{\text{vt}}}\right)^{0.357} \lambda_{\text{vt}}^{0.039} \qquad (15.48)$$

$$W_{\text{fuselage}} = 0.052\, S_f^{1.086} (N_z W_{\text{dg}})^{0.177} L_t^{-0.051}$$
$$\times (L/D)^{-0.072} q^{0.241} + W_{\text{press}} \qquad (15.49)$$

$$W_{\text{main landing gear}} = 0.095 (N_l W_l)^{0.768} (L_m/12)^{0.409} \qquad (15.50)$$

$$W_{\text{nose landing gear}} = 0.125 (N_l W_l)^{0.566} (L_n/12)^{0.845}$$

(reduce total landing gear weight by 1.4% of TOGW if nonretractable) (15.51)

$$W_{\text{installed engine (total)}} = 2.575 W_{\text{en}}^{0.922} N_{\text{en}}$$
(includes propeller and engine mounts) (15.52)

$$W_{\text{fuel system}} = 2.49 V_t^{0.726} \left(\frac{1}{1 + V_i/V_t}\right)^{0.363} N_t^{0.242} N_{\text{en}}^{0.157} \tag{15.53}$$

$$W_{\text{flight controls}} = 0.053 L^{1.536} B_w^{0.371} (N_z W_{\text{dg}} \times 10^{-4})^{0.80} \tag{15.54}$$

$$W_{\text{hydraulics}} = K_h W^{0.8} M^{0.5} \tag{15.55}$$

$$W_{\text{electrical}} = 12.57 (W_{\text{fuel system}} + W_{\text{avionics}})^{0.51} \tag{15.56}$$

$$W_{\text{avionics}} = 2.117 W_{\text{uav}}^{0.933} \tag{15.57}$$

$$W_{\text{air conditioning and anti-ice}} = 0.265\, W_{\text{dg}}^{0.52} N_p^{0.68} W_{\text{avionics}}^{0.17} M^{0.08} \tag{15.58}$$

$$W_{\text{furnishings}} = 0.0582\, W_{\text{dg}} - 65 \tag{15.59}$$

Weights Equations Terminology

A = aspect ratio
B_h = horizontal tail span, ft
B_w = wing span, ft
D = fuselage structural depth, ft
D_e = engine diameter, ft
F_w = fuselage width at horizontal tail intersection, ft
H_t = horizontal tail height above fuselage, ft
H_t/H_v = 0.0 for conventional tail; 1.0 for "T" tail
H_v = vertical tail height above fuselage, ft
I_y = yawing moment of inertia, lb-ft^2 (see Chap. 16)
K_{cb} = 2.25 for cross-beam (F-111) gear; = 1.0 otherwise
K_d = duct constant (see Fig. 15.2)
K_{door} = 1.0 if no cargo door; = 1.06 if one side cargo door; =1.12 if two side cargo doors; = 1.12 if aft clamshell door; =1.25 if two side cargo doors and aft clamshell door
K_{dw} = 0.768 for delta wing; = 1.0 otherwise
K_{dwf} = 0.774 for delta-wing aircraft; = 1.0 otherwise
K_h = 0.05 for low subsonic with hydraulics for brakes and retracts only; = 0.11 for medium subsonic with hydraulics for flaps; = 0.12 for high subsonic with hydraulic flight controls; = 0.013 for light plane with hydraulic brakes only (and use M = 0.1)
K_{Lg} = 1.12 if fuselage-mounted main landing gear; = 1.0 otherwise
K_{mc} = 1.45 if mission completion required after failure; = 1.0 otherwise
K_{mp} = 1.126 for kneeling gear; = 1.0 otherwise
K_{ng} = 1.017 for pylon-mounted nacelle; = 1.0 otherwise
K_{np} = 1.15 for kneeling gear; = 1.0 otherwise
K_p = 1.4 for engine with propeller or 1.0 otherwise

K_r = 1.133 if reciprocating engine; = 1.0 otherwise
K_{rht} = 1.047 for rolling horizontal tail; = 1.0 otherwise
K_{tp} = 0.793 if turboprop; = 1.0 otherwise
K_{tpg} = 0.826 for tripod (A-7) gear; = 1.0 otherwise
K_{tr} = 1.18 for jet with thrust reverser or 1.0 otherwise
K_{uht} = 1.143 for unit (all-moving) horizontal tail; = 1.0 otherwise
K_{vg} = 1.62 for variable geometry; = 1.0 otherwise
K_{vs} = 1.19 for variable sweep wing; = 1.0 otherwise
K_{vsh} = 1.425 for variable sweep wing; = 1.0 otherwise
K_{ws} = $0.75\ [(1 + 2\lambda)/(1 + \lambda)]\ (B_w \tan\Lambda/L)$
K_y = aircraft pitching radius of gyration, ft ($\cong 0.3L_t$)
K_z = aircraft yawing radius of gyration, ft ($\cong L_t$)
L = fuselage structural length, ft (excludes radome cowling, tail cap)
L_a = electrical routing distance, generators to avionics to cockpit, ft
L_d = duct length, ft
L_{ec} = length from engine front to cockpit—total if multi-engine, ft
L_f = total fuselage length
L_m = extended length of main landing gear, in.
L_n = extended nose gear length, in.
L_s = single duct length (see Fig. 15.2)
L_{sh} = length of engine shroud, ft
L_t = tail length; wing quarter-MAC to tail quarter-MAC, ft
L_{tp} = length of tailpipe, ft
M = Mach number
N_c = number of crew (use 0.5 for UAV)
N_{ci} = number of crew equivalents: 1.0 if single pilot;
= 1.2 if pilot plus backseater; = 2.0 pilot and copilot
N_{en} = number of engines
N_f = number of functions performed by controls (typically 4–7)
N_{gen} = number of generators (typically = N_{en})
N_{Lt} = nacelle length, ft
N_l = ultimate landing load factor; = $N_{gear} \times 1.5$
N_m = number of mechanical functions (typically 0–2)
N_{mss} = number of main gear shock struts
N_{mw} = number of main wheels
N_{nw} = number of nosewheels
N_p = number of personnel onboard (crew and passengers)
N_s = number of flight control systems
N_t = number of fuel tanks
N_u = number of hydraulic utility functions (typically 5–15)
N_w = nacelle width, ft
N_z = ultimate load factor; = 1.5 × limit load factor
q = dynamic pressure at cruise, lb/ft^2
R_{kva} = system electrical rating, kV · A (typically 40–60 for transports, 110–160 for fighters and bombers)
S_{cs} = total area of control surfaces, ft^2
S_{csw} = control surface area (wing-mounted), ft^2
S_e = elevator area, ft
S_f = fuselage wetted area, ft^2

S_{fw} = firewall surface area, ft^2
S_{ht} = horizontal tail area
S_n = nacelle wetted area, ft^2
S_r = rudder area, ft^2
S_{vt} = vertical tail area, ft^2
S_w = trapezoidal wing area, ft^2
SFC = engine specific fuel consumption—maximum thrust
T = total engine thrust, lb
T_e = thrust per engine, lb
V_i = integral tanks volume, gal
V_p = self-sealing "protected" tanks volume, gal
V_{pr} = volume of pressurized section, ft^3
V_t = total fuel volume, gal
W = total fuselage structural width, ft
W_c = maximum cargo weight, lb
W_{dg} = flight design gross weight, lb (typically 50–60% of internal fuel for military aircraft)
W_{ec} = weight of engine and contents, lb (per nacelle), $\cong 2.331 W_{engine}^{0.901} K_p K_{tr}$
W_{en} = engine weight, each, lb
W_{fw} = weight of fuel in wing, lb (if zero, ignore this term)
W_l = landing design gross weight, lb
W_{press} = weight penalty due to pressurization, $= 11.9(V_{pr} P_{delta})^{0.271}$, where P_{delta} = cabin pressure differential, psi (typically 8 psi)
W_{uav} = uninstalled avionics weight, lb (typically = 800–1400 lb)
Λ = wing sweep at 25% MAC
λ = taper ratio (wing or tail)

15.4 Additional Considerations in Weights Estimation

These statistical equations are based upon a database of existing aircraft. They work well for a "normal" aircraft similar to the various aircraft in the database. However, use of a novel configuration (canard pusher) or an advanced technology (composite structure) will result in a poor weights estimate when using these or similar equations. To allow for this, weights engineers adjust the statistical-equation results using "fudge factors" (defined as the variable constant that you multiply your answer by to get the right answer!)

Fudge factors are also required to estimate the weight of a class of aircraft for which no statistical equations are available. For example, there have been too few Mach 3 aircraft to develop a good statistical database. Weights for a new Mach 3 design can be estimated by selecting the closest available equations (probably the fighter/attack equations) and determining a fudge factor for each type of component.

This is done using data for an existing aircraft similar to the new one (such as the XB-70 for a Mach 3 design) and calculating its component weights using the selected statistical equations. Fudge factors are then determined by dividing the actual component weights for that aircraft by the calculated component weights.

To estimate the component weights for the new design, these fudge factors are multiplied by the component weights as calculated using the selected statistical equations.

Table 15.4 Weights estimation "fudge factors"

Category	Weight group	Fudge factor (multiplier)
Advanced composites	Wing	0.85–0.90
	Tails	0.83–0.88
	Fuselage/nacelle	0.90–0.95
	Landing gear	0.95–1.0
	Air induction system	0.85–0.90
Braced wing	Wing	0.82
Braced biplane	Wing	0.6
Wood fuselage	Fuselage	1.60
Steel tube fuselage	Fuselage	1.80
Flying boat hull	Fuselage	1.25
Carrier-based aircraft	Fuselage and landing gear	1.2–1.3

Fudge factors for composite-structure, wood or steel-tube fuselages, braced wings, and flying-boat hulls are provided in Table 15.4. These should be viewed as rough approximations only and subject to heated debate. For example, there are those who claim that a properly designed steel-tube fuselage can be lighter than an aluminum fuselage.

One final consideration in aircraft-weights estimation is the weight growth that most aircraft experience in the first few years of production. This growth in empty weight is due to several factors, such as increased avionics capabilities, structural fixes (such as replacing an aluminum fitting with steel to prevent cracking), and additional weapons pylons.

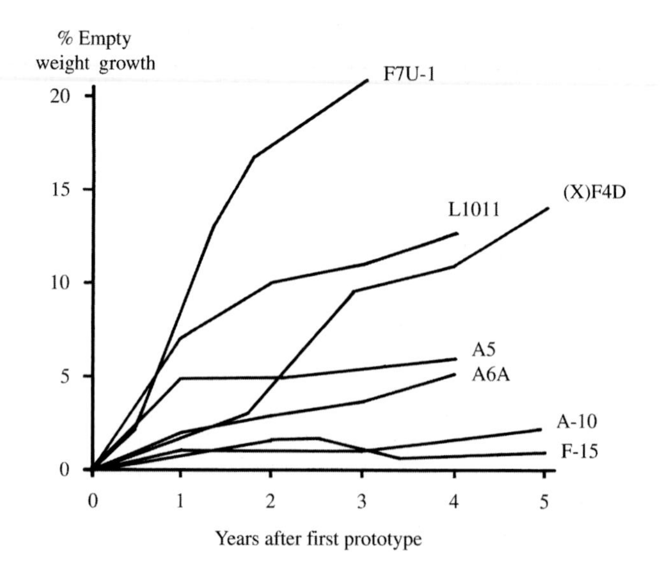

Fig. 15.3 Aircraft weight growth.

Figure 15.3 shows the empty-weight growth of a number of aircraft. In the past, a weight growth of 5% in the first year was common. Today's better design techniques and analytical methods have reduced that to less than 2% in the first year. Still, some specific allowance for weight growth should be made in the conceptual-design weight estimation.

Due to this weight growth and the inaccuracies inherent in any statistical methods, it is wise to increase the calculated empty weight by some arbitrary amount during conceptual and preliminary design. An empty weight margin of 3–5% is reasonable. The U.S. Navy often uses a margin of 5–10% or more.

X-31 Enhanced Fighter Maneuverability Aircraft in flight (NASA photo).

CV-22 Osprey fires counter measures (U.S. Air Force photo).

16
Stability, Control, and Handling Qualities

16.1 Introduction

During early conceptual design, the requirements for good stability, control, and handling qualities are addressed through the use of tail volume coefficients and through location of the aircraft center of gravity (c.g.) at some percent of the wing mean aerodynamic chord (MAC), as discussed in Chapter 6. In larger aircraft companies, the aircraft is then analyzed by the controls experts, probably using a six-degrees-of-freedom (DOF) aircraft-dynamics computer program to determine the required c.g. location and the sizes of the tails and control surfaces.

An understanding of the important stability and control design parameters can be attained through study of simpler methods, which are also suitable for use by homebuilders and designers at smaller companies.

This chapter introduces the key concepts and equations for stability, control, and handling-qualities evaluation. These are based upon classical controls methods, many of which were developed by NACA in the period from 1925–1945. For derivations and additional detail on these methods, see Refs. 7, 37, 65, 66, and especially 67 and 4.

The basic concept of stability is simply that a stable aircraft, when disturbed, tends to return by itself to its original state (pitch, yaw, roll, velocity, etc.). "Static stability" is present if the forces created by the disturbed state (such as a pitching moment due to an increased angle of attack) push in the correct direction to return the aircraft to its original state.

If these restoring forces are too strong, the aircraft will overshoot the original state and will oscillate with greater and greater amplitude until it goes completely out of control. Although static stability is present, the aircraft does not have "dynamic stability."

Dynamic stability is present if the dynamic motions of the aircraft will eventually return the aircraft to its original state. The manner in which the aircraft returns to its original state depends upon the restoring forces, mass distribution, and damping forces. Damping forces slow the restoring rates. For example, a pendulum swinging in air is lightly damped and will oscillate back and forth for many minutes. The same pendulum immersed in water is highly damped and will slowly return to vertical with little or no oscillation.

Figure 16.1 illustrates these concepts for an aircraft disturbed in pitch. In Fig. 16.1a, the aircraft has perfectly neutral stability and simply remains at whatever pitch angle the disturbance produces. While some aerobatic aircraft are

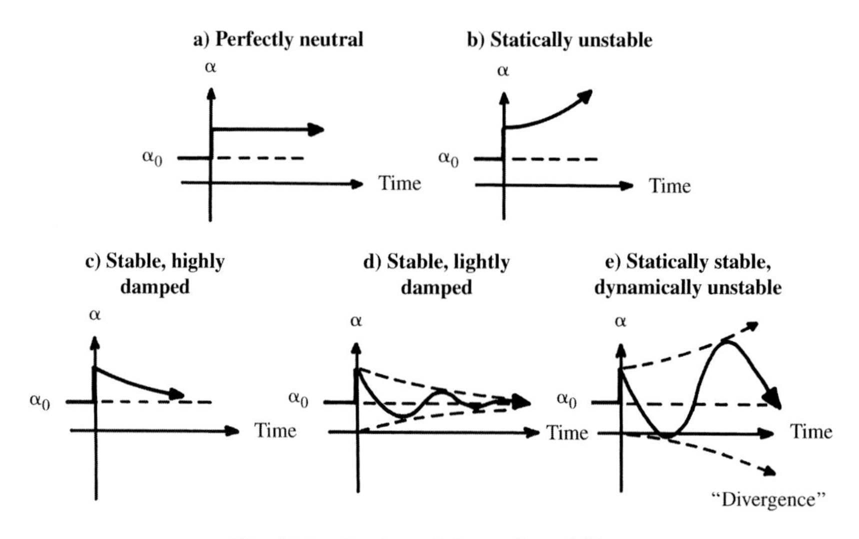

Fig. 16.1 Static and dynamic stability.

nearly neutral in stability, few pilots would care to fly such an aircraft on a long trip in gusty conditions.

Illustration Fig. 16.1b shows static instability. The forces produced by the greater pitch angle actually cause the pitch angle to further increase. Pitchup is an example of this.

In Fig. 16.1c, the aircraft shows static stability with very high damping. The aircraft slowly returns to the original pitch angle without any overshoot.

Illustration Fig. 16.1d shows a more typical aircraft response; the aircraft returns to its original state, but experiences some converging oscillation. This is acceptable behavior provided the time to converge is short.

In Fig. 16.1e, the restoring forces are in the right direction so the aircraft is statically stable. However, the restoring forces are high and the damping forces are relatively low, so the aircraft overshoots the original pitch angle by a negative amount greater than the pitch angle produced by the disturbance. Restoring forces then push the nose back up, overshooting by an even greater amount. The pitch oscillations continue to increase in amplitude until the aircraft "diverges" into an uncontrolled flight mode such as a spin.

Note that instability is not always unacceptable provided that it occurs slowly. Most aircraft have at least one unstable mode, the spiral divergence. This divergence mode is so slow that the pilot has plenty of time to make the minor roll correction required to prevent it. In fact, pilots are generally unaware of the existence of the spiral-divergence mode because the minor corrections required are no greater than the roll corrections required for gusts.

Dynamic-stability analysis is complex and requires computer programs for any degree of accuracy. Most of the stability-analysis methods presented in this chapter evaluate static stability. For conventional aircraft configurations, satisfaction of static-stability requirements will probably give acceptable dynamic stability in most flight modes. Rule-of-thumb methods are presented for stall departure and spin recovery, the dynamic-stability areas of greatest concern.

16.2 Coordinate Systems and Definitions

Figure 16.2 defines the two axis systems commonly used in aircraft analysis. The body-axis system is rigidly fixed to the aircraft, with the X axis aligned with the fuselage and the Z axis upward. The origin is at an arbitrary location, usually the nose. The body-axis system is more "natural" for most people, but suffers from the variation of the direction of lift and drag with angle of attack. (Remember that lift, by definition, is perpendicular to the wind direction.)

The wind axis system solves this problem by orienting the X axis into the relative wind regardless of the aircraft's angle of attack α or sideslip β. The aircraft is not fixed to the axis system, so the axis projections of the various lengths (such as the distance from the wing MAC to the tail) will vary for different angles of attack or sideslip. This variation in moment arms is usually ignored in stability analysis since the angles are typically small.

The stability axis system, commonly used in stability and control analysis, is a compromise between these two. The X axis is aligned at the aircraft angle of attack, as in the wind axis system, but is not offset to the yaw angle. Directions of X, Y, and Z are as in the wind axis system.

Note that the rolling moment is called L. This is easily confused with lift. Also, the yawing moment is called N, which is the same letter used for the normal-force coefficient. (The aerodynamics crowd must have used up all the good letters by the time the stability folks developed their equations!)

Wing and tail incidence angles are denoted by i, which is relative to the body-fixed reference axis. The aircraft angle of attack α is also with respect to this reference axis, so the wing angle of attack is the aircraft angle of attack plus the wing angle of incidence.

Tail angle of attack is the aircraft angle of attack plus the tail angle of incidence, minus the downwash angle (ϵ) which is discussed later. In this

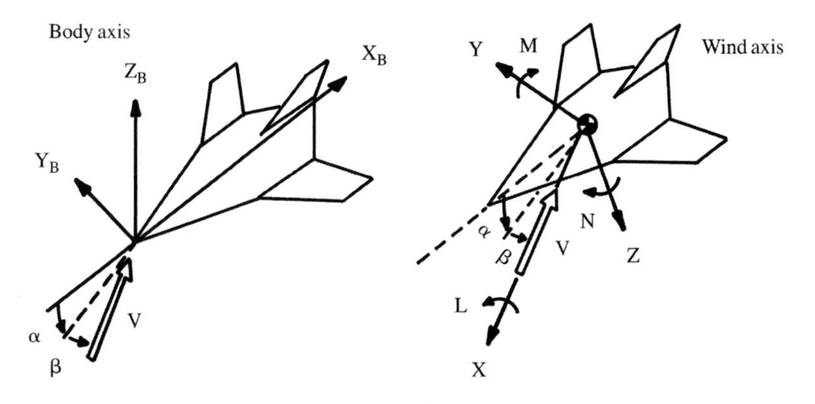

Fig. 16.2 Aircraft coordinate system.

chapter, angles of attack are measured from the zero-lift angle, which was discussed in Chapter 12.

Nondimensional coefficients for lift and drag have been previously defined by dividing by dynamic pressure and wing area. For stability calculations, the moments about the three axes (M, N, and L) must also be expressed as nondimensional coefficients.

Since the moments include a length (the moment arm), they must be divided by a quantity with dimension of length as well as by the dynamic pressure and wing area. This length quantity is the wing MAC chord for pitching moment and the wing span for yawing and rolling moments, as shown in Eqs. (16.1–16.3). Positive moment is nose up or to the right.

$$c_m = M/qS\bar{c} \tag{16.1}$$

$$c_n = N/qSb \tag{16.2}$$

$$c_\ell = L/qSb \tag{16.3}$$

Stability analysis is largely concerned with the response to changes in angular orientation, so the derivatives of these coefficients with respect to angle of attack and sideslip are critical. Subscripts are used to indicate the derivative. For example, $C_{n\beta}$ is the yawing moment derivative with respect to sideslip, a very important parameter in lateral stability.

Similarly, subscripts are used to indicate the response to control deflections, indicated by δ. Thus, $C_{m\delta_e}$ indicates the pitching-moment response to an elevator deflection.

Unless otherwise indicated, all sweep angles in this chapter are quarter-chord sweeps, and all chord lengths c are the wing MAC. Also, all angles are in radians unless otherwise mentioned. Angle terms that are not estimated in radians must be converted to radians before use in stability equations.

16.3 Longitudinal Static Stability and Control

Pitching-Moment Equation and Trim

Most aircraft being symmetrical about the centerline, moderate changes in angle of attack will have little or no influence upon the yaw or roll. This permits the stability and control analysis to be divided into longitudinal (pitch only) and lateral-directional (roll and yaw) analysis.

Figure 16.3 shows the major contributors to aircraft pitching moment about the c.g., including the wing, tail, fuselage, and engine contributions. The wing pitching-moment contribution includes the lift through the wing aerodynamic center and the wing moment about the aerodynamic center. Remember that the aerodynamic center is defined as the point about which pitching moment is constant with respect to angle of attack. This constant moment about the aerodynamic center is zero only if the wing is uncambered and untwisted. Also, the aerodynamic center is typically at 25% of the MAC in subsonic flight.

Another wing moment term is the change in pitching moment due to flap deflection. Flap deflection also influences the wing lift, adding to that term. Flap deflection has a large effect upon downwash at the tail, as discussed later.

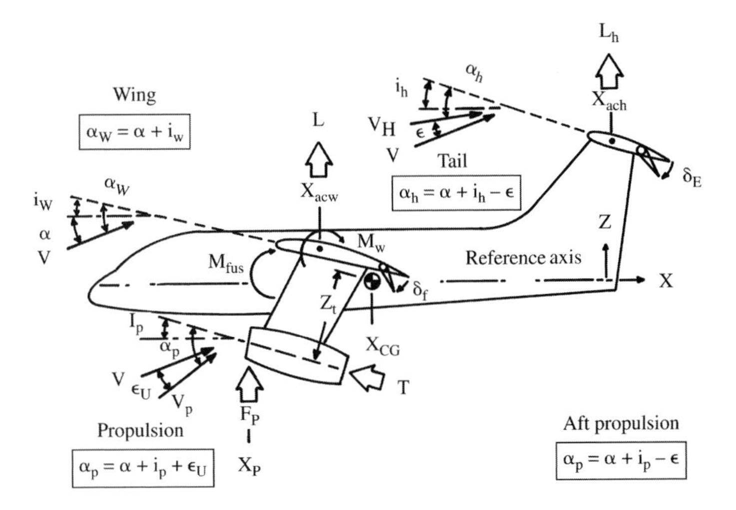

Fig. 16.3 Longitudinal moments.

Drag of the wing and tail produces some pitching moment, but these values are negligibly small. Also, the pitching moment of the tail about its aerodynamic center is small and can be ignored.

On the other hand, the long moment arm of the tail times its lift produces a very large moment that is used to trim and control the aircraft. While this figure shows tail lift upward, under many conditions the tail lift will be downward to counteract the wing pitching moment.

A canard aircraft has a "negative" tail moment-arm that should be applied in the equations that follow. If an aircraft is tailless, the wing flap must be used for trim and control. Because of the short moment arm of such a control, the trim drags will be substantially higher for off-design c.g. locations.

The fuselage and nacelles produce pitching moments that are difficult to estimate without wind-tunnel data. These moments are influenced by the upwash and downwash produced by the wing.

The engine produces three contributions to pitching moment. The obvious term is the thrust times its vertical distance from the c.g. Less obvious is the vertical force F_p produced at the propeller disk or inlet front face due to the turning of the freestream airflow. Also, the propwash or jet-induced flowfield will influence the effective angle of attack of the tail and possibly the wing.

Equation (16.4) expresses the sum of these moments about the c.g. The effect of elevator deflection is included in the tail lift term. Equation (16.5) expresses the moments in coefficient form by dividing all terms by $(qS_w c)$ and expressing the tail lift in coefficient form. Note that, to facilitate understanding, these equations are defined in the body-axis coordinate system rather than the stability-axis system.

$$\begin{aligned} M_{cg} &= L(X_{cg} - X_{acw}) + M_w + M_{w\delta f}\delta_f + M_{fus} \\ &\quad - L_h(X_{ach} - X_{cg}) - Tz_t + F_p(X_{cg} - X_p) \end{aligned} \tag{16.4}$$

This equation in coefficient form has a term representing the ratio between the dynamic pressure at the tail and the freestream dynamic pressure, which is defined in Eq. (16.6) as η_h. This ranges from about 0.85–0.95, with 0.90 as the typical value.

To simplify the equations, all lengths can be expressed as a fraction of the wing mean chord c. These fractional lengths are denoted by a bar. Thus, $\overline{X}_{cg}$ represents X_{cg}/c. This leads to Eq. (16.7).

$$C_{m_{cg}} = C_L\left(\frac{X_{cg} - X_{acw}}{c}\right) + C_{m_w} + C_{m_{w\delta f}}\delta_f + C_{m_{fus}} - \frac{q_h S_h}{q S_w} C_{L_h}\left(\frac{X_{ach} - X_{cg}}{c}\right) - \frac{T z_t}{q S_w c} + \frac{F_p (X_{cg} - X_p)}{q S_w c} \tag{16.5}$$

$$\eta_h = q_h/q \tag{16.6}$$

$$C_{m_{cg}} = C_L(\overline{X}_{cg} - \overline{X}_{acw}) + C_{m_w} + C_{m_{w\delta f}}\delta_f + C_{m_{fus}} - \eta_h \frac{S_h}{S_w} C_{L_h}(\overline{X}_{ach} - \overline{X}_{cg}) - \frac{T}{q S_w}\overline{Z}_t + \frac{F_p}{q S_w}(\overline{X}_{cg} - \overline{X}_p) \tag{16.7}$$

For a static "trim" condition, the total pitching moment must equal zero. For static trim, the main flight conditions of concern are during the takeoff and landing with flaps and landing gear down and during flight at high transonic speeds. Trim for the high-g pullup is actually a dynamic problem (discussed later). Usually the most forward c.g. position is critical for trim. Aft-c.g. position is most critical for stability, as discussed next.

Equation (16.7) can be set to zero and solved for trim by varying some parameter, typically tail area, tail lift coefficient (i.e., tail incidence or elevator deflection), or sometimes c.g. position. The wing drag and tail trim drag can then be evaluated. Methods for the first-order evaluation of the terms of Eq. (16.7) are presented later.

Static Pitch Stability

For static stability to be present, any change in angle of attack must generate moments that oppose the change. In other words, the derivative of pitching moment with respect to angle of attack [Eq. (16.8)] must be negative. Note that the wing pitching moment and thrust terms have dropped out, as they are essentially constant with respect to angle of attack.

Because of downwash effects, the tail angle of attack does not vary directly with aircraft angle of attack. A derivative term accounts for the effects of wing and propeller downwash, as described later. A similar derivative is provided for the propeller or inlet normal-force term F_p.

$$C_{m\alpha} = C_{L\alpha}(\overline{X}_{\text{cg}} - \overline{X}_{\text{acw}}) + C_{m_{\alpha\text{fus}}} - \eta_h \frac{S_h}{S_w} C_{L\alpha_h} \frac{\partial \alpha_h}{\partial \alpha}(\overline{X}_{\text{ach}} - \overline{X}_{\text{cg}}) + \frac{F_{p\alpha}}{qS_w}\frac{\partial \alpha_p}{\partial \alpha}(\overline{X}_{\text{cg}} - \overline{X}_p) \tag{16.8}$$

Equation (16.8) seems to offer no mechanism for stabilizing a tailless aircraft ("flying wing"). In fact, the tailless aircraft must be stabilized in the first term by providing that the wing aerodynamic center is behind the c.g., making the first term negative.

The magnitude of the pitching-moment derivative [Eq. (16.8)] changes with c.g. location. For any aircraft there is a c.g. location that provides no change in pitching moment as angle of attack is varied. This airplane aerodynamic center, or neutral point X_{np}, represents neutral stability (Fig. 16.1a) and is the most-aft c.g. location before the aircraft becomes unstable.

Equation (16.9) solves Eq. (16.8) for the neutral point ($C_{m\alpha} = 0$). Equation (16.10) then expresses the pitching moment derivative in terms of the distance in percent MAC from the neutral point to the c.g. This percentage distance, called the "static margin," is the term in parentheses in Eq. (16.10).

$$\overline{X}_{\text{np}} = \frac{C_{L\alpha}\overline{X}_{\text{acw}} - C_{m_{\alpha\text{fus}}} + \eta_h \frac{S_h}{S_w} C_{L\alpha_h} \frac{\partial \alpha_h}{\partial \alpha}\overline{X}_{\text{ach}} + \frac{F_{p\alpha}}{qS_w}\frac{\partial \alpha_p}{\partial \alpha}\overline{X}_p}{C_{L\alpha} + \eta_h \frac{S_h}{S_w} C_{L\alpha_h} \frac{\partial \alpha_h}{\partial \alpha} + \frac{F_{p\alpha}}{qS_w}} \tag{16.9}$$

$$C_{m\alpha} = -C_{L\alpha}(\overline{X}_{\text{np}} - \overline{X}_{\text{cg}}) \tag{16.10}$$

$$\text{Static Margin (SM)} = (\overline{X}_{\text{np}} - \overline{X}_{\text{cg}}) = -\frac{C_{M\alpha}}{C_{L\alpha}} \tag{16.11}$$

The static margin is the most important term in the longitudinal stability of an aircraft, and a "target" static margin is both a requirement and a key design tool for aircraft designers. Note that static margin can also be calculated as the ratio between pitching-moment derivative and lift coefficient derivative.

If the c.g. is ahead of the neutral point (positive static margin), the pitching-moment derivative is negative, and so the aircraft is stable (this is yet another confusing terminology!) At the most-aft c.g. position, a typical transport aircraft has a positive static margin of 5–10%. General-aviation designs are even more stable—the Cessna 172 has a static margin of about 19%.

Earlier fighters typically had positive static margins of about 5%, but newer fighters such as the F-16 and F-22 are being designed with "relaxed static stability (RSS)" in which a negative static margin (zero to −15%) is coupled with a computerized flight control system that deflects the elevator to provide artificial stability. This reduces trim drag substantially.

It is common to neglect the inlet or propeller force term F_p in Eq. (16.9) to determine "power-off" stability. This removes any strong dependence of X_{np} on velocity (q) in the subsonic flight regime. Power effects are then accounted for

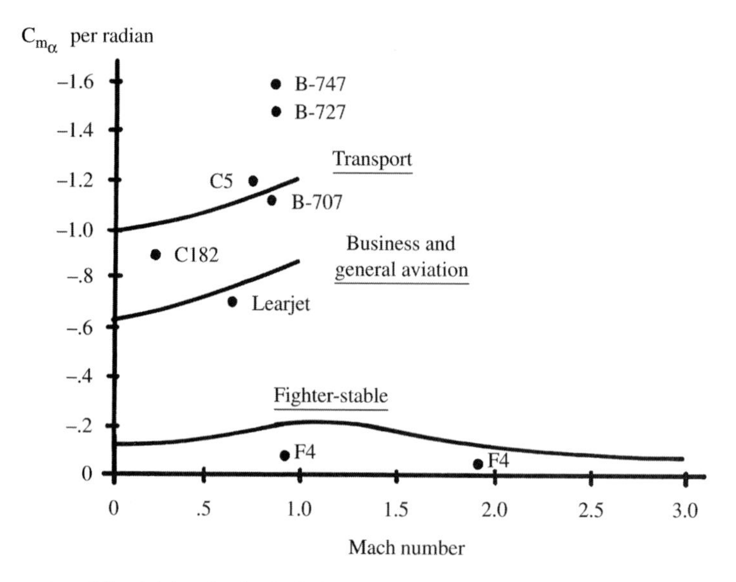

Fig. 16.4 Typical pitching-moment derivative values.

using a static margin allowance based upon test data for a similar aircraft. Typically these allowances for power-on will reduce the static margin by about 1–3% for jets. For propeller-powered aircraft, every one mean-aerodynamic-chord length that the propeller is ahead of the center of gravity will reduce the stability by about 2%.

Figure 16.4 illustrates pitching-moment-derivative values for several classes of aircraft. These may be used as targets for conceptual design. Dynamic analysis during later stages of design may revise these targets.

The evaluation of the terms in Eqs. (16.7–16.9) is difficult without wind-tunnel data. Various semi-empirical methods are presented next, primarily based upon Refs. 4, 37, 67, and 68. Note that these methods are considered crude by the stability and control community, and are only suitable for conceptual design estimates and for student design projects.

Aerodynamic Center

A critical term in Eq. (16.7) is X_{acw}, the location of the wing aerodynamic center. For a high-aspect-ratio wing, the subsonic aerodynamic center will be located at the percent MAC of the airfoil aerodynamic center. For most airfoils this is the quarter-chord point (plus or minus 1%). At supersonic speeds the wing aerodynamic center typically moves to about 45% MAC.

Figures 16.5a–16.5c provide graphical methods for aerodynamic center estimation (Refs. 67 and 37). Note that poor results are obtained at transonic speeds. These methods are also used for estimating the tail aerodynamic center.

A quick approximation of the shift of aerodynamic center with increasing Mach number is given in Eq. (16.12). For a better estimate at supersonic

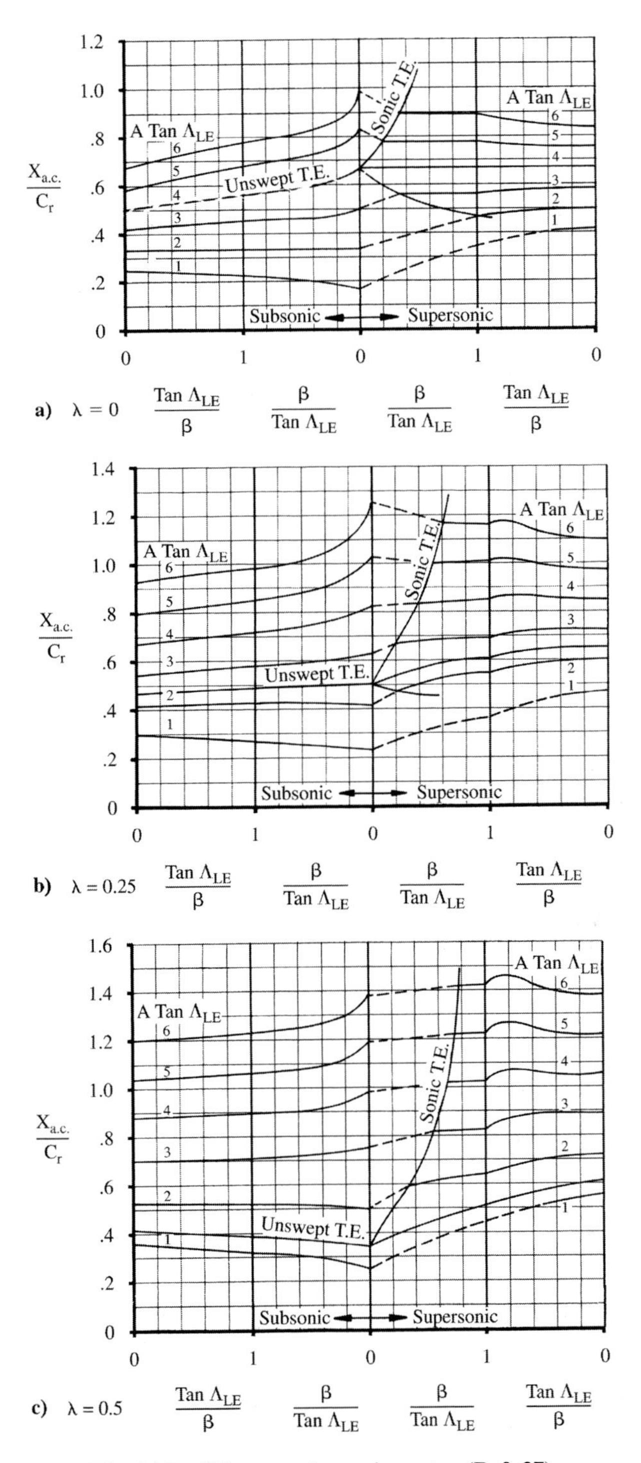

Fig. 16.5 Wing aerodynamic center (Ref. 37).

speeds, aerodynamic panel programs such as the classic Woodward Panel Program can now be run even on personal computers.

$$x_{\text{ac}} = x_{c/4} + \Delta x_{\text{ac}} \sqrt{S_{\text{wing}}} \tag{16.12}$$

where

$$\Delta x_{\text{ac}} = 0.26\,(M - 0.4)^{2.5} \quad (0.4 < M < 1.1)$$
$$\Delta x_{\text{ac}} = 0.112 - 0.004M \quad (M > 1.1)$$

Wing and Tail Lift, Flaps, and Elevators

The lift-curve slopes of the wing and tail are obtained with the methods presented in Chapter 12. The tail lift-curve slope should be reduced about 20% if the elevator gap is not sealed.

The lift coefficients for the wing and tail are simply the lift-curve slopes times the wing or tail angle of attack (measured with respect to the zero-lift angle). These are defined in Eqs. (16.13) and (16.14) based upon the angle-of-attack definitions from Fig. 16.3. Note that for cambered airfoils, the zero-lift angle is a negative value. Also, the tail angle of attack must account for the downwash effect ϵ, which will be estimated later [Eq. (16.24)].

$$\text{Wing:} \quad C_L = C_{L\alpha}(\alpha + i_w - \alpha_{\text{OL}}) \tag{16.13}$$

$$\text{Aft tail:} \quad C_{L_h} = C_{L_{\alpha h}}(\alpha + i_h - \epsilon - \alpha_{\text{OL}_h}) \tag{16.14}$$

where α_{OL} is the angle of attack for zero lift, which is a negative value for a wing or tail with positive camber and/or downward flap/elevation deflection.

The elevator acts as a flap to increase the tail lift. Flap deflection at moderate angles of attack does not change the lift-curve slope, so the lift increment due to flaps can be accounted for by a reduction in the zero-lift angle (i.e., more negative). This reduction in zero-lift angle is equal to the increase in lift coefficient due to flap deflection divided by the lift-curve slope:

$$\Delta\alpha_{\text{OL}} = -\frac{\Delta C_L}{C_{L\alpha}} \tag{16.15}$$

For the complicated high-lift devices seen on most transport wings, the increase in lift coefficient can be approximated using the methods in Chapter 12 or from Fig. 5.3. The change in zero-lift angle can then be determined from Eq. (16.15) and applied to Eq. (16.13).

Plain flaps are used for a modest increase in wing lift and as the control surfaces (elevator, aileron, and rudder) for most aircraft. The change in zero-lift angle due to a plain flap is expressed in Eq. (16.16), where the lift increment with flap deflection is expressed in Eq. (16.17). The 0.9 factor is an approximate

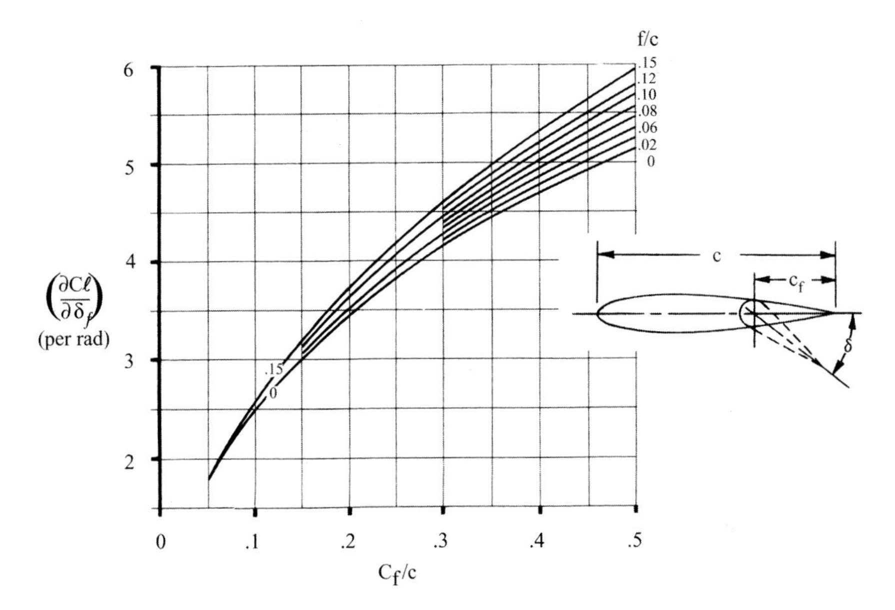

Fig. 16.6 Theoretical lift increment for plain flaps (Ref. 37).

adjustment for flap tip losses.

$$\Delta\alpha_{\text{OL}} = \left(-\frac{1}{C_{L_\alpha}}\frac{\partial C_L}{\partial \delta_f}\right)\delta_f \tag{16.16}$$

where

$$\frac{\partial C_L}{\partial \delta_f} = 0.9K_f\left(\frac{\partial C_\ell}{\partial \delta_f}\right)_{\text{airfoil}}\frac{S_{\text{flapped}}}{S_{\text{ref}}}\cos\Lambda_{\text{H.L.}} \tag{16.17}$$

Figures 16.6 and 16.7 provide the theoretical airfoil lift increment for flaps at small deflections and an empirical adjustment for larger deflections. A typical flap used for control will have a maximum deflection of about 30 deg. Flap deflection must be converted to radians for use in Eq. (16.16). (Note that L is lift in these equations.)

This empirically corrected theoretical method from Ref. 37 sometimes overpredicts the flap or control surface effectiveness, implying that a flap deflection gives even more lift than an equal-incidence deflection of the entire wing or tail. This should not normally occur, and can be avoided by ensuring that the product of the first two terms in Eq. (16.16) is less than 1. Equation (16.18) is a purely empirical expression of data from Ref. 70 that provides a reasonable upper limit on the surface effectiveness terms and can be applied to Eq. (16.15).

$$\begin{aligned}\frac{\Delta\alpha_{\text{OL}}}{\delta_e} &= \frac{1}{C_{L_\alpha}}\frac{\partial C_L}{\partial \delta_f} \\ &= 1.576\,(C_f/C)^3 - 3.458\,(C_f/C)^2 + 2.882\,(C_f/C)\end{aligned} \tag{16.18}$$

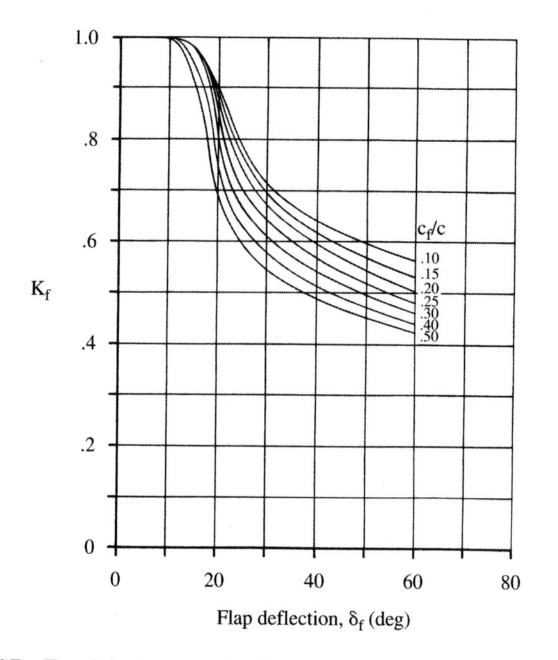

Fig. 16.7 Empirical correction for plain lift increment (Ref. 37).

Figure 16.8 defines the geometry for these equations. H.L. refers to the flap hinge-line sweep, S_{flapped} refers to the portion of the wing or tail area with the flap or control surface. The MAC of the flapped portion of the wing or tail (c') is determined geometrically by considering the flapped portion as a separate surface.

If a flap, elevator, rudder, or aileron has an unsealed hinge gap, the effectiveness will be reduced because of the air leaking through the opening. This reduction will be approximately 15% of the lift increment due to flap deflection.

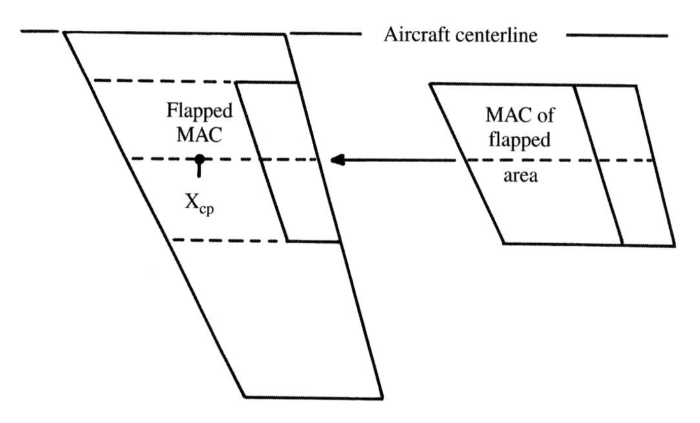

Fig. 16.8 Flapped area and flapped MAC (c′).

These flap lift approximations are reasonable at low Mach numbers. At higher speeds flap lift tends to follow the trends of Fig. 12.5, so as a rough approximation one can adjust flap lift by $C_{L\alpha}$ at the given Mach, divided by $C_{L\alpha}$ at Mach 0.

Wing Pitching Moment

The wing pitching moment about the aerodynamic center is largely determined by the airfoil pitching moment. Equation (16.19) provides an adjustment for wing aspect ratio and sweep for a straight wing or an untwisted swept wing at low subsonic speeds. The wing twist adds an increment of approximately (−0.01) times the twist (in degrees) for a typical swept wing. A more detailed estimation of the wing twist effect is available in Ref. 37. Transonic effects increase the magnitude of the wing pitching moment by about 30% at Mach 0.8.

$$C_{m_w} = C_{m_{0_{\text{airfoil}}}} \left(\frac{A \cos^2 \Lambda}{A + 2 \cos \Lambda} \right) \tag{16.19}$$

The pitching-moment increment due to flap deflection is approximated as the lift increment due to the flap times the moment arm from the center of pressure of the flap lift increment to the c.g. [Eq. (16.20)]. The center of pressure of the flap lift increment (X_{cp}) is determined as a percent of the flapped MAC (c') using Fig. 16.9.

$$C_{m_{w\delta f}} = -\frac{\partial C_L}{\partial \delta_f} (\overline{X}_{\text{cp}} - \overline{X}_{\text{cg}}) \tag{16.20}$$

For a highly swept wing the center of pressure of the flap lift increment can be ahead of the c.g., creating a positive moment increment. This reduces the download required by the tail. Conversely, a canard configuration will put the center of pressure of the flap lift increment well behind the c.g., requiring a huge balancing force.

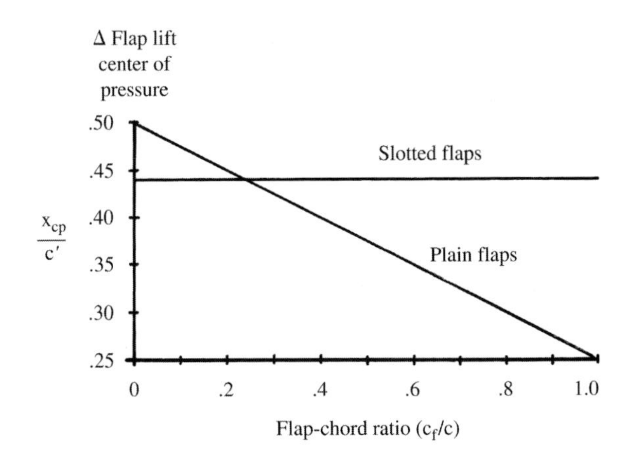

Fig. 16.9 Center of pressure for lift increment due to flaps (after Ref. 37).

Downwash and Upwash

The remaining terms in Eq. (16.7) are strongly influenced by the wing flowfield, as shown in Fig. 16.10. Ahead of the wing, the air in subsonic flight is pulled upward by the reduced pressures above the wing. This upwash pushes upward on the fuselage forebody and also turns the flow prior to reaching a propeller or inlet located ahead of the wing.

Behind the wing, the flow has an initial downward direction theoretically equal to the wing angle of attack. This downwash angle diminishes aft of the wing to a value of approximately half the wing angle of attack at the tail of a typical aircraft. Also, the downwash varies across the span and approaches zero near the wing tips.

The downwash reduces the tail angle of attack and pushes downward on the aft fuselage, contributing to the fuselage pitching moment. Downwash is strongly affected by the propwash.

The upwash-angle (ϵ_u) derivative with respect to wing angle of attack is determined from Fig. 16.11. The downwash angle (ϵ) derivative is determined from Fig. 16.12 at low subsonic speeds (unswept wing). The spanwise variation in downwash behind the wing reduces the average downwash experienced by the tail by approximately 5%. The additional downwash due to flap deflection is determined from Fig. 16.13 in which h is the tail height above the wing.

At transonic speeds (around Mach 0.9) the downwash-angle derivative increases by about 30–40% then reduces at higher speeds. Equations (16.21) provide a rough approximation of the downwash at high subsonic and supersonic speeds.

$$\text{Subsonic:} \quad \frac{\partial \epsilon}{\partial \alpha} = \left(\left. \frac{\partial \epsilon}{\partial \alpha} \right|_{M=0} \right) \left(\frac{C_{L\alpha}}{C_{L\alpha}|_{M=0}} \right) \tag{16.21a}$$

$$\text{Supersonic:} \quad \frac{\partial \epsilon}{\partial \alpha} = \frac{1.62 C_{L\alpha}}{\pi A} \tag{16.21b}$$

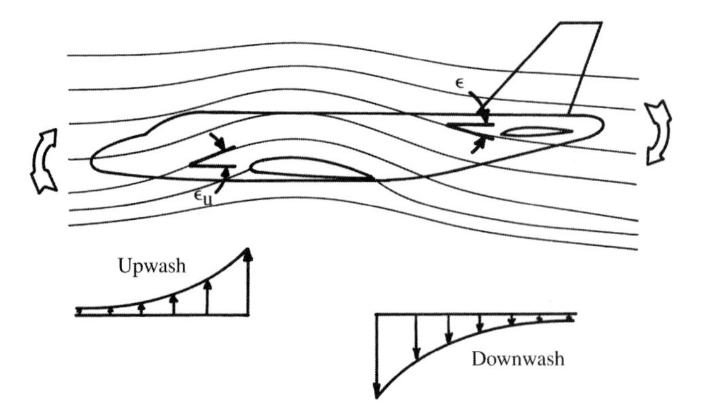

Fig. 16.10 Wing flowfield effect on pitching moment.

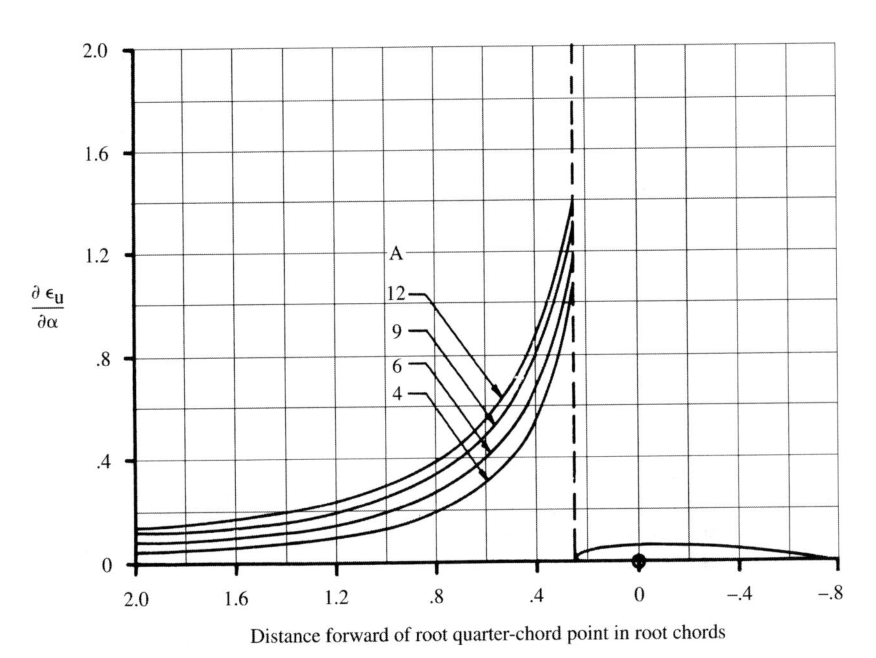

Fig. 16.11 Upwash estimation (subsonic only) (Ref. 37).

The resulting angle of attack considering the effect of upwash or downwash is determined by adding an upwash or subtracting a downwash from the freestream angle of attack. The angle-of-attack derivatives are therefore as expressed in Eqs. (16.22) and (16.23). Equation (16.23) is the tail angle-of-attack derivative from Eq. (16.8), called β in many texts, which is easily confused with yaw angle. The downwash derivative is with respect to the wing angle of attack, so the tail angle of attack can now be determined as shown in Eq. (16.24).

$$\text{Upwash:} \quad \frac{\partial \alpha_u}{\partial \alpha} = 1 + \frac{\partial \epsilon_u}{\partial \alpha} \tag{16.22}$$

$$\text{Downwash:} \quad \frac{\partial \alpha_h}{\partial \alpha} = 1 - \frac{\partial \epsilon}{\partial \alpha} \tag{16.23}$$

$$\alpha_h = (\alpha + i_w)\left(1 - \frac{\partial \epsilon}{\partial \alpha}\right) + (i_h - i_w) + \Delta\alpha_{\mathrm{OL}} \tag{16.24}$$

A canard will obviously experience no downwash from the wing, but its own downwash will influence the wing. The estimation of the effect of canard downwash on the wing is very difficult because the downwash varies across the canard span and because the canard tip vortices actually create an upwash on the wing outboard of the canard.

The effect of canard downwash on the wing may be crudely approximated by assuming that the canard downwash as calculated with these methods uniformly

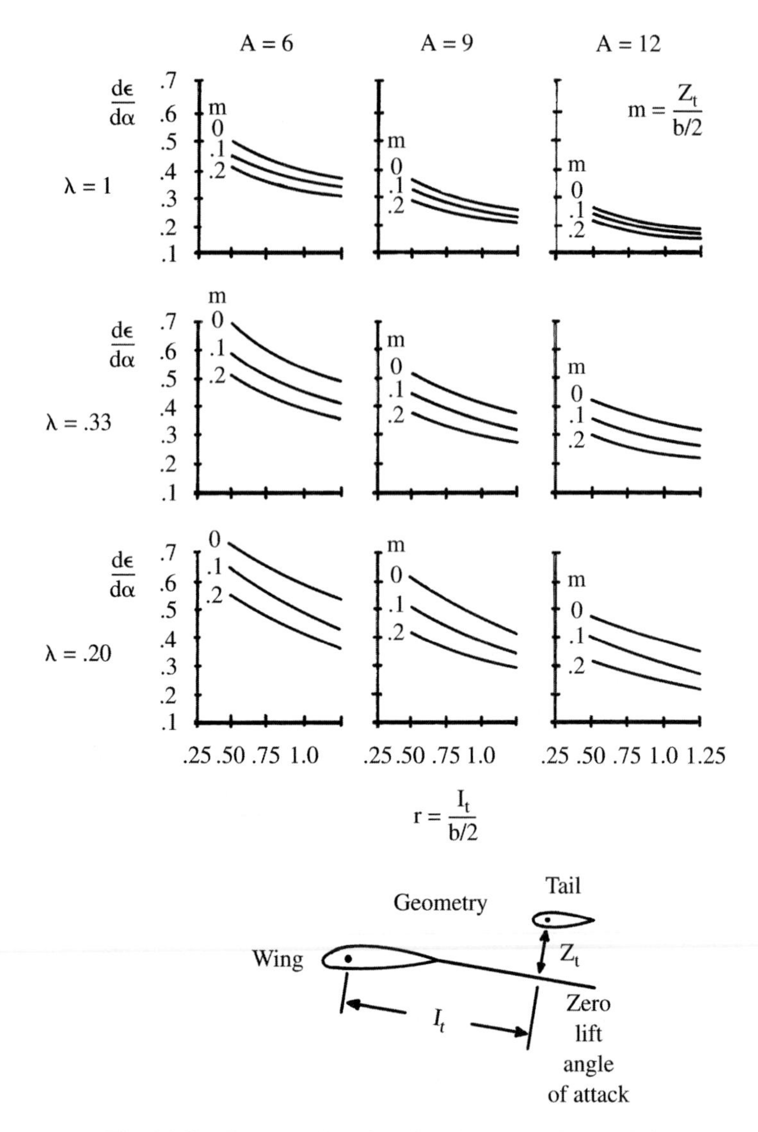

Fig. 16.12 Downwash estimation ($M = 0$) (after Ref. 4).

affects the wing inboard of the canard tips. This reduces the angle of attack at the wing root.

Wing Vertical Position

The vertical position of the wing also has an effect on stability. This easy to visualize—if the nose comes up on a high wing configuration, the wing actually

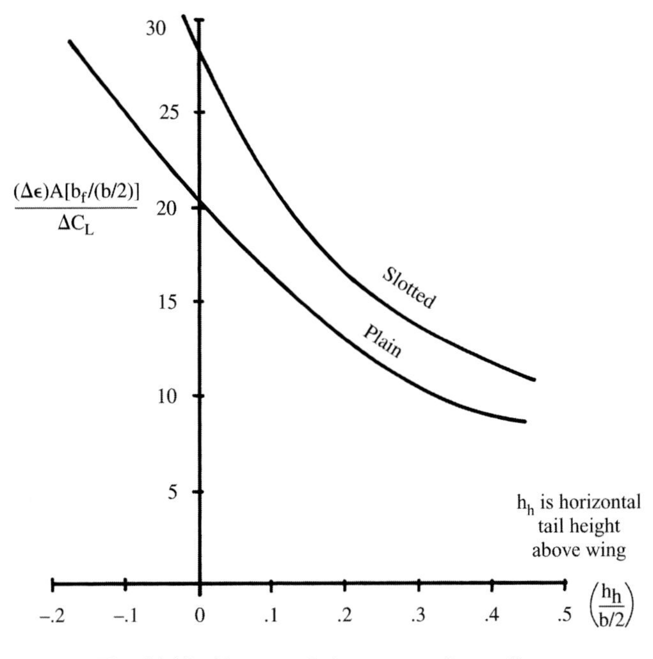

Fig. 16.13 Downwash increment due to flaps.

moves to the rear relative to the center of gravity and thus provides an additional nose-down pitching moment. As a rough approximation, it can be assumed that a high wing increases the static margin by 10% of the vertical distance of the wing above the c.g., divided by wing MAC.

Fuselage and Nacelle Pitching Moment

The pitching-moment contributions of the fuselage and nacelles can be approximated by Eq. (16.23) from NACA TR 711. The W_f is the maximum width of the fuselage or nacelle and L_f is the length. Figure 16.14 provides the empirical pitching-moment factor K_f.

$$C_{m_{\alpha\,\text{fuselage}}} = \frac{K_f W_f^2 L_f}{c S_w}, \quad \text{per deg} \tag{16.25}$$

Thrust Effects

The remaining terms in Eq. (16.7) are thrust effects upon pitching moment. Thrust has three effects, namely, the direct moment of the thrust, the propeller or inlet normal force due to turning of the air, and the influence of the propwash or jet-induced flows upon the tail, wing, and aft fuselage.

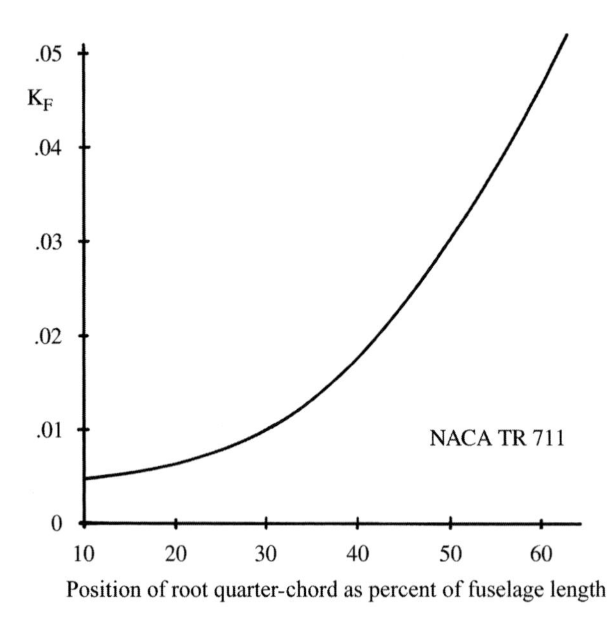

Fig. 16.14 Fuselage moment term.

The direct moment of the thrust is simply the thrust times the moment arm about the c.g., as defined in Eq. (16.7). If the thrust axis passes through or near the c.g., this term may be ignored.

The normal force due to the turning of the air at an inlet front face F_p can be calculated from momentum considerations. This normal force equals the mass flow into the inlet times the change in vertical velocity. Since the angles are small, the change in vertical velocity is approximately the turning angle (α_p—see Fig. 16.3) times the aircraft velocity [Eq. (16.26)]. The engine mass flow can be approximated by assuming a capture-area ratio of one [Eq. (16.27)] if installed engine mass-flow data are unavailable. Note that in British units mass flow is in slugs per second, which equals pounds per second divided by 32.2. The metric system avoids such confusing adjustments!

$$F_p = \dot{m}V \tan \alpha_p \cong \dot{m}V\alpha_p \tag{16.26}$$

$$\dot{m} \cong \rho V A_{\text{inlet}} \tag{16.27}$$

$$F_{p\alpha} = \dot{m}V \tag{16.28}$$

The derivative of the normal force with respect to angle of attack is the mass flow times the velocity [Eq. (16.28)]. The derivative of α_p with respect to angle of attack [see Eq. (16.9)] is the upwash derivative Eq. (16.22) if the inlet is ahead of the wing, and the downwash derivative Eq. (16.23) if the inlet is behind the wing. For an inlet mounted under the wing, the wing turns the flow before it reaches the inlet front face so the normal force is approximately zero.

For a propeller-powered aircraft, a normal force contribution to pitching moment is also produced by the momentum change caused by the turning of the airstream. Unlike the jet inlet, the actual turning angle is not apparent because the propeller does not fully turn the airflow to align with the propeller axis.

Equation (16.29) is an empirical method for estimation of the propeller normal force based upon charts in Ref. 68; N_B is the number of blades per propeller, and A_p is the area of one propeller disk. The derivative term is the normal force exerted by one blade when the propeller is operating at zero thrust, found in Fig. 16.15 as a function of advance ratio. The function $f(T)$ adjusts for nonzero thrust and is found in Fig. 16.16.

$$F_{p\alpha} = qN_B A_p \frac{\partial C_{N_{\text{blade}}}}{\partial \alpha} f(T) \tag{16.29}$$

Note in Eq. (16.7) that a propeller mounted aft of the c.g. is stabilizing. This is one of the advantages of the pusher-propeller configuration.

The propwash affects the downwash seen by the horizontal tail and reduces the tail's effectiveness. Equation (16.30) estimates this propeller downwash effect as a derivative that is added to the wing downwash derivative. The constant terms come from Fig. 16.17.

$$\frac{\partial \epsilon_p}{\partial \alpha} = K_1 + K_2 N_B \frac{\partial C_{N_{\text{blade}}}}{\partial \alpha}\left(\frac{\partial \alpha_p}{\partial \alpha}\right) \tag{16.30}$$

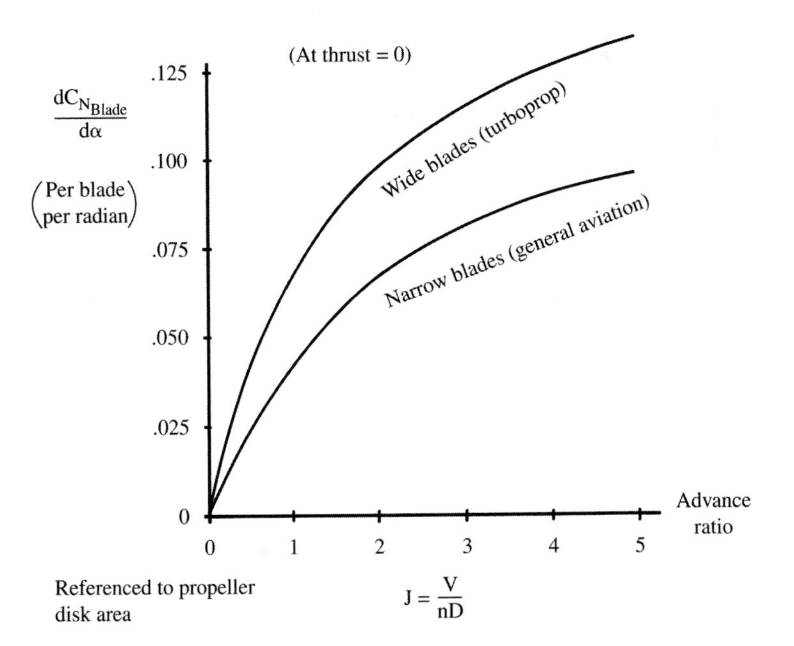

Fig. 16.15 Propeller normal force coefficient (after Ref. 68).

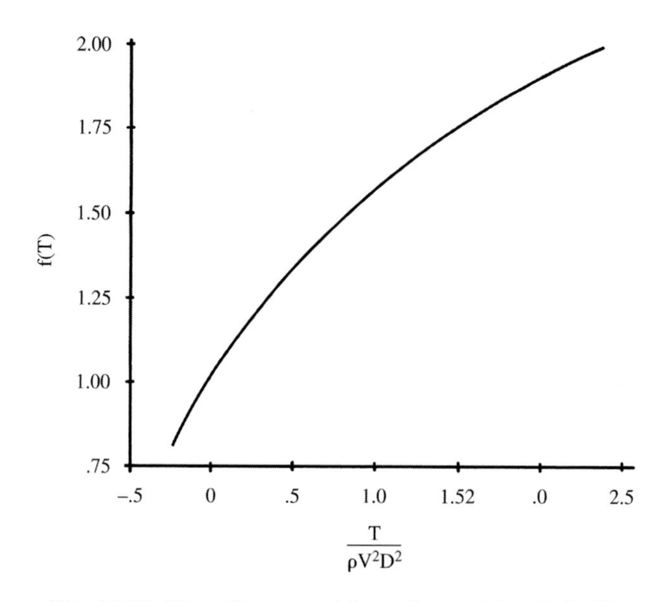

Fig. 16.16 Propeller normal force factor (after Ref. 68).

If largely in the propwash, the tail will experience an increased dynamic pressure, as shown in Eq. (16.31). The tail dynamic pressure ratio η_h for zero thrust is approximately 0.9. If the tail is only partly in the propwash, the right-side term in the parentheses should be reduced proportionately. This term can

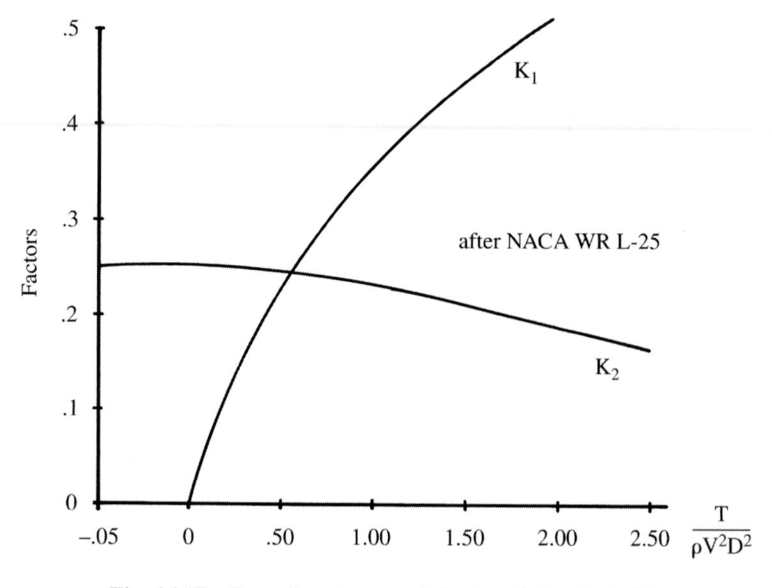

Fig. 16.17 Propeller downwash factors (after Ref. 68).

also be applied to estimate increase in dynamic pressure at the wing, which may especially affect the pitching moment due to flap deflection.

$$\eta_h = \eta_{h_{T=0}}\left(1 + \frac{T}{qA_p}\right) \tag{16.31}$$

The increase in dynamic pressure at the tail will increase the magnitude of the tail lift which, being downward in most cases, causes a nose-up trim change with application of power. It is not uncommon in single-engined propeller aircraft to incline the propeller axis several degrees downward to counteract the power effect upon trim.

Trim Analysis

We now have all the information required for trim analysis. Trim requires that the total moment about the c.g. [Eq. (16.7)] equals zero. For a given flight condition, we can determine the values in the equation and see if they sum to zero. If not, we can vary the tail lift by changing elevator deflection or tail incidence until the total moment is zero.

However, the change in tail lift will change the aircraft total lift, which must equal the weight. Therefore, as the tail lift changes, the aircraft angle of attack must change. This can be solved by a computerized iterative process or by a graphical technique.

For the graphical solution, arbitrarily assumed aircraft angles of attack and elevator deflection angles (δ_E) are used to calculate the total-pitching-moment coefficient ($C_{m_{cg}}$) using Eq. (16.7). Equation (16.32) is used to determine the tail-lift term.

$$C_{L_h} = C_{L_{\alpha h}}\left[(\alpha + i_w)\left(1 - \frac{\partial\epsilon}{\partial\alpha}\right) + (i_h - i_w) - \alpha_{\text{OL}_h}\right] \tag{16.32}$$

$$C_{L_{\text{total}}} = C_{L_\alpha}[\alpha + i_w] + \eta_h \frac{S_h}{S_w} C_{L_h} \tag{16.33}$$

For the arbitrarily assumed angles of attack and elevator deflection, the total lift coefficient $C_{L_{\text{total}}}$ can be estimated using Eq. (16.33). This equation sums the wing and tail lift coefficients, including the effects of dynamic pressure at the tail. Remember that by definition an upload on the tail is positive. If a download exists on the tail, the tail lift reduces the total lift.

The total-pitching-moment coefficient is then plotted vs the total lift coefficient for the various elevator-deflection angles. The elevator deflection for trim is determined by interpolating for zero pitching moment at the required total lift coefficient. This is illustrated in Fig. 16.18.

The total induced drag including trim-drag effects can now be calculated at the trim angle of attack and elevator deflection angle using Eq. (16.34). Note that the term K_h is the drag-due-to-lift factor for the horizontal tail. This is determined using the methods of Chapter 12, treating the horizontal tail as a wing. Since the tail's induced drag is much smaller than the wing induced drag, it is permissible to use the simpler empirical methods for K (or e) rather than the

Calculation table

	$\delta_E = -2°$	$0°$	$2°$
$\alpha = 0°$	$C_{m_{cg}} = 0.033$ $C_{l_{total}} = -0.07$	0.018 −0.05	0.002 −0.03
$\alpha = 5°$	$C_{m_{cg}} = 0.012$ $C_{l_{total}} = 0.53$	−0.004 0.54	−0.02 0.56
$\alpha = 10°$	$C_{m_{cg}} = -0.005$ $C_{l_{total}} = 1.03$	−0.021 1.04	−0.038 1.06

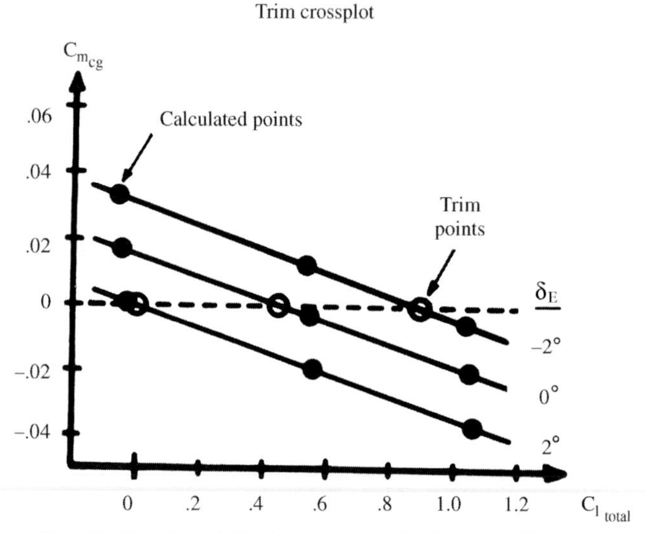

(Note: Positive δ_E as defined produces an upload on the tail.)

Fig. 16.18 Graphical trim analysis.

leading-edge-suction method.

$$C_{D_{i_{\text{trimmed}}}} = K\ [C_{L_\alpha}(\alpha + i_w)]^2 + \eta_h \frac{S_h}{S_w} K_h\ [C_{L_h}]^2 \tag{16.34}$$

For an aircraft with an aft tail, the downwash off the wing has an additional effect on total trimmed drag. The directions of lift and drag of the tail are slightly rotated, since they are always perpendicular and parallel to the local flow direction. The change in drag direction has a trivial effect on total drag, but the change in lift direction may be nontrivial. For a stable aircraft, the tail is often experiencing a download (negative lift) to trim the aircraft, and this downward lift vector is rotated by the wing's downwash so that it has a slight forward component. This is in effect a reduction of trimmed drag, and as a result, the conventional aft tail does not have as much trim drag as might be assumed. On the other hand, if we design so that the aft tail is lifting to minimize trim drag, the lift vector is rotated to the rear. This causes a slight increase in drag that reduces the trim drag savings expected for such an aircraft (which usually requires an unstable aircraft with a computerized flight control system).

This downwash effect on the direction of tail lift can be estimated by determining the downwash angle as already described, then multiplying the lift on the tail by the sine of the angle and adding or subtracting the result to the aircraft's total drag.

Another small trim drag contribution is the parasitic drag of the elevator, if it must be deflected to maintain trim. This drag can be estimated using Eq. (12.37), although test data on a similar configuration are preferred. Avoidance of this drag contribution is one reason that many aircraft have a variable incidence (all-moving) horizontal tail.

For an all-moving tail, the tail incidence angle is varied rather than elevator angle. For a tailless configuration, the wing flap acts as the elevator. Otherwise the procedure is similar.

Due to the amount of computation involved, it is common in early conceptual design to calculate the trim condition without including the thrust effects unless the thrust axis is well above or below the c.g.

(Most stability-and-control textbooks introduce a secondary derivative term $C_{m\delta_E}$ that directly relates the elevator deflection to its influence upon pitching moment. I choose to leave the elevator effect as a change in tail lift to avoid further complexity in terminology and to leave the tail moment as a clearly understood "force-times-distance" term. This understanding is especially important in conceptual design because the designer still has the freedom to change the "distance.")

Ground Effect on Trim Calculation

The trim equation (16.7) is strongly affected by ground effect (Chapter 12). When the aircraft approaches the ground to within about 20% of the span, the wing and tail lift-curve slopes will increase by about 10%. Furthermore, the downwash is reduced to about half of the normal value, which requires a greater elevator deflection to hold the nose up.

The aircraft must have sufficient elevator effectiveness to trim in ground effect with full flaps and full-forward c.g. location, at both power-off and full power. Some additional elevator authority must then be available for control.

Takeoff Rotation

Sometimes the elevator of an aircraft is sized by the requirement for takeoff rotation. For a tricycle-gear aircraft the elevator should be powerful enough to rotate the nose at 80% of takeoff speed with the most-forward c.g. For a taildragger aircraft the elevator should be powerful enough to lift the tail at half the takeoff speed with the most-aft c.g. (Ref. 66).

For rotation analysis, Eq. (16.7) may be employed with the addition of two landing gear terms. The analysis assumes that the nosewheel or tailwheel is just resting on the ground without carrying any of the weight. The weight on the wheels is the aircraft weight minus the total lift at that angle of attack. This exerts a vertical force with a moment arm equal to the distance from the main gear to the c.g. as measured parallel to the ground.

The rolling friction of the mainwheels exerts a rearward force equal to the weight on the wheels times the rolling friction coefficient (0.03 is typical). This rolling friction force acts through a moment arm equal to the vertical height of the c.g. above the ground.

These additional moments due to the vertical and rearward landing gear forces must be converted to moment coefficients by dividing them by $(qS_w c)$.

The previously described changes in lift-curve slopes and downwash angles due to ground effect must be considered in takeoff rotation analysis.

Velocity Stability

This brief discussion of longitudinal stability and control has focused upon the angle-of-attack stability derivatives. The aircraft must also have velocity stability, implying that an increase in velocity must produce forces which slow the aircraft down, usually by raising the nose. For most contributors to pitching moment, angle-of-attack stability implies velocity stability as well.

One additional term that affects velocity stability is the variation in thrust with velocity. For propellers, the thrust reduces with increased aircraft velocity. If the propeller is mounted substantially above the c.g., an increase in velocity will reduce the thrust, causing the aircraft to pitch nose up. This produces a slight climb that will reduce the velocity, so a high-mounted propeller is stabilizing.

Roughly speaking, the apparent static margin increases one-quarter of a percent for every 1% MAC that the thrust axis is above the c.g. Conversely, a propeller mounted below the c.g. is destabilizing by the same amount. However, this apparent stability response only occurs after enough time has passed for the aircraft's velocity to increase or decrease enough to affect the propeller's thrust. This doesn't change the immediate response of the aircraft to a pitch disturbance, so the benefit of a high thrust axis cannot be used to lessen the aircraft's power-off static margin. On the other hand, the velocity stability detriment for a low thrust axis probably should be considered as it will tend to exaggerate the effect of a slight out-of-trim condition over a long period of time. (But how can you put a propeller below the aircraft without having it hit the ground?)

The high-mounted propeller also demands a large trim force required to counter the nose-down pitching moment of the high thrust axis. The high-mounted propeller is usually used only to provide water clearance in a seaplane.

For jet aircraft, the velocity effect upon thrust being negligible, engine vertical position has little effect upon velocity stability.

16.4 Lateral-Directional Static Stability and Control

Yaw/Roll Moment Equations and Trim

In many ways the lateral-directional analysis resembles the longitudinal analysis. However, the lateral-directional analysis really embraces two closely coupled analyses: the yaw (directional) and the roll (lateral).

It is important to realize that both are driven by the yaw angle β, and that the roll angle ϕ actually has no direct effect upon any of the moment terms! Furthermore, the deflection of either rudder or aileron will produce moments in both yaw and roll. (Note: to reduce verbiage, "lateral" is used synonymously with "lateral-directional" in the following discussion.)

The geometry for lateral analysis is illustrated in Fig. 16.19, showing the major contributors to yawing moment N and rolling moment L. By definition, yaw and roll are positive to the right. Note that unlike the longitudinal terms, most of these terms have a zero value when the aircraft is in straight and level flight. Also, by the sign conventions used for β and yaw, a positive value of yawing moment derivative with respect to β is stabilizing. However, a negative value of the rolling moment derivative with respect to β is stabilizing (dihedral effect).

The major yawing moment is due to the lateral lift of the vertical tail, denoted by F_v. This counteracts the fuselage yawing moment, which is generally negative to the sense shown in the figure. Rudder deflection acts as a flap to increase the lateral lift of the vertical tail.

A vertical tail immersed in the propwash experiences an additional force. The air in the propwash has a rotational component caused by the propeller and in the same direction that the propeller rotates. A propeller usually rotates clockwise when seen from behind. For a vertical tail above the fuselage, the propwash

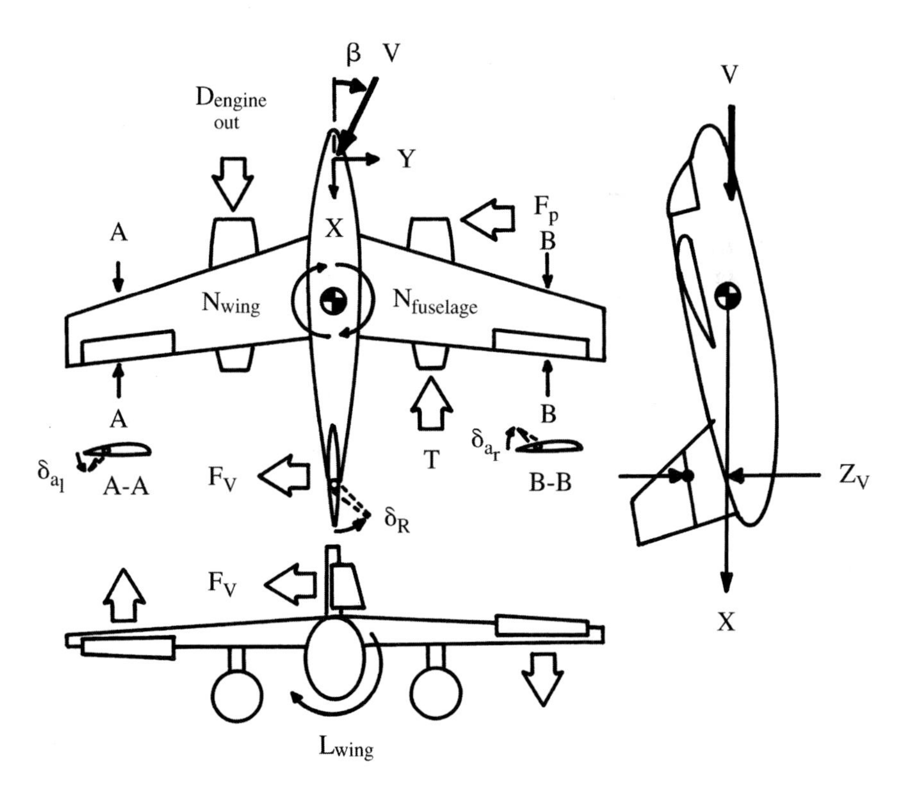

Fig. 16.19 Lateral geometry.

rotational component causes the angle of sideslip at the tail to become more negative, thus slightly yawing the nose of the aircraft to the left.

A stronger yawing moment caused by the propeller occurs when the disk of the propeller is at an angle to the freestream flow, typically during a low-speed climb. The blade going downward has a higher angle of attack and is also at a slightly higher velocity because it is advancing into the relative flow. Therefore it experiences higher thrust, causing the effective thrust axis to move toward that side. This is called "*p*-effect," yawing the nose to the left for a clockwise propeller rotation, and is difficult to predict.

Many single-engined aircraft have 1 or 2 deg of incidence built into the vertical tail to counteract *p*-effect. Alternatively, some aircraft have the propeller axis angled to the right.

The wing yawing moment can be visualized as an increase in drag on the side of the wing that is more nearly perpendicular to the oncoming flow. If the wing is swept aft, this yawing moment is stabilizing as shown.

Another wing yawing moment occurs with aileron deflection. The wing with increased lift due to aileron deflection has more induced drag, so the yawing moment is in the opposite direction from the rolling moment due to the aileron deflection. This is known as "adverse yaw."

The engines have the same three effects upon lateral moments that they have on longitudinal moments (direct thrust, normal force, and propwash or jet-induced flowfield effects). In yaw, the thrust is balanced unless an engine fails. Then the remaining engine(s) create a huge yawing moment that is made worse by the drag of the failed engine.

The inlet front face or propeller disk has the same normal-force term discussed for longitudinal stability. As in pitch, this is destabilizing in yaw if the inlet or propeller is in front of the c.g.

The propwash or jet-induced flowfield effects are generally negligible in yaw unless the vertical tail is in the propwash or near the jet exhaust. In this case the dynamic pressure and angle of sideslip at the tail will be affected much as the horizontal tail is affected by propwash.

In roll, the major influence is the wing rolling moment due to dihedral effect. As discussed in Chapter 4, this rolling moment tends to keep the aircraft level because it sideslips downward whenever a roll is introduced. The dihedral effect rolls the aircraft away from the sideslip direction.

The ailerons, the primary roll-control device, operate by increasing lift on one wing and reducing it on the other. The aileron deflection δ_a is defined as the average of the left and right aileron deflections in the directions shown. (Some texts define aileron deflection as the total of left and right.) Positive aileron deflection rolls the aircraft to the right.

Spoilers are an alternative roll-control device. These are plates that rise up from the top of the wing, usually just aft of the maximum-thickness point. This disturbs the airflow and "spoils" the lift, dropping the wing on that side. Spoiler deflection also increases drag, so the wing yaws in the same direction that it rolls (proverse yaw).

The vertical tail contributes positively to the roll stability because it is above the c.g. Note that the moment arm for the vertical tail roll contribution is from the vertical tail MAC to the *X* axis in the stability (wind) axis system. This *X* axis is through the c.g. and is aligned with the relative wing. Thus, this term changes substantially with angle of attack.

The major thrust effect on static roll moments is the engine-out case. The air in the propwash has higher dynamic pressure and thus produces more lift on the wing. With propwash on only one side of the wing there is a difference in lift between the left and right wing. This can frequently be ignored because the resulting roll moment is so much less than the engine-out yaw moment. The equivalent jet-induced effect on roll is negligible unless the jet exhaust impinges upon the flaps, as in the YC-15.

Propwash can also alter the wing dihedral effect. When the aircraft yaws, one side of the wing gets more propwash than the other, producing a destabilizing roll moment. This is more severe for single-engined aircraft where the propeller is way in front of the wing.

There will be a thrust normal force contribution to rolling moment at angle of sideslip if the engines are substantially above or below the c.g. A high-mounted engine would be stabilizing. This is usually negligible.

$$\begin{aligned} N = {} & N_{\text{wing}} + N_{w\delta_a}\delta_a + N_{\text{fus}} + F_v(X_{\text{acv}} - X_{\text{cg}}) \\ & - TY_p - DY_p - F_p(X_{\text{cg}} - X_p) \end{aligned} \tag{16.35}$$

$$L = L_{\text{wing}} + L_{w\delta_a}\delta_a - F_v(Z_v) \tag{16.36}$$

These yaw and roll moments are summed in Eqs. (16.35) and (16.36) for a twin-engined aircraft with one engine out. Similar equations for other engine arrangements should be obvious from inspection of Fig. 16.19. These are strictly static equations. Dynamic terms will be considered in a later section.

The lateral lift force on the vertical tail appears in both equations. This is much like the horizontal-tail lift, and must be calculated using the local dynamic pressure and angle of sideslip. The local angle of sideslip is less than the free-stream sideslip angle because of a "sidewash" effect largely due to the fuselage. Propwash can also reduce the effective angle of sideslip. Equation (16.37) expresses the lateral lift force on the vertical tail. Note that the tail lateral-lift-force derivative C_{F_β} is equivalent to $C_{L\alpha}$ in longitudinal notation and is calculated the same way.

$$F_v = q_v S_v C_{F_{\beta_v}} \frac{\partial \beta_v}{\partial \beta} \beta \tag{16.37}$$

The yaw- and roll-moment equations are expressed in coefficient form by dividing through by (qS_wb), as shown in Eqs. (16.38) and (16.40). Lengths are expressed as a fraction of wing span using the "bar" notation. Thus, $(\overline{Y})$ denotes (Y/b). The ratio between dynamic pressure at the tail and the free-stream dynamic pressure is denoted by η_v. The vertical-tail contributions to yaw and roll are expressed by the derivatives defined in Eqs. (16.39) and (16.41).

Yaw:

$$C_n = \frac{N}{qS_w b} = C_{n_{\beta_w}}\beta + C_{n_{\delta_a}}\delta a + C_{n_{\beta_{\text{fus}}}}\beta + C_{n_{\beta_v}}\beta - \frac{T\overline{Y}_p}{qS_w} - \frac{D\overline{Y}_p}{qS_w} - \frac{F_p}{qS_w}(\overline{X}_{\text{cg}} - \overline{X}_p) \tag{16.38}$$

where

$$C_{n_{\beta_v}} = C_{F_{\beta_v}} \frac{\partial \beta_v}{\partial \beta} \eta_v \frac{S_v}{S_w}(\overline{X}_{\text{acv}} - \overline{X}_{\text{cg}}) \tag{16.39}$$

Roll:

$$C_\ell = \frac{L}{qS_w b} = C_{\ell_{\beta_w}}\beta + C_{\ell_{\delta_a}}\delta_a + C_{\ell_{\beta_v}}\beta \tag{16.40}$$

where

$$C_{\ell_{\beta_v}} = -C_{F_{\beta_v}} \frac{\partial \beta_v}{\partial \beta} \eta_v \frac{S_v}{S_w}\overline{Z}_v \tag{16.41}$$

Lateral-Trim Analysis

The main static lateral-trim condition of concern is the engine-out case on takeoff. The vertical tail with rudder deflected must produce sufficient yawing moment to keep the aircraft at zero angle of sideslip at takeoff speed (1.1 times the stall speed) with one engine out and at the aftmost c.g. location. Rudder deflection should probably be no more than 20 deg to allow additional deflection for control.

Another lateral-trim condition that should be checked is the crosswind-landing case. The aircraft must be able to operate in crosswinds equal to 20% of takeoff speed, which is equivalent to holding an 11.5-deg sideslip at takeoff speed. Again, no more than 20 deg of rudder should be used.

If the vertical tail cannot provide sufficient force to produce zero yawing moment in Eq. (16.38) for either of these cases, there are several approaches to correct the problem. The brute-force method simply increases the vertical-tail size, but this penalizes aircraft weight and drag.

The rudder chord and/or span can be increased to improve the rudder effectiveness. This can also be increased by using a double-hinged rudder, as seen on the DC-10. An all-moving vertical tail as seen on the F-107 and SR-71 provides the greatest "rudder" control power for a given tail area, but is heavy.

Sometimes the engines may be moved inward to reduce the engine-out moment. However, this increases wing structural weight as discussed.

The rudder deflection and propwash effects for the engine-out case will also cause a rolling moment. Usually this is small enough to be ignored, but a short-coupled aircraft with widely separated engines may require excessive aileron deflections to counter the rolling moments. The adverse yaw of the aileron deflections will then make the yawing situation even worse!

The aileron control authority must also be checked at the 11.5-deg sideslip condition using Eq. (16.40). An aircraft with a large amount of effective dihedral may not have sufficient aileron area to prevent the aircraft from rolling away from the sideslip.

Static Lateral-Directional Stability

The yaw and pitching-moment derivatives with respect to sideslip are provided in Eqs. (16.42) and (16.43). The power-off C_{n_β} is simply the sum of the wing, fuselage, and vertical-tail contributions.

$$C_{n_\beta} = C_{n_{\beta_w}} + C_{n_{\beta_{\text{fus}}}} + C_{n_{\beta_v}} - \frac{F_{p\beta}}{qS_w}\frac{\partial \beta_p}{\partial \beta}(\overline{X}_{\text{cg}} - \overline{X}_p) \tag{16.42}$$

$$C_{\ell_\beta} = C_{\ell_{\beta_w}} + C_{\ell_{\beta_v}} \tag{16.43}$$

It would be possible to solve Eq. (16.42) for the c.g. position for zero yaw stability. This would be the lateral neutral point. This is not usually calculated because it is common practice to determine the most-aft c.g. position from longitudinal considerations and then vary the vertical-tail area until gaining sufficient yaw stability.

Figure 16.20 provides suggested goal values for C_{n_β}. These are somewhat less than those suggested by the NASA curve. C_{ℓ_β} should be of negative sign with magnitude about half that of the C_{n_β} value at subsonic speeds, and about equal to it at transonic speeds.

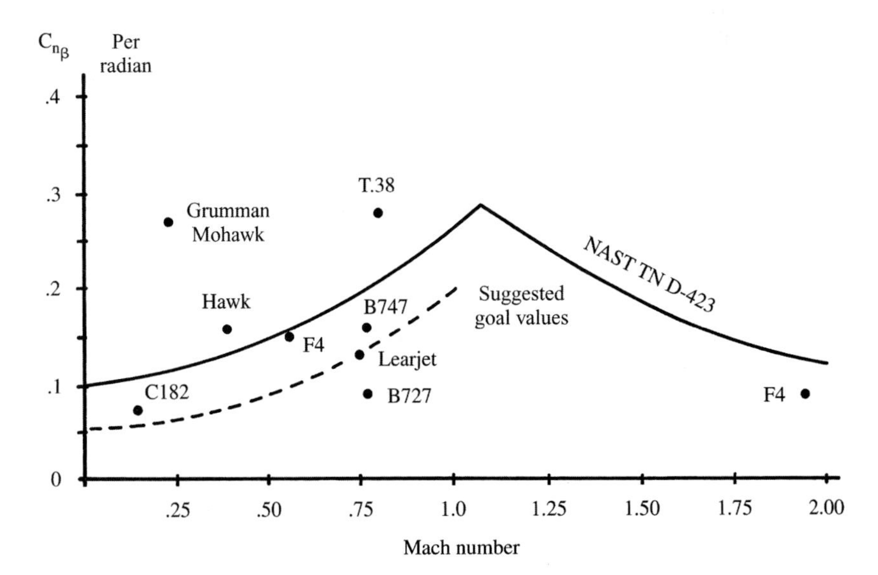

Fig. 16.20 Typical yaw moment derivative values.

Final selection of these values requires dynamic analysis based upon wind-tunnel data, and it is not unheard of for the vertical-tail size or wing dihedral to be changed after the prototype flies (F-100, B-25).

The following sections provide crude estimation procedures for the terms of these lateral equations. Many of these terms are identical to longitudinal terms as previously discussed, and the reader should refer back to that material. These include the tail aerodynamic center, tail-force (lift) curve slope, rudder (flap) effectiveness, and propeller or inlet normal force.

Wing Lateral-Directional Derivatives

Reference 37 provides an empirical expression for the wing yawing moment due to sideslip [Eq. (16.44)].

$$C_{n_{\beta_w}} = C_L^2 \left\{ \frac{1}{4\pi A} - \left[\frac{\tan\Lambda}{\pi A\,(A + 4\cos\Lambda)} \right] \times \left[\cos\Lambda - \frac{A}{2} - \frac{A^2}{8\cos\Lambda} + \frac{6(\overline{X}_{\text{acw}} - \overline{X}_{\text{cg}})\sin\Lambda}{A} \right] \right\} \tag{16.44}$$

The rolling moment due to sideslip, or dihedral effect, is proportional to the dihedral angle but also includes the effects of sweep and wing vertical position on the fuselage; C_{l_β} for a straight wing is approximately 0.0002 times the dihedral angle in deg, so 1 deg of "effective dihedral" is defined to be a C_{l_β} of 0.0002 per deg, or 0.0115 per radian.

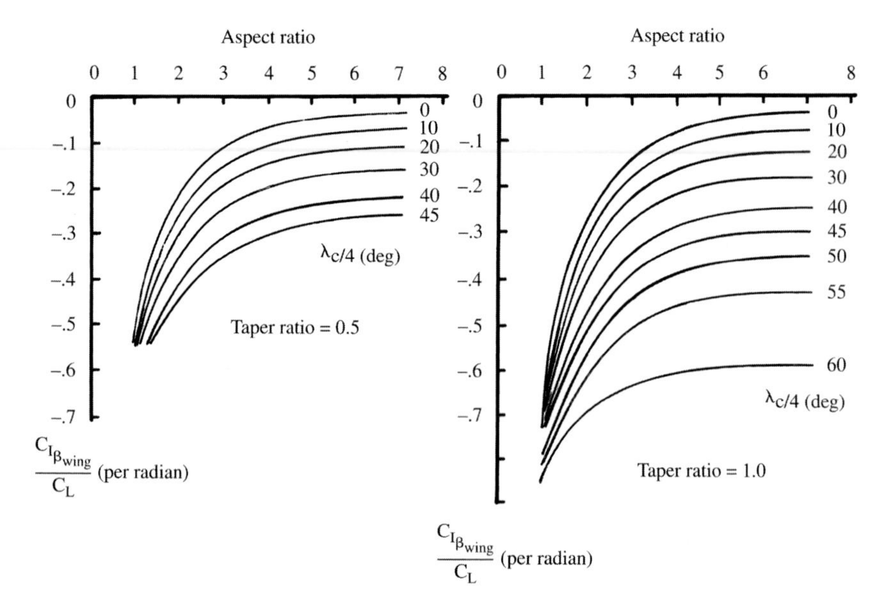

Fig. 16.21 Dihedral effect of sweep (Ref. 10).

Figure 16.21 (replotted from Ref. 10) provides an estimate of the wing dihedral effect due to sweep for a wing with no geometric dihedral. Two taper ratios are provided, requiring interpolation or extrapolation for other taper ratios. The values from the figure are per unit lift coefficient, so the final value is obtained by multiplying by the wing C_L.

Equation (16.45) from Ref. 37 estimates the effect of the geometric dihedral angle (radians). Equation (16.46) from Ref. 68 determines the effect of wing vertical placement on the fuselage; Z_{wf} is the vertical height of the wing above the fuselage centerline, and D_f and W_f are the depth and width of the fuselage.

These two additional dihedral contributions are added to the value from Fig. 16.21, as shown in Eq. (16.47). All terms should be negative except that the wing vertical placement term will be positive (destabilizing) for a low wing.

$$(C_{\ell_\beta})_\Gamma = -\frac{C_{L\alpha}\Gamma}{4}\left[\frac{2(1+2\lambda)}{3(1+\lambda)}\right] \tag{16.45}$$

$$C_{\ell_{\beta_{\text{wf}}}} = -1.2\frac{\sqrt{A}\,Z_{\text{wf}}(D_f + W_f)}{b^2} \tag{16.46}$$

$$C_{\ell_{\beta_w}} = \left(\frac{C_{\ell_{\beta_w}}}{C_L}\right)C_L + (C_{\ell_\beta})_\Gamma + C_{\ell_{\beta_{\text{wf}}}} \tag{16.47}$$

The aileron control power can be approximated using a strip method. The portion of the wing having the aileron is broken into strips as shown in Fig. 16.22. The lift increment due to aileron deflection is estimated as a flap effect using the method presented in Eq. (16.17). This lift increment is then multiplied by the strip's moment arm from the aircraft centerline (Y_1), as shown in Eq. (16.48), where K_f and the lift derivative with flap deflection come from Figs. 16.6 and 16.7. Remember to reduce the aileron effectiveness about 15%

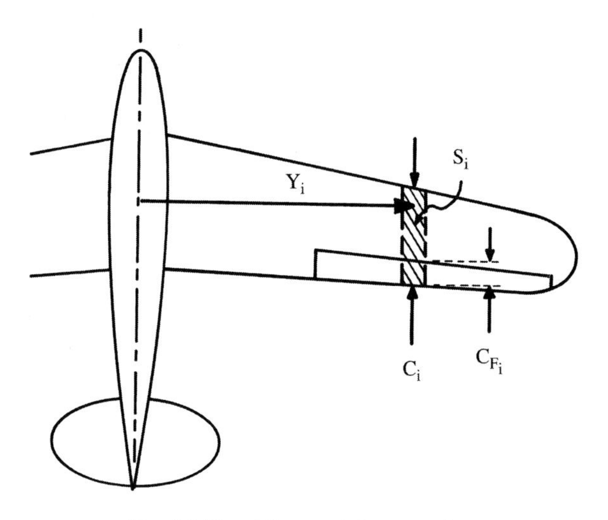

Fig. 16.22 Aileron strip geometry.

if the hinge gap is not sealed.

$$C_{\ell_{\delta_a}} = \frac{2\Sigma K_f \left(\frac{\partial C_L}{\partial \delta_f}\right)' Y_i S_i \cos \Lambda_{\text{H.L.}}}{S_w b} \tag{16.48}$$

$$C_{n_{\delta_a}} = -0.2 C_L C_{\ell_{\delta_a}} \tag{16.49}$$

The yawing moment due to aileron deflection depends upon the spanwise distribution of induced drag with the aileron deflected. This varies with the wing lift coefficient as well as the aileron deflection. Yawing moment due to aileron deflection can be approximated by Eq. (16.49), a simplification of the method from Ref. 37; C_L is the wing lift coefficient.

Fuselage and Nacelle Lateral-Directional Derivatives

The yawing moment due to sideslip is expressed in Eq. (16.50) as a function of the fuselage or nacelle volume, depth, and width. The fuselage contribution to roll is usually negligible except for its influence upon the wing effective dihedral, as previously discussed.

$$C_{n_{\beta_{\text{fus}}}} = -1.3 \frac{\text{volume}}{S_w b} \left(\frac{D_f}{W_f}\right) \tag{16.50}$$

Lateral-Directional Derivatives

The vertical-tail lateral derivatives were expressed in Eqs. (16.39) and (16.41). The lateral lift-curve slope is found using the methods in Chapter 12. The vertical-tail aspect ratio should be increased for the endplate effects of the fuselage and horizontal tail. Typically the effective aspect ratio will be about 55% higher than the actual aspect ratio. Also, the lateral lift-curve slope should be reduced by about 20% if the rudder hinge gap is not sealed.

The remaining unknowns in Eqs. (16.39) and (16.41) are the local dynamic pressure ratio and sideslip derivative. These can be estimated in an empirical Eq. (16.51) from Ref. 37; S'_{vs} is the area of the vertical tail extended to the fuselage centerline.

$$\left(\frac{\partial \beta_v}{\partial \beta} \eta_v\right) = 0.724 + \frac{3.06 \frac{S'_{\text{vs}}}{S_w}}{1 + \cos \Lambda} - 0.4 \frac{Z_{\text{wf}}}{D_f} + 0.009 A_{\text{wing}} \tag{16.51}$$

Thrust Effects on Lateral-Directional Trim and Stability

The thrust effects on the lateral trim and stability are similar to the longitudinal effects. There are direct-thrust moments, normal-force moments, and propwash or jet-induced effects.

When all engines are running, the direct-thrust moments cancel each other. The normal-force moments of the engines are additive.

When one engine fails, the remaining engine(s) produce(s) a substantial yawing moment. Also, the failed engine contributes an additional drag term as previously presented in Chapter 13.

The propwash dynamic pressure effect is estimated using Eq. (16.31). The propwash effect upon sidewash can be estimated using Eqs. (16.30) and (16.23) and then be applied to the result from Eq. (16.51).

(With the exception of rudder sizing for engine-out, the lateral analysis is frequently ignored in early conceptual design. To obtain good lateral results usually requires six-DOF analysis using wind-tunnel data. During early conceptual design, previous aircraft data and rule-of-thumb methods such as the tail volume coefficient are relied upon to select tail areas, dihedral angle, and the rudder and aileron areas.)

16.5 Stick-Free Stability

The preceding analysis has assumed that the control surfaces are rigidly held to the desired deflection. This "stick-fixed" assumption is reasonable for most modern fighters and large transports that employ fully powered flight-control systems.

Many smaller aircraft use purely manual or simply boosted control systems in which the airloads upon the control surfaces cause them to change deflection angle as the angles of attack and sideslip vary. Such a case requires a "stick-free" stability analysis.

A worst-case analysis for stick-free longitudinal stability would assume that the elevator "floats" up so much that it contributes nothing to the tail lift. In this case the percent reduction in total tail lift-curve slope would equal the elevator's area as a percent of total tail area.

This is generally not the case, and the elevator will usually float to a lesser angle depending upon the airfoil pressure distribution and the amount of "aerodynamic balance" (i.e., the portion of the elevator ahead of the hinge line). Data in Ref. 70 indicate that a typical free elevator with aerodynamic balance will reduce the total tail lift-curve slope by approximately 50% of the elevator's area as a percent of total tail area. Thus, a stick-free elevator which is 40% of the total tail area will experience a reduction in the tail slope of the lift curve of about 20%.

In fact, the elevator can be "overbalanced" so that it floats into the relative wind and therefore adds to the stability. However, this may produce unusual control forces. Due to the strong effect of the boundary layer, control-surface float is difficult to predict even with wind-tunnel data.

References 4 and 67 provide detailed methods for analyzing the stick-free stability based upon test data for control surface hinge moments. Typically the stick-free neutral point is 2–5% ahead of the stick-fixed neutral point.

Stick-free directional stability is also reduced as a result of rudder float. This can be approximated using the percent reduction in tail slope of the lift curve just described.

16.6 Effects of Flexibility

The preceding discussion also assumes that the aircraft is rigid. In fact, many aircraft are quite flexible, especially in fuselage longitudinal bending, wing spanwise bending, and wing torsional deflection. These can have a major effect upon the stability characteristics.

If the fuselage is flexible in longitudinal bending, the horizontal-tail incidence angle will reduce when the aircraft angle of attack is increased. This reduces the effectiveness of the tail as a restoring force for pitch stability. The vertical tail experiences the same effectiveness reduction due to lateral fuselage bending.

Similarly, a swept flexible wing will deflect such that the wing tips have a reduced angle of attack compared to the rigid aircraft. This reduces the slope of the lift curve and moves the wing aerodynamic center forward, destabilizing the aircraft. These effects are shown in Fig. 16.23.

A typical swept-wing transport at high subsonic speeds will experience a reduction in wing lift-curve slope of about 20%, a reduction in tail pitching-moment contribution of about 30%, and a reduction in elevator effectiveness of about 50% due to flexibility effects. The wing aerodynamic center will shift forward about 10% MAC due to flexibility.

In addition, the aileron effectiveness may be reduced by 50 to over 100%! At high dynamic pressures the ailerons will produce torsional moments on the wing that twist it in the opposite direction from the aileron deflection. This wing twist produces a rolling moment in the opposite direction from the desired rolling moment.

If the wing twists enough, this effect may overpower the aileron forces, producing "aileron reversal." To retain roll authority, many jet transports lock the outboard ailerons at high speeds and rely upon spoilers or small inboard ailerons.

Figure 16.24 shows the aileron reversal experienced with the B-47, the first transonic jet having thin, highly swept wings. Technically, this aircraft was in many ways the forerunner of today's jet transports. As can be seen, the ailerons had zero roll rate effectiveness at about 470 knots due to flexibility effects. At higher speeds, the ailerons worked backward; an "up" left aileron would twist the wing trailing edge down so much that lift would increase, causing a roll to the right rather than to the left as expected! By adding wing spoilers, roll control was possible at a much higher speed. Pilots were taught that if the spoilers failed to operate at a speed greater than 470 knots, they should simply move the control stick in the opposite direction from the way that they wished to roll.

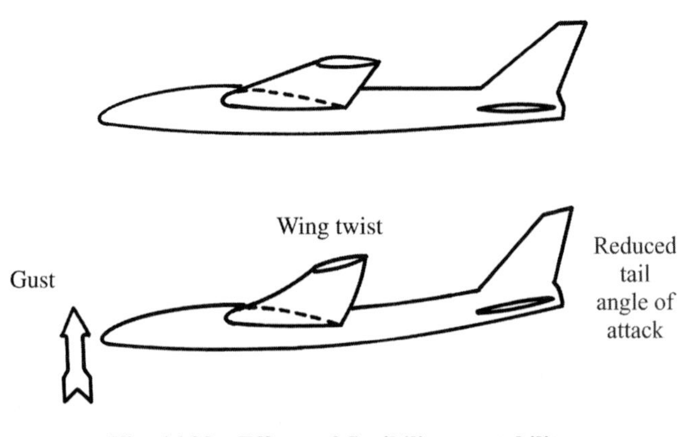

Fig. 16.23 Effects of flexibility on stability.

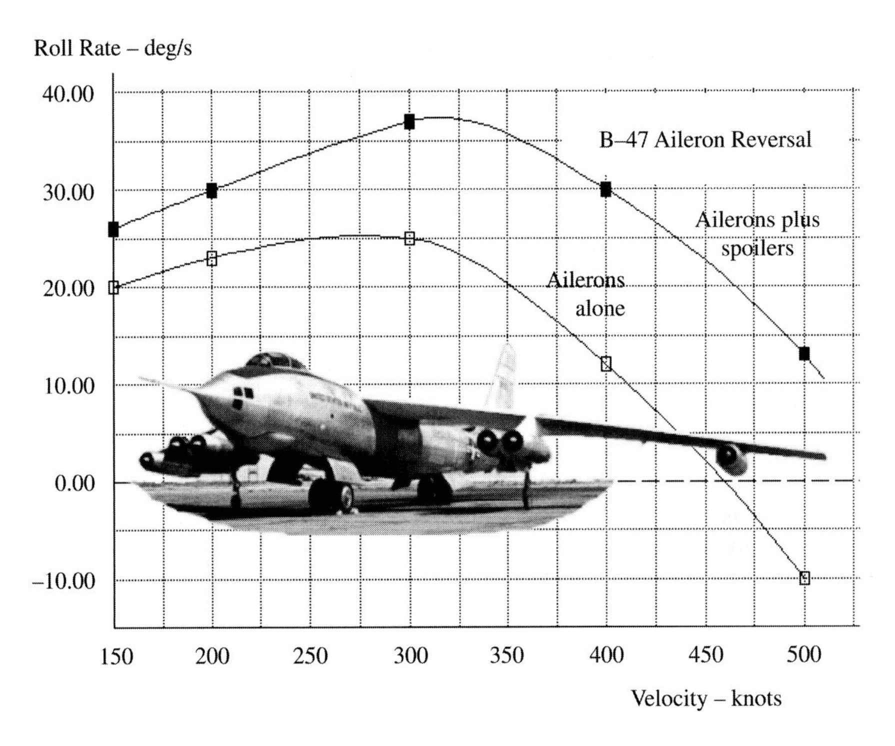

Fig. 16.24 Aileron reversal caused by flexibility effects.

Today, of course, we could use a computerized flight control system to do the same thing without the pilot ever being aware of it. This could save hundreds or even thousands of pounds compared to adding enough structural rigidity to avoid aileron reversal at all speeds.

These effects are functions of dynamic pressure, with the greatest impact seen at the low-altitude, high-speed condition. A "stiffer" aircraft such as a fighter, with a low wing aspect ratio and a short fuselage, will have less impact on its static stability derivatives due to flexibility.

16.7 Dynamic Stability

Dynamic stability concerns the motions of the aircraft, so two new classes of force must be considered: the inertia forces and the damping forces.

Mass Moments of Inertia

Inertia forces derive from the tendency of mass to resist accelerations. The mass for rotational accelerations is represented by "mass moment of inertia" terms, denoted by I. Mass moment of inertia describes a body's resistance to rotational accelerations, and is calculated by integrating the products of mass elements and the square of their distance from the Ref. axis.

For aircraft dynamic analysis, the mass moments of inertia about the three principal axes must be determined: I_{xx} about the roll axis, I_{yy} about the pitch axis, and I_{zz} about the yaw axis.

These can be initially determined using historical data based upon the nondimensional radii of gyration $\overline{R}$, as described in Ref. 11. Equations (16.52–16.54) are used with typical $\overline{R}$ values from Table 16.1.

$$I_{xx} = \frac{b^2 M \overline{R}_x^2}{4} = \frac{b^2 W \overline{R}_x^2}{4g} \tag{16.52}$$

$$I_{yy} = \frac{L^2 M \overline{R}_y^2}{4} = \frac{L^2 W \overline{R}_y^2}{4g} \tag{16.53}$$

$$I_{zz} = \left(\frac{b+L}{2}\right)^2 \frac{M \overline{R}_z^2}{4} = \left(\frac{b+L}{2}\right)^2 \frac{W \overline{R}_z^2}{4g} \tag{16.54}$$

Damping Derivatives

Aerodynamic damping forces resist motion. The rotational damping forces are proportional to the pitch rate Q, roll rate P, and yaw rate R. (Note: Avoid confusing Q with dynamic pressure, q).

Table 16.1 Nondimensional radii of gyration[a]

Aircraft class	$\overline{R}_x$	$\overline{R}_y$	$\overline{R}_z$
Single-engine prop	0.25	0.38	0.39
Twin-engine prop	0.34	0.29	0.44
Business jet twin	0.30	0.30	0.43
Twin turboprop transport	0.22	0.34	0.38
Jet transport—Fuselage-mounted engines	0.24	0.36	0.44
—2 wing-mounted engines	0.25	0.38	0.46
—4 wing-mounted engines	0.31	0.33	0.45
Military jet trainer	0.22	0.14	0.25
Jet fighter	0.23	0.38	0.52
Jet heavy bomber	0.34	0.31	0.47
Flying wing (B-49 type)	0.32	0.32	0.51
Flying boat	0.25	0.32	0.41

[a]Typical values, see Ref. 11 for examples.

These damping forces arise because of a change in effective angle of attack due to the rotational motion, as shown in Fig. 16.25 for the lift on the horizontal tail during a steady pitchup and for the lift on a segment of the wing during a steady roll. The lateral lift on a vertical tail in a steady yawing motion would change similarly to the horizontal tail.

The change in effective angle of attack, and hence the change in lift, is directly proportional to the rotation rate and the distance from the c.g. The moment is proportional to the lift times the distance from the c.g. Rotational damping moment is therefore proportional to the rotational rate and the square of the distance from the c.g.

Equations (16.55) and (16.56) provide first-order estimates of the pitch- and yaw-damping derivatives. The wing drag term in Eq. (16.56) accounts for the yaw-damping effect of the wing. Dynamic pressure ratios (η) for horizontal and vertical tails can be approximated as 0.9.

Roll damping is estimated with Fig. 16.26, based upon data in NACA 1098 covering the lower aspect ratios and NACA 868 covering the higher aspect ratios. The sweep factor is multiplied times the unswept damping derivative.

$$C_{m_Q} = -2.2\eta_h \frac{S_h}{S_w} C_{L_{\alpha h}} \left(\frac{X_{\text{ach}} - X_{\text{cg}}}{c} \right)^2 \tag{16.55}$$

$$C_{n_R} = -2.0\eta_v \frac{S_v}{S_w} C_{F_{\beta v}} \left(\frac{X_{\text{acv}} - X_{\text{cg}}}{c} \right)^2 - \frac{C_{D_{\text{wing}}}}{4} \tag{16.56}$$

There are also "cross-derivative" damping terms. The yaw rate will affect the roll moment, and the roll rate will affect the yaw moment. These are both functions of wing lift coefficient. As a rough approximation, the rolling moment due to yaw rate C_{ℓ_R} is about $C_L/4$ and the yawing moment due to roll rate C_{n_P} is about $-C_L/8$.

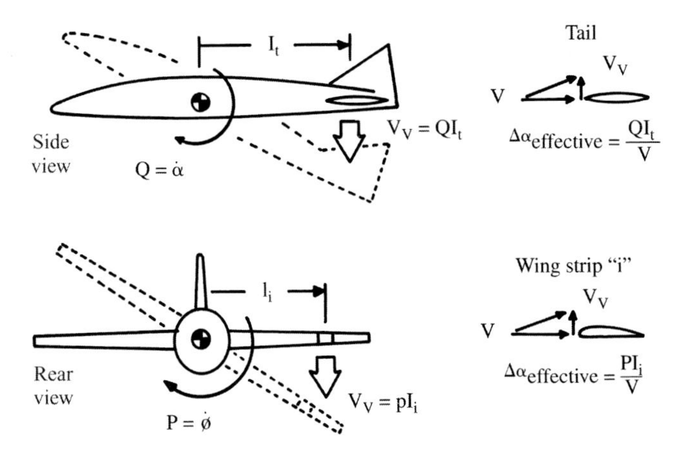

Fig. 16.25 Origin of damping forces.

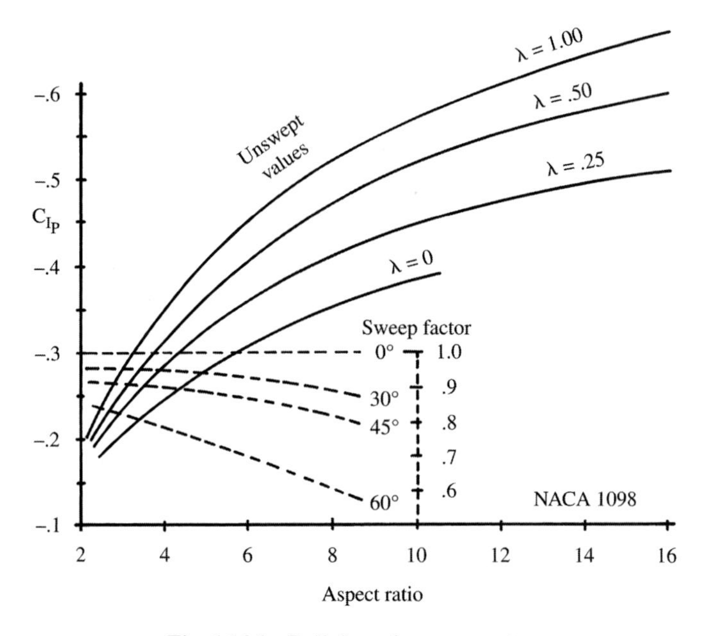

Fig. 16.26 Roll damping parameter.

One-DOF Dynamic Equations

A six-DOF analysis is required to fully evaluate aircraft dynamic stability and control. The six-DOF allows simultaneous rotations in pitch, yaw, and roll, and allows the aircraft velocity to change in the vertical, lateral, and longitudinal directions. All these motions affect each other, requiring a tremendous number of cross derivatives to account fully for all forces and moments. References 37 and 67 are recommended for the equations for six-DOF analysis.

The one-DOF equations may be used for initial analysis of several flight conditions, such as pull-up and steady roll. The one-DOF rotation equations are based upon the fact that the rotational acceleration times the mass moment of inertia equals the sum of the applied moments (which includes the damping moments). Equations (16.57–16.59) provide these:

$$\text{Pitch:}\quad I_{yy}\dot{Q} = qS_w cC_{m_\alpha}\alpha + qS_w cC_{m_Q}Q \tag{16.57}$$

$$\text{Yaw:}\quad I_{zz}\dot{R} = qS_w bC_{n_\beta}\beta + qS_w bC_{n_R}R \tag{16.58}$$

$$\text{Roll:}\quad I_{xx}\dot{P} = qS_w bC_\ell + qS_w bC_{\ell_P}P \tag{16.59}$$

These are second-order differential equations since Q, R, and P are the derivatives with time of pitch, yaw, and roll. Note that there is no first-order term in the roll equation since the roll angle does not affect the roll moments if the sideslip remains zero.

Aircraft Dynamic Characteristics

With proper input data, these one-DOF equations may be solved for time history after a given disturbance. However, the results will be incorrect since real aircraft motions always involve more than one-DOF. Longitudinal analysis requires a minimum of three-DOF to account for the interplay between pitch angle, vertical velocity, and changes in horizontal velocity. An additional equation is required for elevator deflection in a stick-free analysis.

Lateral analysis with stick fixed also requires a minimum of three-DOF, which account for lateral velocity, sideslip angle, and rolling angle. For stick-free lateral analysis, two additional equations are required to account for the aileron and rudder deflections. A full six-DOF (nine-DOF for stick-free) is preferable because of the interplay between lift coefficient and the lateral derivatives, especially at higher angles of attack.

Analytical techniques for three- or six-DOF simulations are beyond the scope of this book, but a few comments on typical results are in order. Longitudinally, there are two oscillatory solutions to the equations of motion. One is a short-period mode, which is typically heavily damped and provides the desired dynamic stability in response to a pitch disturbance. The other solution is a long-period lightly damped mode called the "pitch phugoid." This involves a slow pitch oscillation over many seconds in which energy is exchanged between vertical and forward velocity. Many aircraft have a slight unnoticeable pitch phugoid. An excessive phugoid should be avoided.

The lateral equations of motion yield three solutions to a yaw disturbance. One is the desired heavily damped direct convergence. The spiral divergence mode, another solution, involves an increasing bank angle with the aircraft turning tighter and tighter until control is lost. However, the time to diverge is so long that pilots can easily correct for spiral divergence.

The third lateral solution, a short-period oscillation called Dutch roll, sees the aircraft waddle from side to side, exchanging yaw and roll. If the Dutch roll is excessive, this oscillation will be objectionable to passengers and crew. Dutch roll is largely caused by the dihedral effect.

Dutch roll damping is determined mainly by the size of the vertical tail, and is usually the driving criteria for tail sizing other than engine-out control. For this reason, vertical-tail size should not be reduced below the size indicated by the tail volume coefficient method until a six-DOF analysis has been conducted, preferably with wind-tunnel data for the dynamic derivatives.

Dutch roll is aggravated by flexibility effects at high speeds. Most large, swept-wing aircraft use a powered rudder mechanized with a gyro to deflect into a yaw, thus increasing the effective Dutch roll damping.

16.8 Quasi Steady State

Setting the rotational accelerations in Eqs. (16.57–16.59) to zero yields quasi-steady-state equations. These represent a steady pitch, yaw, or roll rate and are identical to the steady-state trim equations presented earlier, but with the addition of damping terms.

Pull-up

Pull-up is a quasi-steady-state trim condition in which the aircraft accelerates vertically at a load factor n. Level flight implies that $n = 1$. The longitudinal-trim equation previously presented [Eq. (16.7)], with the addition of the pitch damping moment (C_{m_Q} times Q), is solved to provide a total aircraft lift equal to n times the aircraft weight. The required elevator deflection is then determined from the required tail lift. The pitch rate Q is related to the load factor in a pull-up as follows:

$$Q = \frac{g(n-1)}{V} \tag{16.60}$$

Level Turn

A level turn is similar to the pull-up in that the aircraft experiences an increased load factor and a steady pitch rate. Note that the sideslip remains zero during a coordinated turn so that the level turn is strictly a longitudinal problem! The load factor due to a bank angle ϕ is obtained from Eq. (16.61), and the resulting pitch rate is obtained from Eq. (16.62).

$$n = 1/\cos\phi \tag{16.61}$$

$$Q = \frac{g}{V}\left(n - \frac{1}{n}\right) \tag{16.62}$$

Steady Roll

The steady roll is found by setting Eq. (16.59) to zero. Equation (16.40) C_l indicates that the only rolling-moment term that remains when the sideslip equals zero is the roll due to aileron deflection. This leads to Eq. (16.63), which is solved for roll rate (radians) as a function of aileron deflection in Eq. (16.64).

$$I_{xx}\dot{P} = 0 = qS_w bC_{\ell_{\delta_a}}\delta_a + qS_w bC_{\ell P}P \tag{16.63}$$

$$P = -\left(\frac{C_{\ell\delta_a}}{C_{\ell_p}}\right)\delta_\alpha \tag{16.64}$$

For many years the roll-rate requirement was based upon the wing helix angle $Pb/2V$. NACA flight tests (NACA 715) determined that most pilots consider an aircraft to have a good roll rate if the wing helix angle is at least equal to 0.07 (0.09 for fighters).

Military specifications (MIL-F-8785B or Mil Std 1797) require that the aircraft reach a certain roll angle in a given number of seconds, as noted in Table 16.2. These assume that the aircraft is in level flight upon initiation of the roll, so the rotational acceleration should be accounted for. However, aircraft generally reach maximum roll rate quickly; the quasi-steady-state roll rate therefore may be used to initially estimate the time to roll.

16.9 Inertia Coupling

The F-100 prototype, the first fighter capable of level supersonic flight, featured a thin swept wing and long heavy fuselage compared to previous fighters.

Table 16.2 MIL-F-8785 B roll requirements

Class	Aircraft type	Required roll
I	Light utility, observation, primary trainer	60 deg in 1.3 s
II	Medium bomber, cargo, transport, ASW, recce.	45 deg in 1.4 s
III	Heavy bomber, cargo, transport	30 deg in 1.5 s
IV A	Fighter-attack, interceptor	90 deg in 1.3 s
IV B	Air-to-air dogfighter	{ 90 deg in 1.0 s 360 deg in 2.8 s
IV C	Fighter with air-to-ground stores	90 deg in 1.7 s

During flight testing, a series of high-speed rolls suddenly diverged in angle of attack and sideslip, much to the surprise of all concerned. Detailed analysis and simulation discovered the cause to be "inertia coupling."

Figure 16.27 shows a typical fighter in roll. The mass of the forebody and aft-fuselage are concentrated like a barbell for illustrative purposes.

Like all objects, the fighter tends to roll about its principal (longitudinal) axis. However, if the fighter rolled 90 deg about its longitudinal axis, the angle of attack would be exchanged with the angle of yaw, as shown. The C_{n_β} effect of the vertical tail would oppose this increase in yaw angle with roll.

In addition, the aileron rolling moments are about the wind axis. The aircraft thus actually rolls around an axis somewhere between the principal axis and the wind axis.

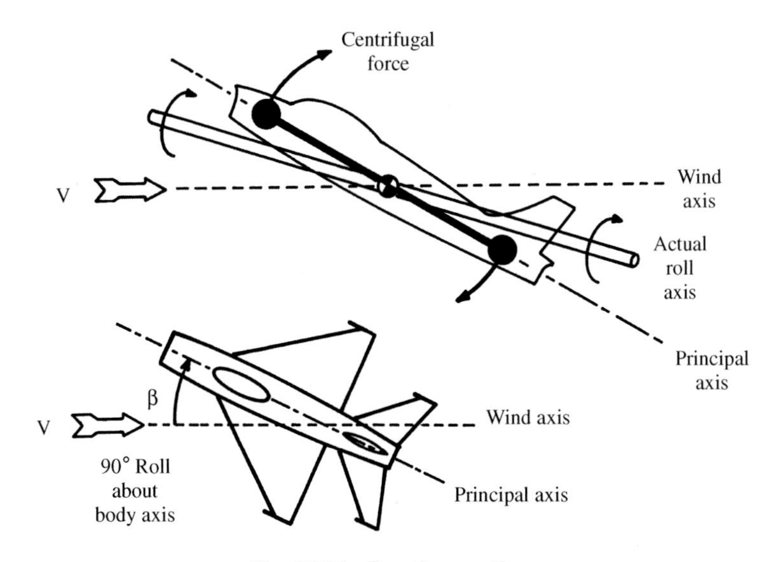

Fig. 16.27 Inertia coupling.

The masses of the forebody and aft-fuselage are above and below this actual roll axis. Centrifugal force tends to pull them away from the roll axis, creating a nose-up pitching moment. The combination of the increase in yaw angle with roll and the nose-up pitching moment due to inertia is called inertia coupling.

Inertia coupling becomes a problem only when the moments produced by the inertia forces are stronger than the aerodynamic restoring moments. This is most likely to happen at high altitudes (lower air density) and at high Mach numbers where the tail loses lift effectiveness.

The solution to inertia coupling in the F-100 was a larger vertical tail. This remains the typical solution. For this reason the vertical-tail area should not be reduced below the statistical tail-volume-method result until a more detailed analysis is available.

16.10 Handling Qualities

Cooper–Harper Scale

Aircraft handling qualities are a subjective assessment of the way the plane feels to the pilot. Few modern pilots fully appreciate the great advances in handling qualities made since the dawn of aviation. Early fighters such as the Fokker Eindecker had handling qualities which were so poor that the pilots felt that without constant attention, the aircraft would "turn itself inside out or literally swap ends" (movie stunt pilot Frank Tallman, quoted from Ref. 71).

A number of "goodness" criteria such as the wing helix angle have already been discussed. It is important that the aircraft have a nearly linear response to control inputs and that the control forces be appropriate for the type of aircraft. The control forces required due to flap deflection or power application should be small and predictable.

These handling-qualities criteria are generally considered later in the design cycle. Figure 16.28 illustrates the Cooper–Harper Handling Qualities Rating Scale, which is used by test pilots to categorize design deficiencies (Ref. 72). Handling qualities are discussed in detail in Ref. 69.

Departure Criteria

One of the most important aspects of handling qualities is the behavior of the aircraft at high angles of attack.

As the angle of attack increases, a "good" airplane experiences mild buffetting to warn the pilot, retains control about all axes, and stalls straight ahead with immediate recovery and no tendency to enter a spin. If a spin is forced, the "good" airplane can be immediately recovered.

A "bad" airplane loses control in one or more axes as angle of attack increases. A typical bad characteristic is the loss of aileron roll control and an increase in aileron adverse yaw. When the aircraft is near the stall angle of attack, any minor yaw resulting from aileron deflection may slow down one wing enough to stall it. With only one wing generating lift, the "bad" airplane will suddenly depart into a spin or other uncontrolled flight mode.

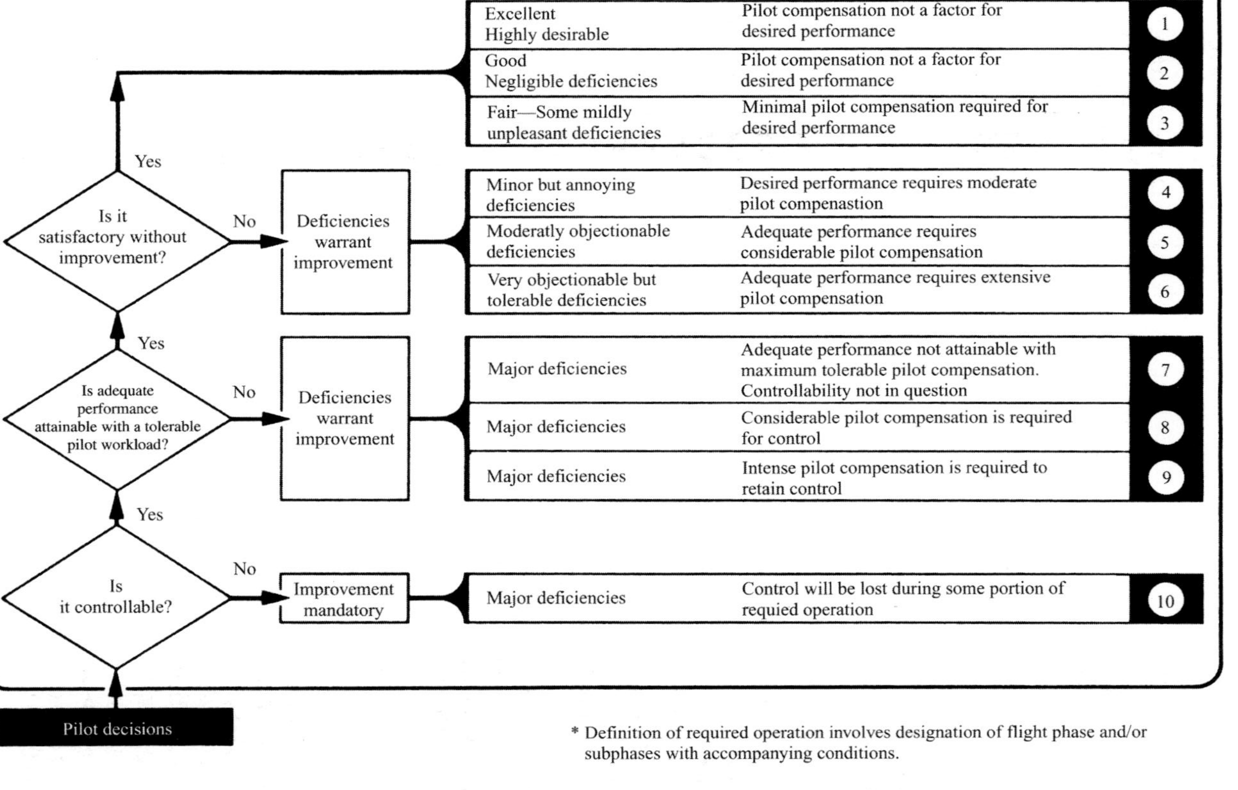

Fig. 16.28 Cooper–Harper Handling Qualities Rating Scale.

Design features for good departure and spin characteristics have been discussed in earlier chapters. There have been many criteria proposed for good departure characteristics. Several aerodynamic coefficients are important to departure characteristics, especially C_{n_β}, $C_{n_{\delta a}}$, C_{ℓ_β}, and $C_{\ell_{\delta a}}$.

These are combined in the lateral control departure parameter (LCDP), sometimes called the lateral control spin parameter or the aileron-alone divergence parameter [Eq. (16.65)]. The LCDP focuses upon the relationship between adverse yaw and directional stability.

Equation (16.66) shows another departure parameter, $C_{n_{\beta\mathrm{dynamic}}}$, which includes the effects of the mass moments of inertia. Both of these parameters should be positive for good departure resistance. A typical goal is to have $C_{n_{\beta\mathrm{dynamic}}}$ greater than 0.004.

$$\mathrm{LCDP} = C_{n_\beta} - C_{\ell_\beta}\frac{C_{n_{\delta a}}}{C_{\ell_{\delta a}}} \tag{16.65}$$

$$C_{n_{\beta\mathrm{dynamic}}} = C_{n_\beta}\cos\alpha - \frac{I_{zz}}{I_{xx}}C_{\ell_\beta}\sin\alpha \tag{16.66}$$

Figure 16.29 shows a crossplot of the LCDP and $C_{n_{\beta\mathrm{dynamic}}}$ with increase in angle of attack. In Ref. 73 the boundaries for acceptable departure resistance were determined from high-g simulator tests using experienced pilots. The earlier Weissman criteria are also shown.

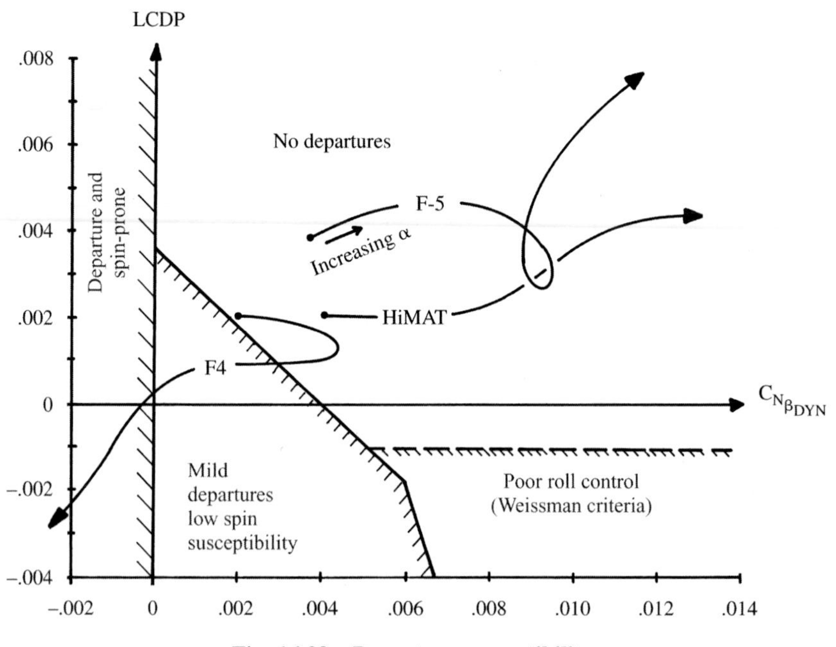

Fig. 16.29 Departure susceptibility.

Note the departure-parameter crossplot for the F-5. This aircraft is widely considered to be one of the best fighters at high angle of attack. Both departure parameters are increasing with angle of attack.

On the other hand, the F-4 has poor departure characteristics. Its departure-parameter crossplot starts in the acceptable zone, but crosses into the unacceptable zone as angle of attack increases.

The HiMat fighter shows that even an advanced supersonic canard configuration can have good departure characteristics. The HiMat has highly cambered outboard wing leading edges and has large twin tails with a substantial portion below the wing.

Unfortunately, the stability derivatives used to calculate these departure parameters become very nonlinear near the stall. First-order estimation techniques used in conceptual design may not give usable results for departure estimation. However, the configuration designer can expect to be instructed to "fix it" when the first wind-tunnel data are available!

There are a few design rules that can be applied during early configuration layout. The fuselage forebody shape has a huge effect upon the stability characteristics at high angles of attack. This is mostly due to a tendency of vortices to form asymmetrically, i.e., stronger on one side than the other thus pulling the nose strongly to one side. An elliptical nose cross section that has width greater than height is desirable. Also, some sort of strake or sharp edge on each side of the nose tends to create symmetric vortices, reducing this problem.

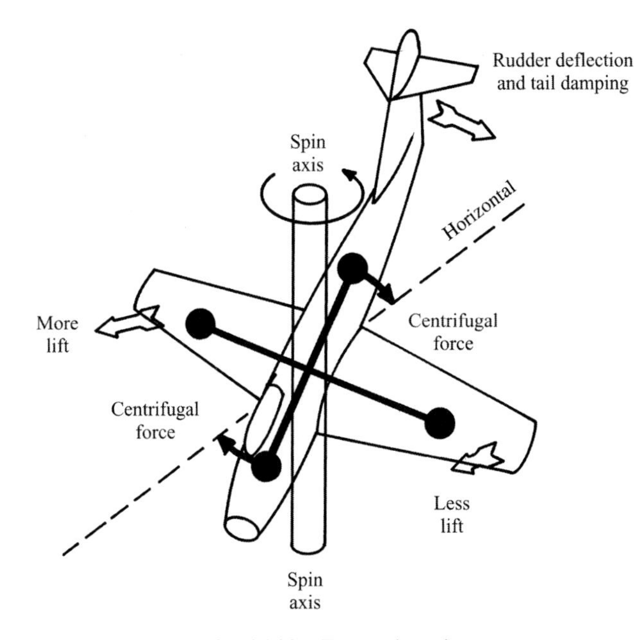

Fig. 16.30 Forces in spin.

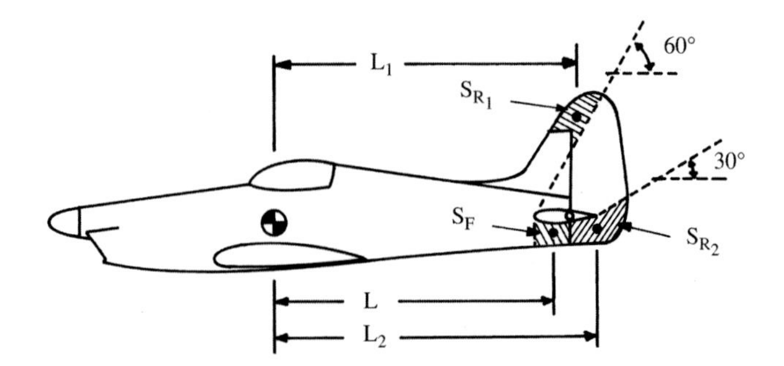

Fig. 16.31 Geometry for spin recovery estimation.

Wing-tip stalling should be prevented by the use of wing twist, fences, notches, or movable leading-edge devices. It is also desirable for departure prevention to have a substantial ventral-tail surface.

Spin Recovery

After stall, a spin will develop in a "bad" airplane or a good airplane severely abused. Figure 16.30 shows the forces acting in a fully developed spin. The fuselage and wing masses are represented by barbells. The centrifugal

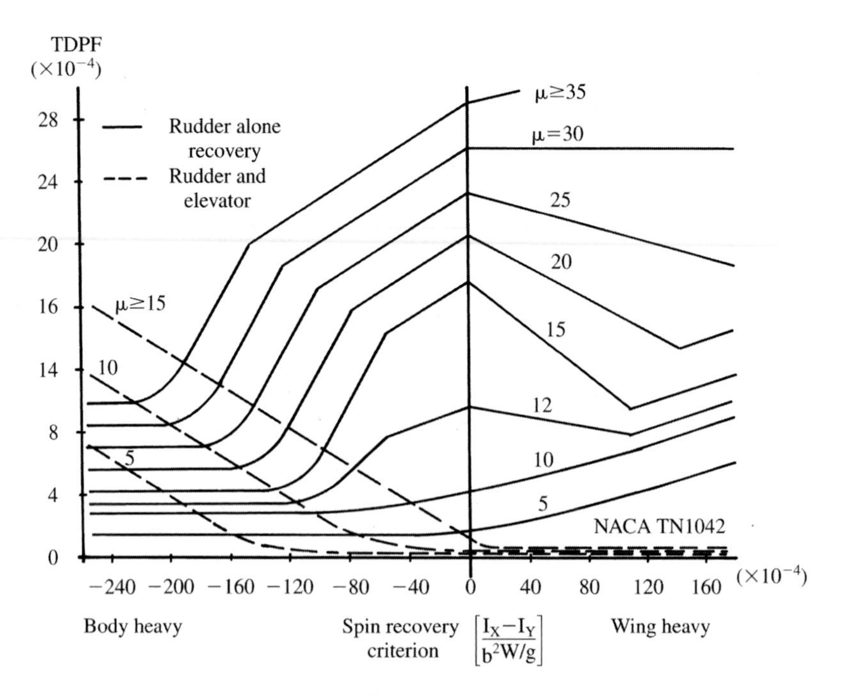

Fig. 16.32 Spin recovery criteria.

forces acting on the fuselage tend to raise the nose, further increasing the wing stall.

The spin is primarily driven by the difference in lift between the outer, faster wing and the inner, slower wing, which is more fully stalled. The spin is opposed by damping forces, primarily from portions of the aft fuselage and vertical tail underneath the horizontal tail (S_F—see Fig. 16.31).

For recovery, the rudder is deflected against the spin. However, only the part of the rudder not blanketed by the stalled air from the horizontal tail will aid the recovery (S_{R_1} and S_{R_2}).

Figure 16.32 presents an empirical estimation of the required tail damping and rudder area for spin recovery for straight-winged aircraft (Ref. 74). This determines the minimum allowable tail-damping power factor (TDPF), defined in Eq. (16.67) where TDR is the tail damping ratio [Eq. (16.68)] and URVC is the unshielded rudder volume coefficient [Eq. (16.69)]. The airplane relative density parameter (μ) is defined in Eq. (16.70).

$$\mathrm{TDPF} = (\mathrm{TDR})(\mathrm{URVC}) \tag{16.67}$$

$$\mathrm{TDR} = \frac{S_F L^2}{S_w (b/2)^2} \tag{16.68}$$

$$\mathrm{URVC} = \frac{S_{R_1} L_1 + S_{R_2} L_2}{S_w (b/2)} \tag{16.69}$$

$$\mu = \frac{W/S}{\rho g b} \tag{16.70}$$

This empirical estimation technique is dominated by the ability of the rudder, vertical tail, and aft fuselage to oppose the aircraft's rotation in the spin. One can also delay spin entry or enhance spin recovery by reshaping the wing leading edges to minimize the lift imbalance, typically with a drooped leading edge near the wing tips. This, however, imposes some drag penalty during regular flight.

Thunderbird F-16 showing strakes (U.S. Air Force photo).

17
Performance and Flight Mechanics

17.1 Introduction and Equations of Motion

The last chapter discussed stability and control, which largely concern the rotational motions of the aircraft in pitch, yaw, and roll. This chapter introduces flight mechanics, the study of aircraft translational motions. The geometry for flight mechanics is shown in Fig. 17.1.

The climb angle γ is the angle between horizontal and the wind (stability) X axis (Xs). The climb gradient (G), the tangent of the climb angle, represents the vertical velocity divided by the horizontal velocity.

Summing forces in the Xs and Zs directions yield Eqs. (17.1) and (17.2). The resulting accelerations on the aircraft in the Xs and Zs directions are determined as these force summations divided by the aircraft mass (W/g):

$$\Sigma F_x = T\cos(\alpha + \phi_T) - D - W\sin\gamma \tag{17.1}$$

$$\Sigma F_z = T\sin(\alpha + \phi_T) + L - W\cos\gamma \tag{17.2}$$

$$\dot{W} = -CT \tag{17.3}$$

$$C = C_{\text{power}}\frac{V}{\eta_p} = C_{\text{bhp}}\frac{V}{550\ \eta_p} \tag{17.4}$$

$$T = P\eta_p/V = 550\ \text{bhp}\ \eta_p/V \tag{17.5}$$

Equation (17.3) defines the time rate of change in aircraft weight as the specific fuel consumption (C) times the thrust. For a piston-propeller engine, Eq. (17.4) determines the equivalent C based upon the piston-engine definition of C_{power} or C_{bhp} (see Chapter 5), and Eq. (17.5) determines the thrust of the propeller.

These simple equations are the basis of the most detailed sizing and performance programs used by the major airframe companies. The angle of attack and thrust level are varied to give the required total lift (including load factor) and the required longitudinal acceleration depending upon what maneuver the aircraft is to perform (level cruise, climb, accelerate, turn, etc.). Angle of attack and lift are restricted by the maximum lift available. The thrust level is restricted to the available thrust, as obtained from a table of installed engine thrust vs altitude and velocity (or Mach number).

What makes the sizing and performance programs complicated is not the actual calculation of the aircraft response to the forces at a given angle of

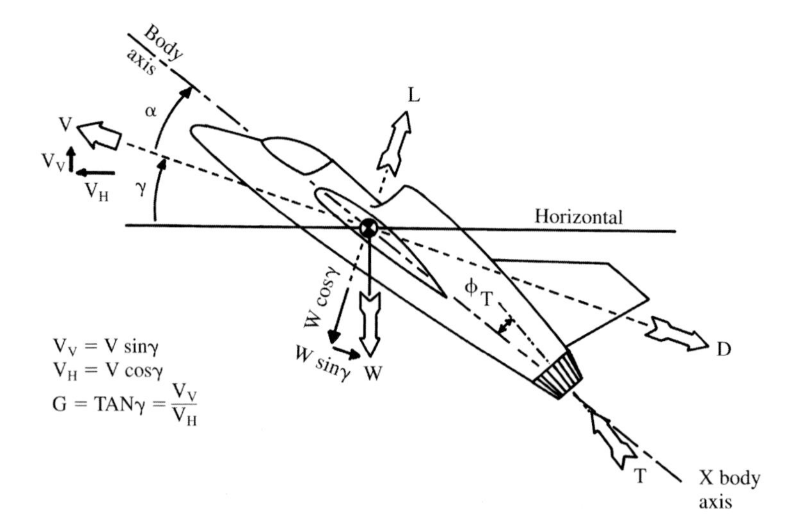

Fig. 17.1 Geometry for performance calculation.

attack and thrust level. The complications arise in determining what the angle of attack and thrust level should be to perform some maneuver.

For example, the rate of climb varies with velocity. What combination of velocity and thrust setting will allow an airliner to climb to cruise altitude with the least fuel consumption over the total mission? This chapter will address such performance issues.

For most aircraft the thrust axis has little incidence with respect to the wind axis under most flight conditions. This is by design, and permits simplifying Eqs. (17.1) and (17.2) to the forms shown in Eqs. (17.6) and (17.7).

$$\Sigma F_x = T - D - W \sin\gamma \tag{17.6}$$

$$\Sigma F_z = L - W \cos\gamma \tag{17.7}$$

A word of caution: Be especially careful with fps (British) units in the performance calculations. Apply each equation to the units of the data you are using to be sure that all units cancel, leaving you with the units of the desired answer. Be wary of equations involving horsepower. Anytime the constant "550" appears in an equation, the other units must be converted to feet, pounds, and seconds (one bhp = 550 ft-lb/s). The more logical metric system avoids such confusion. Another potential source of confusion is the specific fuel consumption C, which is usually given in units of hours^{-1} (actually lb-fuel per hour per lb-thrust!). This must be divided by 3600 to yield seconds^{-1}. Also, see Author's Note on use of metric equivalent units.

17.2 Steady Level Flight

If the aircraft is flying in unaccelerated level flight, then climb angle γ equals zero and the sum of the forces must equal zero. This leads to Eqs. (17.8) and

(17.9), the most simple versions of the translational equations of motion. They state simply that, in level flight, thrust equals drag and lift equals weight. These are expressed using aerodynamic coefficients for the analysis that follows.

$$T = D = qS(C_{D_0} + KC_L^2) \tag{17.8}$$

$$L = W = qSC_L \tag{17.9}$$

$$V = \sqrt{\frac{2}{\rho C_L}\left(\frac{W}{S}\right)} \tag{17.10}$$

From Eq. (17.9), the velocity in level flight can be expressed as a function of wing loading, lift coefficient, and air density [Eq. (17.10)].

These equations imply that the actual T/W in level flight must be the inverse of the L/D at that flight condition [Eq. (17.11)]. The T/W and L/D in level flight can be expressed in terms of the wing loading and dynamic pressure by substituting Eq. (17.9) into Eq. (17.8), as follows:

$$\frac{T}{W} = \frac{1}{L/D} = \frac{qC_{D_0}}{(W/S)} + \left(\frac{W}{S}\right)\frac{K}{q} \tag{17.11}$$

Minimum Thrust Required for Level Flight

From Eq. (17.11) it follows that the condition for minimum thrust at a given weight is also the condition for maximum L/D. To find the velocity at which thrust is minimum and L/D is maximum, the derivative with respect to velocity of Eq. (17.11) is set to zero. This is shown in Eq. (17.12), and solved in Eq. (17.13) for the velocity at which the required thrust is minimum and the L/D is at a maximum.

$$\frac{\partial(T/W)}{\partial V} = \frac{\rho V C_{D_0}}{W/S} - \frac{W}{S}\frac{2K}{\frac{1}{2}\rho V^3} = 0 \tag{17.12}$$

$$V_{\text{min thrust or drag}} = \sqrt{\frac{2W}{\rho S}\sqrt{\frac{K}{C_{D_0}}}} \tag{17.13}$$

$$C_{L\,\text{min thrust or drag}} = \sqrt{\frac{C_{D_0}}{K}} \tag{17.14}$$

Substituting this velocity into Eq. (17.9) yields the lift coefficient for minimum drag in level flight [Eq. (17.14)]. This optimal lift coefficient is only dependent upon the aerodynamic parameters. At any given weight, the aircraft can be flown at the optimal lift coefficient for minimum drag by varying velocity or air density (altitude).

If the lift coefficient for minimum drag is substituted back into the total-drag Eq. (17.8), the induced-drag term will equal the zero-lift drag term. The total drag at the lift coefficient for minimum drag will then be exactly twice the zero-lift drag [Eq. (17.15)].

$$D_{\substack{\text{min thrust}\\ \text{or drag}}} = qS\left[C_{D_0} + K\left(\sqrt{\frac{C_{D_0}}{K}}\right)^2\right] = qS(C_{D_0} + C_{D_0}) \tag{17.15}$$

Minimum Power Required for Level Flight

The conditions for minimum thrust and minimum power required are not the same. Power is force times velocity, which in steady level flight equals the drag times the velocity as shown in Eq. (17.16). Substituting the lift coefficient in level flight from Eq. (17.9) yields Eq. (17.17).

$$P = DV = qS(C_{D_0} + KC_L^2)V = \frac{1}{2}\rho V^3 S(C_{D_0} + KC_L^2) \tag{17.16}$$

$$P = \frac{1}{2}\rho V^3 SC_{D_0} + \frac{KW^2}{\frac{1}{2}\rho VS} \tag{17.17}$$

The velocity for flight on minimum power is obtained by setting the derivative of Eq. (17.17) to zero, as shown in Eqs. (17.18) and (17.19). Substituting this into Eq. (17.9) yields the lift coefficient for minimum power, Eq. (17.20). Substituting this into Eq. (17.8) gives the drag at minimum power required [Eq. (17.21)]:

$$\frac{\partial P}{\partial V} = \frac{3}{2}\rho V^2 SC_{D_0} - \frac{KW^2}{\frac{1}{2}\rho V^2 S} = 0 \tag{17.18}$$

$$V_{\substack{\text{min}\\ \text{power}}} = \sqrt{\frac{2W}{\rho S}\sqrt{\frac{K}{3C_{D_0}}}} \tag{17.19}$$

$$C_{L\,\substack{\text{min}\\ \text{power}}} = \sqrt{\frac{3C_{D_0}}{K}} \tag{17.20}$$

$$D_{\substack{\text{min}\\ \text{power}}} = qS(C_{D_0} + 3C_{D_0}) \tag{17.21}$$

Note that the velocity for minimum power required is approximately 0.76 times the velocity for minimum thrust [Eq. (17.13)]. The aircraft is flying at a lift coefficient for minimum power, which is about 73% higher than the lift coefficient for minimum drag [Eq. (17.14)].

The induced drag at the lift coefficient for minimum power is exactly three times the zero-lift drag, so the total drag is four times the zero-lift drag [Eq. (17.21)]. This drag coefficient is twice as high as at minimum drag [Eq. (17.15)].

Remember that at the minimum-power condition the aircraft is flying at a slower speed (reduced dynamic pressure) than at the minimum-drag condition.

The actual drag increase will thus be less than the factor of two indicated by the drag coefficients. The actual drag increase is 2.0 times the ratio of dynamic pressures (0.76^2), or only 15.5% higher than the total drag at minimum-drag conditions. Thus, the L/D when flying at the velocity for minimum power required is $1/1.155$, or 0.866 times the maximum L/D.

Graphical Analysis for Thrust and Power Required

The analytical optimizations in the last two sections depend upon the assumptions that the zero-lift drag coefficient is constant with velocity, that the drag due to lift follows the parabolic approximation, and that K is constant with velocity. As seen in Chapter 12, these assumptions are not very good other than for an aircraft with a high-aspect-ratio wing that is flying at low Mach numbers.

To determine the actual thrust (or power) required for level flight, the aerodynamic results are plotted vs velocity or Mach number and compared to the engine data.

For piston-powered aircraft, power is virtually constant with velocity. The only power variation with velocity is due to ram pressure in the intake manifold. For jet aircraft, equivalent power varies widely with velocity but thrust is roughly constant with velocity.

It is therefore common practice to graph the propulsive requirements of an aircraft vs velocity (or Mach number), using thrust for jet aircraft and using power for propeller aircraft. These are shown in Fig. 17.2. The power required is found by multiplying the drag by the velocity. The equivalent thrust for the propeller aircraft is also shown for illustration, but is not commonly plotted.

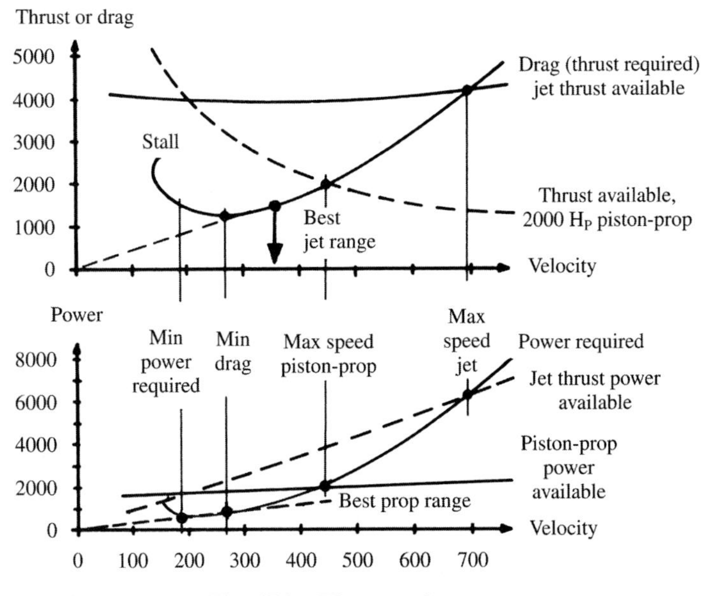

Fig. 17.2 Thrust and power.

The velocities for minimum thrust and minimum power are shown. Note that the minimum-power-required velocity is about 86.6% of the minimum-thrust-required velocity, as predicted in the last section. Also, the superiority of the jet engine for high-speed flight should be clear from this illustration.

The excess thrust at full throttle is determined simply by subtracting the thrust required from the thrust available. This excess can be used to accelerate or climb, as discussed later.

Such a plot of thrust or power vs velocity is different at each altitude.

Range

The range of an aircraft is its velocity multiplied by the amount of time it can remain in the air. Time in the air equals the amount of fuel carried divided by the rate at which the fuel is burned. This in turn is the required thrust multiplied by the specific fuel consumption.

Unfortunately, the simple equation implied by the last paragraph is complicated by the fact that the aircraft weight drops as fuel is burned. This changes the drag, which then changes the thrust required. Net result: the aircraft goes farther but the calculation is more difficult!

However, the "instantaneous range" derivative can be calculated using the simple relationship just described, which is expressed in Eq. (17.22). This describes the additional distance the aircraft will travel with the next incremental amount of fuel burned. This can also be expressed in terms of the L/D and weight, as shown. Instantaneous range is a commonly used measure of merit and is usually discussed in units of nautical miles per pound of fuel.

$$\frac{dR}{dW} = \frac{V}{-CT} = \frac{V}{-CD} = \frac{V(L/D)}{-CW} \tag{17.22}$$

$$R = \int_{W_i}^{W_f} \frac{V(L/D)}{-CW} dW = \frac{V}{C}\frac{L}{D}\ln\left(\frac{W_i}{W_f}\right) \tag{17.23}$$

Integrating the instantaneous range with respect to the change in aircraft weight yields the Breguet range equation [Eq. (17.23)]. This integration assumes that the velocity, specific fuel consumption, and L/D are approximately constant.

These assumptions require that the aircraft hold lift coefficient constant. To hold the lift coefficient constant as the aircraft becomes lighter requires reducing the dynamic pressure. Since velocity is also being held constant, the only way to reduce dynamic pressure is to reduce air density by climbing. This results in a flight path known as the cruise-climb, which maximizes range.

The cruise-climb is not normally permitted for transport aircraft because of the desire by air-traffic control to keep all aircraft at a constant altitude and airspeed. It is possible to develop a rather messy range equation under these assumptions.

However, the Breguet range equation can be applied with little loss of accuracy by breaking the cruise legs into several shorter mission-segments, using the appropriate L/D values as aircraft weight drops.

On a long flight, air traffic control may permit a "stairstep" flight path in which the aircraft climbs to a more optimal altitude several times during the cruise as fuel is burned off.

Range Optimization—Jet

The Breguet range equation can be applied equally well to jets or propeller aircraft, with the use of Eq. (17.4) to determine an equivalent thrust specific fuel consumption for the propeller aircraft. However, the conditions for maximum range differ for jets and props because of the effect of velocity on thrust for the propeller.

The terms in the Breguet range equation that do not involve the weight change [i.e., $(V/C)\,(L/D)$] are known as the "range parameter" and are a measure of the cruising performance. For subsonic jet aircraft the specific fuel consumption is approximately independent of velocity and the range parameter can be expanded as shown in Eq. (17.24).

Setting the derivative of Eq. (17.24) with respect to velocity equal to zero yields Eq. (17.25), the velocity for best range for a jet. The resulting lift coefficient and drag are given in Eqs. (17.26) and (17.27).

$$\frac{V}{C}\left(\frac{L}{D}\right) = \frac{V}{C}\left(\frac{C_L}{C_{D_0} + KC_L^2}\right) = \frac{2W/\rho VS}{CC_{D_0} + (4KW^2C)/(\rho^2V^4S^2)} \tag{17.24}$$

$$V_{\substack{\text{best}\\ \text{range}}} = \sqrt{\frac{2W}{\rho S}\sqrt{\frac{3K}{C_{D_0}}}} \tag{17.25}$$

$$C_{L_{\substack{\text{best}\\ \text{range}}}} = \sqrt{\frac{C_{D_0}}{3K}} \tag{17.26}$$

$$D_{\substack{\text{best}\\ \text{range}}} = qS\left(C_{D_0} + \frac{C_{D_0}}{3}\right) \tag{17.27}$$

Note that the drag coefficient for best range for a jet is 1.33 times the zero-lift drag coefficient. This is a lower drag coefficient than the drag coefficient for best L/D, which was shown to be 2.0 times the zero-lift coefficient. However, when maximizing range, the aircraft flies at a higher velocity [31.6% faster—divide Eq. (17.25) by Eq. (17.13)]. This increases the dynamic pressure, which increases the actual drag magnitude.

As a result, the actual drag while flying at the velocity for best range will be higher than the drag at the velocity for best L/D. The ratio between the drags at the best range velocity and the best L/D velocity is determined as the ratio of drag coefficients (1.33/2.0) multiplied by the ratio of dynamic pressures (1.316^2), or about 1.154.

Since drag is in the denominator of L/D, the L/D at the velocity for best range will be found to be 86.6% of the best L/D ($1/1.154 = 0.866$). This result was presented without proof in Chapter 5.

These range optimization equations were based on the assumption that the range parameter $(V/C)(L/D)$ does not vary with weight as Eq. (17.23) is integrated, which we attempt to provide by holding a constant lift coefficient during cruise. We do this by climbing, but eventually that will change specific fuel consumption (C) because it is a function of altitude for jet and prop engines. Furthermore, our derivation of Eq. (17.24) implicitly assumed that C_{D_0} and K do not vary as velocity changes when we solve for V in Eq. (17.25), which we also know to be only a rough approximation. Thus, Eqs. (17.25–17.27) are not exactly correct in the real world.

A more correct optimum condition for range can be found by exhaustively searching throughout the flight envelope at the current aircraft weight, looking for the place where the range parameter $(V/C)(L/D)$ is at a maximum. This is the method used by the computer programs in the major aircraft companies. The same is true for the following loiter optimization methods.

Range Optimization—Prop

Substituting Eq. (17.4) into Eq. (17.23) yields the Breguet range equation for propeller-powered aircraft [Eq. (17.28)]. The velocity term seen in the jet range equation has disappeared. Since all other terms are constant with respect to velocity, it follows that propeller aircraft range will maximize by flying at the speed and lift coefficient for maximum L/D, as was determined with Eqs. (17.13) and (17.14):

$$R = \frac{\eta_p}{C_{\text{power}}} \frac{L}{D} \ell n \left(\frac{W_i}{W_f} \right) = \frac{550\eta_p}{C_{\text{bhp}}} \frac{L}{D} \ell n \left(\frac{W_i}{W_f} \right) \tag{17.28}$$

Loiter Endurance

The amount of time an aircraft can remain in the air is simply its fuel capacity divided by the rate of fuel consumption (thrust multiplied by specific fuel consumption). The change in weight due to fuel consumption complicates the equation.

The "instantaneous endurance" as defined in Eq. (17.29) is the amount of time the aircraft will remain aloft from the next increment of fuel burned. This can be expanded as shown to express instantaneous endurance in terms of L/D and weight.

$$\frac{\mathrm{d}E}{\mathrm{d}W} = -\frac{1}{CT} = \frac{1}{-CW} \left(\frac{L}{D} \right) \tag{17.29}$$

$$E = \int_{W_i}^{W_f} \frac{1}{-CT} \mathrm{d}W = \int_{W_f}^{W_i} \frac{1}{CW} \left(\frac{L}{D} \right) \mathrm{d}W = \left(\frac{L}{D} \right) \left(\frac{1}{C} \right) \ell n \left(\frac{W_i}{W_f} \right) \tag{17.30}$$

Equation (17.30) integrates for total endurance E. For propeller aircraft, the endurance is obtained by using the equivalent C obtained from Eq. (17.4).

Loiter Optimization—Jet

For jet aircraft the only term in the endurance equation that varies with velocity is the L/D. Therefore, the endurance for jet aircraft is maximized by maximizing the L/D, as determined from Eqs. (17.13) and (17.14).

Loiter Optimization—Prop

Substituting Eq. (17.4) into Eq. (17.30) yields Eq. (17.31), the endurance equation for propeller aircraft. This substitution introduces a velocity term into the loiter endurance equation, so the condition for best prop loiter will not simply be the maximum L/D.

The terms in Eq. (17.31) that vary with velocity are expanded and the derivative with respect to velocity is set to zero in Eq. (17.32). This eventually leads to Eq. (17.33), the velocity condition for maximum loiter time for a propeller aircraft.

$$E = \left(\frac{L}{D}\right)\left(\frac{\eta_p}{C_{\text{power}}V}\right)\ell n\left(\frac{W_i}{W_f}\right) = \left(\frac{L}{D}\right)\left(\frac{550\eta_p}{C_{\text{bhp}}V}\right)\ell n\left(\frac{W_i}{W_f}\right) \tag{17.31}$$

$$\frac{\partial}{\partial V}\left(\frac{L}{DV}\right) = \frac{\partial}{\partial V}\left[\frac{2W/\rho V^3 S}{C_{D_0} + (4KW^2/\rho^2 V^4 S^2)}\right] = 0 \tag{17.32}$$

$$V = \sqrt{\frac{2W}{\rho S}\sqrt{\frac{K}{3C_{D_0}}}} \tag{17.33}$$

This last equation is identical to Eq. (17.19), the velocity condition for minimum power required. The lift coefficient and drag for maximum prop endurance are therefore identical to the minimum-power results defined by Eqs. (17.20) and (17.21). As was shown, the aircraft flies at a velocity that is 76% of the velocity for best L/D. The L/D when flying at the minimum power velocity was shown to be 86.6% of the best L/D.

Relationship Between Loiter and Cruise

In preliminary design studies of derivative aircraft, the available loiter time of existing aircraft is often needed for evaluation of their usability for other missions. As described in a recent paper (Ref. 134), there is a simple relationship between range and endurance based on the Breguet range and loiter equations. Given a known aircraft range and cruise speed, equivalent loiter time can be estimated with reasonable accuracy by

$$E_{\text{loiter}} = 1.14\left\{\frac{R_{\text{cruise}}}{V_{\text{cruise}}}\right\} \tag{17.34}$$

Effects of Wind on Cruise and Loiter

While the design mission for an aircraft often assumes zero wind, the real world is usually not so cooperative. In fact, when you fly east in the morning

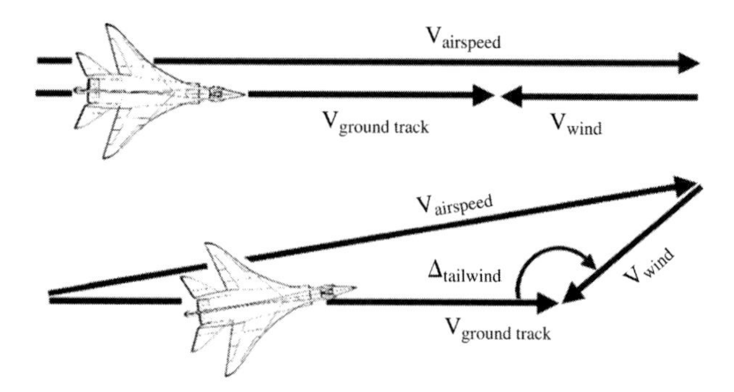

Fig. 17.3 Effects of wind.

and west in the afternoon, you often find a strong headwind both ways! This has a direct effect on the range as calculated in Eq. (17.23). If you have a direct headwind that makes your groundspeed 10% lower than in no-wind conditions, then your range during cruise for a certain amount of fuel will be 10% less. If you are sizing to a required range, you must increase the required cruise range R in the mission segment weight fraction (19.10) by the ratio of velocities, $(V_{\text{airspeed}}/V_{\text{groundspeed}})$, while still using the actual airspeed for V in the equation. If you have a tailwind, the cruise range is improved.

Also, the real world usually offers a wind that is neither a headwind nor a tailwind. You must solve for the groundspeed along the desired flight direction using the Law of Sines and a wind vector diagram as shown at the bottom of Fig. 17.3. Note that the aircraft has its nose pointed to the left of the desired groundtrack to compensate for the wind. If we define the relative wind angle such that a tailwind has angle zero, and a headwind has angle of π radians (180 deg), then we can derive:

$$V_{\text{groundspeed}} = \frac{V_{\text{airspeed}} \sin\{\pi - \Delta_{\text{tailwind}} - \sin^{-1}[V_{\text{wind}}(\sin \Delta_{\text{tailwind}})/V_{\text{airspeed}}]\}}{\sin \Delta_{\text{tailwind}}} \tag{17.35}$$

The traditional pilot's "flight computer" solves this equation graphically, telling where to point the nose and what the resulting groundspeed will be. From the calculated groundspeed, the cruise range or the mission segment weight fraction equation can be adjusted as shown before.

The presence of wind affects the optimal cruising speed for maximizing range. Basically, you should fly faster into a headwind so that you do not fight it as long, and slower if a tailwind is pushing you forward. Unless the wind is very strong, these will only change your airspeed by perhaps 5–10% or so, gaining just a few percent in range over the range if you flew at the no-wind optimal speed. Complicated adjustments can be made to the range optimization equations (see Ref. 113), but as was already discussed, the use of pure equations for optimizing for range is not the preferred method anyway. Instead, we use a computer program that will exhaustively search throughout the flight envelope at the current aircraft weight, looking for the place where the range parameter $(V/C)(L/D)$ is at a maximum. We can adjust the velocity V in the range parameter as just described and use the same search routine to find the best answer for range optimization with winds considered. Then, calculate the range obtained with the velocity adjusted as already described.

The wind has no effect on loiter time or loiter optimization airspeeds, unless somehow the wind speed is greater than your optimum loiter speed and you find you are being blown backward!

17.3 Steady Climbing and Descending Flight

Climb Equations of Motion

Rate of climb is a vertical velocity, typically expressed in feet or meters per minute (which must be converted to feet per second for the following calculations). Climb gradient G is the ratio between vertical and horizontal distance traveled. This is approximately equal to the vertical climb rate divided by the aircraft velocity, or the sine of the climb angle γ.

Equations (17.6) and (17.7) sum the forces depicted in Fig. 17.1 when γ is not zero. Setting the sum of the forces to zero yields the steady climb Eqs. (17.36) and (17.37). Solving for climb angle in Eq. (17.36) produces Eq. (17.38). For normal climb angles (less than 15 deg) the cosine term is approximately one.

The rate of climb, or vertical velocity, is the velocity times the sine of the climb angle [Eq. (17.39)].

$$T = D + W \sin\gamma \tag{17.36}$$

$$L = W \cos\gamma \tag{17.37}$$

$$\gamma = \sin^{-1}\left(\frac{T-D}{W}\right) = \sin^{-1}\left(\frac{T}{W} - \frac{\cos\gamma}{L/D}\right) \cong \sin^{-1}\left(\frac{T}{W} - \frac{1}{L/D}\right) \tag{17.38}$$

$$V_v = V \sin\gamma = V\left(\frac{T-D}{W}\right) \cong V\left(\frac{T}{W} - \frac{1}{L/D}\right) \tag{17.39}$$

The velocity for steady climbing flight can now be derived from Eq. (17.37), as shown in Eq. (17.40).

The thrust-to-weight ratio is no longer the inverse of the lift-to-drag ratio as was the case for level flight. Solving Eq. (17.38) for T/W yields Eq. (17.41), the thrust-to-weight ratio required for a steady climb at angle γ.

$$V = \sqrt{\frac{2}{\rho C_L}\left(\frac{W}{S}\right)\cos\gamma} \tag{17.40}$$

$$\frac{T}{W} = \frac{\cos\gamma}{L/D} + \sin\gamma \cong \frac{1}{L/D} + \sin\gamma = \frac{1}{L/D} + \frac{V_v}{V} \tag{17.41}$$

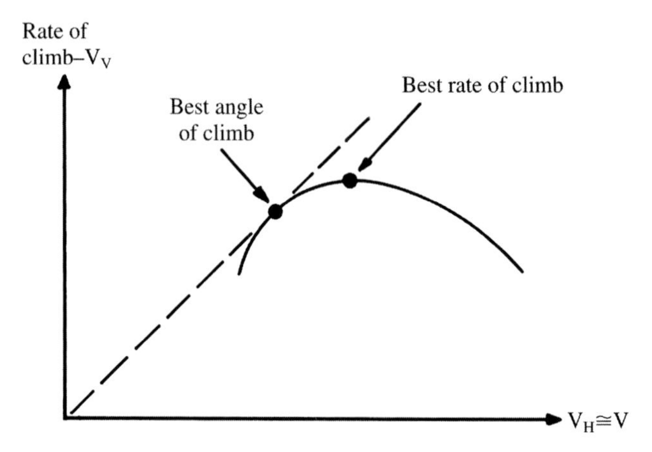

Fig. 17.4 Graphical method for best climb.

Graphical Method for Best Angle and Rate of Climb

Two climb conditions especially concern the aircraft designer: the best rate of climb, which provides the maximum vertical velocity (V_v), and the best angle of climb, which provides a slightly lower vertical velocity but at a reduced horizontal speed, so that the angle of climb is maximized. Therefore, the aircraft gains more altitude for a given horizontal distance, important for clearing mountains!

The most accurate way to determine best rate and angle of climb is to plot the rate of climb vs velocity, using Eq. (17.39), and the actual thrust and drag data as shown in Fig. 17.4. The best rate of climb is obviously the peak of the curve. The best angle of climb is the point of tangency to a line from the origin. The angle of climb is the arctangent of the vertical velocity divided by the horizontal velocity at that point.

Best Angle and Rate of Climb—Jet

Analytical optimization of velocity for best angle and rate of climb can be messy. Graphical analysis is more reliable, but doesn't give an analytical feeling for the key variables.

For a jet aircraft, the thrust is essentially constant with velocity so Eq. (17.38) can be directly maximized for the conditions for best climb angle. Since the T/W term is constant with velocity, the velocity for best L/D should be used to maximize climb angle. This velocity was determined in Eq. (17.13).

To determine the velocity for best rate of climb of a jet aircraft, Eq. (17.39) must be maximized. Equation (17.42) is obtained from Eq. (17.39) by expanding the drag term and assuming that γ is small enough that lift approximately equals weight:

$$V_v = V\left(\frac{T-D}{W}\right) = V\left(\frac{T}{W}\right) - \frac{\rho V^3 C_{D_0}}{2(W/S)} - \frac{2K}{\rho V}\left(\frac{W}{S}\right) \tag{17.42}$$

$$\frac{\partial V_v}{\partial V} = 0 = \frac{T}{W} - \frac{3\rho V^2 C_{D_0}}{2(W/S)} + \frac{2K}{\rho V^2}\left(\frac{W}{S}\right)$$

$$V = \sqrt{\frac{W/S}{3\rho C_{D_0}}[T/W + \sqrt{(T/W)^2 + 12C_{D_0}K}]} \tag{17.43}$$

In Eq. (17.43), the derivative of the vertical velocity with respect to aircraft velocity is set to zero and solved for velocity for best climb.

Note that if the thrust is zero this equation collapses to the equation for the velocity for minimum power required [Eq. (17.19)], which serves as a lower boundary on the solution. The effect of nonzero thrust is a significant increase in the velocity for best climb rate with increasing thrust.

The velocity for best climb rate including the effects of thrust may be on the order of twice the velocity for minimum power. Velocities of 300–500 knots are not uncommon for the best climb speed for a jet. The B-70 has a best climb speed of 583 knots {1080 km/hr}.

This climb optimization will only determine the velocity for the best rate of climb at some altitude. It will not tell you what the complete climb profile should be to minimize time to a given altitude. For many supersonic aircraft, minimizing total time to climb requires leveling off or even diving as the aircraft accelerates through transonic speeds to minimize the time spent at these high-drag conditions. In a later section, the specific excess power method will be presented as a means for determining the climb profile that minimizes total time to climb.

Best Angle and Rate of Climb—Prop

Equation (17.44) expresses the climb angle of a propeller aircraft, as obtained by substituting Eq. (17.5) into Eq. (17.38). This equation can be expanded and the derivative taken with respect to velocity:

$$\gamma = \sin^{-1}\left[\frac{P\eta_p}{VW} - \frac{D}{W}\right] = \sin^{-1}\left[\frac{550\,\text{bhp}\ \eta_p}{VW} - \frac{D}{W}\right] \tag{17.44}$$

However, the theoretical optimal velocities obtained with the resulting equation tend to be too low (sometimes lower than the stall speed) for the parabolic drag approximation to be valid, because of the separation drag at high angles of attack. Also, the thrust no longer follows Eq. (17.5), which implies that thrust is infinite at zero airspeed.

If thrust and drag data are available at low speeds, the graphical method will produce good results. Most propeller aircraft have a best angle-of-climb speed about 85–90% of the best rate-of-climb speed. This can be used for an initial estimate.

Best rate of climb for a propeller aircraft is obtained by substituting Eq. (17.5) into Eq. (17.39). This yields Eq. (17.45), simply the power available minus the power required, divided by aircraft weight. Therefore the best rate of climb occurs at the velocity for minimum power required, as defined in

Eq. (17.19):

$$V_v = V\sin\gamma = \frac{P\,\eta_p}{W} - \frac{DV}{W} = \frac{550\,\text{bhp}\;\eta_p}{W} - \frac{DV}{W} \tag{17.45}$$

Time to Climb and Fuel to Climb

The time to climb to a given altitude is the change in altitude divided by the vertical velocity (rate of climb), as shown in Eq. (17.46) for an incremental altitude change. Fuel burned is the product of the thrust, specific fuel consumption, and time to climb [Eq. (17.47)].

$$\mathrm{d}t = \frac{\mathrm{d}h}{V_v} \tag{17.46}$$

$$\mathrm{d}W_f = -CT\,\mathrm{d}t \tag{17.47}$$

The air density, aircraft weight, drag, thrust, specific fuel consumption, and best climb velocity all change during the climb. A good approximation over small changes in altitude is that the rate of climb at a given weight and constant-thrust setting and constant velocity will reduce linearly with the altitude. This is shown in Eq. (17.48), where the linear constant a is determined from the rates of climb at any two altitudes h_1 and h_2 [Eq. (17.49)]. These two altitudes used to determine a should be near the beginning and ending altitudes of the climb being analyzed, but need not be exactly the same altitudes.

$$V_v = V_{v_i} - a(h_{i+1} - h_i) \tag{17.48}$$

$$a = \frac{V_{v_2} - V_{v_1}}{h_2 - h_1} \tag{17.49}$$

If the climb is broken into short segments (less than 5000 ft {~1500 m} in altitude gain), the fuel burned will be an insignificant portion of the total aircraft weight and can be ignored in the time integration. Substituting Eq. (17.48) into Eq. (17.46) and integrating yields Eq. (17.50), the time to climb from altitude i to altitude $i+1$.

Oddly enough, the change in altitude has dropped out of the equation! However, the change in altitude is implicit in the change in rate of climb (V_v) due to change in altitude. The fuel burned will then be described by Eq. (17.51).

$$t_{i+1} - t_i = \frac{1}{a}\ell n\left(\frac{V_{v_i}}{V_{v_{i+1}}}\right) \tag{17.50}$$

$$\Delta W_{\text{fuel}} = (-CT)_{\text{average}}(t_{i+1} - t_i) \tag{17.51}$$

If desired, the accuracy of Eq. (17.50) can be improved upon by iteration. The rate of climb at the end of the climb segment can be recalculated using the reduced aircraft weight obtained by subtracting the fuel burned [Eq. (17.51)] from the original weight. This revised rate of climb can then be applied back into Eq. (17.50).

17.4 Level Turning Flight

In level turning flight, the lift of the wing is canted so that the horizontal component of the lift exerts the centripetal force required to turn. The total lift on the wing is n times the aircraft weight W, where n is the load factor. Since the vertical component of lift must be W, the horizontal component of lift must be W times the square root of $(n^2 - 1)$. The geometry of a level turn is shown in Fig. 17.5.

$$\dot{\psi} = \frac{W\sqrt{n^2 - 1}}{(W/g)V} = \frac{g\sqrt{n^2 - 1}}{V} \tag{17.52}$$

Turn rate ($\mathrm{d}\psi/\mathrm{d}t$) equals the radial acceleration divided by the velocity, as shown in Eq. (17.52). Turn rate is usually expressed in degrees per second. Equation (17.52) yields radians per second, which must be multiplied by 57.3 to get degrees per second.

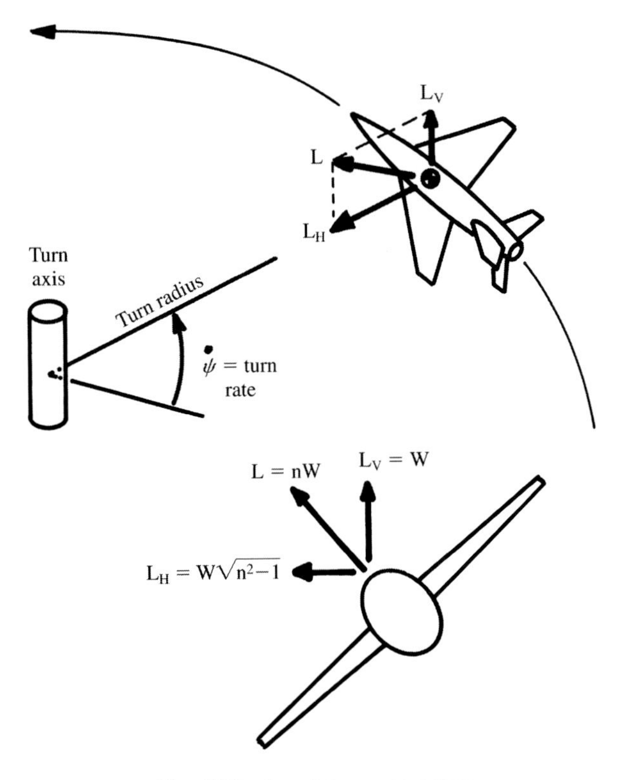

Fig. 17.5 Level turn geometry.

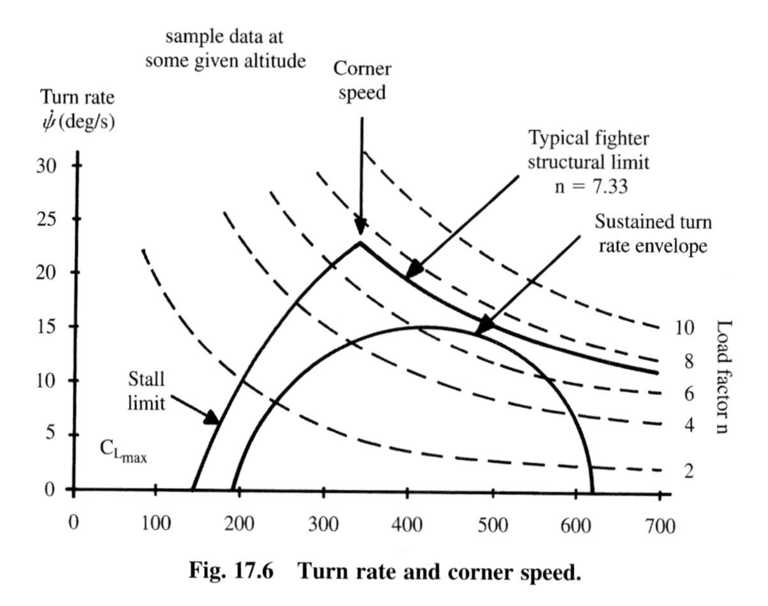

Fig. 17.6 Turn rate and corner speed.

Instantaneous Turn Rate

If the aircraft is allowed to slow down during the turn ("instantaneous turn"), the load factor n will be limited only by the maximum lift coefficient or structural strength of the aircraft. Figure 17.6 shows these stall and structural limits expressed as turn rate vs velocity for a typical fighter aircraft.

The intersection of the stall limit and the structural limit defines the corner speed, which is the velocity for maximum instantaneous turn rate. For a typical fighter, corner speed is about 300–350 knots {560–650 km/hr}. In a classical turning dogfight, opponents will try to get to their own corner speed as quickly as possible.

Sustained Turn Rate

In a sustained turn, the aircraft is not permitted to slow down or lose altitude during the turn. In a sustained turn the thrust must equal the drag and the lift must equal load factor n times the weight. Thus the maximum load factor for sustained turn can be expressed as the product of the thrust-to-weight and lift-to-drag ratios [Eq. (17.53)], assuming that the thrust axis is approximately aligned with the flight direction.

To solve for the sustained load factor in terms of the basic aerodynamic coefficients, the drag is expanded using ($C_L = nW/qS$) and set equal to the thrust. This leads to Eq. (17.54), which defines the maximum available sustained load factor for a given flight condition.

Note that the drag-due-to-lift factor (K) is a function of lift coefficient, as described in Chapter 12. Since n is also a function of lift coefficient, iteration is required to solve Eq. (17.54).

$$n = (T/W)(L/D) \tag{17.53}$$

$$n = \sqrt{\frac{q}{K(W/S)}\left(\frac{T}{W} - \frac{qC_{D_0}}{W/S}\right)} \tag{17.54}$$

Equation (17.53) implies that the sustained-turn load factor can be optimized by flying at the lift coefficient for maximum L/D, which was determined in Eq. (17.14). Using this lift coefficient and setting lift equal to n times W leads to Eq. (17.55). This can be readily solved for either velocity or wing loading to obtain the maximum sustained-turn load factor.

$$L = nW = qS\sqrt{\frac{C_{D_0}}{K}} \tag{17.55}$$

Figure 17.6 shows the sustained turn-rate envelope. This is derived using Eq. (17.52) to determine the turn rates provided by the sustained load factors available at the various flight conditions.

Turn Rate with Vectored Thrust

Vectored thrust offers improved turn performance for future fighters, and is already used in the VSTOL Harrier fighter to maximize turn rate. The direction the thrust should be vectored depends upon whether instantaneous or sustained turn rate is to be maximized.

In a level turn with vectored thrust, the load factor times the weight must equal the lift plus the contribution of the vectored thrust, as shown in Eq. (17.56).

The maximum load factor (and turn rate) is obtained by taking the derivative with respect to vector angle and setting it to zero [Eq. (17.57)]. This yields Eq. (17.58), which states simply that the thrust vector for maximum instantaneous turn rate should be perpendicular to the flight direction.

$$nW = L + T\sin(\alpha + \phi_T) \tag{17.56}$$

$$\frac{\partial n}{\partial \phi_T} = \frac{\partial}{\partial \phi_T}\left[\frac{L}{W} + \frac{T}{W}\sin(\alpha + \phi_T)\right] = \left(\frac{T}{W}\right)\cos(\alpha + \phi_T) = 0 \tag{17.57}$$

$$\phi_T = 90\,\text{deg} - \alpha \tag{17.58}$$

Since none of the thrust is propelling the aircraft forward, it will slow down very rapidly! British pilots in combat have used the 90-deg vectoring of the Harrier to generate a high turn-rate while decelerating, causing pursuing pilots to overshoot.

In a sustained turn with vectored thrust, the drag equals the thrust times the cosine of the total thrust angle, so the load factor n is expressed as in Eq. (17.59). Setting the derivative with respect to thrust-vector angle equal to zero [Eq. (17.60)] yields Eq. (17.61).

$$n = \left[\frac{T\cos(\alpha + \phi_T)}{W}\right]\left(\frac{L}{D}\right) \tag{17.59}$$

$$\frac{\partial n}{\partial \phi_T} = \frac{T}{W}\sin(\alpha + \phi_T)\left(\frac{L}{D}\right) = 0 \tag{17.60}$$

$$\phi_T = -\alpha \tag{17.61}$$

Equation (17.61) implies that the thrust vector for maximum sustained turn rate should be aligned with the flight direction. If the aircraft is at a positive angle of attack, the thrust should be vectored upward (relative to the fuselage axis) to align it with the freestream! However, this calculation ignores the jet flap effect that may produce a drag reduction with slight downward deflection if the nozzles are located near the wing trailing edge.

17.5 Gliding Flight

Straight Gliding Flight

Gliding flight is similar to climbing flight with the thrust set to zero. Equations (17.36) and (17.37) become Eqs. (17.62) and (17.63). The direction of the gliding angle γ is assumed to be reversed from that used for climb.

$$D = W\sin\gamma \tag{17.62}$$

$$L = W\cos\gamma \tag{17.63}$$

$$\frac{L}{D} = \frac{W\cos\gamma}{W\sin\gamma} = \frac{1}{\tan\gamma} \cong \frac{1}{\gamma} \tag{17.64}$$

The lift-to-drag ratio is the inverse of the tangent of the glide angle [Eq. (17.64)]. In sailplane terminology, the "glide ratio" is the ratio between horizontal distance traveled and altitude lost, and is equal to the lift-to-drag ratio. A high-performance sailplane with a glide ratio of 40 will travel over seven statute miles for every thousand feet of altitude lost.

(Cultural note: In sailplane terminology, a "sailplane" is an expensive, high-performance unpowered aircraft. A "glider" is a crude, low-performance unpowered aircraft!)

To maximize range from a given altitude, the glide ratio should be maximized. This requires flying at the velocity for maximum L/D as found in Eq. (17.13), repeated below as Eq. (17.65). The lift coefficient for maximum L/D is repeated as Eq. (17.66). The resulting maximum L/D (glide ratio) is determined from Eq. (17.15), as shown in Eq. (17.67).

$$V_{\max L/D} = \sqrt{\frac{2W}{\rho S}\sqrt{\frac{K}{C_{D_0}}}} \tag{17.65}$$

$$C_{L_{\max L/D}} = \sqrt{\frac{C_{D_0}}{K}} \tag{17.66}$$

$$\left(\frac{L}{D}\right)_{\max} = \frac{1}{2\sqrt{C_{D_0}K}} = \frac{1}{2}\sqrt{\frac{\pi Ae}{C_{D_0}}} \tag{17.67}$$

The time a glider may remain in the air is determined by the "sink rate," the vertical velocity V_v, which is negative in this case. Sink rate is the aircraft velocity times the sine of the glide angle, as expressed in Eq. (17.68).

$$V_v = V\sin\gamma = \sin\gamma\sqrt{\left(\frac{W}{S}\right)\frac{2\cos\gamma}{\rho C_L}} \tag{17.68}$$

$$\sin\gamma = \frac{D}{L}\cos\gamma = \frac{C_D}{C_L}\cos\gamma \tag{17.69}$$

$$V_v = \sqrt{\frac{W}{S}\frac{2\cos^3\gamma C_D^2}{\rho C_L^3}} \cong \sqrt{\frac{W}{S}\frac{2}{\rho(C_L^3/C_D^2)}} \tag{17.70}$$

Equation (17.68) contains both sine and cosine terms. In Eq. (17.69) the sine of the glide angle is expressed in cosine terms to allow substitution into Eq. (17.68), as shown in Eq. (17.70). For typical, small glide angles the cosine term may be ignored.

The lift coefficient for minimum sink rate is solved for by maximizing the term involving C_L and C_D. This is shown in Eq. (17.71), with the result in Eq. (17.72). Note that this is also the lift coefficient for minimum power required, so the velocity can be expressed as in Eq. (17.73). The L/D at minimum sink speed is given by Eq. (17.74).

$$\frac{\partial}{\partial C_L}\left(\frac{C_L^3}{C_D^2}\right) = \frac{\partial}{\partial C_L}\left[\frac{C_L^3}{(C_{D_0} + KC_L^2)^2}\right] = 0 \tag{17.71}$$

$$C_{L_{\text{min sink}}} = \sqrt{\frac{3C_{D_0}}{K}} \tag{17.72}$$

$$V_{\text{min sink}} = \sqrt{\frac{2W}{\rho S}\sqrt{\frac{K}{3C_{D_0}}}} \tag{17.73}$$

$$\left(\frac{L}{D}\right)_{\text{min sink}} = \sqrt{\frac{3}{16KC_{D_0}}} = \sqrt{\frac{3\pi Ae}{16C_{D_0}}} \tag{17.74}$$

The velocity for minimum sink rate is 76% of the velocity for best glide ratio. Sailplane pilots fly at minimum sink speed when they are in "lift" (i.e., in an air

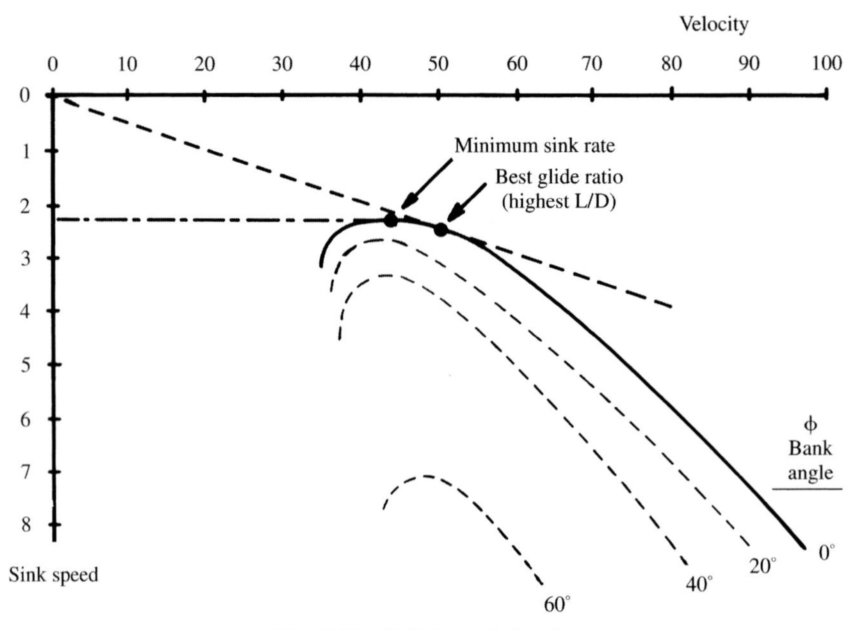

Fig. 17.7 Sailplane sink rate.

mass moving upward). When the lift "dies," they accelerate to the velocity for best glide ratio to cover the most ground while looking for the next lift. An instrument called a "variometer" tells the sailplane pilots when they are in lift.

Figure 17.7 shows a graphical representation of sink rate for a sailplane. This is known as a "speed-polar," or "hodograph," and can be used to graphically determine the velocities for minimum sink rate and best glide ratio.

Turning Gliding Flight

When sailplane pilots find lift, they turn in a small circle to stay within the lifting air mass. Due to the additional wing lift required to turn, the sailplane will experience higher drag and a greater sink rate. Equation (17.63) must be modified to account for the bank angle ϕ [Eq. (17.75)].

$$L\cos\phi = W\cos\gamma \cong W \tag{17.75}$$

Turn rate is equal to the centripetal acceleration divided by the velocity, and is also equal to the velocity divided by the turn radius [Eq. (17.76)]. This allows the centripetal acceleration to be expressed as the velocity squared divided by the turn radius [Eq. (17.77)]. In Eq. (17.78), the turning force due to the lateral component of wing lift is equal to the aircraft mass times the centripetal acceleration.

$$\dot{\psi} = a/V = V/R \tag{17.76}$$

$$a = V^2/R \tag{17.77}$$

$$L\sin\phi = \frac{WV^2}{gR} = W\sqrt{n^2 - 1} \tag{17.78}$$

$$R = \frac{V^2}{g\tan\phi} = \frac{V^2}{g\sqrt{n^2 - 1}} \tag{17.79}$$

Equation (17.78) can be solved for turn radius as expressed in terms of either bank angle or load factor [Eq. (17.79)].

The vertical velocity (sink rate) can be determined by substituting $C_L \cos\phi$ for C_L in Eq. (17.70). This yields Eq. (17.80), which is simply the previous result divided by the cosine of ϕ, raised to the 3/2 power. The radius of the turn is found by substituting Eq. (17.75) into Eq. (17.79), as shown in Eq. (17.81):

$$V_v = \frac{1}{\cos^{3/2}\phi}\sqrt{\frac{W}{S}\frac{2}{\rho(C_L^3/C_D^2)}} \tag{17.80}$$

$$R = \frac{2W}{\rho S C_L g \sin\phi} \tag{17.81}$$

Because the ϕ term in Eq. (17.80) does not vary with velocity, the prior results for the velocities for best glide ratio and minimum sink rate can be applied.

One unique problem for a slow-flying sailplane in a turn is the variation in velocity across the long span of the wing. The wing on the inside of the turn may stall due to the lower velocity. This is shown in Fig. 17.8. The velocity across the span varies linearly with distance from the axis of the turn. Also, the bank angle shortens the wing span when seen from above. These effects are shown in Eq. (17.82):

$$V = V_{cg}\left[1 + \frac{Y}{R}\cos\phi\right] \tag{17.82}$$

$$V_{\text{inner}} = V_{cg}\left[1 - \frac{b}{2R}\cos\phi\right] \tag{17.83}$$

In Eq. (17.83), the velocity at the inner wing tip is shown as a function of wing span, turn radius, and bank angle. In normal flight this velocity difference is easily corrected with a little aileron to increase the lift coefficient on the inner wing. However, when flying near the stall at even a moderate bank angle, this can reduce the velocity of the inner wing tip enough to create a one-wing stall, which leads to a spin.

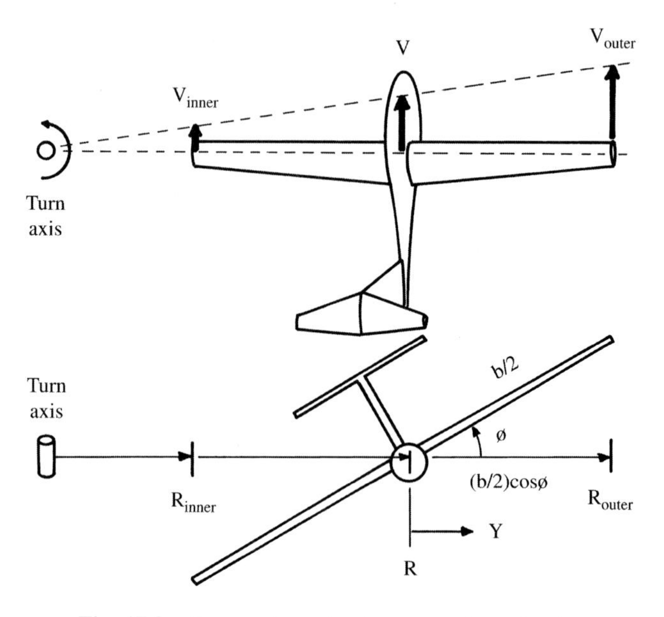

Fig. 17.8 Turn radius effect on wing-tip velocity.

17.6 Energy-Maneuverability Methods

Energy Equations

Fighter pilots have always known that management of energy is critical to survival and success. In World War I the experienced pilots always tried to enter a dogfight from above. They could then exchange the potential energy of altitude for the kinetic energy of speed or turn rate.

Jet-fighter dogfight maneuvers largely rely upon the exchange of potential and kinetic energy to attain a positional advantage. For example, the "high speed yo-yo" maneuver is used when overtaking a slower aircraft in a hard turn. The attacker pulls up, trading kinetic energy for potential energy and slowing to allow a higher turn rate. After turning, the attacker rolls partially inverted and pulls down astern of the opponent, now exchanging potential energy back for speed.

Fighter pilots understand that potential and kinetic energy can be exchanged, and that the sum of the aircraft energy must be managed to attain success. This intuitive measure of goodness can be analytically developed and applied to aircraft design (first defined in Ref. 104).

$$E = Wh + \frac{1}{2}\left(\frac{W}{g}\right)V^2 \tag{17.84}$$

$$h_e = \frac{E}{W} = h + \frac{1}{2g}V^2 \tag{17.85}$$

$$P_{s_{\text{used}}} = \frac{\mathrm{d}h_e}{\mathrm{d}t} = \frac{\mathrm{d}h}{\mathrm{d}t} + \frac{V}{g}\frac{\mathrm{d}V}{\mathrm{d}t} \tag{17.86}$$

At any point in time, the total energy of an aircraft (the "energy state") is the sum of the potential and kinetic energy, as shown in Eq. (17.84). Dividing by aircraft weight gives the "specific energy" [Eq. (17.85)]. Specific energy has units of distance (feet or meters), and is also called the "energy height" (h_e) because it equals the aircraft altitude if the velocity is zero.

Power is the time rate of energy usage, so the specific power $(P_s)_{\text{used}}$ can be defined as the time rate at which the aircraft is gaining altitude or velocity [Eq. (17.86)]. Since specific energy has units of distance, specific power has units of distance per time (ft/s or m/s).

This power being used by the aircraft to gain height or velocity has to come from somewhere. In the discussions of power required vs power available, it was pointed out that the excess power could be used to climb or accelerate. This excess power is the excess thrust $(T - D)$ times the velocity [Eq. (17.87)].

The specific excess power (P_s) is the excess power divided by the weight, and equals the specific power used, as shown in Eq. (17.88).

$$P = V(T - D) \tag{17.87}$$

$$P_s = \frac{V(T - D)}{W} = \frac{\mathrm{d}h}{\mathrm{d}t} + \frac{V}{g}\frac{\mathrm{d}V}{\mathrm{d}t} \tag{17.88}$$

$$P_s = V\left[\frac{T}{W} - \frac{qC_{D_0}}{W/S} - n^2\frac{K}{q}\frac{W}{S}\right] \tag{17.89}$$

Drag, and therefore P_s, is a function of the aircraft load factor. The higher the load factor, the greater the drag, and thus the less excess power available. Equation (17.88) can be expanded in terms of the load factor and the aerodynamic coefficients as shown in Eq. (17.89). Note that T/W and W/S are at the given flight condition, not the takeoff values!

Specific excess power P_s has the same units as rate of climb. In fact, Eq. (17.88) is identical to the rate-of-climb equation if the longitudinal acceleration ($\mathrm{d}V/\mathrm{d}t$) is zero. The P_s at a load factor of one is actually the rate of climb that would be available if the pilot chose to use all of the excess power for climbing at constant velocity.

When P_s equals zero, the drag of the aircraft exactly equals the thrust so there is no excess power. This does not necessarily mean that the aircraft isn't climbing or accelerating. However, if the sum of the energy usage equals zero, then the aircraft must be flying level, or climbing and decelerating, or descending and accelerating.

Equations (17.88) and (17.89) assume that the thrust axis is approximately aligned with the flight direction. If this is not the case, the thrust components in the lift and drag directions yield Eq. (17.90).

$$P_s = V\left\{\frac{T\cos(\alpha + \phi_r)}{W} - q\frac{C_{D_0}}{W/S} - \frac{n^2K}{WqS}[W - T\sin(\alpha + \phi_r)]\right\} \tag{17.90}$$

P_s *Plots*

For any given altitude, P_s can be calculated using Eq. (17.89) for varying Mach numbers and load factors once the aerodynamic coefficients and installed thrust data are available. Design specifications for a new fighter will have a large number of "must meet or exceed" P_s points, such as $P_s = 0$ at $n = 5$ at Mach 0.9 at 30,000 ft {9144 m}.

P_s values are calculated and plotted against Mach number as shown in Fig. 17.9 for a number of altitudes. Computers are especially handy for this "number crunching."

From the P_s charts at the various altitudes (Fig. 17.9), several additional charts can be prepared by cross-plotting.

The level turn rate can be determined for the various load factors at a given altitude and Mach number, and plotted vs P_s (Fig. 17.10). This is compared to the data for a threat aircraft at that altitude and Mach number. With an equivalent P_s at a higher turn rate, the new fighter would always be able to turn inside the opponent without losing relative energy. A turn-rate advantage of 2 deg/s is considered significant.

In Fig. 17.11, $P_s = 0$ contours are plotted for different load factors on a Mach number vs altitude chart. This is a major tool for the evaluation of new fighters, and permits comparisons between two aircraft for all Mach numbers and altitudes on one chart. To win a protracted dogfight, an aircraft should have $P_s = 0$ contours that envelop those of an opponent.

In Fig. 17.12, contour lines of constant P_s at a given load factor are plotted onto a Mach number vs altitude chart. A separate chart is prepared for each load factor. The chart for load factor equals one is especially important because it provides the rate of climb and the aircraft ceiling, and because it is used to determine an optimal climb trajectory.

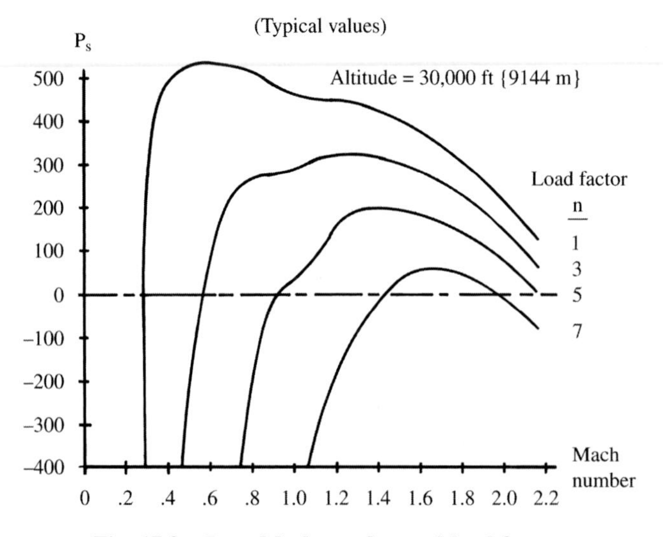

Fig. 17.9 P_s vs Mach number and load factor.

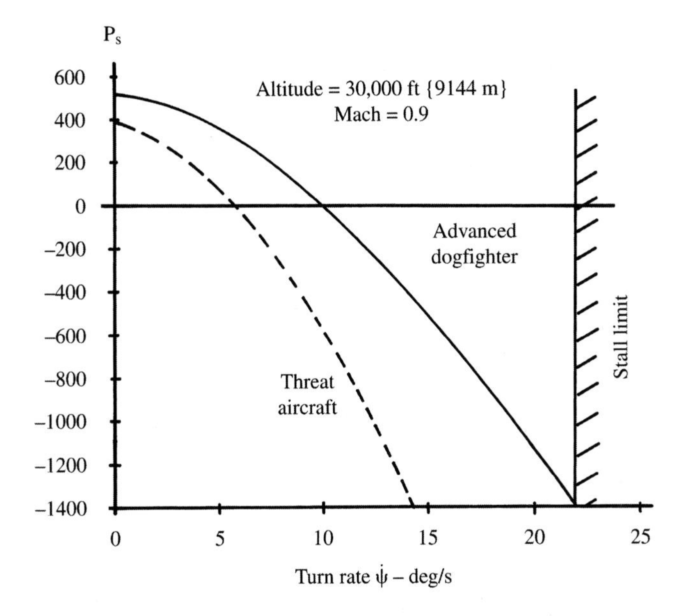

Fig. 17.10 Turn rate vs P_s.

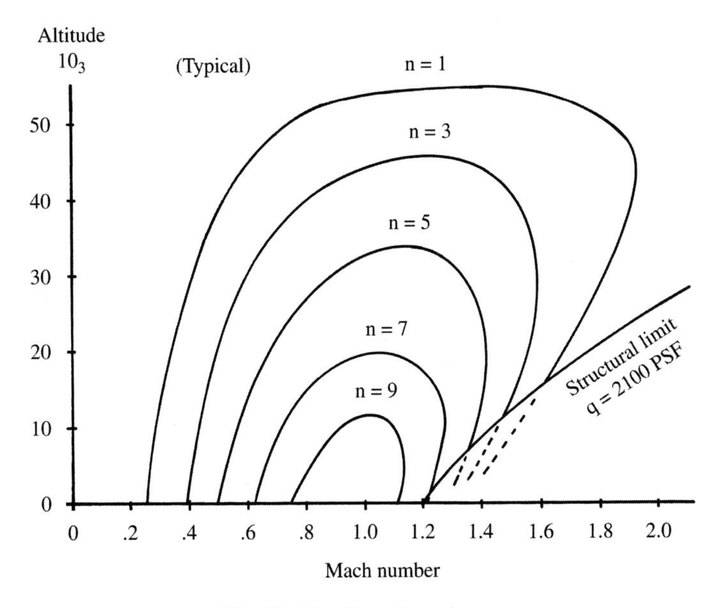

Fig. 17.11 P_s = 0 contours.

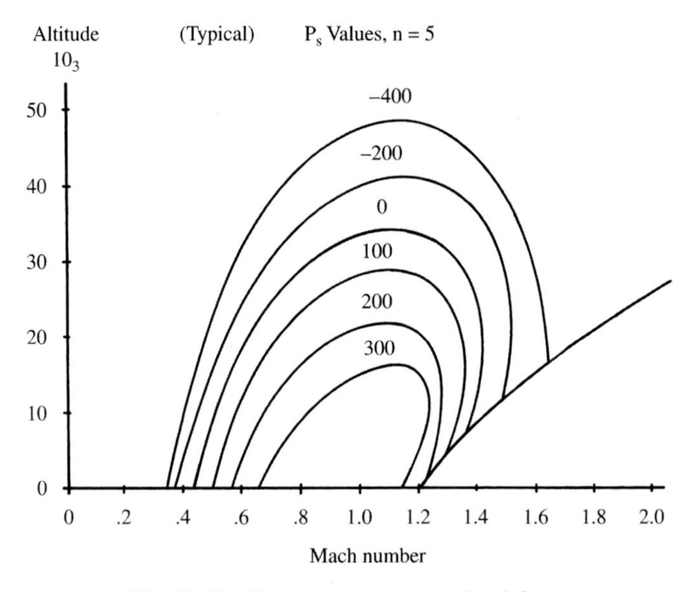

Fig. 17.12 P_s contours, constant load factor.

Minimum Time-to-Climb Trajectory

Figure 17.13 is a plot of energy height vs Mach number and altitude. This is merely a graphical representation of Eq. (17.85), and has nothing to do with the particulars of any one aircraft. An F-16 or a Boeing 747 would have an energy height of 42,447 ft {12,938 m} if flying at Mach 0.9 at 30,000 ft {9144}.

$$dt = \frac{dh_e}{P_s} \tag{17.91}$$

$$t_{1-2} = \int_{h_{e1}}^{h_{e2}} \frac{1}{P_s} dh_e \tag{17.92}$$

Equation (17.86) can be rearranged into Eq. (17.91), which expresses the incremental time to change energy height (h_e) as the change in energy height divided by the P_s at that flight condition. This is then integrated in Eq. (17.92) for the time to change energy height.

Equation (17.92) shows that the time to change energy height is minimized if the P_s is maximized at each energy height. This occurs at those points on the Mach number vs altitude plot of 1-g P_s (Fig. 17.12) where the P_s curve is exactly tangent to an energy-height curve (Fig. 17.13).

In Fig. 17.14, the 1-g P_s curves for a typical current-technology-high-thrust fighter are superimposed on the h_e curves of Fig. 17.13. The trajectory for minimum time to climb is shown as passing through the dots representing the points where the P_s curves are tangent to h_e curves. For such a fighter, the minimum time to climb is obtained by staying low and accelerating to transonic

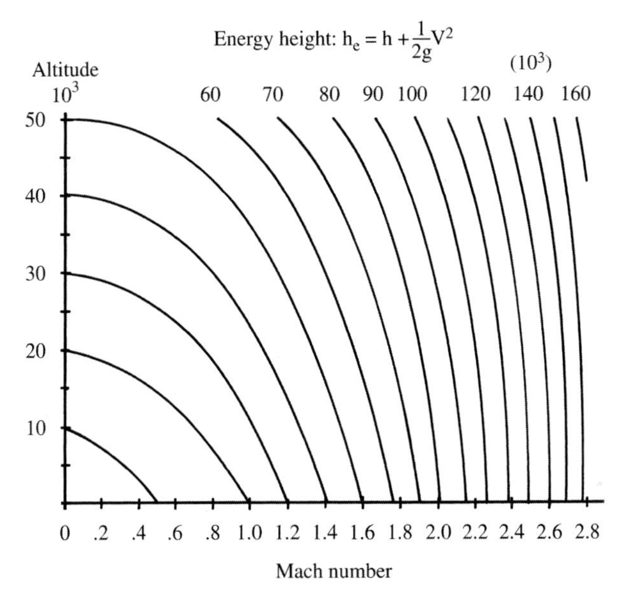

Fig. 17.13 Lines of constant energy height.

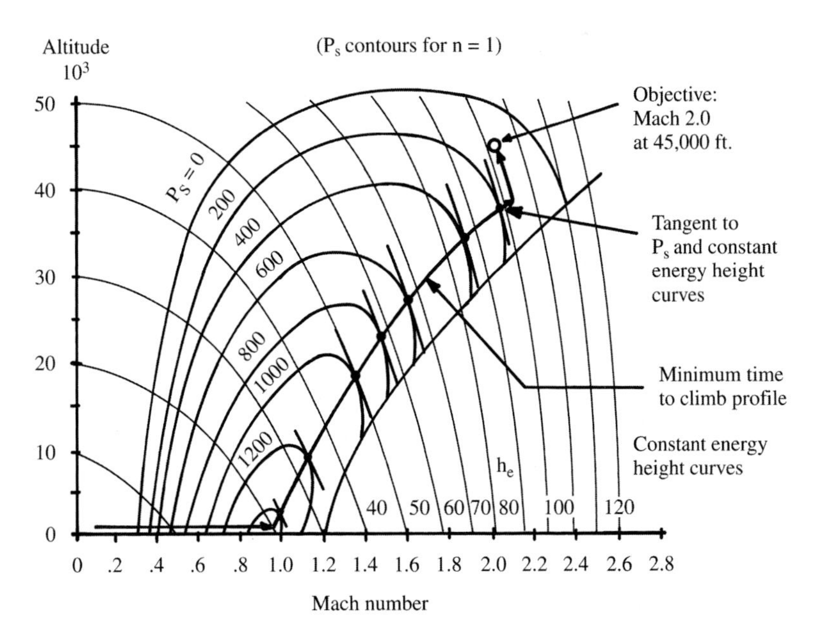

Fig. 17.14 Minimum time-to-climb trajectory, high-thrust fighter.

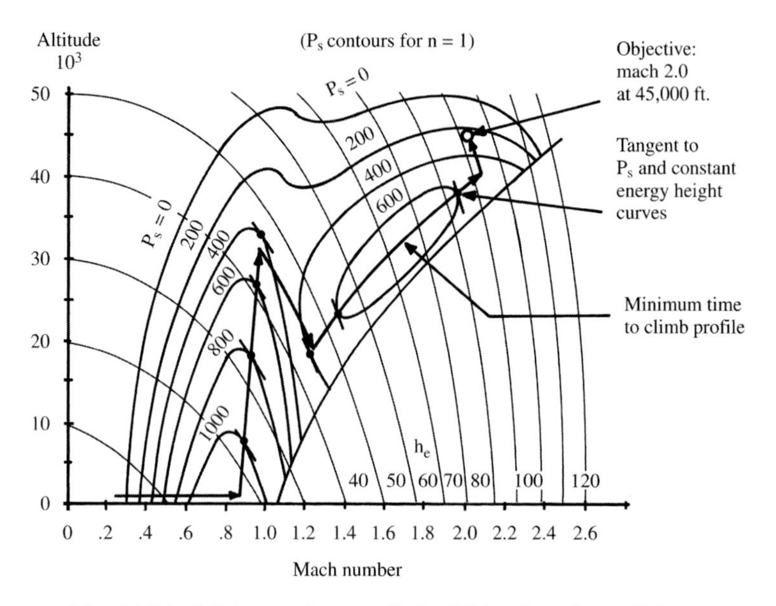

Fig. 17.15 Minimum time-to-climb, SST or low-thrust fighter.

speeds, then pitching up into a steep climb at approximately constant indicated airspeed (i.e., dynamic pressure), as shown by the optimal trajectory.

Figure 17.15 shows the 1-g P_s curves for a typical 1960s era jet fighter. These fighters had significantly less thrust, and suffered a "thrust pinch" at transonic speeds—in which the thrust minus drag would reduce to almost zero. This causes the P_s contours to form "bubbles."

The minimum-time-to-climb trajectory requires jumping from one bubble to the other. This is done by dividing or climbing along lines of constant energy height tangent to P_s lines of the same numerical value for both bubbles, as shown in Fig. 17.15.

Note that Fig. 17.15 requires diving through Mach 1.0 to minimize time to climb for this aircraft. This was common in earlier jets, and makes sense intuitively. Since thrust minus drag is nearly zero at transonic speeds, acceleration will be slow and the aircraft will spend a lot of time in transonic acceleration. Diving reduces this time. The altitude lost is easily regained at higher speeds where the drag is less. Supersonic transports such as Concorde also use this strategy.

This method minimizes time to climb with no constraint on ending velocity. To climb to a given altitude with a specified ending velocity, the optimal trajectory is flown until the aircraft reaches the energy-height curve of the desired ending condition. Then that energy-height curve is followed to the ending altitude and velocity, by either climbing or diving.

$$t_{1-2} \cong \frac{\Delta h_e}{(P_s)_{\text{average}}} \tag{17.93}$$

The actual time to climb is determined by numerically integrating along the optimal trajectory using Eq. (17.92). The time to change energy height is approximately expressed in Eq. (17.93) as the change in energy height divided by the average P_s during the change. As always, accuracy is improved with smaller integration steps.

Note that the time to follow lines of constant energy height is usually negligible for a first-order analysis.

Minimum Fuel-to-Climb Trajectory

The energy equations can be modified to determine the climb trajectory that minimizes fuel consumption. The fuel specific energy (f_s) is defined as the change in specific energy per change in fuel weight. This is shown in Eq. (17.94) to equal the P_s divided by the fuel flow, which is the thrust times the specific fuel consumption.

Like P_s, the f_s values can be calculated and plotted vs Mach number for each altitude and then cross-plotted as contour lines on a Mach number vs altitude chart, as shown in Fig. 17.16.

$$f_s = \frac{\mathrm{d}h_e}{\mathrm{d}W_f} = \frac{\mathrm{d}h_e/\mathrm{d}t}{\mathrm{d}W_f/\mathrm{d}t} = \frac{P_s}{CT} \tag{17.94}$$

$$W_{f_{1-2}} = \int_{h_{e_1}}^{h_{e_2}} \frac{1}{f_s}\mathrm{d}h_e \tag{17.95}$$

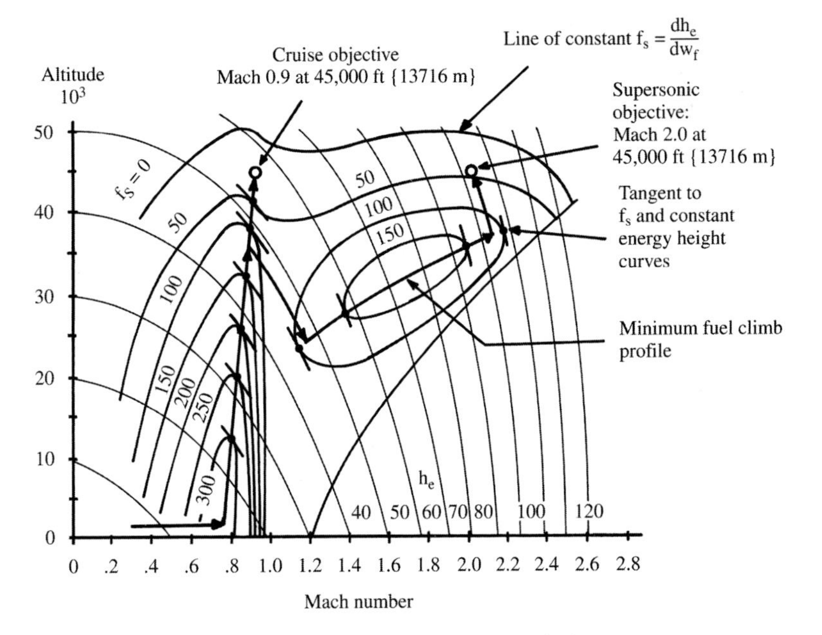

Fig. 17.16 Minimum fuel to climb.

In Eq. (17.95), Eq. (17.94) is rearranged and integrated to yield the change in fuel weight for a change in energy height (h_e). Note that this is minimized when f_s is maximized for each energy height. This implies that the minimum-fuel-to-climb trajectory passes through those points for which f_s contours are exactly tangent to the h_e contours. This is shown in Fig. 17.15, which greatly resembles the chart used to determine the minimum-time-to-climb trajectory.

$$W_{f_{1-2}} \cong \frac{\Delta h_e}{(f_s)_{\text{average}}} \tag{17.96}$$

The fuel consumed during the climb is determined by numerically integrating along the minimum-fuel trajectory, using Eq. (17.96) as an approximation.

Energy Method for Mission-Segment Weight Fraction

Equation (17.97) is an expression of the mission-segment weight fraction for any flight maneuver involving an increase in energy height. This can be used for climbs or accelerations or combinations of the two. Remember that the mission-segment weight fraction expresses the total aircraft weight at the end of the mission segment divided by the total aircraft weight at the beginning of the mission segment. This is used for sizing as discussed in earlier chapters.

$$\frac{W_i}{W_{i-1}} = \exp\left[\frac{-C\Delta h_e}{V(1 - D/T)}\right] = \exp\left[\frac{-C\Delta h_e}{V\{1 - [1/(T/W)(L/D)]\}}\right] \tag{17.97}$$

Unfortunately, a maneuver involving a reduction in energy height cannot create fuel as would be implied by putting a negative value for the change in h_e into Eq. (17.97)!

17.7 Operating Envelope

The aircraft "operating envelope" or "flight envelope" maps the combinations of altitude and velocity that the aircraft has been designed to withstand. The "level-flight operating envelope" has the further restriction that the aircraft be capable of steady level flight.

The operating envelope for a typical fighter is shown in Fig. 17.17. Fighter operating envelopes are the most complicated and contain all the elements of the operating envelopes of other classes of aircraft.

The level-flight operating envelope is determined from the $P_s = 0$ and stall limit lines. The $P_s = 0$ limit is usually shown for both maximum thrust and for military (nonafterburning) thrust.

Since the $P_s = 0$ and stall lines vary with aircraft weight, some assumption about aircraft weight must be made. Typically the operating envelope is calculated at takeoff weight, cruise weight, or combat weight.

The "absolute ceiling" is determined by the highest altitude at which $P_s = 0$. Some small rate-of-climb capability (i.e., P_s) is required at the "service ceiling."

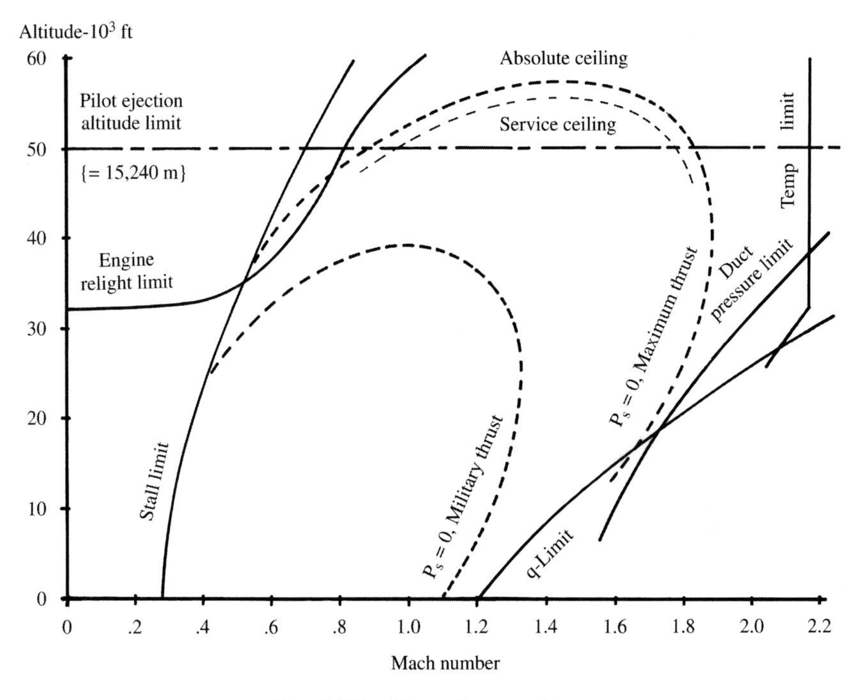

Fig. 17.17 Operating envelope.

FARs require 100 fpm {30.5 mpm} for propeller aircraft and 500 fpm {152 mpm} for jets. Military specifications require 100 fpm {30.5 mpm} at the service ceiling (300 fpm {91 mpm} for U.S. Navy).

For some jet aircraft, the limitation on usable ceiling is the pilot. The odds of surviving an ejection above 50,000 ft {15,240 m} are rather small without an astronaut-type pressure suit or some type of capsule. This limits the usable ceiling as shown.

Another limitation to the level flight envelope of many jet aircraft is the low-q engine operating limit. At low velocities and high altitudes there may not be enough air available to restart the engine in the event of a flameout. It may also be impossible to operate or light the afterburner. These limits are provided by the engine manufacturer.

The remaining limits shown in Fig. 17.17 are structural. The external-flow dynamic pressure q as defined in Eq. (17.98) has a direct impact upon the structural loads. A maximum q limit is specified in the design requirements and used by the structural designers for stress analysis. Typical fighter aircraft have a q limit of 1800–2200 psf {86–105 kN/m^2}. This corresponds to transonic speeds at sea level.

$$q = \frac{1}{2}\rho_\infty V_\infty^2 = 0.7 P_{\text{static}} M^2 \tag{17.98}$$

$$P_{T_0} = P_{\text{static}}[1 + 0.2M^2]^{3.5} \tag{17.99}$$

The airload pressures exerted within the inlet duct are greater than the free-stream pressures because the inlet slows the air down (typically to about Mach 0.4–0.5 at the engine front face). The total pressure of the oncoming air is determined from Eq. (17.99), using the static atmospheric pressure at that altitude from the Standard Atmosphere Table in Appendix B.

The total pressure within the duct equals the outside total pressure times the inlet-duct pressure recovery, as discussed in Chapter 13. Equation (17.99) is used again for the flow within the duct and solved for the static pressure at the Mach number at the engine front face. This is the maximum wall pressure exerted within the inlet duct, and may easily be three times the outside dynamic pressure. As shown in Fig. 17.17, the inlet-duct pressure limit does not follow the same slope as the dynamic-pressure limit.

The remaining operating envelope limit is the temperature limit due to skin aerodynamic heating. This depends upon the selected structural materials. A design chart for skin temperature vs Mach number and altitude was presented in Chapter 14.

17.8 Takeoff Analysis

A widely used empirical chart for determining takeoff distance was presented in Chapter 5. Later in the design process, a more detailed analysis breaks the takeoff into segments for more accurate analysis.

Figure 17.18 illustrates the segments of the takeoff analysis. The ground roll includes two parts, the level ground-roll and the ground roll during rotation to the angle of attack for liftoff. After rotation, the aircraft follows an approximately circular arc ("transition") until it reaches the climb angle.

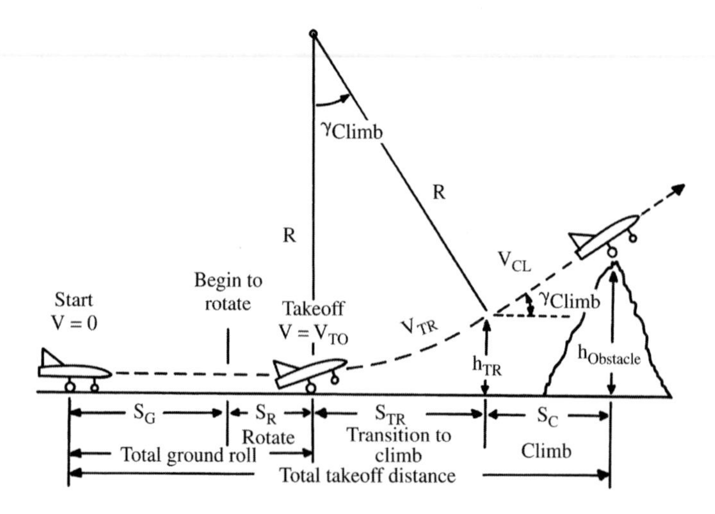

Fig. 17.18 Takeoff analysis.

Table 17.1 Ground rolling resistance

	μ-typical values	
Surface	Rolling (brakes off)	Brakes on
Dry concrete/asphalt	0.03–0.05	0.3–0.5
Wet concrete/asphalt	0.05	0.15–0.3
Icy concrete/asphalt	0.02	0.06–0.10
Hard turf	0.05	0.4
Firm dirt	0.04	0.3
Soft turf	0.07	0.2
Wet grass	0.08	0.2

Ground Roll

During the ground roll, the forces on the aircraft are the thrust, drag, and rolling friction of the wheels, this last being expressed as a rolling friction coefficient μ times the weight on the wheels $(W - L)$. A typical μ value for rolling resistance on a hard runway is 0.03. Values for various runway surfaces are presented in Table 17.1.

The resulting acceleration of the aircraft, as expressed by Eq. (17.100), can be expanded in terms of the aerodynamic coefficients. This requires evaluating the lift and drag of the aircraft in ground effect and with landing gear down and flaps in the takeoff position, as discussed in Chapter 12. The lift coefficient is based on the wing angle of attack on the ground (measured to the zero lift angle), and is typically less than 0.1 unless large takeoff flaps are deployed.

$$a = \frac{g}{W}[T - D - \mu(W - L)] = g\left[\left(\frac{T}{W} - \mu\right) + \frac{\rho}{2W/S}(-C_{D_0} - KC_L^2 + \mu C_L)V^2\right] \tag{17.100}$$

$$S_G = \int_{V_i}^{V_f} \frac{V}{a}\mathrm{d}V = \frac{1}{2}\int_{V_i}^{V_f} \frac{1}{a}\mathrm{d}(V^2) \tag{17.101}$$

The ground-roll distance is determined by integrating velocity divided by acceleration, as shown in Eq. (17.101). Note the mathematical trick that simplifies the integration by integrating with respect to V^2 instead of V.

The takeoff velocity must be no less than 1.1 times the stall speed, which is found by setting maximum lift at stall speed equal to weight and solving for stall speed. The maximum lift coefficient is with the flaps in the takeoff position. Remember that landing gear geometry may limit maximum angle of attack (and hence lift coefficient) for takeoff and landing.

Equation (17.101) is integrated for ground-roll distance from V_{initial} to V_{final} in Eq. (17.102), where the terms K_T and K_A are defined in Eqs. (17.103) and (17.104). K_T contains the thrust terms and K_A contains the aerodynamic terms.

$$S_G = \frac{1}{2g}\int_{V_i}^{V_f} \frac{d(V^2)}{K_T + K_A V^2} = \left(\frac{1}{2gK_A}\right)\ell n\left(\frac{K_T + K_A V_f^2}{K_T + K_A V_i^2}\right) \tag{17.102}$$

$$K_T = \left(\frac{T}{W}\right) - \mu \tag{17.103}$$

$$K_A = \frac{\rho}{2(W/S)}(\mu C_L - C_{D_0} - KC_L^2) \tag{17.104}$$

Equation (17.102) integrates ground roll from any initial velocity to any final velocity. For takeoff, the intial velocity is zero, and the final velocity is V_{TO}. Because the thrust actually varies somewhat during the ground roll, an averaged thrust value must be used. Since we integrate with respect to velocity squared, the averaged thrust to use is the thrust at about 70% (1/square-root 2) of V_{TO}.

For greater accuracy, the ground roll can be broken into smaller segments and integrated using the averaged thrust for each segment in Eq. (17.102). The averaged thrust is the thrust at 70% of the velocity increase for that segment. Also, K may be reduced due to ground effect (Chapter 12).

The time to rotate to liftoff attitude depends mostly on the pilot. Maximum elevator deflection is rarely employed. A typical assumption for large aircraft is that rotation takes 3 s. The acceleration is assumed to be negligible over that short time interval, so the rotation ground-roll distance S_R is approximated by three times V_{TO}. For small aircraft the rotational time is on the order of 1 s, and $S_R = V_{\text{TO}}$.

Transition

During the transition, the aircraft accelerates from takeoff speed (1.1 V_{stall}) to climb speed (1.2 V_{stall}). The average velocity during transition is therefore about 1.15 V_{stall}. The average lift coefficient during transition can be assumed to be about 90% of the maximum lift coefficient with takeoff flaps. The average vertical acceleration in terms of load factor can then be found from Eq. (17.105):

$$n = \frac{L}{W} = \frac{\frac{1}{2}\rho S(0.9\, C_{L_{\max}})(1.15\, V_{\text{stall}})^2}{\frac{1}{2}\rho S C_{L_{\max}} V_{\text{stall}}^2} = 1.2 \tag{17.105}$$

$$n = 1.0 + \frac{V_{\text{TR}}^2}{Rg} = 1.2 \tag{17.106}$$

$$R = \frac{V_{\text{TR}}^2}{g(n-1)} = \frac{V_{\text{TR}}^2}{0.2g} \tag{10.107}$$

The vertical load factor must also equal 1.0 plus the centripetal acceleration required to cause the aircraft to follow the circular transition arc. This is shown in Eq. (17.106), and solved for the radius of the transition arc in Eq. (17.107).

The climb angle γ at the end of the transition is determined from Eq. (17.108). The climb angle is equal to the included angle of the transition arc (see Fig. 17.18), and so the horizontal distance traveled during transition can be determined from Eq. (17.109). The altitude gained during transition is determined from the geometry of Fig. 17.18 to be as indicated in Eq. (17.108).

$$\sin\gamma_{\text{climb}} = \frac{T-D}{W} \cong \frac{T}{W} - \frac{1}{L/D} \tag{17.108}$$

$$S_T = R\sin\gamma_{\text{climb}} = R\left(\frac{T-D}{W}\right) \cong R\left(\frac{T}{W} - \frac{1}{L/D}\right) \tag{17.109}$$

$$h_{\text{TR}} = R(1-\cos\gamma_{\text{climb}}) \tag{17.110}$$

If the obstacle height is cleared before the end of the transition segment, then Eq. (17.111) is used to determine the transition distance.

$$S_T = \sqrt{R^2 - (R - h_{\text{TR}})^2} \tag{17.111}$$

Climb

Finally, the horizontal distance travelled during the climb to clear the obstacle height is found from Eq. (17.112). The required obstacle clearance is 50 ft {15.24 m} for military and small civil aircraft and 35 ft {10.7 m} for commercial aircraft.

$$S_c = \frac{h_{\text{obstacle}} - h_{\text{TR}}}{\tan\gamma_{\text{climb}}} \tag{17.112}$$

If the obstacle height was cleared during transition, then S_c is zero.

Balanced Field Length

The "balanced field length" (discussed in Chapter 5) is the total takeoff distance including obstacle clearance when an engine fails at "decision speed" V_1, the speed at which, upon an engine failure, the aircraft can either brake to a halt or continue the takeoff in the same total distance. If the engine fails before decision speed, the pilot can easily brake to a halt. If the engine fails after decision speed, the pilot must continue the takeoff.

An empirical method for balanced field-length estimation was presented in Chapter 5. A more detailed equation, as developed in Ref. 23, takes this form:

$$\text{BFL} = \frac{0.863}{1+2.3G}\left(\frac{W/S}{\rho g C_{L_{\text{climb}}}} + h_{\text{obstacle}}\right)\left(\frac{1}{T_{\text{av}}/W - U} + 2.7\right) + \left(\frac{655}{\sqrt{\rho/\rho_{\text{SL}}}}\right) \tag{17.113}$$

$$\text{JET: } T_{av} = 0.75\, T_{\substack{\text{takeoff}\\ \text{static}}} \left[\frac{5 + \text{BPR}}{4 + \text{BPR}}\right] \tag{17.114}$$

$$\text{PROP: } T_{av} = 5.75\ \text{bhp}\left[\frac{(\rho/\rho_{SL})N_e D_p^2}{\text{bph}}\right]^{\frac{1}{3}} \tag{17.115}$$

where

BFL = balanced field length (ft)
$G = \gamma_{climb} - \gamma_{min}$
γ_{climb} = arcsine $[(T - D)/W]$, 1-engine-out, climb speed
γ_{min} = 0.024 2-engine; 0.027 3-engine; 0.030 4-engine
$C_{L_{climb}} = C_L$ at climb speed (1.2 V_{stall})
$h_{obstacle}$ = 35 ft commercial, 50 ft military
$U = 0.01\, C_{L_{max}} + 0.02$ for flaps in takeoff position
BPR = bypass ratio
bhp = engine brake horsepower
N_e = number of engines
D_p = propeller diameter (ft)

For a more accurate determination of the balanced field length, the takeoff roll should be integrated with an engine failure at an assumed V_1, and compared with a braking analysis at that V_1 using the methods in the next section. The assumed V_1 should be iterated until the total takeoff distance including a 35-ft-obstacle clearance equals the total distance with braking.

It is usually assumed that the pilot waits one second before recognizing the engine failure and applying the brakes. The use of reverse thrust is not permitted for the balanced field length calculations. To permit positive rate of climb after engine failure, the pilot will not take off at minimum flight speed where the drag due to lift is excessive. Instead takeoff will be delayed until a higher speed, which is calculated to minimize balanced field length. This may be 20–40% higher than minimum takeoff speed.

17.9 Landing Analysis

Landing is much like taking off, only backward! Figure 17.19 illustrates the landing analysis, which contains virtually the same elements as the takeoff. Note that the aircraft weight for landing analysis is specified in the design requirements, and ranges from the takeoff value to about 85% of takeoff weight. Landing weight is not the end-of-mission weight, because this would require dumping large amounts of fuel to land immediately after takeoff in the event of an emergency.

Approach

The approach begins with obstacle clearance over a 50-ft {15.24-m} object. Approach speed V_a is 1.3 V_{stall} (1.2 V_{stall} for military). The steepest approach

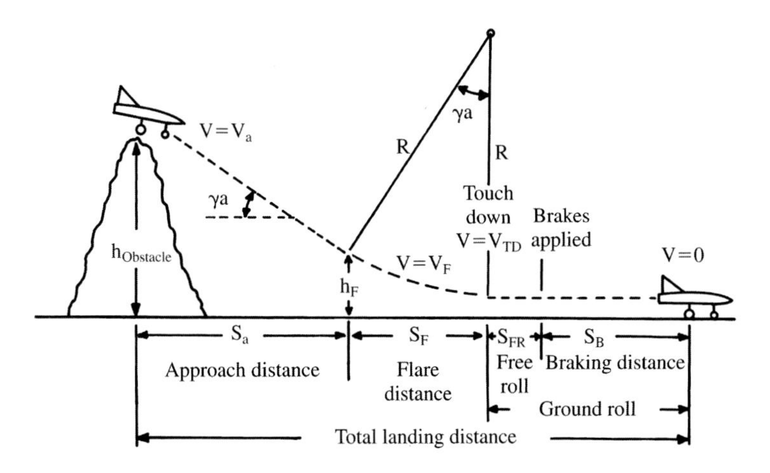

Fig. 17.19 Landing analysis.

angle can be calculated from Eq. (17.108), with idle thrust and drag with full flaps deflected.

For transport aircraft the approach angle should be no steeper than 3 deg (0.052 rad), which may require more than idle thrust. Approach distance is determined from Eq. (17.112) using the flare height h_f.

Flare

Touchdown speed V_{TD} is 1.15 V_{stall} (1.1 V_{stall} for military). The aircraft decelerates from V_a to V_{TD} during the flare. The average velocity during the flare V_f is therefore 1.23 V_{stall} (1.15 V_{stall} for military). The radius of the flare circular arc is found by Eq. (17.107) using V_f, and where $n = 1.2$ for a typical aircraft.

The flare height can now be found from Eq. (17.110), and the horizontal distance during flare can be found from Eq. (17.109).

Although the deceleration from V_a to V_{TD} would imply additional energy and thus additional distance, this is negligible because the pilot usually pulls off all remaining approach power when the flare is begun.

Ground Roll

After touchdown, the aircraft rolls free for several seconds before the pilot applies the brakes. The distance is V_{TD} times the assumed delay (1–3 s).

The braking distance is determined by the same equation used for takeoff ground roll [Eq. (17.102)]. The initial velocity is V_{TD}, and the final velocity is zero.

The thrust term is the idle thrust. If a jet aircraft is equipped with thrust reversers, the thrust will be a negative value approximately equal to 40 or 50% of maximum forward thrust.

Thrust reversers cannot be operated at very slow speeds because of reingestion of the exhaust gases. Thrust reverser "cutoff speed" is determined by the engine manufacturer, and is typically about 50 knots (85 ft/s or 93 km/hr). The ground roll must be broken into two segments using Eq. (17.102) and the appropriate values for thrust (negative above cutoff speed, positive idle thrust below cutoff speed).

Reversible propellers produce a reverse thrust of about 40% of static forward thrust (60% for turboprops), and can be used throughout the landing roll.

The drag term may include the additional drag of spoilers, speed brakes, or drogue chutes. Drogue chutes have drag coefficient of about 1.4 times the inflated frontal area, divided by the wing reference area.

The rolling resistance will be greatly increased by the application of the brakes. Typical μ values for a hard runway surface are about 0.5 for civil and 0.3 for military aircraft. Values for various surfaces are provided in Table 17.1.

The FAA requires that an additional two-thirds be added to the total landing distance of commercial aircraft to allow for pilot technique. Thus the "FAR field length" is equal to 1.666 times the sum of the approach, flare, and total ground roll.

17.10 Other Fighter Performance Measures of Merit

The standard measures of merit for fighter aircraft including turn rate, corner speed, load factor, and specific excess power P_s do not completely distinguish between a good and a not-so-good fighter. For example, two fighters with exactly the same turn rate vs P_s will be widely different in combat effectiveness if one aircraft has unpredictable and uncontrollable behavior at high angle of attack. There is now great interest in defining new fighter measures of merit that can account for such differences.

There are several key deficiencies in current measures of merit. First, they focus on steady-state performance abilities, whereas a real dogfight is characterized by continuous change in aircraft state. In the high-speed yo-yo discussed earlier, the aircraft quickly pitches up, then rolls and turns at approximately corner speed for a few seconds, then rolls to almost inverted flight, pitches up (down) again, and then rolls out and dives.

While turn rate at corner speed is important, the ability to rapidly execute these changes in state is also very important. Furthermore, these changes of state are usually being executed simultaneously such as pitching and rolling at the same time, known affectionately as "yank and bank").

Another deficiency is that the current measures of merit are oriented around the classical gun attack in which a tail chase with your opponent in front is the desired outcome. Modern missiles are getting so good that in combat the first aircraft to point its nose at the opponent will win, regardless of energy state. The most-modern missiles have such good "off-boresight" capability that even rapid nose-pointing may become irrelevant—simply "see-and-shoot"!

It must be remembered, however, that missiles are expensive and that each fighter can only carry a few of them. Future fighters must also have good classical dogfighting abilities.

Current measures of merit also fail to address the importance of what is called "decoupled energy management" to permit nonstandard fighter maneuvers. "Coupled energy management" refers to maneuvers in which potential and kinetic energy are exchanged. In the high-speed yo-yo, kinetic energy is exchanged for potential energy in the initial climb, and the potential energy is then exchanged back for kinetic energy after the turn. This makes the aircraft predictable.

In decoupled energy management, the potential and kinetic energy are changed independently. For example, speed may be reduced rapidly and without gaining altitude by using large speed brakes and/or in-flight thrust reversing.

Figure 17.20 shows the "energy management envelope" measure of merit. In this extended version of energy manueverability, the maximum and minimum (most negative) P_s values obtainable are plotted vs turn rate. If suitable controls over thrust and drag are available, the pilot can manage his energy state by selecting any P_s level within the envelope at a given turn rate. In the traditional evaluation of Fig. 17.10, only the maximum P_s obtainable is considered.

Note from Fig. 17.20 that an aircraft controllable after the stall has the option of developing a tremendous drag force for reduction of energy state. Under certain combat conditions this can be used to force the opponent to overshoot.

Also, turn rate is inversely proportional to velocity. If an aircraft can be momentarily slowed to extremely low speeds, well below stall, the turn rate can greatly exceed that in conventional flight. This may allow a missile first-shot opportunity. This "post-stall maneuvering" (Ref. 75) is employed on the X-31 test aircraft.

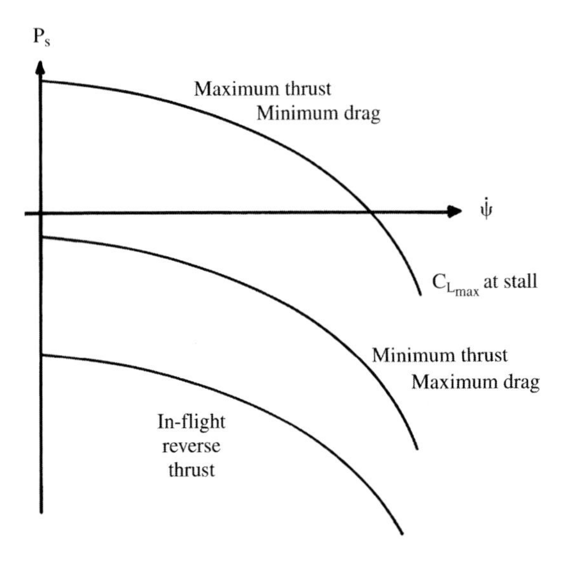

Fig. 17.20 Energy management envelope.

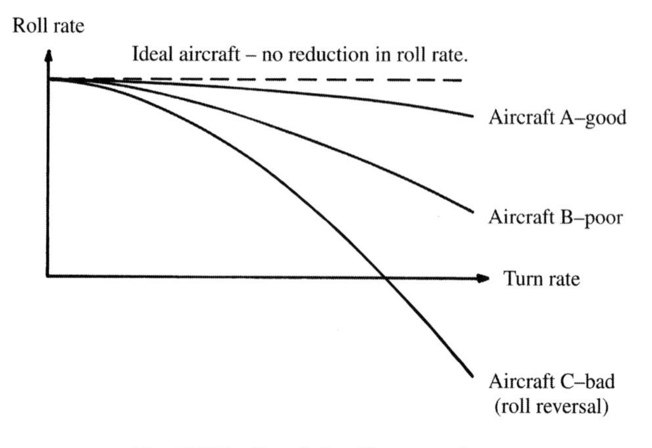

Fig. 17.21 Loaded roll comparison.

The "loaded roll" measure of merit deals with the effect of angle of attack on roll performance. A number of existing fighters lose their roll ability at high load factors due to aeroelastic effects, adverse yaw, and aileron flow separation. An aircraft sluggish in roll during a high-g turn will be at a clear disadvantage. Figure 17.21 illustrates this comparison for a "good" aircraft, a "fair" aircraft, and an aircraft that experiences complete roll reversal.

Reference 76 defines a number of alternative fighter measures of merit.

Supermaneuver and Poststall Maneuver

Fighter capabilities variously called poststall maneuver (PSM), enhanced fighter maneuver, and supermaneuver offer substantial advantages in close-in combat. With the successful flight test of the X-31, and the YF-22s demonstration of 60-deg angle-of-attack operation, these capabilities have finally come of age. A supermaneuver capability allows a fighter to point its nose at an opponent more rapidly, getting the first missile shot in a "face-to-face" dogfight. This is attained primarily by the combination of thrust-induced turning and dynamic turning, usually involving high angles of attack as described next.

Contrary to science-fiction movies, a rocket can turn in space only by thrusting in a direction perpendicular to the flight path (Fig. 17.22). This produces a turn load factor (n) that is the component of thrust perpendicular to the flight path, divided by the weight of the vehicle. Turn rate can then be expressed simply as equal to (gn/V).

Note that at zero velocity, turn rate seems to go to infinity! While limited by pitch rate capability, a rocket could attain extremely high turn rates if it slowed to a very low speed.

An aircraft can also turn using thrust, provided that its thrust can be angled to have a substantial component perpendicular to the flight path. This can be done in three ways.

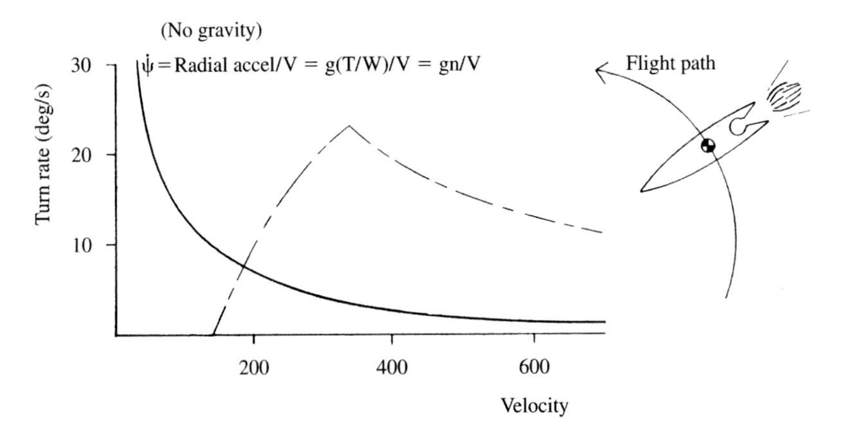

Fig. 17.22 Thrust-induced turning.

Figure 17.23 shows one way to direct the aircraft thrust perpendicular to the flight path, namely by providing thrust-vectoring nozzles at or near the aircraft center of gravity. This allows the pilot to vector thrust at will, without concern for thrust-produced pitching moments. Such vectoring is available on the Harrier, and is proposed for the Reverse-Installation Vectored Engine Thrust ("RIVET") VSTOL concept (Ref. 93).

For such a design, the turn-rate plot and the vectored thrust-induced turn-rate plot are essentially summed. The wing can be kept at the angle of attack for maximum lift, while the nozzles are directed approximately perpendicular to the flight path for maximum instantaneous turn rate as proven in Eq. (17.57).

Note that the wing stall limit line of Fig. 17.23 goes to zero rather than the level-flight stall speed. This indicates that we are momentarily ignoring gravity, going to a 90-deg bank to maximize instantaneous turn rate. Obviously, this can only be done for a few seconds!

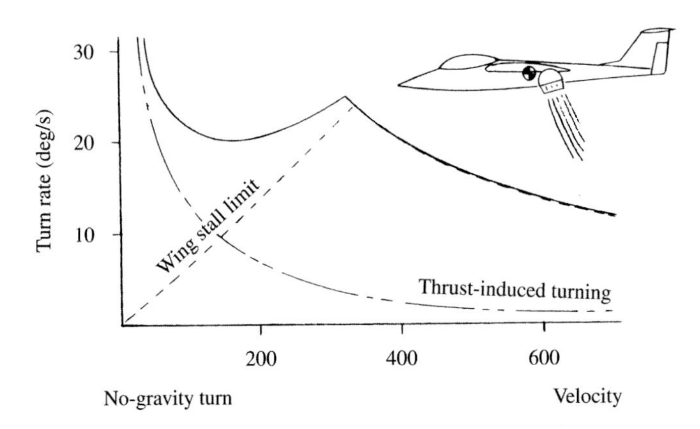

Fig. 17.23 Thrust vectoring at center of gravity.

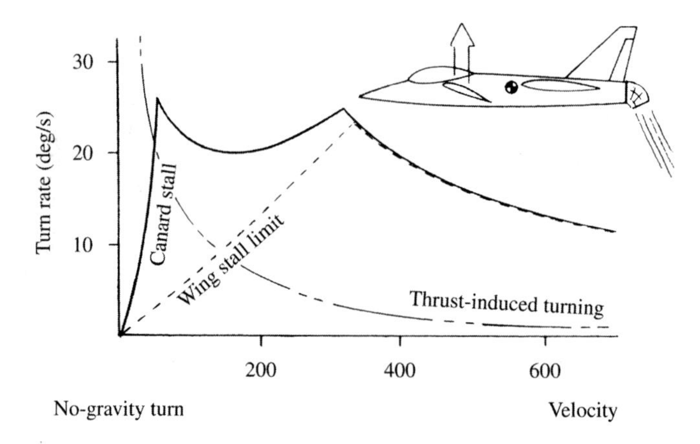

Fig. 17.24 Aft nozzle plus canard.

Another option for vectoring the thrust perpendicular to the flight path is the addition of a thrust-vectored nozzle at the rear. This is a relatively easy feature to add to a design, and in fact the F-22 has such a nozzle already. However, the F-22 nozzles cannot be used for thrust-induced turning because the downward vectoring of thrust produces a large nose-down pitching moment. To be useful for thrust-induced turning, this pitching moment must be balanced by some nose-up moment, which can be attained by the addition of a large canard as seen on the F-15 STOL/Maneuver demonstrator, and an early Lockheed JSF design concept.

This approach, the aft-nozzle-plus-canard, allows the aircraft to retain the full turning ability due to wing lift, plus the additional vectored thrust-induced turning, down to the speed at which the canard stalls (Fig. 17.24). By selecting canard size, the designer can select the lowest turning speed available. However, the larger the canard, the greater the weight and drag impact on the design.

In the third option, the aircraft acts like a rocket, pointing its fuselage at a very high angle to the flight path (Fig 17.25). Note that whereas a nearly 90-deg angle of attack is shown, substantial thrust turning is available at lesser angles.

In this option the aircraft angle of attack is well past the stall angle (hence poststall maneuvering). This clearly requires that the aircraft not just have flying abilities at poststall angles, but that it also retain good controllability and acceptable air quality into the inlet duct so that the engine continues running.

This poststall thrust-induced turning, used by X-31, has several problems. It is difficult for the pilots, because the airplane is flying in a direction downward through the floorboards! The pilot is blind in the direction of flight. A roll about the velocity vector looks like a yaw to the pilot, and so disorientation is very possible.

Also, flight into the poststall region means just that—the wing is stalled, and hence is producing only a fraction of its maximum lift. However, if velocity is slow enough, the jet thrust alone ensures that turn rate will be high anyway.

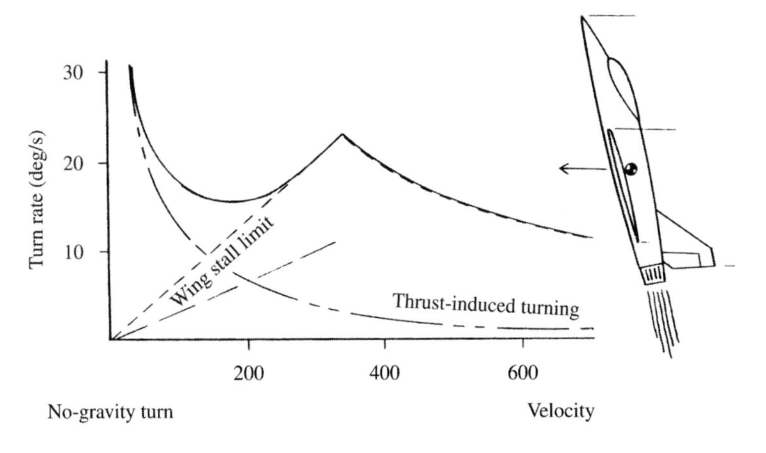

Fig. 17.25 Fuselage pointing.

In any case the drag at extreme angle of attack will be very high, and the thrust component in the flight direction very small, so that the aircraft will decelerate very rapidly and may reach velocities near zero if the pilot is not careful (as this author discovered while flying the X-31 simulator!).

However, this option requires far less compromise to the design than the preceding options, and a substantial number of studies have verified that pilots can be trained to fly and fight at these extreme angles of attack.

Figure 17.26 shows a frustrating moment for the fighter pilot—the opponent is almost, but not quite, in his sight! If he could just lift the nose a little more. . . .

To the pilot, pitch rate looks like turn rate in such a situation. With a normal aircraft the pilot should not pitch up faster than the aircraft turn rate or the aircraft will stall. With a PSM aircraft, however, that is not a problem. The pilot can just pull the nose up past stall, take the shot, and put the nose back down in a quick, smooth motion.

While the nose is coming up, there appears to be a much greater turn rate, equal to the actual aircraft turn rate plus the available pitching rate at that condition. This is shown in Fig. 17.27.

This "dynamic turn" depends only upon the pitch rate capability of the aircraft and can be virtually as rapid as the pilots desire. For a modern, relaxed-stability fighter, the aircraft is always trying to pitch up anyway and the computer must fight to prevent it! Therefore, pitch-up can occur at 90 deg/s or more, and will be limited only to make the airplane less sensitive to fly. However, pitch-down from a high angle of attack is more difficult to obtain and is often the limiting case for control surface sizing.

Dynamic turning is not the result of a turning of the velocity vector and so does not produce increased load factor on the airplane. Thus, the apparent violation of the structural limit line is permissible—the aircraft does not see any additional *g* force.

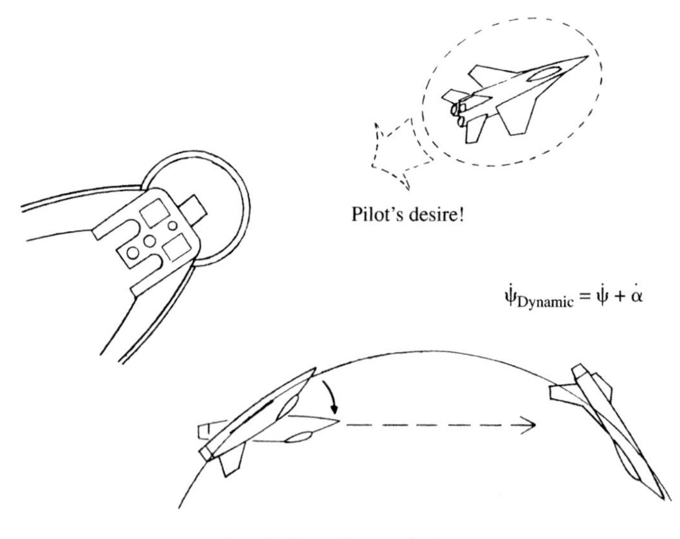

Fig. 17.26 Dynamic turn.

Realize, though, that this dynamic turn can be used for only a brief period of time before the aircraft reaches its maximum angle of attack, even if that limit is 90 deg. Also, as the aircraft pitches past the stall angle of attack, the conventional turn rate will sharply decrease. Drag will go up, and speed and energy will reduce. This dynamic turn maneuver must take place rapidly or the PSM airplane

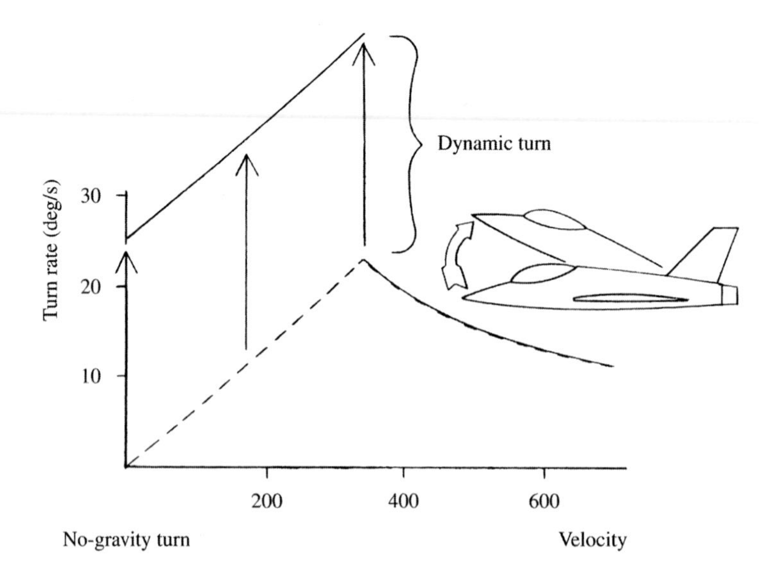

Fig. 17.27 Dynamic turn on the turn-rate plot.

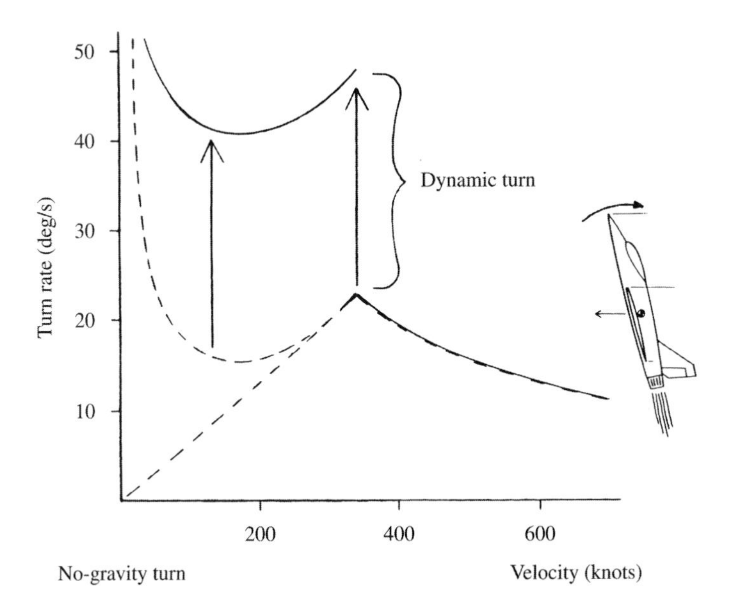

Fig. 17.28 Combined dynamic and PSM turning.

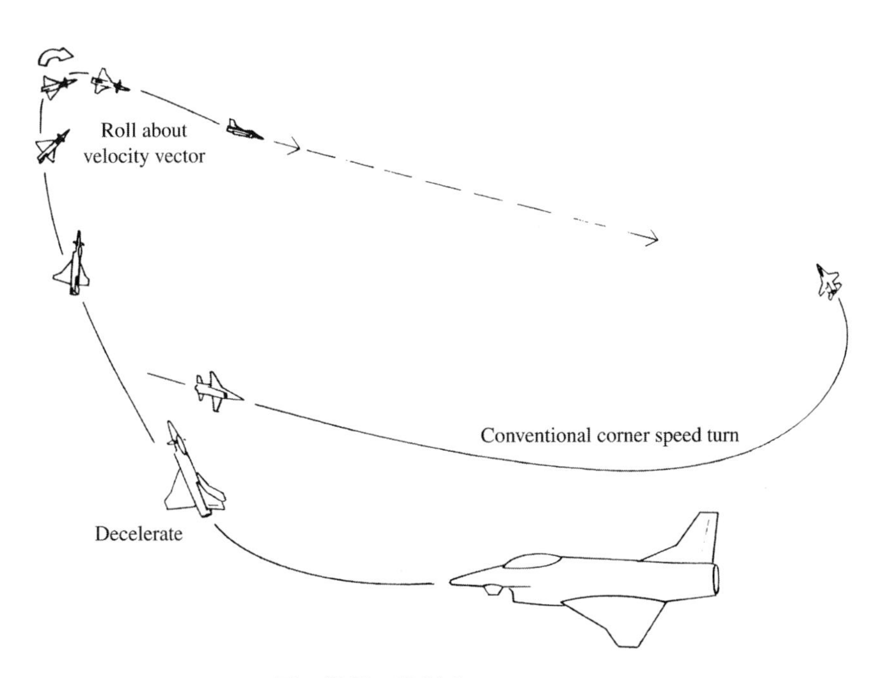

Fig. 17.29 X-31 Supermaneuver.

will wind up at a large energy disadvantage should the opponent fail to die as planned.

Figure 17.28 illustrates the effects of combining poststall dynamic turning with thrust-induced turning. As can be seen, such designs have ample capability to exceed the turn rates predicted by the classic turn-rate plot for a brief period of time (Ref. 94).

Figure 17.29 shows a combined supermaneuver developed for the X-31 program that minimizes total time to reach a shot opportunity. Here the aircraft pitches up and bleeds off speed, slowing to extremely low speeds while initiating the turn.

At the top of the maneuver, the aircraft is rapidly turned using engine thrust until the velocity vector is roughly 90 deg to the target aircraft. Meanwhile, the aircraft is also rolled approximately 90 deg around the velocity vector, which looks like a 90-deg yaw to the pilot due to the high angle of attack. This results in the nose pointing at the target aircraft!

Although the velocity vector is not yet pointed at the target aircraft, the pilot can take the shot, then accelerate out of the high-angle-of-attack condition.

F-14 arrested landing (U.S. Navy photo).

18
Cost Analysis

18.1 Introduction

When the aircraft manufacturers submit their proposals for a new aircraft, the customer faces a problem. All of the proposed aircraft will meet the design requirements! This is ensured by the methods of this book and the much more in-depth methods used by the aircraft companies.

The customer must use some criteria other than aircraft performance to select the best proposal. While there will be some differences in technical credibility, data substantiation, and intrinsic design qualities, the final contractor selection will probably hinge on cost.

Aircraft cost estimation occupies the fuzzy gray area between science, art, and politics. Cost estimation is largely statistical, and in the final analysis we predict the cost of a new aircraft based on the actual costs of prior aircraft. However, it is very difficult to determine how much a prior aircraft really did cost in terms that are meaningful to the next aircraft.

For example, anybody attempting to use the B-1B actual costs to predict the costs of a similar future bomber would have a terrible time establishing a meaningful baseline cost. The B-1B began as the B-1A, which was cancelled before entering production. Rockwell stored a warehouse full of B-1A parts and tooling, some of which (but not all) were usable in the B-1B program when it was started. The new requirements for the B-1B required substantial reengineering, especially in the nacelle and avionics areas.

The cost estimation for a future bomber should not be based upon the sum of all of the costs in the B-1A and B-1B programs. Hopefully, a future aircraft would be designed and produced without the inefficient stop-and-restart experienced by the B-1B. It would be virtually impossible, though, to try to determine an equivalent program cost for the B-1B had it been designed from the ground up.

While the B-1B case is extreme, other aircraft programs pose similar problems in establishing a baseline program cost. For political reasons most military aircraft production programs are stretched out. To reduce "this year's" defense budget, the number of aircraft produced per year may be reduced well below the optimal production rate. In some cases, production rates are less than one per month.

This will greatly increase the cost per aircraft. Should this be included in cost estimation for the next aircraft? Or should the actual cost be adjusted to determine a cost baseline at the optimal production rate? This approach ensures a cost overrun when the new aircraft's production rate is slowed.

In fact, it is very difficult to compare costs for two aircraft that are already in production. Part of the problem hinges upon what type of money to use. Program cost comparisons can be made in "then-year" or "constant-year" dollars. Then-year dollars are the actual dollars spent in each year of the program, past, present, and future. For future program costs, an estimate of the inflation rate must be made.

For comparison of program costs and for establishing a cost baseline for new aircraft cost prediction, constant-year dollars should be used (the actual dollars spent, ratioed by inflation factors to some selected year). However, budgeting in Congress is done in then-year dollars, so most cost data are prepared in then-year dollars.

As an example, Ref. 77 quotes from Congressional testimony the actual costs per aircraft of $17.6 million for the F-15 and $10.8 million for the F-16 (in late-1970s dollars). Considering the far-greater capabilities of the F-15, it would seem a better bargain at only 60% more than the F-16. But, these are in then-year dollars for procurements during different years! In constant 1978 dollars, the costs were $18.8 million for the F-15 and $8.2 million for the F-16, so the F-15s actually cost 130% more in constant dollars.

Another problem in cost comparison is the aircraft production quantity and rates. The more aircraft produced, the more the manufacturer learns and the cheaper the next aircraft can be produced. This is known as the "learning curve" effect.

Roughly speaking, every time the production quantity is doubled the labor cost per aircraft goes down 20% (i.e., an 80% learning curve). Aircraft production typically follows a 75–85% learning curve (see Fig. 18.1.)

Due to the learning curve effect, cost comparisons are not meaningful between a new aircraft just entering production and an old aircraft already produced in the hundreds or thousands.

Still another problem in cost comparison is that different costs are used, frequently without proper identification. Comparing the flyaway cost of one aircraft to the program or life-cycle cost of another is meaningless.

18.2 Elements of Life-Cycle Cost

When you buy a car, the "cost" is what the dealer charges you to drive it home. Most car buyers today are somewhat influenced by the expected cost of ownership (gas mileage and maintenance), but would never consider that as an actual part of the purchase price. However, a typical $15,000 car will cost at least 25 cents per mile to operate, which adds another $25,000 to the probable "life-cycle cost" of the car!

Figure 18.2 shows the elements that make up aircraft life-cycle cost (LCC). The sizes of the boxes are roughly proportional to the magnitude of the costs for a typical aircraft.

"RDT&E" stands for research, development, test, and evaluation, which includes all the technology research, design engineering, prototype fabrication, flight and ground testing, and evaluations for operational suitability. The cost of aircraft conceptual design as discussed in this book is included in the RDT&E cost.

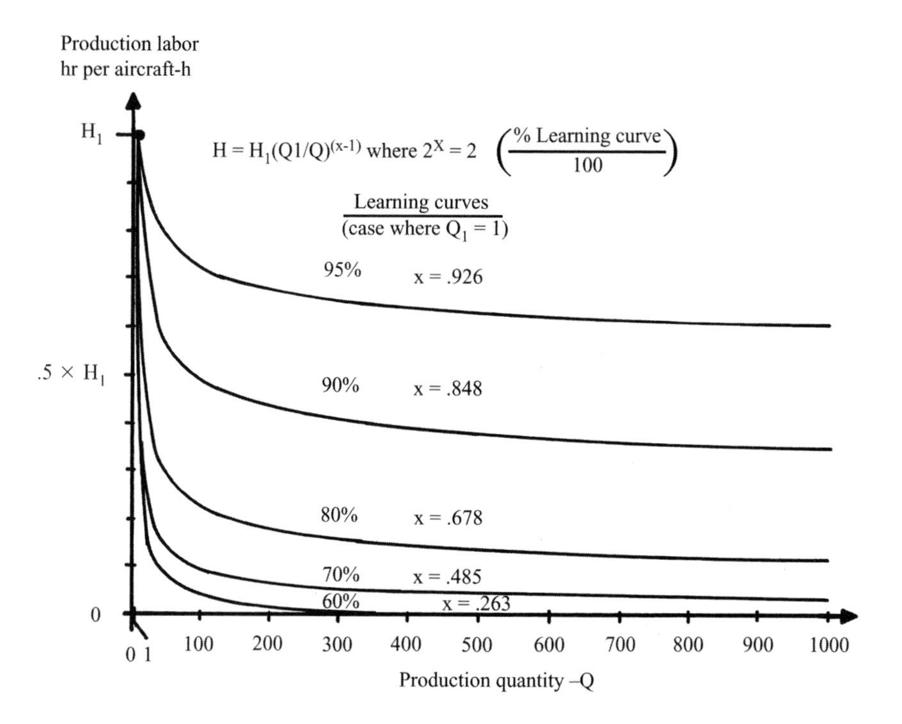

Fig. 18.1 Production learning curve.

RDT&E includes certification cost for civil aircraft. For military aircraft, RDT&E includes the costs associated with the demonstration of airworthiness, mission capability, and compliance with Mil-Specs. RDT&E costs are essentially fixed (nonrecurring) regardless of how many aircraft are ultimately produced. RDT&E, while a lot of money, is typically less than 10% of total life-cycle cost.

The aircraft "flyaway" (production) cost covers the labor and material costs to manufacture the aircraft, including airframe, engines, and avionics. This cost includes production tooling costs. Note that "cost" includes the manufacturer's overhead and administrative expenses. Production costs are recurring in that they are based upon the number of aircraft produced. The cost per aircraft is reduced as more aircraft are produced due to the learning curve effect. Production is typically about half of LCC for military aircraft, but less for commercial aircraft due to the greater number of flying hours in the commercial world.

The purchase price for a civil aircraft is set to recover the RDT&E and production costs, including a fair profit. Since the RDT&E costs are fixed, some assumption must be made as to how many aircraft will be produced to determine how much of the RDT&E costs each sale must recover.

For military aircraft, the RDT&E costs are paid directly by the government during the RDT&E phase, so these costs need not be recovered during production. Military aircraft "procurement cost" (or "acquisition cost") includes the production costs as well as the costs of required ground support equipment, such as flight simulators and test equipment, and the cost of the initial spare

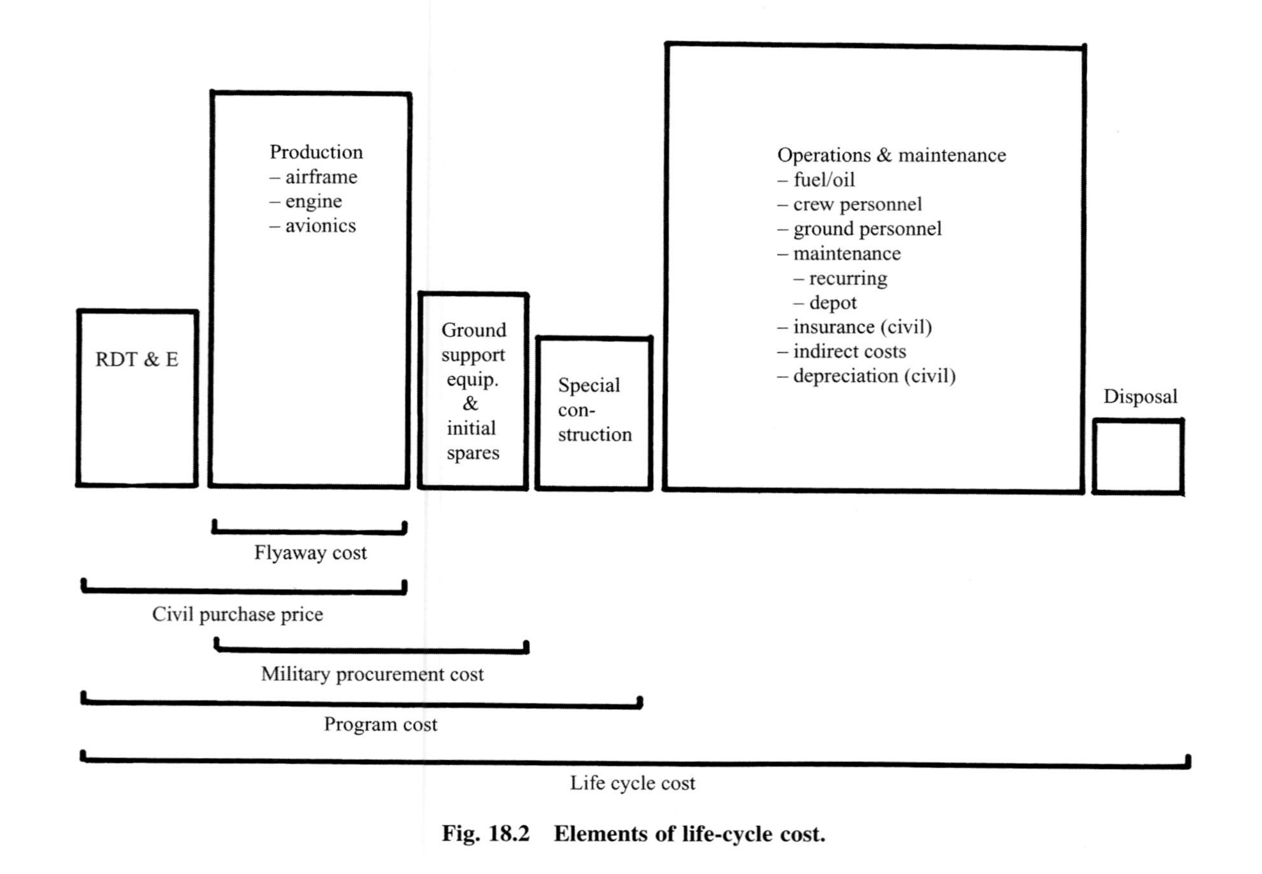

Fig. 18.2 Elements of life-cycle cost.

parts during operational deployment. For civil aircraft, these are normally purchased separately.

One recent trend in military aircraft procurement is called "cost sharing"; the contractor is invited to share some of the RDT&E costs with the expectation of recovering them later during production. It remains to be seen whether future administrations will permit full cost recovery at a later date.

"Program cost" covers the total cost to develop and deploy a new aircraft into the military inventory. Some aircraft require special ground facilities for operational deployment. For example, a fighter/attack aircraft with a large wing span may not fit into the existing bombproof shelters in Europe. The cost of constructing new shelters would be included in the total program cost along with the RDT&E and procurement costs.

"Operations and Maintenance" (O&M) costs (also called Operations and Support, or O&S) are usually much larger than development and production costs for commercial aircraft, and about equal in the military environment. O&M covers fuel, oil, aircrew, maintenance, and various indirect costs. For civil aircraft, insurance will be part of operations cost.

For the operators of commercial aircraft, the depreciation of the aircraft based upon purchase price is also considered to be a part of the operating cost. "Depreciation" is an accounting term that refers to the allocation of the purchase price out over a number of years, using some depreciation schedule.

The simplest depreciation schedule is a straight-line formula, in which each year's depreciation is the purchase price divided by the number of years over which depreciation is spread. Commercial aircraft are usually depreciated over 12–14 years, although they may have a useful life of 20 years or more.

The final element making up the total life-cycle cost concerns "disposal." Obsolete military aircraft are flown one last time to Arizona for "pickling" and storage. The expense of this is not large, so it is frequently ignored in LCC estimation. Civil aircraft have a negative disposal cost because they are worth something on the resale market (typically 10% of purchase price).

18.3 Cost-Estimating Methods

Aircraft, like bologna, are bought by the pound. In 1999, most aircraft cost roughly $200–400 per pound of DCPR or AMPR weight. (DCPR or AMPR weight, defined in Chapter 15, typically equals 60–70% of empty weight). The actual cost varies depending upon the maximum speed, avionics sophistication, production rate, and numerous other factors, but weight remains the most important cost factor within a given class of aircraft.

The cost-estimating methods for a full-scale development proposal are based upon a detailed assessment of the actual tasks to design, test, and produce the aircraft. A "work breakdown structure" (WBS) is prepared. This is an organized tabulation of all of the tasks, and in its most complex form may include hundreds or thousands.

Hours estimates for each task in the WBS are prepared by the appropriate functional groups in the company. This requires estimates for such things as the number of drawings, wind-tunnel tests, tooling fixtures, etc. Other costs such as raw-material purchases, vendor items, computer time, and purchased services are estimated separately. Full-scale-development proposal cost estimation is a massive effort.

Cost estimation during conceptual design is largely statistical. As a simple example, the purchase price data for a number of airliners were plotted vs empty weight where an obvious straight-line trend was noted. This allows prediction of airliner price as simply $2.3 million plus 371 times empty weight in pounds {816 times empty weight in kilograms}, with results in 1995 USD.

To obtain better statistical relationships, cost data for a number of aircraft are broken out by the various cost elements and analyzed using curve-fit programs to prepare cost estimating relationships (CER).

CER input variables include such factors as aircraft DCPR (or AMPR) weight, maximum velocity, and production rate. The output of a CER is either cost or labor hours (engineering, production, etc.), which are converted to cost by multiplying by the appropriate hourly rate.

CERs are developed by the aircraft companies for their own use, and by the various customer organizations for evaluation of proposed aircraft. The U.S. Air Force has developed a complex cost model called Modular Life Cycle Cost Model (MLCCM) that is used for detailed cost estimation. The Rand Corporation has published a number of reports featuring simple CERs for conceptual design, some of which are presented later in the chapter.

Because of the statistical nature of most cost estimation, cost analysts often have to resort to adjustments to the results obtained from equations—in other words, "fudge factors," previously defined with tongue in cheek as "variable constants that you multiply your answer by to get the right answer." These take into account the differences between the underlying assumptions of your design and the realities of the existing aircraft whose cost data were used to obtain the statistical equations.

18.4 RDT&E and Production Costs

RDT&E and production costs are frequently combined to develop CERs. It is difficult to separate clearly the RDT&E from production costs, especially in the areas of engineering and prototype fabrication.

For example, production of long-lead-time items (e.g., landing-gear forgings) is usually initiated before the prototype has flown. The engineering support of these production items should be considered a part of production. It is difficult for the developer of a CER to determine, years later, how many engineering hours during the RDT&E phase were actually spent in support of production.

It is also common in the development of CERs to assume that the prototype aircraft will have a cost based upon the production-cost CER, with the higher cost of prototypes accounted for by the early position on the learning curve. However, prototypes are usually built virtually by hand, with simplified prototype tooling, and may have labor hour costs much greater than accounted for by the learning curve.

The best CERs are those developed using recent aircraft that are highly similar to the new aircraft being analyzed. Because detailed cost data are usually proprietary, this puts the current producers of aircraft at a great advantage when it comes to estimating the cost of a new aircraft. Boeing has no trouble estimating with great accuracy the costs of a new jetliner, using the costs of their current aircraft.

When a detailed cost baseline for a highly similar aircraft is available, even simple CERs can yield great accuracy. Merely multiplying the component weights of the new aircraft times the dollars per pound or hours per pound for a similar baseline aircraft is probably better than a sophisticated CER based upon a number of not-so-similar aircraft.

For example, the selected cost-baseline aircraft may have required 50 hr/lb {110 hr/kg} of manufacturing labor for the fuselage and subsystems and 90 hr/lb {198 hr/kg} for the wings and tails. These typical values are multiplied by the appropriate component weights of the new aircraft to determine hours, which are then multiplied times the manufacturing hourly rate to determine cost.

This technique is especially useful for prototype and flight demonstrator (X-series) aircraft, which are poorly estimated by sophisticated CERs based upon production aircraft. However, it may be difficult to find a recent and similar prototype or demonstrator aircraft to use as a cost baseline.

RAND DAPCA IV Model

A set of CERs for conceptual aircraft design developed by the RAND Corporation (Ref. 78) is known as "DAPCA IV," the Development and Procurement Costs of Aircraft model.

DAPCA is probably not the very best set of CERs for any one class of aircraft, but is notable in that it seems to provide reasonable results for several classes of aircraft including fighters, bombers, and transports.

DAPCA estimates the hours required for RDT&E and production by the engineering, tooling, manufacturing, and quality control groups. These are multiplied by the appropriate hourly rates to yield costs. Development support, flight-test, and manufacturing material costs are directly estimated by DAPCA.

Engineering hours include the airframe design and analysis, test engineering, configuration control, and system engineering. Engineering hours are primarily expended during RDT&E, but there is some engineering effort throughout production. In the estimating equation presented next, the total engineering effort for a 500-aircraft production run is about three times the engineering effort for a one-aircraft "production run."

The engineering effort performed by the airframe contractor to integrate the propulsion and avionics systems into the aircraft is included under engineering hours. However, the actual engineering effort by the propulsion and avionics contractors is not included. Those items are treated as purchased equipment. Engineering support of tooling and production planning are included in those areas instead of in engineering.

Tooling hours embrace all of the preparation for production: design and fabrication of the tools and fixtures, preparation of molds and dies, programming for numerically controlled manufacturing, and development and fabrication of production test apparatus. Tooling hours also cover the ongoing tooling support during production.

Manufacturing labor is the direct labor to fabricate the aircraft, including forming, machining, fastening, subassembly fabrication, final assembly, routing (hydraulics, electrics, and pneumatics), and purchased part installation (engines, avionics, subsystems, etc). The following equation includes the manufacturing hours performed by airframe subcontractors, if any.

Quality control is actually a part of manufacturing, but is estimated separately. It includes receiving inspection, production inspection, and final inspection. Quality control inspects tools and fixtures as well as aircraft subassemblies and completed aircraft.

The RDT&E phase includes development support and flight-test costs. Development-support costs are the nonrecurring costs of manufacturing support of RDT&E, including fabrication of mockups, iron-bird subsystem simulators, structural test articles, and various other test items used during RDT&E. In DAPCA these costs are estimated directly, although some other models separately estimate the labor and material costs for development support.

Flight-test costs cover all costs incurred to demonstrate airworthiness for civil certification or Mil-Spec compliance except for the costs of the flight-test aircraft themselves. Costs for the flight-test aircraft are included in the total production-run cost estimation. Flight-test costs include planning, instrumentation, flight operations, data reduction, and engineering and manufacturing support of flight testing.

Manufacturing materials—the raw materials and purchased hardware and equipment from which the aircraft is built—include the structural raw materials, such as aluminum, steel, or prepreg graphite composite, plus the electrical, hydraulic, and pneumatic systems, the environmental control system, fasteners, clamps, and similar standard parts.

These may be contractor-furnished equipment (CFE) or government-furnished equipment (GFE). Manufacturing materials include virtually everything on the aircraft except the engines and avionics.

The following DAPCA equations have been modified to include the quantity term provided in an appendix to Ref. 78.

DAPCA assumes that the engine cost is known. A turbojet-engine cost estimation equation from Ref. 79 has been included for use where the engine cost is unknown. For a turbofan engine, cost should be increased 15–20% higher than predicted with this equation. Note that the equation does not include the cost to develop a new engine.

Modified DAPCA IV Cost Model (costs in constant 1999 dollars):

$$\text{Eng hours} = H_E = 4.86 W_e^{0.777} V^{0.894} Q^{0.163} \quad \text{(fps)}$$

$$= 5.18 W_e^{0.777} V^{0.894} Q^{0.163} \quad \text{(mks)} \tag{18.1}$$

$$\text{Tooling hours} = H_T = 5.99 W_e^{0.777} V^{0.696} Q^{0.263} \quad \text{(fps)}$$

$$= 7.22 W_e^{0.777} V^{0.696} Q^{0.263} \quad \text{(mks)} \tag{18.2}$$

$$\text{Mfg hours} = H_M = 7.37 W_e^{0.82} V^{0.484} Q^{0.641} \quad \text{(fps)}$$

$$= 10.5 W_e^{0.82} V^{0.484} Q^{0.641} \quad \text{(mks)} \tag{18.3}$$

$$\text{QC hours} = H_Q = 0.076 \text{ (mfg hours) if cargo airplane}$$

$$= 0.133 \text{ (mfg hours) otherwise} \tag{18.4}$$

$$\text{Devel support cost} = C_D = 66.0 W_e^{0.630} V^{1.3} \quad \text{(fps)}$$

$$= 48.7 W_e^{0.630} V^{1.3} \quad \text{(mks)} \tag{18.5}$$

$$\text{Flt test cost} = C_F = 1807.1 W_e^{0.325} V^{0.822} \text{FTA}^{1.21} \quad (\text{fps})$$
$$= 1408 W_e^{0.325} V^{0.822} \text{FTA}^{1.21} \quad (\text{mks}) \tag{18.6}$$
$$\text{Mfg materials cost} = C_M = 16 W_e^{0.921} V^{0.621} Q^{0.799} \quad (\text{fps})$$
$$= 22.6 W_e^{0.921} V^{0.621} Q^{0.799} \quad (\text{mks}) \tag{18.7}$$
$$\text{Eng production cost} = C_{\text{eng}} = 2251.0[0.043 T_{\max} + 243.25 M_{\max} + 0.969 T_{\text{turbine inlet}} - 2228] \quad (\text{fps})$$
$$= 2251[9.66 T_{\max} + 243.25 M_{\max} + 1.74 T_{\text{turbine inlet}} - 2228] \quad (\text{mks}) \tag{18.8}$$
$$\text{RDT\&E} + \text{flyaway} = H_E R_E + H_T R_T + H_M R_M + H_Q R_Q + C_D + C_F + C_M + C_{\text{eng}} N_{\text{eng}} + C_{\text{avionics}} \tag{18.9}$$

where

W_e = empty weight (lb or kg)
V = maximum velocity (knots or km/hr)
Q = lesser of production quantity or number to be produced in five years
FTA = number of flight-test aircraft (typically 2–6)
N_{eng} = total production quantity times number of engines per aircraft
$T_{\max}$ = engine maximum thrust (lb or kN)
$M_{\max}$ = engine maximum Mach number
$T_{\text{turbine inlet}}$ = turbine inlet temperature (°R or K)
C_{avionics} = avionics cost

The hours estimated by DAPCA are based upon the design and fabrication of an aluminum aircraft. For aircraft that are largely fabricated from other materials, the hours must be adjusted to account for the more difficult design and fabrication. Based upon minimal information, the following "fudge factors" are recommended:

aluminum 1.0
graphite–epoxy 1.1–1.8
fiberglass 1.1–1.2
steel 1.5–2.0
titanium 1.1–1.8

Reference 103 offers a more complete discussion and methodology for adjusting cost CERs for the use of advanced materials, but all such adjustment factors are highly debatable.

The hours estimated with this model are multiplied by the appropriate hourly rates to calculate the labor costs. These hourly rates are called "wrap rates" because they include the direct salaries paid to employees as well as the employee benefits, overhead, and administrative costs. Typically the employee salaries are a little less than half the wrap rate. Average wrap rates were presented in Ref. 78, as follows (adjusted to 1999):

engineering .... \$86 = R_E
tooling................ \$88 = R_T

quality control .... $81 = R_Q$
manufacturing .... $73 = R_M$

Predicted costs are then ratioed by some inflation factor to the selected year's constant dollar. Aircraft costs do not all follow the same inflation factor. For example, the salaries of the engineers may increase at a slower rate than the raw-material cost for aluminum.

"Economic escalation factors" for the various cost elements are based upon the actual and predicted cost-inflation for the more important cost drivers. One such factor, the "Federal Price Deflator for the Aircraft Industry," is derived from an in-depth analysis of the costs of items used for aircraft production.

For initial estimates and student design projects, the Consumer Price Index (CPI) may be used as an approximate economic escalation factor. The CPI is the purchasing value of the dollar expressed as a percentage of some chosen base year (changed occasionally to avoid large CPI numbers). The past and projected CPI is published by the government and is readily available.

DAPCA does not estimate avionics costs. They must be estimated from data on similar aircraft or from vendors' quotations. Avionics costs range from roughly 5–25% of flyaway cost depending upon sophistication, or can be approximated as $3000–$6000 per pound ($7–$13 per gram) in 1999 dollars.

DAPCA does not include an allowance for the cost of interiors for passenger aircraft, such as seats, luggage bins, closets, lavatories, insulation, ceilings, floors, walls, and similar items. Reference 11 suggests that cost per aircraft be increased by $2500 per passenger for jet transport, $1250 for regional transports, or $625 for general-aviation aircraft (adjusted to 1999 dollars).

DAPCA IV was developed from a statistical data base of nonstealth, noncomposites fighters, trainers, transports, and bombers. It does not handle the most advanced designs very well, and it is recommended that hours and cost estimates be increased by about 20–40% for such aircraft. DAPCA tends to overpredict commercial aircraft development costs by roughly 10%.

Predicted aircraft costs will be multiplied by an "investment cost factor" to determine the purchase price to the customer. The investment cost factor includes the cost of money and the contractor profit; it is considered highly proprietary by a company. Investment cost factor may be roughly estimated as 1.1–1.4.

Initial spares will add perhaps 10–15% to an aircraft's purchase price.

18.5 Operations and Maintenance Costs

O&M costs are determined from assumptions as to how the aircraft will be operated. The main O&M costs are fuel, crew salaries, and maintenance. For a typical military aircraft, the fuel totals about 15% of the O&M costs, the crew salaries about 35%, and the maintenance most of the remaining 50%. Over one-third of U.S. Air Force manpower is dedicated to maintenance.

For commercial aircraft (which fly many more hours per year), the fuel totals about 38% of O&M costs, the crew salaries about 24%, and the maintenance about 25%. The depreciation of the aircraft purchase price is about 12% of total O&M costs, and the insurance is the remaining 1%. Commercial aircraft also pay landing fees that add about 2% to operating costs. (These are rough numbers: actual values vary widely.)

Fuel and Oil Costs

When flying the design mission, the aircraft burns all of the available fuel except what will be required for loiter and for reaching an alternate airport. However, the actual missions will rarely resemble the design mission. Most of the time the aircraft will land with substantial fuel in the tanks that can be used on the next flight.

To estimate yearly fuel usage, a typical mission profile is selected, and the total duration and fuel burned are used to determine the average fuel burned per hour. This is multiplied by the average yearly flight hours per aircraft, which must be assumed based upon typical data for that class of aircraft. Table 18.1 provides some rough guidelines for flight hours per year and other LCC parameters.

Finally, the total amount of fuel burned per year of operation is multiplied by the fuel price as obtained from petroleum vendors, ratioed to the appropriate year's dollar. Fuel prices can change rapidly, going from about 80 cents per gallon in 1998 to over $1.80 per gallon in 2006. Note that oil costs average less than half a percent of the fuel costs, and can be ignored.

Crew Salaries

Crew expenses for military and civil aircraft are calculated differently. The cost of a civil-aircraft crew (including flight and cabin crew) can be statistically estimated based upon the yearly "block hours."

Block hours measure the total time the aircraft is in use, from when the "blocks" are removed from the wheels at the departure airport to when they are placed on the wheels at the destination. Block hours therefore include taxi time, ground hold time, total mission flight time, airborne holding time, extra time for complying with air-traffic-control approach instructions, and time spent on the ground waiting for a gate.

Block speed (V_B)—the average block velocity, i.e., the trip distance divided by the block time—will be substantially less than the actual cruise velocity.

Reference 52 provides detailed formulas prepared by the Airline Transport Association of America for airline block-time estimation. Block time can be approximated by the mission flight time plus 15 min for ground maneuver and 6 min for air maneuver.

Table 18.1 LCC parameter approximations

Aircraft class	FH/YR/AC	Crew ratio	MMH/FH
Light aircraft	500–1000	—	1/4–1
Business jet	500–2000	—	3–6
Jet trainer	300–500	—	6–10
Fighter (modern)	300–500	1.1	10–15
Bomber	300–500	1.5	25–50
Military transport	700–1400	1.5 if FH/YR < 1200 2.5 if $1200 <$ FH/YR < 2400 3.5 if $2400 <$ FH/YR	20–40
Civil transport	2500–4500	—	5–15

Note that the mission distance will not simply be the straight-line distance between the two airports. Airliners must follow federal airways, which may not directly connect the two airports. The additional distance will be approximately 2% of distances over 1400 miles, and $(0.015 + 7/D)\%$ for shorter trips.

The block hours per year can be determined from the ratio between block hours and flight hours for the selected mission, times the total flight hours per year per aircraft (Table 18.1). For a long-range aircraft, the block hours equal approximately the flight hours; but for short-range aircraft with average trip times under an hour, the block time can be substantially greater than the flight time.

Crew cost per block hour can be estimated using Eqs. (18.10) and (18.11). These were provided in Ref. 52 from Boeing data (converted to 1999 dollars).

$$\text{Two-man crew cost} = 51\left(V_c\frac{W_0}{10^5}\right)^{0.3} + 122\,(\text{fps})$$

$$= 54\left(V_c\frac{W_0}{10^5}\right)^{0.3} + 122\,(\text{mks}) \tag{18.10}$$

$$\text{Three-man crew cost} = 68\left(V_c\frac{W_0}{10^5}\right)^{0.3} + 172\,(\text{fps})$$

$$= 72\left(V_c\frac{W_0}{10^5}\right)^{0.3} + 172\,(\text{mks}) \tag{18.11}$$

where

V_c = cruise velocity in knots or km/hr
W_0 = takeoff gross weight, lb or kg

Costs are estimated in 1999 dollars per block hour.

These equations must be viewed as rough approximations only. The current turmoil in the airline industry has created a wide variation in crew costs. The B-747 crew costs per block hour in 1987 ranged from \$1013 for an old established airline to \$189 for a new low-fare airline!

For military aircraft, crew costs are determined by estimating how many flight-crew members will have to be kept on the active-duty roster to operate the aircraft. This is the number of aircraft times the number of crew members per aircraft, times the "crew ratio."

Military pilots no longer get their own airplane as in the movies. There are always more pilots and other crew members than the number of aircraft. The crew ratio defines the ratio of aircrews per aircraft. It ranges from 1.1 for fighters to 3.5 for transports that are flown frequently. Typical crew ratios are provided in Table 18.1.

The average cost per crew member, as is obtained from military sources, varies depending upon airplane type. As in civilian life, the cost is much greater than the salaries alone, to cover benefits and overhead. In the absence of better data, the engineering hourly wrap-rates times 2080 hr per year may be used for initial trade studies and student design projects.

Maintenance Expenses

Unscheduled maintenance costs depend upon how often the aircraft breaks and the average cost to fix it.

Scheduled maintenance depends upon the number of items requiring regularly scheduled maintenance and the frequency and cost of the scheduled maintenance. Maintenance is usually scheduled by accumulated flight hours. For example, light aircraft require a complete inspection every 100 hours. For commercial aircraft, there are also maintenance activities that are scheduled by the number of flights ("cycles").

Maintenance activities are lumped together under Maintenance Man hours per Flight Hour (MMH/FH). This is the primary measure of maintenance "goodness." MMH/FHs range from well under 1.0 for small private aircraft to over 100 for certain special-purpose aircraft. Typical values are shown in Table 18.1.

Reducing MMH/FH is a key goal of aircraft design, as discussed earlier. MMH/FH is roughly proportional to weight because the parts count and systems complexity go up with weight.

MMH/FH is also strongly affected by the aircraft utilization. An aircraft that is constantly flying will receive more scheduled maintenance per year and will be maintained by more experienced mechanics. For example, the DC-9 has a MMH/FH of about 6.4 in civilian operation. The same plane in military service (C-9), flying only about half as many hours per year, has a MMH/FH of about 12.

From the MMH/FH and flight hours per year, the maintenance manhours per year can be estimated. The maintenance labor cost can then be determined from the labor wrap-rate obtained from airline or military sources. In the absence of better data, the labor cost can be approximated by the manufacturing wrap-rate presented earlier.

Materials, parts, and supplies used for maintenance will approximately equal the labor costs for military aircraft.

For civil aircraft, Ref. 52 presents the following rough equations for materials cost per flight hour and per cycle (adjusted to 1999 dollars). The number of cycles per year is estimated by determining the total yearly block time divided by the block time per flight. The total materials cost is the cost per flight hour times the flight hours per year, plus the cost per cycle times the cycles per year.

$$\frac{\text{material cost}}{\text{FH}} = 3.3\left(\frac{C_a}{10^6}\right) + 10.2 + \left[58\left(\frac{C_e}{10^6}\right) - 19\right]N_e \qquad (18.12)$$

$$\frac{\text{material cost}}{\text{cycle}} = 4.0\left(\frac{C_a}{10^6}\right) + 6.7 + \left[7.5\left(\frac{C_e}{10^6}\right) + 4.1\right]N_e \qquad (18.13)$$

where

C_a = aircraft cost less engine
C_e = cost per engine
N_e = number of engines

Costs are in 1999 dollars per flight hour or cycle.

Depreciation

For commercial aircraft, the depreciation is considered a part of the operating expenses. Depreciation is really the allocation of the purchase price over the operating life of the aircraft. While complicated depreciation formulas are used by accountants, a simple straight-line schedule provides a reasonable first estimate. The airframe and engine have different operating lives, and so must be depreciated separately.

The airframe yearly depreciation is the airframe cost less the final resale value, divided by the number of years used for depreciation. If the resale value is 10% of purchase price and the depreciation period is 12 years, the yearly airframe depreciation is the airframe cost times 0.9 divided by 12. (Here airframe cost refers to the total cost minus the total engine costs.)

Engine resale value can be neglected for initial analysis. If the engine is depreciated over four years, the yearly depreciation cost per engine is the engine purchase price divided by 4.

Insurance

Insurance costs for commercial aircraft add approximately 1–3% to the cost of operations.

18.6 Cost Measures of Merit (Military)

Once the cost is estimated, it is incorporated into several cost-effectiveness measures of merit. For military aircraft (fighters and bombers), the ultimate measure of merit is the cost to "win the war" (or at least avoid losing it!).

This is determined through parametric variations in sophisticated "campaign models" that simulate in great detail the conduct of a postulated war. The improvement in the outcome of the war is compared to the total LCC to develop the new aircraft and operate it for (typically) 20 years.

Other cost-effectiveness measures of merit in common use include the cost per weapon pound delivered and the cost per target killed. These require detailed analysis of the sortie rate, survivability, and weapons effectiveness that goes beyond the scope of this book.

Trade studies are conducted to determine the variation in these measures of merit with design changes such as payload and turn rate. Conceptual designers in military aircraft companies become very familiar with these measures of merit.

In some military aircraft procurements, cost alone becomes the driving measure of merit. The aircraft must cost less than some stated value regardless of performance and range requirements.

If the aircraft as designed to meet the stated performance and range requirements costs more than the design-to-cost, then either performance or range must be sacrificed. At this point the designers sincerely hope that no other company has succeeded in designing an aircraft in full compliance with performance, range, and cost requirements. This is discussed further in Chapter 19.

18.7 Aircraft and Airline Economics

DOC and IOC

Cost effectiveness for an airliner is purely economic. The aircraft must generate sufficient revenue in excess of operating costs that the purchase investment is more profitable than investing the same amount of money elsewhere.

Airline operating costs are divided into direct operating costs (DOC) and indirect operating costs (IOC). DOC costs concern flight operations as discussed earlier, namely, fuel, oil, crew, maintenance, depreciation, and insurance.

Another significant expense for airliners is the landing fee now charged by commercial airports. Normally landing fees are proportional to landing weight, and are increasingly seen as a major source of revenue by local governments. In some cases the landing fee can nearly equal the fuel cost for the flight, although a more typical value is about one-third of the fuel cost. Landing fees can be found by calling airport operations offices.

DOC costs for economic analysis are expressed as cost per seat-mile flown, where the seat-miles are equal to the number of seats on the aircraft times the statute miles flown. DOC per seat-mile is frequently used to compare aircraft and is used as the measure of merit for design trade studies. Wide-body transports average DOC of about 3–4 cents per seat-mile.

IOC costs, the remaining costs to run an airline, include the depreciation costs of ground facilities and equipment, the sales and customer service costs, and the administrative and overhead costs.

IOC costs are difficult to estimate because they are based on how the airline chooses to run its operations. Certain companies such as Southwest Airlines are famous for holding these costs to a bare minimum, whereas other airlines pride themselves on more of a "full service" operation. IOC costs do not lend themselves to statistical analysis and depend very little upon the aircraft design itself. Typically, the yearly IOC costs will range from a third of to about equal to the DOC costs, but the variation is so great that the only way to obtain reliable IOC costs for economic analysis is from the airlines themselves. Reference 137 is recommended for an in-depth discussion of airliner design and costs analysis.

Airline Revenue

Airline revenue comes primarily from ticket sales. Ticket prices are approximately proportional to trip distance, but are higher per mile for shorter distances. Tickets are sold in four classes: first class, business class, coach (tourist) class, and excursion.

Roughly speaking, first class costs twice as much as coach, business class costs 1.5 times coach, and excursion fares are 50–90% of coach fares. However, there is tremendous variation. As this is written, one airline is quoting a higher Los Angeles to Sacramento excursion fare than its coach fare!

For revenue estimation, a phone call to the local airline ticket office will provide current fares over selected routes. For future fares, the current fares can be ratioed by the assumed inflation.

The number of tickets sold in these various classes must be estimated. For the North Atlantic routes, tickets sold are typically 5% in first class, 15% in business class, 10% in coach, and 70% in excursion. If a weighted average of the different fares is calculated, it turns out that the average fare paid is approximately the coach fare.

The remaining parameter to be determined for revenue estimation, the "load factor," measures how full the aircraft is. Load factor equals the seats sold divided by the total seats available. Current load factors range from 60–70%. Load-factor data are provided occasionally in the trade magazine *Aviation Week and Space Technology* (along with other airline operating-cost data).

Thus, the revenue per seat-mile flown can be determined as the average fare sold over that route (approximately the coach fare) times the load factor.

Breakeven Analysis

The term "breakeven" refers to a time or occurrence in which revenues have equaled expenses, and is a common calculation in business circles. For aircraft design, there are two breakeven conditions of special interest.

For the aircraft manufacturer, breakeven is said to occur when enough aircraft have been sold to pay off the development costs. A sales price per aircraft is determined as the cost to develop and certify the aircraft, plus the cost to produce each aircraft, plus the "cost of money," which is an accounting for the use of the funds during the development of the aircraft. To this the company adds, as we say, "all the market will bear," to determine a sales price. The difference between the cost to produce an aircraft and what we sell that aircraft for is the "contribution margin," because it contributes to paying off the development cost. Eventually, the contribution margin allows for some company profit.

For example, if a new business jet costs $400 million to develop and certify (including cost of money), and we can sell it for $2 million more than the actual production cost, then we will break even at aircraft number 200. Before that, we really won't have made any actual profit except in, perhaps, an accounting sense. After aircraft 200, we will make a profit of $2 million per aircraft.

We must set our sales price carefully—too high, and we won't make many sales. But if we set the price too low, the breakeven point may not ever be reached.

The other breakeven calculation of interest is related to airline operations. The airline will estimate the breakeven load factor to determine, among other measures of merit described next, whether our aircraft will be a good addition to their fleet, and whether they should buy our aircraft.

The load factor for breakeven can be calculated as the cost per seat-mile divided by the average fare per seat-mile. The operating-cost breakeven analysis uses the DOC per seat-mile. This is the load factor at which the passengers pay just enough to fly the airplane, with no excess for covering indirect costs or to provide the airline any profit.

The total-cost breakeven analysis uses the DOC plus IOC per seat-mile. The IOC per seat-mile is determined by dividing the airline's total yearly indirect operating costs by the total number of seat-miles flown by the airline each year. As a rough approximation, the IOC per seat-mile approximately equals the DOC per seat-mile.

Investment Cost Analysis

The ultimate decision by the airline as to whether or not to buy a particular aircraft is based upon an investment cost analysis that takes the "net present value" of the revenue minus cost over the useful life of the aircraft and compares it to the investment cost (purchase price).

Net present value (NPV) is an economic valuation based on the concept that money in the hand today is more valuable than money received in the future. At the very least, the money in hand today could be drawing interest in the bank. Even better, money in hand today could be invested in some reasonably safe business venture and draw a higher yearly return.

The "net present value" of future money is the amount of money in hand today which would yield the given future amount of money if invested at a "normal" rate of return. For example, \$110 to be received a year from today would have a net present value of \$100 if a normal investment returns 10% interest per year.

Equation (18.14) determines the future value V_n after n years of an initial investment of value V_0, given an interest rate r. In Eq. (18.15) this is solved for the required investment today to yield a given future value. V_0 is therefore the net present value V_{np} as just described. The interest rate r is known as the "discount factor" in net-present-value calculations.

$$V_n = V_0(1 + r)^n \tag{18.14}$$

$$V_0 = V_{\mathrm{np}} = \frac{V_n}{(1 + r)^n} \tag{18.15}$$

The NPV of an airliner is the total of the net present values of all of the yearly operating profits during the life of the aircraft (usually taken to be the depreciation period). The yearly operating profits are the yearly revenues minus the DOC and IOC, not including depreciation. Depreciation is not included in NPV calculation because it is the yearly apportionment of the purchase price.

The NPV is determined by estimating the revenues and costs for each year of operation, including the effects of the estimated inflation. The yearly operating profit is then converted to NPV using Eq. (18.15). Finally, the NPVs of all of the years of operation are summed. To this is added the NPV of the salvage value of the aircraft at the end of its life (typically equal to 10% of purchase price).

The total NPV must be greater than the purchase price of the aircraft, or the investment will not return the expected normal rate of return, i.e., the discount factor r.

Selection of the appropriate discount factor is critical to the NPV calculation. The selected discount factor should be greater than the interest received from extremely safe investments such as government bonds, but should be less than the return from risky investments such as volatile stocks. The selected discount rate should probably be no less than the real rate of return on the airline company's stock, which equals the yearly dividends plus the increase in stock value, divided by the stock purchase price.

Alternatively, the discount factor can be solved for the value for which the investment just barely breaks even. The discount factor r for which the NPV exactly equals the investment is called the "internal rate of return"; it represents the equivalent interest rate returned by the airline investment. This can be compared to the expected rate of return on other investments to determine if the new airliner is a good buy.

Titan IV Centaur (U.S. Air Force photo).

19
Sizing and Trade Studies

19.1 Introduction

We have come full circle in the design process. We began with a rough conceptual sketch and a first-order estimation of the T/W and W/S to meet the performance requirements. A "quick and dirty" sizing method was used to estimate the takeoff weight and fuel weight required to meet the mission requirements.

The results of that sizing were used to develop a conceptual design layout that incorporated considerations for the real world, including landing gear, structure, engine installation, etc. The design layout was then analyzed for aerodynamics, weights, installed-engine characteristics, structures, stability, performance, and cost.

The as-drawn aircraft might or might not actually meet all of the performance and mission requirements. The refined estimates for the drags, weights, and installed engine characteristics are all somewhat different from our earlier crude estimates. Therefore, the selected T/W and W/S are probably not optimal. The same is true for the aspect ratio, sweep, taper ratio, and other geometric parameters. Also, the as-drawn weights are probably wrong.

Now we are ready to revisit the sizing analysis using our far-greater knowledge about the aircraft. Refined trade-study methods will allow us to determine the size and characteristics of the optimal aircraft, which meets all performance and mission requirements.

19.2 Detailed Sizing Methods

Equations (17.1) and (17.2) [repeated as Eqs. (19.1) and (19.2)] define the sum of the forces on the aircraft in the X_s and Z_s directions. The resulting accelerations on the aircraft are determined as these force summations divided by the aircraft mass (W/g):

$$\Sigma F_x = T\cos(\alpha + \phi_T) - D - W\sin\gamma \tag{19.1}$$

$$\Sigma F_z = T\sin(\alpha + \phi_T) + L - W\cos\gamma \tag{19.2}$$

$$\dot{W} = -CT \tag{19.3}$$

Equation (19.3) defines the time rate of change in aircraft weight as the specific fuel consumption c times the thrust. Equations (19.4) and (19.5) determine the equivalent c and thrust for a piston-engine aircraft (see Chapter 5):

$$C = C_{\text{power}} \frac{V}{\eta_P} = C_{\text{bhp}} \frac{V}{550\eta_P} \tag{19.4}$$

$$T = \frac{P\,\eta_P}{V} = \frac{550 \text{ bhp } \eta_P}{V} \tag{19.5}$$

These equations are the basis of the highly detailed sizing programs used by the major airframe companies. In these programs the fuel weight is actually calculated by determining the required thrust level and resulting fuel flow during each segment of the mission.

The angle of attack and thrust level are varied to give the required total lift and the required longitudinal acceleration depending upon what maneuver the aircraft must perform (level cruise, climb, accelerate, turn, etc.). Angle of attack and lift are restricted by the maximum lift available. The thrust level is restricted to the available thrust obtained from a table of installed-engine thrust vs altitude and velocity (or Mach number).

To improve the accuracy, the mission is broken into a large number of very short segments that may be less than one minute in duration. The reduction in the aircraft weight during each of these short mission segments is determined by calculating the actual fuel burned based upon the required thrust setting.

The computer iterates for sized takeoff weight by varying the assumed takeoff weight until the ending empty-weight fraction matches the empty-weight fraction determined by the detailed weight estimation. More sophisticated sizing programs will use statistical weights equations to automatically recalculate the allowable empty weight for the sizing variations in takeoff weight, wing area, thrust, aspect ratio, and other trade parameters.

Such methods go beyond the scope of this book. Those who take jobs as sizing and performance specialists in major aircraft companies will find that these computer programs are so large today that they are programmed by a team.

19.3 Improved Conceptual Sizing Methods

Review of Sizing Method

For sizing and trade studies during conceptual design, an improved version of the method presented in Chapter 6 is adequate. Remember that the aircraft was sized iteratively by assuming a takeoff weight. A statistical method was used to determine the empty weight for this assumed takeoff weight.

The fuel used was determined by breaking the mission into mission segments, numbered from 1 to x. For each mission segment, the change in aircraft weight was calculated as either a mission-segment weight fraction (W_i/W_{i-1}) due to fuel burned, or as a discrete change in weight due to payload dropped.

Starting with the assumed takeoff weight, the aircraft weight was reduced for each mission segment either by subtracting the discrete weight or by multiplying by the mission-segment weight fraction. The fuel burned during each mission segment was totalled throughout the mission to determine the total fuel burned. A 6% allowance was added to the mission fuel to account for reserve and trapped fuel.

The aircraft takeoff weight was then calculated by summing the payload, crew, fuel, and empty weight. This calculated takeoff weight was compared to the assumed takeoff weight. A new assumed takeoff weight was selected somewhere between the two, and the sizing process was iterated toward a solution.

This same sizing process can be employed for sizing the as-drawn aircraft, but the method can be improved based upon our greater knowledge of the design.

In the initial sizing before the aircraft design layout was prepared, the mission fuel was determined using simplified equations and statistical estimates of the aerodynamic properties and installed-engine characteristics. The empty weight was determined from statistical equations based only upon the takeoff weight.

At this later stage in the design process, we can calculate better estimates for the fuel used during each mission segment, and we have a better estimate of the empty weight based upon a detailed analysis of the as-drawn aircraft. These improved methods are presented next.

Many of these methods rely upon calculating, by the methods of the performance chapter, the duration of time to perform the mission segment. The fuel burned during a duration of d at a given thrust T and specific fuel consumption C is then determined by Eq. (19.6). The mission-segment fuel fraction is solved for in Eq. (19.7), where C and $(T/W)_i$ are the average actual values during mission segment i:

$$W_{f_i} = CTd \tag{19.6}$$

$$\frac{W_i}{W_{i-1}} = 1 - Cd\left(\frac{T}{W}\right)_i \tag{19.7}$$

Note that if $(T/W)_i$ remains essentially constant during the iterations for takeoff weight, the result of Eq. (19.7) can be used unchanged for each iteration. This is the case for "rubber engine" sizing.

For "fixed-engine" sizing, Eq. (19.7) would have to be recalculated for each iteration step because the T/W for a fixed thrust changes as the weight is changed. Alternatively, Eq. (19.6) can be used to calculate the actual weight of the fuel burned by that fixed-size engine. The fuel burned is then treated as a weight drop in the sizing iterations.

(A word of caution: Mission-segment weight fractions should range between about 0.9 and 1.0. If a mission-segment weight fraction is less than 0.9, the accuracy should be improved by breaking that mission segment into two or more smaller segments. If the mission-segment weight fraction is calculated to be greater than 1.0, you have probably used the wrong units somewhere or have forgotten the negative sign on an exponent!)

Engine Start, Warmup, and Taxi

In the initial sizing method, the mission-segment weight fraction for engine start, warmup, and taxi was lumped with the takeoff, and assumed to be 0.97–0.99.

A better estimate for the fuel used during engine start, warmup, and taxi uses the actual engine characteristics to calculate the fuel burned by the engine in a certain number of minutes at some thrust setting. Typically this would be 15 min at idle power. Equation (19.7) is used to determine the resulting mission-segment weight fraction.

Takeoff

The takeoff distance was broken into segments and calculated in Chapter 17. The time duration d of those segments is approximately the segment distance divided by the average velocity during the segment. Equation (19.7) can then be used to calculate the mission-segment weight fraction using the appropriate average takeoff thrust and fuel consumption.

Sometimes the design requirements may lump together the engine start, warmup, taxi, and takeoff into a single requirement based upon some amount of time at a given thrust setting. For military combat aircraft this is usually 5 min at maximum dry power. For transports and commercial aircraft, 14 min at ground idle plus 1 min at takeoff thrust have often been specified.

Climb and Acceleration

The energy methods of Chapter 17 provided Eq. (17.96), repeated as Eq. (19.8), for the mission-segment weight fraction for a change in altitude and/or velocity. The average values of C, V, D, and T should be used. A long climb or large change in velocity should be broken into segments such that the quantity $C/[V(1 - D/T)]$ is approximately constant.

$$\frac{W_i}{W_{i-1}} = \exp\left[\frac{-C\Delta h_e}{V(1 - D/T)}\right] \tag{19.8}$$

$$\Delta h_e = \Delta\left(h + \frac{1}{2g}V^2\right) \tag{19.9}$$

The distance travelled during climb is usually "credited" to the cruise segment that follows, i.e., that distance is subtracted from the required cruise range. Distance travelled during climb is calculated as average velocity times the time to climb, which equals $\Delta h_e/P_s$.

Cruise and Loiter

In Chapter 17, methods for determining the optimal velocities and altitudes for cruise and loiter were presented, and the Breguet equations for cruise and loiter were derived. Solving these for mission segment weight fraction yields

Eqs. (19.10) and (19.11), where R is the range and E is the endurance time.

$$\text{Cruise:}\quad \frac{W_i}{W_{i-1}} = \exp\left[\frac{-RC}{V(L/D)}\right] \tag{19.10}$$

$$\text{Loiter:}\quad \frac{W_i}{W_{i-1}} = \exp\left[\frac{-EC}{L/D}\right] \tag{19.11}$$

Equation (19.10) provides the mission segment weight fraction for a cruise-climb, as discussed in Chapter 17. For a constant-airspeed, constant-altitude cruise, the cruise must be broken into shorter segments and the L/D revised as the weight changes.

If your sizing mission specifies some headwind, you must increase the required cruise range R in the mission segment weight fraction equation by the ratio of velocities, $(V_{\text{airspeed}}/V_{\text{groundspeed}})$, while still using the actual airspeed for V in the equation (see Chapter 17). Loiter is not affected by wind.

Combat and Maneuver

Fighter aircraft are sized with a requirement for air-combat time. This may be explicitly stated, such as "5 min at maximum thrust at 30,000 ft at 0.9 Mach number." Alternatively, a certain number of turns at combat conditions may be specified. In that case, the time to perform the turns is determined from the performance methods of Chapter 17.

Once the combat time is known, Eq. (19.7) can be used.

Descent

Descent was statistically estimated in the initial sizing method, and no range credit was taken for the horizontal distance travelled during descent. A more accurate calculation will probably yield a small improvement in sized takeoff weight.

$$V_v = V\left(\frac{T}{W}\right) - \frac{\rho V^3 C_{D_0}}{2(W/S)} - \frac{2K}{\rho V}\left(\frac{W}{S}\right) \tag{19.12}$$

Descent is a negative climb, i.e., thrust less than the drag. The climb equation developed in Chapter 17 is repeated as Eq. (19.12), in which V_v is vertical velocity or rate of descent. Descent is usually flown at cruise velocity and idle power setting, unless this produces an extreme descent angle (arcsine V_v/V).

The time to descend is determined from the vertical velocity, and the mission-segment weight fraction is determined from Eq. (19.7). A long descent should be broken into segments for greater accuracy. Also, credit should be taken for the distance travelled unless the mission requirements specifically exclude range credit.

{The detailed calculation of descent fuel is probably more trouble than it is worth for quick studies and student design projects. The earlier historical method [Eq. (6.22)] is usually good enough.}

Landing

Landing was previously approximated by a small W_i/W_{i-1} fraction (0.992–0.997). This is probably good enough even for more refined sizing.

From obstacle clearance height to full stop takes less than one minute, and is usually flown at idle power. Even if thrust reversers are employed, the impact upon total fuel weight is small because the thrust reversers are operated for only about 10 s.

If more accuracy is desired, the fuel for landing can be calculated by determining the time to land from the distances calculated in Chapter 17, using the average velocity for each landing segment. Then Eq. (19.7) can be employed.

Empty-Weight Estimation and Refined Sizing

Previously the empty weight was estimated statistically using the takeoff weight. Now that we have a design layout, the methods of Chapter 15 can be used to calculate the empty weight for the as-drawn aircraft by a detailed estimation of the weight of each major component of the aircraft.

During the first refined sizing iteration, the assumed takeoff weight is the as-drawn takeoff weight. The empty weight is the as-drawn empty weight. The fuel required is calculated using the refined methods just presented, plus an allowance for reserve and trapped fuel (6%).

Unless the designer has been very lucky, the takeoff weight calculated from the refined estimate of fuel burned and the as-drawn empty weight will not equal the as-drawn takeoff weight. The as-drawn takeoff weight was based upon initial sizing with limited information about the aircraft, and cannot be expected to be very accurate.

Since the calculated takeoff weight does not equal the as-drawn takeoff weight, the designer must iterate by assuming a new takeoff weight. The empty weight must then be determined for the new assumed takeoff weight.

It would be possible to go back to the detailed weight equations of Chapter 15 and recalculate the empty weight by summing the component weights. Without the aid of a sophisticated computer program, however, the time involved would be prohibitive if this were done for each step of the sizing iteration.

An approximate noncomputerized method relies upon the statistical data from Chapter 3 to adjust the as-drawn empty weight based upon the new assumed takeoff weight. Remember that Fig. 3.1 showed the trend of the empty weight ratio W_e/W_0 decreasing with increasing takeoff weight. A good approximation for the new empty weight would be found by adjusting the as-drawn empty weight ratio along the slope shown in Fig. 3.1 for that class of aircraft. The empty weight for the new assumed takeoff weight can therefore be estimated by adjusting the as-drawn empty weight for the new takeoff weight, as shown in Eq. (19.13). The value of C (not to be confused with SFC) represents the slope of the empty-weight-ratio trend line and is taken from Table 3.1.

$$W_e = W_{e_{\text{as drawn}}} \left[\frac{W_0}{W_{0_{\text{as drawn}}}} \right]^{1+c} \tag{19.13}$$

The value c typically equals (-0.1), so $(1 + c)$ equals about 0.9. This indicates that the empty weight as a fraction of takeoff weight will reduce as the assumed takeoff weight is increased.

A statistical value of c (such as -0.1) may be used, or c may be calculated from your aircraft concept. Make an arbitrary change in W_0, say a 10% increase, and recalculate W_e with all effects considered. These include changes in wing and tail areas, increased fuselage size, heavier landing gear, and larger engine. Then, with the new values as W_0 and W_e, solve for c in Eq. (19.13).

At this point, sufficient information is available to size the aircraft using the sizing method of Chapter 6 with the improved estimates for fuel burned and empty weight.

If the resulting sized-aircraft weight substantially differs from the as-drawn weight, the results should be considered suspicious and the aircraft redrawn, re-analyzed, and resized. "Substantially different" is a matter of opinion, but this author gets nervous at a takeoff-weight difference greater than about 30% of the as-drawn weight.

19.4 Sizing Matrix and Carpet Plots

Sizing Matrix

The sizing procedure just described ensures that the as-drawn aircraft, scaled to the sized takeoff weight, will meet the required mission range. However, there is no assurance that it will still meet the numerous performance requirements such as turn rate or takeoff distance.

The configuration geometry was initially selected to meet these requirements based upon assumptions as to lift, drag, thrust, etc. The as-drawn aircraft will have different characteristics and may no longer meet all requirements, or it may exceed all of them, indicating that it has been overdesigned and is not the lightest possible design.

Sufficient information is now available on the as-drawn aircraft to analyze its performance vs the requirements. If it falls short in some performance area, the thrust or wing area could be changed to attain the desired performance. Rather than this time-consuming "hit or miss" method, the designer can apply the "sizing matrix" method.

In the sizing-matrix method, the thrust-to-weight ratio T/W and wing loading W/S are arbitrarily varied from the as-drawn baseline values (typically plus and minus 20%).

Each combination of T/W and W/S produces a different airplane, with different aerodynamics, propulsion, and weights. These different airplanes are separately sized to determine the takeoff weight of each to perform the design mission.

	$W/S = 50 (\text{lb/ft}^2)$	$W/S = 60$	$W/S = 70$
$T/W = 1.1$	1 $W_0 = 56{,}000$ lb $P_s = 700$ fps ($M0.9$, 30k ft, 5g's) $S_{to} = 340$ ft $a = 46$ s	2 $W_0 = 49{,}000$ lb $P_s = 330$ fps $S_{to} = 430$ ft $a = 42$ s	3 $W_0 = 46{,}000$ lb $P_s = 30$ fps $S_{to} = 660$ ft $a = 39$ s
$T/W = 1.0$	4 $W_0 = 48{,}500$ lb $P_s = 430$ fps $S_{to} = 450$ ft $a = 50.5$ s	Resized baseline 5 $W_0 = 43{,}700$ lb $P_s = 30$ fps $S_{to} = 595$ ft $a = 47$ s	6 $W_0 = 42{,}000$ lb $P_s = -190$ fps $S_{to} = 800$ ft $a = 45$ s
$T/W = 0.9$	7 $W_0 = 44{,}000$ lb $P_s = 140$ fps $S_{to} = 670$ ft $a = 56$ s	8 $W_0 = 39{,}000$ lb $P_s = -230$ fps $S_{to} = 810$ ft $a = 53$ s	9 $W_0 = 36{,}000$ lb $P_s = -320$ fps $S_{to} = 1070$ ft $a = 51$ s

Require: $P_s \geq 0$ at $M0.9$, 30k ft {9144 m}, 5g's
$S_{to} \leq 500$ ft {152 m}
a 50 s from $M0.9$ to $M1.5$

Fig. 19.1 Sizing matrix.

They are also separately analyzed for performance. If the T/W and W/S variations are wide enough, at least one of the aircraft will meet all performance requirements, although it will probably be the heaviest airplane when sized to perform the mission.

Figure 19.1 shows an example of a sizing matrix for a small fighter. Nine T/W–W/S variations of the aircraft have been sized and analyzed for takeoff distance, P_s, and acceleration time. Performance requirements for this example are a takeoff distance under 500 ft {152 m} zero P_s at Mach 0.9/5 g/30,000 ft {9144 m} and an acceleration time under 50 s from Mach 0.9–1.5.

From the data in the matrix it can be seen that the as-drawn baseline (number 5) exceeds the requirements, as do numbers 1, 2, and 6. Number 3 greatly exceeds the requirements but is very heavy. Numbers 4, 7, 8, and 9 are deficient in some requirement but lighter in weight.

The important question becomes: "What combination of T/W and W/S will meet all of the requirements at a minimum weight?"

Sizing Matrix Plot

Optimization of T/W and W/S requires crossplotting the sizing-matrix data, as shown in Fig. 19.2. For each value of thrust-to-weight ratio, the sized takeoff gross weight, P_s, and takeoff distance are plotted vs wing loading. The data points from the sizing matrix in Fig. 19.1 are shown as numbered black dots. (The acceleration data points were plotted in a similar fashion, but not shown.)

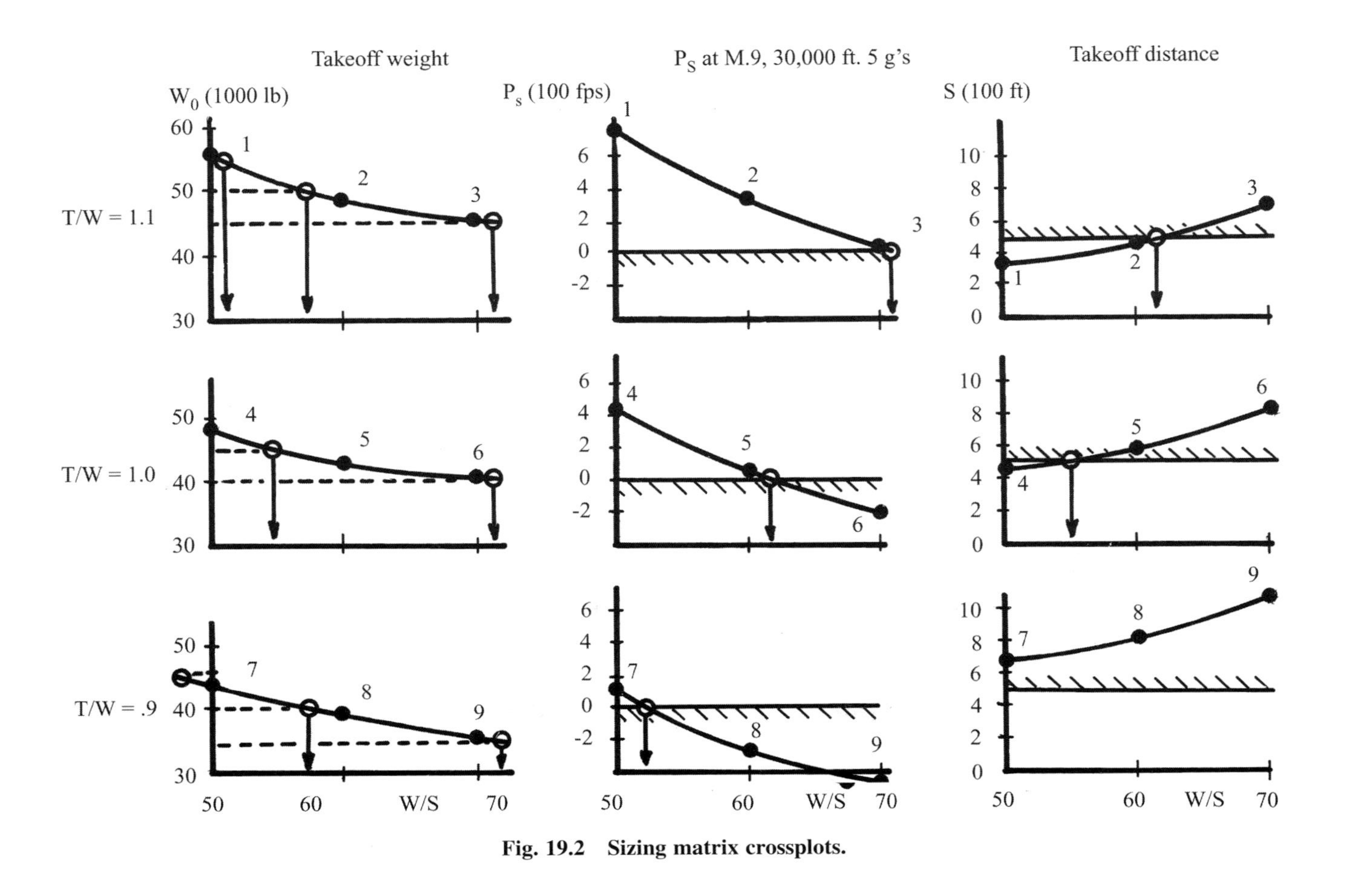

Fig. 19.2 Sizing matrix crossplots.

From the takeoff-weight graphs in Fig. 19.2, the wing loadings corresponding to regularly spaced arbitrary gross weights are determined. For this example, gross weights at 5000-lb increments were selected. For these arbitrary weight increments, the corresponding W/S values are shown as circles on Fig. 19.2.

The W/S and T/W values for the arbitrary gross-weight increments are transferred to a T/W–W/S graph as shown in Fig. 19.3. Smooth curves are drawn connecting the various points that have the same gross weight to produce lines of constant-size takeoff gross weight (Fig. 19.3). From these curves one can readily determine the sized takeoff weight for variations of the aircraft with any combination of T/W and W/S.

Next, the W/S values that exactly meet the various performance requirements are obtained from the performance plots for different T/W values (right side of Fig. 19.2). These values are again shown as circles.

These combinations of W/S and T/W that exactly meet a performance requirement are transferred to the T/W–W/S graph and connected by smooth curves, as shown in Fig. 19.4. Shading is used to indicate which side of these "constraint lines" the desired answer must avoid.

The desired solution is the lightest aircraft that meets all performance requirements. The optimum combination of T/W and W/S is found by inspection, as shown in Fig. 19.4, and usually will be located where two constraint lines cross.

This is a simple example with only three performance constraints. In a real optimization, a dozen or more constraint lines may be plotted. While it is not necessary to include every performance requirement in the sizing matrix plot, all those that the baseline aircraft does not handily exceed should be included.

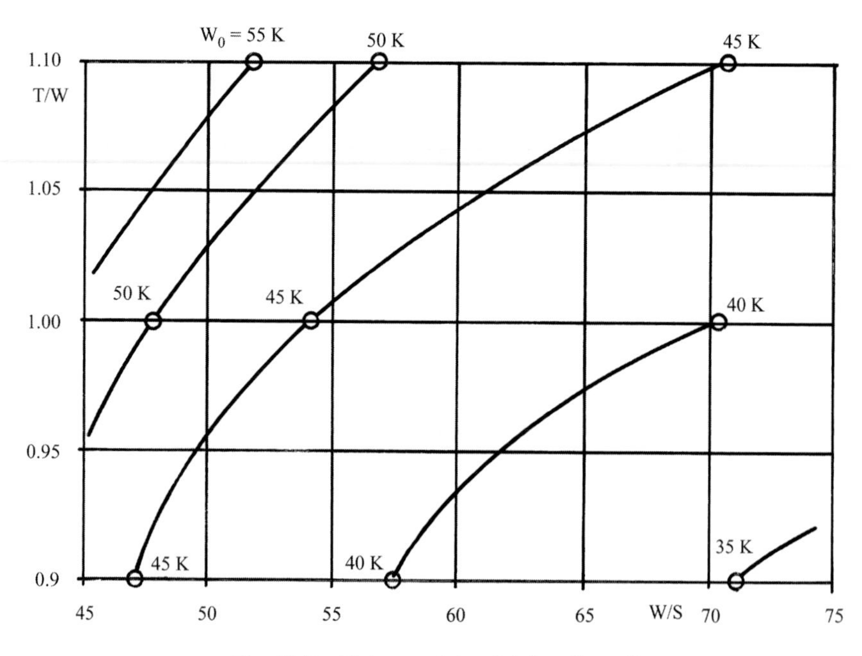

Fig. 19.3 Sizing matrix plot (continued).

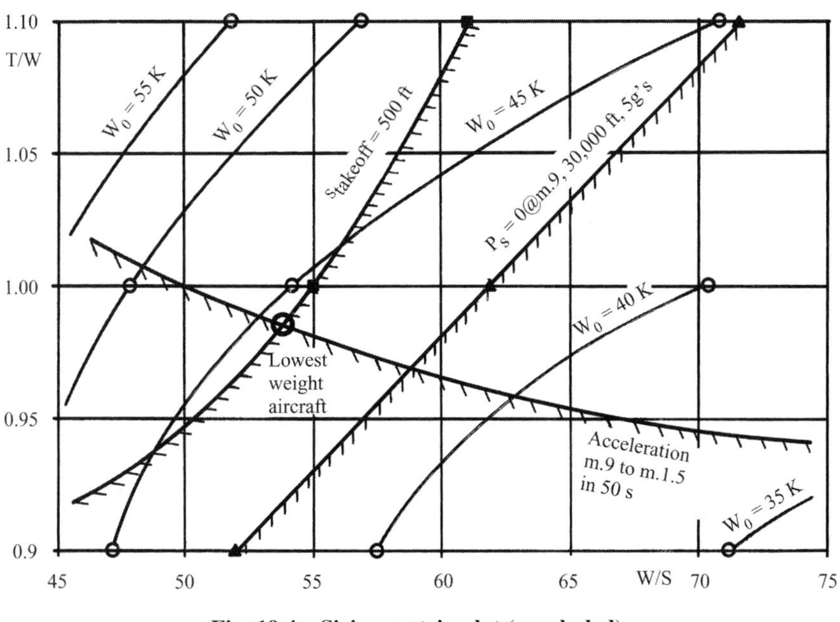

Fig. 19.4 Sizing matrix plot (concluded).

This example showed only a 3 × 3 sizing matrix. For better accuracy, 5 × 5 and larger sizing matrices are used at the major aircraft companies.

Carpet Plot

Another presentation format for the sizing matrix, the so-called carpet plot, is based upon superimposing the takeoff weight plots from Fig. 19.2.

In Fig. 19.5, the upper-left illustration from Fig. 19.2 is repeated showing a plot of sized takeoff gross weight W_0 vs W/S for a T/W of 1.1. The points labeled 1, 2, and 3—data points from the matrix (Fig. 19.1)—represent wing loadings of 50, 60, and 70 psf {244, 293, and 342 kg/sqm}.

The next illustration of Fig. 19.5 superimposes the next W_0 vs W/S plot from Fig. 19.2. This plot represents a T/W of 1.0. The data points labeled 4, 5, and 6 again represent wing loadings of 50, 60, and 70.

To avoid clutter, the horizontal axis has been shifted to the left some arbitrary distance. This shifting of the axis is crucial to the development of the carpet-plot format.

In the lower illustration of Fig. 19.5, the third curve of W_0 vs W/S has been added, again shifting the horizontal axis the same increment. The points labeled 7, 8, and 9 again represent wing loadings of 50, 60, and 70.

Now these regularly spaced wing-loading points on the three curves can be connected, as shown. The resulting curves are said to resemble a carpet, hence the name. The horizontal axis can be removed from the carpet plot because one can now read wing loadings by interpolating between the curves.

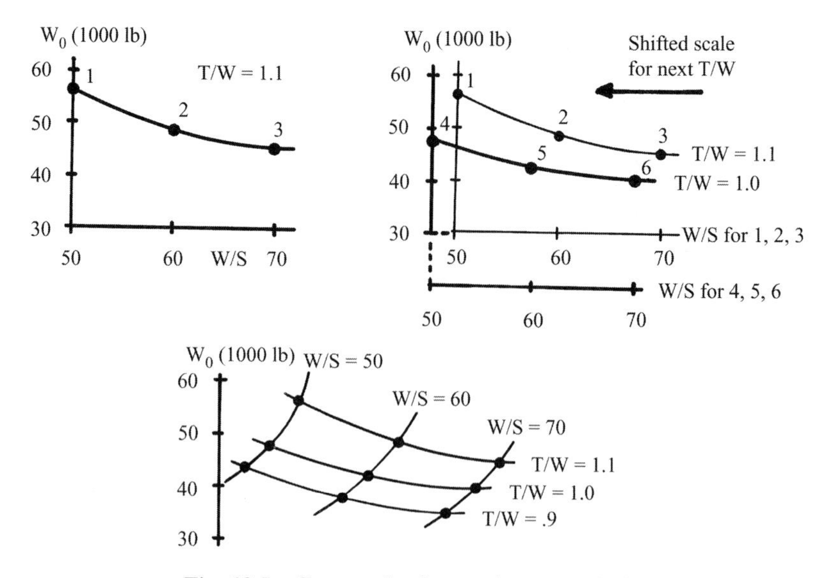

Fig. 19.5 Carpet plot format (same results!).

In Fig. 19.6, the wing loadings that exactly meet the takeoff, P_s, and acceleration requirements (from Fig. 19.2) have been plotted onto the carpet plot and connected with constraint lines.

The optimal aircraft is found by inspection as the lowest point on the carpet plot that meets all constraints. This usually occurs at the intersection of two constraint curves.

The carpet plot and the sizing-matrix crossplot format give the same answer. Some people prefer the carpet-plot format because the "good" direction for

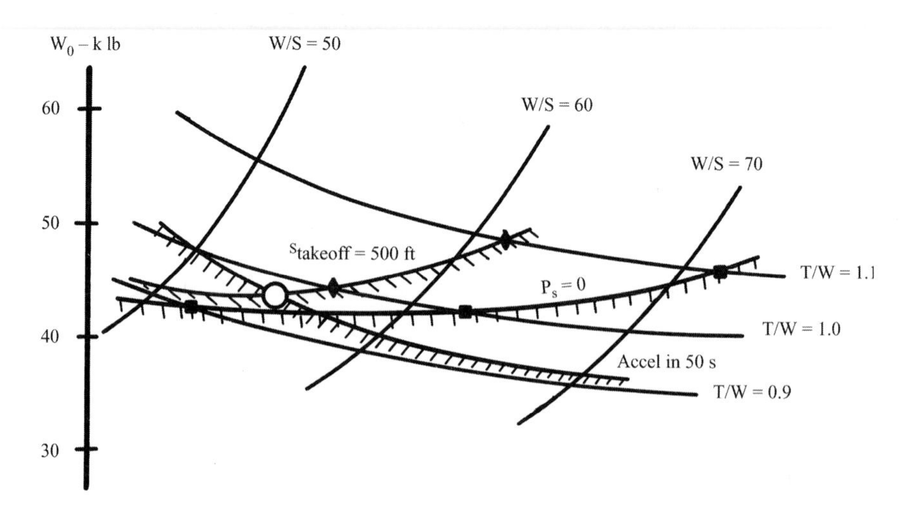

Fig. 19.6 Completed carpet plot.

minimum weight is obvious (down). Others prefer the sizing-matrix crossplot format because it is easier to read the optimal thrust-to-weight ratio and wing loading once they are found. Note that both formats are commonly referred to as carpet plots.

It is also possible to create sizing plots in which the measure of merit is cost rather than weight. The plotting procedure is the same except that cost values are used rather than weight values in the development of the sizing plot. However, for most aircraft types the minimization of weight will also minimize cost for a given design concept.

Sizing-Matrix Data Approximations

A massive amount of work would be required to analyze fully the impact of variations in T/W and W/S on the aerodynamic, propulsion, and weight data required to develop a carpet plot. A variation in T/W affects the thrust and fuel flow, but also affects the wetted area and wave drag due to the change in nacelle size.

A change in W/S affects the wetted area and wave drag. Additionally, changing W/S affects the drag-due-to-lift (K) factor because the fuselage covers up more or less of the wing span. Note that, while the total parasite drag usually increases as the wing size increases, the drag coefficient may drop because it is referenced to the wing area!

At the major aircraft companies, sophisticated modules for analyzing the effects of the parametric variations of T/W and W/S are incorporated into the sizing programs.

For initial studies and student designs, this analysis can be approximated by ratioing the baseline analysis for the affected parts of the airplane.

The change in zero-lift drag can be assumed to be proportional to the change in wetted area due to the wing-area and nacelle-size variations. Wing wetted area varies approximately directly with wing area. Nacelle wetted area varies roughly with the variation in thrust.

For a supersonic aircraft the wave drag should be recalculated. The wing cross-sectional area varies directly with a change in wing area. This is used to determine the new total cross-sectional area that is used to approximate the wave drag.

The variation in K due to relative fuselage size, being small, may be ignored for initial studies. If the wing area is changed, however, then the aircraft will fly at different lift coefficients.

The statistical equations in Chapter 15 show that the wing and tail component weights vary approximately by the 0.7 power of the change in wing area. The engine itself varies in weight by the 1.1 power of a change in thrust.

Installed propulsion performance can be assumed to ratio directly with the thrust.

These and similar, reasonable approximations can be used to estimate the revisions to aerodynamic, weight, and propulsion data for sizing analysis and carpet plotting.

19.5 Trade Studies

Trade Study Categories

Trade studies produce the answers to design questions beginning with "What if . . . ?" Proper selection and execution of the trade studies is as important in aircraft design as a good configuration layout or a correct sizing analysis. Only through the trade studies will the true optimum aircraft emerge.

The "granddaddy" of all trade studies is the $T/W-W/S$ carpet plot. This is such an integral part of aircraft analysis that it is not usually even thought of as a trade study. A $T/W-W/S$ carpet plot in good measure determines the minimum-weight aircraft that meets all performance requirements.

Table 19.1 shows a number of the trade studies commonly conducted in aircraft design. These are loosely organized into design trades, requirements trades, and growth sensitivities.

Design trades reduce the weight and cost of the aircraft to meet a given set of mission and performance requirements. These include wing-geometry and propulsion variations as well as configuration arrangement trades.

Requirements trades determine the sensitivity of the aircraft to changes in the design requirements. If one requirement forces a large increase in weight or cost, the customer may relax it.

Growth-sensitivity trade studies determine how much the aircraft weight will be impacted if various parameters such as drag or specific fuel consumption should increase. These are typically presented in a single graph, with percentage

Table 19.1 Typical trade studies

Design trades	Requirements trades	Growth sensitivities
T/W and W/S	Range/payload/passengers	Dead weight[a]
A, Λ	Loiter time	C_{D_0} and K
		$C_{D\text{wave}}$
t/c, λ	Speed	$C_{L_{\text{max}}}$
Airfoil shape and camber	Turn-rate, P_s, n_{max}	Installed thrust and
High-lift devices	Runway length	SFC
Fuselage fineness ratio	Time-to-climb	Fuel price
BPR, OPR, TIT, etc.	Signature level	
Propeller diameter	Design-to-cost	
Materials		
Configuration		
tail type		
variable sweep		
number and type of engines		
maintainability features		
observables		
passenger arrangement		
Advanced technologies		

[a]Dead weight is a catch-all phrase for "the airplane might get heavier by X pounds." It might come about because the empty weight gets higher, or because the payload has been increased, or perhaps because the airplane was out of balance and ballast had to be added. We use the concept of dead weight in trade studies to calculate how sensitive the design will be to any weight increase later in development. When the actual aircraft is put up on scales, there is no "dead weight." By that time, any dead weight has turned into actual weight increases to various components—hopefully not too much!

change of the various parameters on the horizontal axis and percentage change in takeoff weight on the vertical axis.

Be aware of an important consideration in all of these trade studies: the realism factor. There is an unfortunate tendency to minimize redesign effort, especially for yet another boring trade study! If asked to study the impact of carrying two more internal missiles, the designer may find a way to "stuff them in" without changing the external lines of the aircraft.

This might completely invalidate the results of the trade study. If there were sufficient room in the baseline to fit two more missiles internally, then the baseline was poorly designed. If the baseline was already "tight," then the revised layout must be a fake!

The best way to avoid such problems is to insist that all redesigned layouts used for trade studies be checked to maintain the same internal density as the baseline, calculated as takeoff weight divided by internal volume.

The other design trade studies shown in Table 19.1 must be calculated using a complete T/W–W/S carpet plot for each data point. For example, to determine the optimal aspect ratio the designer might parametrically vary the baseline aspect ratio up and down 20%.

For each aspect ratio, a T/W–W/S carpet plot would be used to determine the minimum-weight airplane. These minimum weights would then be plotted vs aspect ratio to find the best aspect ratio. If the designer wished to optimize for, say, aspect ratio and sweep, a matrix of parametric variations of these would have to be defined. At a minimum, this would consist of three variations of each, or a total of nine variations. For each variation, a T/W–W/S carpet plot optimization would be done, each with nine variations of T/W and W/S, to find the best possible aircraft for the specified combination of aspect ratio and sweep. These nine resulting "bests" would be used to make an aspect ratio-sweep carpet plot to finally find the "best-best." Notice that 9×9, or 81 parametric variations of the aircraft, would have to be defined and analyzed as to aerodynamics, propulsion, weights, sizing, and mission performance. This allows a "multivariable optimization" of aspect ratio, sweep, T/W, and W/S.

But what about taper ratio? And t/c? And fineness ratio, and bypass ratio, and . . . ?

Multivariable/Multidisciplinary Design Optimization

As this example shows, the workload for multivariable optimization trade studies can rapidly exceed manual capabilities or even the capabilities of most computer programs used for conceptual design analysis. To optimize T/W, W/S, aspect ratio, taper ratio, sweep, and thickness (the basic set of design parameters) requires a minimum of 3^6, or 729 data points (5^6, or 15,625 data points would be better). Each data point represents a different airplane and requires full analysis for aerodynamics, propulsion, weights, sizing, and performance. Also, you need a technique to find the best aircraft by interpolating between those 729 cases. How do you draw a six-dimensional carpet plot?

To truly optimize an aircraft, even more design parameters from Table 19.1 such as fuselage fineness ratio and engine bypass ratio or propeller diameter should be included in a simultaneous optimization. In fact, one could attempt to simultaneously optimize all of these and many more, and also have the computer optimally change the actual shape of the design including wing planform breaks, nacelle locations, and tail locations, and perhaps optimize the airfoils and the APU installation at the same time.

Such "everything optimization" is neither feasible nor desirable. After a certain point, excessive time spent on defining, executing, and understanding an optimization method or computer program is just time taken away from other pressing design tasks.

All optimization methods must revolve around one or more measures of merit, which implies that we know exactly how the aircraft will be operated. In the history of aviation there has probably never been a case of an aircraft flying its "design mission," i.e., the exact same mission that was used for sizing and optimization during its conceptual design. Even if a pilot looked up the original design mission and tried to duplicate it in flight, it could not be done unless the pilot could find a perfect standard day, and happened to have a perfect, nominal engine and an aircraft whose design was not changed or compromised during development.

Even more important, most aircraft are converted to missions that were never anticipated during their design. The F-4, one of the most successful fighters of all time, was designed for a supersonic, deck-launched interception mission totally unrelated to its widespread use as a multirole fighter-bomber. The F-16, in use around the world, was conceived, sized, and optimized as a lightweight dogfighter with the designers' battle cry "not a pound for air-to-ground." It is now the U.S. Air Force's main ground-attack fighter (but is still a potent air-to-air machine).

Another problem is that aircraft optimization is, by definition, making changes to the shape of the aircraft. After nearly 600 pages of aircraft design methods emphasizing the actual conceptual layout, the reader should instantly scream, "but how does the computer know if the landing gear fits, and the radar fits, and the passengers fit, and the fuel tanks are big enough, and the overnose vision angle is still correct, and. . . ."

Of course, each of these and many more could be programmed into the optimizer, but the time to develop the inputs for an optimization model must be considered against the time constraints of conceptual design.* Furthermore, once a time-consuming optimization model is developed for a certain design approach, there will be an understandable reluctance to look at totally different design approaches that are not represented by the model. This could serve as a dampener on the essence of aircraft conceptual design.

However, if we are careful to use optimization in a balanced fashion, with experienced designers always "in the loop," it can be a very powerful tool for improving our design. In this author's opinion, it is best used when based on

*One approach, called "Net Design Volume," seeks to ensure that the modified design will have sufficient internal volume to hold the payload, fuel, and aircraft internal components. See Raymer (Ref. 135) for its description, equations, and application if you are developing an aircraft optimization program.

analysis of a realistic and complete aircraft conceptual design layout, and when its goal is to quickly tell the aircraft designer how to change the design layout to make it better, and is used in the next design iteration as only one of many "inputs," as described in Chapter 2.

There are many mathematical techniques for multivariable optimization, including the repetitive use of carpet plots as already described. Better, the multivariable parametric data as already discussed can be fit to an approximating multidimensional surface equation called a "response surface," which can then be mathematically or numerically solved for an optimum.

A concept called "Latin squares" has been used at a number of companies including Boeing and can be viewed as a mathematical approximation for reducing the number of data points needed to be calculated. Essentially, Latin squares tells you which data points to skip, and how to approximate the results that the skipped points would have provided. It is analogous to the old sizing expert's trick—surprisingly good—of drawing a family of curves from five data points.

Another technique for multivariable optimization uses a "finite difference" approach. Small parametric changes are made to the aircraft one at a time, and the change in the measure of merit (such as sized takeoff weight) is used to define a slope (first derivative) of the "system response" to a change in that variable. These derivatives are then used to predict the optimum solution, and iteration is used to drive out the obvious linearization errors.

This author has had good results with exhaustive searching by a simple gradient method to simultaneously optimize an aircraft for the six basic design parameters just described (Ref. 108). Each variable is parametrically varied by plus and minus some selected "step size," and the resulting aircraft are all analyzed for aerodynamics, weights, sizing, cost, and performance. The "best" variant, that with the lowest value of the selected measure of merit, which also meets all performance requirements, is remembered, and when all parametric variations about the initial baseline are exhausted, becomes the centerpoint baseline for the next iteration loop. This continues until no better variant is found, then the stepping distance is shortened and the process repeated until some desired level of resolution is obtained.

Multidisciplinary design optimization (MDO) carries multivariable optimization to the next stage: the optimization of systems across widely different functional areas. J. Sobieski of NASA Langley Research Center defines MDO as "a methodology for design of complex engineering systems that are governed by mutually interacting physical phenomena and made up of distinct interacting subsystems" and goes on to explain MDO as suitable for systems for which "in their design, everything influences everything else" (Ref. 108).

That is, in fact, a pretty good description of aircraft conceptual design, and the various multivariable optimizations just described can be viewed as MDO—even the simple sizing carpet plots that, after all, optimize over disparate disciplines of aerodynamics, weights, propulsion, sizing, and performance.

MDO, though, seeks to carry the level of analysis to a much higher level without losing the interconnectivity of the different functional areas. Ideal for engineering systems for which no single mathematical model is possible, it permits assembling mathematical models from different functional disciplines such as aerodynamics and structures and then deriving system optimizations.

MDO methods include the finite difference technique just discussed, as well as more exotic techniques. The Implicit Function Theorem method differentiates the various governing equations to obtain sensitivity equations. These are used to set up simultaneous linear algebraic equations, which are then solved for an optimal solution.

"Decomposition" works by partitioning a large engineering design optimization problem into a number of smaller, solvable problems ("submodules"). During execution of the optimizer, top-level routines pass data between the submodules in a structured manner that retains their coupling and accommodates the defined system constraints. For example, a wing analysis decomposition may have an aerodynamics module that knows how to calculate drag and airloads if it knows the wing shape, and a structures module that knows how to calculate weight and structural deflections if it knows the airloads. Each executes separately, passing their results to the other until they converge at an optimum for the measure of merit such as weight or drag, or a blended bit of both.

The "genetic algorithm" approach works by applying a process of "survival of the fittest." While Darwin is not normally associated with aircraft design, the modeling of aircraft characteristics as "genes" of design variables shows much promise. The design variables are coded into binary strings such that a collection of 1s and 0s defines a particular aircraft, at least as regards the design variables being optimized (Ref. 109).

Rather than starting with a single "baseline" design and trying to improve upon it, the genetic algorithm starts with a number of random collections of 1s and 0s defining some initial "population" of designs. Those are analyzed and evaluated as to "fitness," based on the measure(s) of merit, and the most fit are most likely to be permitted to "reproduce." Reproduction occurs by breaking apart genes and combining them randomly with others. The "child" might be able to say "I got my large engine from my father, and my area ruling from my mother." The next generation is evaluated as to "fitness," and the process continues until the population all resemble each other. This is presumed to represent an optimum (but occasionally it doesn't—the subject of much research today).

A detailed overview of MDO methods and especially genetic algorithms as applied to aircraft conceptual design optimization can be found in this author's 2002 Ph.D. dissertation, "Enhancing Aircraft Conceptual Design Using Multidisciplinary Optimization" (Ref. 136). This also includes a discussion of the variables and constraints most suitable and useful to aircraft conceptual design projects and methods for automatically revising the vehicle geometry to enhance realism of the optimization results.

Cost as the Measure of Merit

It has been assumed in the preceding discussion that the measure of merit for trade studies and optimization is the sized takeoff gross weight. In an actual design competition or sale, cost will probably be the final deciding criteria. Using weight as the measure of merit is usually a good approximation to minimizing acquisition cost because cost is so strongly driven by the weight (especially empty weight) for a given design approach. However, if you are doing trade studies of alternative technologies, engines, avionics, manufacturing methods, or similar items, then weight is a poor approximation to cost. Also, life-cycle cost is largely driven by fuel costs, which may not be minimized by finding the minimum weight airplane. A higher-aspect-ratio wing is heavier but saves fuel. If you are designing a commercial transport, the airlines will be more interested in their return on investment and even the net present value, as explained in Chapter 18.

It is a fairly simple matter to use purchase price, fuel cost, operating cost, life-cycle cost, return on investment, or net present value as the measure of merit for carpet plots. Estimate the desired cost value for each parametric design variation from its sized empty weight, and use the cost rather than weight on the carpet plot. The same can be done with multivariable optimizers and even with MDO.

Estimating cost from changes in sized weight can be readily done with the DAPCA cost model described earlier, where empty weight is a key analysis input. Other costs items such as avionics acquisition will be insensitive to aircraft sized weight and must be estimated in some other manner. The fuel carried by the sized aircraft can be used to ratio fuel usage for operations costing.

It is more difficult to use cost as estimated by the detailed WBS methods described in Chapter 18. The relationship between sized aircraft weight and the number of hours to perform the various tasks is not clear or easy to define, and the number of inputs and assumptions overwhelms the optimization process. For these reasons, most companies use DAPCA or an in-house equivalent for conceptual design trade studies and optimizations, then use a detailed WBS method for the final contract pricing.

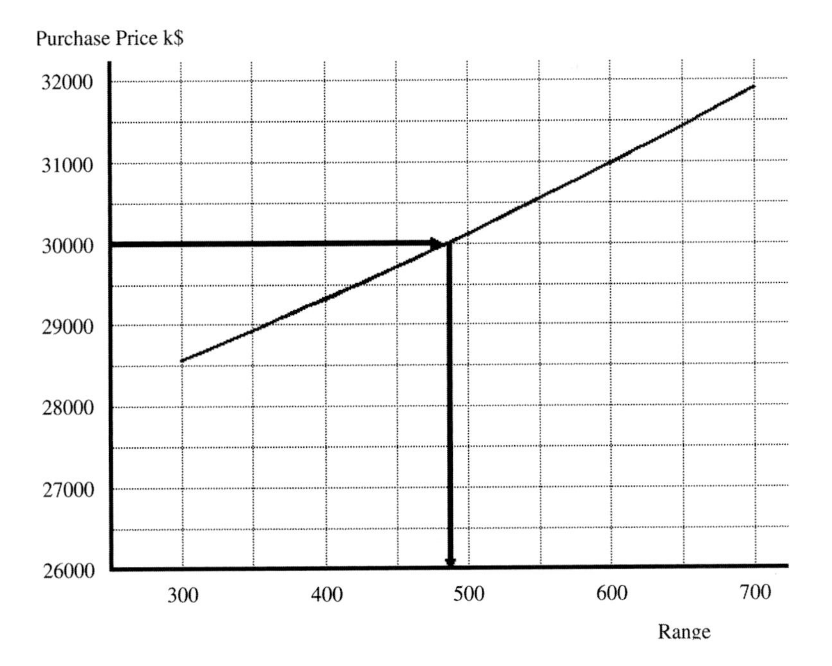

Fig. 19.7 Cost-driven range trade.

As was stressed in Chapters 2 and 3, a key part of the early conceptual design process is the design of the requirements. The aircraft designers work with the intended customers to understand the requirements and to change them as required to provide the best combination of aircraft capabilities and cost. Sometimes, the cost part is clear. The customer simply doesn't have the ability to pay more than a certain amount.

"Design-to-cost" is the term historically used to describe a design process wherein a cost target cannot be exceeded. A more recent term, "cost as an independent variable," or CAIV, says it more strongly—the plane has to cost "X"—now tell me what you can give me for that!

CAIV forces early cost-based trade studies of requirements, technologies, and concepts. At its simplest, a CAIV study could involve a parametric variation of mission range, calculating sized takeoff weight and hence acquisition cost. From the cost the customer will pay, we can read off the range of the aircraft he or she can afford (Fig. 19.7).

CAIV goes far beyond that, though, following throughout the design and development cycle and permeating all design decisions. The cost impact of any change, whether to fix a problem or to add a new capability, is assessed and used to bound the options. Management and the technical staff commit to keeping within the cost bounds and driving cost downwards at every chance. Finally, the customers commit to work with the contractors to keep costs down rather than assuming a traditional adversarial role.

NASA X-Wing Hybrid Helicopter (NASA photo).

20
Vertical Flight—Jet and Prop

20.1 Introduction

"Vertical take-off is necessary only in a circus."—P. Dementyev, Chairman of State Committee of Aircraft Technology, USSR

To most people the operational benefits of an ability to take off and land vertically are self-evident. Conventional aircraft must operate from a relatively small number of airports or airbases with long paved runways. For commercial transportation, the airport is rarely where you actually wish to go and is usually crowded, causing delays in the air and on the ground.

The military airbase is vulnerable to attack, and during a wartime situation the time expended cruising to and from the in-the-rear airbase increases the required aircraft range and also increases the amount of time it takes for the aircraft to respond to a call for support.

The first type of vertical takeoff heavier-than-air aircraft was the helicopter, which was conceived by Leonardo da Vinci but not regularly used until shortly after World War II. The helicopter rapidly proved its worth for rescue operations and short-range point-to-point transportation, but its inherent speed and range limitations restricted its application.

For propeller-powered aircraft, the tilt-rotor concept seems to offer the best compromise between helicopter-like vertical flight and efficient wing-borne cruise. This is the basis of the V-22 Osprey now entering service.

For jet aircraft, a clear "best" solution for vertical lift has yet to emerge. Instead, there are many different vertical-lift concepts, some tested and some not, available for incorporation into a new design. Selection of a best concept depends upon the intended mission and operational environment as well as the technical details of the selected lift concept. Ultimately, system-level trade studies should be used to select the best approach for a new project.

20.2 Jet VTOL

Introduction

Vertical takeoff and landing (VTOL) capability was pursued almost as soon as the jet was invented. Its high thrust-to-weight ratio seems to lend itself to vertical flight, and both military and commercial customers would be delighted to be able to take off and land without the need for long runways.

To date there have only been two operational jet VTOL designs—the British Harrier and the Russian YAK 38. These are both subsonic aircraft. It is especially difficult to design a VTOL aircraft with supersonic capabilities. Although the VTOL Mirage III-V flew at Mach 2 back in 1966, it was not considered practical enough for operational development. Similarly, the Russian YAK-141 flew at M1.7 but was cancelled as impractical and too expensive. In the many years since, a supersonic VTOL aircraft has not entered service.

This is largely because of the increased internal volume required for the vertical-lift apparatus and vertical-flight fuel. Also, most concepts for vertical lift tend to increase the aircraft's cross-sectional area near the aircraft's wing, and that increases the supersonic wave drag. In addition, the engine modifications and extra equipment for VTOL flight impose an excessive weight penalty on VTOL designs that adds to the aircraft empty weight—and that imposes an even greater penalty as a result of the leverage effect on sized TOGW. It has simply not been possible up to now to provide both vertical flight and supersonic forward flight in an operational aircraft of any usable range.

However, the overall level of aircraft, engine, and VTOL technology is advancing rapidly. The supersonic F-35 fighter is in development at this time and is to have a VTOL version based on the Lockheed shaft-driven lift fan concept, described next. As for commercial jet VTOL applications—that still seems to be far away.

VTOL Terminology

VTOL refers to a capability for Vertical TakeOff and Landing, as opposed to Conventional Takeoff and Landing (CTOL).

An aircraft that has the flexibility to perform either vertical or short takeoffs and landings is said to have Vertical or Short TakeOff and Landing (VSTOL) capability. An aircraft that has insufficient lift for vertical flight at takeoff weight but that can land vertically at landing weight is called a Short TakeOff and Vertical Land (STOVL).

The "tail-sitter" or Vertical Attitude TakeOff and Landing (VATOL) aircraft cannot use its vertical lift capability to shorten a conventional takeoff or landing roll. In contrast, a Horizontal Attitude TakeOff and Land (HATOL) concept can usually deflect part of its thrust downward while in forward flight enabling it to perform a Short TakeOff and Landing (STOL).

Fundamental Problems of VTOL Design

A number of unique problems characterize the design and operation of jet VTOL aircraft. Two fundamental problems stand out because they tend to have the greatest impact upon the selection of a VTOL propulsion concept and upon the design and sizing of the aircraft: balance and thrust matching.

Modern supersonic jet fighters have a T/W exceeding 1.0, so it would seem fairly easy to point the jet exhaust downward and attain vertical flight. Unfortunately, this is complicated by the balance problem.

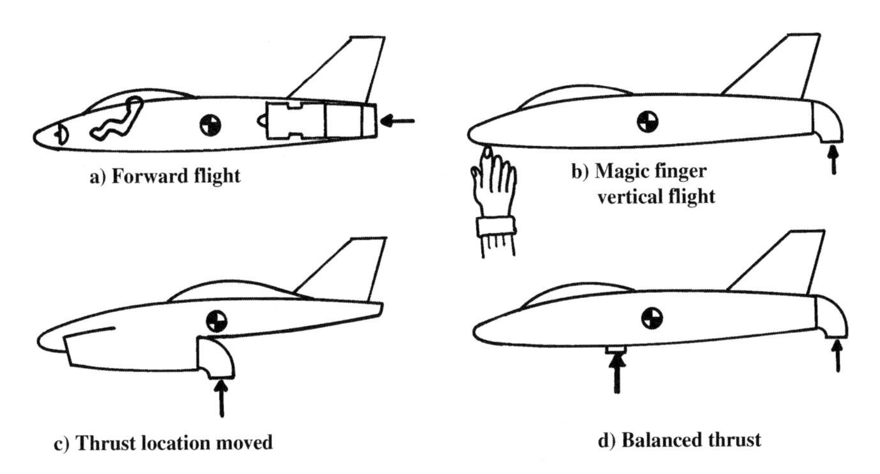

Fig. 20.1 The balance problem.

Many subsonic jets and virtually all supersonic jets are designed with the engine at the rear, the cockpit and avionics at the nose, and the payload and fuel near the center of the aircraft. This traditional layout places the expendables on the c.g., collocates the parts of the aircraft requiring cooling (crew and avionics), and keeps the avionics away from the hot and vibrating engine.

Figure 20.1a illustrates this traditional (and usually optimal) layout. If the aircraft's thrust exceeds its weight, vertical flight could be obtained simply by deflecting the thrust downward, as shown in Fig. 20.1b. However, a "magic finger" must hold up the nose to balance the vertical thrust force at the tail. This balance problem is possibly the single most important driver of the design of the VTOL jet fighter.

There are really only two conceptual approaches to solving the balance problem. Either the thrust can somehow be moved to the c.g. (Fig. 20.1c), or an additional thrust force can be located near the nose (Fig. 20.1d). Both of these approaches will tend to compromise the aircraft away from the traditional and usually optimal layout.

For cruise-dominated VTOL aircraft such as transports, a more severe problem involves thrust matching. If the thrust required for vertical flight is provided by the same engines used for cruise, the engines will be far too large for efficient cruise.

As an example, imagine designing a VTOL transport using four of the TF-39 engines used in the C-5. These produce about 40,000 lb {178 kN} of thrust at sea-level static conditions, or 160,000 lb {712 kN} altogether. If the aircraft is to have a typical 30% thrust surplus for vertical flight ($T/W = 1.3$), then the aircraft can weigh no more than 123,077 lb {55,827 kg} at takeoff. Note that this is far less than the C-5 at 764,000 lb {346,544 kg}.

Assuming a typical cruise L/D of 18 yields a required T/W during cruise of about 1/18, or 0.056. If the aircraft weight at the beginning of cruise is about 95% of the takeoff weight, then the total thrust required during cruise is only 6496 lb ($123{,}077 \times 0.95 \times 0.056$) {29 kN}.

This is only 1624 lb {7 kN} of thrust per engine, which is about 18% of the available thrust for that engine at a typical cruise altitude of 35,000 ft {10,668 m}. It is doubtful that the engine would even run at that low of a thrust setting.

At 35,000 ft and Mach 0.9, the best SFC for this engine would be about 0.73 at a thrust of 9000 lb {40 kN} per engine. The SFC at the 50% throttle setting is about 1.2 {34 mg/Ns}. This is 64% worse than the SFC at the higher thrust setting. If the engine would run at only 18% of its available thrust, its SFC would be even worse than the 1.2 value.

Aircraft range is directly proportional to SFC. The mismatch between thrust for vertical flight and thrust for cruise will produce a tremendous fuel consumption and range penalty for a cruise-dominated design that uses only the vectored thrust of its cruise engines for vertical flight. For this reason many conceptual VTOL transport designs incorporate separate "lift engines" used during vertical flight.

If three of the TF-39 engines in the preceding example could be turned off during cruise (without a drag penalty), the remaining engine could be operated at a 72% thrust setting where it gets an SFC of about 0.8 {23 mg/Ns}. This is a big improvement over all engines being used for both lift and cruise. However, the use of separate lift engines introduces additional problems, as discussed later.

There are numerous other problems associated with VTOL aircraft design including transition, control, suckdown, hot gas ingestion, FOD, inlet flow matching, and ground erosion. These are discussed next following a brief discussion of the various VTOL jet propulsion options that are currently available to the designer.

VTOL Jet Propulsion Options

Broadly speaking, jet VTOL concepts can be divided into those that utilize fairly conventional engines and those that use engines modified so that the fan and core air are split, with the fan air ducted and exhausted from some place separate from the core air.

The conventional-engine VTOL concepts that do not use additional lift engines for vertical flight must have a net takeoff T/W in excess of 1.0. If the jet exhaust is not diverted to some other location for vertical flight, the aircraft must either be a tail sitter (VATOL), or have the engine exhaust at the aircraft c.g. and capable of vectoring downward for vertical flight. This can be accomplished by using a vectoring nozzle or nacelles which tilt (Fig. 20.2).

The YAK-36 and the X-14 research aircraft had vectoring nozzles at the c.g., with the engines out in front. This is probably not a good arrangement for most applications because the cockpit winds up in back, for balance, and thus does not provide acceptable visibility for the pilot. Also, in forward flight the jet exhaust scrubs alongside the fuselage, causing thermal and acoustic problems.

An alternative approach is to place the nozzles at the center of gravity but put the engine in the rear fuselage as on a regular aircraft, but installed backward! This "Reverse Installation Vectored Engine Thrust" (RIVET) concept offers design simplicity, reduced weight, ease of transition, and inherent vectoring in forward flight (VIFF). However, inlet duct losses of 5% or more will be

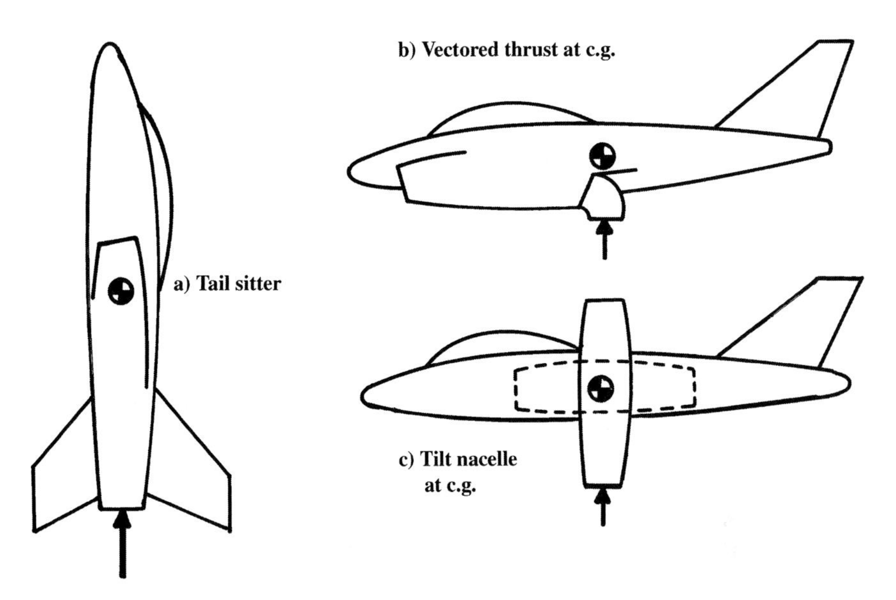

Fig. 20.2 Conventional engine, no lift engine, no flow diversion.

caused by the 180-deg bend required to supply air to a backwards engine. Sizing studies (Ref. 93) indicate that despite these duct losses, RIVET offers a viable option for supersonic V/STOL.

Tilt nacelles, although heavy, may be the best compromise for some applications. Grumman pursued a tilt-nacelle concept for naval applications for a number of years, even flying a subscale model.

Some VTOL concepts provide a means of diverting the exhaust flow to gain vertical lift. This is generally done by a retracting blocker device in the engine that shuts off the flow through the rearward-facing nozzle. The flow is then diverted forward through internal ducting (Fig. 20.3).

The diverted flow can be exhausted directly downwards, or it can be "augmented" by either a tip-driven fan or an ejector. Both of these can actually increase the thrust obtained from the diverted flow by using the energy of the exhaust flow to accelerate a larger mass of air. This augments thrust by increasing the propulsive efficiency.

The tip-driven fan (a ducted fan) is turned by turbine blades at its tip. The diverted engine exhaust is passed over the turbine blades to spin the fan. The Ryan XV-5A used such fans and attained an "augmentation ratio" of almost three. (The lifting thrust attained with the tip-driven fans was almost three times the thrust produced by the jet engines during normal forward flight.)

The ejector makes use of the viscosity of the air. Any exhaust jet will "drag" along adjacent air molecules, accelerating the free air in its vicinity. The ejector consists of a short duct with an exhaust stream blowing down it. Additional air is pulled by viscosity into the duct, accelerated, and ejected through a nozzle. This produces thrust greater than the thrust due to the jet exhaust alone.

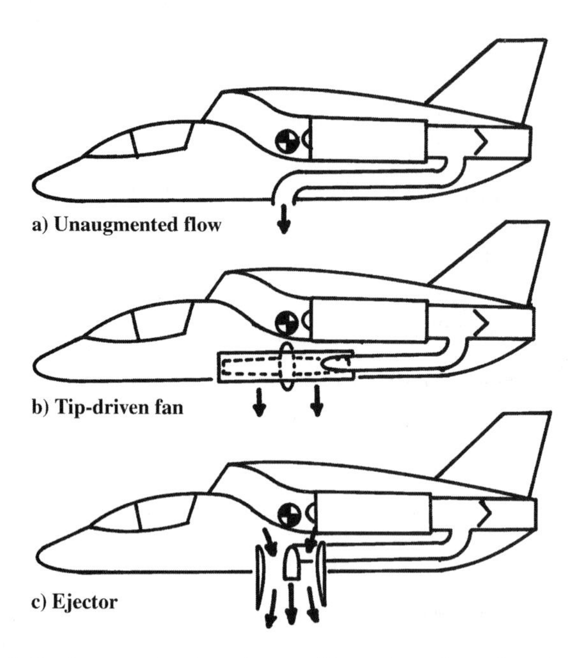

Fig. 20.3 Conventional engine, no lift engines, flow diversion used.

While ejectors promise theoretical augmentation ratios of 3 or more, a more realistic value ranges from about 1.3 to perhaps 2.2. The Rockwell XFV-12A featured ejectors along the entire span of the wing and canard. It was expected to produce a high value of augmentation ratio. The actual value achieved was only about 1.5, and it never flew.

Both ejectors and tip-driven fans are heavy and tend to chop up the aircraft structure. Also, the internal ducting is bulky and poses a thermal problem. However, ejectors and tip-driven fans tend to reduce the thrust-matching problem since the engines do not have to be sized to lift the aircraft by jet thrust alone. The resulting improvement in cruise fuel consumption may offset the weight of the ejector or tip-driven fan.

One of the simplest ways of providing VTOL capability is to add lift engines to an essentially conventional aircraft (Fig. 20.4). This brute-force approach was used in the Mirage III-V. Obviously, the separate lift engines add considerable weight and volume to the design, but the forward-flight engine can be sized for efficient cruise, thus solving the thrust-matching problem.

Since the lift engines are designed for a single operating condition, they can be highly optimized for that condition. Existing lift engines have uninstalled engine T/W on the order of 15, compared to about 6–8 for a typical forward-flight engine. Future lift engines are expected to have engine T/W of 25 or more. Installation including doors and a vectoring lift nozzle will roughly double the weight.

A more subtle approach than the use of separate engines for lift and cruise is to size the forward-flight engine for efficient cruise, but also provide a means of

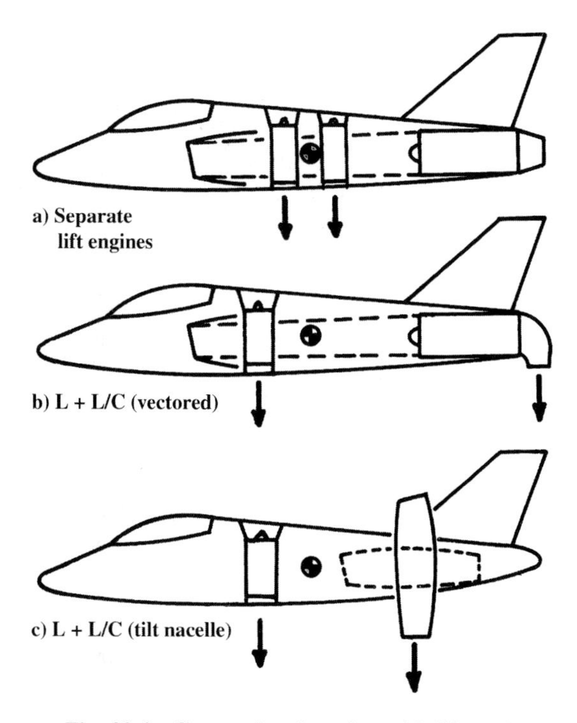

Fig. 20.4 Conventional engine with lift engines.

vectoring its thrust downward for vertical flight. The vertical-thrust shortfall is made up by the addition of lift engines. This is known as a "lift plus lift/cruise ($L + L/C$)" approach, and was used on the YAK-38 and YAK-41.

Because the forward-flight engine is providing some vertical thrust, the thrust required from the lift engines is reduced. The forward-flight thrust can be vectored by a vectoring nozzle as in the YAKS or by tilt nacelles.

A major problem with the $L + L/C$ approach is the transition from vertical to forward flight. During the transition period, the lift/cruise engine thrust is being vectored rearward, decreasing the vertical component of thrust. Since the lift/cruise engine is at the back of the aircraft, additional thrust is required to avoid a nose-up pitching moment. Alternatively, vectoring added to the lift engine will improve transition but add weight and complexity.

If a lift engine should fail during vertical flight or transition, the aircraft would instantly pitch nose-down. The Yakovlev designs have an automatic ejection seat to save the pilot in this event.

The $L + L/C$ approach is especially poor for providing vectoring in forward flight (VIFF). The pilot would have to start up the lift engines before the selection of in-flight vectoring.

Another problem is that the aircraft operators would rather not have to provide tools, spare parts, and trained mechanics for two types of engine in one aircraft.

One possible benefit for the $L + L/C$ concept is the ability to use the lift engine to return to base in the event that the cruise engine fails. This requires some aft vectoring ability for the lift engine, which is desirable anyway to aid in transition.

Related to the $L + L/C$ concept is the "shaft-driven lift fan" (SDLF). In the SDLF concept, a driveshaft runs from the engine to a separate lift fan positioned where the lift engine on the $L + L/C$ concept is located. The driveshaft is powered by the main engine through a modified or supplemental turbine, and spins the lift fan to provide vertical thrust. This avoids the need to develop a complete new lift engine, although the fan, driveshaft, gearbox, and turbine must be developed. Also, SDLF has a cooler front exhaust since the forward lift exhaust is not combusted, and the core exhaust from the main engine is cooled down when power is extracted to run the lift fan. SDLF does give up the return-to-base capability of the $L + L/C$. Figure 20.5 shows an early Lockheed SDLF concept (Ref. 93) with stealth shaping.

A number of VTOL propulsion concepts are based upon a "split-flow" modification to the turbofan engine. The airflow from the fan is split away from the core airflow and used in some fashion to address the balance and/or thrust-matching problems.

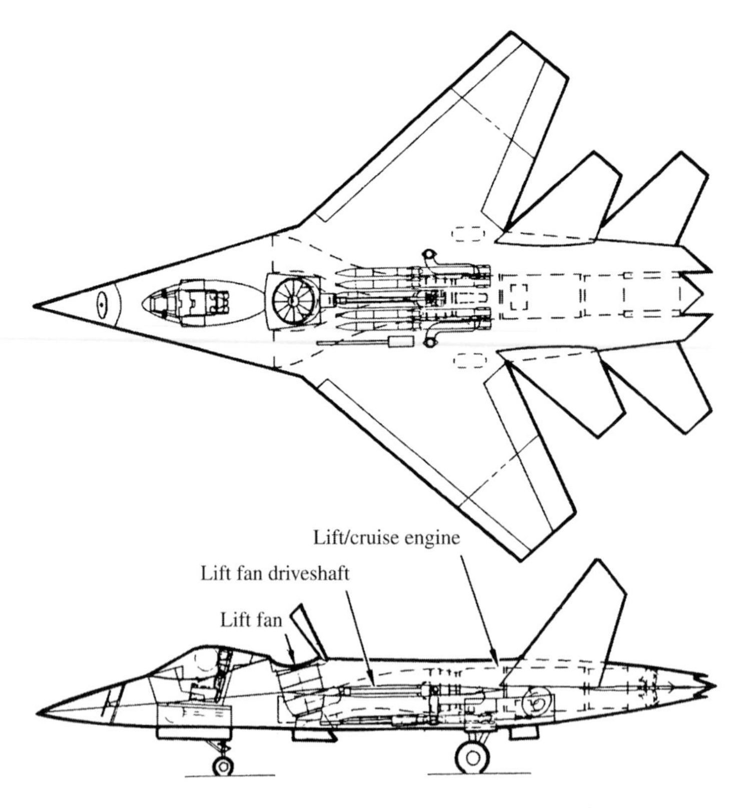

Fig. 20.5 Early Lockheed shaft-driven lift fan STOVL fighter concept.

One such approach exhausts the fan air separately and provides a means for vectoring it downward for vertical flight (Fig. 20.6a). The AV-8 Harrier uses the high-bypass Pegasus engine in which the fan air and core air are each separately vectored through "elbow" nozzles (described later). This permits nearly instantaneous vectoring of thrust with no mode changes (such as starting a lift engine or diverting air into an ejector). This approach also simplifies transition and enhances maneuverability.

On the negative side, the Pegasus-type engine suffers the thrust-matching problem since the engine thrust must provide all of the required lifting force. Also, the engine must straddle the aircraft c.g. This increases the aircraft's cross-sectional area right at the wing location, and thus increases supersonic wave drag (the Harrier is subsonic).

It is possible to augment the thrust of such an engine by essentially providing an "afterburner" for the fan and core airflows in so-called plenum-chamber burning (PCB). There is considerable debate about the desirability of such high exhaust temperatures for VTOL operation.

Another means of providing afterburning to the Pegasus-type split flow engine is to duct the fan and core exhausts together during forward flight, and provide a conventional afterburner that is only used in forward flight (i.e., for combat). This avoids extreme ground footprint temperatures during vertical operation, and is a simple extrapolation of current technology and hardware. However, this requires that the engine be sized for vertical flight without afterburning, leading to combat thrust-to-weight ratios of 1.5 and more! Clearly, this imposes fuel efficiency, weight, and cost penalties.

A close relative of the vectored-thrust split-flow engine is the tandem fan (Fig. 20.6b), a dual-cycle engine that features an additional fan ahead of the regular one. During forward flight, the front fan acts to "supercharge" the rear fan and engine core.

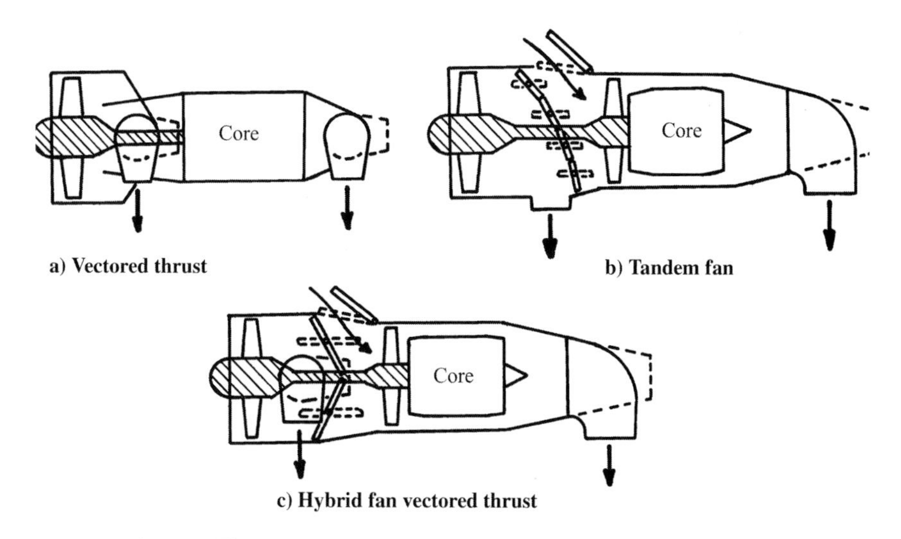

Fig. 20.6 Split-flow engines (vectored fan air).

For vertical flight the flow from the front fan is diverted by blocker doors and exits from a downward-facing front nozzle. Auxiliary doors open to provide air to the rear fan and engine core. This has the effect of increasing the total amount of air that the engine acts upon, and therefore increasing the thrust efficiency.

The tandem fan is fairly heavy compared to a normal engine, but does provide a clever means of augmenting the vertical lift and of moving the center of vertical lift forward. A potential problem for the tandem-fan concept involves the transition from vertical to forward flight. The core engine may surge or stall because of the distortion introduced when the blocker doors and auxiliary inlet are being opened or closed.

The tandem fan can be considered a dual-cycle engine. The engine has a much higher effective bypass ratio when operating in vertical than in forward flight. A modified version of the tandem fan called the hybrid fan (Fig. 20.6c) permits the high-bypass mode to be used in forward flight.

In the hybrid fan, the nozzles for the front fan have a full vectoring capability much like the conventional-cycle vectored thrust engine (Pegasus-type). This permits both vertical lift and highly efficient cruise operation with the front fan flow exiting through the front nozzles.

Note that the hybrid-fan engine requires an auxiliary inlet capable of efficient operation in forward flight. In the hybrid fan, the auxiliary inlets are used in forward flight to provide air for the rear fan and engine core. In the tandem fan, the auxiliary inlets are used only for vertical flight and do not have to be efficient in forward flight.

For high-speed flight, the hybrid-fan blocker doors are opened and the front nozzles and auxiliary inlet are closed. The front fan acts to supercharge the rear fan and engine core.

The hybrid fan, like the tandem fan, is somewhat heavy and complicated. The dual-mode capability for forward flight may provide cruise and loiter fuel-efficiencies that are substantially better than a conventional engine's. This could more than make up for the engine weight penalty.

In the remaining class of jet VTOL propulsion concepts, the engine airflow is split and the fan air is ducted away from the core air (Fig. 20.7). The core air is exited through a vectoring nozzle, and is deflected downward for vertical flight.

In forward flight the fan air goes out an aft-facing nozzle, while in vertical flight the fan air is ducted forward for balance. The fan air is usually augmented in some fashion to increase total thrust in vertical flight.

The remote augmented lift system (RALS) acts like an afterburner. Fuel is added to the fan air and burned before exiting through a front nozzle.

Tip-driven fans and ejectors can also be used to augment the fan-air thrust. These act in the manner previously described.

These concepts are poor at vectoring in forward flight because of the significant mode change from forward to vertical flight.

In another variant of the split-flow concept, a valve is provided that mixes the fan and core air during forward flight.

In addition to the concepts just described, there are many possible combinations of these basic VTOL propulsion schemes. For example, an advanced Harrier-like concept was once proposed that incorporated a PCB Pegasus-type engine as well as separate lift engines.

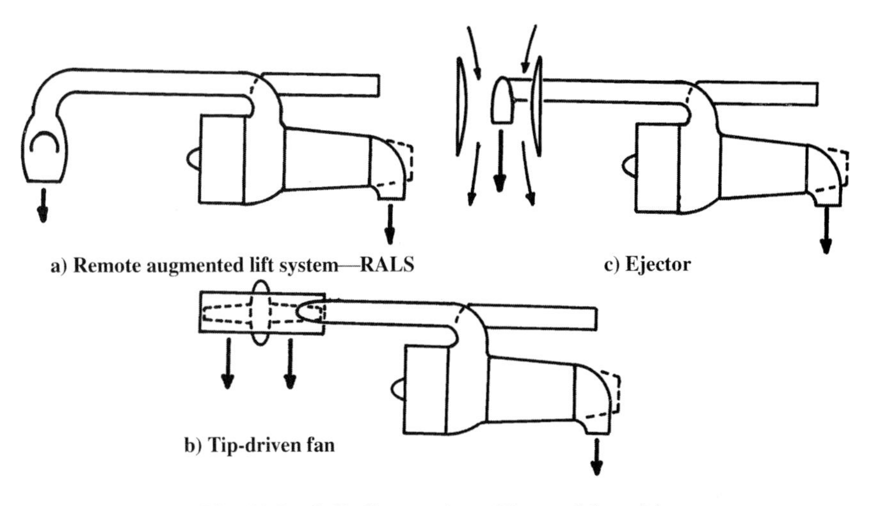

Fig. 20.7 Split flow engines (diverted fan air).

Vectoring-Nozzle Types

The selection of the type of nozzle-vectoring mechanism is almost as important to a VTOL aircraft design as the type of propulsion system. The ideal vectoring mechanism would weigh little more than a conventional nozzle and would provide continuous vectoring from 0 to over 90 deg with negligible thrust loss. Such a nozzle has yet to be designed.

The common types of VTOL nozzle are shown in Fig. 20.8. The most obvious means of vectoring the thrust, vectoring flaps (Fig. 20.8a), deflect the engine flow much as wing flaps deflect the external airflow. These vectoring flaps can be an integral part of the nozzle system, as shown. This type of vectoring system introduces a thrust loss of roughly 3–6% when vectored 90 deg.

The vectoring flaps can also be external to the nozzle as a part of the wing flap system. This approach was used on the XC-15 transport prototype. Although this was not a VTOL aircraft, its wing flap system was able to turn the engine flow more than 60 deg for STOL landings. This, combined with a landing gear that permits a 30-deg nose-up position, would provide the required 90 deg of total thrust vectoring for vertical flight.

The bucket vectoring mechanism (Fig. 20.8b) is similar to the commonly used clamshell thrust reverser. The great advantage of this concept is that the flow-turning forces are all carried through the hingeline; thus, the actuator can be fairly small. Also, the bucket can be designed with a smooth turning surface to raise the turning efficiency. A bucket vectoring nozzle can be designed to have a thrust loss of only about 2–3% when vectored 90 deg.

Figure 20.8c shows an "axisymmetric" vectoring system like that used on the Yak-41. The tailpipe is broken along slanted lines into three pieces, as shown. The three pieces are connected with circular rotating-ring bearings so that the middle (shaded) piece can be rotated about its longitudinal axis while the other

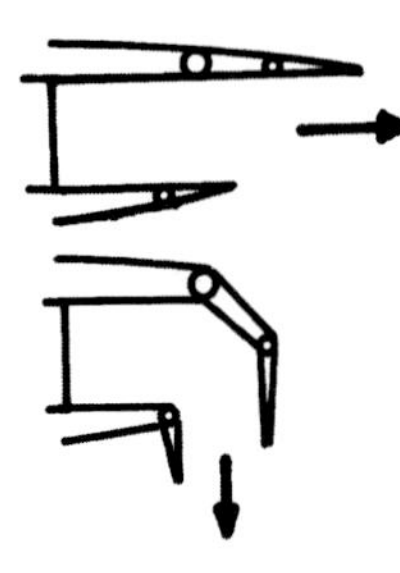

a) Vectoring flaps

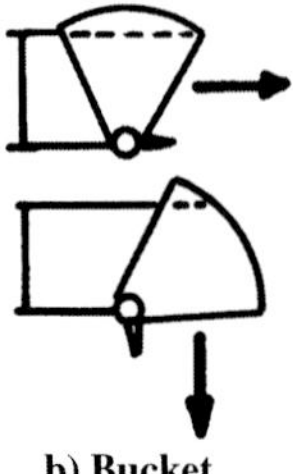

b) Bucket

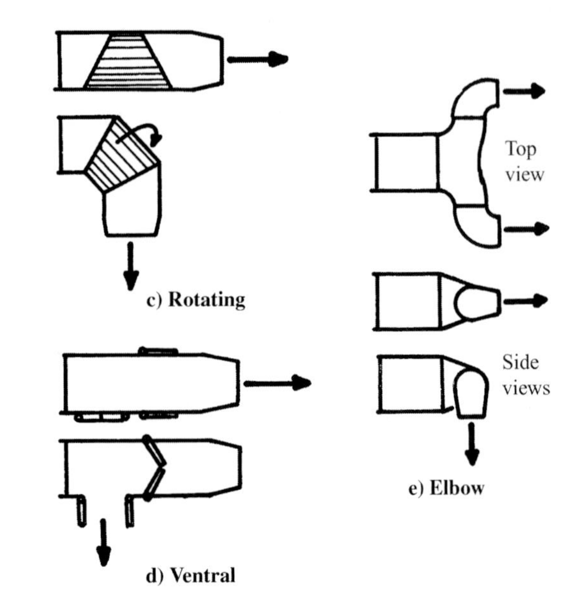

c) Rotating

d) Ventral

e) Elbow

Fig. 20.8 Vectoring nozzles.

parts remain unrotated. This causes the middle and end pieces of the tailpipe to vector downward as shown.

The rotating-ring bearings must be circular in shape, so the tailpipe must have a circular cross section along the slanted lines shown. For this to occur the perpendicular cross-sectional shape of the tailpipe must be an ellipse, so this nozzle system is not truly axisymmetric, despite its name. This type of vectoring nozzle has roughly a 3–5% thrust loss when vectored 90 deg.

The ventral nozzle (Fig. 20.8d) is simply a hole in the bottom of the tailpipe leading to a downward-facing nozzle. The flow out the regular nozzle is blocked off with some type of door.

To reduce hot-gas ingestion and damage to the runway, an afterburner is not usually used for vertical lift. A ventral nozzle can therefore be placed upstream (forward) of the afterburner. This moves the vertical thrust substantially forward compared to a vectoring nozzle at the end of the entire engine. That helps the balance problem.

The ventral nozzle has a thrust loss on the order of 3–6% when vectored 90 deg.

The "elbow" nozzle is used on the Pegasus engine in the highly successful AV-8 Harrier. In the elbow nozzle the flow is turned 90 deg outboard (see top view in Fig. 20.8e). A circular ring bearing connects to the movable part of the nozzle which turns the flow 90 deg back to the freestream direction. To vector the flow downward, the ring bearing is rotated 90 deg, as shown.

The elbow nozzle is simple and lightweight, and requires a minimum of actuator force for vectoring. However, the flow is always being turned through a total of 180 deg, even in forward flight. Because of this, the engine is always suffering a thrust loss of approximately 6–8%. All the other types of vectoring nozzle only impose a thrust loss during vertical flight.

To reduce this thrust loss, the elbow nozzle can be designed using turns of less than 90 deg by canting the ring bearings downward and rearward. This can reduce the total turning angle to about 110 deg, which reduces the thrust loss to about 4–6%. However, the nozzles will then yaw inward during transition from horizontal to vertical thrust. This reduces the total usable thrust during transition and also increases the exhaust impingement upon the fuselage.

Another alternative is to provide elbow nozzles that are only used during vertical flight. A blocker door like that used for the ventral nozzle can divert the airflow from a conventional nozzle to the elbow nozzles. Like the ventral nozzle, the "part-time" elbow nozzles can be located forward of the afterburner for balance. The use of a conventional nozzle in forward flight saves fuel during cruise. This can more than compensate for the extra nozzle weight.

There is another important factor in the selection of nozzle type. This is the effect of the vectoring mechanism on the resultant thrust magnitude during transition.

For the elbow-type nozzle as used in the Harrier, the vector angle has virtually no effect upon the thrust magnitude. For the vectoring flap, bucket, and rotating segment-type nozzles the only effect is the previously mentioned loss of thrust due to turning. This thrust loss is zero at 0 deg of deflection and gradually approaches the maximum value as the thrust deflection approaches 90 deg.

For the ventral nozzle, another factor causes a further reduction in thrust during transition. To transition from forward to vertical flight, the flow out the rear nozzle is gradually blocked off and the ventral nozzle is gradually opened.

The net thrust direction during transition is the vector sum of the thrusts produced by the aft and ventral nozzles. For example, if the exhaust mass flow out the aft nozzle and the ventral nozzle are equal, then each nozzle has a thrust approximately equal to half the total engine thrust. The direction of the net resultant thrust is therefore 45 deg downward.

The magnitude of the net resultant thrust is found by vector addition to be 0.707 times the total engine thrust (square root of 0.5^2 plus 0.5^2). Thus the magnitude of the thrust is reduced by almost 30% when the net resultant thrust is vectored 45 deg during transition! To avoid this thrust loss during transition, some designs use vanes in the ventral nozzle to deflect the thrust rearward for transition—but this adds weight and complexity.

Suckdown and Fountain Lift

The VTOL aircraft in hover is not in stagnant air. The jet exhaust that supports the aircraft also accelerates the air mass around it. This entrainment is due to viscosity and is strongest near the exhaust plume, producing a downward flowfield about the aircraft (Fig. 20.9a).

This downward flowfield pushes down on the aircraft with a "vertical drag" force equivalent to a loss of typically 2–6% of the lift thrust. The magnitude of this vertical drag force depends largely upon the relative locations of the exhaust nozzles and the wing. If the nozzles are right under the wing, the entrained airflow will exert a large downward force.

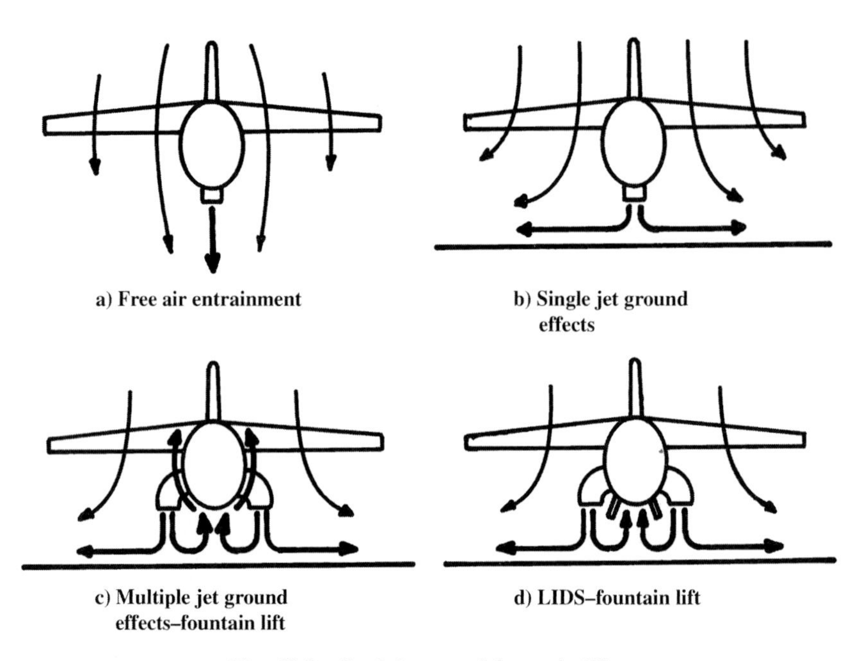

Fig. 20.9 Suckdown and fountain lift.

Unfortunately, the nozzles and the wing are both near the c.g. for most VTOL concepts. The use of a tandem wing, forward-swept wing, or joined wing may reduce the entrained download by separating the wing away from nozzles, which are located at the c.g.

Figure 20.9b shows the effect of the ground on the entrained flowfield. The jet exhaust strikes the ground and spreads outward. This increases the mixing between the jet exhaust and the adjacent air, which increases the entrainment effect. The entrained download (or "suckdown") therefore increases as the ground is approached.

A single-jet VTOL concept can experience a 30% reduction in effective lift due to suckdown. Furthermore, the suckdown increases as the ground is approached—a very undesirable handling quality!

Figure 20.9c shows a VTOL concept with widely separated multiple nozzles near the ground. The jet exhausts strike the ground and spread outward. The exhausts meet in the middle. Since there is nowhere else to go, they merge and rise upward, forming a "fountain" under the aircraft.

This fountain pushes upward on the aircraft with a magnitude that will often cancel the suckdown force. The strength of the fountain lift depends upon the exact arrangement of the nozzles and the shape of the fuselage. Lower-fuselage shaping that makes it more difficult for the fountain to flow around the fuselage will increase the fountain effect. For example, square lower corners are better than round ones.

Fountain lift increases as the ground is approached. This desirable handling quality counters the undesirable effect of suckdown.

The fountain lift can be increased even more by the use of lift improvement devices (LIDS) (called cushion augmentation devices in Britain). These are longitudinal strakes located along the lower fuselage corners which capture the fountain (Fig. 20.9d). LIDS added to the AV-8B increased the net vertical lift over 6%.

Note that multiple nozzles near to each other may not produce a fountain effect because the exhaust plumes may merge into a single jet, producing a flowfield more like that shown in Fig. 20.9a.

Recirculation and Hot-Gas Ingestion

A VTOL aircraft hovering near the ground tends to "drink its own bathwater." The hot exhaust gases find their way back into the inlet, causing a significant reduction in thrust. Also, this "recirculated" air can include dirt and other erosion particles that can damage or destroy the engine. In some cases the dirt kicked up by a hovering VTOL aircraft can completely obscure the pilot's vision.

Figure 20.10 shows the three contributors to exhaust recirculation: buoyancy, fountain, and relative wind. Buoyancy refers to the natural tendency of hot gases to rise. The jet exhaust mixes with the ambient air and slows down as it moves farther away from the airplane. Eventually it has slowed enough that the outward momentum becomes negligible and the buoyancy effect takes over. The now-warm air rises up around the aircraft and can eventually be drawn back into the inlet.

The buoyancy effect takes time. It takes about 30 s in hover for the air around the Harrier to heat up by 5°C. This 5°C increase in air temperature entering the inlet reduces the engine thrust by about 4%.

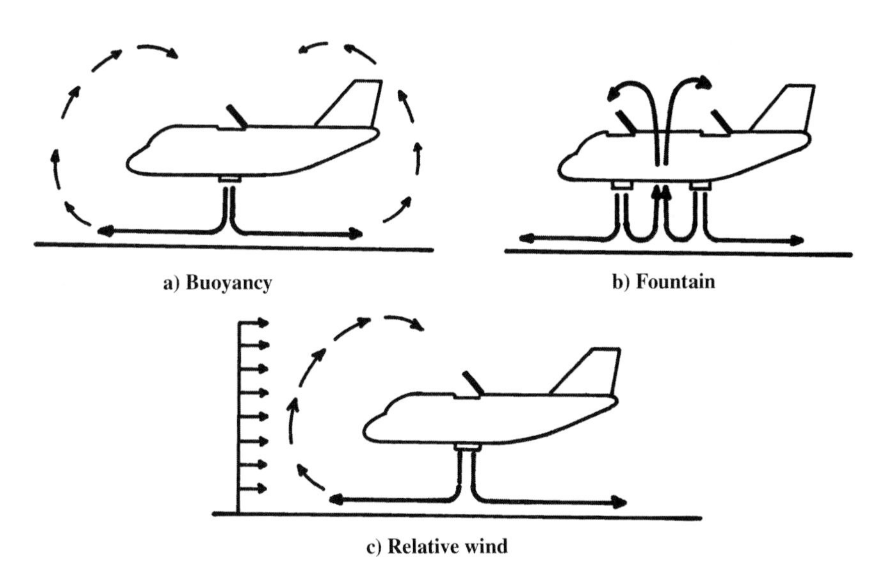

Fig. 20.10 Recirculation.

If the nozzle arrangement produces a fountain, the recirculation will be greatly increased. This causes additional hot-gas ingestion (HGI) in addition to the buoyancy effect. Unlike the buoyancy effect, the fountain effect takes little time to increase the temperature of the air entering the inlet.

The Harrier experiences a 10°C temperature rise due to the fountain effect. This reduces thrust by about 8%.

The third contributor to recirculation, relative wind, can be due to atmospheric wind or to aircraft forward velocity. Essentially, the relative wind pushes back on the spreading exhaust gases, forcing them up. At some combination of relative wind and exhaust velocity, the hot gases will wind up back in the inlet.

Hot-gas ingestion is typically limited to speeds below about 50 knots {93 km/hr}. If the nozzles can rapidly vector from full-aft to a downward angle, a rolling takeoff can be used to minimize HGI problems. The pilot starts the takeoff with the nozzles fully-aft and quickly accelerates to about 50 knots. Then the nozzles are vectored downward and the aircraft takes off.

VTOL Footprint

The "footprint" of a VTOL aircraft refers to the effect of the exhaust upon the ground. This is largely determined by the dynamic pressure and temperature of the exhaust flow as it strikes and flows along the ground. Even a helicopter cannot operate from a very loose surface such as fine sand or dust. The exhaust of a turbojet VTOL aircraft can be of such high pressure and temperature that it can erode a concrete landing pad if the aircraft is hovered in one spot for too long.

No exact method exists to determine the acceptable exhaust pressures and temperatures for VTOL operation off of a given surface. Roughly speaking, a turbojet exhaust is marginal for operation off concrete and is too hot and high pressure for asphalt. The front-fan exhaust of a split-flow turbofan is generally acceptable for concrete, asphalt, and dense sod. However, the core-flow exhaust of the turbofan may be too hot and high pressure for asphalt and sod.

Ejectors and tip-driven fans substantially reduce the exhaust temperature and pressure, allowing operation from regular sod and perhaps hard-packed soil.

In general, the nozzles should be as far above the ground as possible. The ground temperatures due to a turbofan will be reduced by about 30% if the nozzles are five nozzle diameters above the ground. This suggests that a pair of side-mounted elbow nozzles are preferable to a single ventral nozzle because they are higher off the ground and have less diameter for the same total airflow.

VTOL Control

The VTOL aircraft in hover and transition must be controlled by some form of thrust modulation. Most VTOL concepts use a reaction control system (RCS), in which high-pressure air is ducted to the wing tips and the nose and/or tail. This high-pressure air can be expelled through valve-controlled nozzles to produce yaw, pitch, and roll control moments.

The high-pressure air for the RCS is usually bled off the engine compressor, causing a reduction in thrust. The Harrier loses roughly 10% of its lift thrust due to RCS bleed air.

Bleed-air RCS systems can be light in weight. For the Harrier, the whole system only weighs about 200 lb {91 kg}. However, the RCS ducting occupies a significant volume in the aircraft. Also, RCS ducting is hot and cannot be placed too near the avionics.

If a VTOL concept has three or more lift nozzles placed well away from the c.g., modulation of the lift thrusts can be used for control in vertical flight. For example, if the thrust from the forward nozzle is reduced, the nose will pitch down. Vectoring the left-side nozzles forward and the right-side nozzles rearward will cause the nose to yaw to the left.

In addition to three-axis control (roll, pitch, and yaw), a VTOL needs vertical-velocity control ("heave" control). This is done by varying the lifting thrust. For an aircraft with fixed nozzle-exit area (such as the Harrier), the lifting thrust is varied by engine throttle setting.

An engine with variable nozzle-exit area can change its lifting thrust more rapidly by changing exit area, leaving the throttle setting unchanged. Provision of acceptable heave control generally adds about 5% to the required hover T/W.

A multi-engine aircraft should remain under control following the loss of an engine. This common requirement is far more difficult for a VTOL aircraft to meet than for a conventional aircraft. For example, if a VTOL aircraft requires two engines to hover, a third engine of the same thrust would be required to assure hover ability after loss of an engine. Not only that, but the engines must be arranged so that their combined thrust passes through the c.g. with all engines running and with any one engine failed.

The more engines a VTOL concept has, the smaller the impact of adding one extra engine for engine-out hover. However, the increased pilot workload for operating multiple engines and the additional maintenance makes this option less attractive. Also, the more engines, the greater the probability of an engine failure.

Another technique studied for engine-out control involves cross-shafting the engine fans so that the fans of all the engines can be driven from the cores of the other engines. This minimizes the asymmetric thrust loss from the failure of one engine core. However, the weight impact of the cross-shafting mechanisms must be considered.

Some multi-engine VTOL concepts have been designed with several jet engines operating together through some form of augmentation devices. For example, the Ryan XV-5A had two jet engines that were diverted to three tip-driven fans. Either engine could drive all three fans.

VTOL Propulsion Consideration

Thrust matching has already been discussed as one of the key problems facing VTOL designers. Inlet matching presents a similar problem. For efficient jet-engine operation at zero airspeed, the inlet should look much like a bellmouth as seen on jet-engine test stands. The inlet should have a large inlet area and generous inlet-lip radii. These features cause unacceptable drag levels during high-speed flight.

As a compromise, inlets can be sized for cruise operations and provided with auxiliary doors for VTOL operation. For reasonable low-speed efficiency these auxiliary doors must be very large compared with typical auxiliary doors as seen on a CTOL aircraft.

Another propulsion consideration is the amount of vertical thrust required for vertical flight. As a minimum, the net T/W for vertical flight must obviously exceed 1.0. For acceptable response in heave (vertical acceleration), the net T/W should equal or exceed 1.05.

The net thrust available for vertical lift will be reduced by suckdown, hot-gas ingestion, and RCS bleed. Because of these factors, the required T/W for vertical flight will greatly exceed the 1.05 value required merely to hold the airplane up and provide heave control.

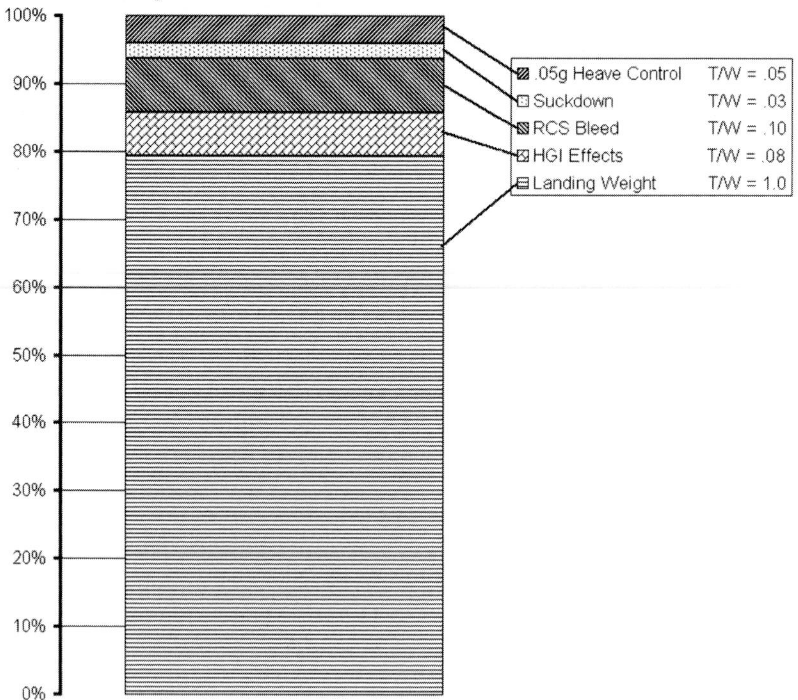

Fig. 20.11 Typical jet hover T/W breakdown.

For most types of VTOL aircraft, the overall installed T/W for vertical flight ranges between about 1.2 and 1.5, with 1.3 being a typical value. Figure 20.11 shows a typical breakdown of contributors to required T/W.

Weight Effects of VTOL

It is difficult to assess statistically the impact of VTOL on aircraft weights using design data from existing aircraft. VTOL designs are so strongly driven by weight considerations that the designers will push much harder to reduce weight than in a normal CTOL design.

For example, the Harrier was designed so that it requires removing the wing to remove the engine. This would be considered a fatal design flaw in a CTOL aircraft but is tolerated in the Harrier because of the weight savings compared to the immense doors that would otherwise be required to remove the engine.

Because of such design compromises, the Harrier has an empty-weight fraction W_e/W_0 of only 0.48, whereas a statistical approach based upon similar CTOL designs would indicate that the Harrier should have an empty-weight fraction of about 0.55. By way of reference, the A-4M, which performs a similar mission, has an empty-weight fraction of 0.56.

A VTOL aircraft designed to the same ground rules as a CTOL aircraft will always be heavier in two areas, propulsion and control systems.

The propulsion system will be heavier due to the compromises just described for solving the balance and/or thrust-matching problems. The various VTOL propulsion concepts all incorporate some additional features such as vectoring nozzles, extra internal ducting, tilt nacelles, or lift engines. These add weight.

Reference 87 compares CTOL and VTOL versions of a carrier-based utility aircraft (similar to the S-3). The CTOL version's propulsion-system weighs 8% of the takeoff weight. The VTOL version's tilt-nacelle propulsion system weighs 20% of the takeoff weight.

Data from Refs. 87–89 indicate that a typical supersonic CTOL fighter design may have a propulsion-system weight about 16–18% of the takeoff weight. An equivalent VTOL design would have a propulsion-system weight about 18–22% of the takeoff weight.

The far-greater propulsion-system weight for the cruise-dominated utility aircraft reflects the fact that the fighter concept already requires large engines for supersonic flight.

Control-system weights are increased about 50% for most VTOL designs. This is due to the ducting, nozzles, and valves of the typical RCS. However, the total control-system weight is only a small fraction of the takeoff weight (2% for a typical CTOL design), so the impact is slight.

For most VTOL designs the landing gear will weigh the same as for a CTOL design. Carrier-based aircraft may show reduced landing gear weight with VTOL.

The landing gear of carrier-based aircraft are substantially heavier than the landing gear of other aircraft because of the extremely high sink-rates during landing and because of the catapult and arresting-hook loads. These can increase the landing gear weight from about 4% to about 6% of the aircraft takeoff gross weight.

A VTOL aircraft designed for carrier operation need not incorporate the heavier landing gear. This represents a weight savings compared to the CTOL carrier-based aircraft.

As mentioned, it is difficult to provide an estimate for the total impact of VTOL on W_e/W_0 based upon statistics. However, data in Refs. 87–89 indicate that a fighter aircraft will experience an increase in W_e/W_0 of roughly 4% if designed to the same groundrules as an equivalent CTOL aircraft. Similarly, a transport/utility aircraft will have an increase in W_e/W_0 of about 7%. These estimates should be considered to be extremely crude.

Sizing Effects of VTOL

The final sized weight of a VTOL aircraft will be increased by the empty-weight effects just described. Also, a thrust mismatch between vertical flight and cruise may force the engine to be operated well away from the optimal thrust setting for cruise efficiency. This increases fuel consumption, which increases sized aircraft weight.

These factors will clearly increase the sized aircraft takeoff weight if a VTOL aircraft is flown over the same mission as an equivalent CTOL. In some cases, though, the mission requirements can be reduced for the VTOL aircraft with no real loss in operational effectiveness.

For example, a Close-Air Support (CAS) aircraft like the A-10 should be based as near as possible to the ground troops being supported. VTOL ability may permit basing literally at the forward lines, thus reducing the mission range requirements while providing better operational effectiveness. The VTOL aircraft can also "loiter" on the ground, unlike the CTOL aircraft, which may have a requirement for a one-hour loiter on a CAS mission.

VTOL greatly simplifies instrument landings. Helicopters can "feel their way around" in foggy conditions that ground all CTOL aircraft. A VTOL capability should therefore reduce landing reserves for loiter or diversion to alternate airports.

On the other hand, the fuel burned by a vertical landing can be substantial, whereas a CTOL aircraft uses virtually no fuel in landing.

Another favorable effect of a VTOL capability comes in the optimization of wing loading. For many aircraft the wing loading will be determined by either the takeoff or landing requirements. A VTOL capability removes this consideration, possibly permitting a smaller wing, which in turn reduces weight and fuel usage.

Taken altogether, these factors indicate that the jet VTOL aircraft will usually be heavier than an equivalent CTOL design. Based upon the data in the references, an increase in sized takeoff weight of about 10–20% can be expected for a fighter design. A transport/utility design will typically size to a weight roughly 30–60% greater than the CTOL design. As before, these are very crude trends, not estimates for any particular design.

20.3 Prop VTOL and Helicopter

"Helicopters really can't fly—they're just so ugly that the Earth repels them."—Anon.

Introduction

Jet VTOL, for all the progress being made, is still in its infancy. As a measure of this, Harriers and Yak-38s can put on a great airshow simply by taking off, straight up. The crowd loves it, but nobody is impressed when a helicopter does the same thing—it happens every day.

Vertical flight via some sort of "airscrew" was conceived of by da Vinci, but he failed to anticipate blade aerodynamics, power-to-weight requirements, or controllability issues. It wasn't until Sikorsky's "VS-300" that a helicopter truly controllable in hover and forward flight would be demonstrated (1940). This was such an advance on prior attempts that its successor, the R-4 "Hoverfly," was immediately ordered into production and saw combat service just four years later (earning the nickname "Eggbeater," still applied to helicopters).

Fundamentally, the reason that helicopters are routinely successful at vertical flight while jet VTOLs are "putting on a show" every time they do it is because helicopters make better use of Eq. (13.4). This proved that efficient thrust is obtained by applying the power of the engine to a large cross section (S) of air, which is accelerated ($V - V_0$) by a relatively small amount. The large rotor "gently" accelerates a large disk of air, compared to the small exhaust flow "screaming" out of a jet engine at near-sonic speeds. Quite simply, the helicopter can hover on a much smaller power-to-weight ratio than can the jet.

This subchapter presents an overview and key concepts for design of helicopter and other propeller VTOL aircraft, including first-order analysis methods, with emphasis on how their design differs from and is similar to the design of other types of aircraft as discussed in this book. Specialized helicopter textbooks should be referred to for the details of blade aerodynamics, rotor analysis, power estimation, vehicle dynamics, and range and performance analysis (Refs. 68, 114, and 115 are recommended).

Two fundamental differences between helicopter design and the design of wing-borne aircraft should be noted. First, for helicopters there is nothing equivalent to the Breguet range equation [Eq. (3.5)], i.e., a simple equation relating the fuel burned to the range. This greatly complicates the calculation of range and the sizing of the helicopter to a range requirement as in Chapter 3.

Second, the rotor blade aerodynamics dominate even the earliest design studies. A helicopter's range and flight performance depend so much on the rotor analysis that the helicopter designers almost immediately perform in-depth rotor calculations using blade element or numerical calculations. Helicopter designers simply don't spend much time doing top-level, order-of-magnitude conceptual trade studies.

A note on terminology: a helicopter "rotor" is like a variable-pitch "propeller," but is much smarter—it can vary its pitch all at once ("collective") like the propeller, but can also vary its pitch as the blade goes around through 360 deg of rotation ("cyclic"). In other words, the pitch can vary as the rotor "cycles" around. Also, a helicopter is a "rotary-wing aircraft," and helicopter people sometimes don't like it when nonhelicopter people use the word "aircraft" as synonymous with "fixed-wing aircraft."

Helicopter Design Concepts

There have been many different approaches to the basic helicopter idea. All must provide lift at the center of gravity and have good control about three axes while avoiding unneeded complexity and weight. Figure 20.12 illustrates the most common approaches.

The simplest helicopter concept uses just a single main rotor located at the center of gravity (c.g.). The vast majority of helicopters in production have a single main rotor. This can be called the "conventional" helicopter approach, and as with aircraft tail arrangements, "conventional" usually means best for most design applications. This provides the greatest disk area for a given weight of rotor system, and by helicopter standards, is fairly simple as to control mechanisms (described next). However, the application of power to the single main rotor will cause a strong torque that must be countered in some fashion, as will be discussed later. Also, the large diameter of the single main rotor may be a disadvantage for high-speed flight since it causes the advancing tip to have a higher relative velocity and will reach Mach effects sooner.

To avoid the torque problem, various versions of co-axial counter-rotating rotors have been used. The Kamov Design Bureau in Russia is well known for this approach, which has also been employed by U.S. helicopter designers.

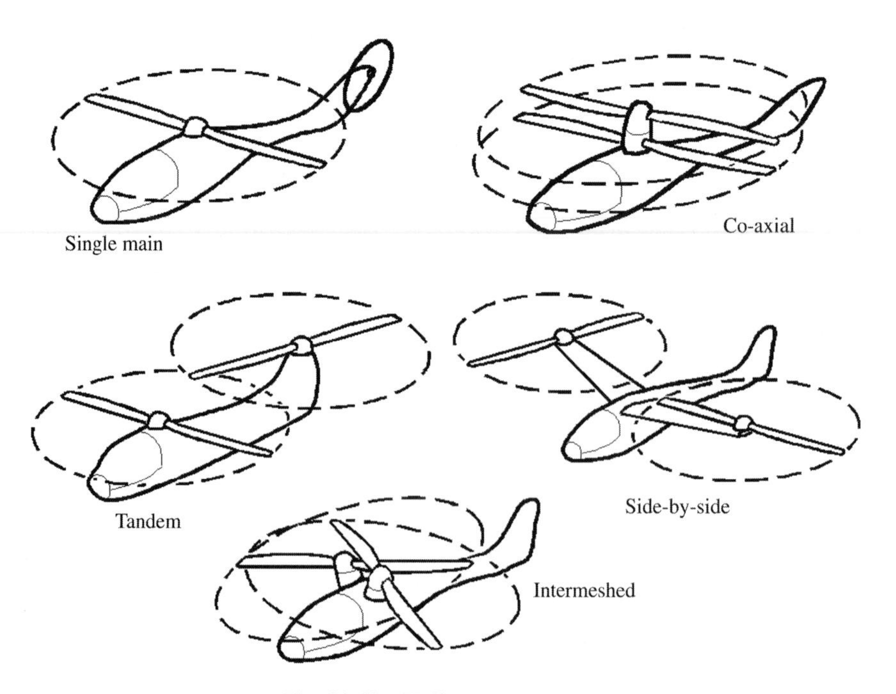

Fig. 20.12 Helicopter concepts.

Counter-rotating propellers in general have a slight advantage in propeller efficiency resulting from the vector direction of the local flow as seen by the second propeller. This in effect makes the blade lift more directly in the vertical direction, and can be visualized as "taking out the swirl" of the first propeller (swirl represents wasted energy).

Co-axial rotors have several disadvantages. The "mast" (vertical post on which the rotor is mounted) must be quite tall to provide sufficient separation between the blades, because it would be catastrophic for them to strike each other. This mast height, roughly 0.3 times the rotor radius, adds drag and weight. Also, the mechanization is quite complex. A counter-rotating gearbox must convert the engine's power onto two concentric shafts, with suitable bearings. The pilot's control inputs must somehow be passed through the plane of the lower rotor to reach the upper rotor, which of course is rotating in the opposite direction. For a military helicopter, all of this mechanization adds to the vulnerable area.

To avoid the complexity of concentric shafts and passing of control inputs, the "intermeshed" rotor was developed. It has two counter-rotating rotors set at outward-tilting angles, driven by a single gearbox that ensures that the rotors *just* miss each other! This concept was pioneered by Kaman Corporation, which produced a number of such designs primarily used for U.S. Air Force search and rescue.

The tandem helicopter is used to provide a wide center-of-gravity range for a cargo helicopter, such as the Boeing CH47 or the classic Vertol H-21 Flying Banana. Lift can be shifted to the front or rear just by changing collective on the rotors. Also, placing the lift at the ends of the fuselage somewhat reduces its structural weight. Such designs obtain yaw control by "flying" the rotors to opposite roll angles (seen from the front). The tandem helicopters suffer from interference effects between the two rotors that reduce efficiency and cause strange flying characteristics unless the controls are artificially augmented.

The side-by-side helicopter is used for really large helicopters such as the Russian Mil V-12, which lifted a record-setting 68,410 lb {31,030 kg} payload. This monster actually has four turboshaft engines of 6500 hp {4847 kW} each, and uses the rotor/gearbox/engine package from two Mi-6 helicopters, one on each wing tip. This arrangement suffers from extra structural weight resulting from the aircraft being suspended from the tips of its "wings," but avoids the interference problems of the tandem helicopter and may benefit in forward flight from an apparent doubling of the "aspect ratio." (In forward flight the rotor disk acts like a wing, as will be discussed next.)

The basic mechanization for a rotor is almost scary to fixed-wing airplane designers and pilots. Control (and life) depend on the proper operation of an assembly of small mechanical parts rotating, vibrating, and rubbing on each other, and if one thing goes wrong . . . ! However, it all works with a surprising degree of reliability, and most helicopter crashes, like most aircraft crashes, involve a perfectly good machine flying into something hard.

Figure 20.13 shows the fundamental helicopter control mechanism, providing roll, pitch and heave (vertical acceleration). Each rotor blade is free to independently pivot in pitch, which is controlled by a rod that is linked to a rotating "swashplate" placed around the mast. This swashplate rotates with the mast and blades, and is connected to a nonrotating, or fixed swashplate such that when the fixed swashplate is moved up or down or is tilted, the rotating

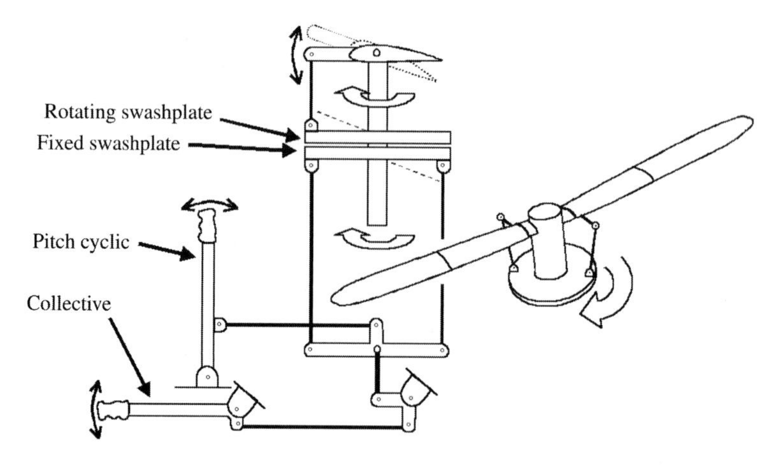

Fig. 20.13 Helicopter cyclic and collective controls.

swashplate moves too. The pilot's "collective" control moves the swashplates up and down, changing blade pitch to change total lift. The pilot's "cyclic" control tilts the swashplates. When the swashplates are tilted, the rotor blades go from a higher pitch at one blade rotation angle to a lower pitch angle when that blade is on the opposite side. This causes the plane of the rotor disk to tip, as the pilot desired. Because of gyroscopic lag, the controls are mechanized such that the most-positive blade pitch occurs 90 deg before the desired tilt.

The rotor blades are hinged as shown in Fig. 20.14 to permit them to "flap" up and down to facilitate this tilting of the rotor plane. Even more important, since the blades are pivoted at the root there cannot be any root bending moment. This allows the blades to be much lighter than if they were rigid (although "rigid rotor" helicopters have been tested and do provide better maneuverability).

Such flapping also solves an obvious problem—in forward flight the advancing blade should get more lift because it sees a higher relative velocity, causing the helicopter to roll on its back.

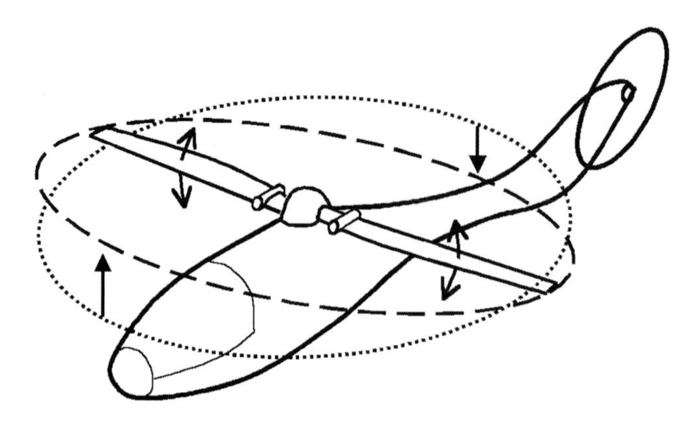

Fig. 20.14 Rotor blade flapping.

If the blades are permitted to flap as shown, the advancing blade will flap upward. This upward motion reduces its lift, while the retreating blade flaps downward, increasing its lift. The rotor flapping reaches equilibrium when the lift is balanced from side to side. This results in the plane of the rotor disk, defined by the track of the blade tips, being tipped backward relative to the actual plane of rotation.

As a result of this motion, the blade tip is accelerating and decelerating in its rotational motion around the helicopter causing in-plane structural stresses. To avoid these stresses, yet another pivot is added, with a vertical shaft. This "lead-lag" hinge eliminates in-plane stresses at the blade root. All together, a rotor blade is "attached" to the fuselage through four separate pivots—the rotor shaft, the pitch pivot, the flapping pivot, and finally the lead-lag pivot.

In the single main rotor configuration, some form of anti-torque device is required (Fig. 20.15). For most helicopters this is provided by a tail rotor, which is driven by a shaft linked to the main rotor. Tail rotors are typically about 15–20% of the main rotor's diameter. The pilot's rudder pedals control the tail rotor's blade pitch, causing the yaw to change. For a helicopter without an augmented control system, the pilot must learn to "dance" on the rudder pedals, continuously making small corrections as every change in speed, altitude, power setting, cyclic control, and wind gust affects the yaw. The tail rotor, present on the first successful helicopter and on most helicopters since, is efficient and responsive, but has a few problems. It is a significant source of noise and vibration. It also adds drag in forward flight.

When walking to or from a helicopter with the rotor turning, a basic human instinct forces most of us to duck down, even if the rotor is at twice our height. For some reason, though, every year a few people try to run around the back of the helicopter and run into the forgotten, spinning tail rotor. Putting a shroud around the rotor minimizes this possibility. Also, the shroud can reduce

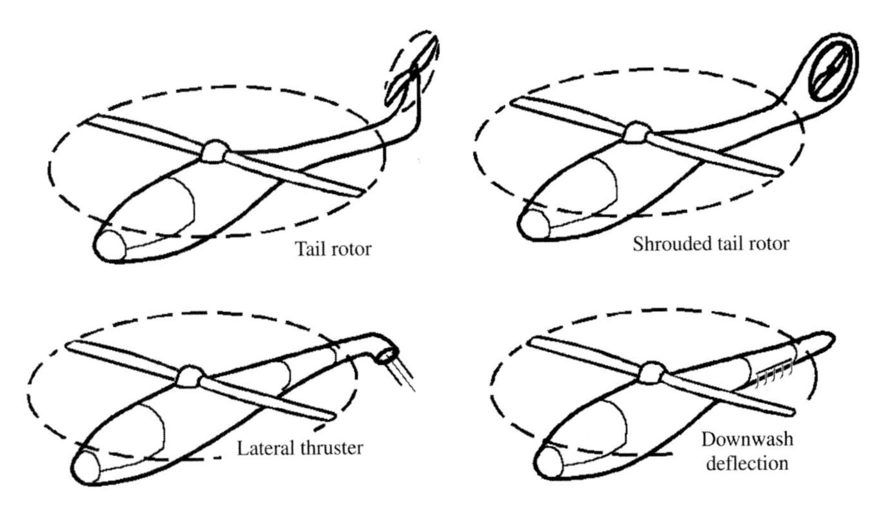

Fig. 20.15 Anti-torque devices.

noise and reduce drag in forward flight. The RAH-66 Comanche uses a shroud to reduce noise and to minimize its radar cross section from the front.

A lateral thruster, really a ducted fan inside the aft fuselage, can be used for anti-torque. While inefficient for hover, it may offer drag advantages in high-speed flight.

Various concepts for deflecting the rotor downwash to provide a sideways lift force for anti-torque have been attempted. The obvious use of a large airfoil vane has not been successful. However, the use of forced circulation by blowing has been. Several models of McDonnell (Hughes) helicopters use what they call NOTAR (no tail rotor) in which the aft end of the fuselage has a round cross section and a fore-and-aft slot. An internal fan blows air out of the slot, tangent to the surface, forcing circulation around the aft fuselage that produces a sideways lift force for anti-torque. NOTAR reduces noise and drag in forward flight, and cannot injure people on the ground, but it consumes more power and is heavier than a conventional tail rotor.

Helicopters have a big problem—they are inherently slow, for a fundamental reason. The advancing blade has an airspeed equal to the helicopter's airspeed plus the rotational velocity of the blade. The retreating blade has the helicopter's airspeed minus the blade's rotational velocity—and because the retreating blade has its trailing edge pointing the wrong way, the net airspeed must be a negative number. In other words, the rotor must spin fast enough that the retreating blade has a rearward rotational tip speed substantially greater than the helicopter's forward speed if it is to develop any lift.

At a minimum, for the retreating tip to have zero net airspeed the advancing blade must have double the helicopter's airspeed. To generate any lift on the retreating blade, the advancing blade must go perhaps three times the helicopter's airspeed, which means that the advancing blade tip may approach sonic speeds when the helicopter approaches just 200 knots {370 km/hr}.

This fundamental helicopter speed limit can be circumvented in several ways, although all have their own penalties. One approach, the "compound helicopter," has a wing and an extra forward propulsion system (jet, prop, or ducted fan). For high-speed operation, the rotor blades "unload," going to a flat pitch and generating little lift. This results in a compromised fixed-wing aircraft attached to a compromised helicopter—capable of vertical takeoff and fast forward flight but not very practical as to range and payload.

Figure 20.16 shows other approaches to beating the helicopter "speed limit." The "Advancing Blade" concept uses counter-rotating rotors, unloading the retreating blades and flying only on the lift of the two advancing blades. This requires a sophisticated blade pitch control system and usually entails an additional thrust system for high-speed flight. Also, the blades must be much stronger and therefore heavier than on a normal helicopter.

The "Stopped Rotor" system does just that—the rotor is stopped for high-speed flight, and acts as a wing. In the "X-wing" concept illustrated in Fig. 20.16, the rotor has four blades arranged as an "X" and acts as a strange tandem wing when stopped. The rotors must be rigid and with all the strength of wings to carry the lift loads when stopped. Another concept being investigated uses a two-bladed rotor/wing, with a large canard that helps control the aircraft during the transitional stopping of the rotor.

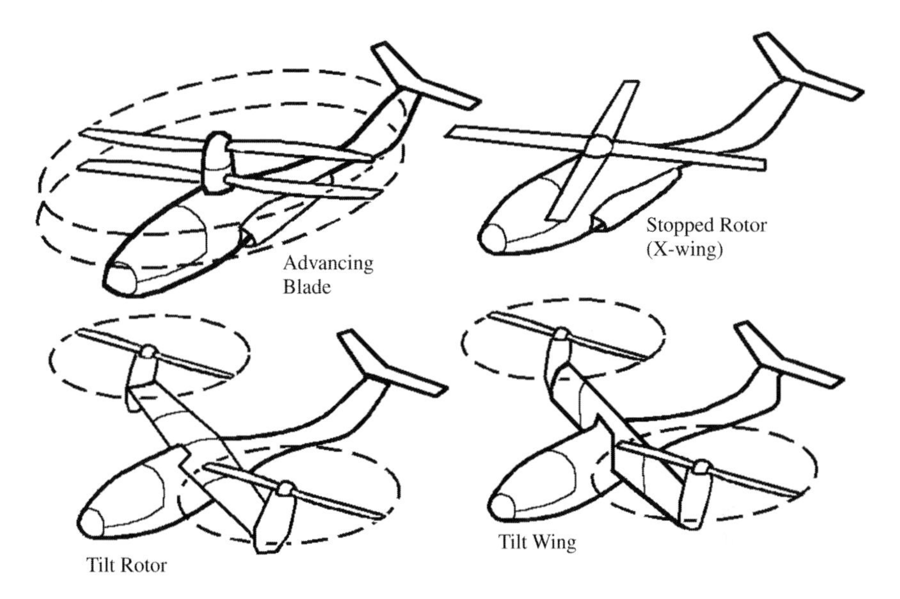

Fig. 20.16 High-speed helicopters and prop VTOLs.

The oddest aspect of stopped rotor concepts is that the airfoil on the retreating blade side of the aircraft will be in reverse flow when stopped— the "pointy end" will be forward. A circular-arc, sharp leading-edge airfoil can be used that will have acceptable aerodynamics no matter which end is forward.

In a more sophisticated approach, a "circulation-control wing" can be used in which the trailing and leading edges are round, with spanwise slots out of which air can be blown. This blowing forces circulation so that the airfoil generates lift, and the greater the blowing, the more the lift. Thus, blowing can be used instead of changing blade pitch for control during helicopter flight. In forward flight, the blowing on the retreating blade side of the aircraft can be switched to the opposite end of the airfoil, making a good forward flight wing. However, this system is quite complex and to date has not been used for a production aircraft.

The other types of high-speed helicopters really are not helicopters in forward flight. Instead, the aircraft takes off as a helicopter but rotates its rotors (or propellers) so that they provide forward thrust, while lift is provided by conventional wings. The "Tilt-Rotor" was first demonstrated in the Bell XV-3, followed by the highly successful XV-15. Today, the V-22 is entering production and the civil Bell-Augusta 609 is under development. The tilt-rotor is fast. While few regular helicopters can exceed 170 knots {320 km/hr}, the tilt-rotors exceed 270 knots {500 km/hr}.

For the tilt-rotor, the engine nacelles rotate out at the wing tips, requiring separate pivots and actuators plus some mechanism to guarantee that both move at the same time. In the "Tilt-Wing," the entire wing rotates to tilt the rotors from vertical to horizontal, which seems intuitively simpler. The nacelles can be firmly attached to the wing, and all that is needed to pivot the entire wing/nacelle/rotor assembly is a hinge and an actuator at the wing root. The tilt-wing should be lighter, simpler, more reliable, and since the wing is aligned with the propwash in hover, should attain a higher net lift. However, it has one large problem. During transition, the wing is at an extreme angle of attack and so it stalls. This makes controllability difficult unless the wing is almost entirely within the propwash and the engines are kept at a high power setting—difficult to do when you want to slow down.

The tilt-wing concept was first tested in the Boeing Vertol VZ-2. This had rotors and used cyclic control for pitch, differential collective for roll, and had a tail rotor for yaw. It flew quite well provided that transition was always done at a high power setting.

Tilting wings or nacelles can also be used with propellers (i.e., blades without a cyclic capability), or with ducted fans. In either case, differential blade pitch can be used for roll control but some other means of yaw and pitch control must be provided. The propellers or ducted fans typically have a lesser diameter compared to tilting rotors and so require more power for hover (see the following for power calculations).

The Chance-Vought XC-142 demonstrated tilt-wing with propellers, and had impressive performance. Weighing 41,500 lb {18,824 kg}, it could fly from 30 kn {56 km/hr} backward to 350 knots {643 km/hr} forward, and had range of 710 n miles {1320 km}. However, in production it would have cost far more than a STOL aircraft carrying a similar payload.

Helicopter Design Parameters and Blade Airfoil Selection

When designing a helicopter or another type of propeller VTOL aircraft, two design parameters dominate, namely, "Power Loading" (W/P) and "Disk Loading" (W/S). These are similar in importance and effect to the T/W and W/S for a fixed-wing aircraft, and together they largely determine the helicopter's hover, climb, speed, range, and autorotate capabilities. (W is the takeoff gross weight. P is maximum engine power. S is the rotor disk area, not to be confused with the actual area of the blades themselves.

The definition of power loading is identical to that of propeller-powered fixed-wing aircraft, and has a similar reverse connotation—a *big* number implies a *small* engine relative to the size of the helicopter. In fact, typical helicopter power loadings are about the same as those of high-powered propeller aircraft, roughly 4–8 lb/hp {2.4–4.9 kg/kW}. Table 20.1 provides typical power loadings for various classes of rotary-wing aircraft.

As can be seen, there is a lot of "scatter" in the data, and this author has not found a reliable statistical correlation between the desired helicopter speed and its power loading as was presented in Table 5.4 for fixed-wing aircraft.

Disk loading (W/S) is the equivalent of wing loading for a fixed-wing aircraft, and the disk area is the same as the "S" of Eq. (13.4). This proved that higher thrust efficiency is obtained by applying the power of the engine to a large cross section of air. Therefore, the lower the disk loading, the smaller the engine required to hover or climb (i.e., the *larger* the power loading permitted). However, a lower disk loading implies a larger rotor blade that has more weight,

Table 20.1 Helicopter power loadings

	Typical W/P	
Aircraft type	lb/hp	kg/kW
Scout/attack helicopter	3–5	1.8–3.1
Transport helicopter	5–7	3.1–4.3
Civil/utility helicopter	3–8	1.8–4.9
Tilt rotor	4–5	2.4–3.1
Tilt wing (propeller)	~3.4	~2.1

more drag in forward flight, and a greater tendency to encounter shocks on the advancing blade. Thus, for high speed the disk loading should be higher, but as will be shown next, the vertical sink speed in a power-off autorotate is proportional to the square root of the disk loading, putting an upper limit on W/S.

Typical helicopter disk loadings are provided in Table 20.2. These data also have a lot of scatter, and an actual W/S for a similar helicopter should be used where possible.

As with T/W and W/S for fixed-wing aircraft, the helicopter's power and disk loadings must be determined simultaneously using performance calculations based on design requirements. Approximate methods are presented next, but for any serious helicopter design work, detailed analysis methods should be used.

Another key parameter in the design of a helicopter rotor is the "solidity," or σ. This is the ratio of the total blade area to the total disk area, and, like activity factor [Eq. (13.15)], is a measure of how much power can be put into the rotor. A high disk loading, high-powered helicopter needs a high solidity to absorb all the power and convert it into lift and thrust. Otherwise the blades will stall before full power is reached.

Airfoil selection for a helicopter rotor blade is similar to the selection of wing airfoils, but has several key differences. Low drag at the design lift coefficient is,

Table 20.2 Helicopter disk loadings

	Typical W/S	
Aircraft type	lb/ft^2	kg/m^2
Scout/attack helicopter	8–10	39–49
Transport helicopter	6–15	29–73
Civil/utility helicopter (low speed)	4–6	20–29
Civil/utility helicopter (high speed)	6–10	29–49
Tilt rotor	15–25	73–122
Tilt wing (propeller)	~50	~245

Low speed ~<150 knots {280 km/hr}

of course, good, as is a high drag-divergent Mach number, to delay formation of shocks on the advancing blade. A high maximum lift is also good, to avoid blade stall that usually limits the helicopter's hover ceiling. Unfortunately, many of the airfoils that are "good" for wings in terms of maximum lift or shock-delaying characteristics are not good for rotors because their shape creates an excessive pitching moment. This causes torsional twisting on the rotor. Because of its extreme span relative to its chord length, a rotor blade is torsionally very weak. For this reason many rotor blade airfoils are symmetric, or nearly so.

To avoid torsional flutter, it is necessary to mass-balance a rotor blade everywhere along its span such that the c.g. is at the airfoil's aerodynamic center. Thus, it is good to select an airfoil whose aerodynamic center is more toward the rear, minimizing the amount of weight required to balance.

A good blade airfoil is thick enough for structural depth, and has a simple shape for ease of manufacture.

Momentum Theory for Hover and Vertical Climb

In Chapter 13, the thrust of an aircraft propeller was calculated by defining an efficiency parameter equal to the thrust power obtained in forward flight ($T \times V$) divided by the power put into the propeller (P). Because this method breaks down at zero speed, empirical static thrust data were applied and visually faired to the forward flight results. This suffices for aircraft performance estimation where the static thrust is of concern only for the start of the takeoff roll. For helicopters, the ability to hover is crucial to the determination of power requirements, and so a more sophisticated thrust method must be employed.

In static conditions, a rotor (or propeller) does not actually experience zero airflow velocity. The rotor induces a velocity, pulling the air into itself. This is shown in Fig. 20.17 where the rotor disk area is S. V_0 is the velocity high

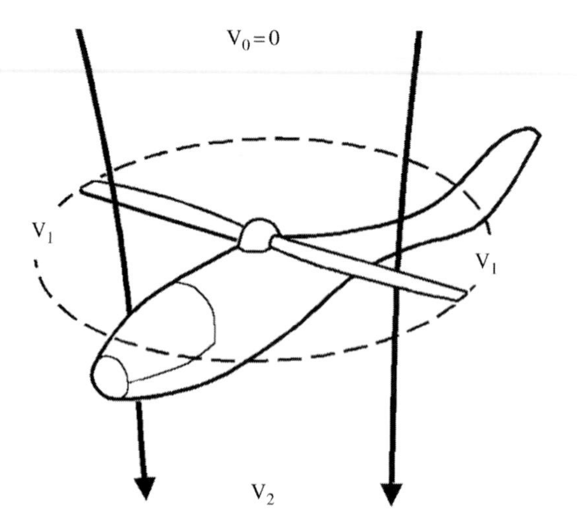

Fig. 20.17 Helicopter in hover.

above the rotor, and since the helicopter is hovering, $V_0 = 0$. The velocity right at the rotor disk is V_1, and the downwash velocity below the helicopter is V_2.

Hover Momentum Theory is derived by equating the power inherent in the induced velocity at the rotor disk V_1 [Eq. (20.2)] with the increase in kinetic energy in the downwash V_2 [Eq. (20.3)]. Equations (20.4–20.7) equate the two and solve for the induced velocity at the plane of the rotor disk in terms of the thrust disk loading T/S.

$$T = \dot{m}\Delta V = (\rho V_1 S)(V_2 - V_0) = \rho V_1 S V_2 \tag{20.1}$$

At 1:

$$P = TV = \rho V_1^2 S V_2 \tag{20.2}$$

At 2:

$$P = \Delta \text{ Kinetic Energy} = \Delta(1/2\dot{m}V^2) = 1/2\rho V_1 S V_2^2 \tag{20.3}$$

so

$$\rho V_1^2 S V_2 = 1/2\rho V_1 S V_2^2 \tag{20.4}$$

$$V_1 = V_2/2 \tag{20.5}$$

$$T = 2\rho V_1^2 S \tag{20.6}$$

$$V_1 = \sqrt{(T/S)/2\rho} \tag{20.7}$$

The induced velocity determined in Eq. (20.7) can then be applied to the definition of power to determine the "induced" or "ideal" power [Eq. (20.8)]. This can be solved for the "ideal thrust" of a rotor in hover as a function of power and disk loading [Eq. (20.9)]. This assumes that thrust disk loading T/S equals the weight disk loading W/S, but actually we should add roughly 3% for the force of the downwash blowing on the fuselage, or $(T/S) = 1.03(W/S)$.

$$P = TV_1 = T\sqrt{(T/S)/2\rho} \tag{20.8}$$

$$T_{\text{ideal}} \cong P\sqrt{2\rho/(W/S)} = 550\,\text{hp}\sqrt{2\rho/(W/S)} \tag{20.9}$$

Inherent in momentum theory are a number of assumptions including uniform flow throughout the rotor disc and an instantaneous, "magical" imparting of energy to the airflow. It also ignores airfoil profile drag losses, tip losses, and residual rotational velocities. Actual helicopter losses include roughly 6% for nonuniform inflow, up to 30% for airfoil profile drag, about 3% for tip losses, and less than 1% for slipstream effects. Together, the net thrust is typically 83% or less of the theoretical ideal thrust.

An empirical "measure of merit" (M) is used to adjust the momentum theory's estimation of power required, much as propeller thrust is adjusted by the propeller efficiency parameter. To avoid the "$V = 0$" problem just discussed, the measure of merit [Eq. (20.10)] is defined as the ratio between the ideal power [Eq. (20.8)] and the actual power required. Typically, $M = 0.6$ to 0.8, and is used to estimate actual power required via Eq. (20.11) [Eq. (20.12) in British units]. (Do not confuse measure of merit M with Mach number).

Define:

$$M = P_{\text{ideal}}/P_{\text{actual}} \tag{20.10}$$

$$P_{\text{actual}} = \frac{T}{M}\sqrt{\frac{T/S}{2\rho}} \tag{20.11}$$

$$\text{hp}_{\text{acutal}} = (\text{T}/550\,M)\sqrt{(T/S)/2\rho} \quad \text{(fps)} \tag{20.12}$$

$$P_{\text{total}} = (P_{\text{rotor}} + P_{\text{tail rotor}})/\eta_{\text{mechanical}} = \frac{P_{\text{rotor}}\left(1 + \dfrac{P_{\text{tail rotor}}}{P_{\text{rotor}}}\right)}{\eta_{\text{mechanical}}} \tag{20.13}$$

where

$\eta_{\text{mechanical}} \cong 0.97$
$P_{\text{tail rotor}}/P_{\text{rotor}} \cong 0.14$ to 0.22

This estimate of actual power required by the main rotor for hover must be further adjusted to account for the power required to drive the tail rotor and for mechanical losses, as shown in Eq. (20.13).

Another consideration in hover is the ground effect, which has the same beneficial effect on helicopters as was described in Chapter 12 for fixed-wing aircraft. By constraining the downwash, ground effect increases efficiency, which reduces the power (required for hover). At half the rotor's diameter above the ground, a helicopter gains about 5% in thrust, and at a height of 20% of the rotor's diameter, thrust increases by about 18%. This allows helicopters to take off and land when in mountainous terrain at an altitude well above the free-air hover ceiling.

Momentum theory can be extended to analysis of the vertical climb of a helicopter. One might assume that the additional power to climb would equal the theoretical time derivative of the increase in potential energy, i.e., the weight times the climb speed. This is pessimistic because the climb speed favorably affects the thrust equation. Climb momentum theory is based on Fig. 20.17, setting V_0 equal to climb speed V_C. Repeating the derivations of Eqs. (20.1–20.8) will derive the conclusion that the additional power to climb is only half the time derivative of the increase in potential energy, that is, the additional power required to climb is approximately half of the helicopter's weight times the climb speed. This is added to the hover power requirement to determine total power to climb.

Combining these equations yields Eq. (20.14), which can be used to calculate power required for vertical climb or hover (setting climb rate to zero).

Hover or vertical climb:

$$P_{\text{climb}} = \left[\left(\frac{fW}{M} \sqrt{\frac{fW/S}{2\rho}} \right) + \frac{WV_{\text{climb}}}{2} \right] \left[\frac{(1 + P_{\text{tail rotor}}/P_{\text{rotor}})}{\eta_{\text{mechanical}}} \right] \tag{20.14}$$

where

W = helicopter weight
S = rotor disk area
M = measure of merit
V_{climb} = climb speed (=0 for hover)
f = adjustment for downwash on fuselage (typically $f = 1.03$)
(in fps units, divide by 550 to yield horsepower)

Another important consideration is the requirement for "autorotation." When a helicopter's engine fails, it does not immediately fall out of the sky. The rotor will turn of its own accord if it is set to a lower pitch. When a rotor is autorotating, by definition, its power requirement is zero. Referring to Fig. 20.17 and Eq. (20.2), it is clear that since power equals thrust times velocity and the power during autorotation is zero, the induced velocity through the rotor disk must also be zero. This implies that the rotor is acting like a parachute, creating vertical drag by preventing airflow through its disk.

If we assume an ideal parachute with drag coefficient of 1.0, and set vertical drag equal to weight, we can solve for descent velocity and find that it equals twice the induced velocity in hover ($2V_1$), as determined by Eq. (20.7). This simply derived approximation is actually reasonably close to the correct answer.

Power Required for Forward Flight

Proper analysis of the helicopter in forward flight requires blade element or numerical methods as described next. For initial analysis, we can roughly analyze power requirements by treating the rotor like a wing. Actually, in forward flight the rotor does act like a wing—both turn the flow inducing a downwash, both form trailing vortices, both experience a drag due to lift, and both have roughly an elliptical lift distribution. If the rotor were a round wing, it would have an aspect ratio of $[d^2/\pi\,(d^2/4)] = [4/\pi]$. Empirical data suggest that, operating as a wing, the rotor has an equivalent Oswald's efficiency factor (e) of 0.5 to 0.8, which can be used in Eq. (12.48) to estimate induced drag.

Parasitic drag can be estimated using the methods of Chapter 12, plus some helicopter-specific drag data as provided in Table 20.3. Not having a wing, the drag of a helicopter is normally given in terms of drag area (D/q). The data in Table 20.3, multiplied by frontal area of the component, give D/q.

Note that a well-streamlined helicopter fuselage could be analyzed using the form factors and skin friction values estimated in Chapter 12, but most helicopters have a fuselage of such irregular shape that the drag is better estimated using these D/q data. Better yet would be the use of wind-tunnel data on a similar configuration, ratioed by fuselage frontal area.

Table 20.3 Helicopter drag data

Component D/q	(Per unit frontal area)
Fuselage	0.07–0.10
Tubular landing skid	1.01
Streamlined landing skid	0.40
Unfaired rotor hub	1.0–1.4
Faired rotor hub	0.5–0.8
Downwash interference drag (per unit fuselage frontal area)	0.02
Leakage and protuberance drag	10–20% added to parasitic drag

The rotor also provides the forward propulsion of a helicopter, so it can be analyzed as if it were an aircraft propeller. Empirical data indicate a propeller efficiency η_p of 0.6 to 0.85, applied to Eq. (13.17), gives a reasonable approximation of the effectiveness of the rotor for forward thrust.

Setting thrust equal to drag [Eq. (13.17) = Eq. (12.4)] and solving for power yields the power required for the rotor. To this we add the adjustments above for the tail rotor and mechanical losses, yielding Eq. (20.15). Note that "S" is the rotor disk area, and the rotor disk aspect ratio $[4/\pi]$ is already included in the equation.

Level forward flight:

$$P_{\text{level}} = \frac{V}{\eta_{\text{p}}}\left\{q(D/q) + \frac{W^2}{4eqS}\right\}\left(\frac{1 + P_{\text{tail rotor}}/P_{\text{rotor}}}{\eta_{\text{mechanical}}}\right) \tag{20.15}$$

Climbing forward flight:

$$P_{\text{climb}} = \frac{V}{\eta_{\text{p}}}\left\{q(D/q) + \frac{W^2}{4eqS} + W\sin\gamma\right\}\left(\frac{1 + P_{\text{tail rotor}}/P_{\text{rotor}}}{\eta_{\text{mechanical}}}\right) \tag{20.16}$$

(in fps units, divide by 550 to yield horsepower)

As was presented in Eq. (17.35), an aircraft in a climb has a projected weight contribution in the drag direction, based on the climb path angle γ. This can be added to the drag term in forward flight resulting in Eq. (20.16). [We neglect the slight reduction in required lift implied by Eq. (17.36).]

Usually, the power required for climb at a moderate forward speed is substantially less than the power required for a vertical climb as expressed in Eq. (20.14). For this reason, helicopter pilots often lift the helicopter a few feet off the ground, accelerating forward while staying in ground effect, and begin to climb only when a substantial forward speed is reached.

Blade Element Theory and Numerical Methods

At helicopter design organizations, momentum theory/measure of merit methods are used only for the earliest rough calculations. Almost as soon as a design project is begun, computerized rotor analysis methods are employed to optimize W/A and W/P, select solidity and the blade airfoil, and determine the blade planform and twist. These computer programs are based on either blade element theory or numerical methods.

In blade element theory, the blade is broken into chordwise strips from root to tip, and the angle-of-attack of each blade element is described in equations as a function of helicopter forward velocity, rotational velocity, local induced velocity, blade twist, cyclic control input, radial position, azimuth position, and blade flapping. The analysis must also consider advancing blade compressibility, retreating blade stall, and tip losses. Once local angle of attack is determined considering the preceding, the lift and drag can be integrated over the blade elements and resolved into the thrust and torque directions, then summed. Blade element theory is very complex mathematically, with multipage equations, but "canned" computer programs are readily available.

Numerical methods have largely supplanted blade element methods because, once someone else has written the computer program, they are no more difficult to use and give better results. Numerical methods use various aerodynamics analysis techniques ranging from relatively simple linearized panel codes to Navier–Stokes CFD codes (see Chapter 12). The rotor blade is divided into panels, and the flowfield is gridded if CFD is employed. The analysis can include all the considerations just described as well as the effects of unsteady flow, nonuniform induced velocities, regions of reversed flow, dynamic stall, blade flapping, and even dynamic blade bending and twisting. Numerical methods typically follow a blade around one rotation, integrating the forces and moments acting on the blade to calculate the vertical position of the blade after one complete revolution. The computer program iterates until the calculated end position equals the start position, as it must in the real world. Then thrust and torque can be summed.

Figure 20.18 illustrates research at NASA Ames Research Center into the application to helicopters of unstructured grid, Reynolds-averaged Navier–Stokes (NS) methods (see Chapter 12). The gridding of the flowfield around a rotor is illustrated, along with calculated pressure contours on a Comanche helicopter from a different analysis. Currently, full NS analysis of the spinning rotor blades and the complete helicopter together is not accomplished, but research is progressing rapidly.

Helicopter Range Analysis

As mentioned in the introduction to this section, there is no equation for helicopters equivalent to the Breguet range equation [Eq. (3.5)], which for aircraft directly relates the fuel burned to the range obtained. Calculation of helicopter range or the sizing of the helicopter to a range requirement must be done by a method similar to that used in the most sophisticated aircraft range calculations programs. The actual drag is calculated at the current flight condition, then the

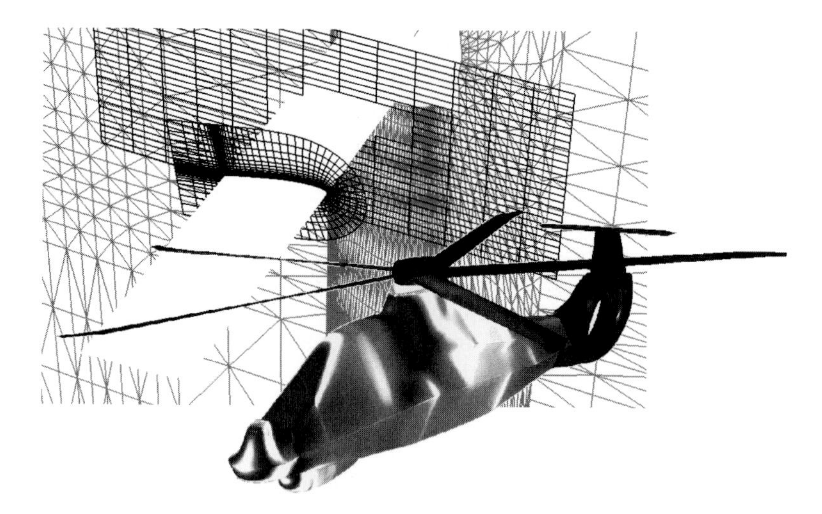

Fig. 20.18 Helicopter CFD numerical methods (NASA Ames Research Center).

power setting required to overcome that drag is calculated, and then the engine's fuel consumption data are used to determine fuel flow.

For helicopters, the fuel used during a cruise mission segment is estimated as follows:

1) Assume a helicopter weight (between the start weight and the end weight of cruise).

2) Calculate power required using the preceding equations, or by iterating power until the desired velocity is obtained using more sophisticated analysis techniques.

3) Calculate or look up the fuel flow at that power setting.

4) Calculate the specific range (distance traveled per unit fuel used) as velocity divided by fuel flow.

5) Iterate back to step 1, assuming another helicopter weight. Then, plot specific range vs helicopter weight. The total range is found by graphical integration, i.e., the area under the curve.

This is actually similar to the derivation of the Breguet range equation where we integrated a specific range equation with respect to a change in aircraft weight. The reason that we cannot derive a similar direct equation for helicopters is that there is nothing equivalent to L/D since the rotor provides both lift and thrust.

For loiter, the same method is followed, but the specific loiter (time per unit fuel used) is graphically integrated rather than the specific range.

Helicopter Initial Sizing

To determine a first estimate of helicopter weight and fuel weight to perform the required mission, we use the aircraft sizing equation [Eq. (3.4)], repeated next. Fuel fraction is found, not from a mission segment fuel fraction based on

Table 20.4 Helicopter empty weight fractions

Aircraft type	Typical W_e/W_0
Scout/attack helicopter—light armor and weapons	0.5–0.6
Scout/attack helicopter—heavy armor and weapons	0.6–0.8
Transport helicopter	0.45–0.55
Civil/utility helicopter	0.45–0.6
Tilt rotor	0.55–0.7

Breguet, but from the "known-time fuel burn" equation [Eq. (19.6)] that is modified to use power specific fuel consumption and the helicopter power loading in Eq. (20.18). The total mission duration is assumed from the desired range and cruise speed, plus an allowance for takeoff and climb. This assumes that the helicopter flies at nearly full power throughout the mission, ignoring the effect of the weight reduction as fuel is burned, so it is probably conservative, and we can safely ignore the time spent in descent and landing. Helicopters typically require a 5% margin on engine fuel consumption plus a 10% fuel reserve, or an allowance equal to 20–30 min of flight at best loiter speed.

$$W_0 = \frac{W_{\text{crew}} + W_{\text{payload}}}{1 - (W_f/W_0) - (W_e/W_0)} \tag{20.17}$$

$$\frac{W_i}{W_{i-1}} = 1 - \frac{c_{\text{power}} d}{W/P} \tag{20.18}$$

Empty weight fraction is determined historically—typical data are in Table 20.4, but actual data for a similar helicopter should be used where available. Note that the empty weight fractions in the table are not given as functions of takeoff weight as they were for fixed-wing aircraft. The historical data do not support such a trendline. This does simplify the solution of Eq. (20.17)—unlike the case with fixed-wing aircraft, no iteration is required!

Helicopter Design Process

Helicopter design is similar to the steps described in the Intermission between Chapters 11 and 12, but with certain key differences. As with aircraft, you must have design requirements to begin, including payload and/or number of passengers, range, rate of climb, and certain flight speeds especially maximum and cruise. For a helicopter you also need allowable autorotate descent speed and required hover ceiling (in- or out-of-ground effect). As with an aircraft you must gather a lot of data such as internal component geometries and weights, and should also identify some candidate engines and obtain geometric, weights, and performance data.

You may wish to develop design sketches of alternative configuration concepts, including different options for rotor configuration and anti-torque technique. The next task is to select initial values for W/A and W/P, and perform initial sizing calculations to estimate design takeoff gross weight and fuel weight.

A fixed-wing aircraft designer would continue on developing an initial configuration layout based on these initial estimates. However, at this time the helicopter designer will probably run a computer program to better calculate and optimize the rotor parameters including details as to solidity, blade shape, airfoil, twist, and similar parameters, and determine in more detail the optimal disk loading and the required power.

Then, the actual design layout can be developed as described in the Intermission.

STS 114-S-037 Space Shuttle launch (NASA photo).

21
Extremes of Flight

21.1 Introduction

The preceding chapters cover the design of aircraft in the "normal" speed range, from low subsonic to about Mach 2.2. This chapter discusses excursions in both directions—much faster and much slower.

Weirdly enough, there are some similarities in these extremes. For example, rocket launch vehicles and airships are both dominated in their design by the volume of their "propellants," if you consider an airship's lifting gas as propelling it upward. In fact, hydrogen is the ideal airship lifting gas and also the ideal rocket propellant, if practical considerations are ignored. Rockets and airships are both far more sensitive to vehicle empty weight than a normal aircraft. For different reasons, neither can use the Breguet range equation that we rely upon so much for other aircraft. Another similarity is that these extremes of flight are both a lot of fun to design.

21.2 Rockets, Launch Vehicles, and Spacecraft

These days the line between aircraft and spacecraft design has gotten blurry. Many aircraft designers including this author are spending more and more time in the design of rockets, launch vehicles, and spacecraft.* The design of these vehicles is mostly similar to aircraft design as described throughout this book. There is sizing, preliminary layout, design analysis, and performance calculations, and design iteration is an important part of the process. There are also important differences, which are discussed in the following.

The history of rockets and spacecraft is well known and will not be repeated here. Suffice it to say that after countless ages of staring up at the skies and wishing to understand the mysterious moon and the fixed and moving lights seen overhead, mankind has finally developed the science to comprehend and the technology to reach at least some of them. We can even make our own and can put them up there for all to see.

The design of launch vehicles and spacecraft is largely driven by propulsion. The acceleration requirements to do any useful mission in space are so prodigious

---

*To some, the term "spacecraft" refers to satellites and planetary probes, not to the complete powered vehicle. Here it is used as the space equivalent of "aircraft," and satellites and such are considered the spacecraft's payload.

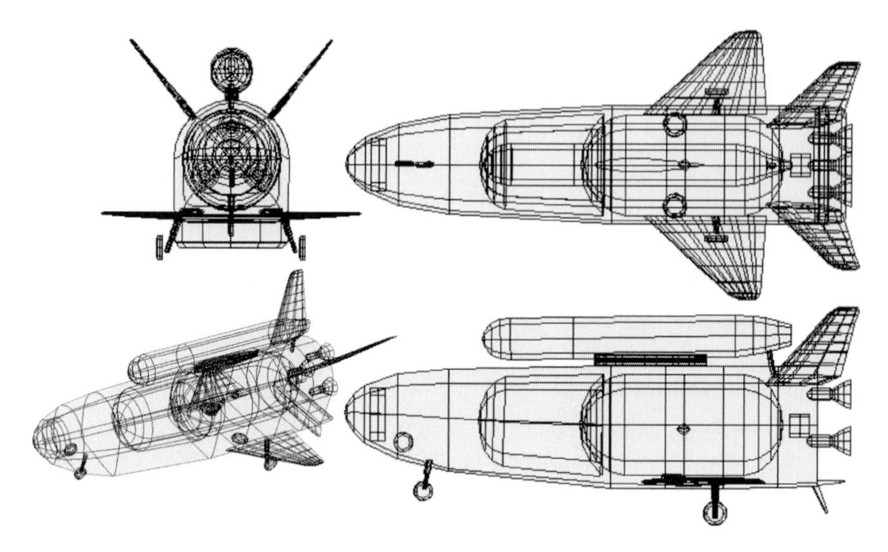

Fig. 21.1 Reusable first-stage launch-vehicle system (Ref. 145).

that the propellant is around 9/10 of the vehicle mass. This compares to a typical aircraft where it is one-third of the mass. The current record for aircraft fuel fraction is the GlobalFlyer, where it is 83% of total mass—a rather poor rocket value.

Although we would like to grip the fabric of space–time and pull ourselves up to space, most forms of spacecraft propulsion yet devised by mankind involve some sophisticated version of throwing rocks out of the back of a canoe. The more rocks you have and the faster you can throw them, the faster your canoe can go.

Such rock-throwing propulsion is called a reaction drive, and there are two categories. In most rocket propulsion* the energy that accelerates the propellant is actually contained in that propellant as chemical energy—the fuel throws itself out the back in a chemical reaction that produces heat and pressure in the combustion chamber.

In the other category of reaction drive, the energy source is separate from the "rocks," typically being stored as electricity, nuclear power, or solar-energy collection. In this case, the best rock to throw is hydrogen because it has the lowest atomic number. For chemical rockets we must compromise between low atomic number of the exhaust and the reaction energy content of the propellants.

Whatever the source of the energy, the thrust of a reaction drive is found from Newton's Third and Second Laws. The Third Law says that for every reaction there is an equal and opposite reaction, so that if the rocket pushes the propellant out the back, the propellant pushes the rocket forward. The Second Law allows us to calculate this force as being equal to the change in momentum per change in

*"Rocket" is used as a generic name for a reaction drive engine, but is also used to describe the entire vehicle. "Motor" generally refers to a solid rocket engine, whereas "engine" refers to a liquid rocket engine.

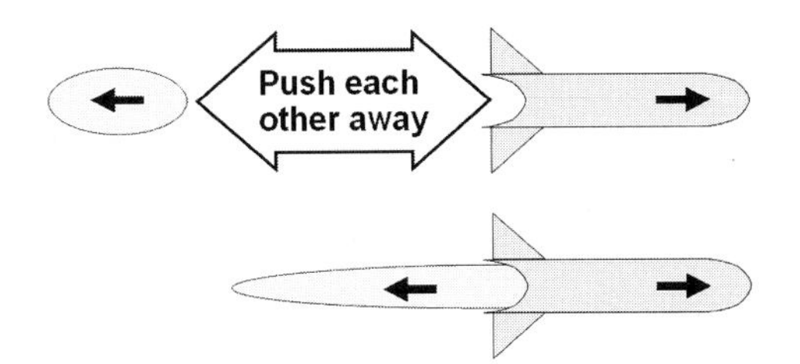

Fig. 21.2 Discrete and continuous propellants pushing a rocket.

time. Its full form includes both discrete masses (rocks) and ejected fluid mass flows, as follows (V_{exhaust} is relative to the vehicle) (Fig. 21.2).

Newton's Second Law:

$$F = ma + \dot{m}V_{\text{exhaust}} \tag{21.1}$$

The total "impulse" applied to the rocket is this force* times its duration, or by integration (assuming fluid mass flow, not thrown rocks),

Total Impulse:

$$I_t = \int_0^t F\,\mathrm{d}t = \int_0^t (\dot{m}V_{\text{exhaust}})\,\mathrm{d}t \tag{21.2}$$

For the following performance calculations we will need to know the impulse obtained per unit mass of propellant consumed, that is, the specific impulse. However, to make the units come out nicely, we normally use unit propellant weight instead of mass, as follows.

Specific Impulse:

$$I_{\text{sp}} = \frac{\text{Total Impulse}}{\text{Fuel burned}} = \frac{\int_0^t (\dot{m}V_{\text{exhaust}})\,\mathrm{d}t}{g_0 \int \dot{m}\,\mathrm{d}t} = \frac{V_{\text{exhaust}}}{g_0} \tag{21.3}$$

The last equality assumes that the exhaust velocity is constant for the duration of the burn, which is reasonable in most cases. Note that g_0 is the Earth-standard acceleration constant but is used even if the spacecraft is far from Earth because it is being used to convert a mass to an equivalent weight force. Adjusting this term for altitude effects is a common student error.

With this term properly employed, I_{sp} becomes the ratio of the thrust force obtained per unit propellant mass flow in weight force equivalent. Thus, the

*There is also a pressure thrust contribution equal to the nozzle exit area times the difference between the exhaust pressure and the ambient atmospheric pressure. It is for this reason that rockets have different thrust at different altitudes. Oddly enough, the maximum thrust is obtained when this pressure thrust is zero because this allows all of the energy available to be used to accelerate the exhaust.

force terms cancel, and the I_{sp} units become "seconds," whether British or metric units are employed. If we chose to use mass flow rather than "weight flow" of propellant, the units of I_{sp} would be the same as velocity and, in fact, I_{sp} would be identical to the effective exhaust velocity. This more correct definition is sometimes used especially in Europe, but notice that the numerical value changes in different systems of units.

Specific impulse I_{sp} is obviously related to the equivalent aircraft parameter, specific fuel consumption (SFC). SFC is defined "upside down." In British units SFC is given in pounds weight of fuel per hour per pound of thrust force generated. We cancel the pounds, leaving units of per hour. In British units, SFC and I_{sp} are therefore the inverse of each other, except that one is in hours and the other is in seconds. Thus, I_{sp} is 3600 divided by SFC and vice versa. Also note that for SFC, "big is bad," whereas for I_{sp} "big is good."

In metric units, the SFC is more properly defined as propellant mass flow per hour divided by thrust force produced, and so the conversion of I_{sp} to SFC requires use of the gravitational acceleration constant (see Appendix A).

One might ask, why don't rocket scientists use the same specific fuel consumption definition that aircraft designers have always used? Perhaps the answer is historical. When Tsiolkovsky derived the Rocket Equation in 1895 (Ref. 143), the aircraft term SFC had not yet been invented. Neither had aircraft. Another possibility was embarrassment. A bad jet engine has an SFC of 2. A good chemical rocket has an SFC of 10. The comparison is less obvious if we say instead that $I_{sp} = 360$ s. This is not their fault. A rocket must carry its oxygen along as a propellant, whereas a jet gets it for free (actually not so free—an air-breathing turbojet engine is 5–10 times heavier for the thrust it produces, mostly because of the difficulty in capturing and using that "free" oxygen).

The efficiency of a rocket engine changes with the exhaust velocity. This is similar to what was found in Chapter 13 for aircraft propulsion, with one important difference: the aircraft is accelerating an oncoming fluid, namely, air—the fuel being a negligible portion of the accelerated mass flow. The rocket must accelerate fluid that it carries along.

Rocket propulsion efficiency is formally defined as the ratio between the power inherent in the vehicle itself (thrust times velocity) divided by the total power inherent in the vehicle plus the residual kinetic energy of the exhaust, or as follows.

Propulsive Efficiency:

$$\eta_p = \frac{FV}{FV + \frac{1}{2}(\dot{m})(V_{\text{exhaust}} - V)^2} = \frac{2V/V_{\text{exhaust}}}{1 + (V/V_{\text{exhaust}})^2} \tag{21.4}$$

where

F = thrust force
V = vehicle velocity
V_{exhaust} = effective exhaust velocity relative to vehicle
V_{exhaust} = thrust/mass flow = $g_0 \, I_{sp}$*

*Note that "real" rocket scientists use C for V_{exhaust}, whereas for aircraft C is a shortened form of SFC and appears in the Breguet equation, among others.

This is maximized when $V_{exhaust} = V$, yielding perfect efficiency (=1.0). This is the same analytical result as was obtained for aircraft, but has a different impact. For aircraft, ideal efficiency produces zero thrust because this implies that there is no acceleration imparted to the oncoming flow. For rockets, this is not the case because thrust is produced by accelerating the exhaust from its relative stationary condition inside the vehicle to its exhaust velocity.

Another interesting observation is that $V_{exhaust} = V$ implies that the exhaust ends up with no velocity relative to an outside observer. If the exhaust really were rocks, it would appear that the spacecraft flew by and "laid" stationary rocks along its path. This is, in fact, optimal because residual velocity in the exhaust takes energy to produce, but does not help to push the rocket.

Yet another observation is the following: although $V_{exhaust} = V$ provides maximum efficiency in terms of energy usage, it does not provide maximum thrust per propellant expended. If energy is readily available, by all means use the highest exhaust velocity technically possible. When the energy source is separate from the exhausted propellant, such as in a nuclear-thermal rocket, a trade study should be conducted to see if the mission performance is maximized with a more powerful but heavier energy source, or with additional propellant mass.

Typical values of I_{sp} are provided in Table 21.1.

Although I_{sp} is very important to rocket performance, there is another aspect of propulsion that must be considered. Different propellants have different densities, and this affects the size and empty weight of the vehicle. Hydrogen, which is excellent in terms of I_{sp}, has a very low density, and so its tanks must be large, leading to a larger, heavier airframe. This is also a serious problem whenever hydrogen is studied as an aircraft propellant (see Fig. 22.7). Solid propellants have worse I_{sp} values than hydrogen but are very dense, leading to smaller stages.

Density impulse is a useful parameter that designers sometimes use to make comparisons between propellant options. Density impulse is calculated as the product of propellant specific gravity (density) and I_{sp}. A larger value implies a better vehicle design will result. However, density impulse is only a guide, and a detailed design trade study is still required to determine the best propellant for a given mission.

Table 21.1 Typical specific impulse for rockets

Rocket type	Typical I_{sp}, s
Chemical, liquid propellant	
LOX-Hydrogen	360–450
LOX-Methane	270–350
LOX-RP (kerosene)	250–330
Chemical, solid propellant	180–220
Nuclear thermal	800–2000
Nuclear pulse (Orion)	4000 +
Electrothermal	400–2000
Ion	4000–25,000
Solar heating	400–700

Spacecraft propulsion not involving the "throwing of rocks" is called a reactionless drive. There are few viable candidates, and practical application seems beyond current technology—but that might change.

Solar sails use photonic pressure to create free thrust, but require sail areas measured in units more typical for real estate than for flight vehicles. The sail weight must be impossibly light, and the problems of unfurling and controlling the sail have yet to be resolved. Research is continuing, and solar sails might prove applicable for certain missions in the near future.

Other candidate reactionless drives include tethers, space elevators, electrodynamic, and more. A number of the exotic possibilities for future spacecraft propulsion were described by science fiction author and noted research scientist Dr. Robert L. Forward in the aptly titled book *Indistinguishable from Magic* (Ref. 144).

There are two practical ways of changing the velocity of a spacecraft without propulsion—gravity assist and aerodynamic assist. Gravity assist involves a flyby path near a planet and is similar to a bicyclist grabbing a passing car. If properly done, it will change the vehicle's speed and direction, while minutely doing the reverse to the planet!

Aeroassist involves the use of aerodynamic drag for deceleration at a destination planet possessing an atmosphere. The returning Apollo capsules used aeroassist to dissipate the velocity acquired by "falling" from the orbit of the moon, thus saving a lot of rocket propellant.

For the selected propulsion system, we now need to determine the propellant required to obtain the desired mission capabilities. In aircraft design, we establish a mission range requirement at the start of a project and then use equations such as those of Breguet to determine the required fuel fraction, which becomes a target for the design effort.

We do something similar in the design of rockets. Rather than a range requirement, we determine an equivalent parameter called Delta-V. Rather than the Breguet equation, we use Tsiolkovsky's Rocket Equation. These are described next.

Delta-V is exactly what the name implies—a change in velocity. It applies regardless of propulsion type and is calculated from the overall mission objective. For example, to fly from Earth orbit to Mars orbit takes a certain total change in velocity, roughly 38,000 fps {12,000 mps}.

This seems odd—surely the important parameter is acceleration. After all, that is what the rocket engine provides. This is true, but the purpose of the acceleration is to change the velocity. The time of the rocket burn is normally so short compared to the transit time that it can be ignored for initial design purposes.

The Delta-V obtained by a rocket burn is used to place us in a different orbit—one that takes us from where we are to where we wish to go. So, the calculation of the required Delta-V is actually an exercise in orbital mechanics.

When an object is in a circular orbit around some body, its horizontal speed is great enough that the centrifugal force equals the weight. Actually, centrifugal force is a fraudulent term—there is no such thing. What is really happening is that the object is falling, but moving forward fast enough that the ground is falling away at the same rate. But, centrifugal force is a convenient engineering fiction. It is calculated as mass times velocity squared, divided by the distance R from the center of the body being orbited. Setting this equal to the weight gives

Centrifugal force = weight:

$$\frac{mv_s^2}{R} = mg \tag{21.5}$$

From Newton's Law of Gravitation, we know that the acceleration due to gravity reduces as distance increases, giving

Gravitational Acceleration:

$$g = g_0\left(\frac{R_0}{R_0 + h}\right)^2 \tag{21.6}$$

We can substitute this into Eq. (21.5) and obtain

Required orbital velocity:

$$v_s = R_0\sqrt{\frac{g_0}{R_0 + h}} \tag{21.7}$$

where

h = height above ground
g_0 = acceleration at planet's surface
= 32.1727 $\mathrm{f/s^2}$ {9.8062 $\mathrm{m/s^2}$} (Earth)
R_0 = planet's radius
= 20,925,646 ft {6,378,137 m} (Earth)
(Data for other planets are in Table 21.2.)

This equation tells us the velocity needed for orbit at a particular height. If you can get to that height, once there the Delta-V to enter orbit is just this required velocity minus your current velocity.

If trying to reach a due-East Earth orbit from the ground, the rotational velocity of your starting point on the ground will help so that your Delta-V requirement can be reduced. This assistance equals the Earth's rotational speed at the equator (1542 fps or 470 mps) adjusted for latitude (multiply by the cosine of the latitude). If launching into a polar orbit, the Earth's rotation does not help, and if you try to launch towards the West you need additional Delta-V. Also note that the closer the launch site is to the equator, the easier it is to reach orbit. It is for this reason that the Soviet Union chose Kazakhstan rather than a site in Russia for its launch complex.

You must also fight both gravity and aerodynamic drag on the way up. These add roughly 6000 fps {1830 mps} to the Delta-V required to reach Earth orbit. The energy height methods of Chapter 17 can be used to approximate the velocity equivalent of the altitude to be gained. For a better solution, time-stepping simulation programs are commonly used.

To travel from planet to planet around the sun, or from Earth orbit to the moon, we first need to get out of the gravity well of the planet we are near. Escape velocity is the speed at which, for the current altitude, the kinetic energy of the vehicle equals the work needed to overcome gravity all of the way out to infinity. This equals the orbital velocity times the square root of two. From this we

Table 21.2 Data for heavenly bodies (after Ref. 146)

Name	Orbit radius, (mil st. miles)	Period of revolution about sun	Mean diameter, km	Relative mass (Earth = 1.0)	Specific gravity (1 = water)	Acceleration of gravity at surface, m/s^2	Escape velocity at surface, m/s
Sun	—	—	1,393,000	332,000	1.41	273.4	616,000
Moon	0.238	27.3 days	3,475	0.012	3.34	1.58	2,380
Mercury	35.96	87.97 days	4,990	0.053	5.30	3.60	4,200
Venus	67.20	224.7 days	12,200	0.815	4.95	8.50	10,300
Earth	92.90	365.256 days	12,755	1.00	5.52	9.806	11,179
Mars	141.6	686.98 days	6,760	0.107	3.95	3.749	5,000
Jupiter	483.3	11.86 yr	14,000	318.4	1.33	26.0	61,000
Saturn	886.2	29.46 yr	125,000	95.2	0.69	13.7	36,600
Uranus	1783	84.0 yr	47,600	14.5	1.56	9.39	21,900
Neptune	2794	164.8 yr	44,700	17.2	2.27	14.9	25,000
Pluto	3670	248.4 yr	14,000	0.90	4.00	7.62	10,000

calculate a "planet exit" Delta-V. If we are already in orbit, we can credit the orbital velocity we already possess.

Next, we need to find the Delta-V to travel from the orbital radius we are in to the orbital radius of the target, around the sun (or Earth, for travel from Earth to moon). There are several strategies we could follow, including just pointing where we want to go and firing up our science-fiction "warp drive" engines. Given the limited capabilities of actual rocket engines, we prefer to follow a minimum-fuel trajectory called a Hohmann transfer orbit. This follows an elliptical orbit that is exactly tangent to the starting and ending orbital radii. Hohmann transfer orbit analysis can be found in Bate et al. (Ref. 147) among others. Typical results from Earth orbit are summarized in Table 21.3.

Although you can always do a Hohmann transfer from one planet's orbital radius to another, it is not always the case that the planet is there when you arrive! Minimum fuel "windows" are times when the start and ending planet are in the correct locations such that, after a Hohmann transfer flight, the target planet is there. If you cannot launch during such a window, a less optimal trajectory will be required, and more propellant will be needed.

Once at the target planet, we find that our velocity is not exactly the same as the orbital velocity at that radius, and so we make a second Delta-V burn to "circularize" the orbit. If we are going to land, though, we can take advantage of the gravity well of the target planet to pull us in. In fact, we will have to somehow counter the extra velocity we pick up in falling from "infinity" to that planet's surface, which equals the escape velocity just discussed. If the planet has an atmosphere, aeroassisted braking can be used.

Calculation of Delta-V to perform the required mission is, of course, more complicated than this brief overview. This is especially the case where maneuvers like gravity assist are employed. To really get the correct answer, even relativity must be considered.

Once the design mission has been analyzed to determine the total Delta-V required, the amount of propellant that must be carried by the vehicle to obtain that Delta-V can be found. For this, we use the famous Rocket Equation.

Table 21.3 Hohmann transfer orbit results

Planet	Minimum launching velocity, mps	Transit time
Mercury	13,411	110 days
Venus	11,582	150 days
Mars	11,582	260 days
Jupiter	14,021	2.7 years
Saturn	14,935	6 years
Uranus	15,545	16 years
Neptune	15,850	31 years
Pluto	16,154	46 years

The Rocket Equation is very much like the Breguet equation in that it relates propellant consumption to the spaceflight equivalent of range, namely, Delta-V. The Rocket Equation can be derived by starting with the top illustration in Fig. 21.2. This shows a rocket pushing out a discrete "blob" of propellant—the "rock" referred to earlier. By conservation of momentum, we know that the momentum before the blob is pushed out has to equal the momentum afterwards, or

Momentum:

$$\text{Before:} \quad (m_{\text{final}} + m_{\text{propellant}})V_0 \tag{21.8}$$

$$\text{After:} \quad m_{\text{final}}(V_0 + \Delta V) + m_{\text{propellant}}(V_0 - V_{\text{exhaust}}) \tag{21.9}$$

where

V_{exhaust} = relative to the vehicle
ΔV = increase in velocity

Equating and solving for ΔV gives

$$\Delta V = \frac{m_{\text{propellant}} V_{\text{exhaust}}}{m_{\text{final}}} \tag{21.10}$$

So the resulting change in vehicle velocity is just the relative exhaust velocity, ratioed by the propellant mass vs final mass.

Replacing the discrete propellant blob by a continuous mass flow and integrating from an initial mass to a final mass gives one form of the Rocket Equation. Substituting from Eq. (21.3) gives a more useful form.

Rocket Equation using V_{exhaust}:

$$\Delta V = V_{\text{exhaust}} \ln\left(\frac{m_i}{m_f}\right) \tag{21.11}$$

Rocket Equation using I_{sp}:

$$\Delta V = g_0 I_{\text{sp}} \ln\left(\frac{m_i}{m_f}\right) \tag{21.12}$$

Rocket Equation–Mass Ratio:

$$\frac{m_i}{m_f} = e^{\frac{\Delta V}{g_0 I_{\text{sp}}}} = e^{\frac{\Delta V}{V_{\text{exhaust}}}} \tag{21.13}$$

An even more useful form for designers is found by solving for the required mass ratio, shown in Eq. (21.13). This is in the form of a mission segment weight fraction as derived in Chapter 3, allowing the calculation of the propellant mass required to obtain the required Delta-V.

One more consideration—staging. The idea of stacking rockets on top of each other to improve performance was suggested in 1650 and theoretically analyzed by Tsiolkovsky, who developed a modified version of Rocket Equation to analyze staging, shown in the following.

The basic idea of staging is simple—improve performance by cutting away parts of the launch vehicle when they are no longer needed. Why carry big empty tanks to orbit? Throw them away as soon as they are empty. Why carry the big engines needed to lift off the ground because the vehicle will be much lighter when most of its propellant is gone? Throw them away too.

There are a number of staging geometries that can be considered, shown in Fig. 21.3. Most staged rockets, including the one conceived in 1650, are stacked vertically requiring a sequential burn. The top-stage engines do not operate at liftoff and must be started immediately when the lower stage is dropped off. This seems like a bad idea, not using all engines on liftoff. Actually, engines that are designed to operate only at high altitudes will have higher thrust and I_{sp} than engines that can also run well at sea level, so that it might be better if the upper-stage engines are not operating at liftoff. Also, vertical stacking will have less drag and probably less weight than the other approaches.

The next two geometries allow parallel burn—all engines are operating at liftoff. This maximizes initial thrust and hence increases allowable gross liftoff weight (GLOW). The extra strap-on boosters or engines are normally used just for initial portions of the flight and are dropped at a fairly low altitude. Parallel burn is normally combined with a third stage designed for high-altitude operation.

The final geometry is used by the space shuttle. Propellant is carried in a fairly cheap external tank much like the throw-away drop tanks long used in military aircraft. The expensive stuff (main engines, avionics, etc.) is located in the recovered portion of the shuttle. The shuttle also incorporates strap-on parallel burn boosters to increase GLOW.

The Rocket Equation is modified for a staged design just by adding up the Delta-V contribution of each stage (Eq. 21.14). The second equation details the individual

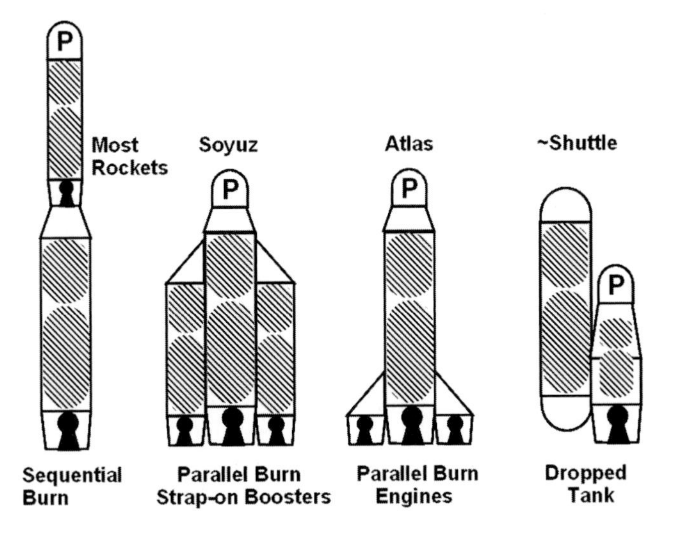

Fig. 21.3 Staging geometries.

stages' Delta-Vs. In the third equation, these are combined by assuming that all stages have the same I_{sp}. This can be approximately true if the same engine type and propellants are used, but the altitude effect on engine performance must be considered. In the last equation, it is further assumed that the empty weight of all but the last stage can be ignored. This is dubious, but does illustrate how the mass ratio appears much better for a staged rocket than for a single stage.

Staged Rocket Equation:

$$\Delta V_f = \sum_{1}^{n} \Delta V = \Delta V_1 + \Delta V_2 + \Delta V_3 + \cdots \tag{21.14}$$

$$\Delta V_f = g_0 \left\{ \begin{array}{c} I_{sp1} \ell n(m_{i1}/m_{f1}) + I_{sp2} \ell n(m_{i2}/m_{f2}) \\ + I_{sp3} \ell n(m_{i3}/m_{f3}) + \cdots \end{array} \right\} \tag{21.15}$$

$$\Delta V_f = g_0 I_{sp} \ell n \left[\left(\frac{m_{i1}}{m_{f1}} \right) \left(\frac{m_{i2}}{m_{f2}} \right) \left(\frac{m_{i3}}{m_{f3}} \right) \cdots \right] \tag{21.16}$$

$$\Delta V_f = g_0 I_{sp} \ell n \left[\frac{m_{i1}}{m_{f(\text{last stage})}} \right] \tag{21.17}$$

Another launch vehicle issue is receiving a lot of attention these days. Existing launch vehicles are largely thrown away during each flight. The launch customer must buy a new flight vehicle for each flight—imagine doing that with airliners! Even the space shuttle, developed for reusability, throws away its huge expendable propellant tank. It also drops its two solid boosters, which are recovered but require complete disassembly and remanufacture before they can be used again.

Launch-vehicle reusability should reduce operational costs, but it adds to the system development cost and weight. The booster must survive the heating and loads of reentry and must be capable of landing in some fashion, preferably at a preselected land location rather than parachuted into the ocean. This might require some sort of "flyback" capability using turbojets, rockets, aerial towing, or efficient gliding. This can also impose horizontal flight stability requirements, which might be difficult to obtain because of the aft center of gravity typical for an empty booster stage. The booster must be capable of operation for a greater flight duration, which increases subsystems requirements. Finally, the booster must be sized to its mission including the empty weight impact of the preceding, and boosters have extremely high sizing growth factors.

All of these add to the weight and cost of a reusable system. The question is—are those costs so large that it remains cheaper to throw the whole thing away after each flight? The jury is still out, but the potential payoff is great, and so a lot of research and design effort is going into reusable launch systems such as that depicted in Fig. 21.1.

This brief introduction has barely touched on the subject of the preliminary design of rockets, launch vehicles, and spacecraft. Other sources should be consulted for detailed descriptions of orbital mechanics, launch analysis, and rocket thrust calculations as previewed here, plus the numerous subjects untouched including structural design, weight estimation, subsystems, avionics, communications, thermal management, guidance, control, payload, and others.

21.3 Hypersonic Vehicles

After a checkered history of promises and problems, hypersonic flight is undergoing a resurgence. "Hypersonic" is roughly described as Mach 5 or higher (the Mach 3 SR-71 seems stationary by comparison), but is actually defined by the presence of the following flow characteristics not found at lower speeds:

1) The shock angles are so steep that they lie close to the surface forming a "shock layer" and causing a strong interaction between the boundary layer and the shocks. Because of this interaction, the boundary layer is one to two orders of magnitude thicker than at lower speeds, and creates an apparent body around the actual body. This, among other things, causes the nose to appear blunt to the freestream flow no matter how pointed its actual shape may be. Note that the shock–boundary-layer interaction violates a common assumption in CFD codes, so specialized hypersonic codes must be used.
2) Extreme flow heating causes molecular excitation and even dissociation of the air into ions, so the air is not really air anymore! Again, regular aerodynamics codes must be modified to account for this, and even the full NS equations do not include this possibility.
3) If the hypersonic flight is at high altitude, the low density–high speed causes the usual "no-slip" assumption to be untrue. The molecules right at the surface have a tangential velocity, unlike the case in slower flows where they are "stuck" to the surface. Regular CFD codes assume "no-slip."
4) Forces and moments change in a significantly nonlinear fashion with respect to angle of attack.

At hypersonic speeds, a first approximation of the pressures exerted upon the vehicle can be found from the Newtonian impact theory. Newton thought that fluid flow could be modeled as a stream of pellets hitting the surface, which proved to be very wrong at subsonic speeds but reasonably correct at Mach 5 or higher. The analysis assumes that air particles hitting the vehicle are turned parallel to the surface, and that the perpendicular component of the air's momentum is exerted as pressure on the vehicle.

From the Newtonian assumption, the center of lift of a hypersonic vehicle can be roughly approximated as the geometric centroid of the total planform area (including fuselage). The design must be configured so that this centroid is relatively close to the c.g. for hypersonic flight. The normal design requirements as to stability must be accommodated in subsonic flight, which may place the wing farther back than desirable for hypersonic balance. This often leads to the use of a strake or double-delta arrangement.

A key issue for hypersonic vehicle design is thermal management. Supersonic aircraft like the SR-71 use their fuel supply as a heat sink, routing the fuel through heat exchangers to absorb the heat generated at the nose and leading edges, plus the excess engine heat. The black paint on the SR-71 is specifically formulated to radiate heat. Flight profiles are actually limited by heat absorption capabilities—when you cannot absorb any more heat, you have to slow down.

The problems are even worse at hypersonic speeds, and thermal analysis must be incorporated from the earliest phases of design. The Space Shuttle experiences maximum surface temperatures of over 3000°F {1650°C}. Extreme thermal loads limit the minimum radius on the nose of a hypersonic reentry vehicle to perhaps 1–2 ft {30–60 cm}, and the nose behind this nose cap should not be sloped less than about 15–20 deg from horizontal. Heating also limits the leading-edge radius for the wing and tail to a minimum of roughly 1–2 in. {3–5 cm} unless very exotic materials plus active cooling are employed.

Despite the high temperature environment, the Space Shuttle and various other hypersonic designs use a conventional aluminum structure. This is protected by thermal tiles or blankets plus exotic materials such as carbon–carbon composite or ceramics in the regions of highest heating (nose and leading edges). The thickness and weight of such a "thermal protection system (TPS)" must be included in the earliest design stages, and TPS experts must select and analyze the materials and provide data to the configuration designer. As a first approximation, an allowance of about 1–2 in. {3–5 cm} on the bottom, and less than 1 in. on the top should suffice. Advanced TPS coverings weigh roughly 0.5–1.0 lb/ft^2 {2.4–5 kg/m^2} of surface area, whereas the Space Shuttle tile TPS averages about 1.6 lb/ft^2 {7.8 kg/m^2}. To this must be added about 0.25–1.0 lb/ft^2 for attachments, usually a bonding agent and a strain isolation pad.

Alternatively, exotic materials such as Inconel can be employed without TPS, but numerous trade studies indicate that the total weight is usually heavier. Final selection depends upon maximum Mach number, duration of high-speed flight, and availability of fuel for cooling.

Some hypersonic vehicles such as the Space Shuttle and various cruise missiles have fairly normal fuselage–wing configurations, with wing planform selected as a compromise between landing speed and high-speed drag. For a reentry vehicle, the reentry *g*-loading often sets a limit on wing loading.

For efficient hypersonic cruise flight, a concept similar to the compression lift described in Chapter 8 offers promise. The "Hypersonic Waverider" is basically a highly swept flying wing configured such that the shocks are defined and constrained by the leading edges, and the vehicle flies on top of the shock waves it creates. Early waverider concepts resembled swept triangular wedges with negative dihedral, but later analysis including the effects of viscosity determined a revised optimal shape. The upper part of Fig. 21.4 shows a typical viscous-optimized Hypersonic Waverider shape, looking like a thumbnail in top view and a downward-pointing bow in cross section. This offers a hypersonic L/D substantially better than a simple wing–fuselage arrangement, but the integration of the optimal shape into a reasonable design, with engines, landing gear, cockpit, and payload, is left to the designer! The lower illustration in Fig. 21.4 shows a notional vehicle design done at the University of Maryland for a slightly different waverider geometry.

The Space Shuttle and ICBM reentry vehicles are actually unpowered hypersonic gliders. Obtaining positive net thrust from an airbreathing engine is quite difficult at hypersonic speeds (especially over about Mach 8). Net thrust for any airbreathing engine is found as gross thrust minus engine-related drag. The drag includes the momentum drag of slowing the air down enough to mix it with fuel and get it to burn. If we slow the air down enough to make that easy, the momentum drag is so large that net thrust is virtually impossible. If we do not slow it down very much, it is very difficult to get the air mixed and burned

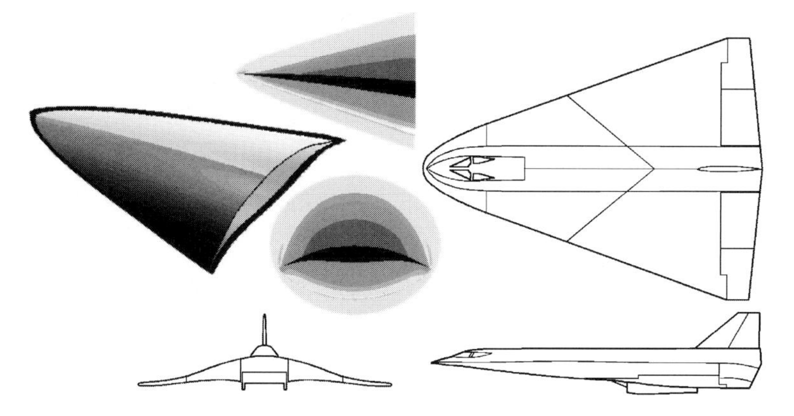

Fig. 21.4 Hypersonic Waverider (NASA Langley/University of Maryland).

in the fraction of a second that it is inside the engine. Even if we can, the net thrust will be found as the difference between two very large numbers, gross thrust and drag. If either of those changes by a small percentage from our expectations, the net thrust will be negative.

Regular turbojets require slowing the air down to about Mach 0.4–0.5, which raises temperature as well as pressure. At speeds much over Mach 3, the temperature of the air itself is already almost too hot for turbine blades—and if we add fuel and combust the mixture, we will burn off any turbine blades. The Air Turbo Rocket concept avoids this by *not* passing the outside air through the turbine blades. Instead, a rocket motor drives the turbine that in turn drives the compressor. The rocket is run fuel-rich, which keeps the temperature down, and provides leftover fuel to be burned when the rocket exhaust is mixed with the compressor air downstream of the turbine. Note that, even though this is an airbreathing engine, some amount of oxidizer must be carried for the rocket motor at its core.

The ramjet concept avoids the turbine blade heating problem by not having any. Compression is provided by the inlet system alone. Fuel is added, burned, and exited through a nozzle to provide thrust. Of course, this does not work at low speeds so some other propulsion device such as a rocket booster must be used for takeoff. The Air Turbo Rocket does provide thrust at zero speed, one of its benefits, but the ramjet is more efficient at higher speeds. The hybrid Air Turbo Rocket–Ramjet (or just Air Turbo Ramjet) diverts its flowpaths to form a ramjet at higher speeds.

The ramjet slows the air down to subsonic speeds. At hypersonic velocities the momentum drag losses become excessive. The Supersonic Combustion Ram Jet, or scramjet, attempts to mix and burn the air at supersonic internal speeds. While this has been demonstrated in the laboratory, an operational scramjet engine has not been developed yet. Figure 21.5 shows the bottom of an airframe-integrated scramjet configuration, illustrating CFD analysis done at NASA Langley Research Center (Ref. 128). Note that the vehicle's forebody forms the inlet ramps, whereas the afterbody forms a nozzle expansion surface.

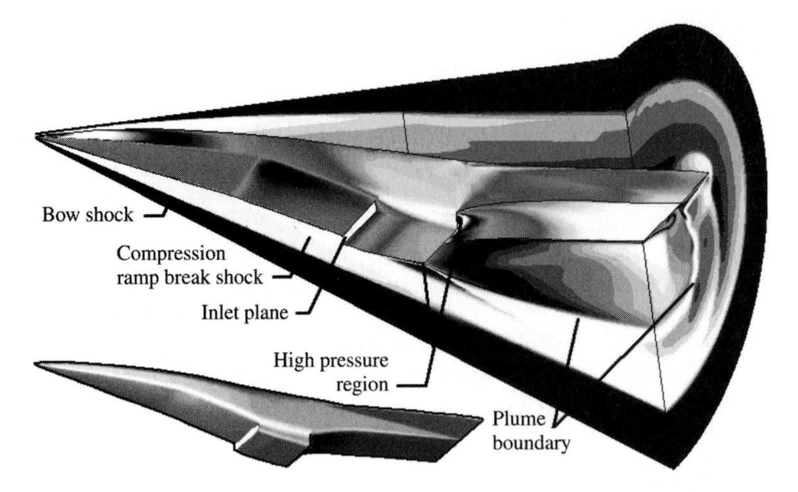

Fig. 21.5 Scramjet CFD analysis (NASA Langley Research Center).

NASA's Hyper-X research vehicle (X-43), similar to Fig. 21.5, successfully flight tested its hypersonic scramjet engines at a speed of Mach 9.6 in 2004. This proved that it is possible to obtain fuel mixing, combustion, and net thrust with internal flow speeds well above the speed of sound. However, this vehicle did not take off and fly to that test condition—it was boosted by a rocket stage many times larger than itself. Also, its test time was just a few seconds, and so we still have a long way to go until we can use scramjets to climb and accelerate to hypersonic speeds, or to cruise a substantial distance. But, Hyper-X was a giant step forward!

References 117 and 119 are recommended for aerodynamics, propulsion, and flight mechanics of hypersonic and reentry vehicles.

21.4 Lighter-Than-Air

Wilbur and Orville were not the first to fly. This brash statement depends, of course, upon the definition of the words "to fly." If your definition matches the ancient dream of mankind—to ascend into the air, go where you wish under power and control, and safely return to Earth when and where you choose—then the first to truly fly was probably Alberto Santos-Dumont in his 1898 gasoline-powered airship (Ref. 138). In his first flight he reached 1300 ft {400 m} above Paris, circling about with ease before an astounded and ecstatic crowd.[§] The landing was less than perfect because of a problem with the ballonet, but, as they would later say, any landing you can walk away from is a good one.

Said Santos-Dumont of that flight, in words that ring true to any pilot today, "I cannot describe the delight, the wonder, and the intoxication of this free diagonal

[§]Others had flown variously powered airships before, but their designs were heavy and clumsy, their flights were not repeated, and they made no further developments.

movement onward and upward, or onward or downward, combined at will with brusque changes of direction horizontally when the airship answers to the touch of the rudder!" (Ref. 139).

In 1901, Santos-Dumont won the Deutsch Prize and worldwide fame for a flight along a specified seven-mile {11-km} course in less than 30 min including a rounding of the Eiffel Tower. By 1903, Santos-Dumont was flying almost daily in his ninth design, landing on the roofs of his favorite Parisian restaurants and allowing an acquaintance to become the first woman to fly a powered flying machine—solo. And these fellows from Ohio say they flew how far . . .?

Count Ferdinand von Zeppelin, whose name has become synonymous with "airship," launched his first design in 1900. Unlike Santos-Dumont's balloon-like designs, the LZ-1 featured a rigid outer structure with the lifting gas held in separate internal cells. It was huge—420 ft long {128 m} with a hydrogen lifting gas capacity of 399,000 ft^3 {11,300 m^3}. On its first flight it carried five people and, coincidentally, also reached an altitude of 1300 ft {400 m}. Von Zeppelin, a retired general, immediately began to explore military applications and by WWI the "zeppelin" had evolved into a feared long-range bomber despite its explosive vulnerability.

Following the war, the Zeppelin Company developed successful passenger airships that flew regularly scheduled trans-Atlantic service at a time when heavier-than-air airliners were in their infancy. The 1928 Graf Zeppelin flew over a million miles without incident and made an aerial circumnavigation of the globe—carrying passengers.

However, the fragility of the airship's lightweight structure coupled with the explosive properties of hydrogen led to numerous and highly public disasters. Other than the Graf Zeppelin, most of the famous airships eventually crashed. This culminated in the 1937 Hindenburg disaster, caught on newsreel footage and endlessly replayed in theaters around the world ("Oh, the humanity!"). This ended public faith in airships at about the same time that airplanes became practical for passenger service, thus ending the "golden age of airships."

Lighter-than-air flight continued in the military. During WWII, U.S. Navy blimps were successful escorts for shipping convoys because they could fly long distances at slow ship speeds and could spot and attack the enemy submarines. The last were retired in 1962, after which powered lighter-than-air flight was mostly limited to advertising, notably the famous "Goodyear Blimps."

Today there is renewed interest in airships for various applications including sightseeing, carriage of bulky cargo, border patrol, and special missions such as logging and missile defense. DARPA has recently begun an airship project called WALRUS, described as a global reach air vehicle. WALRUS is intended to allow rapid deployment of massive amounts of troops and equipment directly from rear area and U.S. locations to unprepared sites in or near a war zone. Two winning contractors are in early design definition at this time (2005) and hope to construct demonstrator vehicles in the near future.

Another interesting application of airships is the "poor-man's satellite," where lighter-than-air technology is used to hover indefinitely at high altitude over a single spot, with applications such as communications or cell phone relay, wireless internet networking, and real-time sensing. Perhaps the most ambitious idea is the "orbital airship," where a high-altitude airship climbs to 200,000 ft

{61,000 m} using buoyancy and then boosts itself to orbital speeds using electric ion propulsion.

There are three major types of airships, or "dirigibles." The word dirigible is not a corruption of "rigid" as some suppose, but instead comes from the French *dirigeable*, that is, "steerable."

Nonrigid airships are basically streamlined balloons. They require a slight overpressure to hold their shape against the aerodynamic "dishing" likely to occur from dynamic pressure at the nose. The overpressure required is small, typically about 5 psf {0.24 $\mathrm{kN/m^2}$}, so that a puncture only results in a slow leak unless the hole is towards the top. Nonrigid airships are often called "blimps," a word said to come from a British military designation "Type B-Limp." Others believe it comes from the odd sound made when you tap on the envelope.

Semirigid airships are blimp-like and require internal pressure to hold their shape, but add an internal or external structure to distribute loads to the fabric. Most of Santos-Dumont's designs were of semirigid construction, with the engine and cockpit attached to a braced frame slung below the envelope.

Rigid airships, generically called "zeppelins," have an external structure that holds its shape without the need for internal pressurization. Most have fabric-covered skins, but some have been built with metal skins. Rigid airships usually hold their lifting gas in separate cells for redundancy.

An interesting airship variation is the *hybrid*, one that gets part of its lift from aerodynamics such as a lifting hull, airplane-like wings, or even a helicopter rotor. The airship portion can be nonrigid, semirigid, or rigid. The hybrid airship notion is not new—Santos-Dumont flew an airship with wings in 1903. However, the 1927 *Airship Design* textbook by C. Burgess (Ref. 140) flatly declared that such a craft "combines the disadvantages, and loses the merits of both types."

This is probably true in terms of weight and drag, but ignores the operational advantages of a hybrid vehicle over a traditional airship. The conventional airship is, by definition, as light as the air and thus is difficult to handle on the ground. To land, its lines must be caught by a ground crew or machine and then attached to a mooring mast. There the airship must be free to pivot with the wind and must be monitored or tied to make sure that a change in atmospheric density or a wind gust cannot drive the tail into the ground, or lift it embarrassingly and dangerously vertical. With payload and fuel removed, the conventional airship experiences excess buoyancy and so must be ballasted before unloading. When fuel is burned during flight, it becomes too light to land, requiring the valving off of lifting gas, or the use of a mechanism to recover water vapor from the engine exhaust.

The hybrid airship avoids these problems because it is actually heavier than air. With only partial hydrostatic lift, it has a substantial download when sitting on the ground and can land and taxi like a normal airplane. The hydrostatic lift reduces the need for aerodynamic lift, which gives a high effective lift-to-drag ratio at low speeds. Analytical studies by this author indicate that a 50-50 split between aerodynamic and hydrostatic lift provides the best balance between cruising efficiency and ground handling (Ref. 141). Note that this form of hybrid airship cannot hover for cargo loading unless a powered VTOL mechanism is added.

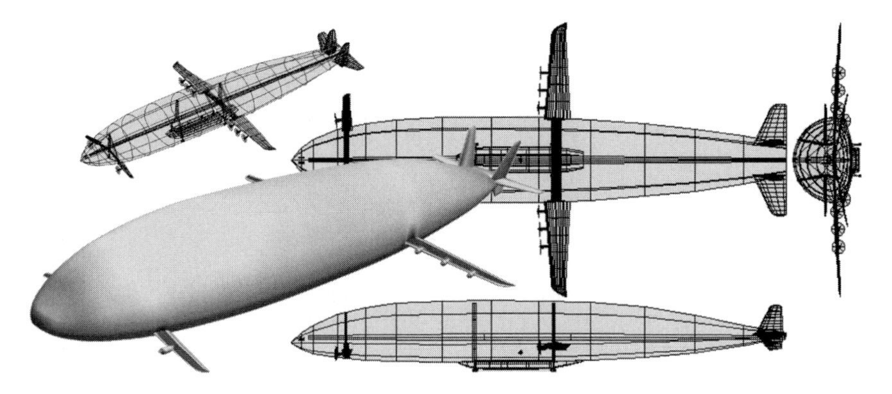

Fig. 21.6 Ohio airship "Dynalifter" hybrid airship (D. Raymer, 2001).

The design of an airship has many similarities to the design of any flying machine, but has one obvious difference—the provision and accounting for hydrostatic lift. The concept is simple—the lifting gas has less density than the air it displaces, and so an upward force is exerted. We merely need to provide sufficient internal gas volume and some form of containment, ensure that the lift loads are transferred to the structure of the vehicle, and calculate the weight of the components and the lift of the gas.

Airship lift is fundamentally determined by Archimedes Principle, which states that the gross buoyancy is determined by the weight of the displaced gas, namely, air. So, before we consider the lifting gas we must look at the air to be displaced. Then we will subtract the weight of the lifting gas used to displace the air to get the net lift.

As found in Appendix A, air has a sea-level, standard-day (59°) density of 0.0023769 slugs/ft^3 {1.225 kg/m^3}, or a weight force equivalent (multiply by g) of 0.0765 lbs/ft^3 {12.01 N/m^3}. This represents the weight force of the air that is displaced by the lifting gas, hence the gross lift.

Air density is affected by altitude, temperature, barometric pressure, and humidity. Atmosphere tables such as Appendix A give the altitude variations. The effects of temperature can be found using Charles' Law (volume varies directly with absolute temperature if pressure is held constant). The effects of barometric pressure can be found using Boyle's Law (volume varies inversely with pressure if temperature is held constant).

Moist air is less dense than dry air because water vapor is less dense than the air it displaces. The water-carrying capability of air depends upon the temperature—hotter air can hold more. At freezing temperature (32°F, 0°C), fully saturated air weighs only 1/2% more than dry air, but at 90°F {32°C} saturated air weighs 5.2% more than dry air. The density variation is approximately linear with percent saturation, and 50% saturated air is often used as a design assumption.

There are three candidate lifting gases—hydrogen, helium, and hot air. Hydrogen is desirable for several reasons. It is common, being an element of water, and can be produced anywhere by electrolysis or various chemical reactions. During the U.S. Civil War, army balloonists traveled with wagons full of acid to make hydrogen. It also has a greater lifting capacity than any other gas. It only has one problem—it is highly flammable and explosive under the right circumstances. For this reason it is now forbidden to use hydrogen in passenger-carrying airships.

Helium has about 10% less lifting capacity and is more expensive. Being an inert element, helium does not combine to form chemical compounds and thus cannot be "broken" out of readily obtainable substances. In fact, helium was unknown to mankind until spectral analysis of sunlight revealed a new element that was wrongly thought to be a metal, hence the name ($\sim$sun + metal). Luckily, helium is found blended with natural gas in underground wells and can be purified out for a reasonable price.

As a conservative rule of thumb, for each 1000 ft^3 of gas hydrogen will lift roughly 68 lb, whereas helium will lift about 60 lb. A similar metric rule of thumb is, for each 1000 liters (cubic meter) of gas, hydrogen will lift about 1.1 kg, whereas helium will lift about 1.0 kg.

Hot air is widely used for recreational balloons and sometimes airships, usually heated with propane burners. The air must be heated for the duration of the flight, thus limiting flight time. Depending upon temperature, hot air can lift roughly 20 lb per 1000 ft^3 {0.300 kg per m^3}.

To properly calculate an airship's net lifting force, the gross lifting force of the displaced air is reduced by the weight force of the displacing gas. Under sea-level, standard-day (59°) conditions, hydrogen has a weight force density of 0.00532 lb/ft^3 {0.836 N/m^3), and helium has a weight force density of 0.01056 lb/ft^3 {1.66 N/m^3).

Density and hence lifting capacity are affected by the purity of the lifting gas. Although it should be nearly pure (98% or better) when first loaded into the airship's gas cells, there will be a slight leakage with time that allows lifting gas to exit and air to enter and dilute the mixture. Golden Age airships would lose purity at a rate of about 2–3% per year. Today it should be half that. When the lifting gas purity gets too low, the gas needs to be replaced or "scrubbed." The density of the impure gas mixture is found simply as the weighted average of the lifting gas and the air.

For calculation of hydrostatic lift under nonstandard conditions, we would consider the density variations in the lifting gas resulting from changes in altitude and temperature. However, these affect the lifting gas density by the same amount that they affect the density of the air being displaced. If the gas bag is free to expand or contract, the net lift will be unchanged.

This seems odd but is true—as you climb higher, the gas bag expands, but the lifting force remains unchanged. Balloons designed for extreme altitude flight are huge and mostly empty when launched, resembling a tower of plastic with a jellyfish at the top, but they swell up to a nearly round shape at their design altitude. If they go higher than the design altitude, the maximum volume is exceeded, and if the gas cannot be quickly vented, the balloon will burst. If the gas is vented, the lift is reduced, which can be a real problem when you later try to land!

This maximum altitude is characteristic of all lighter-than-air vehicles and is called the "pressure height." It is selected by the designer, and hull volume is calculated at pressure height making it much bigger than the volume of the required lifting gas at sea level. If a higher pressure height is selected, the gas bag will have to be made even larger, resulting in more weight, cost, and aerodynamic drag. If a lower pressure height is selected, the airship will be more limited in its flight routes and more susceptible to bad weather.

The external shape of an airship cannot be allowed to change as the lifting gas bags swell and shrink with altitude. For a rigid airship, we simply make the hull large enough so that the fully expanded gas bags will fit. At altitudes below pressure height, external air enters the hull and presses the gas bags up towards the top of the hull (remember—they float).

For semirigid and nonrigid airships, the problem is especially severe. If the excess gas is vented while ascending, upon descent the envelope will be only partially full and will tend to collapse. A clever solution (previously invented) was used on Santos-Dumont's first airship—the ballonet. This is a "balloon within a balloon" affixed to the inside of the airship hull. During ascent, this air-filled balloon is allowed to collapse as the lifting gas expands into its space. Air, not irreplaceable lifting gas, is vented overboard. During descent, the ballonet is reinflated by an air fan, pressing the lifting gas into the top of the hull. Alternatively, ram scoops behind the propellers can be used to reinflate the ballonet on descent.

On Santos-Dumont's first flight the air fan (actually a pump) was not powerful enough to quickly reinflate the ballonet on descent. The weight of the engine and pilot pulled the cigar-shaped envelope into a dangerous "V" shape that could have torn the suspension lines loose, high over Paris. He was lucky to survive and devised more powerful air fans for his later designs.

The calculation of the extra volume required within the hull can be made by determining the percent fullness ($\%F$) at sea level based on the change in lifting gas density, using Charles' and Boyle's Laws as follows:

Percent Fullness:

$$\%F = \frac{\rho_H}{\rho_{\mathrm{SL}}} = \frac{P_H T_{\mathrm{SL}}}{P_{\mathrm{SL}} T_H} \tag{21.18}$$

P_H and T_H are the atmospheric pressure and absolute temperature at the desired pressure height. SL denotes the desired sea-level conditions, which might or might not be standard conditions. This equation determines how full the lifting gas volume can be at sea level if the desired pressure height is to be reached without venting gas. To calculate the total required volume, find the required lifting gas volume at sea level and divide it by the percent fullness.

For example, to allow reaching an altitude of 10,000 ft without venting any lifting gas requires a sea-level percent fullness of about 0.74. Thus, the hull volume must be 1.35 times the volume of lifting gas required at sea level, and 26% of this hull volume will be air when the vehicle is at sea level. For a nonrigid or semirigid design, this air will be contained in ballonets.

Some high-altitude balloons are of the "superpressure" type, where the internal pressure can be substantially greater than the external pressure. This relaxes the requirements for percent fullness—the envelope is made strong enough that at high altitude it can withstand the lifting gases' attempt to expand without valving off the excess gas. This is especially important for long-duration balloons where the heating during the day can overexpand the gas leading to valving off. This is also being considered for long-duration airships, although the weight penalty is obvious.

For lift calculations of a hot air balloon, Charles' Law is used to find the volume of the heated air. This determines density, hence lift. Normally, the internal temperatures of a hot air balloon envelope are about 250°F {120°C}.

Although much of airship design is similar to aircraft design, there are important differences. For example, airship designers cannot use wing area as the reference for aerodynamic coefficients because there is not a wing. Frontal area is sometimes used, especially in drag tables. Normally though, the reference area used by airship designers is the total hull volume, raised to the $\frac{2}{3}$ power to get units of area.

Airship tail sizing is also based on $V^{0.66}$, often estimated as 13% times this parameter. Most of the old airships were actually unstable in yaw, but the time to diverge was so long that a good "helmsman" had no trouble compensating. In pitch, the low center of gravity relative to the hull's hydrostatic lift provides additional stability to counter the aerodynamic instability.

Another interesting parameter in use is the standard displacement (D), which is the volume of the hull times the air's sea-level standard-day density. In other words, D is the weight of air displaced by the hull.

The optimal airship hull fineness ratio was discussed in Chapter 6 and should probably be somewhere between six and eight for best aerodynamics. However, the structural weight is critical for airships, and a higher fineness ratio hull will be heavier. Recent studies indicate that nonrigid and semirigid airships should probably have a fineness ratio of about four. Rigid airships should be about six.

Note that the parasitic drag analysis methods described in Chapter 12 work fairly well for airships. To estimate hull drag, a 0.85 adjustment factor should be applied to the body drag equation to account for beneficial scale effects.

Another difference is that the Breguet Range Equation does not apply to airships. Breguet integrates for range using the aerodynamic lift-to-drag ratio as the vehicle weight changes. A conventional airship has little or no aerodynamic lift. Therefore, drag does not change as vehicle weight changes, and Breguet is therefore not applicable. Instead, range is found simply as engine run time multiplied by speed.

Because of the large size of airships, the engines must accelerate not just the vehicle's mass but also the "apparent mass." This is the outside air that, through viscosity, clings to the vehicle and therefore must also be accelerated. An old rule of thumb (NACA TR 117) suggests using 2.5% of the total hull volume, times the density of air, as an apparent mass for airship acceleration calculations. Apparent mass also exists for airplanes, but we normally ignore it. Also, for an airship the mass of the lifting gas must be accelerated even though the gas seems to have a negative weight!

For a hybrid airship in which substantial aerodynamic lift is used during cruise, the aircraft performance equations including Breguet can generally be used after one important substitution—all vehicle weight terms W should be reduced by the hydrostatic lift. This reduces the weight being carried by the wings and hence reduces drag due to lift. This simple analysis adjustment does not work for performance calculations involving accelerations, such as takeoff.

The engines must accelerate the full mass of the vehicle. The hydrostatic reduction in the downward weight force does not reduce this. Also, do not forget the apparent mass and the mass of the lifting gas.

Weights analysis of airships is similar to that of aircraft. Statistical and analytical methods can be used, but unfortunately most of the available statistical methods are approaching the century mark in age and are more rules of thumb than sophisticated statistical equations. For example, one old approximation says that fixed weight excluding powerplant is 30% of the hull's standard displacement D (just defined). Crew, ballast, and stores weights were estimated together as 5.5% of D.

To get better weights estimates, structural analysis methods have to be applied along with information from vendors of the various subsystems, skin coverings, and gas cell materials used in the design. Note that some of the structural loads that must be considered are unique to airships, and to save weight airship structure is designed to be as flimsy as possible. The possibility of an in-flight breakup is very real—be careful!

Airship design criteria can be found in FAA Document P-8110-2. Type certification requirements are described in FAA Document AC 21.17-1A.

NASA Schweizer 1-36 Deep Stall Research Aircraft (NASA photo).

Transonic Target Drone (Courtesy Composite Engineering, Inc.)

22
Design of Unique Aircraft Concepts

22.1 Introduction

"They laughed at the Wright Brothers!" Throughout the history of aviation, this refrain has been quoted by thousands of wacky inventors with ridiculous ideas. This author had the dubious privilege at two major aircraft companies of being the person who got to read, evaluate, and reply to the unsolicited inventions and design concepts that would arrive. And, if I see one more flying saucer proposal, . . .!!

But, sometimes wacky ideas work. At least, sometimes they work for a particular application or requirement. And sometimes, today's wacky idea is tomorrow's normal design practice—such as sweptback wings, canard pushers, helicopters, and airplanes made of strings and glue (i.e., composites). But usually, today's wacky idea is tomorrow's wacky idea, and they keep on coming back!

Following are some unique aircraft concepts that seem to have merit. Some of them might, in this author's opinion, eventually find widespread operational use—perhaps the blended wing body and the asymmetrical aircraft. Others might not, but who can tell which? These unique concepts are presented here mainly to discuss how their design differs from the design of normal aircraft as described in the preceding chapters and to provide some specific design guidelines and analysis methods and data for those attempting to design such aircraft.

Be advised that others, especially proponents of a particular unique concept, might hotly disagree with some of this author's opinions and data. Also, no claim is made as to the absolute correctness of this information. It might be too pessimistic, reflecting the author's "show-me" engineering mentality, or it might be too optimistic, reflecting this author's love of novel and creative engineering approaches. (After all, this author has the patent on the "RIVET" VSTOL fighter, the one with the engine mounted backward!)

22.2 Flying Wing, Tailless, Lifting Fuselage, and Blended Wing Body

The pure flying wing, with neither fuselage nor tails, is the "ultimate airplane" in the minds of many. Flying wing advocates point out that all an aircraft really needs is lift and thrust, and that a fuselage, tails, nacelles, and other components just add weight and drag! All else being equal, they are right, but practical problems overcome the theoretical advantages for many applications.

Early pioneers of the pure flying wing were Reimar and Walter Horten of Germany, and Jack Northrop of the United States. The Hortens flew their first powered all-wing design, the H IIm "D-Habicht" in 1935. With the pilot lying prone, the only "bump" from the pure wing geometry was the landing gear and the faired shaft for the propeller. Over the next 10 years the Hortens flew dozens of designs culminating in the first-ever turbojet-powered flying wing, the Ho IX, which flew in 1945. This pure flying wing had a span of 52.5 ft {16 m} and was capable of 470 knots {870 km/hr}. As mentioned in Chapter 8, this advanced design used RAM and configuration shaping for stealth. Its successor, the never-flown Ho229, can be seen in pieces at the Smithsonian Paul Garber Facility, awaiting restoration.

The Hortens employed a design philosophy of reducing the lift at the wing tips to nearly zero, twisting the wing to generate most of the lift on the inboard part of the wing. This allowed moving the c.g. forward and created a design very stable in pitch, and, with proper sweepback, in yaw and roll. The Horten wings did not require any vertical tails nor the negative-dihedral "crank" seen on the outboard panels of some flying wings. The inefficiency of the unusual lift distribution was corrected with increased aspect ratio. Notionally, you can think of the Horten wings as conventional aft-tailed aircraft but with those aft tails stuck out on the wing tips.

Jack Northrop's first flying wing flew in 1940. Like the Horten wings it was virtually pure, with only the canopy and propeller shaft violating the wing contours. The N-1M originally had the negative-dihedral crank previously mentioned, but it was removed after initial testing. After flying a one-third scale prototype and several related designs, Northrop began work on the huge XB-35 flying wing bomber that first flew in 1946. It was converted to jet power and redesignated YB-49, flying in 1948. Much bigger than the Horten jet, this had a span of 172 ft {52.4 m}, a weight of 196,193 lb {88,990 kg}, and a speed of 430 knots {800 km/hr} (Ref. 120).

Northrop preferred to avoid excess twist, and his designs required some vertical tail surface. In the propeller versions the fairing for the propeller shaft and the stabilizing effect of the pusher-propellers themselves provided all the needed directional stability. When converted to a jet, the YB-49 required small vertical tails to replace the lost contribution of the propellers. Flying wing fanatics are still arguing as to whether the several crashes of Northrop flying wings, including the YB-49, were caused by pitch instability or by more mundane causes such as structural or hydraulic system failures. Whatever the cause, the flight performance of the YB-49 was outstanding for its day. The only technical difficulty this author is aware of, a tendency to "hunt" in yaw making for a poor bombing platform, could have been solved a few years later with the active yaw damper developed for the B-47.

Northrop was vindicated late in life when his company won the contract to build the B-2 "stealth bomber." While technically not directly related to the YB-49, its outstanding aerodynamics and low observability have taken the flying wing out of the "oddity" category and into the "viable option" category.

Flying wing design is similar to the design of other aircraft, with a few key differences. Obviously, the planform wing geometry, twist, and airfoil shaping must be carefully considered and analyzed as quickly as possible, and a detailed stability and control analysis should be done early in the project. Center-of-gravity location is critical. Use of sweepback and twist to attain pitch stability has been discussed. Alternatively, a "reflexed" airfoil can be selected, having the trailing edge lifted slightly to provide a naturally stable airfoil. Such airfoils tend to be less efficient, and are typically limited to slower aircraft.

The relaxed-stability, active flight controls developed for fighter aircraft permit the B-2 and other modern flying wings to be more optimized for aerodynamics with less of a compromise for stability and control. This is especially beneficial because the use of longitudinal instability allows the flying wing to take off and land with its trailing-edge surfaces angled downwards like flaps, rather than upward (see Chapter 16).

The flying wing requires special attention as to yaw control. Northrop's wings, including the B-2, rely on wing-tip-mounted surfaces that split open, creating drag that gives a yaw force. This has a very nonlinear response. When first opened, nothing happens. When it is opened more, it suddenly "catches" the air creating a large yawing moment. For this reason the B-2 has a "pilot comfort" mode wherein both of these drag rudders are cracked open just to the point of "catching," allowing the flight control system to fine tune the aircraft's dynamic motion and dampen any "hunting" in yaw. This can be seen in most pictures of the B-2 in flight.

Other possible wing-tip drag rudders are plates that extend up and down near the middle of the airfoil, and clamshell-like devices at the leading edge. Less "pure" flying wings have used vertical tails, often at the wing tips. Conventional rudders can then be used. If they cannot provide enough yawing moment, they can be mechanized to increase drag as they open outward, magnifying their effect.

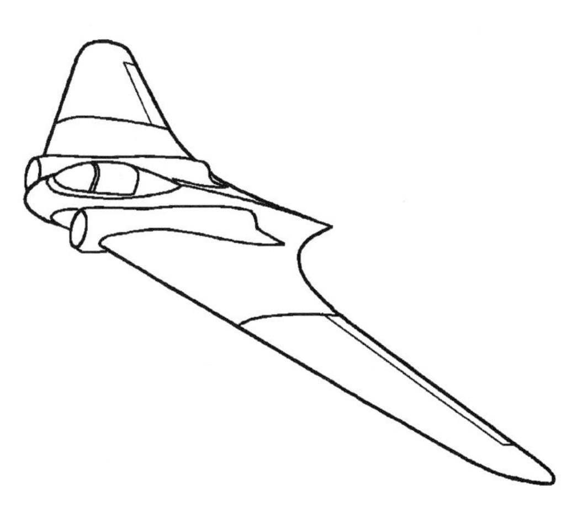

Fig. 22.1 Horten Jet flying wing.

Roll and pitch control for flying wings is usually done with conventional trailing-edge surfaces. It is best to combine these functions as "elevons"—elevator-ailerons—so that the nose-up deflection, which is trailing edge up, delays stall at the wing tips and enhances rather than reverses the wing twist effect. Several early Horten designs had ailerons outboard and elevators inboard, and were almost unflyable.

Properly done, the flying wing should obtain reduced wetted area compared to a conventional design, and should also have a lower structural weight. This is due to the reduced number of components, and also due to the "spanloading" effect discussed in Chapter 8. In fact, the "spanloaded flying wing" has been proposed as a massive cargo aircraft. Designs large enough for a root-to-tip cargo bay capable of holding U.S. Air Force outsized cargo within the airfoil contours have been proposed. Nobody has determined where to land such a monster, however!

As a rough estimate, the wing weight adjustment for delta wings (0.768—see Chapter 15) can be applied to flying wings if they are well designed and reasonably spanloaded. Reference 121 is recommended for further information on flying wings and tailless aircraft.

The lifting fuselage grows out of the desire to make the best possible use of all aircraft components. Rather than allow the fuselage to go "along for the ride," Burnelli and others have designed aircraft where the fuselage is shaped like a wing so that it can contribute to the lift. Burnelli envisioned a wide fuselage with an airfoil shape from front to rear, resembling an untapered wing of aspect ratio 0.4 or less, with conventional wings attached. This would have a high structural weight, probably overcoming any potential aerodynamic benefits.

However, virtually all airliners incorporate the lifting fuselage principle to some extent. By designing so that the fuselage is at a small angle of attack during cruise, a little bit of lift is generated for free, and a dip in the spanwise lift distribution is avoided. Also, the positive pressures that this causes under the fuselage help to turn the flow at the back of the aircraft, reducing separation drag. This was overdone on the Lockheed L-1011, and flight attendants have been complaining ever since that they have to push the carts uphill.

It is generally assumed that the next airliner is going to look just like the last airliner—a tubular fuselage with low wings, a conventional tail, and nacelles either under the wings or on the aft fuselage. This may not always be true, and Boeing is currently investigating the radical "blended-wing-body" (BWB) concept that may provide a revolutionary improvement in subsonic airliner efficiency (Fig. 22.2).

The BWB is basically a flying wing with a delta-shaped wing/fuselage in the center, large enough for a passenger cabin. In some sense it is related to the Burnelli configuration just described, but the center section is blended into the wing panels. This concept reduces the total wetted area of the airplane and, with its deep center section, improves structural efficiency. The BWB has about half of the root bending stresses of a conventional configuration. The wing-tip-mounted vertical tails also act as winglets to reduce drag due to lift. BWB requires relaxed static stability and an automated flight control system to fly efficiently, optimize span loading, and avoid the need for a tail. The BWB optimizes at a wing loading of about 100 psf {488 kg/sqm}, much less than

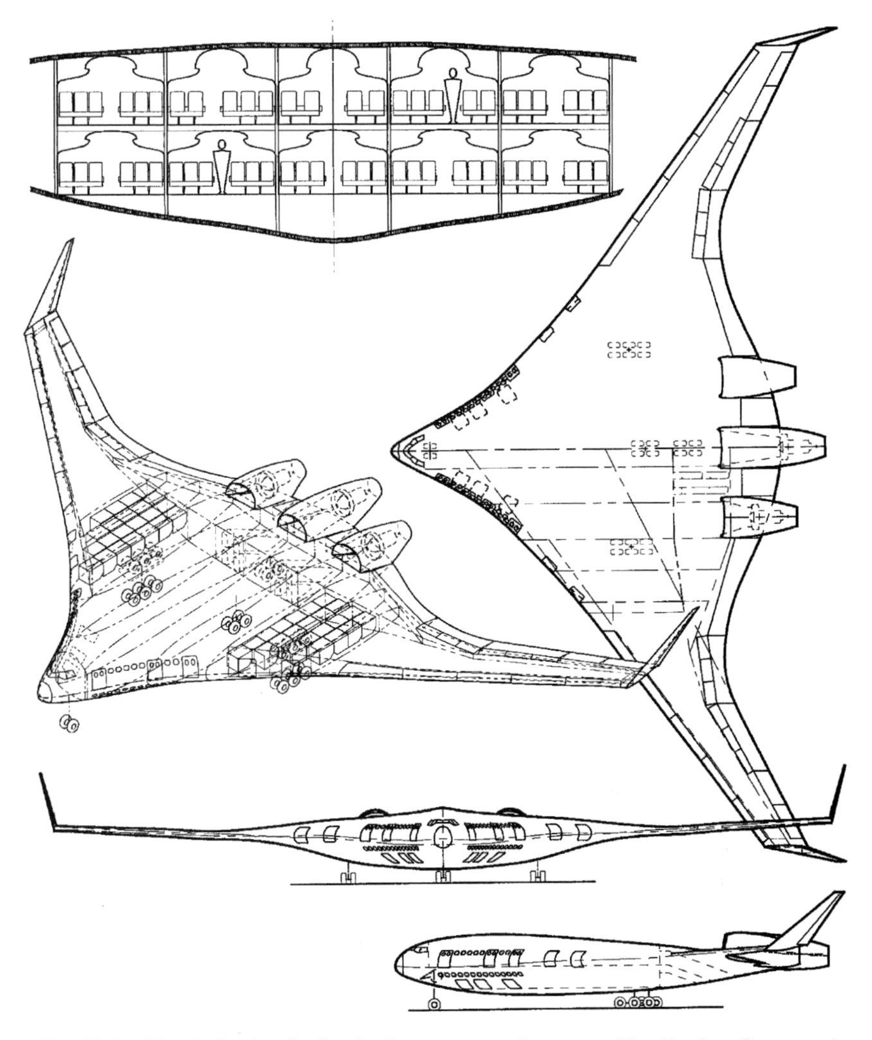

Fig. 22.2 Blended-wing-body airplane concept (courtesy The Boeing Company).

the 160 psf {781 kg/sqm} of most airliners. This low wing loading permits the elimination of high-lift flaps, and only a leading-edge slat on the outboard wing is needed in addition to the wing trailing-edge controls.

Boeing studies predict, compared with an equivalent conventional configuration, a 15% reduction in sized takeoff weight, a 20% improvement in L/D, and a 27% reduction in fuel usage (Ref. 122). A crucial problem to solve is the attainment of a cabin pressure vessel without a huge weight penalty, since the cabin is not a capped cylinder as in a conventional airliner. Boeing plans to carry both pressure and bending loads by a 5-in.-thick {13-cm} composite structural sandwich or a deep hat stringer structural shell. Also, for packaging reasons it seems that BWB is most suitable for a very large airliner (800 passengers), and there is concern that the dearth of windows may be claustrophobic to some passengers.

22.3 Delta and Double-Delta Wing

The delta wing configuration offers certain advantages, especially for high-speed flight. Extensive research led by Dr. Alexander Lippisch showed that the true delta (straight trailing edge) or near-delta planform offers benefits in wing structural weight, increased internal volume, and transonic and supersonic drag.

While tailless flying-wing delta designs are more "pure," Lippisch favored the inclusion of vertical tails as seen on his Me-163 Comet. He stated that "without these vertical surfaces it is impossible to obtain a degree of directional stability comparable to the normal aircraft" (Ref. 123). The A-12 carrier-based attack aircraft was to be a pure tailless flying-wing design, relying on modern computerized flight control systems to obtain the stability that Lippisch could not. It was cancelled mostly because of weight growth—always a problem, especially when a design has to comply with carrier launch and recovery requirements (see Appendix F).

Delta wings usually offer structural weight savings compared to a conventional swept wing because, with the delta, the internal structure need not be swept. Typically, a delta wing has its spars going out perpendicular to the fuselage, and the load path from tip to tip is a straight line. The statistical weight equations of Chapter 15 suggest a 0.768 wing weight adjustment for delta wings, and a further weight adjustment to the fuselage of 0.774.

The high-speed drag reductions are obvious from the sweep of the delta wing. Since the wing has relatively small aspect ratio and a near-zero taper ratio, the wing root chord is very large, and so the root thickness is very deep. This reduces structural bending loads and provides extra room for fuel, landing gear, and structure.

This long wing root can be a disadvantage. Sometimes there is no room left for a horizontal tail, forcing the use of a cantilevered structure, or the use of a canard, or the use of a tailless design approach. Because of the high sweep and low aspect ratio, deltas often require a lower wing loading.

During advanced bomber design studies at Rockwell North American Aviation (1977), a delta-shaped stealth flying wing incorporating the span loading philosophy was conceived called the Delta Spanloader (Fig. 22.3). Extensive analysis indicated a substantial structural weight savings, and RCS test results were quite good (Ref. 124). Like the B-2, this concept used relaxed static stability to minimize trim drag and permit operation of flaps for takeoff and landing. This concept yielded a sized takeoff gross weight a full 30% lower than a conventional bomber design with the same technologies and design mission.

As described in the preceding section on hypersonics, the center of lift of a high-speed vehicle can be roughly approximated as the geometric centroid of the total planform area. The proper location of the wing for subsonic stability

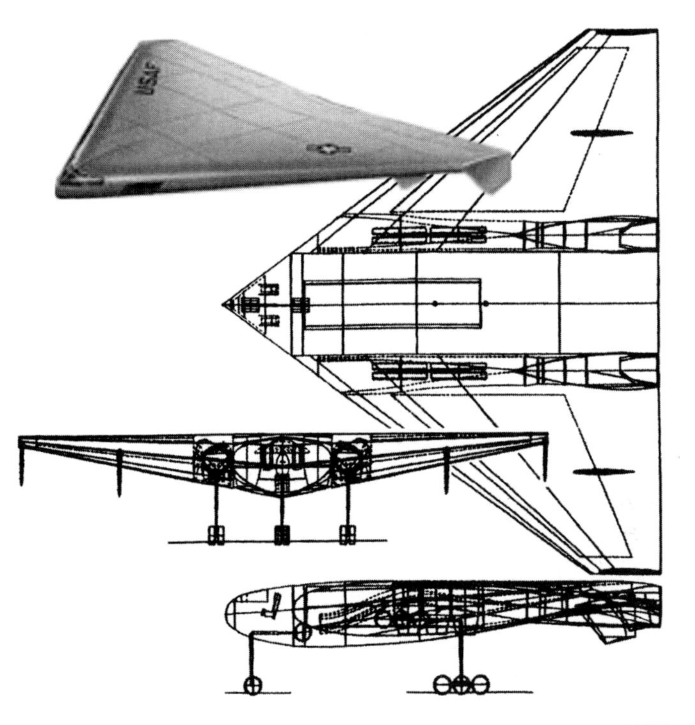

Fig. 22.3 Delta Spanloader stealth flying wing (D. Raymer, 1977).

is usually farther back than desirable for supersonic stability. Adding some sort of highly swept lifting surface forward of the main wing can solve this problem, producing little lift at low speeds but producing increasing lift forward of the wing at higher speeds. One way to accomplish this is the use of the "Double-Delta," an efficient planform for high-speed aircraft. This consists of a delta wing with a "kink" in the leading edge, and a more highly swept delta planform inboard of the kink. The double-delta configuration is seen in the Swedish Draken fighter (Fig. 7.19), which was the first double-delta fighter, flying in 1955. The Lockheed SR-71 has such extensive fuselage chines that it is technically a double-delta, and when the engineers in the "nonblack" side of Lockheed were developing their supersonic transport design, their counterparts from the Skunkworks gave them a sketch of the double-delta arrangement and said "use it—it works—but don't ask how we know!"

22.4 Forward-Swept Wing

The forward-swept wing offers the same transonic and supersonic drag reduction of the aft-swept wing, and has an added benefit in stall. For an aft-swept wing the spanwise flow resulting from the sweep causes boundary-layer thickening and separation out at the tips, leading to early tip stall. For the forward-swept wing, this phenomenon affects the roots instead, leaving the tips with good airflow up to high angles of attack, improving maximum lift and retaining aileron control.

Forward sweep has been incorporated in a number of aircraft designs, for various reasons. The German Hansa Executive jet was forward swept just so that the wing box could pass behind the passenger cabin rather than below the fuselage, reducing frontal area and drag. While a clever idea, the Hansa was heavy and ultimately unsuccessful due to the problem of structural divergence.

A wing with a conventional structural arrangement will, when experiencing a lift load, bend about an axis perpendicular to the direction of the wing box structure. This is illustrated in the top example of Fig. 22.4, where this perpendicular direction is indicated by lines on the wing box. In effect, the wing bends as if those lines are hinges. This is no problem for an aft-swept wing.

When a forward-swept wing bends upward about those "hinges," the tips are deflected such that they experience a higher angle of attack. This increases lift, causing more bending, causing a higher tip angle of attack. At some aircraft speed, the wing will "diverge," and break. The structure must be strong enough that this divergence occurs at a speed higher than the aircraft's maximum placard speed, often resulting in a substantial weight penalty.

Composite materials offer a way to build a forward-swept wing with minimal weight penalty. Composites are typically stiffer than aluminum, so the divergence speed is higher. Many modern general-aviation aircraft including Rutan's Boomerang (see what follows) use composites to allow a modest amount of forward sweep. This can result in better stall characteristics and provides more design freedom as to the location of the wing carrythrough structure.

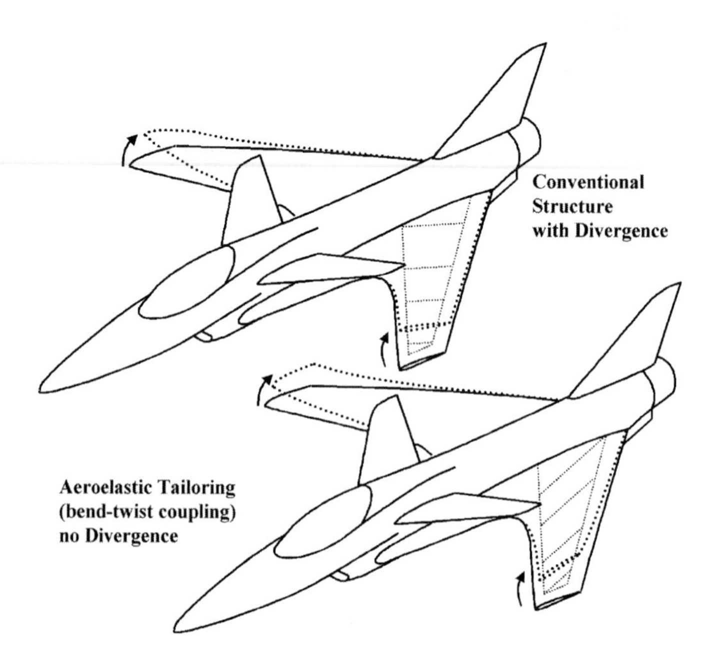

Fig. 22.4 Forward-swept wing—effect of aeroelastic tailoring.

Composites can be "tailored" to allow even more forward sweep, and the proper composite ply orientation can virtually eliminate the divergence problem. This is done by orienting a number of plies in the direction shown in the bottom illustration of Fig. 22.4. If enough plies are placed this way, the wing must bend about those plies—so the wing acts as if it is "hinged" along those lines. When it bends up under an air load, the "bend-twist coupling" caused by this "aeroelastically tailored" structure prevents the tip from experiencing a higher angle of attack, so there is no divergence.

Properly done, this has very little weight penalty. However, forward sweep is not without tradeoffs. While it can result in a higher maximum lift coefficient, that only occurs at very high angles of attack and may not be useful for takeoff and landing. Early claims that forward sweep yielded lower supersonic drag were often based on the assumption that the wing could be reduced in size—not true if landing speed sizes the wing. For a same-sized wing, this author's experience in numerous design studies is that forward sweep, integrated into a real aircraft design, usually has higher supersonic drag.

Forward sweep also places the wing trailing edge more to the rear of the aircraft, creating balance problems when flaps are deflected. Also, a forward-swept geometry causes the flap hinge lines to be more highly swept, which was shown in Chapter 12 to reduce the lift obtained. Together, these limit maximum lift coefficient and may actually require a larger wing. Finally, forward sweep causes problems with radar cross section due to the radar energy bouncing off the wings and onto the fuselage.

22.5 Canard-Pusher

The best aircraft design practice is when the components of the aircraft serve multiple purposes, and when they work together in an interconnected fashion. An example of this is the combination of pusher-propeller, canard horizontal tail, and winglets. While there have been many aircraft incorporating one or two of these concepts, most were inferior to the conventional state of the art despite the claims of their promoters.

The canard pusher with winglets was developed for the homebuilt market by Burt Rutan, and revolutionized and revitalized the entire homebuilding community. His "VariEze" (Fig. 22.5) blended these three technologies and incorporated the then-novel foam-and-fiberglass construction technique to create a fast, safe, and attractive design that could literally be built by anyone, in a garage.

The advantages and disadvantages of the canard were discussed in Chapter 4. The primary advantage of the canard is that, by proper shaping, it can be made to stall before the main wing resulting in a stall-proof aircraft. The aerodynamic advantage of the canard "tail" producing an upload instead of a download is countered by the difficulty of creating high lift for takeoff and landing, often leading to an excessively large wing. But, the canard can readily be used on a pusher-propeller design where there may not be room for a conventional aft tail.

The pusher-propeller was discussed in Chapter 10 where it was noted that the aerodynamic advantage of the aircraft not flying in the propeller wake was countered by the fact that the propeller must operate in the aircraft's wake. But, the pusher-propeller provides excellent visibility and reduces engine noise and

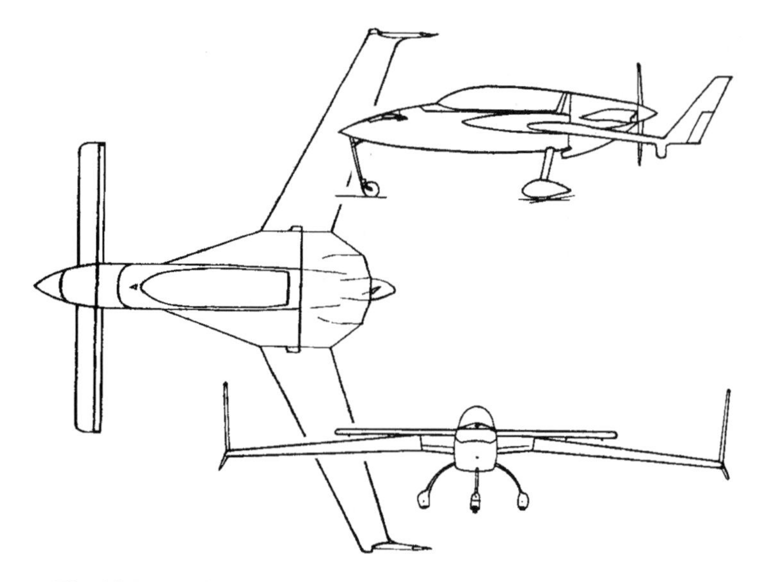

Fig. 22.5 VariEze canard-pusher homebuilt (courtesy E. Rutan).

vibration in the cabin (although propeller noise is increased since it is operating in the aircraft's wake). Also, the pusher-propeller permits a short fuselage, which reduces wetted area—if you can find a place for a horizontal tail.

The pusher-propeller moves the c.g. to the rear, requiring substantial wing sweep for balance. With the wing swept so much, the wing tips are far enough behind the c.g. that the vertical tails can be located on the wing tips. If they are configured as winglets (Chapter 4), they will reduce the induced drag.

Any one or two of these features would probably not be an optimal design, but by cleverly combining all three, a good airplane results, which cannot be made to stall or spin.

Note that the power lost to cooling is often higher for a pusher, and that much of the aerodynamic efficiency of the VariEze comes from its small frontal area compared to other two-seat aircraft. Also, its composite construction results in smooth skins giving it more laminar flow than production aluminum aircraft, further reducing drag. These features, incorporated in a conventional design layout such as the various Lancair models or the production Cirrus SR20, also yield outstanding aerodynamic efficiency. Is the canard-pusher the best configuration for your new design? Only a design trade study will answer that question!

22.6 Multi-Fuselage

Occasionally, the design requirements or the desire to reuse existing components leads to the notion of a multi-fuselage aircraft. Such a design has two or more distinct fuselage components that may be similar or different. The classic example is the North American Aviation F-82 Twin-Mustang, which resembled

Fig. 22.6 Multi-fuselage C-5 derivative.

a pair of P-51s joined together, although there was actually little part commonality with the P-51. This design was developed to provide long-range bomber escort, using a two-man crew to avoid fatigue. The F-82 flew the longest-ever nonstop distance by a propeller-powered fighter, Hawaii to New York.

A multi-fuselage C-5 derivative was once seriously proposed, as shown in Fig. 22.6. Landing fields for this beast would have been difficult to find!

Other multi-fuselage designs have been proposed to allow carrying a large volume of fuel when a single fuselage would have to be as big as a blimp. A Rockwell North American Aviation design study on the use of low-density hydrogen fuel investigated a triple-fuselage bomber, as shown in Fig. 22.7

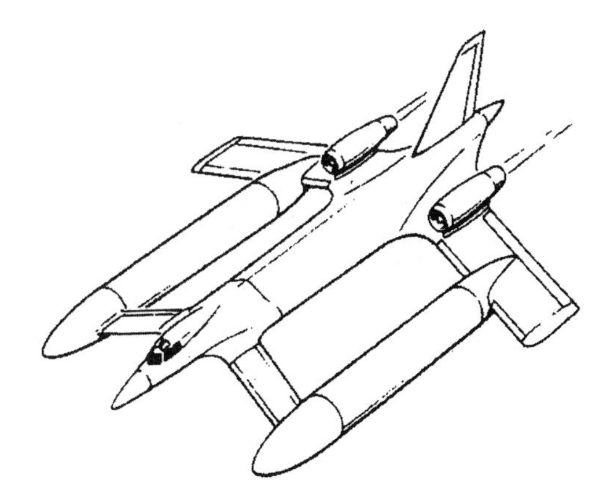

Fig. 22.7 Hydrogen-fueled multi-fuselage bomber proposal (D. Raymer, 1978).

(1977). Analysis indicated that the extra wetted area overly penalized this approach for the required high-speed flight (Ref. 124).

The multi-fuselage approach was successfully used on the around-the-world Rutan Voyager. With a fuel fraction of 72%, it was found that a multi-fuselage approach with "outrigger" fuel tank bodies was more efficient than a single huge fuselage. This approach permitted the application of the spanloading concept as described in Chapter 8, reducing structural weight enough to overcome the wetted area penalty. Also, the joining of the three bodies via a canard provided torsional stiffening to the design, further reducing weight.

Even stranger multi-fuselage concepts are described in the next section.

22.7 Asymmetrical

Most aircraft design is done assuming a plane of symmetry—we usually design just the left half of the aircraft, and assume that the other side is a mirror image. This simplifies design, but even more important, it simplifies lateral-directional stability and handling qualities. Specifically, with a symmetric design we can usually assume that the rolling and yawing moments are zero when the aircraft is at zero sideslip angle regardless of angle of attack, and that these moments increment the same whether a right or left sideslip occurs.

This is generally true for gliders and jets, although the asymmetric formation of vortices on the nose is always a potential problem (Chapter 8). For propeller-powered aircraft, this assumption is often untrue or irrelevant.

A tractor-propeller aircraft is flying in its own propwash, and that prop-wash includes a rotational motion. Thus, the aircraft can have a symmetric flowfield only if it has an even number of propellers and they rotate in opposite directions on opposite sides of the aircraft. Otherwise, a physically symmetric prop-powered aircraft flies in a nonsymmetric flowfield, so it acts like a nonsymmetric aircraft. Furthermore, the "*p*-effect" described in Chapter 16 includes the fact that the center of thrust of a propeller moves laterally when the disk of the propeller is at an angle of attack to the flow, as during a climb. This causes a yawing moment, even for a symmetric aircraft.

Most designers, faced with these problems, design a physically symmetric aircraft that will fly symmetrically when power is turned off, and then try to fix these power-induced effects with trim and by using tricks such as angling the vertical tail or the engine thrust axis. Other designers, though, have reasoned, "since it isn't going to fly symmetrically anyway, why do I have to make it *look* symmetric?" They have even thought of ways to use nonsymmetric geometry to make it fly more symmetrically!

An early example was the German Blohm-Voss Bv-141, flown in 1938 (Fig. 22.8). By placing the pilot and tail gunner in a pod offset from the main fuselage containing the engine and tails, the gunner was afforded a clear field of fire while the pilot got unsurpassed downward visibility for a single-engine attack aircraft. The *p*-effect moved the thrust axis to the right, closer to the centerline, and made the flying characteristics far more symmetrical than the aircraft's appearance would indicate. This strange-looking aircraft attained a range double that of the more-conventional Fw-189 that won the contract.

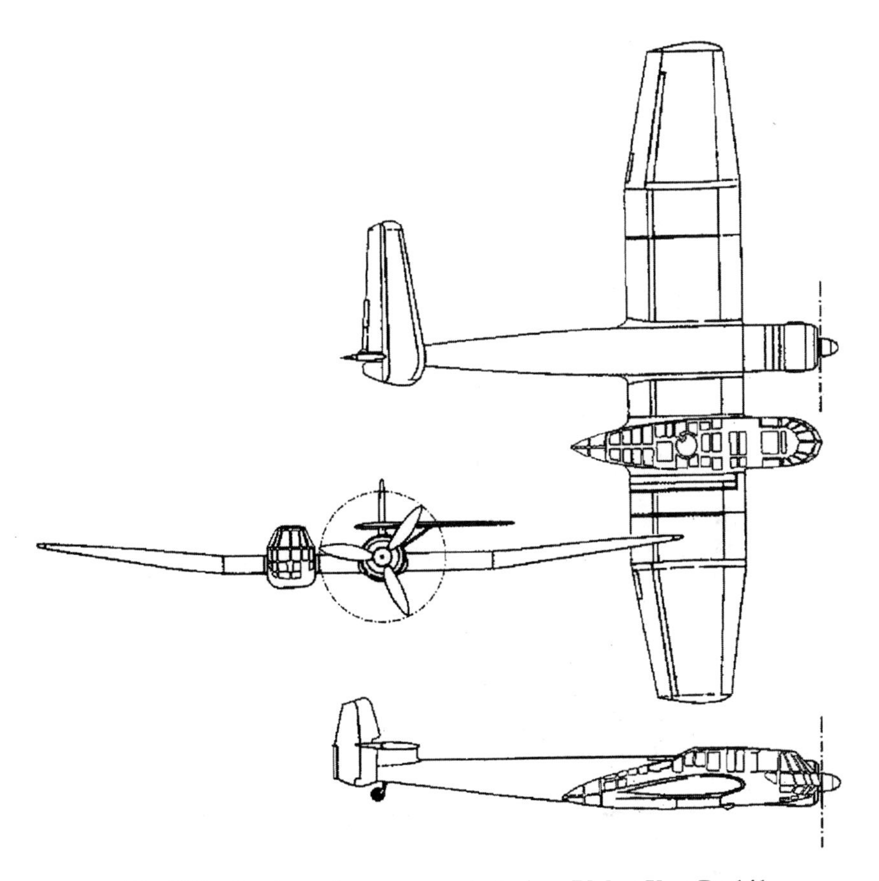

Fig. 22.8 Single-engine asymmetric design: Blohm-Voss Bv-141.

Figure 22.9 shows Burt Rutan's Boomerang, designed for his own personal transportation but now under consideration for production. While the designer of the Bv-141 kept a symmetrical wing and carefully balanced the aircraft masses on each side of its center, the Boomerang is so wildly nonsymmetric that one suspects Rutan was just showing off, until you follow through his logic and realize that everything on the design has a reason.

First, Boomerang had to be twin-engined for safety, but the actual statistics indicate that a twin is no safer than a single-engine aircraft. While having an extra engine can save you if you lose an engine over the Alps at midnight, you also get a doubled risk of having one engine fail. In a conventional design the loss of an engine on takeoff or go-around requires expert piloting to avoid a crash. In fact, most twins have a minimum engine-out flying speed that is higher than the stall speed, and if the pilot allows the aircraft to fall below that speed the running engine will drag the aircraft over on its back or into a spin.

P-effect makes this worse. If the running engine has its downward-traveling blade away from the fuselage (right-hand engine for most aircraft), the *p*-effect

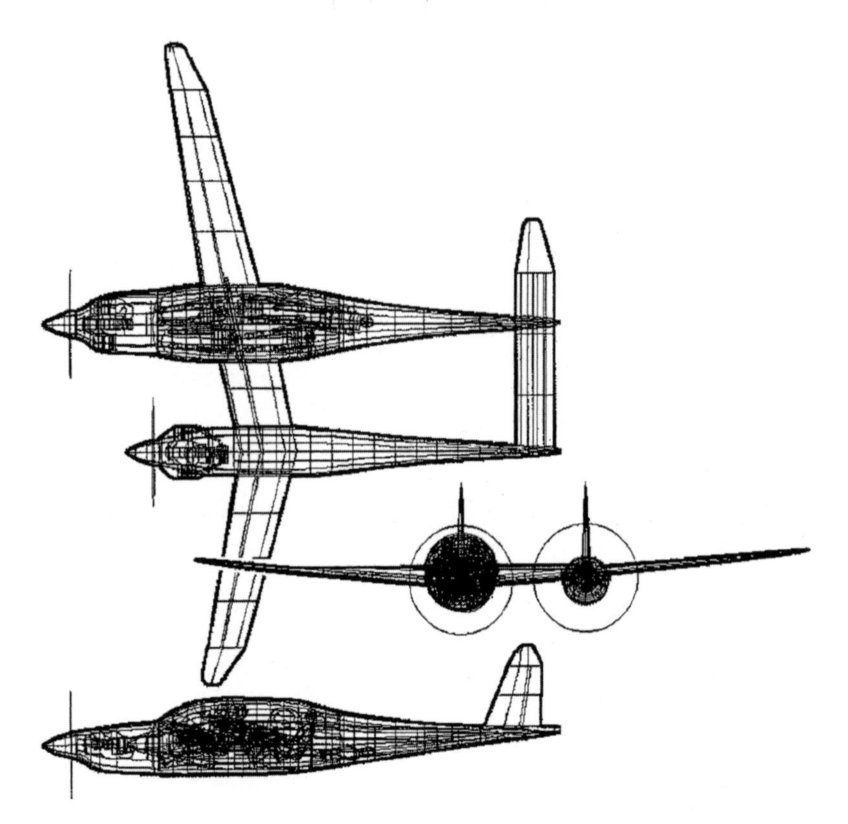

Fig. 22.9 Twin-engine asymmetric design: Rutan Boomerang original design layout (courtesy E. Rutan).

moves the thrust axis even farther out, increasing the yawing moments. Many aircraft have been built with counter-rotating engines so that both propellers have the blade nearest to the fuselage be the downward-traveling one. This adds cost and complexity, and limits choice of engines.

Other designs such as the Cessna Skymaster and Rutan's own Defiant have one pusher and one tractor propeller. These suffer from the inefficiencies and long landing gear of the rear-mounted propeller installation, and have the cabin noise and loss of forward visibility of a single-engine tractor-prop aircraft.

The way to minimize the engine-out yawing is to get the two propellers as close to each other and the centerline of the aircraft as is possible. A normal, fuselage-plus-two-nacelle layout has a minimum propeller separation equal to the width of the fuselage plus a reasonable allowance for propeller clearance. This is too much to avoid a yawing problem. Instead, Boomerang is designed starting with a single fuselage with an engine in front, as if it were to be a single-engine aircraft. A second engine is added alongside, not in a "copycat" fuselage as with the F-82, but in a much smaller engine nacelle. Of course, this makes the design nonsymmetrical—but who cares?! The propeller tips are laterally separated by only 1 ft {30 cm}, and the aircraft lateral center of gravity winds up somewhere between the two bodies. With the engines so close together, the plane flies with almost no rudder input no matter which engine fails (Ref. 125).

To reduce propeller mutual interference and to better balance the design, the extra engine is moved to the rear as far as possible without having the plane of the propeller disk intersect the passenger compartment. This reduces noise and avoids the risk of a thrown blade cutting someone in half. This also moves the extra nacelle far enough back that, with a slight stretch, it can be attached to the horizontal tail to increase torsional rigidity and save some structural weight—the same trick used on the Voyager.

Wings are swept forward to improve stalling characteristics and, probably more important, to keep the wing root behind the pilot for visibility and away from the left propeller to avoid interference. The dihedral break must be at about the middle of the lifting span to avoid nonsymmetric dihedral effect, but nothing says the planform break has to occur at the same place as the dihedral break—so it doesn't! Instead, it is placed within the left nacelle rather than the fuselage to keep the wing farther away from the left propeller.

Rutan says

> *P*-effect makes this aircraft (fly) symmetrical at low speeds, as the thrust lines move right. The combination of having half of the lateral arm for engine-out and about twice the directional stability of conventional twins (long tail arms and both tails in prop wash) gives a very dramatic benefit–this aircraft can be flown at full aft stick and you can feather one engine and apply full power to the other under complete control while not touching the rudder pedals! In this condition it takes 1 lb (1.5 deg) of aileron to hold heading. You can also do aggressive turning maneuvering at full aft stick using only the stick (Ref. 126).

The asymmetry of the Boomerang extends to the cabin. The seats are staggered front to rear rather than being lined up, shoulder-to-shoulder. This allows a narrower fuselage, and thus reduced frontal area yet provides plenty of shoulder room for all.

Another variant of asymmetric design offers the theoretical minimum drag, maximum efficiency supersonic design. The oblique wing is a swept wing with one side swept forward and the other swept to the rear, that is, a single straight wing skewed with respect to the fuselage. This is usually pivoted at the center to allow variable sweep, which for this concept is easy to mechanize with little weight penalty.

Since the wing structural box is straight, the oblique wing is much lighter in weight than other swept wings. It obtains the same transonic and supersonic benefits as other swept wings, and is no worse than the aft-swept wing as to stall angle. Since the aft-swept wing will likely stall first, a stall limiter of some sort is desirable.

The main advantage, though, is the supersonic wave drag. Since the left wing and right wing are not at the same fuselage station (distance from the nose), they do not add to each other in the aircraft's volume distribution. The wing's volume is "spread" in the flight direction, reducing the aircraft's maximum cross section area—and wave drag varies by the square of the maximum cross-section area [Eq. (12.45)].

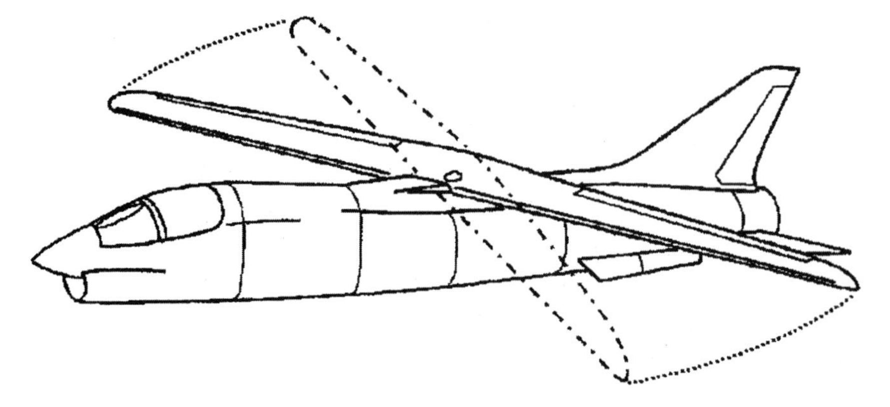

Fig. 22.10 Proposed NASA oblique wing research aircraft (modified F-8).

The oblique wing has yet to be used on a production aircraft, but seems most suitable for a supersonic transport. It was under serious consideration for a Navy attack fighter, but, like forward sweep, it is difficult to make such a design truly stealthy. Figure 22.10 illustrates a proposed NASA research F-8 modified with an oblique wing.

22.8 Joined Wing

A number of aircraft have had biplane wings "joined" in some fashion, at or near the tips. Often such joining was in the form of vertical endplate or similar panel, and had little or no structural implications. Julian Wolkovitch proposed a different and more sophisticated concept in which triangulation is used to enhance structural strength (Ref. 127). The "joined wing" has two wings, a front wing that is swept to the rear, and a back wing that is swept toward the front. The back wing is mounted at the top of the vertical tail and extends downward at a substantial anhedral angle (negative dihedral), meeting and attaching to the front wing. From the front this has a triangular shape, whereas from the top, a diamond shape. Area ruling is employed at the place of attachment, where the back wing begins behind the maximum thickness point of the front wing.

Unlike normal biplanes, such an arrangement can have good transonic and aerodynamic characteristics. Additional tails are probably not required, and trailing-edge surfaces on both wings provide pitch and roll and can provide direct lift and side force, if desired. The main benefit, though, is the substantial reduction in wing structural weight that is achievable, on the order of 30%. This results from the triangulation, with the back wing acting as strut for the front wing. Also, the actual net force vector of a wing is upward (lift) and to the rear (drag), so this arrangement is actually more triangulated than it appears—the net force vector is aligned with the plane of the two wings better than if the front wing were above the back wing as in some other "joined" concepts.

With the loads in the plane of triangulation carried by that triangulation, only the out-of-plane loads are carried by normal wing bending structure. Here an additional structural benefit is obtained—the out-of-plane loads are partly carried

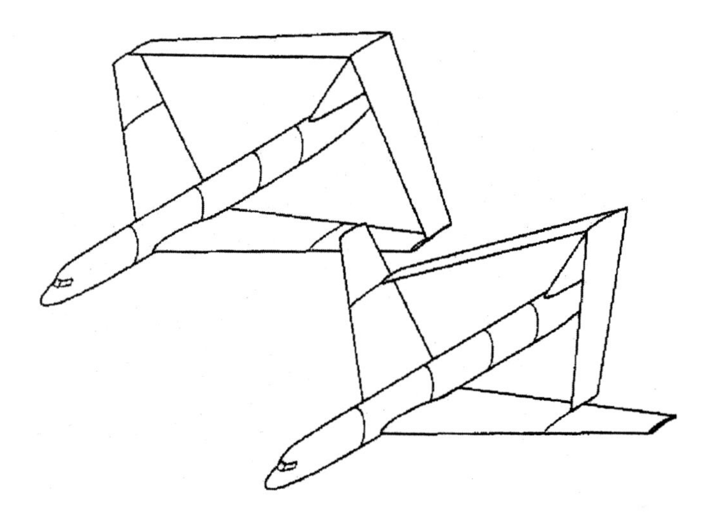

Fig. 22.11 Joined wing concepts.

by the vertical distance between the wing panels' front and rear spars, measured in the direction perpendicular to the plane of triangulation. This increases the effective wing-box depth for structural purposes (Fig. 22.12).

On the negative side, it is difficult to get a trimmed maximum lift coefficient equal to a normal wing–tail configuration, and there can be excess wetted area and interference drag with so many component intersections.

The joined wing configuration has been proposed by Boeing as a carrier-based surveillance aircraft, with antennas on all four wing panels to provide coverage in all directions. This author did an unpublished design study at Rockwell where the joined wing was used for VTOL because without a big wing root at the center of gravity, it may reduce the suckdown effect. Figure 22.11 illustrates two notional joined wing concepts. The lower version, where the wings join somewhere in the middle of the front wing's span, seems preferable according to most studies.

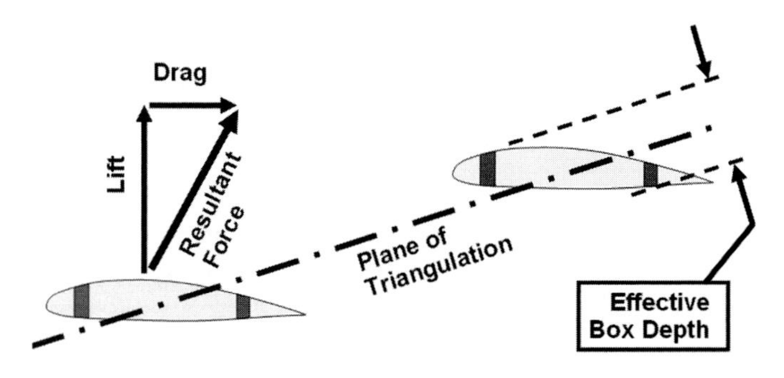

Fig. 22.12 Joined wing structural benefits.

22.9 Wing-in-Ground-Effect

Equation (12.61) is tantalizing. If we could fly very close to the ground, say, at a height equal to one-twentieth of the wing span, we would have an induced drag that is only 27% of normal. Furthermore, the lift coefficient increases near the ground allowing us to fly with a smaller wing—if we can stay at such low heights.

It would be difficult to imagine flying from Los Angeles to Chicago at a height of 5 ft {1.5 m}, but Los Angeles to Tokyo might be feasible! An aircraft designed like a flying boat but intended to keep within ground effect offers fantastic L/D. Called "wing-in-ground-effect (WIG)," such designs were built and flown by Lippisch and others, most notably the Russians. They call such aircraft "Ekranoplanes" and have built huge ones, both propeller and jet powered.

During the Cold War, satellite photos detected something immense flying very fast over the Caspian Sea, so it was nicknamed the "Caspian Sea Monster." It turned out to be the Bartini-Beriev VVA-14 or KM (Russian abbreviation for prototype ship). Depicted in Fig. 22.13, it was designed by the Central Hydrofoil Design Bureau and first flown in 1966. Weighing 1.1 million lb {500,000 kg}, it has eight turbojets mounted on a stub pylon at the front of the fuselage. Their exhausts can be deflected down to the water in front of the wing to create an air cushion for added lift (a technique called "power-augmented ram," or PAR). The KM has two more turbojets mounted on the vertical tail for extra thrust during takeoff, and was said to cruise at 230 knots {430 km/hr}. Unfortunately, the only KM ever built crashed during takeoff in 1980.

The WIG concept works best for a very large vehicle because the height required to miss the waves is a smaller percentage of the wing span for a large vehicle. Also, the WIG is almost by definition a flying boat, and needs a seaplane hull as described in Section 11.7. This will cause a substantial weight and drag penalty. If a relatively small wing is used to minimize drag, then getting out of the water becomes a real problem. Use of PAR and extra engines as just described can help, but results in additional weight and complexity penalties. Finally, operation considerations must be addressed early in the design. Where will it dock? How will passengers and payload be loaded? How will maintenance be done? What new infrastructure will be required? These must all be determined before a design can be finalized and assessed as to economic feasibility.

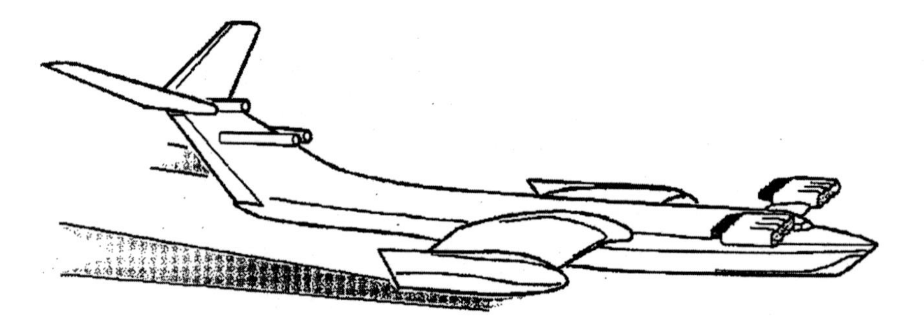

Fig. 22.13 Russian "Caspian Sea Monster" WIG.

22.10 Unmanned/Uninhabited

Unmanned aircraft have been a part of aviation since before the Wright Brothers. In fact, several unmanned aircraft made credible flights before 1903, although without the means of control that separated the Wright Brothers from other early aviation pioneers. Today, the vast computational capabilities that can be put into a tiny chip are bringing about a renaissance in the field of unmanned aircraft.

"Unmanned" implies that a person is not flying the aircraft, which is true for some autonomous aircraft. Usually, an unmanned aircraft actually has a pilot/operator, so a more-correct term is "uninhabited"—nobody is inside, although somebody somewhere is directing and perhaps flying the aircraft. Prior terminology ["drone," "pilotless," and "remotely-piloted vehicle (RPV)"] has fallen out of favor, and most people today refer to such designs as "UAVs," unmanned or uninhabited air vehicles.

There are many UAVs of various types in operation around the world. Many are target drones, such as the Mach 4 AQM-37. A number are for surveillance or are multipurpose, such as the Israel Aircraft Industry Hunter. Most UAV designs range in size and technology from the equivalent of large model airplanes to an uninhabited general-aviation aircraft, although a few UAVs are highly sophisticated.

The General Atomics Predator is typical of prop-powered, low-speed UAVs. It is used for surveillance in trouble spots around the world. Predator is a pusher-propeller aircraft constructed much like a composite homebuilt aircraft. Weighing about 1900 lb {862 kg}, it flies at about 110 knots {204 km/hr} and can operate for a total of over 40 hr. Under continuous operator control, Predator carries visual light and infrared television cameras and can also carry a ground-looking radar.

The much-larger, jet-powered Global Hawk is built by Northrop-Grumman (Ryan) and is somewhat similar in design and mission to an unmanned U-2. Cruising at 65,000 ft {19,812 m} with a total range of 16,566 n miles {30,680 km}, Global Hawk has a takeoff gross weight of 25,600 lb {11,612 kg}. It carries a payload of electronic sensors including electro-optical, infrared, and synthetic aperture radar with moving target indicator.

Design of UAV is much like the design of any other aircraft, but with special considerations for takeoff and landing and for provision of avionics and systems for uninhabited flight. Options for takeoff are many, including conventional wheeled takeoff and landing, use of a launching rail, air- or car-launch, boosted vertical launch, and others. For landing, options include wheels, skids, parachutes, airbags, and none—that is, keep it flying forever or just let it crash! Trade studies should be done to compare these alternatives.

The removal of the aircrew from the design requirements does not save as much in weight or cost as some may assume. The requirements for cooling air are often based on avionics, not crewmembers, and a UAV probably will require more avionics to do a given mission. While it is tempting to assume that we can remove redundancy, reliability, and structural strength margins from an aircraft that has no crewmembers, we must consider the safety of people on the ground. A UAV that is not permitted to operate over land has little operational utility.

Fig. 22.14 Boeing Educational Project UAV (D. Raymer and R. Dellacamera, Instructor/Design Consultants, 2005).

A detailed weights and sizing analysis should reveal a reduction in sized takeoff gross weight of perhaps 10–30% compared to an "inhabited" aircraft. The design layout should benefit from the absence of a cockpit and canopy. Depending on the takeoff and landing gear chosen, further savings in volume and weight may be obtained. Adjustments to structural weights can be made by reducing the required ultimate load factor, assuming that it will not fly in extreme weather, or if it does, a few crashes are OK! In the weights equations of Chapter 15, the use of 0.5–0.7 for the number of crewmembers will approximate the systems that remain even when an aircraft has no crewmembers aboard.

With modern computer and guidance capabilities, it has become feasible to use unmanned systems for a variety of tactical applications including dropping bombs, and even air-to-air dogfighting. The cruise missiles used in Desert Storm were really nonreusable tactical UAVs. Advanced tactical unmanned aircraft could readily attack known-location targets using a variety of conventional payloads such as "smart" bombs, programmed to hit a particular GPS (global positioning system) coordinate. With a more aggressive application of technology, even relocatable targets and targets of opportunity could be attacked by a UAV. If sufficient sensors and target identification capabilities can be provided, an air-to-air combat UAV is conceivable and could have a g-limit double the human tolerance or more.

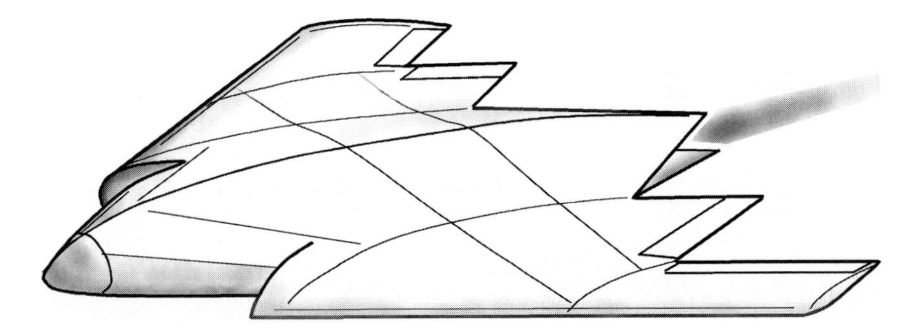

Fig. 22.15 Tactical UAV concept (courtesy Conceptual Research Corp.).

Figure 22.15 shows a notional design concept for a tactical unmanned/uninhabited air vehicle done at Conceptual Research Corporation for investigation of alternative and novel concepts for inlets and exhaust nozzles for such aircraft. While this concept was conceived to simply drop a smart bomb on a predetermined GPS coordinate location, it could also carry surveillance, reconnaissance, and communications relay equipment for alternative missions.

MIT Daedalus Human-Powered Aircraft (NASA photo).

Author with his 1981 Rockwell Advanced Tactical Fighter design.

23
Conceptual Design Examples

23.1 Introduction

This final chapter offers two design examples that illustrate the concepts and methods presented in this book. The design examples are a single-seat aerobatic homebuilt airplane and a lightweight supercruise dogfighter.

These examples illustrate the steps and thought processes used in conceptual design, covering the extremes from propeller-powered, fixed-size engine design to "rubber-engine" supersonic design. The differences and surprising similarities between these extremes of design are shown.

Design requirements for these aircraft were assumed based upon data for similar aircraft. These design requirements were then treated as if they were mandated by some customer and used as the starting point for the design effort.

The examples are designed and analyzed using the methods presented in the book. Every effort has been made by the author to develop credible, realistic designs and to analyze and optimize them properly. However, no claim is made that these are optimal designs or even good designs or that all calculations are correct!

Furthermore, the examples are incomplete in that only the more important analysis areas are presented due to space limitations. Were the author to grade himself in a college design course, these examples would rate at most a "B." The "A" students would conduct far more analysis (structures, roll rate, c.g. envelopes, etc.) and would ultimately redraw the as-optimized aircraft to ensure that the analysis assumptions were realistic.

The first example is designed and analyzed completely by hand and nonprogrammable pocket calculator to illustrate and prove that this initial design process is, in fact, fully understandable. The only exception to the by-hand rule is that a very simple computer program was used for sizing iterations, although even that could have been done by hand using the methods of Chapter 3. This simple program is available (free) at the author's Aircraft Design Web site, *http://www.aircraftdesign.com.*

The second example uses hand calculations for all prelayout activities including initial sizing and selection of T/W and W/S. For the rest of the example, the author's RDS-Student program was used. RDS-Student was written to permit students to analyze and, most important, to optimize their aircraft design in a one-semester class. A user-friendly implementation of the methods already presented is available from AIAA along with this book. Further information about

RDS-Student and the enhanced RDS-Professional can be found at the author's Aircraft Design Web site just mentioned.

The author strongly recommends that students do their initial sizing and other prelayout activities by hand as shown here before being permitted to use RDS-Student or any other "canned" design program. Then, use the computer for the laborious "number-crunching" of sizing, performance, and trade studies.

23.2 Single-Seat Aerobatic

This design represents an aircraft that the author hopes to build and fly some day. It would provide fun weekend aerobatic flying for the occasional pilot and would offer better performance than the Great Lakes Biplane but without the touchy handling qualities of the Pitts Special.

The design is fairly classical in layout but is based around the more recent techniques for quick fabrication using moldless foam–fiberglass sandwich construction. This permits rapid "garage" fabrication of a one-of-a-kind aircraft. Also, the selected engine is already set up for aerobatic flight, which should minimize the installation effort.

One interesting result of the sizing and optimization presented next is that the low-wing loading required for good aerobatic capability has strongly biased the aspect-ratio optimization, leading to a lower-than-expected optimal aspect ratio. In fact, without the addition of a maneuver-related performance constraint, the "optimal" aspect-ratio approaches zero!

(Special note to homebuilders concerning the first example—DON'T BUILD IT!!! This is a first-pass conceptual design only. It would take at least a man-year of design and analysis effort by an experienced designer before this concept could be built and safely flown.)

DR-1 SINGLE SEAT AEROBATIC HOMEBUILT

ENGINE: LYCOMING O-320-A2B (FROM CITABRIA)

150 Hp AT 2700 RPM, $C_{bhp} = .5$ (Assumed)
272 LBS DRY
30" LENGTH 32" WIDTH 23" HEIGHT

DESIGN GOALS: RAPID FABRICATION (FOAM & FIBERGLASS)

PERFORMANCE: BETWEEN PITTS S-1S AND GREAT LAKES

$V_{max} \geq 130$ KTS $V_{stall} \leq 50$ KTS

Takeoff ≤ 1000 FT over 50'

Rate of $\text{Climb}_{S.L.} \geq 1500$ fpm

RANGE ≥ 280 nm (no reserves) AT $V_{cruise} = 115$ kts

$n = +9/-6$ g's

$W_{crew} = 220$ Lb (includes parachute)
-Optional open cockpit! (Used for analysis)

HANDLING QUALITIES:

- SLIGHTLY STABLE (LIKE A FIGHTER)
- GOOD SPIN RECOVERY UPRIGHT AND INVERTED

MISSION

Cruise
280 nm
2 3
Climb
h=8,000 ft
$V_{cruise} = 115$ kts
Takeoff 0 1
Land
4

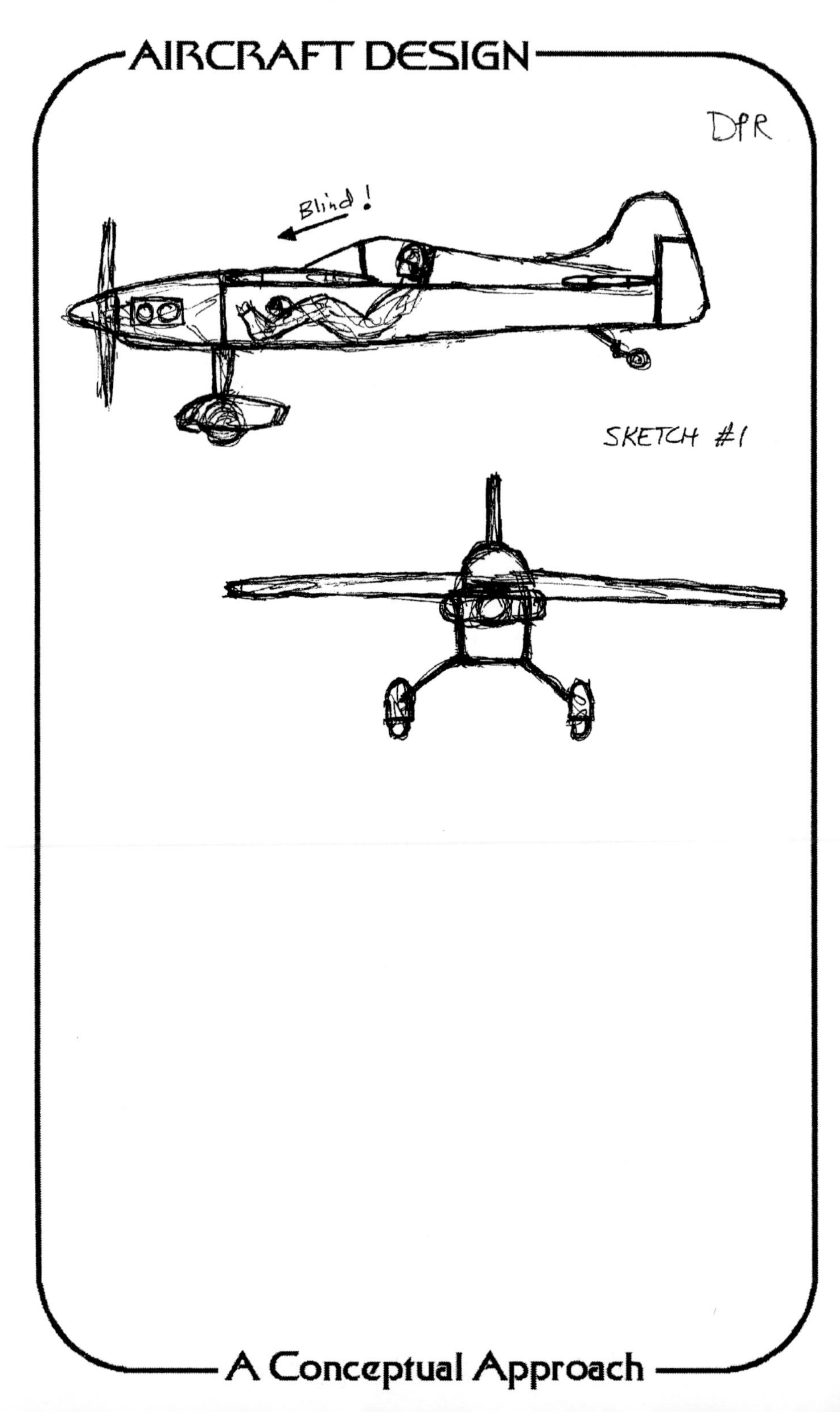
AIRCRAFT DESIGN
DPR
Blind!
SKETCH #1
A Conceptual Approach

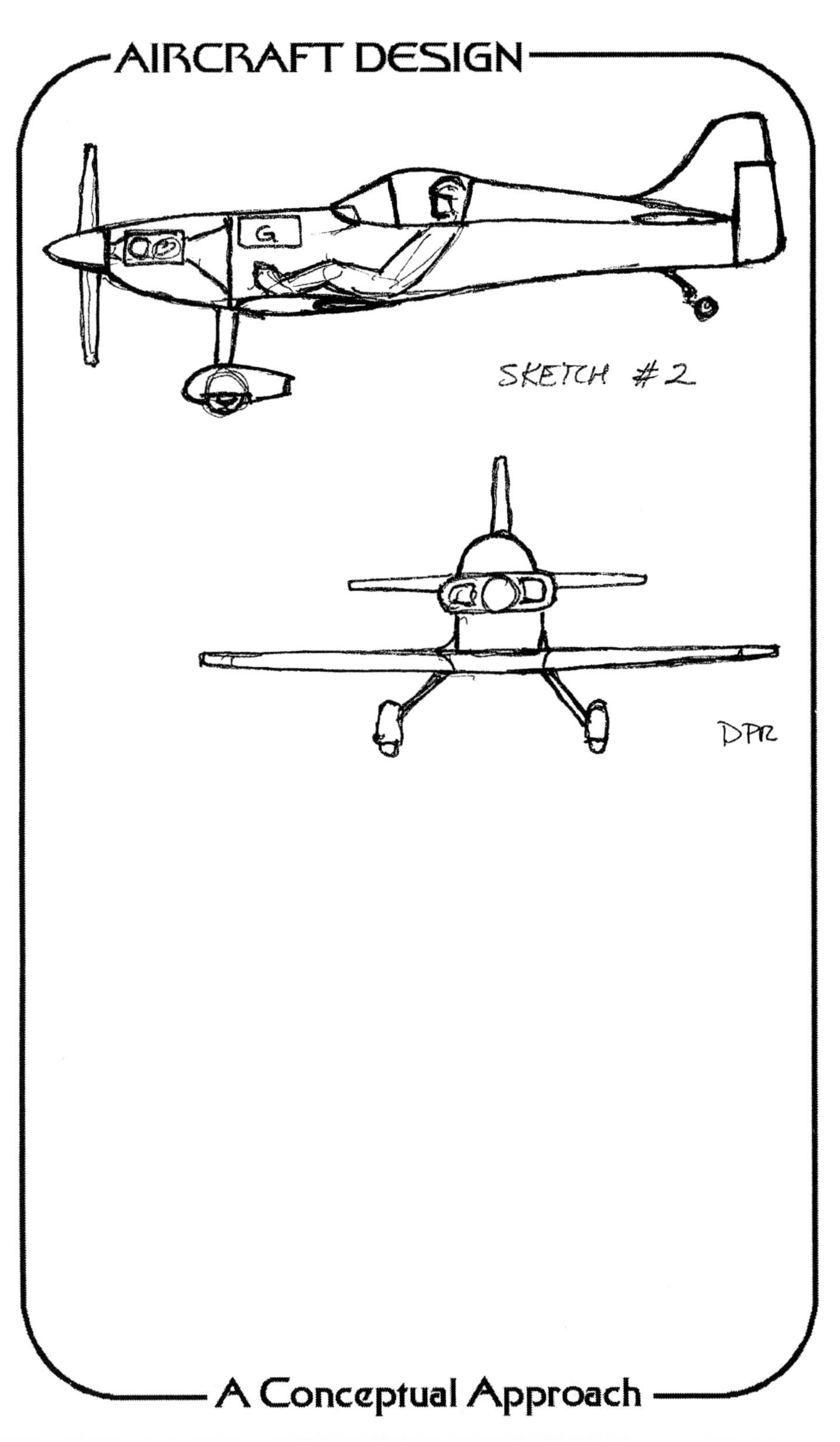
AIRCRAFT DESIGN
G
SKETCH #2
DPR
A Conceptual Approach

AIRCRAFT DESIGN

WING GEOMETRY SELECTION

(From Chapter 4 Charts & Tables)

$A=6$ $\lambda=.4$ $\Lambda_{\bar{c}/4}=0$ $\Gamma=3°$ (0° Effective)

AIRFOIL : NACA 63_2015 (tip)
NACA 63_2012 (root)
(t/c: last two digits)

Higher t/c at tip to prevent tip stall

No twist (to avoid inverted tip stall)

HORIZONTAL TAIL

$A=4$ $\lambda=.4$ NACA 0012

VERTICAL TAIL

$A=1.5$ $\lambda=.4$ NACA 0012

A Conceptual Approach

HP/W RATIO: PITTS: $W/H_P = 6.4$; GREAT LAKES: $W/H_P = 10$ } TENTATIVELY SELECT: $W/H_P = 8$

STATISTICAL (Table 5.4)

$$H_P/W = .004\,V_{max_{mph}}^{.57} = .004(130 \times 1.151)^{.57} = .069$$

or $W/H_P = 14.4$

(but statistics are for cruising aircraft, not aerobatics!)

SO INITIALLY SELECT

★ $W/H_P = 8$ and $W_0 = 8 \times 150 = 1200$ Lb

WING LOADING

HISTORICAL: PITTS: $W/S = 11.7$
GREAT LAKES: $W/S = 9.6$
STEVENS ACRO: $W/S = 13.0$

STALL: No flaps, so $C_{L_{max}} \cong .9\, C_{\ell_{max}} \cong 1.2$

eq 5.6) $W/S \le \frac{1}{2}(.00238)(50 \times 1.689)^2(1.2) = 10.2 \text{ Lb/ft}^2$

TAKEOFF: 1,000 Ft. TAKEOFF $\Rightarrow$ TAKEOFF PARAMETER $= 120$

$$C_{L_{takeoff}} = C_{L_{max}}\left(\frac{V_{stall}}{V_{takeoff}}\right)^2 = 1.2\left(\frac{1}{1.1}\right)^2 = .99$$

eq 5.8)
$$T.O.P. = 120 = \frac{W/S}{(1)(.99)(1/8)} \; ; \; W/S \le 14.9$$

CLIMB: Assume climb speed $= 70$ kts, so $q = 16.6 \text{ Lb/ft}^2$

$$G = \frac{V_v}{V} = \frac{1500/60}{70 \times 1.689} = .212$$

$$T/W = \frac{550 \times 0.8}{70 \times 1.689}\left(\frac{1}{8}\right) = .465$$

Assume $e = .8$

From sketch, $\frac{S_{wet}}{S_{ref}} \cong 3.5$ so $C_{D_0} \cong 3.5 \times .0055 = .02$ (eq 12.23)

eq 5.30)
$$W/S \le \left[(.465 - .212) + \sqrt{(.465 - .212)^2 - \frac{4(.02)}{\pi \times 6 \times .8}}\right] \Big/ \left[\frac{2}{16.6\pi \times 6 \times .8}\right] = 62$$

CRUISE: $q_{cruise} = 35 \text{ Lb/ft}^2$

$$W/S = 35\sqrt{\pi \times 6 \times .8 \times .02} = 20$$

☆ SELECT LOWEST: $W/S = 10.2$; $S = \frac{1200}{10.2} = 118 \text{ FT}^2$

INITIAL SIZING

EMPTY WEIGHT FRACTION:

Table 6.2) $\frac{W_e}{W_0} = .59\, W_0^{-.1} (6)^{.05} (1/8)^{.1} (10.2)^{-.05} (130)^{.17} = 1.093\, W_0^{-.1}$

But this is for cruising, not aerobatic aircraft. We adjust the equation using the Stevens Acro:

Stevens Acro: $W_0 = 1300$ $W_e = 950$ $W_e/W_0 = .73$

Our equation gives: $\frac{W_e}{W_0} = 1.093(1300)^{-.1} = .533$ (too low!)

Use a fudge-factor to adjust the equation:

$$\frac{W_e}{W_0} = \left(\frac{.73}{.533}\right) 1.093\, W_0^{-.1} = 1.495\, W_0^{-.1}$$

At $W_0 = 1200$ Lb, $W_e = 883$ Lb

MISSION SEGMENT WEIGHT FRACTIONS

Table 3.2) $\frac{W_1}{W_0} = .97$ $\frac{W_2}{W_1} = .985$

CRUISE: $W/S = (10.2)(.97)(.985) = 9.7$; $q = 35$ Lb/ft²

eq 6.13) $$L/D = \frac{1}{\frac{35 \times .02}{9.7} + \frac{9.7}{35 \times \pi \times 6 \times 0.8}} = 11.04$$

eq 6.12) $$\frac{W_3}{W_2} = e^{\frac{-(280 \times 6076)(.5/3600)}{11.04 \times 550 \times 0.8}} = .953$$

Table 3.2) $\frac{W_4}{W_3} = .995$

$$\frac{W_4}{W_0} = .97 \times .985 \times .953 \times .995 = .906$$

eq 3.11) $\frac{W_f}{W_0} = 1.06(1 - .906) = .0997$

$W_f = 1200 \times .0997 = 120$ Lb (total)

$W_{f\,usable} = 120/1.06 = 112$ Lb

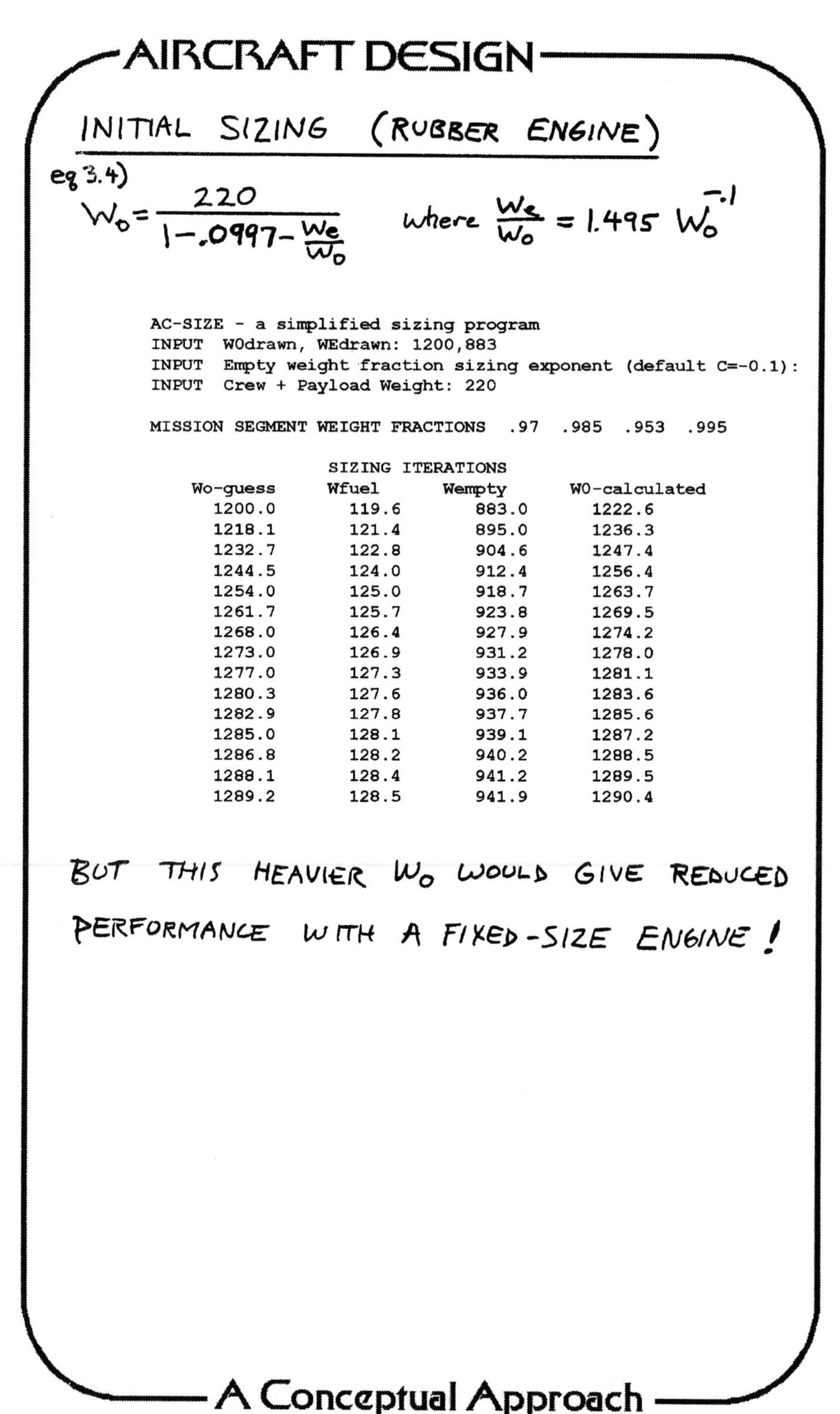
AIRCRAFT DESIGN
INITIAL SIZING (RUBBER ENGINE)
eq 3.4)
$W_0 = \frac{220}{1 - .0997 - \frac{W_e}{W_0}}$ where $\frac{W_e}{W_0} = 1.495\, W_0^{-.1}$
AC-SIZE - a simplified sizing program
INPUT WOdrawn, WEdrawn: 1200,883
INPUT Empty weight fraction sizing exponent (default C=-0.1):
INPUT Crew + Payload Weight: 220
MISSION SEGMENT WEIGHT FRACTIONS .97 .985 .953 .995
SIZING ITERATIONS
Wo-guess Wfuel Wempty W0-calculated
1200.0 119.6 883.0 1222.6
1218.1 121.4 895.0 1236.3
1232.7 122.8 904.6 1247.4
1244.5 124.0 912.4 1256.4
1254.0 125.0 918.7 1263.7
1261.7 125.7 923.8 1269.5
1268.0 126.4 927.9 1274.2
1273.0 126.9 931.2 1278.0
1277.0 127.3 933.9 1281.1
1280.3 127.6 936.0 1283.6
1282.9 127.8 937.7 1285.6
1285.0 128.1 939.1 1287.2
1286.8 128.2 940.2 1288.5
1288.1 128.4 941.2 1289.5
1289.2 128.5 941.9 1290.4
BUT THIS HEAVIER W_0 WOULD GIVE REDUCED PERFORMANCE WITH A FIXED-SIZE ENGINE!
A Conceptual Approach

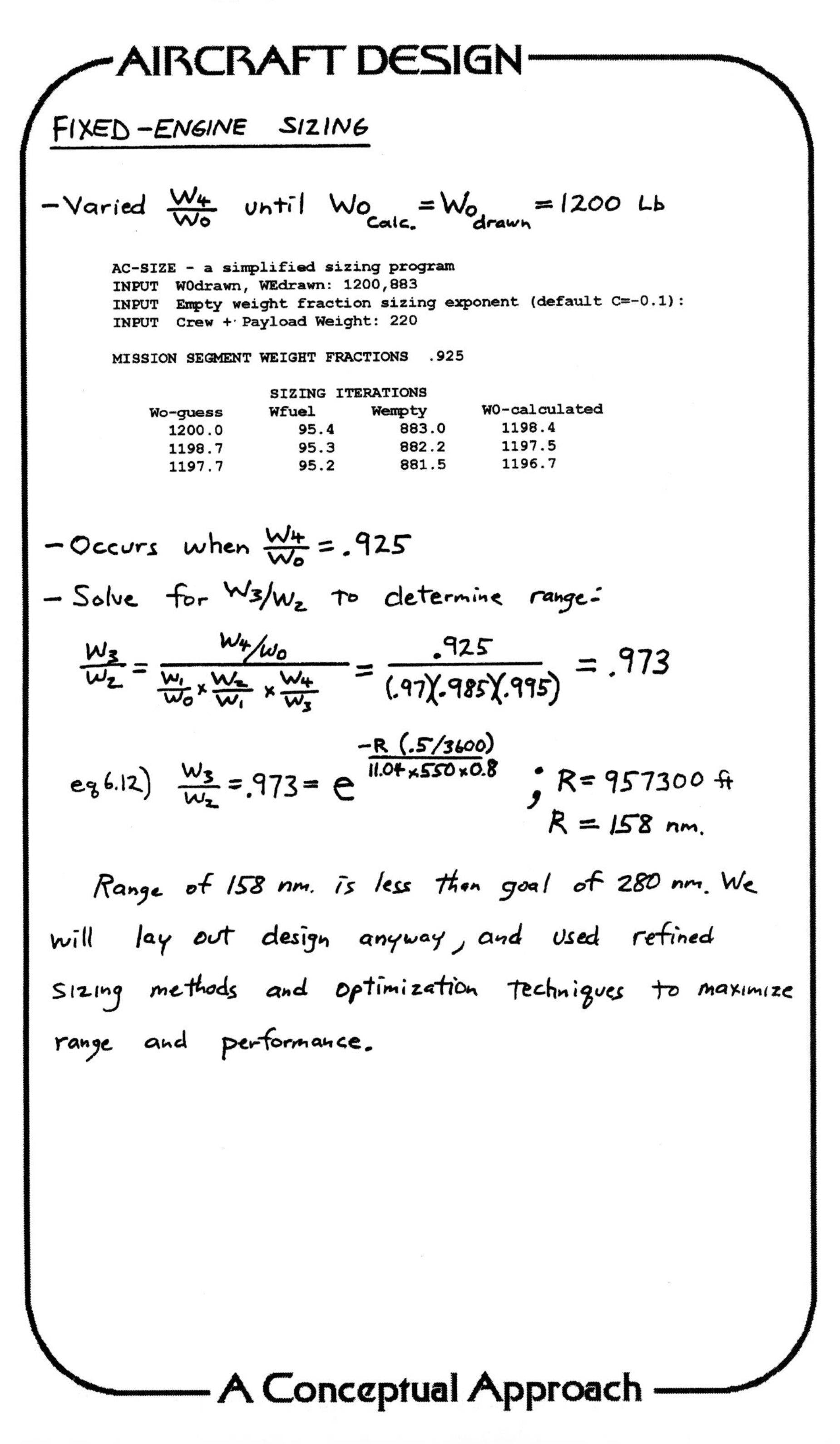

AIRCRAFT DESIGN

FIXED-ENGINE SIZING

– Varied $\frac{W_4}{W_0}$ until $W_{0_{calc.}} = W_{0_{drawn}} = 1200$ Lb

```
AC-SIZE - a simplified sizing program
INPUT  W0drawn, WEdrawn: 1200,883
INPUT  Empty weight fraction sizing exponent (default C=-0.1):
INPUT  Crew + Payload Weight: 220

MISSION SEGMENT WEIGHT FRACTIONS  .925
```

SIZING ITERATIONS

Wo-guess	Wfuel	Wempty	W0-calculated
1200.0	95.4	883.0	1198.4
1198.7	95.3	882.2	1197.5
1197.7	95.2	881.5	1196.7

– Occurs when $\frac{W_4}{W_0} = .925$

– Solve for W_3/W_2 to determine range:

$$\frac{W_3}{W_2} = \frac{W_4/W_0}{\frac{W_1}{W_0} \times \frac{W_2}{W_1} \times \frac{W_4}{W_3}} = \frac{.925}{(.97)(.985)(.995)} = .973$$

$$\text{eq 6.12)} \quad \frac{W_3}{W_2} = .973 = e^{\frac{-R\,(.5/3600)}{11.04 \times 550 \times 0.8}} \; ; \; R = 957300 \text{ ft}$$

$$R = 158 \text{ nm.}$$

Range of 158 nm. is less than goal of 280 nm. We will lay out design anyway, and used refined sizing methods and optimization techniques to maximize range and performance.

A Conceptual Approach

LAYOUT DATA

$W_o = 1200$ Lb $W_e = 883$ Lb $W_f = 120$ Lb (113 Usable)

WING: $S = 118$ Ft^2 $A = 6$ $\lambda = .4$

eq. 7.5) $b = 26.6$ ft $= 319.3$ in
eq. 7.6) $C_r = 75.6$ in
eq 7.7) $C_t = 30.4$ in
eq 7.8) $\bar{C} = 56$ in
eq 7.9) $\bar{Y} = 68.4$ in

FUSELAGE   table 6.3) $L \cong 3.5(1200)^{.23} = 18$ ft
$L_{tail\ arm} \cong 60\%\ L = 10.8$ ft

VERTICAL TAIL   eq 6.26) $C_{VT} \cong .04 = \frac{10.8\ S_{VT}}{26.6 \times 118}$; $S_{VT} = 11.6$ ft^2

$A = 1.5$; $\lambda = 0.4$ so $b = 4.1$ ft $C_r = 4.0$ ft $C_t = 1.6$ ft

HORIZONTAL TAIL   eq 6.27) $C_{HT} \cong .5 = \frac{10.8\ S_{HT}}{4.7 \times 118}$; $S_{HT} = 25.5$ ft^2

$A = 4.0$; $\lambda = 0.4$ so $b = 10.1$ ft $C_r = 3.6$ ft $C_t = 1.4$ ft

FUEL TANK $W_f = 120$ Lb $= 20$ gallons $= 2.7$ ft^3

TIRE SIZE   table 11.1) $D = 11$ in.; $W = 5.3$ in.

PROP DIAMETER   eq 10.23) $d = 22\sqrt[4]{150} = 77$ in (?)

Check tip speed: $V = 115$ kts $= 194$ ft/sec
$n = 2700$ rpm $= 45$ rev/sec

eq. 10.21 and 10.22) $\sqrt{194^2 + (\pi n d)^2} \leq 850$ ft/sec (Wood Prop.)

so $d \leq 5.85$ ft $= 70$ in

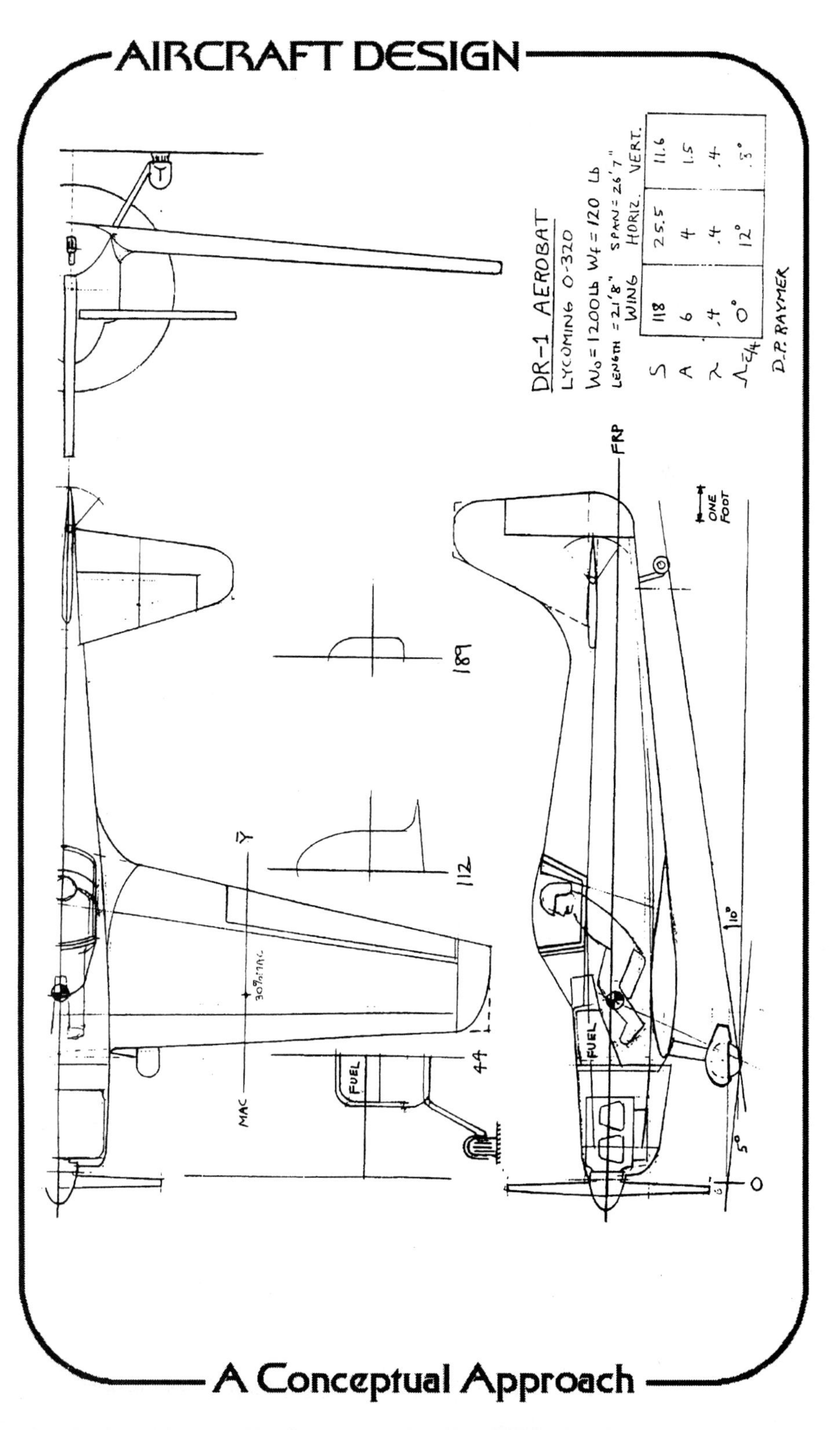
AIRCRAFT DESIGN
DR-1 AEROBAT
LYCOMING O-320
W0 = 1200 Lb Wf = 120 Lb
LENGTH = 21'8" SPAN = 26'7"
WING HORIZ. VERT.
S 118 25.5 11.6
A 6 4 1.5
λ .4 .4 .4
Λc/4 0° 12° 3°
D.P. RAYMER
FRP
ONE FOOT
189
112
Y
30% MAC
MAC
FUEL
44
FUEL
10°
5°
0
A Conceptual Approach

AERODYNAMICS (Wetted and Exposed areas from drawing)

MAXIMUM LIFT: $C_{L_{max}} \cong .9\, C_{\ell_{max}} = (1.35)(0.9) = 1.2$ (eq 12.15)

LIFT CURVE SLOPE:

eq 12.6) $$C_{L_\alpha} = \frac{2\pi \times 6 \times \frac{S_{exp}}{S_{ref}} F}{2 + \sqrt{4 + \frac{6^2}{.95^2}\left(1 + \frac{\tan^2 0}{1}\right)}} = 4.37 \times .837 \times 1.33$$

$C_{L_\alpha} = 4.85$ per radian $= .085$ per degree

PARASITE DRAG (Assuming fully turbulent flow)

Use $V = 100$ kts $= 169$ ft/s, $h =$ sea level to determine friction C_f
so $M = .15$; $\mu = 0.37 \times 10^{-6}$

FUSELAGE $\ell = 22$ ft $\ell/d \cong 6.38$

eq 12.26) Reynolds #: $R = 23{,}916{,}000$

eq 12.28) $R_{cutoff} = 84{,}158{,}000$ (using smooth paint)

eq 12.27) $$C_f = \frac{.455}{(\log_{10} 23{,}916{,}000)^{2.58}\left(1 + .144(.15)^2\right)^{.65}} = .0026$$

eq 12.31) $$FF = 1 + \frac{60}{6.38^3} + \frac{6.38}{400} = 1.25$$

$S_{wet} = 164$ ft^2

$$C_{D_{0\,fuselage}} = .0026 \times 1.25 \times 164/118 = .0046$$

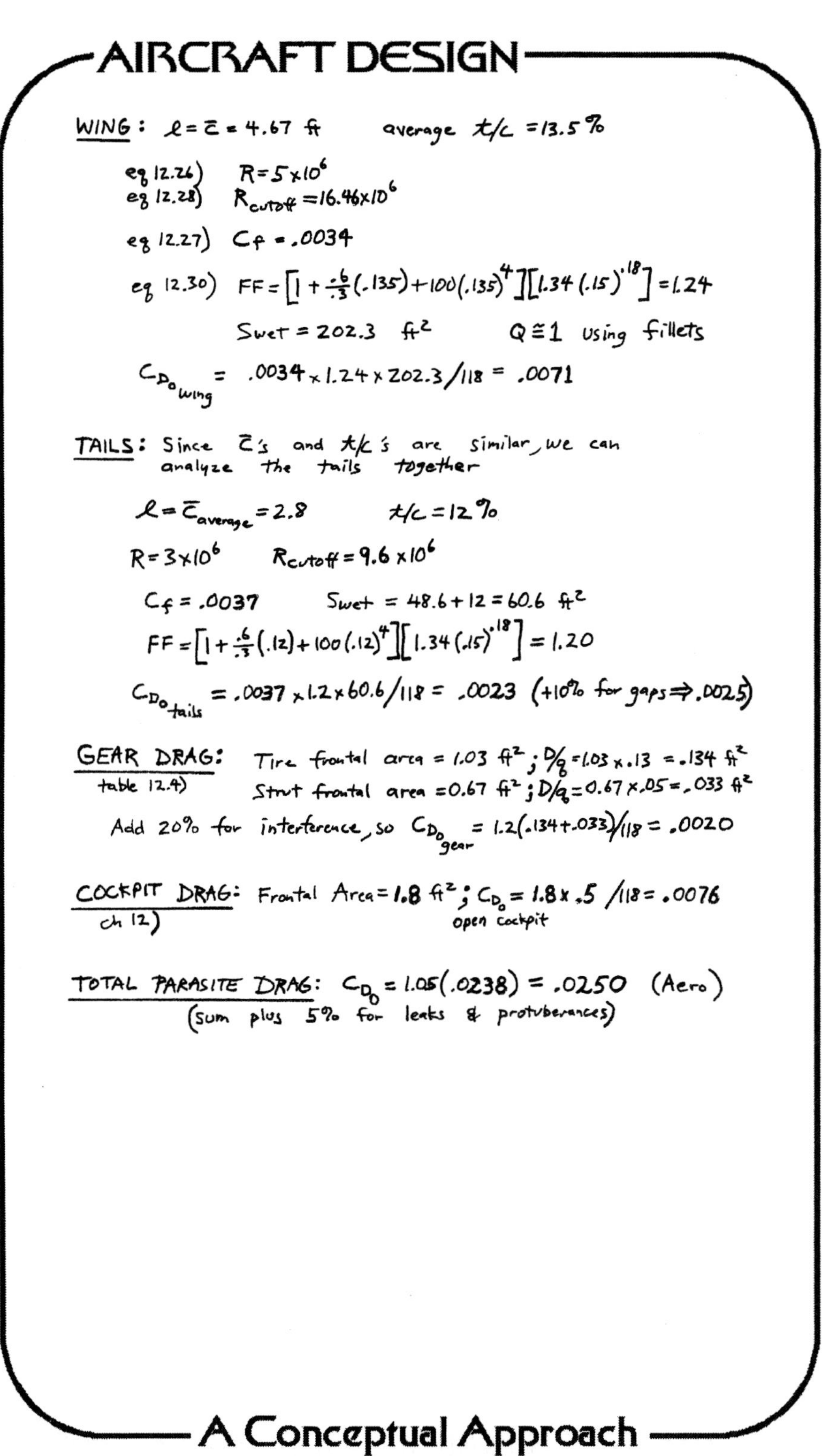

AIRCRAFT DESIGN

WING: $\ell = \bar{c} = 4.67$ ft average $t/c = 13.5\%$

eq 12.26) $R = 5 \times 10^6$
eq 12.28) $R_{cutoff} = 16.46 \times 10^6$

eq 12.27) $C_f = .0034$

eq 12.30) $FF = \left[1 + \frac{.6}{.3}(.135) + 100(.135)^4\right]\left[1.34(.15)^{.18}\right] = 1.24$

$S_{wet} = 202.3$ ft^2 $Q \cong 1$ using fillets

$C_{D_{0_{wing}}} = .0034 \times 1.24 \times 202.3/118 = .0071$

TAILS: Since $\bar{c}$'s and t/c's are similar, we can analyze the tails together

$\ell = \bar{c}_{average} = 2.8$ $t/c = 12\%$

$R = 3 \times 10^6$ $R_{cutoff} = 9.6 \times 10^6$

$C_f = .0037$ $S_{wet} = 48.6 + 12 = 60.6$ ft^2

$FF = \left[1 + \frac{.6}{.3}(.12) + 100(.12)^4\right]\left[1.34(.15)^{.18}\right] = 1.20$

$C_{D_{0_{tails}}} = .0037 \times 1.2 \times 60.6/118 = .0023$ (+10% for gaps ⇒ .0025)

GEAR DRAG: Tire frontal area = 1.03 ft^2; $D/q = 1.03 \times .13 = .134$ ft^2
table 12.4) Strut frontal area = 0.67 ft^2; $D/q = 0.67 \times .05 = .033$ ft^2

Add 20% for interference, so $C_{D_{0_{gear}}} = 1.2(.134 + .033)/118 = .0020$

COCKPIT DRAG: Frontal Area = 1.8 ft^2; $C_{D_0} = 1.8 \times .5/118 = .0076$
ch 12) open cockpit

TOTAL PARASITE DRAG: $C_{D_0} = 1.05(.0238) = .0250$ (Aero)
(sum plus 5% for leaks & protuberances)

A Conceptual Approach

COOLING DRAG:

eq 13.21) $D/q_{cooling} = (4.9 \times 10^{-7}) \frac{150(519)^2}{V} = \frac{19.8}{V}$ ft²

This is for good cowling; triple for light aircraft!

so $C_{D_{0_{cooling}}} \cong .0024$ at V=115 kts (small, so ignore change at other speeds)

MISC. ENGINE DRAG:

eq 13.22) $D/q_{misc} = (2 \times 10^{-4}) 150 = .03$ ft²

$C_{D_{0_{misc}}} = .03/118 = .0003$

TOTAL PARASITE AND ENGINE DRAG:

$$C_{D_0} = .0250 + .0024 + .0003 = .0277$$

INDUCED DRAG:

eq. 12.49) $e = 1.78\left[1 - 0.045(6)^{.68}\right] - 0.64 = 0.87$

$$K = \frac{1}{\pi(6)(.87)} = 0.061$$

PROPULSION

- Custom-design wooden 2-bladed, fixed-pitch propeller
  D = 70 in

ON-DESIGN:

$$J = \frac{V}{nD} = \frac{115 \times 1.689}{(2700/60)(70/12)} = .74$$

$$C_P = \frac{550(150)}{\rho n^3 D^5} = 0.06$$

From fig. 13.12: $\eta_p = .84$ $\theta_{.75} = 20°$

- But wooden propeller reduces η_p by 10%, while 2-bladed is about 3% better than 3-bladed data provided by figure 13.12

Thus; $\eta_p = 0.9 \times 1.03 \times (.84) = .78$ (on-design at 115 kts)

OFF-DESIGN: Use figure 13.13 to adjust η_p (see plot)

STATIC THRUST: From fig. 13.11; $C_T/C_P = 2.5$

eq 13.17) $T_{Static} = 2.5 \dfrac{550(150)}{nD} = 786$ Lb (assuming 3-bladed)

- 2-bladed propeller has 5% less static thrust, thus:

$$T_{static} = .95 \times 786 = 747 \text{ Lb}$$

- But this is for a variable pitch propeller, which changes to a flat blade angle at static conditions.
- Instead, assume the static thrust equals the highest forward thrust value found above (see plot)

(Crude assumption - better to use fixed-pitch data!)

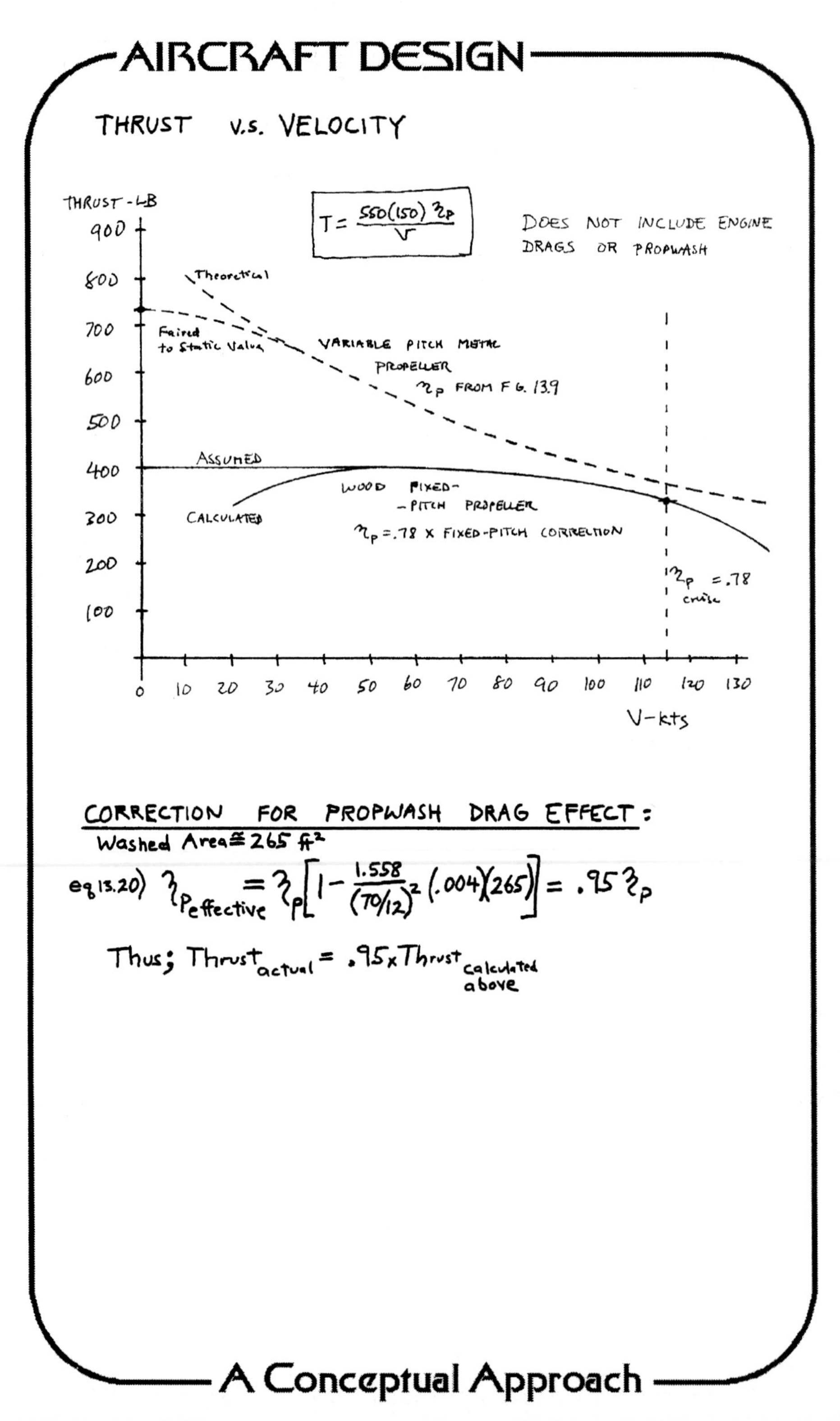
AIRCRAFT DESIGN
THRUST v.s. VELOCITY
THRUST - LB
T = 550(150) ηp / V
DOES NOT INCLUDE ENGINE DRAGS OR PROPWASH
Theoretical
Faired to Static Value
VARIABLE PITCH METAL PROPELLER
ηp FROM FIG. 13.9
ASSUMED
CALCULATED
WOOD FIXED-PITCH PROPELLER
ηp = .78 X FIXED-PITCH CORRECTION
ηp cruise = .78
V - kts
900
800
700
600
500
400
300
200
100
0
10
20
30
40
50
60
70
80
90
100
110
120
130
CORRECTION FOR PROPWASH DRAG EFFECT:
Washed Area ≅ 265 ft²
eq 13.20) ηp effective = ηp [1 − 1.558/(70/12)² (.004)(265)] = .95 ηp
Thus; Thrust actual = .95 x Thrust calculated above
A Conceptual Approach

WEIGHTS

eq 15.46) $W_{wing} = .036(118)^{.758}\left(\frac{6}{\cos^2 0}\right)^{.6}(45)^{.006}(.4)^{.04}\left(\frac{100(.135)}{\cos 0}\right)^{-.3}(9 \times 1200)^{.49} = 168$ Lb

eq 15.47) $W_{ht} = .016(9 \times 1200)^{.414}(45)^{.168}(25.5)^{.896}\left(\frac{100(.12)}{\cos 10}\right)^{-.12}\left(\frac{4}{\cos^2 10}\right)^{.043}(.4)^{-.02} = 20$ Lb

eq 15.48) $W_{vt} = .073(1+.2(0))(9 \times 1200)^{.376}(45)^{.122}(11.6)^{.873}\left(\frac{100(.12)}{\cos 15}\right)^{-.49}\left(\frac{1.5}{\cos^2 15}\right)^{.357}$

$\times (.4)^{.039} = 11$ Lb

eq 15.49) $W_{fus} = .052(164)^{1.086}(9 \times 1200)^{.177}(140)^{-.051}(190/25)^{-.072}(45)^{.241} = 115$ Lb

eq 15.50) $W_{main\ gear} = .095(3 \times 1200)^{.768}(25/12)^{.409} = 69$ Lb

eq 15.52) $W_{installed\ engine} = 2.575(272)^{.922} = 452$ Lb — This seems too high; table 15.2 gives $W_{inst.\ eng} = 1.3(272) = 380$ Lb

eq 15.53) $W_{fuel\ system} = 2.49(20)^{.726}\left(\frac{1}{1+0/1}\right)^{.363} = 22$ Lb

eq 15.54) $W_{flight\ controls} = .053\left(\frac{190}{12}\right)^{1.536}(26.6)^{.371}(9 \times 1200 \times 10^{-4})^{.8} = 13$ Lb

eq 15.56) $W_{electrical} = 12.57(22 + W_{av})^{.51} = 73$ Lb
↳ from next equation

eq 15.57) $W_{av} = 2.117\, W_{uav}^{.933} = 2.117(5)^{.933} = 9.5$ Lb
↳ From data in Ref. 11

$W_{furnishings} \cong 20$ Lb (From data in Ref. 11)

Sharp-eyed readers have found two minor errors here—the limit load factor (9 g) was used rather than the ultimate load factor ($9 \times 1.5 = 13.5\ g$); also the Table 15.2 engine weight factor for fighters and transports (1.3) was used rather than the (1.4) value for general-aviation aircraft. The total increase is 78 lb, which is about the same as the extra weight margins added in the following two pages, and so the total is unchanged, but the margin has been used up already.

AIRCRAFT DESIGN

WEIGHTS BY OTHER METHODS

CESSNA METHODS (Ref. 11)

$$W_{wing} = .047\, W_0^{.397}\, S^{.36}\, n^{.397}\, A^{1.712} = 225 \text{ Lb}$$

$$W_{ht} = .055\, W_0^{.887}\, S_h^{.101}\, A_h^{.138}\, t_{root}^{-.223} = 60 \text{ Lb}$$

$$W_{vt} = .108\, W_0^{.567}\, S_v^{.125}\, A_v^{.482}\, t_{root}^{-.747} (\cos \Lambda_{c/4})^{-.882} = 17.7 \text{ Lb}$$

Ref 10 METHOD

$$W_{fus \text{ w/o nacelle}} = 200 \left[\left(\frac{W_0 n}{10^5}\right)^{.286} \left(\frac{L}{10}\right)^{.857} \left(\frac{W+D}{10}\right) \left(\frac{V_e}{100}\right)^{.338} \right]^{1.1} = 114 \text{ Lb}$$

$$W_{nacelle} = 2.5 \sqrt{H_p} = 31 \text{ Lb}$$

COMPARISON TO ACTUAL DATA (Ref. 11)

$$W_{electrical} \cong 40 \text{ Lb}$$

$$W_{gear} = \left(\frac{W_{gear}}{W_0}\right) W_0 \cong .054(1200) = 64 \text{ Lb}$$

→ Average of C-180 and L-19A values

A Conceptual Approach

AIRCRAFT DESIGN

WEIGHTS ADJUSTMENTS & BALANCE

Sandwich construction, foam-and-fiberglass homebuilts have reduced weight due to design differences, not due to composite construction. Nonetheless, we will use the factors in table 15.4 to estimate the weight savings for each component.

COMPONENT	FUDGE FACTOR	ADJUSTED WEIGHT: Ch15/Other Methods	SELECTED WEIGHT	DISTANCE TO DATUM *
FUSELAGE	.90	104 / 128	130 Lb	115 in
WING	.85	143 / 175	150	70
HOR. TAIL	.83	17 / 45	40	210
VERT. TAIL	.83	9 / 13	15	225
ENGINE	—	452 / 380	380	16
GEAR	.95	66 / 57	60	45
FUEL SYS.	—	22	22	50
FL. CONTROLS	—	13	15	80
ELECTRICAL	—	73 / 40	40	40
AVIONICS	—	9.5	10	60
FURNISHINGS	—	20	20	100

ΣW_e = 882 Lb @ 59.5 in

PILOT & CHUTE	220	85
FUEL (Available, if W_0 = 1200 Lb)	98	50

ΣW_0 = 1200 Lb @ 63.3 in

MOST-AFT C.G IS NO-FUEL: $W_{e+PILOT}$ = 1102 Lb @ 64.5 in

* Measured from back of spinner - see drawing

A Conceptual Approach

STABILITY & CONTROL

$\bar{c} = 56$ in

MOST-AFT C.G. IS AT 64.5 in

$\bar{X}_{cg} = \frac{64.5}{56} = 1.15$

WING AERO. CENTER AT 62 in so $\bar{X}_{acw} = \frac{62}{56} = 1.107$

eq 12.6) $C_{L_\alpha} = 4.85$ per radian

$C_{m_w} = 0$ (since airfoil is symmetric)

FUSELAGE

eq 16.24) $C_{m_{\alpha_{fus}}} = \frac{(.008)(35)^2(234)}{(56)(118 \times 12^2)} = .002$ per deg.

$= .12$ per rad.

TAIL AERO. CENTER AT 206 in so $\bar{X}_{ach} = \frac{206}{56} = 3.68$

eq 12.6) $C_{L_{\alpha_h}} = 3.77$ per radian

DOWNWASH fig 16.12)

$r = \frac{144}{319.3/2} = .9$

$m = \frac{28}{319.3/2} = .2$

$\frac{d\epsilon}{d\alpha} = .38$

eq 16.22) $\frac{d\alpha_h}{d\alpha} = 1 - .38 = .62$

Assume: $\eta_h = q_h/q = .9$ (eq 16.6)

POWER-OFF NEUTRAL POINT (STICK-FIXED)

eq 16.9) $$\bar{X}_{np} = \frac{(4.85)(1.107) - .12 + .9\left(\frac{25.5}{118}\right)(3.77)(.62)(3.68)}{4.85 + .9\left(\frac{25.5}{118}\right)(3.77)(.62)} = 1.27$$

So $X_{np} = 1.27 \times 56 = 71.2$ in. ; which is at 41 % of $\bar{C}_{wing}$

$$\text{STATIC MARGIN} = \frac{71.2 - 64.5}{56} = .12 \quad \text{(ie, 12\% STABLE)}$$

eq 16.10) $$C_{m_\alpha} = -4.85\left(\frac{71.2 - 64.5}{56}\right) = -.58 \quad \text{(very stable)}$$

STICK-FREE (Ch. 16.5) Elevator area ≅ 40% of Tail area,

so $C_{L_{\alpha_h}} \cong .8\ C_{L_{\alpha_h}}$ (fixed)

eq 16.9) $$\bar{X}_{np} = \frac{(4.85)(1.107) - .12 + .8\left[.9\left(\frac{25.5}{118}\right)(3.77)(.62)(3.68)\right]}{4.85 + .8\left[.9\frac{(25.5)}{118}(3.77)(.62)\right]} = 1.26$$

$X_{np} = 1.26 \times 56 = 70.7$ in

$$\text{STATIC MARGIN} = \frac{70.7 - 64.5}{56} = .111 \quad \text{(11.1\% STABLE)}$$

eq 6.10) $C_{m_\alpha} = -4.85\left(\frac{70.7 - 64.5}{56}\right) = -.54$ Nice for weekend pilots, may be too stable for competition.

AIRCRAFT DESIGN

TRIM ANALYSIS

MOMENTS:

eq 16.7) $C_{m_{cg}} = 0 = 4.85\alpha(1.15-1.107) + 0 + 0 + 0.12\alpha - \left[.9\left(\frac{25.5}{118}\right) C_{L_h} (3.68-1.15)\right]$

or $C_{m_{cg}} = 0 = .329\alpha - .492\, C_{L_h}$

eq 16.31) $C_{L_h} = C_{L_{\alpha_h}} \left[.62\alpha + (0-0) - \Delta\alpha_{OL}\right]$

→ due to flap (ie, elevator)

eq 16.15 & 16.16) $\Delta\alpha_{OL} = \frac{.9}{(.95)2\pi}(5.3)(1)(1)\, K_f \delta_e = -.8\, \delta_e$ (small δ_e)

so $C_{m_{cg}} = .329\alpha - .492\,[3.77][.62\alpha + .8\,\delta_e]$

or $C_{m_{cg}} = -0.821\alpha - 1.48\,\delta_e$

LIFT:

eq 16.31 & 16.32) $C_{L_{total}} = 4.85\alpha + .9\left(\frac{25.5}{118}\right)(3.77)[.62\alpha + .8\,\delta_e]$

(Note: α and δ_e in radians !)

A Conceptual Approach

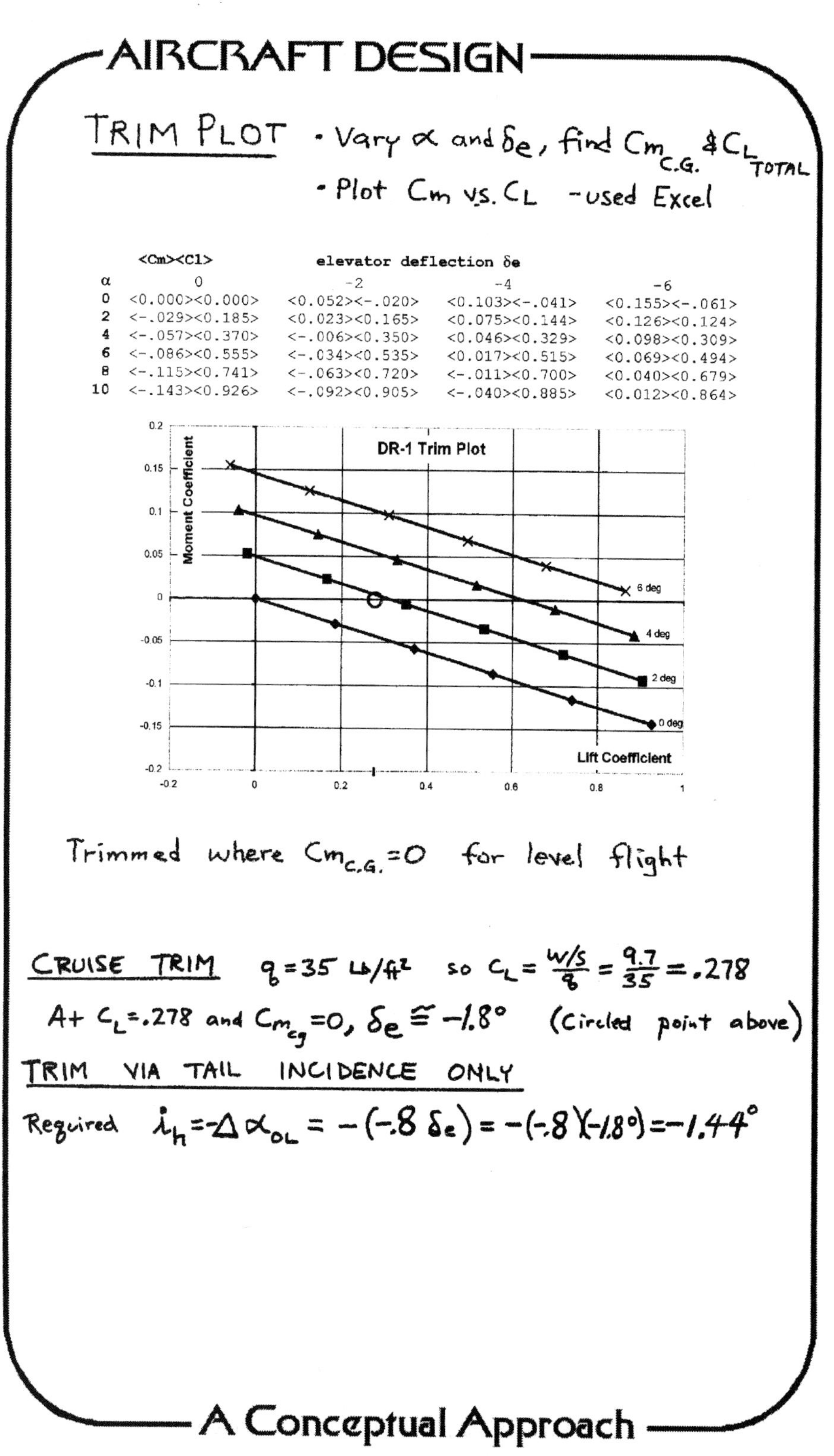

AIRCRAFT DESIGN

TRIM PLOT
- Vary α and δ_e, find $Cm_{C.G.}$ & $C_{L_{TOTAL}}$
- Plot Cm vs. C_L – used Excel

<Cm><Cl> elevator deflection δe

α	0	-2	-4	-6
0	<0.000><0.000>	<0.052><-.020>	<0.103><-.041>	<0.155><-.061>
2	<-.029><0.185>	<0.023><0.165>	<0.075><0.144>	<0.126><0.124>
4	<-.057><0.370>	<-.006><0.350>	<0.046><0.329>	<0.098><0.309>
6	<-.086><0.555>	<-.034><0.535>	<0.017><0.515>	<0.069><0.494>
8	<-.115><0.741>	<-.063><0.720>	<-.011><0.700>	<0.040><0.679>
10	<-.143><0.926>	<-.092><0.905>	<-.040><0.885>	<0.012><0.864>

Trimmed where $Cm_{C.G.} = 0$ for level flight

CRUISE TRIM $\quad q = 35$ Lb/ft² so $C_L = \frac{W/S}{q} = \frac{9.7}{35} = .278$

At $C_L = .278$ and $Cm_{cg} = 0$, $\delta_e \cong -1.8°$ (Circled point above)

TRIM VIA TAIL INCIDENCE ONLY

Required $i_h = -\Delta\alpha_{0L} = -(-.8\,\delta_e) = -(-.8)(-1.8°) = -1.44°$

A Conceptual Approach

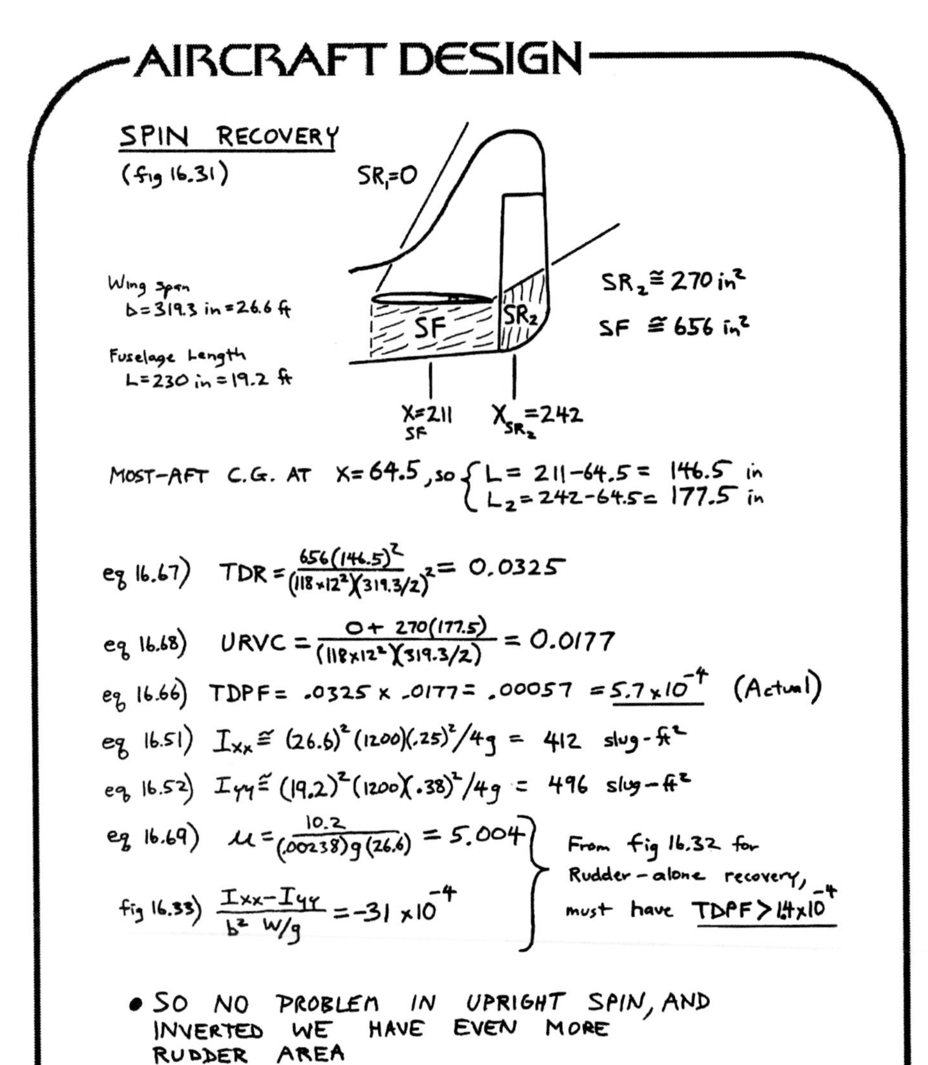

AIRCRAFT DESIGN

SPIN RECOVERY
(fig 16.31)

$SR_1 = 0$

Wing span
$b = 319.3$ in $= 26.6$ ft

Fuselage Length
$L = 230$ in $= 19.2$ ft

SF SR₂

$SR_2 \cong 270\ \text{in}^2$

$SF \cong 656\ \text{in}^2$

$X_{SF} = 211$ $X_{SR_2} = 242$

MOST-AFT C.G. AT $X = 64.5$, so $\begin{cases} L = 211 - 64.5 = 146.5 \text{ in} \\ L_2 = 242 - 64.5 = 177.5 \text{ in} \end{cases}$

eq 16.67) $TDR = \dfrac{656(146.5)^2}{(118 \times 12^2)(319.3/2)^2} = 0.0325$

eq 16.68) $URVC = \dfrac{0 + 270(177.5)}{(118 \times 12^2)(319.3/2)} = 0.0177$

eq 16.66) $TDPF = .0325 \times .0177 = .00057 = 5.7 \times 10^{-4}$ (Actual)

eq 16.51) $I_{xx} \cong (26.6)^2(1200)(.25)^2/4g = 412$ slug-ft²

eq 16.52) $I_{yy} \cong (19.2)^2(1200)(.38)^2/4g = 496$ slug-ft²

eq 16.69) $\mu = \dfrac{10.2}{(.00238)g(26.6)} = 5.004$

fig 16.33) $\dfrac{I_{xx} - I_{yy}}{b^2\ W/g} = -31 \times 10^{-4}$

From fig 16.32 for Rudder-alone recovery, must have $TDPF > 1.4 \times 10^{-4}$

- SO NO PROBLEM IN UPRIGHT SPIN, AND INVERTED WE HAVE EVEN MORE RUDDER AREA

AIRCRAFT DESIGN

AS-DRAWN PERFORMANCE

STALL: $V_{Stall} = \sqrt{\frac{2\,W/S}{\rho C_{L_{max}}}} = 84.5$ ft/sec $= 50$ kts

TAKEOFF:
from eq 5.8) $TOP = \frac{10.2}{(.99)(1/8)} = 82.4$ Fig 5.4) $S_{TO} \cong 900$ ft

V_{max} and RATE OF CLIMB: Use a graph

HORSEPOWER (AND THRUST) ADJUSTMENT FOR ALTITUDE:

eq 13.10) $bhp = bhp_{SL}\left(\rho/\rho_0 - \frac{1-\rho/\rho_0}{7.55}\right) = 0.76\ bhp_{SL}$ (h=8000 ft)

EQUATIONS: $C_L = \frac{W/S}{q} = \frac{1\ .2}{q}$

$C_D = 0.0277 + 0.061\,C_L^2$ $D = 118\,q\,C_D$

eq 17.38) $V_V = V\left(\frac{T-D}{W}\right)$; where T is from Thrust graph prepared earlier, times 0.95 for Propwash effect

V_{KTS}	q - Lb/ft² S.L./8000 ft	C_L S.L./8000	C_D S.L./8000	D - Lb S.L./8000	V_V - ft/s SL/8000
60	12/9.4	.85/1.09	.07/.10	99/111	23/15
80	22/17	.46/.60	.04/.05	104/100	30/22
100	34/27	.30/.38	.033/.036	132/115	20/18
120	49/38	.21/.27	.030/.032	173/143	9/15
140	67/53	.15/.19	.029/.03	230/187	—

A Conceptual Approach

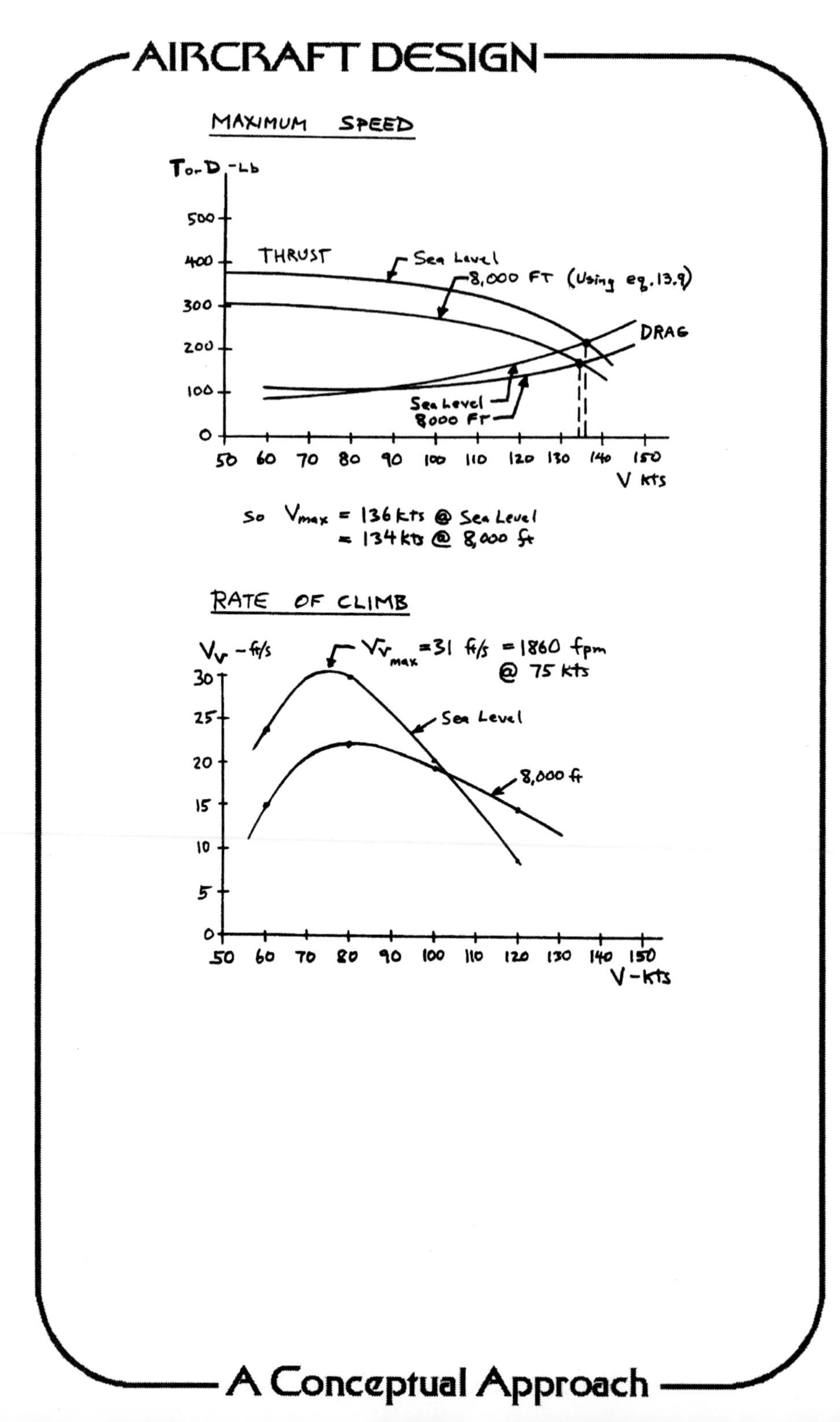
AIRCRAFT DESIGN
MAXIMUM SPEED
T or D -Lb
500
400
300
200
100
0
THRUST
Sea Level
8,000 FT (Using eq. 13.9)
DRAG
Sea Level
8000 FT
50 60 70 80 90 100 110 120 130 140 150
V kts
So Vmax = 136 kts @ Sea Level
= 134 kts @ 8,000 ft
RATE OF CLIMB
Vv - ft/s
Vvmax = 31 ft/s = 1860 fpm
@ 75 kts
30
25
20
15
10
5
0
Sea Level
8,000 ft
50 60 70 80 90 100 110 120 130 140 150
V - kts
A Conceptual Approach

REFINED SIZING

$C_{D_0} = .0277$ $e = 0.87$ $C_{bhp} = .5$ per hour $T = .95\,T_{(\text{from graph})}$ (sea level)

$W_{0_{\text{as drawn}}} = 1200$ $W_{e_{\text{as drawn}}} = 883$ Lb $W_{crew} = 220$ Lb

MISSION SEGMENT WEIGHT FRACTIONS

Warmup & Takeoff: Previously-used $\left\{\frac{W_1}{W_0} = .97\right\}$ seems excessive compared to cruise fraction $\left\{\frac{W_3}{W_2} = .953\right\}$!

Try assuming 5 minutes at maximum power:

$$W_f = C_{bhp}\,(Bhp)(5/60) = .5 \times 150 \times 5/60 = 6.25 \text{ Lb}$$

$$\text{At } W_0 = 1200\ ;\ \frac{W_1}{W_0} = \frac{1200 - 6.25}{1200} = \underline{.995}$$

Use $\frac{W_1}{W_0} = .995$, ignore small change as W_0 changes

Climb & Acceleration: Based on climb chart, climb at 80 kts to 8,000 ft then accelerate to 115 kts for cruise.

eq 19.9) $\Delta h_e = \left(8000 + \frac{1}{2g}(115 \times 1.689)^2\right) - \left(0 + \frac{1}{2g}(80 \times 1.689)^2\right) = 8302$ ft

equiv. $C = (.5)\dfrac{80 \times 1.689}{550(.78)(.95)} = .17$ per hr

From chart; $D/T = 100/360 = .278$ at sea level; $D/T = 105/290 = .362$ at 8,000 ft; Average $D/T = .32$

$$\text{eq 19.8)}\quad \frac{W_2}{W_1} = e^{\frac{-(.17/3600)(8302)}{(80 \times 1.689)(1-.32)}} = \underline{.996}$$

REFINED SIZING, Cont.

Cruise: 280 nm at 115 kts at 8000 ft, so $q = 35$

$$(W/S)_{cruise} = (W/S)_{Takeoff} \times .995 \times .996 = 10.1$$

$$(L/D)_{cruise} = \frac{1}{\frac{35(.0277)}{10.1} + \frac{10.1}{35\pi(6)(.87)}} = 8.804$$

$$\frac{W_3}{W_2} = e^{\frac{-(280 \times 6076)(.5/3600)}{(8.8)(550)(.78 \times .95)}} = \underline{.93624}$$

Descent & Land: Use $\frac{W_4}{W_3} = \underline{.995}$

FUEL FRACTION

$$\frac{W_4}{W_0} = .995 \times .996 \times .936 \times .995 = .92319$$

$$\frac{W_f}{W_0} = 1.06(1 - .923) = \underline{.08142}$$

At $W_0 = 1200$ Lb, $W_{f_{required}} = 1200 \times .08142 = \underline{97.7 \text{ Lb}}$

By pure luck this is close to the as-drawn fuel available, but we resize to show the method.
Assume $W_0 = 1199$ eq 19.13) $We = 882\left(\frac{1199}{1200}\right)^{(1-.1)} = 881.4$
eq 6.3) $W_{0_{calculated}} = 220 + (1199)(.08142) + 881.4 = \underline{1198.97 \text{ Lb.}}$

CLOSE ENOUGH!

W/S – ASPECT RATIO OPTIMIZATION

Since Hp is fixed, we cannot perform Hp/W – W/S optimization. Instead we will optimize W/S and A, holding Range = 280 nm.

- Vary W/S ± 20%, so W/S = 8.16; 10.2; and 12.24
- Vary A ± 33%, so A = 4; 6; and 8

This defines 9 different aircraft, as follows:

A \ W/S:	8.16	10.2	12.24
4	①	②	③
6	④	⑤	⑥
8	⑦	⑧	⑨

⑤ IS THE AS-DRAWN BASELINE

EFFECTS OF (W/S) AND (A) VARIATIONS

WING AREA: $S_W \propto \left(\frac{1}{W/S}\right)$ {"$\propto$" indicates "proportional to"}

TAIL AREA: $S_{tails} \propto \left(\frac{1}{W/S}\right)^{3/2}$ Holding Volume Coefficient constant

(ignore effect of (A) on tail size since $\propto \sqrt{A}$)

WEIGHTS:
- eq 15.46) $W_W \propto S_W^{.758} A^{.6}$
- eq 15.47) $W_{ht} \propto S_{ht}^{.896}$
- eq 15.48) $W_{vt} \propto S_{vt}^{.873}$

Use these to ratio the selected as-drawn component weights.

AERO:
- eq 12.49 used to recompute (e) for each (A)
- Assume wetted areas vary by wing/tail areas
- Must ratio resulting C_{D_0} to new reference area, where needed.

① $W/S = .8 \times 10.2 = 8.16$ $A = .66 \times 6 = 4$

$$S'_{tails} = \left(\frac{1}{.8}\right)^{3/2} S_{tails} = \frac{1}{.72} S_{tails}$$

$$W_w = 130 \left(\frac{1}{.8}\right)^{.758} (.666)^{.6} = 121 \text{ Lb} \qquad \Delta = -9 \text{ Lb}$$

$$W_{ht} = 40 \left(\frac{1}{.72}\right)^{.896} = 54 \text{ Lb} \qquad \Delta = 14 \text{ Lb}$$

$$W_{vt} = 15 \left(\frac{1}{.72}\right)^{.873} = 20 \text{ Lb} \qquad \Delta = 5 \text{ Lb}$$

$$\Delta = 10 \text{ Lb}$$

so $W_e = 882 + 10 = \underline{892 \text{ Lb}}$

Wing: $C_{D_{0_w}} = \left(\frac{1}{.8}\right) .0092 = .0115$ $\Delta = .0023$

tails: $C_{D_{0_t}} = \left(\frac{1}{.72}\right) .0033 = .0045$ $\Delta = .0012$

so $C_{D_0} = .0277 + .0023 + .0012 = .0312$ (ref. to old S_{ref})

$C_{D_0} = .0312(.8) = \underline{.0250}$ (ref. to new S_{ref})

$e = \underline{.93}$ for $A = 4$

MISSION SEGMENT WEIGHT FRACTIONS:

Cruise: $W/S = 8.16\,(.995)(.996) = 8.09$

$$L/D = \frac{1}{\dfrac{35(.0250)}{8.09} + \dfrac{8.09}{35\pi(4)(.93)}} = 7.8$$

$$\frac{W_3}{W_2} = e^{\dfrac{-(280 \times 6076)(.5/3600)}{(7.8)(550)(.78 \times .95)}} = \underline{.928}$$

All OTHER MISSION SEGMENT WEIGHT FRACTIONS ARE ESSENTIALLY UNCHANGED

① Continued

SIZING: USING SAME METHOD BUT WITH

$We_{as\text{-}drawn} = 892$ and $\frac{W_3}{W_2} = .928$

```
INPUT  WOdrawn, WEdrawn: 1200,892
INPUT  Empty weight fraction sizing exponent (default C=-0.1):
INPUT  Crew + Payload Weight: 220

MISSION SEGMENT WEIGHT FRACTIONS  .995  .996  .928  .995
```

SIZING ITERATIONS

Wo-guess	Wfuel	Wempty	W0-calculated
1200.0	108.0	892.0	1220.0
1216.0	109.5	902.7	1232.2
1229.0	110.6	911.3	1242.0
1239.4	111.6	918.3	1249.9
1247.8	112.3	923.9	1256.2
1254.5	112.9	928.4	1261.4
1260.0	113.4	932.0	1265.5
1264.4	113.8	935.0	1268.8
1267.9	114.1	937.3	1271.4
1270.7	114.4	939.2	1273.6
1273.0	114.6	940.7	1275.3
1274.9	114.8	941.9	1276.7
1276.3	114.9	942.9	1277.8
1277.5	115.0	943.7	1278.7

SO THE MODIFIED AIRCRAFT WITH $(W/S) = 8.16$ AND ASPECT RATIO = 4 MUST BE SIZED UP TO $W_0 = 1278$ Lb TO PERFORM THE DESIGN MISSION

The other eight parametric designs that were defined in the array in the preceding two pages are analyzed in a similar manner, with results as shown on the next page.

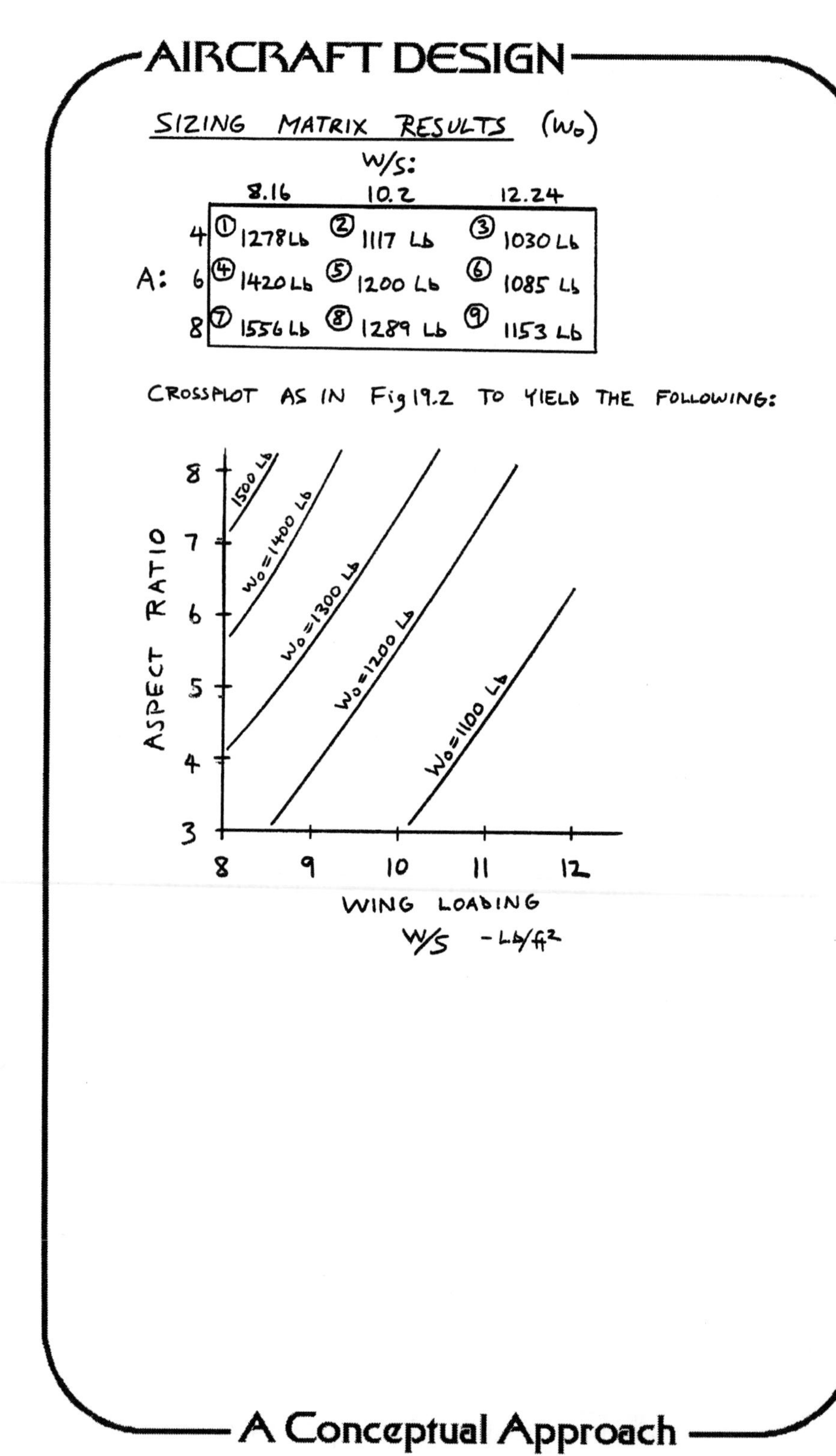
AIRCRAFT DESIGN
SIZING MATRIX RESULTS (Wo)
W/S:
8.16
10.2
12.24
A:
4
6
8
① 1278 Lb
② 1117 Lb
③ 1030 Lb
④ 1420 Lb
⑤ 1200 Lb
⑥ 1085 Lb
⑦ 1556 Lb
⑧ 1289 Lb
⑨ 1153 Lb
CROSSPLOT AS IN Fig 19.2 TO YIELD THE FOLLOWING:
1500 Lb
Wo=1400 Lb
Wo=1300 Lb
Wo=1200 Lb
Wo=1100 Lb
ASPECT RATIO
3
4
5
6
7
8
9
10
11
12
WING LOADING
W/S - Lb/ft²
A Conceptual Approach

PERFORMANCE CONSTRAINT LINES

STALL: As before, $V_{stall} = 50$ requires $W/S \leq 10.2$

RATE OF CLIMB: $V_v \geq 1500$ ft/min at Sea Level

Assume $V = 75$ kts (best R.O.C. for baseline)

so $q = 19.1$ $T = 360$ Lb (from graph)

$$D = qS\left(C_{D_0} + \frac{(W/S/q)^2}{\pi A e}\right) \qquad V_v = \left(\frac{T-D}{W}\right)V$$

① $W = 1278$ Lb $S = 1278/8.16 = 156.6$ ft²

$$D = 156.6 \times 19.1\left(.0250 + \frac{(8.16/19.1)^2}{\pi\, 4\, (.93)}\right) = 122 \text{ Lb}$$

$$V_v = 75 \times 1.689\left(\frac{360-122}{1278}\right) = 23.6 \text{ ft/sec} = 1415 \text{ ft/min}$$

② $W = 1117$ Lb $S = 1117/10.2 = 109$ ft²

$$D = 109 \times 19.1\left(.0277 + \frac{(10.2/19.1)^2}{\pi\, 4\, (.93)}\right) = 108.5 \text{ Lb}$$

$$V_v = 75 \times 1.689\left(\frac{360-108.5}{1117}\right) = 28.5 \text{ ft/sec} = 1711 \text{ ft/min}$$

③ $W = 1030$ Lb $S = 1030/12.24 = 84.1$

$$D = 84.1 \times 19.1\left(.0305 + \frac{(12.24/19.1)^2}{\pi\, 4\, (.93)}\right) = 105.4 \text{ Lb}$$

$$V_v = 75 \times 1.689\left(\frac{360-105.4}{1030}\right) = 31.3 \text{ ft/sec} = 1878 \text{ ft/min}$$

SIMILARLY:

④ $V_v = 1284$ ft/min
⑤ $V_v = 1636$ ft/min
⑥ $V_v = 1863$ ft/min
⑦ $V_v = 1154$ ft/min
⑧ $V_v = 1529$ ft/min
⑨ $V_v = 1774$ ft/min

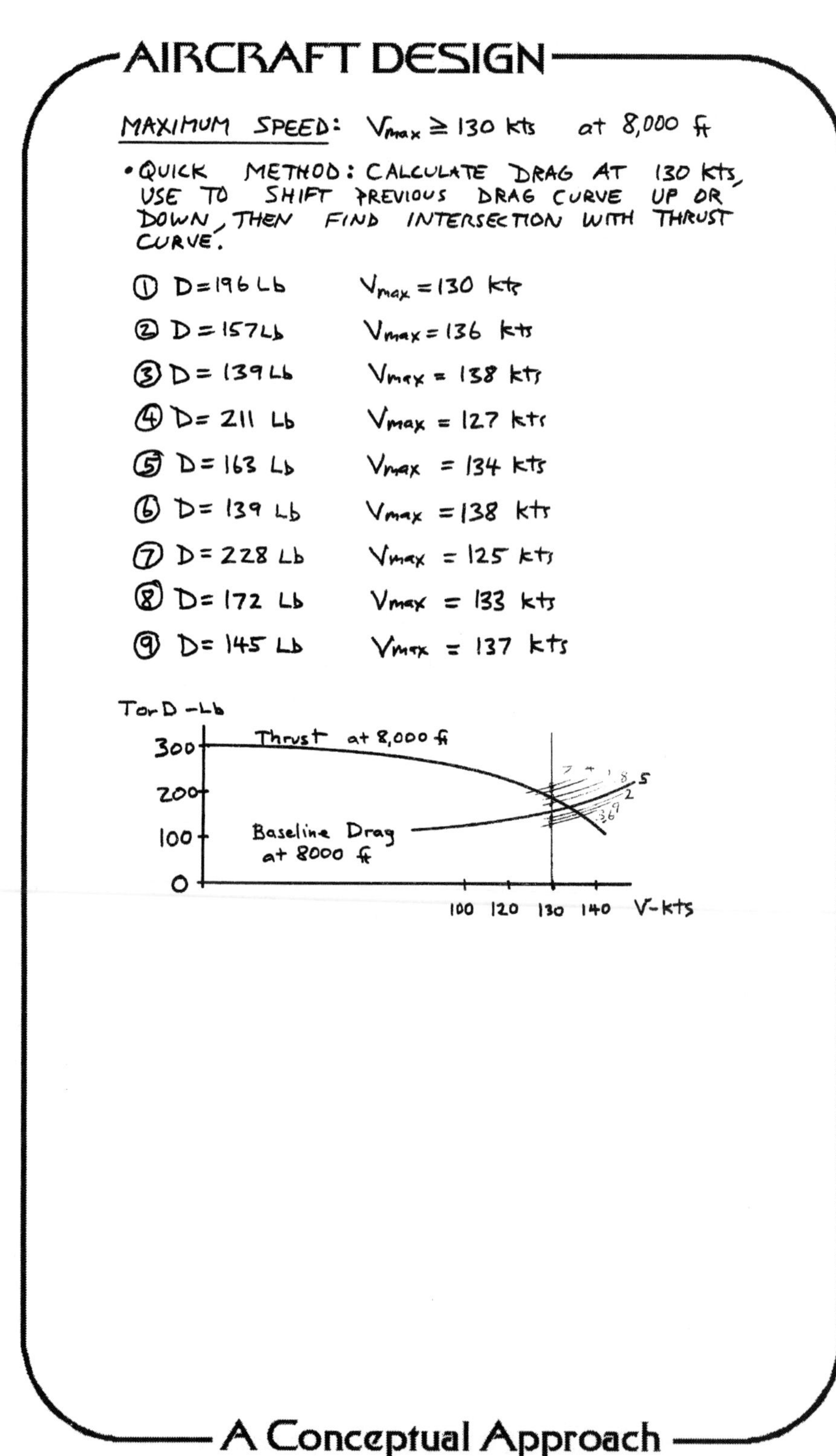
AIRCRAFT DESIGN
MAXIMUM SPEED: Vmax ≥ 130 kts at 8,000 ft
• QUICK METHOD: CALCULATE DRAG AT 130 KTS, USE TO SHIFT PREVIOUS DRAG CURVE UP OR DOWN, THEN FIND INTERSECTION WITH THRUST CURVE.
① D=196 Lb Vmax = 130 kts
② D = 157 Lb Vmax = 136 kts
③ D = 139 Lb Vmax = 138 kts
④ D = 211 Lb Vmax = 127 kts
⑤ D = 163 Lb Vmax = 134 kts
⑥ D = 139 Lb Vmax = 138 kts
⑦ D = 228 Lb Vmax = 125 kts
⑧ D = 172 Lb Vmax = 133 kts
⑨ D = 145 Lb Vmax = 137 kts
T or D – Lb
300
200
100
0
Thrust at 8,000 ft
Baseline Drag at 8000 ft
100 120 130 140 V-kts
A Conceptual Approach

CROSSPLOTTING THE STALL, RATE-OF-CLIMB, AND V_{max} REQUIREMENTS ONTO THE SIZING GRAPH GIVES NO LOWER LIMIT ON ASPECT RATIO! AT VERY LOW ASPECT RATIOS, THE INDUCED DRAG WOULD BECOME EXCESSIVE DURING MANEUVERS. THEREFORE, WE NEED SOME REQUIREMENT BASED ON MANEUVERING.

DEFINE A NEW PERFORMANCE REQUIREMENT BASED ON SUSTAINED TURN:

$\dot{\psi} \geq 30$ °/sec SUSTAINED, AT 100 kts, S.L.

eq (17.51) $\dot{\psi} = 30\,°/\text{sec} = .5236\ \text{rad/sec} = \dfrac{g\sqrt{n^2-1}}{100 \times 1.689}$

so $n \geq 2.92$

T=345 Lb, from graph

eq 17.53) $n = \sqrt{\dfrac{34\pi A e}{W/S}\left(\dfrac{345}{W} - \dfrac{34\,C_{D_0}}{W/S}\right)}$

Check if we are close to stall (which reduces e):

$C_L = \dfrac{n\,W/S}{q} = .88$ for baseline
$= 1.056$ for $W/S = 12.24$

Prior e estimates should be approximately correct.

USING PRIOR DATA:

① $n = 2.8$
② $n = 2.9$
③ $n = 2.8$
④ $n = 3.1$
⑤ $n = 3.3$
⑥ $n = 3.3$
⑦ $n = 3.2$
⑧ $n = 3.4$
⑨ $n = 3.5$

NOTE: Large values of n are incorrect because they imply $C_L > C_{L_{stall}}$. However, we can use those values to crossplot for $n = 2.92$ which is below stall.

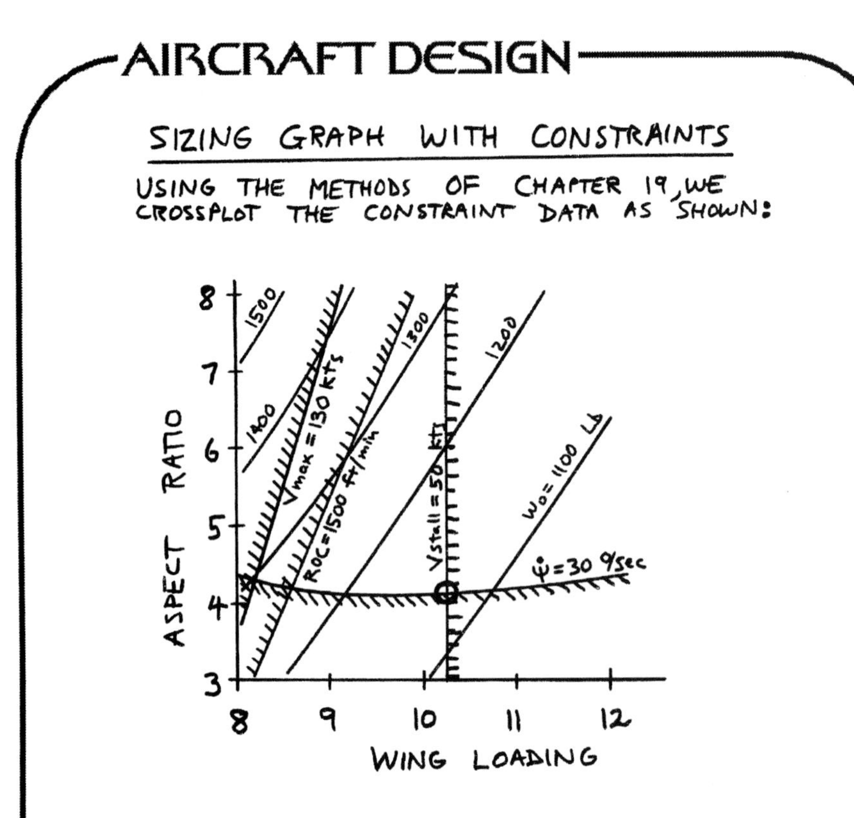

SO THE OPTIMAL AIRCRAFT FOR THE GIVEN REQUIREMENTS OCCURS AT $\{W/S = 10.2\}$ AND $\{A = 4.2\}$ AND HAS $W_0 = 1130$ LB. THE NEXT STEP IN THE DESIGN PROCESS IS TO REDRAW THE AIRCRAFT AND ANALYZE IT IN DETAIL.

23.3 Lightweight Supercruise Fighter

The U.S. Air Force currently operates the F-15 and F-16 as a "high–low" mix of dogfighters. The F-15 has greater range, avionics, and weaponry but is too costly to fill the entire inventory requirement. The F-16, with less capability and cost, rounds out the total required dogfighter inventory and also serves in an air-to-ground role.

The U.S. Air Force is developing the F-22 as a replacement for the F-15. The next fighter after F-22 is then likely to be a replacement for the F-16, which is almost as old as the F-15. This new fighter would be the "low" end of a high–low mix with F-22.

This design example presents such an F-16 follow-on design. Design requirements are based upon assumed improvements to published F-16 capabilities, with the addition of a required capability for sustained supersonic cruise ("supercruise") on dry power. Also, relatively short takeoff and landing requirements are imposed.

The selected design incorporates one unproven technology, the variable dihedral vertical tail. This patented concept purports to control the rearward shift in aerodynamic center as the aircraft accelerates to supersonic flight by converting from a "V" tail subsonically to upright vertical tails supersonically. This should reduce trim drag and enhance maneuverability.

Such a technology study is very common in early conceptual design. As will be seen, the impact of such a technology on aerodynamics, weights, propulsion, etc., is estimated as best as possible, and the aircraft is sized and optimized. The resulting aircraft is then compared to a baseline design that does not incorporate the new technology to determine if the new idea should be pursued.

(Homebuilders: don't build this one either!)

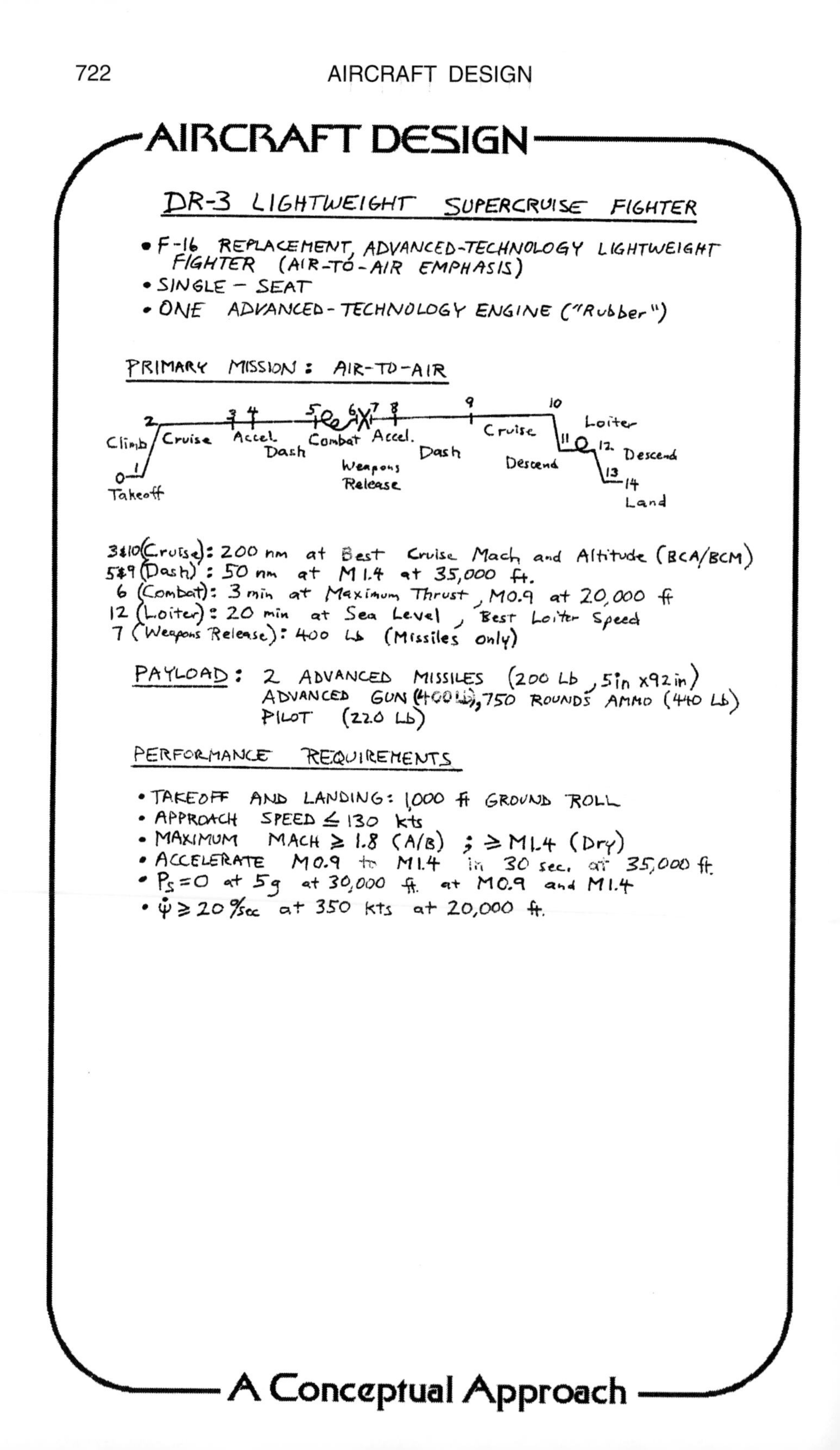

AIRCRAFT DESIGN

DR-3 LIGHTWEIGHT SUPERCRUISE FIGHTER

- F-16 REPLACEMENT, ADVANCED-TECHNOLOGY LIGHTWEIGHT FIGHTER (AIR-TO-AIR EMPHASIS)
- SINGLE – SEAT
- ONE ADVANCED-TECHNOLOGY ENGINE ("Rubber")

PRIMARY MISSION : AIR-TO-AIR

3&10 (Cruise): 200 nm at Best Cruise Mach and Altitude (BCA/BCM)
5&9 (Dash) : 50 nm at M 1.4 at 35,000 ft.
6 (Combat): 3 min at Maximum Thrust, M0.9 at 20,000 ft
12 (Loiter): 20 min at Sea Level, Best Loiter Speed
7 (Weapons Release): 400 Lb (Missiles only)

PAYLOAD: 2 ADVANCED MISSILES (200 Lb, 5 in x 92 in)
ADVANCED GUN (400 Lb), 750 ROUNDS AMMO (440 Lb)
PILOT (220 Lb)

PERFORMANCE REQUIREMENTS

- TAKEOFF AND LANDING: 1,000 ft GROUND ROLL
- APPROACH SPEED $\leq$ 130 kts
- MAXIMUM MACH $\geq$ 1.8 (A/B) ; $\geq$ M1.4 (Dry)
- ACCELERATE M0.9 to M1.4 in 30 sec. at 35,000 ft.
- $P_s = 0$ at 5g at 30,000 ft. at M0.9 and M1.4
- $\dot{\psi} \geq 20$ °/sec at 350 kts at 20,000 ft.

A Conceptual Approach

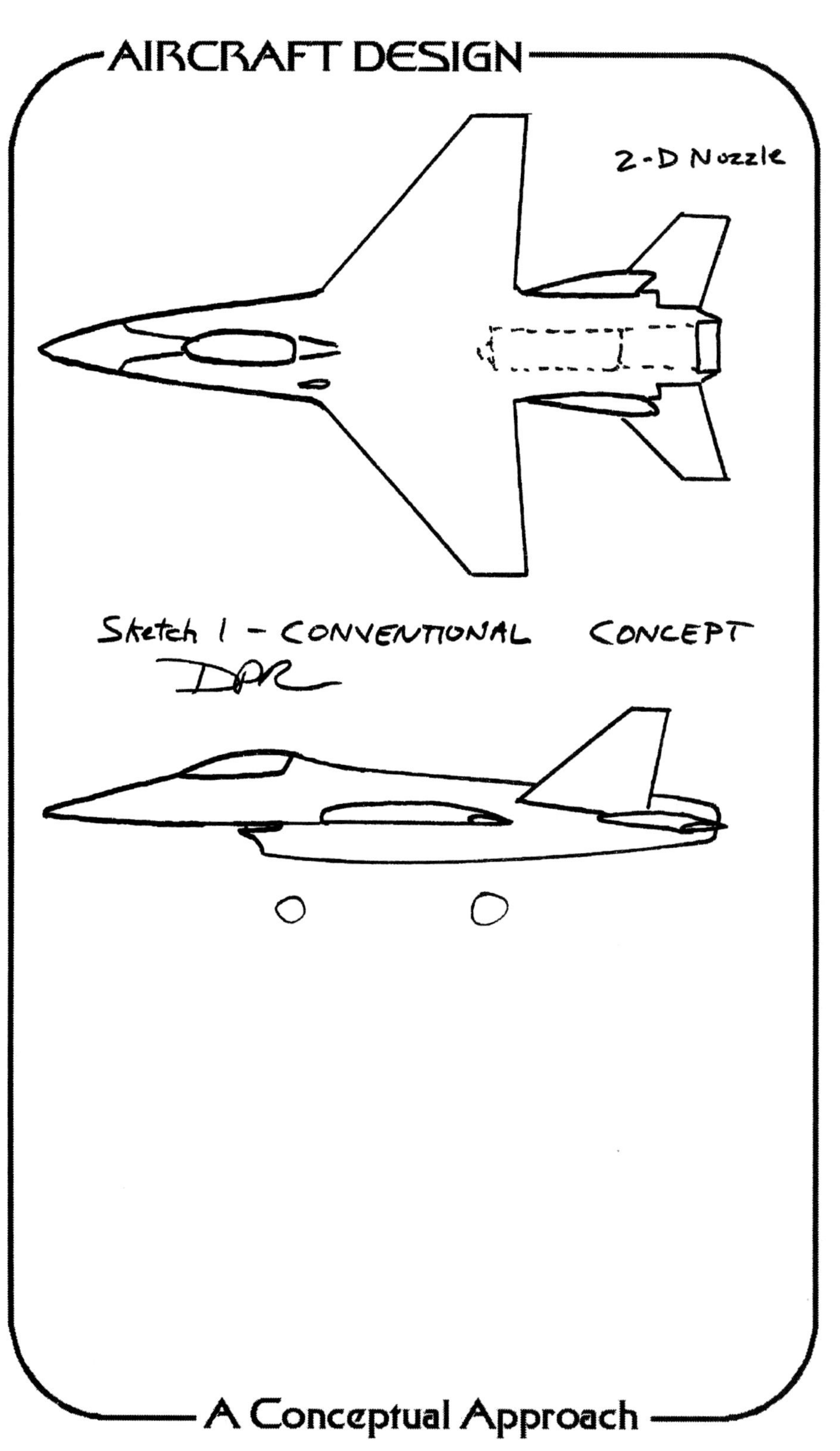
AIRCRAFT DESIGN
2-D Nozzle
Sketch 1 - CONVENTIONAL CONCEPT
DPR
A Conceptual Approach

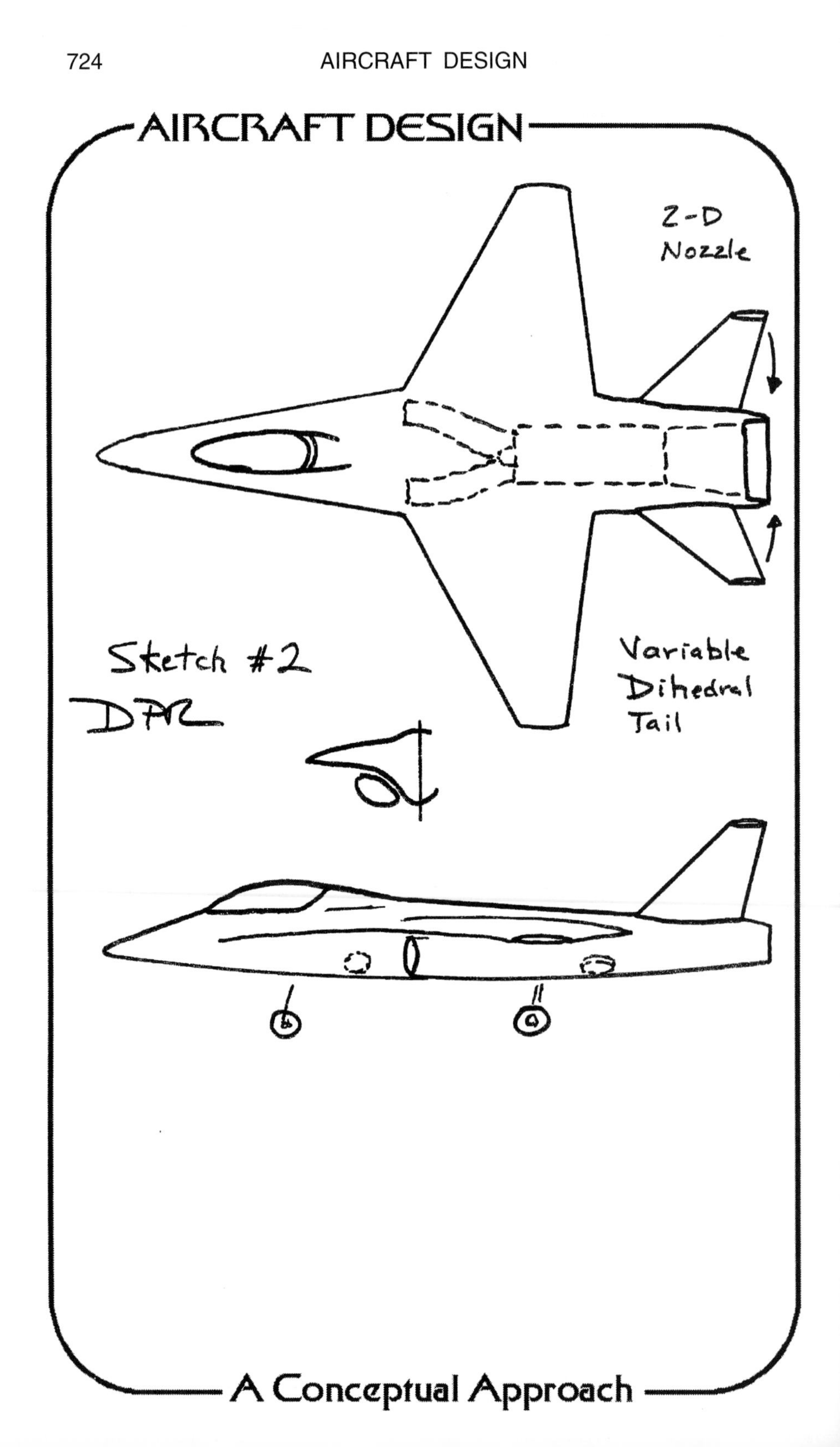
AIRCRAFT DESIGN
2-D
Nozzle
Sketch #2
DPR
Variable
Dihedral
Tail
A Conceptual Approach

WING GEOMETRY

$A = 5.416\,(1.8)^{-.622} = 3.8$

$\Lambda_{LE} = 48°$ $(\Lambda_{c/4} \cong 40°)$

CHECK V.S. FIG. 4.21 INDICATES TRANSONIC PITCHUP! CHANGE TO: $\{A = 3.5,\ \Lambda_{c/4} = 30°\}$ $(\text{so } \Lambda_{LE} \cong 40°)$

SELECT: $\lambda = 0.25$

$t/c = 6\%$

Airfoil: 64A006 (Initially)

ENGINE: POST-2000 "RUBBER" ENGINE. APPROXIMATE WITH APPEN. A.4-1 ENGINE WITH 20% SFC REDUCTION.

T/W

table 5.3) $T/W_{takeoff} = 0.648\,(1.8)^{.594} = .92$ (Use Initially)

W/S

Stall: $V_{approach} \leq 130$ KTS $= 220$ ft/sec

$V_{stall} \leq V_{approach}/1.2 = 183$ ft/sec

$W/S \leq q\,C_{L_{max}}$ at stall

fig 5.3) $C_{L_{max}} \cong 1.5 + .3$ (L.E. FLAP) $\cong 1.8$

so $W/S \leq 72$ Lb/ft² (at Sea Level)

W/S (continued)

Landing:

from eq. 5.11) $S_{\text{landing ground roll}} = 80 \frac{W}{S}\left(\frac{1}{\sigma C_{L_{max}}}\right) \leq 1{,}000$

so $W/S \leq 22.5$ (!)

(MUCH TOO LOW FOR A FIGHTER! WE WILL IGNORE THIS INITIALLY AND USE THRUST REVERSING TO LAND.)

Takeoff:

fig 5.4) $TOP \cong 80$

eq 5.9) $W/S = TOP\left(\frac{C_{L_{max}}}{1.21}\right) T/W = 80\left(\frac{1.8}{1.21}\right)(.92) = 109$

Cruise: table 12.3) $C_{fe} = .0035$

assume $S_{wet}/S_{ref} \cong 4$, so $C_{D_0} \cong .014$ (eq 12.23)

eq 12.50) $e = 4.61\left(1 - .045(3.5)^{.68}\right)(\cos 40°) - 3.1 = 0.86$

At M.9 and 35,000 ft (assumed BCM/BCA), $q = 284 \text{ Lb/ft}^2$

so

$$\left(W/S\right)_{\text{opt. Cruise}} = 284\sqrt{\frac{\pi \times 3.5 \times 0.86 \times 0.014}{3}} = 59.6$$

$$\left(W/S\right)_{\text{takeoff}} \cong \frac{59.6}{.97 \times .977} = 62.9$$

(Using typical values for $\frac{W_1}{W_0}$ and $\frac{W_2}{W_1}$)

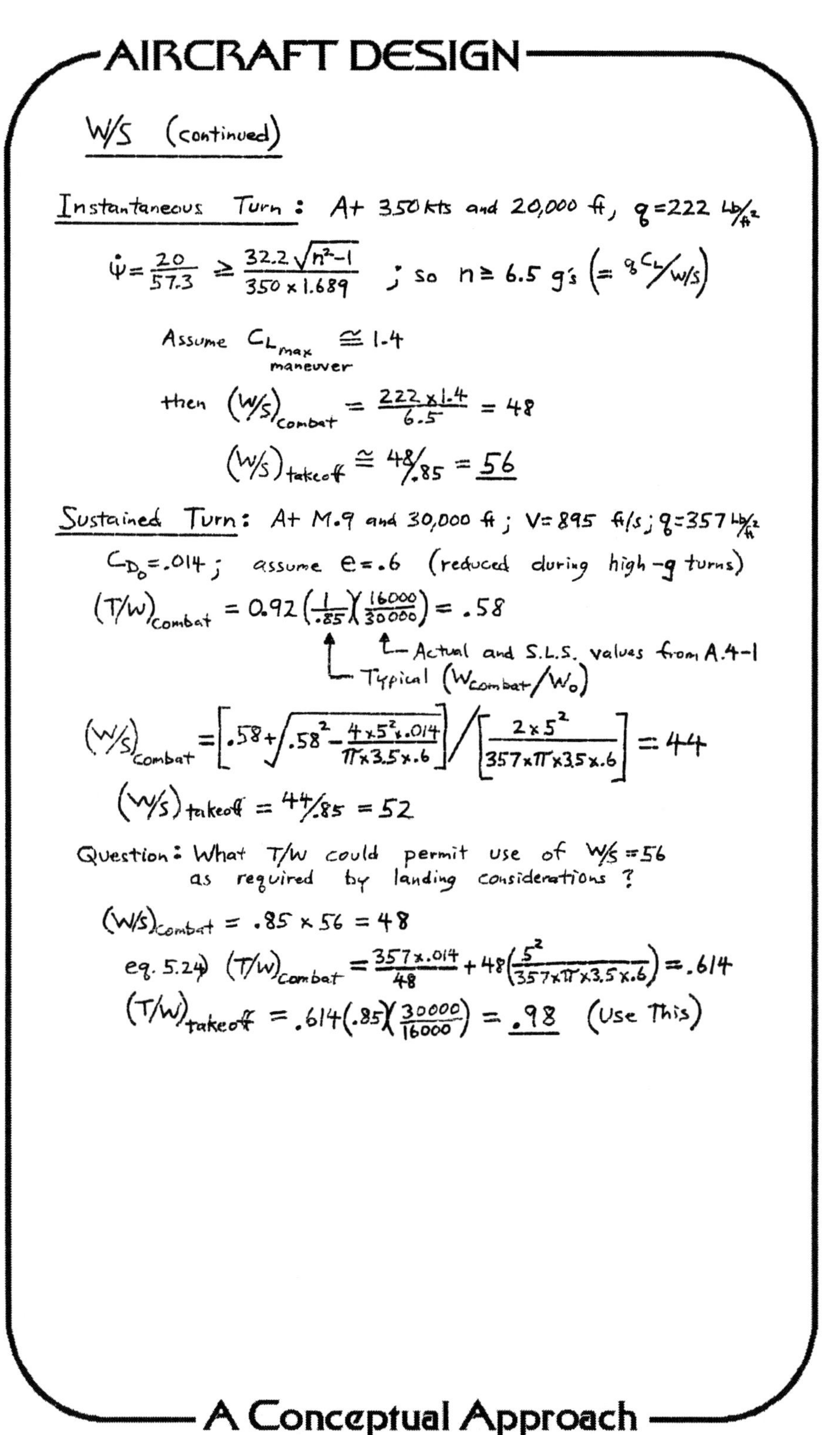

AIRCRAFT DESIGN

W/S (continued)

Instantaneous Turn: At 350 kts and 20,000 ft, $q = 222\ \text{lb/ft}^2$

$$\dot{\psi} = \frac{20}{57.3} \geq \frac{32.2\sqrt{n^2-1}}{350 \times 1.689} \quad ; \text{ so } n \geq 6.5\ g\text{'s} \left(= qC_L/(W/S)\right)$$

Assume $C_{L_{max\ maneuver}} \cong 1.4$

then $(W/S)_{combat} = \dfrac{222 \times 1.4}{6.5} = 48$

$(W/S)_{takeoff} \cong 48/.85 = \underline{56}$

Sustained Turn: At M.9 and 30,000 ft; V = 895 ft/s; $q = 357\ \text{lb/ft}^2$

$C_{D_0} = .014$; assume $e = .6$ (reduced during high-g turns)

$(T/W)_{combat} = 0.92\left(\dfrac{1}{.85}\right)\left(\dfrac{16000}{30000}\right) = .58$

↑ Actual and S.L.S. values from A.4-1

↑ Typical (W_{combat}/W_0)

$$(W/S)_{combat} = \left[.58 + \sqrt{.58^2 - \frac{4 \times 5^2 \times .014}{\pi \times 3.5 \times .6}}\right] \Bigg/ \left[\frac{2 \times 5^2}{357 \times \pi \times 3.5 \times .6}\right] = 44$$

$(W/S)_{takeoff} = 44/.85 = 52$

Question: What T/W could permit use of W/S = 56 as required by landing considerations?

$(W/S)_{combat} = .85 \times 56 = 48$

eq. 5.24) $(T/W)_{combat} = \dfrac{357 \times .014}{48} + 48\left(\dfrac{5^2}{357 \times \pi \times 3.5 \times .6}\right) = .614$

$(T/W)_{takeoff} = .614(.85)\left(\dfrac{30000}{16000}\right) = \underline{.98}$ (Use This)

A Conceptual Approach

AIRCRAFT DESIGN

INITIAL SIZING

Empty Weight Fraction: (Assume composite structure)

table 6.1) $$\frac{W_e}{W_o} = \left[-.02 + 2.16\, W_o^{-.1} \times 3.5^{.2} \times .98^{.04} \times 56^{-.1} \times 1.8^{.08}\right] \times 0.9$$

$$\frac{W_e}{W_o} = 1.75\, W_o^{-.1} - .018$$

Mission Segment Weight Fractions:

Warmup & Takeoff: eq 6.8) $\frac{W_1}{W_o} = .98$

Climb: eq 6.9) $\frac{W_2}{W_1} = 1.0065 - .0325(.9) = .977$

Cruise: (Assume M.9 at 35,000 ft. for BCA/BCM)
V= 876 ft/sec ; q = 283 Lb/ft2

$$W/S = 56 \times .98 \times .977 = 54$$

eq 6.13) $$L/D = \frac{1}{\frac{283 \times .014}{54} + \frac{54}{283\pi \times 3.5 \times .86}} = 10.7$$

SFC:

- A.4-1, Partial Power at M.9 ; 36,000 ft gives: C = 1.07
- Increase this 10% to approximate installation: C = 1.18
- Reduce this 20% for advanced technology: C = 0.94

so;

eq 6.11) $$\frac{W_3}{W_2} = e^{-\frac{(200 \times 6076)(.94/3600)}{876 \times 10.7}} = .967$$

A Conceptual Approach

Acceleration:

eq 6.10) $\left(\frac{W_4}{W_3}\right)_{M.1 \to 1.4} = .9616$

eq 6.9) $\left(\frac{W_4}{W_3}\right)_{M.1 \to .9} = .9773$

$$\left(\frac{W_4}{W_3}\right)_{M.9 \to 1.4} = \frac{.9616}{.9773} = .984$$

Dash: M1.4 at 35,000 ft.; V = 1362 ft/sec; q = 685 Lb/ft²

$W/S = 56 \times .98 \times .977 \times .967 \times .984 = 51$

Looking at fig. 12.31, we roughly estimate;

$$\left(C_{D_0}\right)_{M1.4} \cong 2 \times \left(C_{D_0}\right)_{subsonic} = .028 \quad \text{(crude !)}$$

eq 12.52) $$(K)_{M1.4} \cong \left(\frac{3.5\,(1.4^2-1)}{4 \times 3.5\sqrt{1.4^2-1} - 2}\right) \cos 40^\circ = .22$$

or $e = \frac{1}{\pi A K} = .414$

$$L/D = \frac{1}{\frac{685 \times .028}{51} + \frac{51 \times .22}{685}} = 2.55 \quad (!)$$

SFC:

A.4-1, Military (Max. Dry) Power — $C = 1.2$
Increase 10% (Installation)
Reduce 20% (Advanced Technology) } $C = 1.06$

eq. 6.11) $$\frac{W_5}{W_4} = e^{-\frac{(50 \times 6076)(1.06/3600)}{1362 \times 2.55}} = .975$$

AIRCRAFT DESIGN

Combat $d = 3$ min.

$$(T/W)_{combat} = .98 \times \left(\frac{16000}{30000}\right) \Big/ (.98 \times .977 \times .967 \times .984 \times .975) = .588$$

SFC: A.4-1 for Max. thrust at M.9 at 20,000 ft : $C = 1.78$
Increase 10% for installation
Reduce 20% for advanced technology } $C = 1.57$

$$\frac{W_6}{W_5} = 1 - [(1.57/3600)(.588)(3 \times 60)] = .954$$

Weight Drop: Ignore for initial sizing

Accelerate: $\frac{W_8}{W_7} \cong \frac{W_4}{W_3} = .984$

Dash: $\frac{W_9}{W_8} \cong \frac{W_5}{W_4} = .975$

Cruise: $\frac{W_{10}}{W_9} \cong \frac{W_3}{W_2} = .967$

Descent: Ignore, assuming range credit

A Conceptual Approach

AIRCRAFT DESIGN

Loiter: E = 20 min ; Sea Level

$$(W/S)_{loiter} = 56 \times .98 \times .977 \times .967 \times .984 \times .975 \times .954 \times .984 \times .975 \times .967 = 44$$

$$\text{eq 17.13)} \quad V_{\substack{best\\loiter}} = \sqrt{\frac{2 \times 44}{\rho}\sqrt{\frac{1}{.014 \times \pi \times 3.5 \times .86}}} = 319 \text{ ft/sec}$$

so $q = 121$ Lb/ft^2

$$L/D = \frac{1}{\frac{121 \times .014}{44} + \frac{44}{121\pi \times 3.5 \times .86}} = 13$$

SFC: Adjusted from A.4-1 : C = .906

$$\frac{W_{12}}{W_{11}} = e^{-\frac{(20 \times 60)(.906/3600)}{13}} = .977$$

Descent for Landing: $\frac{W_{13}}{W_{12}} = .993$

Land: $\frac{W_{14}}{W_{13}} = .995$ (eq 6.23)

TOTAL MISSION WEIGHT FRACTION

$$\frac{W_{14}}{W_0} = .98 \times .977 \times .967 \times .984 \times .975 \times .954 \times .984 \times .975 \times .967 \times .977 \times .993 \times .995 = .7586$$

FUEL FRACTION

$$\frac{W_f}{W_0} = 1.06(1 - .7586) = .256$$

A Conceptual Approach

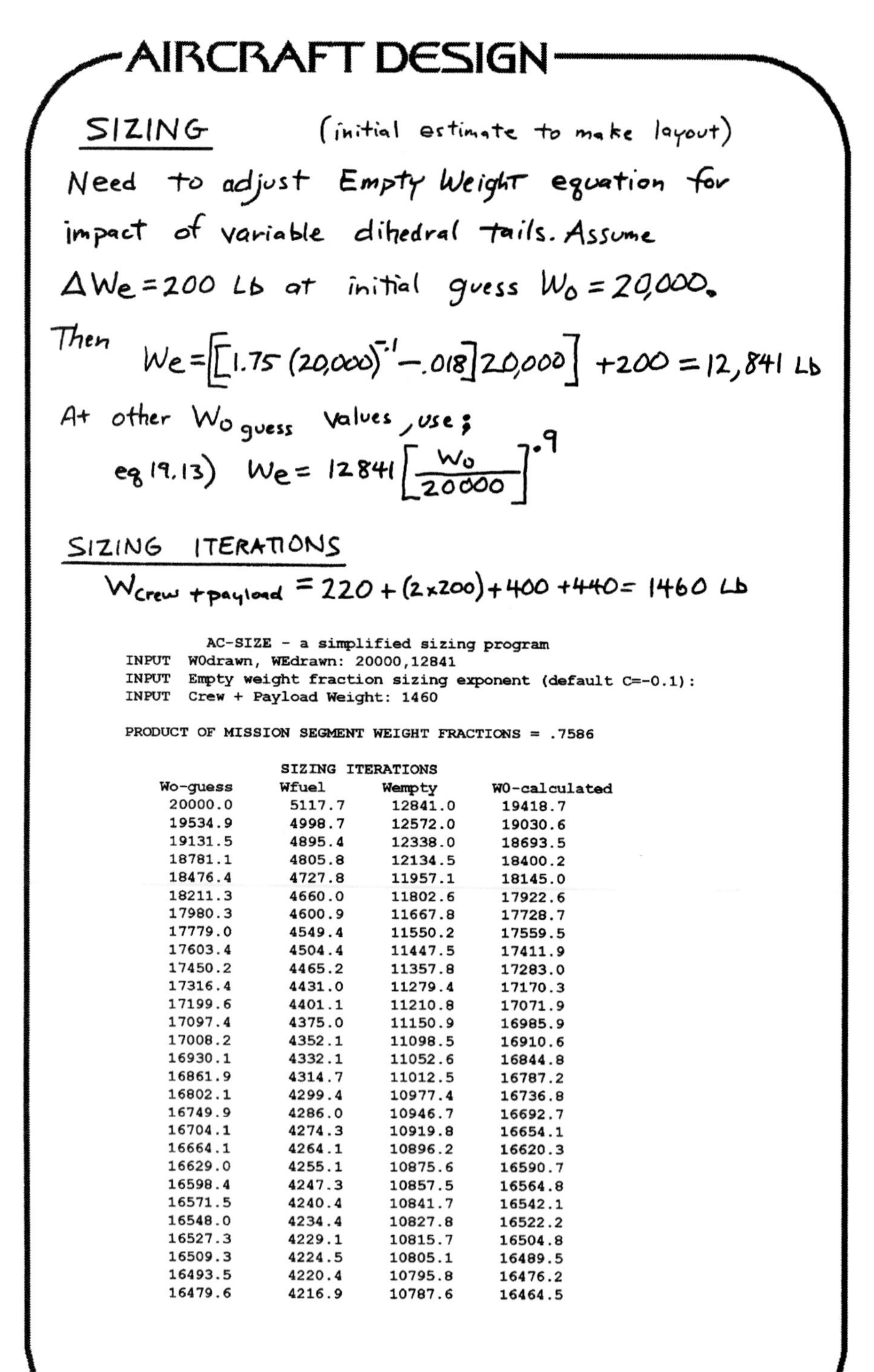

AIRCRAFT DESIGN

SIZING (initial estimate to make layout)

Need to adjust Empty Weight equation for impact of variable dihedral tails. Assume $\Delta W_e = 200$ Lb at initial guess $W_0 = 20{,}000$.

Then $$W_e = \left[\left[1.75\,(20{,}000)^{-.1} - .018\right] 20{,}000\right] + 200 = 12{,}841 \text{ Lb}$$

At other $W_{0\,guess}$ values, use;

$$\text{eq 19.13)} \quad W_e = 12841 \left[\frac{W_0}{20000}\right]^{.9}$$

SIZING ITERATIONS

$$W_{crew+payload} = 220 + (2 \times 200) + 400 + 440 = 1460 \text{ Lb}$$

```
        AC-SIZE - a simplified sizing program
INPUT  W0drawn, WEdrawn: 20000,12841
INPUT  Empty weight fraction sizing exponent (default C=-0.1):
INPUT  Crew + Payload Weight: 1460

PRODUCT OF MISSION SEGMENT WEIGHT FRACTIONS = .7586
```

SIZING ITERATIONS

Wo-guess	Wfuel	Wempty	W0-calculated
20000.0	5117.7	12841.0	19418.7
19534.9	4998.7	12572.0	19030.6
19131.5	4895.4	12338.0	18693.5
18781.1	4805.8	12134.5	18400.2
18476.4	4727.8	11957.1	18145.0
18211.3	4660.0	11802.6	17922.6
17980.3	4600.9	11667.8	17728.7
17779.0	4549.4	11550.2	17559.5
17603.4	4504.4	11447.5	17411.9
17450.2	4465.2	11357.8	17283.0
17316.4	4431.0	11279.4	17170.3
17199.6	4401.1	11210.8	17071.9
17097.4	4375.0	11150.9	16985.9
17008.2	4352.1	11098.5	16910.6
16930.1	4332.1	11052.6	16844.8
16861.9	4314.7	11012.5	16787.2
16802.1	4299.4	10977.4	16736.8
16749.9	4286.0	10946.7	16692.7
16704.1	4274.3	10919.8	16654.1
16664.1	4264.1	10896.2	16620.3
16629.0	4255.1	10875.6	16590.7
16598.4	4247.3	10857.5	16564.8
16571.5	4240.4	10841.7	16542.1
16548.0	4234.4	10827.8	16522.2
16527.3	4229.1	10815.7	16504.8
16509.3	4224.5	10805.1	16489.5
16493.5	4220.4	10795.8	16476.2
16479.6	4216.9	10787.6	16464.5

A Conceptual Approach

AIRCRAFT DESIGN

LAYOUT DATA

$W_0 \cong 16480$ Lb. $W_f = .256 \times W_0 \cong 4220$ Lb $= 94$ ft^3

Fuselage: table 6.3) $L \cong .93\,(16480)^{.39} = 41$ ft $= 492$ in

Wing: $S = 16480/56 = 294$ ft^2

$A = 3.5$ $\lambda = 0.25$ $\Lambda_{\bar{c}/4} = 30°$

eq 7.5) $b = \sqrt{3.5 \times 294} = 32$ ft $= 384$ in

eq 7.6) $C_{root} = \dfrac{2 \times 294}{32\,(1 + .25)} = 14.7$ ft $= 176$ in

eq 7.7) $C_{tip} = 176 \times .25 = 44$ in

eq 7.8) $\bar{C} = 123$ in

eq 7.9) $\bar{Y} = 76.8$ in

Tails: (see ch.4) Lay out "V" tail such that the total tail area equals the sum of the required vertical and horizontal tail areas determined by the volume coefficient method. (Assume $L_t \cong 200$ in)

Vertical Tail: $S_{vt} = .07 \dfrac{b_w S_w}{L_t} = 39$ ft^2
Horizontal Tail: $S_{ht} = .4 \dfrac{\bar{c}_w S_w}{L_t} = 72$ ft^2
Sum $= 111$ ft^2

- If the tails met at the aircraft centerline, $\Gamma_t = \tan^{-1}\left(\frac{39}{72}\right) = 28.4°$
- We will lay out the projected planform for top view as a horizontal equivalent, using $\Gamma_t = 28.4°$ and $S_{t_{true}} = 111$ ft^2.

$(S_{ht})_{projected} = 111 \cos 28 = 97.6$ ft^2
Using $A = 3.5$; $\lambda = .25$; $\Lambda_{\bar{c}/4} = 30°$

$b_h = 18.5$ ft $= 222$ in
$C_{root} = 8.4$ ft $= 101$ in
$C_{tip} = 25.3$ in

- True tail geometry will be graphically determined.

A Conceptual Approach

ENGINE: $T=(T/W)W_0 = .98 \times 16480 = 16150.4$ Lb (SLS)

A.4-1 ; 100%-Sized Engine: $T = 30,000$ Lb
$L = 160$ in
$D = 44$ in
$W = 3,000$ Lb

so Scale Factor: $SF = \frac{16150.4}{30,000} = .538$

$L = 160\,(.538)^{.4} = 125$ in
$D = 44\,(.538)^{.5} = 32$ in
$W = 3000\,(.538)^{1.1} = 1517$ Lb

} Engine with conventional nozzle

TO PROVIDE PITCH CONTROL AT SUPERSONIC SPEEDS (WHEN THE TAILS ARE NEAR-VERTICAL), WE WILL USE A TWO-DIMENSIONAL VECTORING NOZZLE WITH THRUST REVERSING TO SHORTEN THE LANDING. THIS REQUIRES A CIRCLE-TO-SQUARE ADAPTER WHICH LENGTHENS THE ENGINE.

CAPTURE AREA SIZING

From A.4-1 at M1.8 at 30,000 ft, $\dot{m} = 270$ Lbm/s (mass flow)
Scale by Scale Factor: $\dot{m} = .538 \times 270 = 145.3$ Lbm/s

Fig 10.16) $A_c/\dot{m} = 3.8$ at M1.8, so $A_c = 3.8 \times 145.3 = \underline{552 \text{ in}^2}$

LANDING GEAR $W_w \cong .9 \times \frac{16480}{2} = 7416$

MAIN: $D = 1.59\,(7416)^{.302} = 23$ in
$W = .098\,(7416)^{.467} = 6.3$ in

NOSE: $D = 18$ in
$W = 5$ in
} 80% of main gear

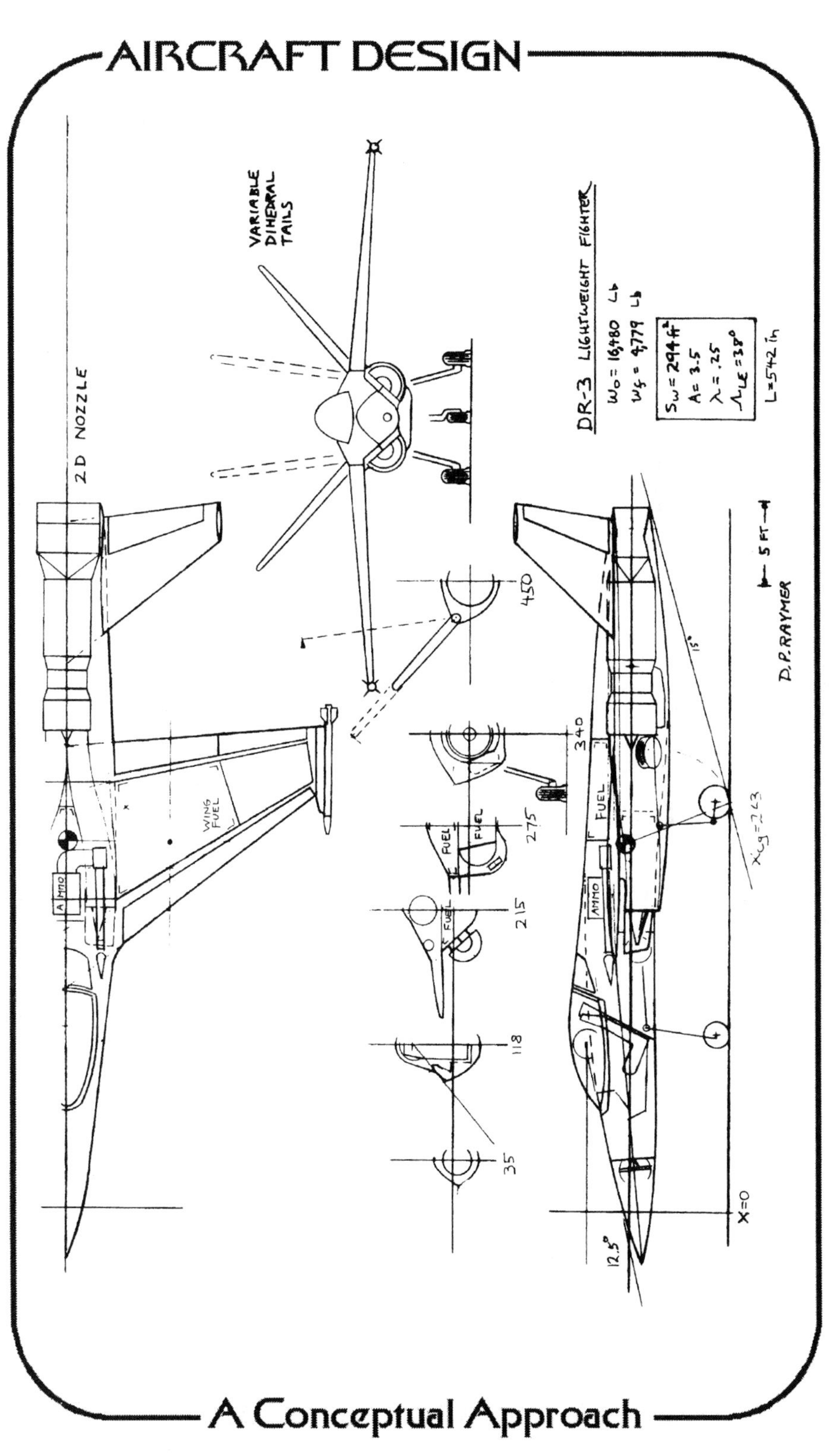
AIRCRAFT DESIGN
2 D NOZZLE
VARIABLE DIHEDRAL TAILS
DR-3 LIGHTWEIGHT FIGHTER
Wo = 16,480 Lb
Wf = 4,779 Lb
Sw = 294 ft²
A = 3.5
λ = .25
ΛLE = 38°
L = 542 in
5 FT
D.P. RAYMER
WING FUEL
AMMO
FUEL
450
340
275
215
118
35
X=0
12.5°
15°
A Conceptual Approach

FUEL TANKS

Required: W_f = .256 × 16480 ≅ 4220 Lb

Assume: Integral wing tanks (85% Usable Volume)

Bladder fuselage tanks (83% Usable Volume)

Volumes Measured from Drawing:

WING { Total: 61 ft^3 at X = 275 in
Usable: 61 × .85 = 52 ft^3 = 2334 Lb

FORWARD FUSELAGE { Total: 38 ft^3 at X = 240 in
Usable: 38 × .83 = 28 ft^3 = 1257 Lb

AFT FUSELAGE { Total: 32 ft^3 at X = 295 in
Usable: 32 × .83 = 26.5 ft^3 = 1189 Lb

Fuel C.G. (desire near aircraft X_{cg} = 265 in)

Wing	2334	@	275 in	
Fwd fus.	1257	@	240 in	
Aft fus.	1189	@	295 in	
Total:	4780 Lb	@	271 in	Too much! Too far aft!

Reduce fuel in Aft fuselage tank:

Wing	2334 Lb	@	275 in	
Fwd fus.	1257	@	240 in	
Aft fus.	629	@	295 in	
Total:	4220 Lb	@	267 in	(Good!)

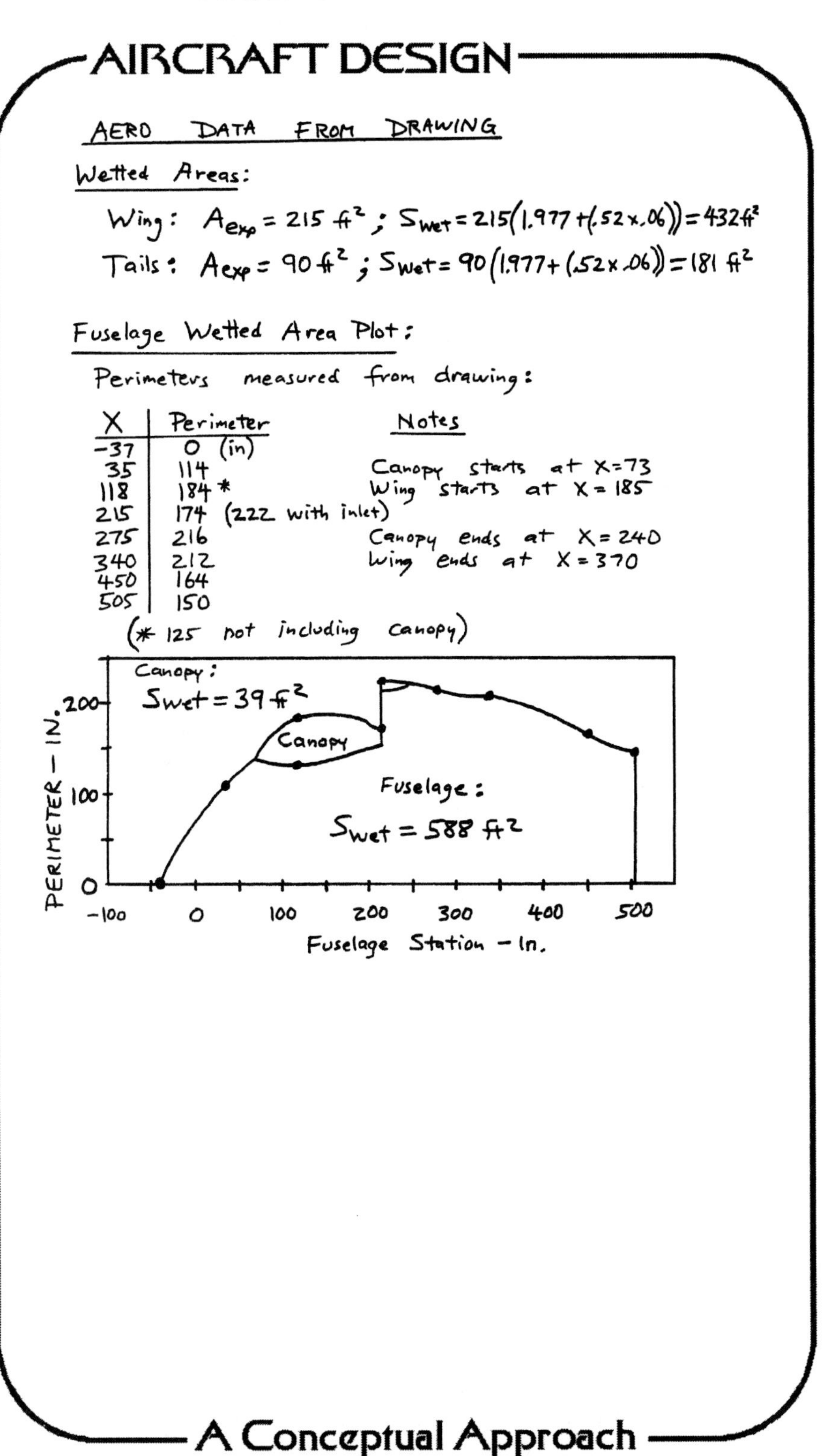

AIRCRAFT DESIGN

AERO DATA FROM DRAWING

Wetted Areas:

Wing: $A_{exp} = 215\ ft^2$; $S_{wet} = 215(1.977 + (.52 \times .06)) = 432\ ft^2$

Tails: $A_{exp} = 90\ ft^2$; $S_{wet} = 90(1.977 + (.52 \times .06)) = 181\ ft^2$

Fuselage Wetted Area Plot:

Perimeters measured from drawing:

X	Perimeter
−37	0 (in)
35	114
118	184 *
215	174 (222 with inlet)
275	216
340	212
450	164
505	150

(* 125 not including canopy)

Notes

Canopy starts at X = 73
Wing starts at X = 185
Canopy ends at X = 240
Wing ends at X = 370

A Conceptual Approach

Design Analysis

From the dimensions and areas measured off the design layout, this initial design layout of the DR-3 aircraft was analyzed using the methods presented in this book. Analysis as described next included aerodynamics, weights, propulsion, sizing, performance, and cost. Also, a *T/W-W/S* carpet plot was prepared to optimize the design. The author's RDS-Student computer program was used, which automates the "number-crunching" of these methods. RDS-Student is available from AIAA along with this textbook, and includes the complete DR-3 sample design files as described next. RDS-Student and the more powerful RDS-Professional are further described at the author's website, http://www.aircraftdesign.com, where additional RDS data files are available for download.

This analysis could have been done by hand like the first design example. Students should be capable of analysis using only a pocket calculator before they are permitted to use RDS or any other "canned" computer program.

Aerodynamic Lift and Drag

Aerodynamic analysis inputs to RDS, including surface areas and geometry for wings, tails, fuselage, canopy, and boundary-layer diverter, are shown on the next two pages. Fuselage and canopy equivalent diameters were determined from maximum cross-section areas. Skin friction drag analysis assumed fully turbulent flow over camouflage paint. *D/q* for the missile was input based upon the AIM-9 data in Fig. 12.24, and a cannon port *D/q* of 0.2 was input as a constant value from zero to Mach 2. Leakage and protuberance drag of 6% was assumed.

For wave drag analysis, the maximum total cross-section area was estimated at 20.9 ft^2 {1.94 sq.m.}, less 3.83 for capture area {0.36 sq.m.}, or a net of 17.07 ft^2 {1.58 sq.m.}. A supersonic wave drag empirical factor (E_{wd}) of 2.0 was assumed, typical of an aircraft designed with some attention to area ruling.

Maximum lift was estimated by adjusting the airfoil maximum lift for the effects of the assumed automatic leading- and trailing-edge maneuver flaps. For a 64-series airfoil, $C_{l\text{-max}}$ is about 0.82, and from Table 12.1, delta-Y is about 1.28. Using Table 12.2 the lift adjustment for trailing-edge plain flaps is about 0.9, and for leading-edge flaps, about 0.3. With hinge line angles of 10 and 39 deg, Eq. (12.21) gives a delta $C_{l\text{-max}}$ of about 0.82, giving an adjusted airfoil $C_{l\text{-max}}$ of 1.64. For landing, a $C_{L\text{-max}}$ value of 1.8 was assumed based on data for modern fighters with leading-edge flaps.

The following pages include a sample of the parasite drag calculation for one altitude and velocity, the total parasite drag as a function of speed and altitude, the parameters determined for calculation of drag-due-to-lift factor (K), a plot of the resulting K vs speed and lift coefficient, and drag polars and *L/D* ratios for the DR-3.

AERODYNAMIC INPUTS: FILE DR3.DAA
AIRCRAFT TYPE : SUPERSONIC, THIN WING

AERODYNAMIC DATA	fps	mks
Max V or M#	2.000	2.000
Max Altitude	50000.	15240.
% Laminar	0.000	0.000
k/10^5 ft	3.330	1.015
%Leak&Protub	6.000	6.000
Amax-aircrft	17.070	1.586
length-eff	45.200	13.777
Ewd	2.000	2.000
CL-cruise	0.210	0.210
WING		
# Componts	1.000	1.000
Sref-wing	294.000	27.313
Sexp-wing	215.000	19.974
A true	3.500	3.500
A effective	3.500	3.500
Lambda=Ct/Cr	0.250	0.250
Sweep-LE	38.000	38.000
t/c average	0.060	0.060
Delta Y	1.280	1.280
Q (interfer)	1.000	1.000
CL-design	0.400	0.400
CLmx-airfoil	1.640	1.640
HORIZONTAL TAIL		
# Componts	1.000	1.000
S-tail	92.000	8.547
Sexp-tail	92.000	8.547
A true	4.000	4.000
A effective	4.000	4.000
Lambda=Ct/Cr	0.340	0.340
Sweep-LE	30.000	30.000
t/c average	0.060	0.060
Delta Y	1.280	1.280
Q (interfer)	1.000	1.000
Dihedral	28.400	28.400

FUSELAGE		
# Componts	1.000	1.000
Swet	588.000	54.627
length	45.200	13.777
diam-effctiv	5.500	1.676
Q (interfer)	1.000	1.000

SMOOTH CANOPY or BLISTER or FAIRING		
# Componts	1.000	1.000
Swet	39.000	3.623
length	13.900	4.237
diam-effctiv	2.000	0.610
Q (interfer)	1.000	1.000

BOUNDARY LAYER DIVERTER		
# Componts	1.000	1.000
# Wedges	2.000	2.000
l	4.200	1.280
d	2.830	0.863
thickness	0.330	0.101

MISC D/q v.s. MACH #:

Mach Number	D/q sq-ft	sq-m
0.000	0.120	0.011
0.500	0.120	0.011
0.980	0.120	0.011
1.100	0.270	0.025
1.200	0.290	0.027
2.000	0.300	0.028

MISC D/q v.s. MACH #:

Mach Number	D/q sq-ft	sq-m
0.000	0.200	0.019
2.000	0.200	0.019

AIRCRAFT DESIGN

SAMPLE AERODYNAMIC RESULTS: FILE DR3.DAA

Altitude = 30000. ft Mach = 0.40

	R# (10^6)	Cf	FF	S-wet	Cdo
WING	11.715	28.859	1.190	431.8	53.5
HORZ TAIL	5.916	32.235	1.202	184.8	25.8
FUSELAGE	51.597	23.046	1.129	588.0	55.1
CNPY/FAIR	15.867	27.516	1.196	39.0	4.6
BL DIVRTR	4.794	33.384	1.674	2.8	0.6
Misc D/q vs M					4.327
Misc D/q vs M					7.211
TOTAL PARASITE DRAG COEFFICIENT				Cdo =	151.131

Altitude = 40000. ft Mach = 1.60

	R# (10^6)	Cf	FF	S-wet	Cdo
WING	31.421	20.520	1.000	431.8	31.9
HORZ TAIL	15.867	22.773	1.000	184.8	15.2
FUSELAGE	138.393	16.590	1.000	588.0	35.2
CNPY/FAIR	42.559	19.618	1.000	39.0	2.8
BL DIVRTR	12.860	23.535	1.000	2.8	0.2
Misc D/q vs					10.636
Misc D/q vs					7.211
Wave Drag coeff.				Cdw =	122.0
TOTAL PARASITE DRAG COEFFICIENT				Cdo =	225.126

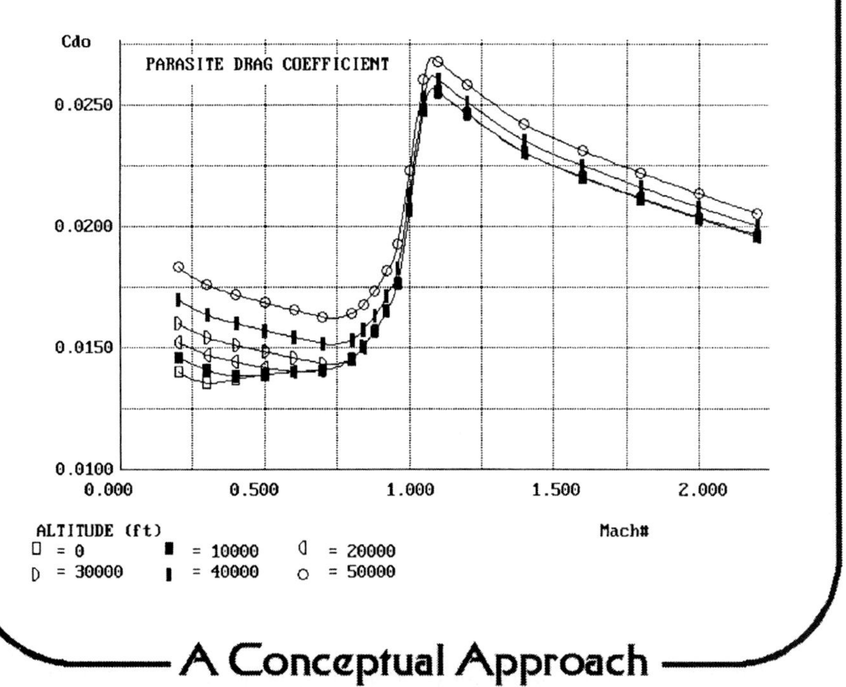

A Conceptual Approach

DRAG-DUE-TO-LIFT (K) FACTOR: REFERENCE AREA = 294

Aspect Ratio = 3.50
Leading edge sweep = 38.00
Sweep of maximum thickness line = 23.67
Sexposed/Sref = 0.73
Fuselage lift factor F = 1.47

Mach number	CL-ALPHA	1/CL-ALPHA
0.2000	3.6717	0.2724
0.3000	3.7163	0.2691
0.4000	3.7821	0.2644
0.5000	3.8729	0.2582
0.6000	3.9951	0.2503
0.7000	4.1587	0.2405
0.8000	4.3809	0.2283
0.8400	4.4925	0.2226
0.8800	4.6215	0.2164
0.9200	4.7722	0.2095
0.9600	4.9507	0.2020
1.0000	5.1222	0.1952
1.0500	5.2966	0.1888
1.1000	5.3923	0.1855
1.2000	5.2005	0.1923
1.4000	4.1591	0.2404
1.6000	3.3519	0.2983
1.8000	2.8183	0.3548
2.0000	2.4413	0.4096
2.2000	2.1628	0.4624

Lift Coeff.	% Suction
0.1500	0.6100
0.2500	0.8100
0.3500	0.9400
0.4500	0.9400
0.5500	0.8700
0.6500	0.7200
0.8000	0.5100
1.0000	0.3300
1.2000	0.2400
1.4000	0.0000

K-100% = 1/PIxAspect Ratio = 0.0909
M# for Sonic leading edge = 1.2691
CLmax at Mach 0.2 = 1.7942

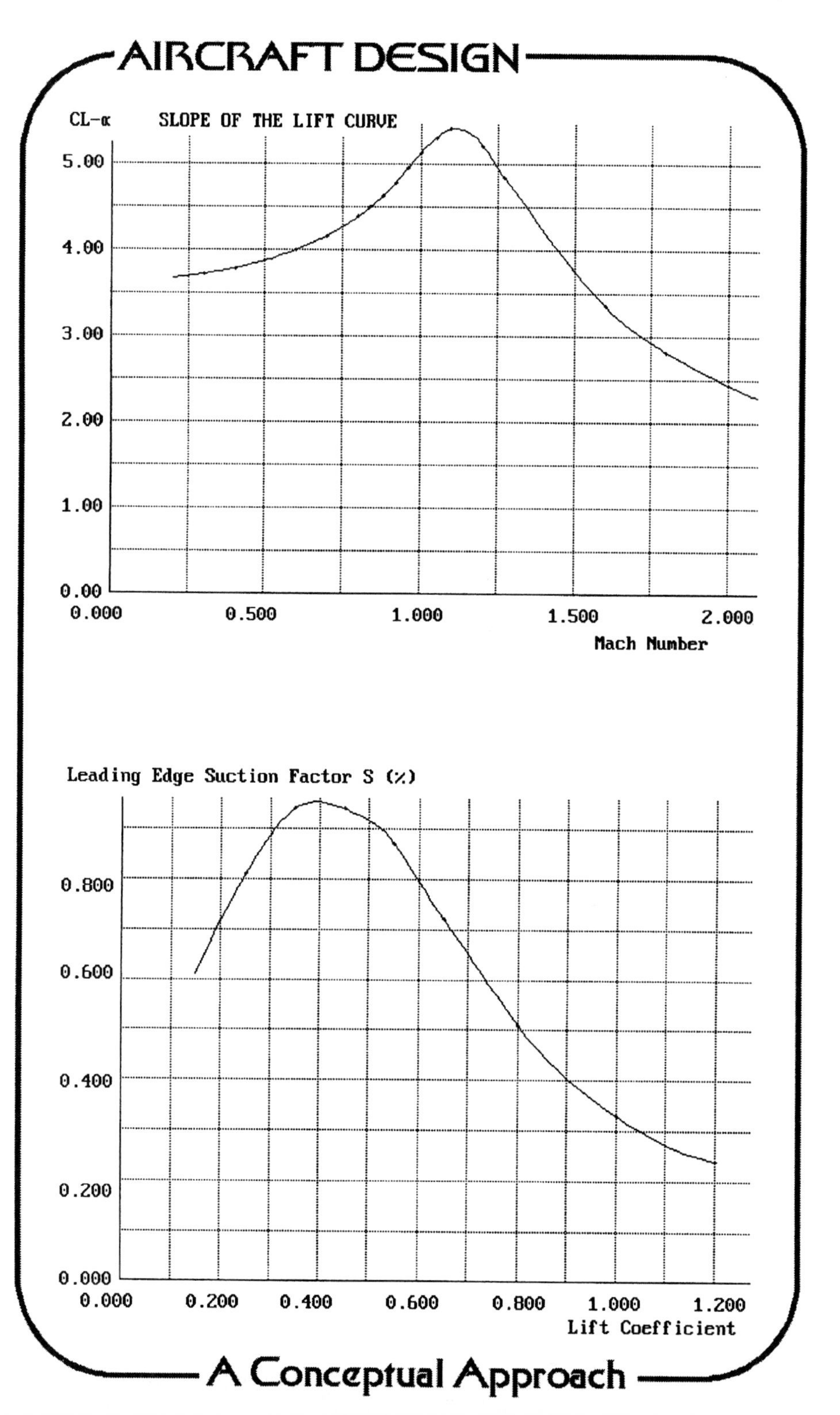
AIRCRAFT DESIGN
CL-α
SLOPE OF THE LIFT CURVE
5.00
4.00
3.00
2.00
1.00
0.00
0.000
0.500
1.000
1.500
2.000
Mach Number
Leading Edge Suction Factor S (%)
0.800
0.600
0.400
0.200
0.000
0.000
0.200
0.400
0.600
0.800
1.000
1.200
Lift Coefficient
A Conceptual Approach

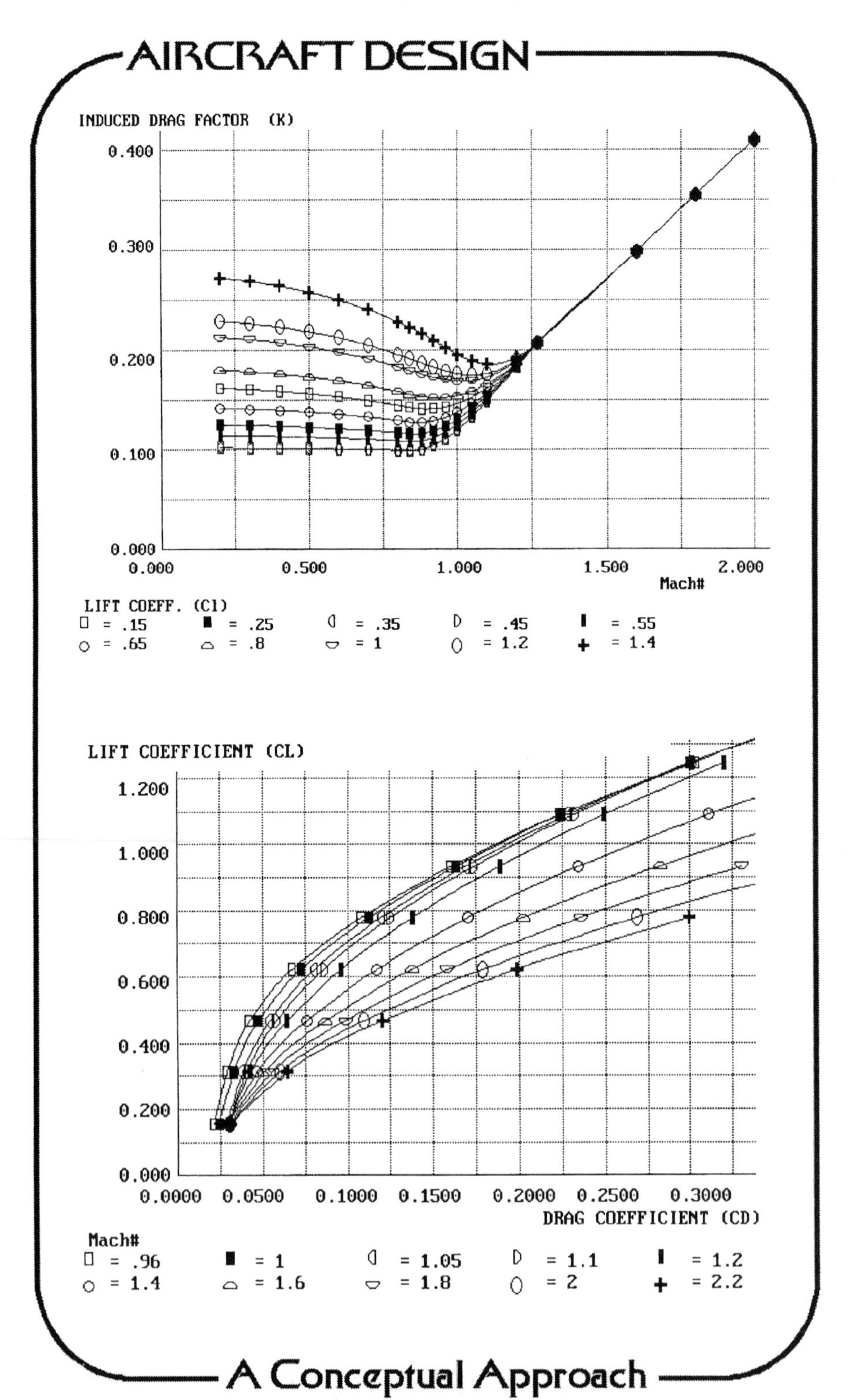
AIRCRAFT DESIGN
INDUCED DRAG FACTOR (K)
0.400
0.300
0.200
0.100
0.000
0.000
0.500
1.000
1.500
2.000
Mach#
LIFT COEFF. (Cl)
= .15
= .25
= .35
= .45
= .55
= .65
= .8
= 1
= 1.2
= 1.4
LIFT COEFFICIENT (CL)
1.200
1.000
0.800
0.600
0.400
0.200
0.000
0.0000
0.0500
0.1000
0.1500
0.2000
0.2500
0.3000
DRAG COEFFICIENT (CD)
Mach#
= .96
= 1
= 1.05
= 1.1
= 1.2
= 1.4
= 1.6
= 1.8
= 2
= 2.2
A Conceptual Approach

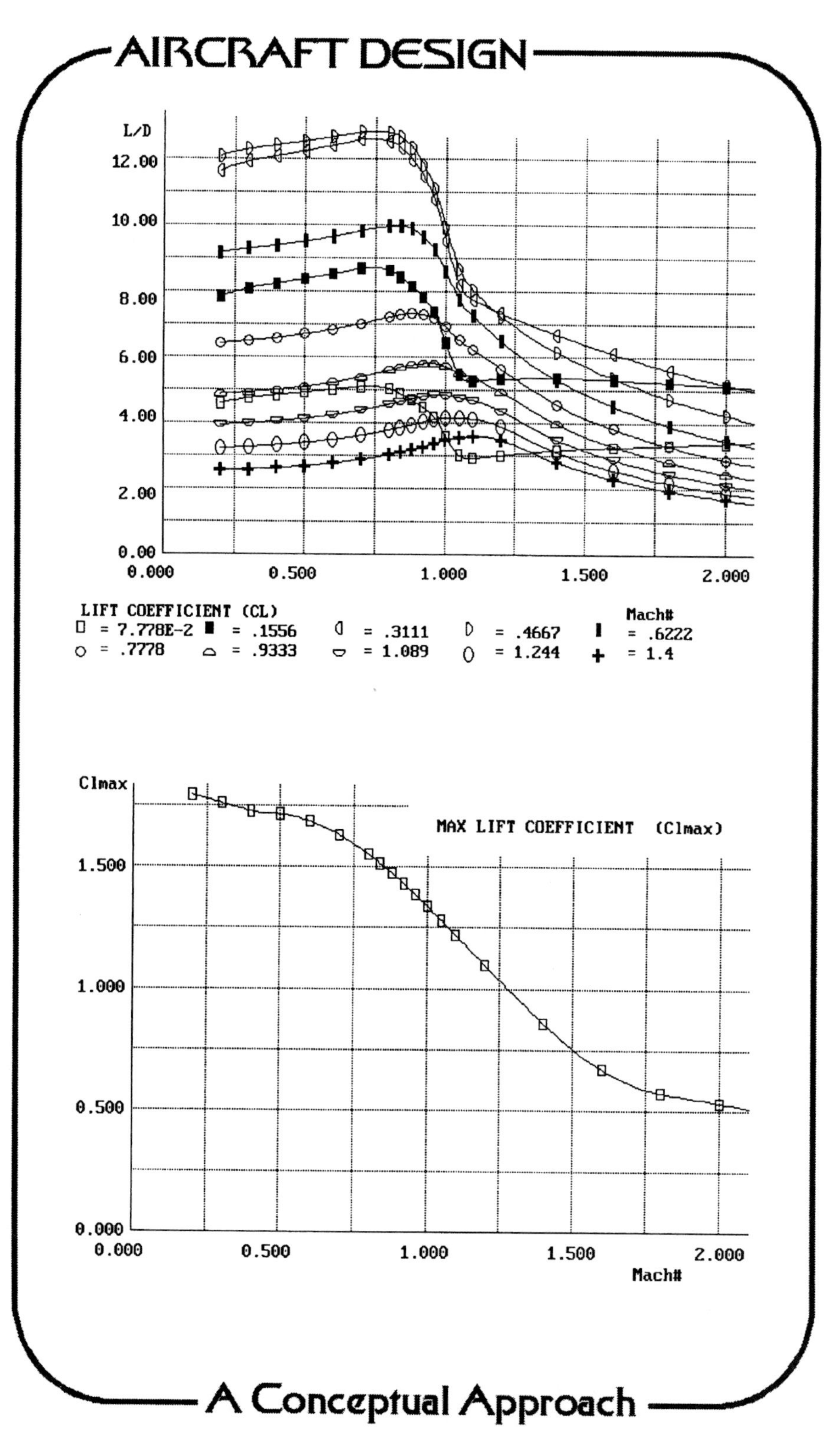
AIRCRAFT DESIGN
L/D
12.00
10.00
8.00
6.00
4.00
2.00
0.00
0.000
0.500
1.000
1.500
2.000
LIFT COEFFICIENT (CL)
Mach#
= 7.778E-2
= .1556
= .3111
= .4667
= .6222
= .7778
= .9333
= 1.089
= 1.244
= 1.4
Clmax
1.500
1.000
0.500
0.000
0.000
0.500
1.000
1.500
2.000
Mach#
MAX LIFT COEFFICIENT (Clmax)
A Conceptual Approach

ADDITIONAL DRAGS FOR TAKEOFF AND LANDING

FLAPS: (Only used for landing)

- Assume $\delta_{flap} = 60°$; from drawing $\left\{\frac{\text{flap span}}{\text{wing span}} = .43\right\}$

- Assume aileron used as flap also, with $\delta_a = 30°$

  from drawing $\left\{\frac{\text{aileron span}}{\text{Wing span}} = .35\right\}$

eq 12.37) $\Delta C_{D_0} = .0023((.43 \times 60) + (.35 \times 30)) = .0835$

LANDING GEAR: (Using values from table 12.4)

	Frontal Area	×	Table 12.4 factor	=	D/q
Main Wheels:	2.1 ft²		.25		.53
Main Struts:	5.8 ft²		.30		1.74
Nose Wheel:	.7 ft²		.25		.18
Nose Strut:	3.2 ft²		.30		.96
			Subtotal		3.41

Increase 20% for interference
Increase 7% for open wheelwells } $D/q = 4.38$

$$\Delta C_{D_{0_{gear}}} = \frac{4.38}{294} = .0149$$

Weights

Weights analysis for the DR-3 was done using the fighter equations in Chapter 15, with adjustments for composite material usage as in Table 15.4. The as-drawn takeoff weight of 16,480 lb {7475 kg} was used as the structural design weight, with ultimate load factor of Nz=7.33x1.5, or 11. Required dimensions and areas, such as the control surface area for the wing, were measured from the drawing.

The V-tail was analyzed with the horizontal tail statistical equation, because a V-tail is loaded for trim and maneuvers much like a horizontal tail. The analysis was based on the true area and aspect ratio of the surface, i.e., 90 ft^2 {8.4 sq.m}, and $A=6.5$ (equivalent). To account for the variable dihedral mechanism and structural requirements, a 200-lb {90.7-kg} weight penalty was added as a part of the miscellaneous empty weight.

For the landing gear, it was assumed that landing weight equals design takeoff weight, and gear load factor was assumed to be 4.

For engine cooling it was assumed that a shroud covers the entire engine, so the shroud length is 14 ft {4.3 m}. Engine control length was estimated from the drawing as 18.3 ft {5.6 m}. To allow for the extra weight of the 2-D vectoring nozzle, an additional 400 lb {181.4 kg} was added to the misc. empty weight.

In the absence of better data, installed avionics weight of 990 lb {449 kg} was guessed using Table 11.6. However, the RDS program requires uninstalled avionics weight, which it uses to estimate installed avionics weight, so Eq. (15.21) was used to back out an uninstalled weight of 727 lb {330 kg}.

The gun was assumed to always stay with the aircraft and so was treated as an addition to the misc. empty weight of 400 lb {181.4 kg}.

The following pages provide the complete weights assumptions and inputs, followed by the resulting summary weights statement. The empty weight for the as-drawn takeoff weight of 16,480 lbs {7475 kg} was determined to be 10,947.2 lb {4965.6 kg}, somewhat above the preliminary prediction of 10,788 lb {4893 kg} used for initial sizing.

Weights estimation inputs and results are tabulated in the following.

AIRCRAFT DESIGN

WEIGHTS INPUTS : FILE DR3.DWT
AIRCRAFT TYPE : FIGHTER/ATTACK

AIRCRAFT DATA	fps	mks
Wo (TOGW)	16480.000	7475.196
Wdg (flight)	16480.000	7475.196
Nz (ultimate)	11.000	11.000
Sw	294.000	27.313
M design max	1.800	1.800
#Engines	1.000	1.000
WING		
X-Location	23.300	7.102
Kdw DeltaWng	1.000	1.000
Kvs VarSweep	1.000	1.000
A	3.500	3.500
t/c	0.060	0.060
lambda Ct/Cr	0.250	0.250
sweep c/4	30.000	30.000
Scsw	72.000	6.689
HORIZONTAL TAIL		
X-Location	39.200	11.948
Fw FusW@Tail	4.700	1.433
Ah	6.500	6.500
Sht	90.000	8.361
FUSELAGE		
X-Location	21.700	6.614
Kdwf DeltWng	1.000	1.000
L FusLength	39.000	11.887
D FusDepth	4.000	1.219
W FusWidth	5.400	1.646
LANDING GEAR		
X-Location	23.800	7.254
KcbCrossBeam	1.000	1.000
Ktpg Tripod	1.000	1.000
Wl LandWt	16480.000	7475.196
Nl n-Land	4.000	4.000
Lm (in)	46.000	1.168
Ln (in)	52.000	1.321
Nnw #nosewhl	1.000	1.000
NoseGear Loc	13.000	3.962

A Conceptual Approach

NACELLE

X-Location	33.300	10.150
T per engine	16150.400	71.840
Sfw Firewall	52.000	4.831
Wengine	1517.000	688.099
Kvg VarGeom	1.000	1.000
Ld DuctLngth	10.700	3.261
Kd DuctConst	1.310	1.310
Ls NoSplitLn	2.000	0.610
De EngDiam	2.700	0.823

ENGINE INSTALLATION

X-Location	33.300	10.150
Ltp Tailpipe	0.000	0.000
Lsh shroud	14.000	4.267
Lec Eng-Ckpt	18.300	5.578
Wt-Oil	50.000	22.680

FUEL SYSTEM gal, lbs-m

X-Location	22.250	6.782
Vt TotalVol	703.300	2662.286
Vi Integral	389.000	1472.528
Vp Protected	314.300	1189.757
Nt #tanks	3.000	3.000
SFC (max T)	1.900	53.804

CONTROLS & INSTRUMENTS

X-Location	21.700	6.614
Scs Controls	94.000	8.733
Ns #systems	4.000	4.000
Nc #crew	1.000	1.000
Nci #crewEqv	1.000	1.000

HYDRAULICS & ELECTRICS

X-Location	21.700	6.614
Kvsh VarSwep	1.000	1.000
Nu #HydrUtil	10.000	10.000
Kmc MsnCmplt	1.450	1.450
Rkva ElectRt	120.000	120.000
La ElectRout	25.000	7.620
Ngenerators	1.000	1.000

```
AIR CONDITIONING & FURNISHINGS
     X-Location          8.300     2.530

LOADS; MISC WT; & AVIONICS
     Wuav              727.000   329.761
     -Location          10.000     3.048
     Wcrew             220.000    99.790
     -Location          10.000     3.048
     Wpayload          840.000   381.017
     -Location          21.700     6.614
     Wmisc(empty)     1000.000   453.592
     -Location          31.800     9.693
```

WEIGHTS RESULTS

FIGHTER/ATTACK GROUP WEIGHT STATEMENT: FPS Units

STRUCTURES GROUP	**4526.2**	**EQUIPMENT GROUP**	**3066.7**
Wing	1459.4	Flight Controls	655.7
Horiz. Tail	280.4	Instruments	122.8
Vert. Tail	0.0	Hydraulics	171.7
Fuselage	1574.0	Electrical	713.2
Main Lndg Gear	631.5	Avionics	989.8
Nose Lndg Gear	171.1	Furnishings	217.6
Engine Mounts	39.1	Air Conditioning	190.7
Firewall	58.8	Handling Gear	5.3
Engine Section	21.0	**MISC EMPTY WEIGHT**	**1000.0**
Air Induction	291.1	**TOTAL WEIGHT EMPTY**	**10947.2**
PROPULSION GROUP	**2354.3**	**USEFUL LOAD GROUP**	**5532.8**
Engine(s)	1517.0	Crew	220.0
Tailpipe	0.0	Fuel	4422.8
Engine Cooling	172.0	Oil	50.0
Oil Cooling	37.8	Payload	840.0
Engine Controls	20.0	Passengers	0.0
Starter	39.5	Misc Useful Load	0.0
Fuel System	568.0	**TAKEOFF GROSS WEIGHT**	**16480.0**

EMPTY CG= 23.8 LOADED-NO FUEL CG= 23.4 GROSS WT CG= 23.1

FIGHTER/ATTACK GROUP WEIGHT STATEMENT: MKS Units

Item	Weight	Item	Weight
STRUCTURES GROUP	2053.1	EQUIPMENT GROUP	1391.0
Wing	662.0	Flight Controls	297.4
Horiz. Tail	127.2	Instruments	55.7
Vert. Tail	0.0	Hydraulics	77.9
Fuselage	713.9	Electrical	323.5
Main Lndg Gear	286.4	Avionics	448.9
Nose Lndg Gear	77.6	Furnishings	98.7
Engine Mounts	17.7	Air Conditioning	86.5
Firewall	26.7	Handling Gear	2.4
Engine Section	9.5	MISC EMPTY WEIGHT	453.6
Air Induction	132.0	TOTAL WEIGHT EMPTY	4965.6
PROPULSION GROUP	1067.9	USEFUL LOAD GROUP	2509.6
Engine(s)	688.1	Crew	99.8
Tailpipe	0.0	Fuel	2006.1
Engine Cooling	78.0	Oil	22.7
Oil Cooling	17.2	Payload	381.0
Engine Controls	9.1	Passengers	0.0
Starter	17.9	Misc Useful Load	0.0
Fuel System	257.6	TAKEOFF GROSS WEIGHT	7475.2

EMPTY CG= 7.2 LOADED-NO FUEL CG= 7.1 GROSS WT CG= 7.0

Propulsion

The DR-3 uses the afterburning turbofan engine in Appendix E.1, with an assumed reduction in specific fuel consumption (SFC or C) of 20% to adjust for "advanced technology." The uninstalled engine, which already includes reasonable allowances for the installation effects of bleed and power extraction, was loaded into RDS data arrays of maximum afterburning thrust and SFC, and maximum dry power thrust and SFC. Then installation values for inlet recovery, inlet drag, and nozzle drag were input and used to calculate installed thrust and SFC throughout the flight envelope.

The engine was assumed to be "rubber," i.e., it could be scaled to virtually any size and thrust to meet the thrust needs of the sized aircraft. This is common in early design studies when we can assume that a new engine will be designed and built for our aircraft design. Note that one should normally scale a "rubber" engine by no more than about 20–30%, and this example required far greater scaling (downward in size). In a real design study, learning that the proposed engine was far too large would lead to selection of a different engine, not to a scaling of that engine by a factor of 50% or more!

Following are the input assumptions, analysis calculations, and graphs of the installed thrust and SFC for the unscaled engine. Note that scaling of the engine is done during mission sizing for this design example.

INLET RECOVERY

- E.1 incorporates $\{P_1/P_0 = .97\}$ subsonic, and $\{.97\}$ times the recovery schedule of Mil E 5008B for supersonic operation. Our inlet is F-104-like, but we assume advanced technology and auxiliary intakes to improve performance at lower speeds.

INLET DRAG (fig. 13.6, axisymmetric inlet)

$$\left.\begin{array}{l} A_c = 552 \text{ in}^2 = 3.83 \text{ ft}^2 \\ S_{ref} = 294 \text{ ft}^2 \end{array}\right\} D_{inlet} = 3.83\,q \left(\frac{D/q}{A_c}\right)_{\text{fig. 13.6}}$$

AIRCRAFT DESIGN

PROPULSION INPUTS: FILE DR3.DPR
AIRCRAFT TYPE : JET PROPELLED

PROPULSION DATA	fps	mks
Thrust-net	16150.400	71.840
SFC Fudge	0.800	0.800
Acapture	3.830	0.356
C-bleed	0.000	0.000
bleed ratio	0.000	0.000
Nozzle Cd	0.015	0.015
Amax-nacelle	16.900	1.570

Mach Number	P1/PoREF	Mach Number	P1/PoACT
0.200	0.970	0.200	0.880
0.400	0.970	0.400	0.940
0.600	0.970	0.600	0.965
0.800	0.970	0.800	0.970
1.000	0.970	1.000	0.970
1.200	0.962	1.200	0.968
1.400	0.949	1.400	0.960
1.600	0.933	1.600	0.945
1.800	0.916	1.800	0.912
2.000	0.897	2.000	0.830
2.200	0.877	2.200	0.720
2.400	0.855	2.400	0.600

Mach Number	Ram Factor	Mach Number	InletDrag
0.200	1.350	0.200	0.010
0.400	1.350	0.400	0.025
0.600	1.350	0.600	0.040
0.800	1.350	0.800	0.070
1.000	1.350	1.000	0.110
1.200	1.320	1.200	0.140
1.400	1.290	1.400	0.145
1.600	1.260	1.600	0.135
1.800	1.230	1.800	0.120
2.000	1.200	2.000	0.080
2.200	1.170	2.200	0.060
2.400	1.140	2.400	0.040

A Conceptual Approach

PROPULSION ANALYSIS: TURBOFAN FPS Units

M#	Inlet loss	Bleed loss	Inlet D/q	Nozzle D/q (sq-ft)
0.4	0.0405	0.0000	0.0957	0.2535
0.6	0.0068	0.0000	0.1532	0.2535
0.8	0.0000	0.0000	0.2681	0.2535
1.0	0.0000	0.0000	0.4213	0.2535
1.2	-0.0083	0.0000	0.5362	0.2535
1.4	-0.0143	0.0000	0.5553	0.2535
1.6	-0.0145	0.0000	0.5171	0.2535
1.8	0.0051	0.0000	0.4596	0.2535
2.0	0.0807	0.0000	0.3064	0.2535
2.2	0.1836	0.0000	0.2298	0.2535
2.4	0.2912	0.0000	0.1532	0.2535

PROPULSION ANALYSIS: TURBOFAN MKS Units

M#	Inlet loss	Bleed loss	Inlet D/q	Nozzle D/q (sq-m)
0.4	0.0405	0.0000	0.0089	0.0236
0.6	0.0068	0.0000	0.0142	0.0236
0.8	0.0000	0.0000	0.0249	0.0236
1.0	0.0000	0.0000	0.0391	0.0236
1.2	-0.0083	0.0000	0.0498	0.0236
1.4	-0.0143	0.0000	0.0516	0.0236
1.6	-0.0145	0.0000	0.0480	0.0236
1.8	0.0051	0.0000	0.0427	0.0236
2.0	0.0807	0.0000	0.0285	0.0236
2.2	0.1836	0.0000	0.0213	0.0236
2.4	0.2912	0.0000	0.0142	0.0236

Note that suitable bleed losses were already included in uninstalled data.

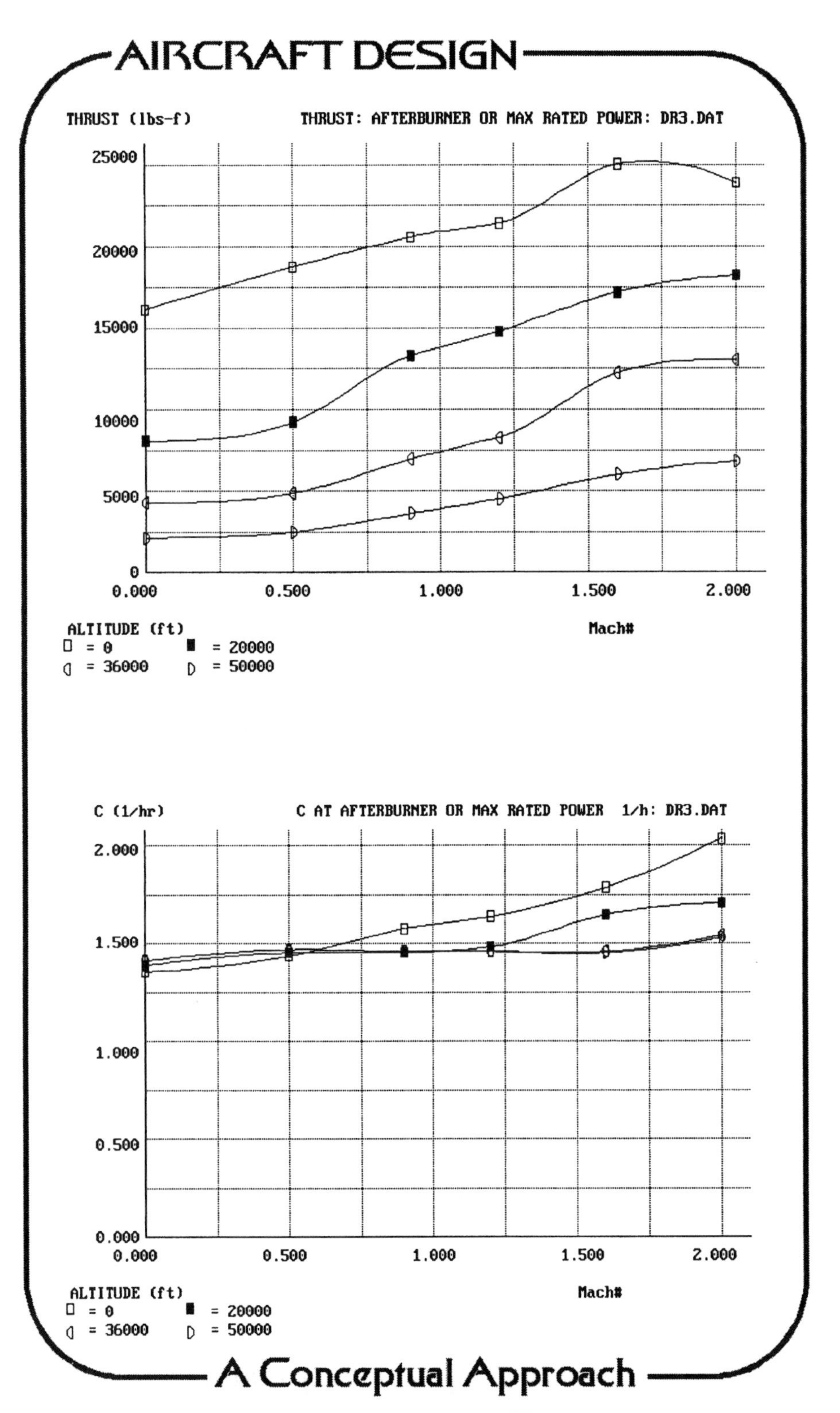

AIRCRAFT DESIGN
THRUST (lbs-f)
THRUST: AFTERBURNER OR MAX RATED POWER: DR3.DAT
25000
20000
15000
10000
5000
0
0.000
0.500
1.000
1.500
2.000
Mach#
ALTITUDE (ft)
= 0
= 20000
= 36000
= 50000
C (1/hr)
C AT AFTERBURNER OR MAX RATED POWER 1/h: DR3.DAT
2.000
1.500
1.000
0.500
0.000
0.000
0.500
1.000
1.500
2.000
Mach#
ALTITUDE (ft)
= 0
= 20000
= 36000
= 50000
A Conceptual Approach

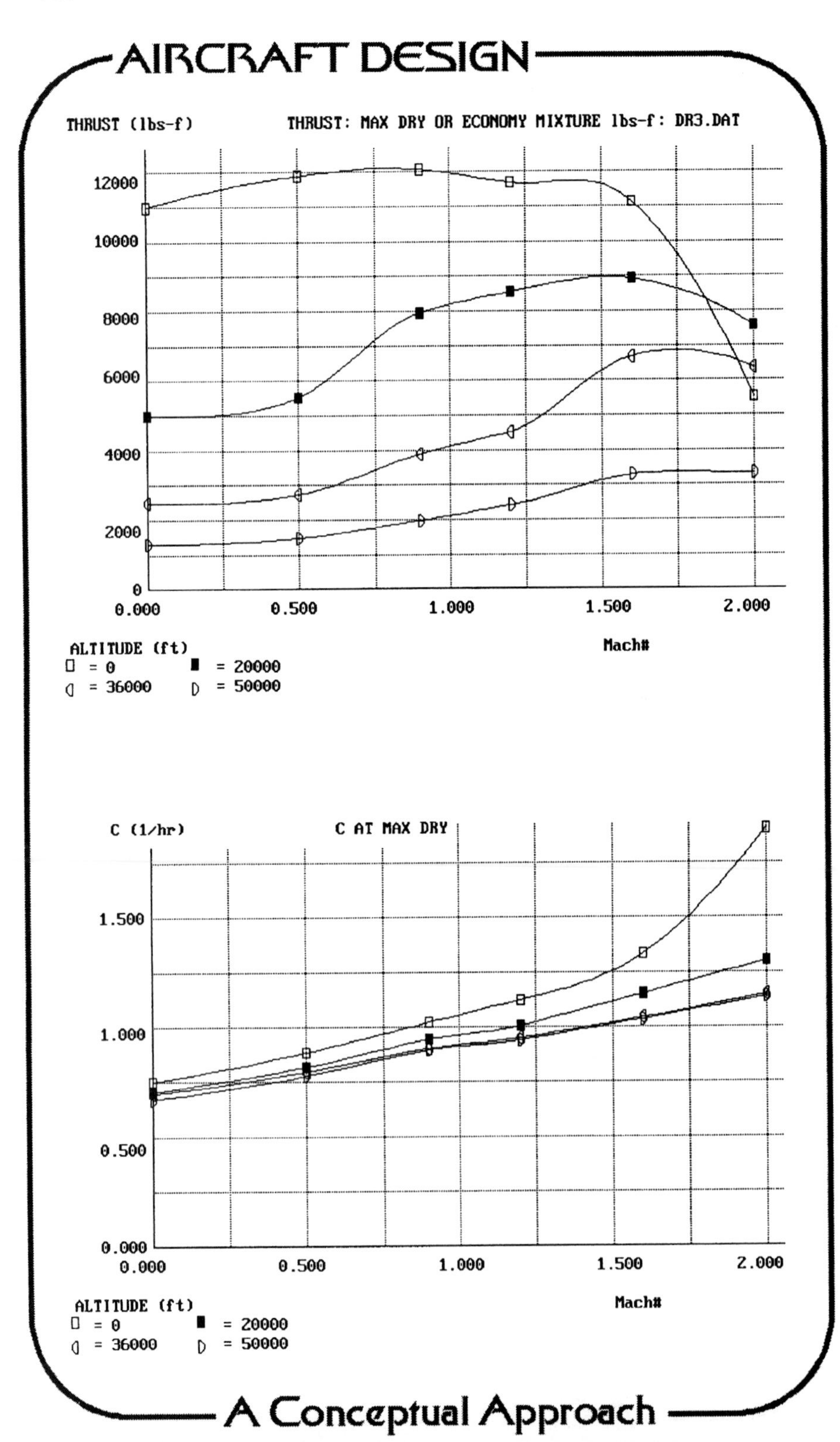
AIRCRAFT DESIGN
THRUST (lbs-f)
THRUST: MAX DRY OR ECONOMY MIXTURE lbs-f: DR3.DAT
12000
10000
8000
6000
4000
2000
0
0.000
0.500
1.000
1.500
2.000
Mach#
ALTITUDE (ft)
= 0
= 20000
= 36000
= 50000
C (1/hr)
C AT MAX DRY
1.500
1.000
0.500
0.000
0.000
0.500
1.000
1.500
2.000
Mach#
ALTITUDE (ft)
= 0
= 20000
= 36000
= 50000
A Conceptual Approach

Stability and Control

An advanced fighter concept such as this would use relaxed static stability with an active, fully computerized flight control system as described in Chapters 11 and 16. For such a concept, early first-order stability and control calculations are not so critical because the aircraft's handling qualities will be determined largely by the computer programming, provided that basic requirements including tail sizes and center-of-gravity location are met and the aircraft has adequately sized control surfaces.

Tails were sized using historical volume coefficients (Chapter 6) and should prove sufficient, pending detailed calculations and/or wind-tunnel testing. The wing location was set to give approximately the desired stability level, in this case slightly negative at subsonic speeds and not too positive (i.e., relaxed stability) at supersonic speeds. From the center of gravity estimated previously, we can use the first-order methods of Chapter 16 to estimate neutral point and hence, static margin. This was done using RDS-Student, with results at M0.6 as detailed next and as graphed next as a function of Mach number. As can be seen, this design is just slightly unstable at subsonic speeds—good for minimizing trim drag. Supersonically, the design is too stable, but this analysis does not take into account the variable dihedral tails. When brought to an upright position, they will destabilize the aircraft preventing a trim drag penalty—exactly as intended.

Following the static margin is a subsonic trim plot showing a slight instability. Note that it takes only a small elevator deflection to trim.

```
WING:                 S-ref =   294.00
          Sexposed/Sref =     0.73
           Aspect Ratio =     3.50
       Mean aero. chord =   10.26
     Leading edge sweep  =    38.00
     Sweep-max thickness =    23.67
           Cl-α (1/rad) =     3.99
                   Xmac =    23.30
TAIL:
                  S-ref =    92.00
           Aspect Ratio =     4.00
     Leading edge sweep  =    30.00
     Sweep-max thickness =    18.32
           dε-tail / dα =     0.46
           Cl-α (1/rad) =     3.12
                   Xmac =    39.20

         FUSELAGE Cm-α =    0.81
       X-NEUTRAL POINT =   23.02
                  X-cg =   23.10
     STATIC MARGIN (%) =   -0.78
                  Cm-α =  0.0310

 TRIM POINT:  CRUISE CL =  0.21
                     α =   2.71 deg
                    δe =   0.11 deg
```

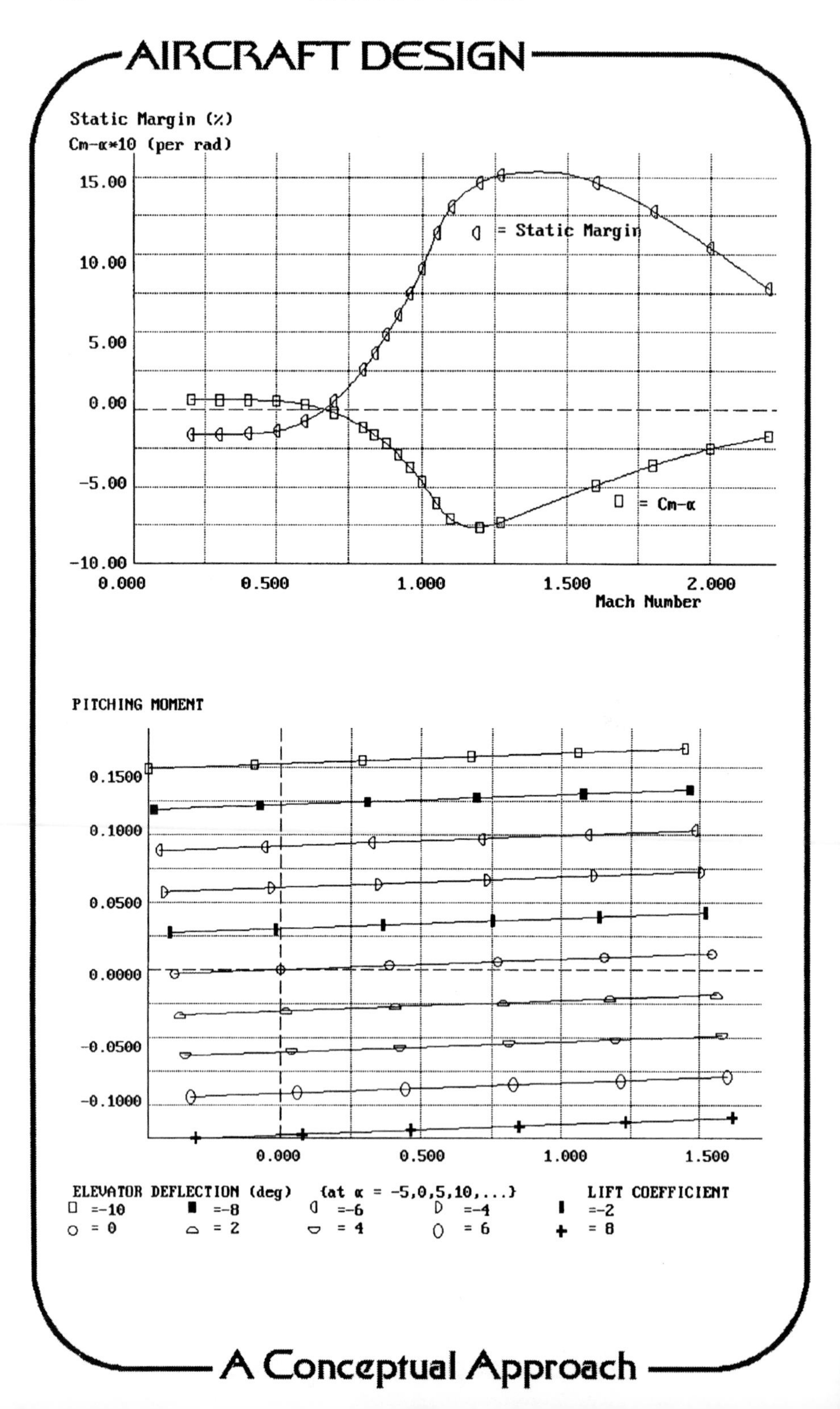
AIRCRAFT DESIGN
Static Margin (%)
Cm-α*10 (per rad)
15.00
10.00
5.00
0.00
-5.00
-10.00
0.000
0.500
1.000
1.500
2.000
Mach Number
= Static Margin
= Cm-α
PITCHING MOMENT
0.1500
0.1000
0.0500
0.0000
-0.0500
-0.1000
0.000
0.500
1.000
1.500
ELEVATOR DEFLECTION (deg) {at α = -5,0,5,10,...}
LIFT COEFFICIENT
=-10
=-8
=-6
=-4
=-2
= 0
= 2
= 4
= 6
= 8
A Conceptual Approach

Refined Sizing and Performance

Next the DR-3 was sized to the design mission, and the required performance values were calculated, using the methods of the book as implemented in the RDS-Student computer program. A detailed mission model was developed in RDS matching the exact mission profile given with the design requirements. Several assumptions and estimations were used. Takeoff was assumed to consist of five minutes at Military (dry) power. Cruise was performed at M0.85 at 45,000 ft {13716 m}, determined from the range optimization plot to maximize range per unit fuel weight used. Best loiter speed at sea level was determined to be M0.29, found using the loiter optimization plot, which calculates seconds of loiter per unit fuel weight used. A 50-n miles {92.6-km} range credit was assumed for the descent from high-altitude cruise down to sea level.

Mission sizing, as detailed next, resulted in a resized takeoff gross weight of 17,062 lb {7739 kg} vs our original as-drawn weight of 16,480 lb {7475 kg}. This is closer than one would usually hope for and probably reflects luck more than skill!

After the mission sizing results are two trade studies, one of the sizing effect of a change in range requirement and one of the effect of an increase or decrease in engine fuel consumption. These were automatically done with the RDS-Professional program. However, these could have easily been done with RDS-Student by parametrically changing range or SFC and rerunning the sizing calculation, recording the resulting sized takeoff gross weights, and plotting the answers in a format similar to these.

Performance calculations were done at a combat weight of 0.89 times takeoff weight, which is the weight at the end of cruise (beginning of combat). The RDS program was again used to perform the calculations and make the graphs, but the same results would be obtained using a pocket calculator and the methods of this book (and a lot of time!).

Performance calculations follow the sizing results, including takeoff, landing, Ps, turn rate, and acceleration. Also included are plots of flight envelope, specific range, rate of climb, Ps, and turn rate.

Note that while most performance requirements were met, the requirement for acceleration from M.9 to M1.4 in 30 s was ***not met*** by the baseline DR-3 design! Either the thrust will have to be increased, or the requirement will have to be relaxed. This is further discussed in the carpet plot optimization at the end of this design example.

AIRCRAFT DESIGN

Sizing Calculations (fps):

```
SEGMENT  1   : TAKEOFF
    W/S = 56.05   T/W =0.666   C = 0.7518   E = 0.0830
MISSION SEGMENT WEIGHT FRACTION = 0.9584

SEGMENT  2   : CLIMB/ACCELERATE
    START ALT =      0.0    END ALT = 45000.0
    START VEL =    0.200    END VEL =    0.850
    W/S =  53.72    T/W = 0.378
     CL = 0.2017    CD0 = 0.0142    K = 0.1349   L/D =10.2636
     C  = 0.8634     Ps = 11411.14  fpm
                  TIME TO CLIMB =     4.92 min
              DISTANCE TRAVELED =    32.9061 nmi
MISSION SEGMENT WEIGHT FRACTION = 0.9736

SEGMENT  3   : CRUISE
    SEGMENT RANGE = 200.0      CLIMB/DESCENT RANGE CREDIT =    32.91
    CRUISE SPEED   = 487.2 kts = Mach 0.850    AT ALTITUDE = 45000.0 ft
    T/W AVAILABLE = 0.167  T/W REQUIRED = 0.083  THRUST SETTING USED =
49.7 %
    W/S = 52.30         CLmax(usable) = 1.5036
    CL  = 0.3358   CD0 = 0.0164    K = 0.1016   L/D =12.0539   C = 0.9948
  SPECIFIC RANGE (nmi/lb) = 0.3842
MISSION SEGMENT WEIGHT FRACTION = 0.9721

SEGMENT  4   : CLIMB/ACCELERATE
    START ALT = 45000.0    END ALT = 35000.0
    START VEL =    0.850    END VEL =    1.400
    W/S =  50.84    T/W = 0.504
     CL = 0.1208    CD0 = 0.0248    K = 0.1977   L/D = 4.3604
     C  = 1.4602     Ps = 19741.05  fpm
                  TIME TO CLIMB =     0.41 min
              DISTANCE TRAVELED =     4.8060 nmi
MISSION SEGMENT WEIGHT FRACTION = 0.9950

SEGMENT  5   : CRUISE
    SEGMENT RANGE =  50.0      CLIMB/DESCENT RANGE CREDIT =     4.81
    CRUISE SPEED   = 806.4 kts = Mach 1.400    AT ALTITUDE = 35000.0 ft
    T/W AVAILABLE = 0.390  T/W REQUIRED = 0.333  THRUST SETTING USED =
85.3 %
    W/S = 50.59         CLmax(usable) = 0.8587
    CL  = 0.0741   CD0 = 0.0233    K = 0.2429   L/D = 3.0059   C = 1.0156
  SPECIFIC RANGE (nmi/lb) = 0.1606
MISSION SEGMENT WEIGHT FRACTION = 0.9813

SEGMENT  6  KNOWN TIME FUEL BURN
    W/S = 49.64   T/W AVAILABLE =0.910   THRUST SETTING USED =100.0 %
      C = 1.4538    E = 0.0500
MISSION SEGMENT FRACTION OR FUEL BURNED =     0.9339

SEGMENT  7   : WEIGHT DROP
WEIGHT DROPPED =    400.0
    SEGMENT IS TURN AROUND POINT FOR THE MISSION - UNUSED CLIMB RANGE
CREDIT IS CANCELLED

SEGMENT  8   : CLIMB/ACCELERATE
    START ALT = 20000.0    END ALT = 35000.0
    START VEL =    0.900    END VEL =    1.400
    W/S =  45.00    T/W = 0.919
     CL = 0.0582    CD0 = 0.0242    K = 0.2020   L/D = 2.3363
     C  = 1.4847     Ps = 37144.35  fpm
                  TIME TO CLIMB =     0.89 min
              DISTANCE TRAVELED =    11.0734 nmi
MISSION SEGMENT WEIGHT FRACTION = 0.9800
```

A Conceptual Approach

```
SEGMENT  9  : CRUISE
    SEGMENT RANGE =  50.0      CLIMB/DESCENT RANGE CREDIT =   11.07
    CRUISE SPEED  = 806.4 kts = Mach 1.400    AT ALTITUDE = 35000.0 ft
    T/W AVAILABLE = 0.447  T/W REQUIRED = 0.377  THRUST USED =  84.2 %
    W/S = 44.10          CLmax(usable) = 0.8587
    CL  = 0.0645   CD0 = 0.0233    K = 0.2429   L/D = 2.6546   C = 1.0176
  SPECIFIC RANGE (nmi/lb) = 0.1624
MISSION SEGMENT WEIGHT FRACTION = 0.9817

SEGMENT  10  : CLIMB/ACCELERATE
    START ALT = 35000.0    END ALT = 45000.0
    START VEL =   1.400    END VEL =   0.850
    W/S =  43.29   T/W = 0.507

  NEGATIVE DELTA ENERGY HEIGHT - CHECK YOUR INPUTS !  SEGMENT IGNORED
     CL = 0.1544   CD0 = 0.0222    K = 0.1470   L/D = 6.0063
     C  = 1.4604    Ps = 19985.68  fpm
                 TIME TO CLIMB =    0.00 min
             DISTANCE TRAVELED =     0.0000 nmi
MISSION SEGMENT WEIGHT FRACTION = 1.0000

SEGMENT  11  : CRUISE
    SEGMENT RANGE = 200.0      CLIMB/DESCENT RANGE CREDIT =   50.00
    CRUISE SPEED  = 487.2 kts = Mach 0.850    AT ALTITUDE = 45000.0 ft
    T/W AVAILABLE = 0.201  T/W REQUIRED = 0.090  THRUST USED =  44.6 %
    W/S = 43.29          CLmax(usable) = 1.5036
    CL  = 0.2779   CD0 = 0.0164    K = 0.1113   L/D =11.1152   C = 1.0420
  SPECIFIC RANGE (nmi/lb) = 0.4086
MISSION SEGMENT WEIGHT FRACTION = 0.9716

SEGMENT  12  : DESCENT
    DISTANCE   TRAVELED =    50.0000
MISSION SEGMENT WEIGHT FRACTION = 0.9900

SEGMENT  13  :LOITER
    LOITER SPEED = 191.6 kts = Mach 0.290    AT ALTITUDE =   200.0 ft
   T/W AVAILABLE = 0.935  T/W REQUIRED = 0.076   THRUST USED =   8.1 %
    W/S = 41.64          CLmax(usable) = 1.7640
    CL  = 0.3367   CD0 = 0.0136    K = 0.1047   L/D =13.2014   C = 1.2412
MISSION SEGMENT WEIGHT FRACTION = 0.9692

SEGMENT  14  : LANDING
MISSION SEGMENT WEIGHT FRACTION = 0.9950

RESERVE & TRAPPED FUEL ALLOWANCE =1.060
```

Sizing Calculations (mks):

```
SEGMENT  1  : TAKEOFF
    W/S =273.68   T/W =0.666   C =21.2894   E = 0.0830
MISSION SEGMENT WEIGHT FRACTION = 0.9584

SEGMENT  2  : CLIMB/ACCELERATE
    START ALT =     0.0     END ALT = 13716.0
    START VEL =   0.200     END VEL =    0.850
    W/S = 262.30   T/W = 0.378
     CL = 0.2017   CD0 = 0.0142    K = 0.1349   L/D =10.2636
     C  =24.4503     Ps = 3478.116  mpm
                  TIME TO CLIMB =     4.92 min
              DISTANCE TRAVELED =    60.9420 km
MISSION SEGMENT WEIGHT FRACTION = 0.9736

SEGMENT  3  : CRUISE
    SEGMENT RANGE = 370.4      CLIMB/DESCENT RANGE CREDIT =   60.94
    CRUISE SPEED  = 902.3 km/h = Mach 0.850    AT ALTITUDE = 13716.0 m
    T/W AVAILABLE = 0.167  T/W REQUIRED = 0.083  THRUST USED =  49.7 %
    W/S =255.36          CLmax(usable) = 1.5036
    CL  = 0.3358   CD0 = 0.0164    K = 0.1016   L/D =12.0539   C =28.1713
  SPECIFIC RANGE (km/kg) = 1.5686
MISSION SEGMENT WEIGHT FRACTION = 0.9721

SEGMENT  4  : CLIMB/ACCELERATE
    START ALT = 13716.0     END ALT = 10668.0
    START VEL =   0.850     END VEL =    1.400
    W/S = 248.24   T/W = 0.504
     CL = 0.1208   CD0 = 0.0248    K = 0.1977   L/D = 4.3604
     C  =41.3499     Ps = 6017.072  mpm
                  TIME TO CLIMB =     0.41 min
              DISTANCE TRAVELED =     8.9006 km
MISSION SEGMENT WEIGHT FRACTION = 0.9950

SEGMENT  5  : CRUISE
    SEGMENT RANGE =  92.6      CLIMB/DESCENT RANGE CREDIT =    8.90
    CRUISE SPEED  =1493.5 km/h = Mach 1.400    AT ALTITUDE = 10668.0 m
    T/W AVAILABLE = 0.390  T/W REQUIRED = 0.333  THRUST USED =  85.3 %
    W/S =247.01          CLmax(usable) = 0.8587
    CL  = 0.0741   CD0 = 0.0233    K = 0.2429   L/D = 3.0059   C =28.7585
  SPECIFIC RANGE (km/kg) = 0.6557
MISSION SEGMENT WEIGHT FRACTION = 0.9813

SEGMENT  6  KNOWN TIME FUEL BURN
    W/S =242.38   T/W AVAILABLE =0.910   THRUST SETTING USED =100.0 %
      C =41.1686   E = 0.0500
MISSION SEGMENT FRACTION OR FUEL BURNED =     0.9339

SEGMENT  7  : WEIGHT DROP
WEIGHT DROPPED =    181.4
    SEGMENT IS TURN AROUND POINT FOR THE MISSION - UNUSED CLIMB RANGE
CREDIT IS CANCELLED

SEGMENT  8  : CLIMB/ACCELERATE
    START ALT =  6096.0     END ALT = 10668.0
    START VEL =   0.900     END VEL =    1.400
    W/S = 219.71   T/W = 0.919
     CL = 0.0582   CD0 = 0.0242    K = 0.2020   L/D = 2.3363
     C  =42.0424     Ps = 11321.6   mpm
                  TIME TO CLIMB =     0.89 min
              DISTANCE TRAVELED =    20.5080 km
MISSION SEGMENT WEIGHT FRACTION = 0.9800
```

```
SEGMENT  9  : CRUISE
    SEGMENT RANGE =  92.6      CLIMB/DESCENT RANGE CREDIT =   20.51
    CRUISE SPEED  =1493.5 km/h = Mach 1.400    AT ALTITUDE = 10668.0 m
    T/W AVAILABLE = 0.447  T/W REQUIRED = 0.377  THRUST USED =  84.2 %
    W/S =215.31         CLmax(usable) = 0.8587
    CL  = 0.0645   CD0 = 0.0233    K = 0.2429   L/D = 2.6546   C =28.8172
  SPECIFIC RANGE (km/kg) = 0.6629
MISSION SEGMENT WEIGHT FRACTION = 0.9817

SEGMENT  10  : CLIMB/ACCELERATE
    START ALT = 10668.0    END ALT = 13716.0
    START VEL =   1.400    END VEL =   0.850
    W/S = 211.36   T/W = 0.507

  NEGATIVE DELTA ENERGY HEIGHT - CHECK YOUR INPUTS !  SEGMENT IGNORED
     CL = 0.1544   CD0 = 0.0222    K = 0.1470   L/D = 6.0063
     C  =41.3550    Ps = 6091.637  mpm
                  TIME TO CLIMB =    0.00 min
              DISTANCE TRAVELED =     0.0000 km
MISSION SEGMENT WEIGHT FRACTION = 1.0000

SEGMENT  11  : CRUISE
    SEGMENT RANGE = 370.4      CLIMB/DESCENT RANGE CREDIT =   92.60
    CRUISE SPEED  = 902.3 km/h = Mach 0.850    AT ALTITUDE = 13716.0 m
    T/W AVAILABLE = 0.201  T/W REQUIRED = 0.090  THRUST USED =  44.6 %
    W/S =211.36         CLmax(usable) = 1.5036
    CL  = 0.2779   CD0 = 0.0164    K = 0.1113   L/D =11.1152   C =29.5061
  SPECIFIC RANGE (km/kg) = 1.6685
MISSION SEGMENT WEIGHT FRACTION = 0.9716

SEGMENT  12  : DESCENT
    DISTANCE   TRAVELED =    92.6000
MISSION SEGMENT WEIGHT FRACTION = 0.9900

SEGMENT  13  :LOITER
    LOITER SPEED = 354.8 km/h = Mach 0.290    AT ALTITUDE =    61.0 m
   T/W AVAILABLE = 0.935  T/W REQUIRED = 0.076   THRUST USED =   8.1 %
    W/S =203.30         CLmax(usable) = 1.7640
    CL  = 0.3367   CD0 = 0.0136    K = 0.1047   L/D =13.2014   C =35.1495
MISSION SEGMENT WEIGHT FRACTION = 0.9692

SEGMENT  14  : LANDING
MISSION SEGMENT WEIGHT FRACTION = 0.9950

RESERVE & TRAPPED FUEL ALLOWANCE =1.060
```

AIRCRAFT DESIGN

Sizing Results

	MISSION SEGMENT	MISSION SEGMENT WEIGHT FRACTION OR DROPPED WEIGHT	Wi/WO
1	TAKEOFF SEGMENT	0.9584	0.9584
2	CLIMB and/or ACCELERATE	0.9736	0.9331
3	CRUISE SEGMENT	0.9721	0.9071
4	CLIMB and/or ACCELERATE	0.9950	0.9025
5	CRUISE SEGMENT	0.9813	0.8856
6	KNOWN TIME FUEL BURN SEGMENT	0.9339	0.8271
7	WEIGHT DROP SEGMENT	400.0000	0.8036
8	CLIMB and/or ACCELERATE	0.9800	0.7875
9	CRUISE SEGMENT	0.9817	0.7731
10	CLIMB and/or ACCELERATE	1.0000	0.7731
11	CRUISE SEGMENT	0.9716	0.7511
12	DESCENT SEGMENT	0.9900	0.7436
13	LOITER SEGMENT	0.9692	0.7207
14	LANDING SEGMENT	0.9950	0.7171

Results (fps units):

```
Seg. 3   CRUISE : 487.2 kts at  45000.0 ft  RANGE =   200.0 nmi
Seg. 5   CRUISE : 806.4 kts at  35000.0 ft  RANGE =    50.0 nmi
Seg. 9   CRUISE : 806.4 kts at  35000.0 ft  RANGE =    50.0 nmi
Seg. 11  CRUISE : 487.2 kts at  45000.0 ft  RANGE =   200.0 nmi
Seg. 13  LOITER : 191.6 kts at    200.0 ft  ENDUR =     0.3 hrs

 TOTAL RANGE =       500.0        TOTAL LOITER TIME =      0.33
 FUEL WEIGHT =      4693.0             EMPTY WEIGHT =   11258.2
LOAD(less Wf)=      1110.0    AIRCRAFT GROSS WEIGHT =   17061.2
```

Results (mks units):

```
Seg. 3   CRUISE : 902.3 km/h at 13716.0 m   RANGE =   370.4 km
Seg. 5   CRUISE :1493.5 km/h at 10668.0 m   RANGE =    92.6 km
Seg. 9   CRUISE :1493.5 km/h at 10668.0 m   RANGE =    92.6 km
Seg. 11  CRUISE : 902.3 km/h at 13716.0 m   RANGE =   370.4 km
Seg. 13  LOITER : 354.8 km/h at    61.0 m   ENDUR =     0.3 hrs

 TOTAL RANGE =       926.0        TOTAL LOITER TIME =      0.33
 FUEL WEIGHT =      2128.7             EMPTY WEIGHT =    5106.6
LOAD less Wf)=       503.5    AIRCRAFT GROSS WEIGHT =    7738.8
```

A Conceptual Approach

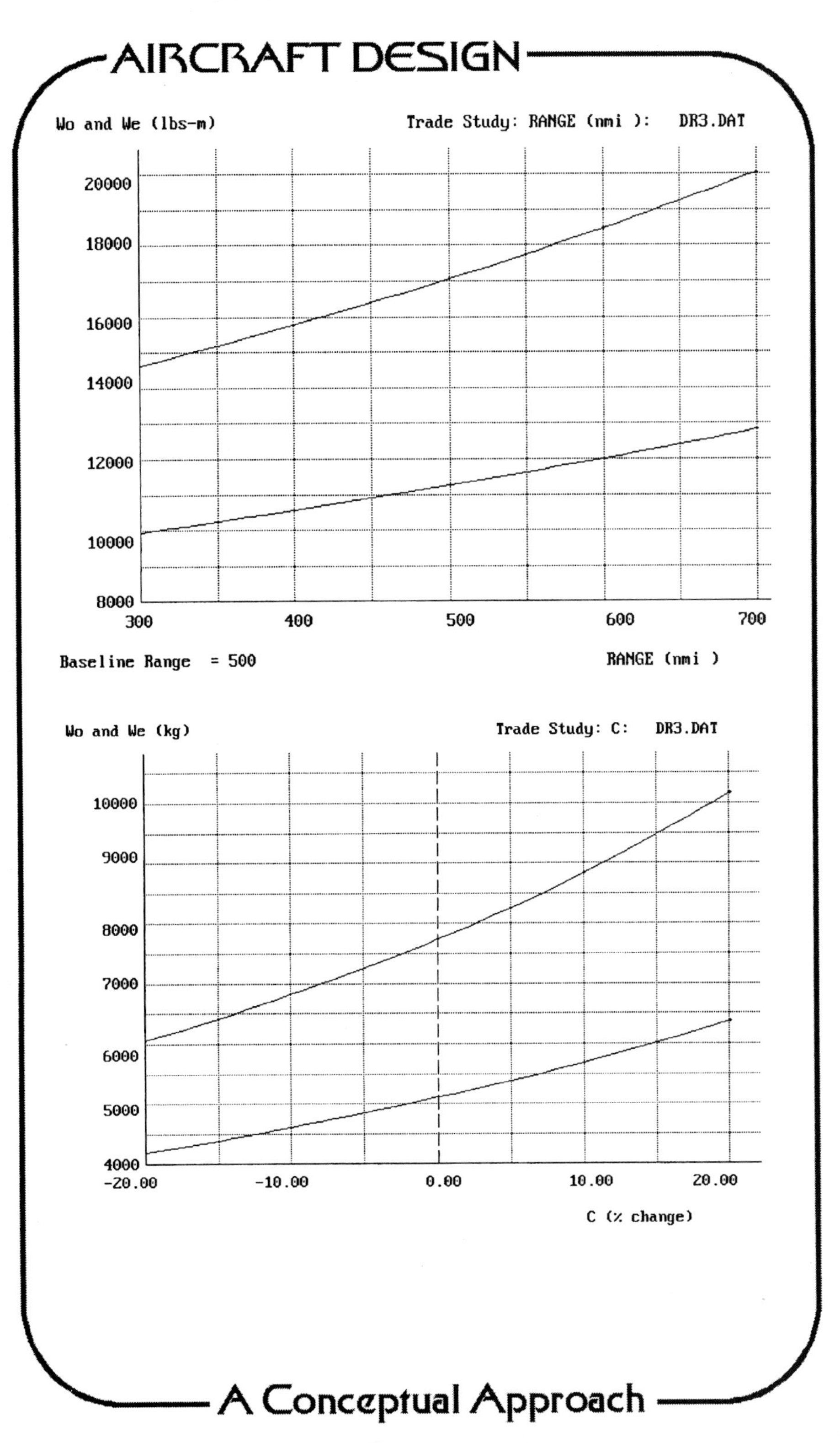
AIRCRAFT DESIGN
Wo and We (lbs-m)
Trade Study: RANGE (nmi): DR3.DAT
20000
18000
16000
14000
12000
10000
8000
300
400
500
600
700
Baseline Range = 500
RANGE (nmi)
Wo and We (kg)
Trade Study: C: DR3.DAT
10000
9000
8000
7000
6000
5000
4000
-20.00
-10.00
0.00
10.00
20.00
C (% change)
A Conceptual Approach

TAKEOFF: DR3

```
AIRCRAFT OPERATING WEIGHT  (Wi) =  16480.0 lb {7475.2 kg}
 OPERATING WEIGHT RATIO (Wi/WO) =  1.000
 THRUST-TO-WEIGHT RATIO    (T/W) =  0.980
     THRUST  (START OF TAKEOFF) =  16150.4 lb {71.8 kN}
     TAKEOFF WINGLOADING   (W/S) =  56.05        {273.68}
                          Vstall =  99.80 kts {184.8 km/h}
                        Vtakeoff =  109.8 kts {203.3 km/h}
  CLIMB ANGLE = 44.97  (deg)  CLIMB  CD0 =  0.0289
           CL =  1.49                   K =  0.2609
                              CLIMB  L/D =  3.07

           GROUND ROLL DISTANCE =    538.2 {164.0}
                ROTATE DISTANCE =    185.4 {56.5}
     TOTAL GROUND ROLL DISTANCE =    723.6 {220.6}
            TRANSITION DISTANCE =    761.6 {232.1}
                 CLIMB DISTANCE =      0.0 {0.0}
         TOTAL TAKEOFF DISTANCE =   1485.2 {452.7}
   FAR PART 25 TAKEOFF DISTANCE =   1707.9 {520.6}
```

LANDING: DR3

```
AIRCRAFT OPERATING WEIGHT  (Wi) =  16480.0 lb {7475.2 kg}
 OPERATING WEIGHT RATIO  Wi/WO) =  1.000
 ROLLOUT THRUST-TO-WEIGHT (T/W) = -0.392
     LANDING WINGLOADING   (W/S) =  56.05        {273.68}
                          Vstall =  95.84 kts {177.5 km/h}
                      Vtouchdown = 115.01 kts {213.0 km/h}

                 APPROACH ANGLE = -3.00  (deg)
                  APPROACH  CD0 =  0.1124
                             CL =  1.62
                              K =  0.2724
                   APPROACH L/D =  2.53

             APPROACH DISTANCE =    773.5 { 235.8}
                FLARE DISTANCE =   2733.1 { 833.1}
         FREE GROUND ROLL DIST =    194.2 { 59.2}
              BRAKING DISTANCE =    796.1 { 242.7}
    TOTAL GROUND ROLL DISTANCE =    990.4 { 301.9}
     NO-FLARE LANDING DISTANCE =   1944.5 { 592.7}
        TOTAL LANDING DISTANCE =   4497.0 { 1370.7}
  FAR PART 25 LANDING DISTANCE =   7495.0 { 2284.5}
```

Ps, TURN, & CLIMB: DR3

```
M = 0.90    ALT = 30000 ft {9144 m}
Wi/WO = 0.872   Wi =   14370.6
W/S =  48.88 {238.65}
T/W = 0.649 THRUST =   9322.7 {41.5kN}
Turn Radius =  3426. Ft {1044 m}

  AT LOAD FACTOR  n = 1
CD0 = 0.0161  K = 0.1411  CL = 0.14  Ps = +458.34 {139.7}
Rate of Climb = 27500fpm {8382mpm} Climb Gradient = 0.51
  AT LOAD FACTOR  n = 2
CD0 = 0.0161  K = 0.1138  CL = 0.27  Ps = +419.78 {127.9}
TURN RATE = 3.57 (deg/sec)
  AT LOAD FACTOR  n = 3
CD0 = 0.0161  K = 0.1022  CL = 0.41  Ps = +362.72 {110.6}
TURN RATE = 5.83 (deg/sec)
   AT LOAD FACTOR  n = 4
CD0 = 0.0161  K = 0.1103  CL = 0.55  Ps = +258.88 {78.91}
TURN RATE = 7.98 (deg/sec)
   AT LOAD FACTOR  n = 5
CD0 = 0.0161  K = 0.1340  CL = 0.69  Ps = +64.17  {19.56}
TURN RATE =10.10 (deg/sec)
   AT LOAD FACTOR  n = 6
CD0 = 0.0161  K = 0.1553  CL = 0.82  Ps = -210.93 {-64.3}
TURN RATE =12.19 (deg/sec)
   AT LOAD FACTOR  n = 7
CD0 = 0.0161  K = 0.1699  CL = 0.96  Ps = -546.41 {-167.}
TURN RATE =14.28 (deg/sec)
```

ACCELERATION: DR3.DAT

```
Wi/WO = 0.872   Wi =   14370.6   W/S =  48.88    Altitude =  35000.0

AT V =  538.80   T/W = 0.522   CD0 = 0.0173    K = 0.1397    CL = 0.1603
     Ps(ft/sec)=    +356.79     dV/dt = 12.6250      DELTA TIME =   3.85
AT V =  567.61   T/W = 0.538   CD0 = 0.0199    K = 0.1450    CL = 0.1444
     Ps(ft/sec)=    +363.57     dV/dt = 12.2120      DELTA TIME =   3.98
AT V =  596.41   T/W = 0.554   CD0 = 0.0239    K = 0.1508    CL = 0.1308
     Ps(ft/sec)=    +354.10     dV/dt = 11.3196      DELTA TIME =   4.30
AT V =  625.21   T/W = 0.569   CD0 = 0.0256    K = 0.1589    CL = 0.1190
     Ps(ft/sec)=    +354.35     dV/dt = 10.8059      DELTA TIME =   4.50
AT V =  654.01   T/W = 0.585   CD0 = 0.0255    K = 0.1703    CL = 0.1088
     Ps(ft/sec)=    +367.15     dV/dt = 10.7032      DELTA TIME =   4.54
AT V =  682.82   T/W = 0.601   CD0 = 0.0250    K = 0.1826    CL = 0.0998
     Ps(ft/sec)=    +382.78     dV/dt = 10.6881      DELTA TIME =   4.55
AT V =  711.62   T/W = 0.629   CD0 = 0.0246    K = 0.1966    CL = 0.0919
     Ps(ft/sec)=    +412.42     dV/dt = 11.0495      DELTA TIME =   4.40
AT V =  740.42   T/W = 0.662   CD0 = 0.0242    K = 0.2111    CL = 0.0849
     Ps(ft/sec)=    +449.13     dV/dt = 11.5650      DELTA TIME =   4.21
AT V =  769.22   T/W = 0.696   CD0 = 0.0238    K = 0.2250    CL = 0.0786
     Ps(ft/sec)=    +487.31     dV/dt = 12.0784      DELTA TIME =   4.03
AT V =  798.02   T/W = 0.729   CD0 = 0.0234    K = 0.2388    CL = 0.0731
     Ps(ft/sec)=    +526.94     dV/dt = 12.5892      DELTA TIME =   3.86

ACCEL TIME FROM    0.900 TO   1.400 IS   42.2 (sec)   DISTANCE =  7.8 nmi
```

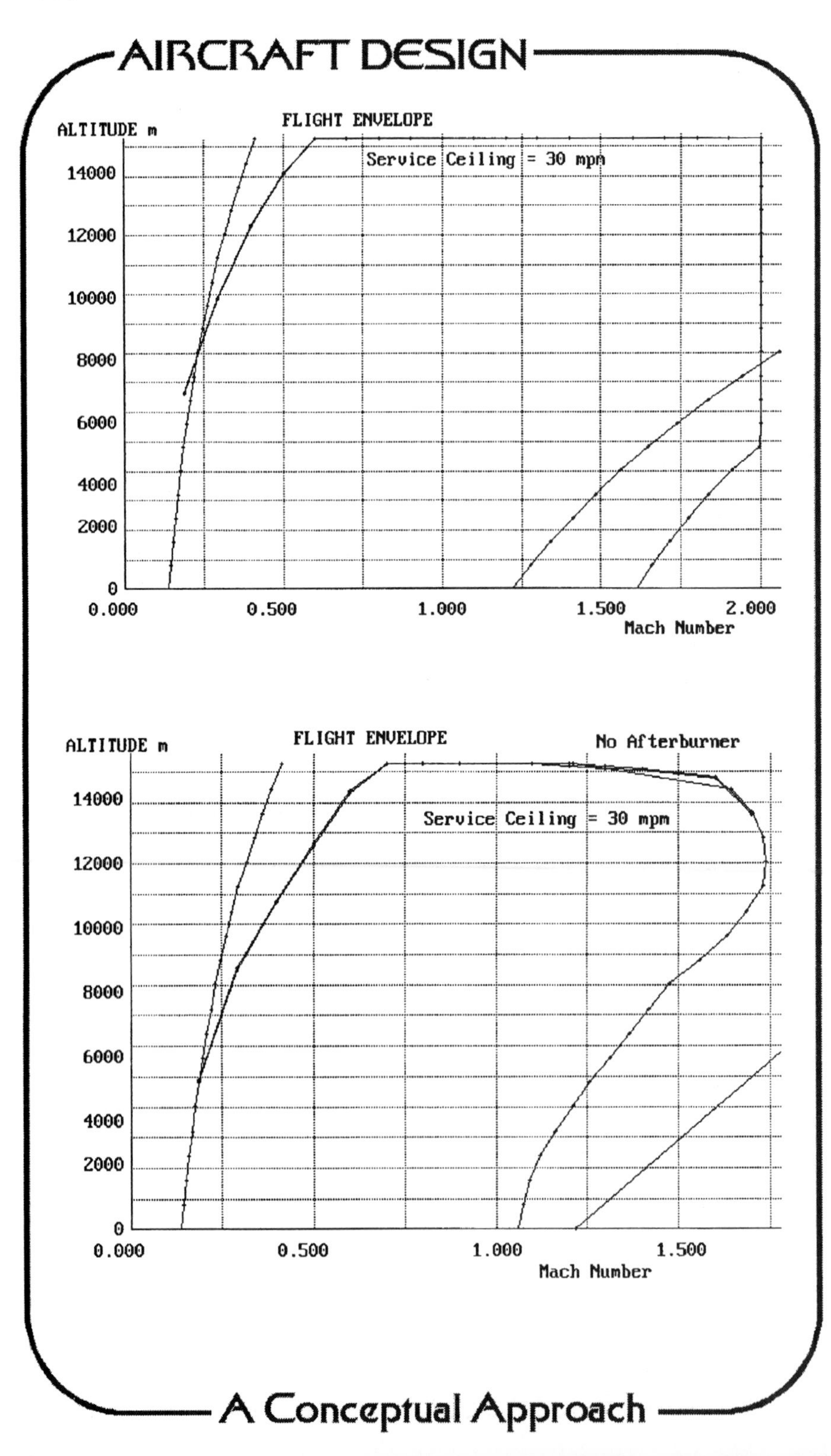
AIRCRAFT DESIGN
FLIGHT ENVELOPE
ALTITUDE m
Service Ceiling = 30 mpm
14000
12000
10000
8000
6000
4000
2000
0
0.000
0.500
1.000
1.500
2.000
Mach Number
FLIGHT ENVELOPE
No Afterburner
ALTITUDE m
Service Ceiling = 30 mpm
14000
12000
10000
8000
6000
4000
2000
0
0.000
0.500
1.000
1.500
Mach Number
A Conceptual Approach

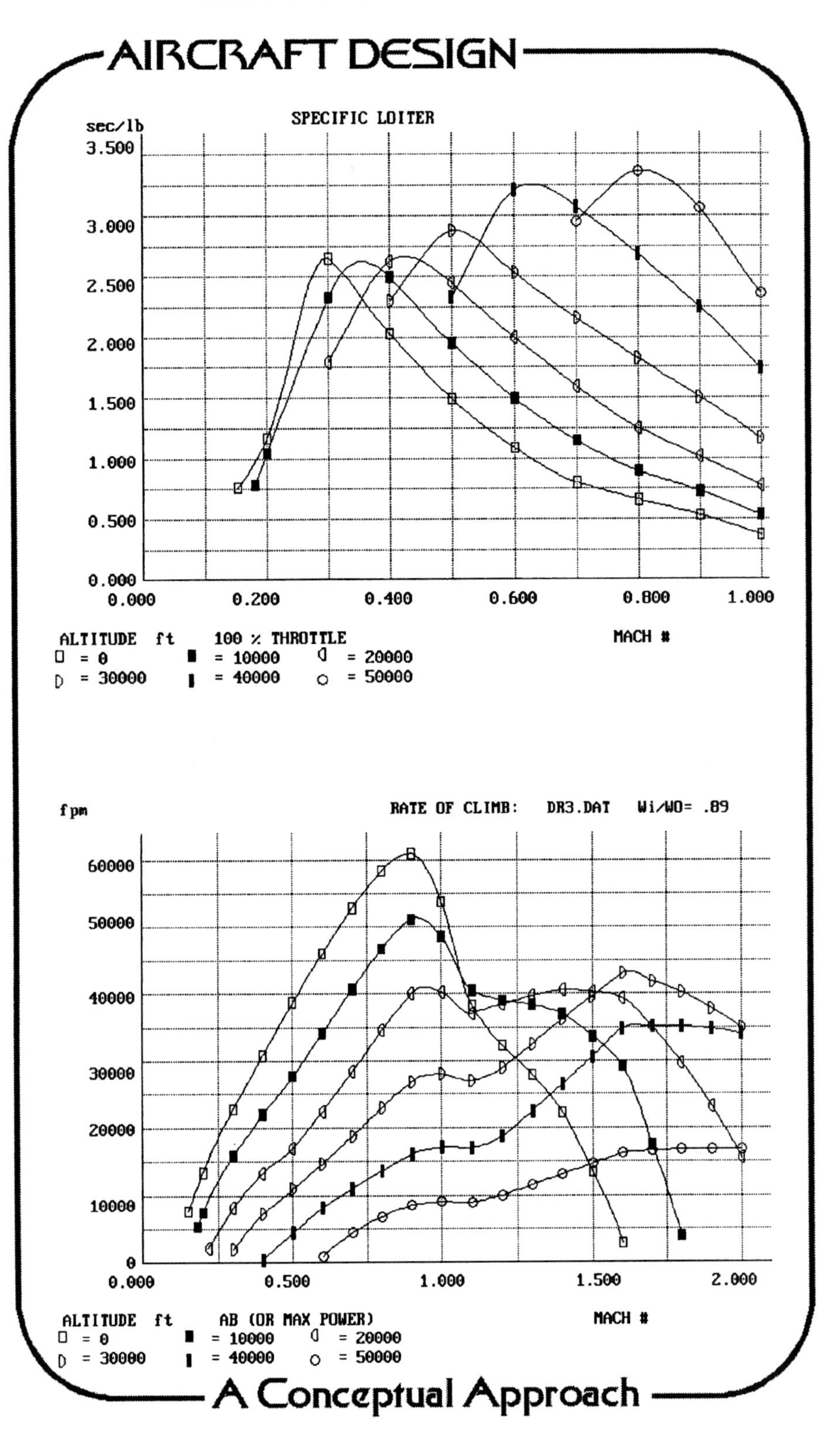
AIRCRAFT DESIGN
SPECIFIC LOITER
sec/lb
3.500
3.000
2.500
2.000
1.500
1.000
0.500
0.000
0.000
0.200
0.400
0.600
0.800
1.000
ALTITUDE ft 100 % THROTTLE
MACH #
= 0
= 10000
= 20000
= 30000
= 40000
= 50000
fpm
RATE OF CLIMB: DR3.DAT Wi/WO= .89
60000
50000
40000
30000
20000
10000
0
0.000
0.500
1.000
1.500
2.000
ALTITUDE ft AB (OR MAX POWER)
MACH #
= 0
= 10000
= 20000
= 30000
= 40000
= 50000
A Conceptual Approach

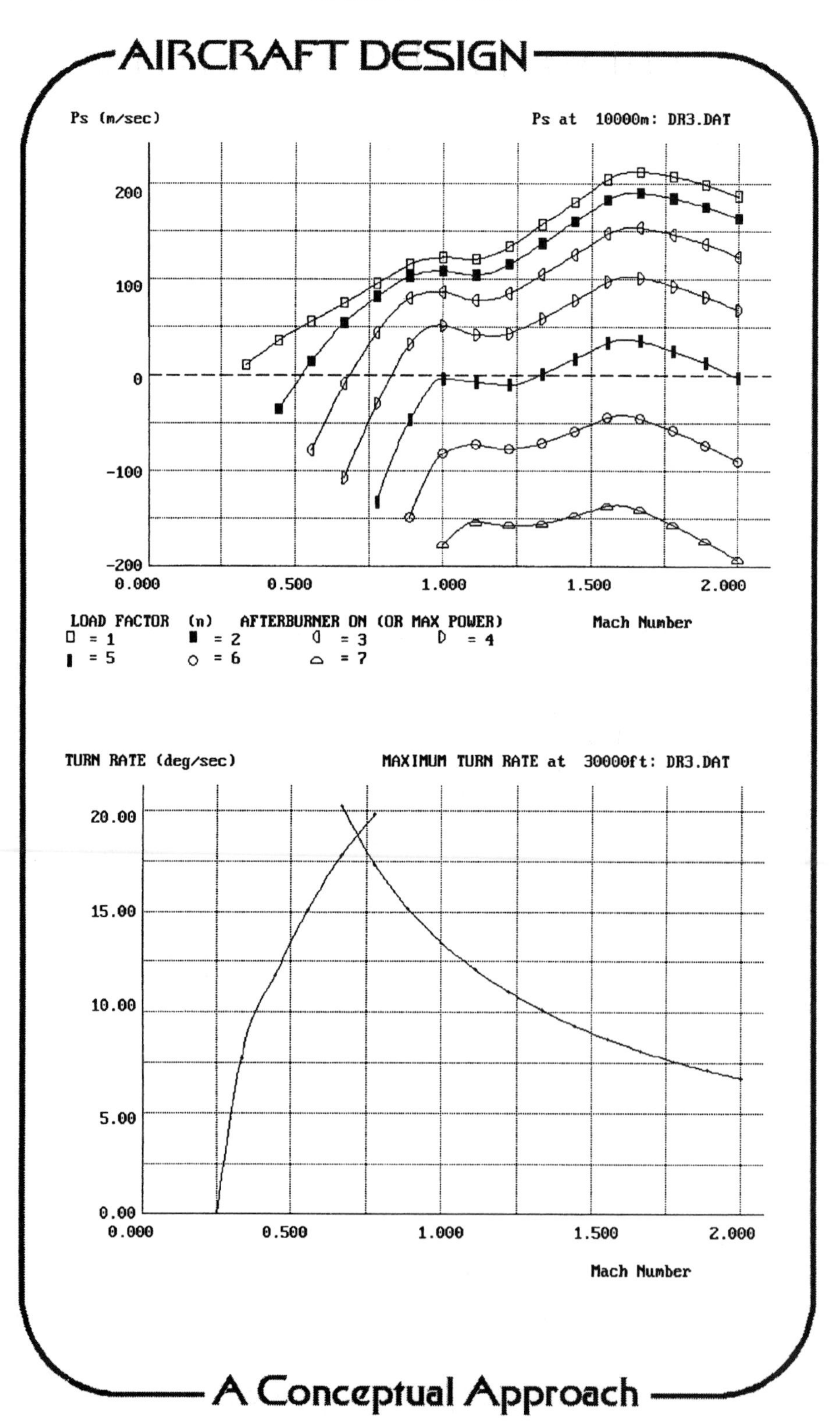
AIRCRAFT DESIGN
Ps (m/sec)
Ps at 10000m: DR3.DAT
200
100
0
-100
-200
0.000
0.500
1.000
1.500
2.000
LOAD FACTOR (n) AFTERBURNER ON (OR MAX POWER)
Mach Number
= 1
= 2
= 3
= 4
= 5
= 6
= 7
TURN RATE (deg/sec)
MAXIMUM TURN RATE at 30000ft: DR3.DAT
20.00
15.00
10.00
5.00
0.00
0.000
0.500
1.000
1.500
2.000
Mach Number
A Conceptual Approach

T/W - W/S Optimization

Optimization of thrust-to-weight ratio and wing loading of the DR-3 was done using the carpet plot methods of Chapter 19. Parametric variations of plus-and-minus 10 and 20% were made about the as-drawn baseline of $T/W = 0.98$ and $W/S = 56$ lb/ft^2 {273.1 kg/sq.m}. RDS was again used.

For variations in *T/W*, it was assumed that engine thrust varies directly with *T/W*. Uninstalled engine weight was assumed to vary by the 1.1 power of the relative change in thrust [see Eq. (10.3)]. The aerodynamic effect of a change in *T/W*, largely the increase in nacelle wetted area, was assumed to be proportional to engine flowpath area and therefore proportional to *T/W*.

The aerodynamic and weight impacts of variations in *W/S* were determined assuming that wing area varies inversely with wing loading. Tail area was varied by the 3/2 power of the relative change in wing area to keep tail volume constant [see Eqs. (6.28) and (6.29)]. The change in wave drag due to the change in wing size was estimated by changing the maximum cross-sectional area (Amax) proportional to the change in wing area, weighted to the wing's baseline percentage of total cross-section area.

For each combination of parametric variations in *T/W* and *W/S* (25 in all), the RDS input matrices were revised. The aircraft was reanalyzed for aerodynamics and weights, and the engine thrust was scaled. Then the sizing, cost, and performance analysis were rerun as tabulated in the following. A carpet plot was prepared using the RDS-Professional program to save time, but could have been done by hand.

The optimum airplane that meets all requirements is shown on the carpet plot at the intersection of the lines labeled "2-Landing" and "7-Acceleration" (30 s). It weighs 19,300 lb {4218 kg} and has a *T/W* of 1.1 and a *W/S* of 59 {288}. This weight is 17% greater than the as-drawn DR-3 weight, but remember that the baseline couldn't meet the acceleration requirement. By relaxing the requirement for acceleration from Mach 0.9 to 1.4 from 30 to 50 s, the weight of an optimal aircraft fully meeting all other requirements drops to 15,600 lb {7076 kg}. Also, this changes the optimal *T/W* to 0.9 and wing loading to 54 {264} and is shown at the intersection of the lines "2-Landing" and "8-Acceleration" (50 s). This is a 19% reduction in sized takeoff gross weight from the aircraft that can meet the acceleration requirement. This will produce a considerable cost savings for the relaxation of that one requirement.

Just for fun, the RDS-Professional Multivariable Optimizer was run to determine what additional weight savings could be obtained by optimizing for sweep, aspect ratio, taper ratio, and thickness as well as *T/W* and *W/S*. This produced an optimized weight of 15,242 lb {6914 kg}, 2% less than the optimum found with a *T/W-W/S* carpet plot. This indicates that the wing planform chosen for the DR-3 using the methods of the book was fairly close to optimal already. However, this further 2% weight savings is obtained for free, resulting just from slight changes to the wing geometry.

AIRCRAFT DESIGN

RDS CARPET PLOT RESULTS: DR3

Var.#	W/S	T/W	Wo	We	Wf
1,	44.843,	0.7840,	15470.,	10308.,	4051.4
2,	50.449,	0.7840,	14303.,	9516.1,	3676.8
3,	56.054,	0.7840,	13478.,	8952.9,	3415.3
4,	61.659,	0.7840,	12928.,	8564.6,	3253.8
5,	67.265,	0.7840,	12532.,	8278.2,	3144.0
6,	44.843,	0.8820,	17381.,	11557.,	4713.9
7,	50.449,	0.8820,	16030.,	10641.,	4279.0
8,	56.054,	0.8820,	15068.,	9985.8,	3972.5
9,	61.659,	0.8820,	14423.,	9533.3,	3779.5
10,	67.265,	0.8820,	13960.,	9201.0,	3649.5
11,	44.843,	0.9800,	19811.,	13115.,	5586.2
12,	50.449,	0.9800,	18200.,	12029.,	5061.1
13,	56.054,	0.9800,	17060.,	11257.,	4692.7
14,	61.659,	0.9800,	16272.,	10712.,	4450.3
15,	67.265,	0.9800,	15716.,	10316.,	4289.7
16,	44.843,	1.0780,	22901.,	15065.,	6725.9
17,	50.449,	1.0780,	20928.,	13748.,	6070.4
18,	56.054,	1.0780,	19528.,	12809.,	5608.8
19,	61.659,	1.0780,	18561.,	12149.,	5301.6
20,	67.265,	1.0780,	17884.,	11674.,	5099.6
21,	44.843,	1.1760,	26811.,	17496.,	8204.8
22,	50.449,	1.1760,	24363.,	15881.,	7372.3
23,	56.054,	1.1760,	22636.,	14736.,	6789.1
24,	61.659,	1.1760,	21414.,	13917.,	6387.5
25,	67.265,	1.1760,	20573.,	13335.,	6127.4

PERFORMANCE RESULTS

Var.#	Takeoff	Landing	Ps@n=5	Ps@n=5	Ps@n=1	Ps@n=1	Accel
1,	712.36,	887.21,	69.818,	-55.65,	219.61,	-112.1,	70.633
2,	789.13,	986.63,	11.604,	-66.96,	277.23,	-82.76,	66.849
3,	865.12,	1085.4,	-51.97,	-83.90,	322.24,	-59.97,	64.226
4,	940.42,	1183.6,	-121.4,	-104.6,	358.25,	-41.85,	62.341
5,	1015.1,	1281.3,	-191.4,	-128.0,	387.64,	-27.17,	60.927
6,	648.88,	846.98,	127.89,	64.593,	400.80,	-46.39,	54.792
7,	717.89,	941.41,	69.679,	53.278,	458.42,	-17.05,	52.530
8,	786.14,	1035.2,	6.0952,	36.346,	503.43,	5.7424,	50.923
9,	853.73,	1128.4,	-63.32,	15.608,	539.44,	23.865,	49.749
10,	920.73,	1221.2,	-133.3,	-7.770,	568.83,	38.547,	48.858
11,	598.61,	811.06,	185.96,	184.84,	581.99,	19.319,	44.825
12,	661.48,	901.02,	127.75,	173.52,	639.61,	48.667,	43.319
13,	723.61,	990.36,	64.169,	156.59,	684.62,	71.460,	42.231
14,	785.09,	1079.1,	-5.252,	135.85,	720.63,	89.583,	41.429
15,	846.01,	1167.4,	-75.30,	112.47,	750.02,	104.26,	40.815
16,	557.82,	778.78,	244.04,	305.08,	763.18,	85.038,	37.956
17,	615.70,	864.73,	185.82,	293.77,	820.80,	114.38,	36.879
18,	672.87,	950.06,	122.24,	276.84,	865.81,	137.17,	36.094
19,	729.40,	1034.8,	52.822,	256.10,	901.82,	155.30,	35.510
20,	785.38,	1119.1,	-17.23,	232.72,	931.21,	169.98,	35.061
21,	524.05,	749.61,	302.11,	425.33,	944.37,	150.75,	32.927
22,	577.81,	831.94,	243.90,	414.02,	1002.0,	180.10,	32.118
23,	630.87,	913.64,	180.31,	397.09,	1047.0,	202.89,	31.524
24,	683.31,	994.82,	110.89,	376.35,	1083.0,	221.01,	31.080
25,	735.20,	1075.5,	40.843,	352.97,	1112.4,	235.70,	30.737
Required Performance Values							
	1000.0,	1000.0,	0.0000,	0.0000,	0.0000,	0.0000,	50.000

A Conceptual Approach

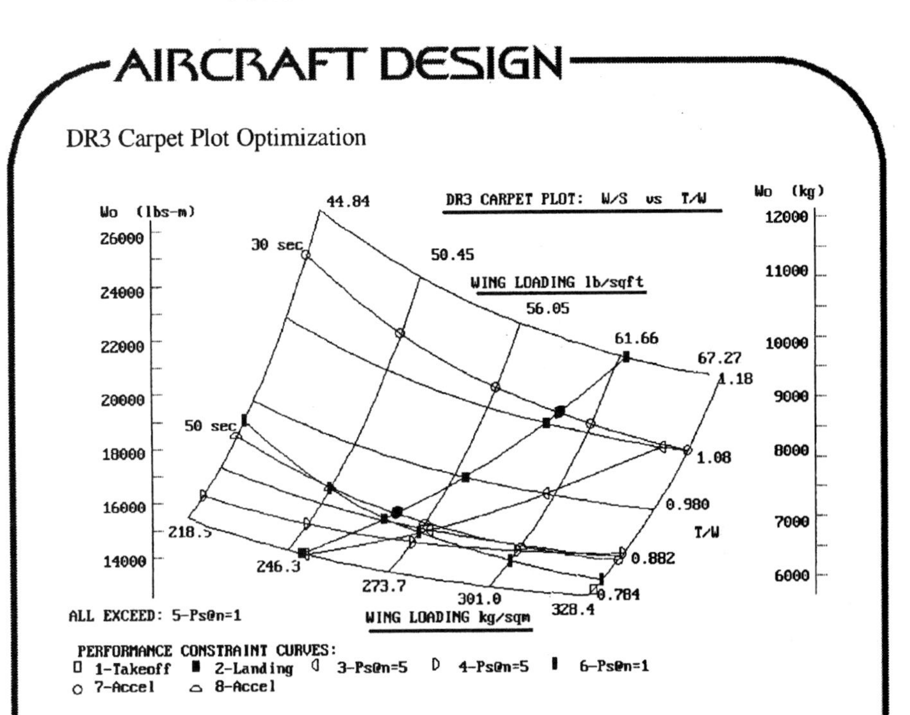

DR-3 MULTIVARIABLE OPTIMIZATION

MEASURE OF MERIT: Wo

	BASELINE	BEST
T/W	0.980	0.919
W/S	56.1	52.6
ASPECT RATIO	3.500	2.800
SWEEP	38.0	34.7
TAPER RATIO	0.250	0.200
WING t/c	0.060	0.068
Sized Wo	17060.2	15242.2
Sized We	11257.5	9925.5
Sized Wf	4692.7	4206.7

	REQUIRED	BASELINE	BEST
Takeoff	1000.0	723.6	720.0
Landing	1000.0	990.4	960.4
Ps@n=5	0.0	64.2	1.7
Ps@n=5	0.0	156.6	62.0
Ps@n=1	0.0	684.6	515.7
Ps@n=1	0.0	71.5	0.1
Accel	50.0	42.2	49.4

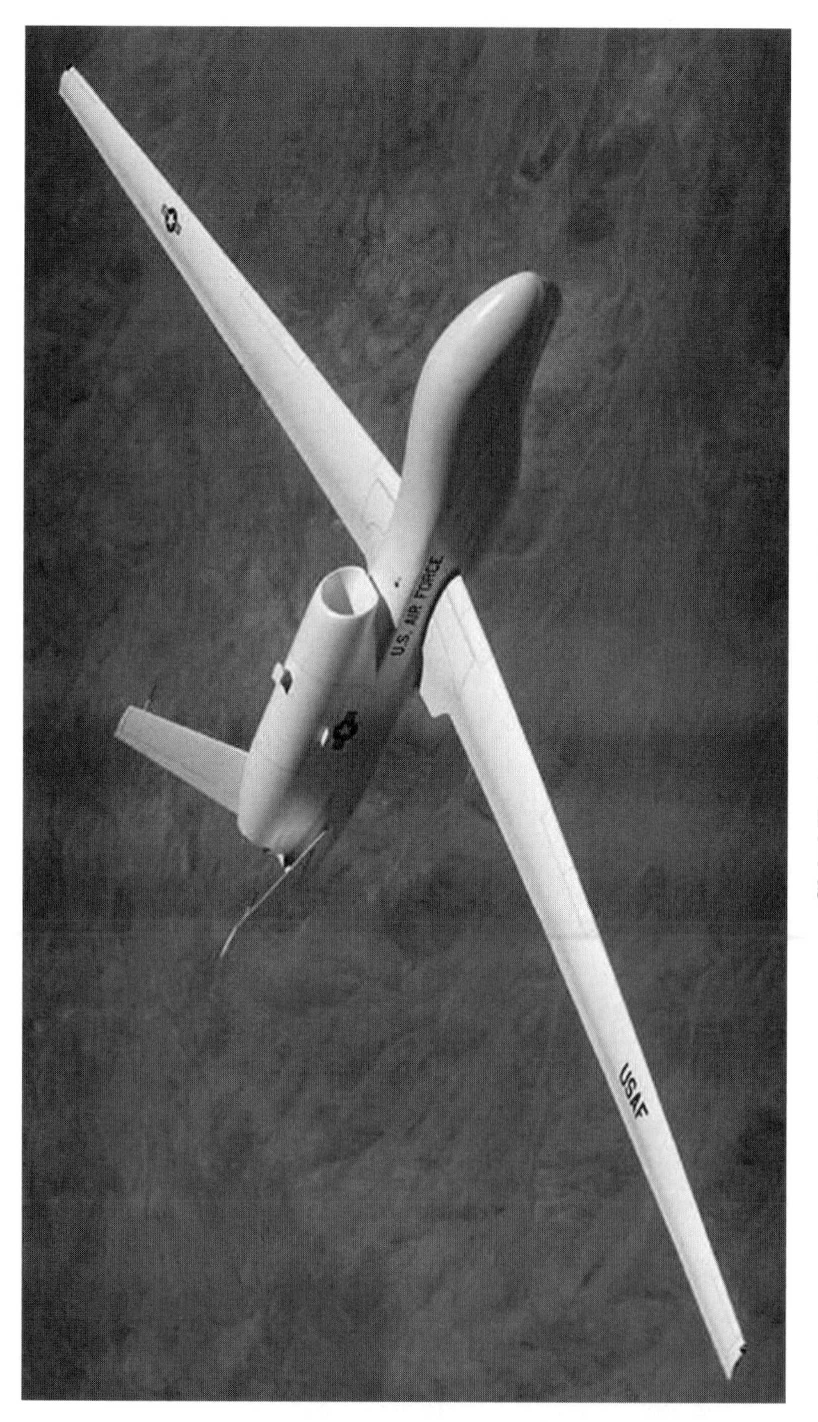

Global Hawk (U.S. Air Force photo)

Appendix A: Unit Conversion

Table A.1 FPS to MKS conversions

given fps units	multiply	by	to obtain
Length & Range	ft	0.30480	m
	in	2.54000	cm
	ft	0.00016	nmi
	nmi	6076.10000	ft
	ft	12.00000	in
	in	0.08333	ft
	nmi	1.85200	km
	nmi	1.15128	st. miles
	st. miles	0.86860	nmi
Area	sq-ft	0.09290	sq-m
	sq-in	6.45200	sq-cm
Volume	gal	3.78542	liter
	cubic ft	28.31700	liter
	Imperial Gal	1.20100	Gallon
	Gallon	0.83264	Imperial Gal
Velocity, Climb & Ps	kts (knots)	1.85201	km/h
	ft/sec	0.30480	m/sec
	fpm	0.30480	mpm
	kts	1.15100	mile per hr
	mile per hr	0.86881	kts
Mass (& Weight)	lbs-m	0.45359	kg
	ounce	28.34467	gram
	slug	14.59000	kg
	ounce	0.00445	lbs-f
	lbs-f	16.00000	ounce
Force & Weight	lbs-f	0.00445	kN
Pressure	lb-f/sqft	0.04788	kN/sqm
Density	slug/ft^3	515.21399	kg/m^3
	lb/ft^3	16.02000	kg/m^3
Weight Loading	lb/sqft	4.88242	kg/sqm
Power	Hp	0.74570	kWatt*
	Hp	550.00000	ft-lb/sec
	ft-lb/sec	0.00182	Hp
	Hp	0.70690	BTU/sec
	BTU/sec	1.41463	Hp
Spec Fuel Consumpt	lb/ lb/hr	28.31800	mg/Ns
Power SFC (C Bhp)	lb/hr/bhp	0.16897	mg/Watt-s
Fs - Fuel Specific Energy	ft/lb	0.67197	m/kg
Range Parameter	nmi/lb	4.08296	km/kg
Loiter Parameter	sec/lb	2.20463	sec/kg
Temperature	deg R	0.55556	deg K

(deg C = deg K - 273.15) (deg F = deg R - 459.67)

interactive version available at www.aircraftdesign.com

*Watt = Joule/sec = N-m/sec

Shaded terms are not properly mass units but their weight equivalents assuming standard acceleration due to gravity—use with caution!

Table A.2 MKS to FPS conversions

given mks units	multiply	by	to obtain
Length & Range	m	3.28084	ft
	cm	0.39380	in
	km	0.53996	nmi
Area	sq-m	10.76391	sq-ft
	sq-cm	0.15499	sq-in
Volume	liter	0.26417	gal
	liter	0.03531	cubic ft
	liter	1000.00000	cubic cm
	cubic cm	0.00100	liter
Velocity, Climb & Ps	km/h	0.53995	kts
	m/sec	3.28084	ft/sec
	mpm	3.28084	fpm
Mass (& Weight)	kg	2.20462	lbs-m
	gram	0.03528	ounce
	kg	0.06854	slug
Force & Weight	kN	224.81003	lbs-f
	kg-f	9.80700	Newton
	Newton	0.10197	kg-f
Pressure	kN/sqm	20.88555	lb-f/sqft
Density	kg/m^3	0.00194	slug/ft^3
	kg/m^3	0.06242	lb/ft^3
Weight Loading	kg/sqm	0.20482	lb/sqft
Power	kWatt*	1.34102	Hp
Spec Fuel Consumpt	mg/Ns	0.03531	lb/ lb/hr
Power SFC (C Bhp)	mg/Watt-s	5.91821	lb/hr/bhp
Fs - Fuel Specific Energy	m/kg	1.48816	ft/lb
Range Parameter	km/kg	0.24492	nmi/lb
Loiter Parameter	sec/kg	0.45359	sec/lb
Temperature	deg K	1.79999	deg R

(deg C = deg K - 273.15) (deg F = deg R - 459.67)

interactive version available at www.aircraftdesign.com

*Watt = Joule/sec = N-m/sec

Shaded terms are not properly mass units but their weight equivalents assuming standard acceleration due to gravity—use with caution!

Appendix B: Standard Atmosphere

Table B.1 Characteristics of the standard atmosphere in British units (fps)

h, ft/10^3	T, °R	p, psf	ρ, sl/ft^3	μ, sl/ft-s	a, ft/s	ν, ft^2/s
0	518.69	2116.2	0.23769 − 2	0.37373 − 6	1116.4	0.15723 − 3
1	515.1	2041	0.2308	0.3717	1112.6	0.1611
2	511.6	1963	0.2241	0.3697	1108.7	0.1650
3	508.0	1897	0.2175	0.3677	1104.9	0.1691
4	504.4	1828	0.2111	0.3657	1101.0	0.1732
5	500.9	1761	0.2043	0.3637	1097.1	0.1776
6	497.3	1696	0.1987 − 2	0.3616 − 6	1093.2	0.1820 − 3
7	493.7	1633	0.1927	0.3596	1089.3	0.1866
8	490.2	1572	0.1869	0.3576	1085.3	0.1914
9	486.6	1513	0.1811	0.3555	1081.4	0.1963
10	483.1	1456	0.1756	0.3534	1077.4	0.2013
11	479.5	1400	0.1701 − 2	0.3514 − 6	1073.4	0.2066 − 3
12	475.9	1346	0.1648	0.3493	1069.4	0.2120
13	472.4	1294	0.1596	0.3472	1065.4	0.2175
14	468.8	1244	0.1546	0.3451	1061.4	0.2233
15	465.2	1195	0.1496	0.3430	1057.4	0.2293
16	461.7	1148	0.1448 − 2	0.3409 − 6	1053.3	0.2354 − 3
17	458.1	1102	0.1401	0.3388	1049.2	0.2418
18	454.6	1058	0.1355	0.3367	1045.1	0.2484
19	451.0	1015	0.1311	0.3346	1041.0	0.2553
20	447.4	9733 − 1	0.1267	0.3325	1036.9	0.2623
21	443.9	9333 − 1	0.1225 − 2	0.3303 − 6	1032.8	0.2697 − 3
22	440.3	8946	0.1184	0.3282	1028.6	0.2772
23	436.8	8572	0.1144	0.3260	1024.5	0.2851
24	433.2	8212	0.1104	0.3238	1020.3	0.2932
25	429.6	7863	0.1066	0.3217	1016.1	0.3017
26	426.1	7527 − 1	0.1029 − 2	0.3195 − 6	1011.9	0.3104 − 3
27	422.5	7203	0.9931 − 3	0.3173	1007.7	0.3195
28	419.0	6890	0.9580	0.3151	1003.4	0.3289
29	415.4	6588	0.9239	0.3129	999.1	0.3387
30	411.9	6297	0.8907	0.3107	994.8	0.3488
31	408.3	6016 − 1	0.8584 − 3	0.3085 − 6	990.5	0.3594 − 3
32	404.8	5746	0.8270	0.3063	986.2	0.3703
33	401.2	5485	0.7966	0.3040	981.9	0.3817
34	397.6	5235	0.7670	0.3018	977.5	0.3935
35	394.1	4993	0.7382	0.2995	973.1	0.4058
36	390.5	4761 − 1	0.7103 − 3	0.2973 − 6	968.7	0.4158 − 3
(a)	390.0	4727	0.7061	0.2969	968.1	0.4205
37	390.0	4539	0.6780	0.2969	968.1	0.4379
38	390.0	4326	0.6463	0.2969	968.1	0.4594
39	390.0	4124	0.6161	0.2969	968.1	0.4820

(Continued)

Table B.1 Characteristics of the standard atmosphere in British units (fps) (continued)

h, ft/10^3	T, °R	p, psf	ρ, sl/ft^3	μ, sl/ft-s	a, ft/s	ν, ft^2/s
40	390.0	3931	0.5873	0.2969	968.1	0.5056
41	390.0	3743 − 1	0.5598 − 3	0.2969 − 6	968.1	0.5304 − 3
42	390.0	3572	0.5337	0.2969	968.1	0.5564
43	390.0	3405	0.5087	0.2969	968.1	0.5837
44	390.0	3246	0.4849	0.2969	965.1	0.6123
45	390.0	3095	0.4623	0.2969	968.1	0.6423
46	390.0	2950 − 1	0.4407 − 3	0.2969 − 6	968.1	0.6738 − 3
47	390.0	2812	0.4201	0.2969	968.1	0.7068
48	390.0	2681	0.4005	0.2969	968.1	0.7415
49	390.0	2556	0.3818	0.2969	968.1	0.7778
50	390.0	2436	0.3639	0.2969	968.1	0.8159
52	390.0	2214 − 1	0.3307 − 3	0.2969 − 6	968.1	0.8978 − 3
54	390.0	2012	0.3006	0.2969	968.1	0.9879
56	390.0	1829	0.2732	0.2969	968.1	0.1087 − 2
58	390.0	1662	0.2482	0.2969	968.1	0.1196
60	390.0	1510	0.2256	0.2969	968.1	0.1316
62	390.0	1373 − 1	0.2051 − 3	0.2969 − 6	968.1	0.1448 − 2
64	390.0	1243	0.1864	0.2969	968.1	0.1593
66	390.0	1134	0.1694	0.2969	968.1	0.1753
68	390.0	1031	0.1540	0.2969	968.1	0.1929
70	390.0	9367 − 2	0.1399	0.2969	968.1	0.2122
72	390.0	8514 − 2	0.1272 − 3	0.2969 − 6	968.1	0.2335 − 2
74	390.0	7739	0.1156	0.2969	968.1	0.2568
76	390.0	7035	0.1051	0.2969	968.1	0.2826
78	390.0	6394	0.9552 − 4	0.2969	968.1	0.3108
80	390.0	5813	0.8683	0.2969	968.1	0.3420
(b)	390.0	5193 − 2	0.7764 − 4	0.2969 − 6	968.1	0.3824 − 2
85	394.3	4533	0.6771	0.2997	973.4	0.4426
90	402.5	3629	0.5253	0.3048	983.5	0.5803
95	410.6	2888	0.4097	0.3099	993.4	0.7565
100	418.8	2309	0.3211	0.3150	1003.2	0.9809
110	435.1	1495 − 2	0.2001 − 4	0.3250 − 6	1022.5	0.1624 − 1
120	451.4	9837 − 3	0.1270	0.3348	1041.5	0.2637
130	467.6	6574	0.8190 − 5	0.3444	1060.1	0.4206
140	483.9	4455	0.5364	0.3539	1078.3	0.6598
150	500.1	3060	0.3564	0.3632	1096.3	0.1019 + 0
(c)	508.8	2515 − 3	0.2880 − 5	0.3682 − 6	1105.7	0.1278 + 0
160	508.8	2125	0.2433	0.3682	1105.7	0.1513
170	508.8	1479	0.1693	0.3682	1105.7	0.2175
(d)	508.8	1218	0.1395	0.3682	1105.7	0.2640
180	499.0	1027	0.1200	0.3626	1095.0	0.3023
190	473.0	7047 − 4	0.8589 − 6	0.3505	1071.7	0.4081
200	457.0	4754	0.6061	0.3381	1047.9	0.5580

Symbols: h = geo. altitude, ρ = density, a = sound speed, ν = kinematic viscosity, T = temperature, μ = viscosity, and p = pressure.
Single digit preceded by plus or minus indicates power of 10 (i.e., 0.23769 − 2 = 0.0023769)
Altitudes of temperature profile discontinuity:
(a) 36,152 ft, (b) 82,346 ft, (c) 155,348 ft, and (d) 175,344 ft.
Data from "US Extension of the ICAO Standard Atmosphere," 1958.

Table B.2 Characteristics of the standard atmosphere in SI units (mks)

Altitude h, km	Temperature T, K	Pressure P, N/m^2	Density ρ, kg/m^3	Speed of sound a, m/s	Viscosity μ, kg/m-s
0.0	288.16	101,325	1.225	340.3	1.79E−05
0.5	284.91	95,461	1.1673	338.4	1.77E−05
1.0	281.66	89,876	1.1117	336.4	1.76E−05
1.5	278.41	84,560	1.0581	334.5	1.74E−05
2.0	275.16	79,501	1.0066	332.5	1.73E−05
2.5	271.92	74,692	0.95696	330.6	1.71E−05
3.0	268.67	70,121	0.90926	328.6	1.69E−05
3.5	265.42	65,780	0.86341	326.6	1.68E−05
4.0	262.18	61,660	0.81935	324.6	1.66E−05
4.5	258.93	57,752	0.77704	322.6	1.65E−05
5.0	255.69	54,048	0.73643	320.5	1.63E−05
5.5	252.44	50,539	0.69747	318.5	1.61E−05
6.0	249.2	47,217	0.66011	316.5	1.6E−05
6.5	245.95	44,075	0.62431	314.4	1.58E−05
7.0	242.71	41,105	0.59002	312.3	1.56E−05
7.5	239.47	38,299	0.55719	310.2	1.54E−05
8.0	236.23	35,651	0.52578	308.1	1.53E−05
8.5	232.98	33,154	0.49575	306.0	1.51E−05
9.0	229.74	30,800	0.46706	303.9	1.49E−05
9.5	226.5	28,584	0.43966	301.7	1.48E−05
10.0	223.26	26,500	0.41351	299.6	1.46E−05
10.5	220.02	24,540	0.38857	297.4	1.44E−05
11.0	216.78	22,700	0.3648	295.2	1.42E−05
11.5	216.66	20,985	0.33743	295.1	1.42E−05
12.0	216.66	19,399	0.31194	295.1	1.42E−05
12.5	216.66	17,934	0.28837	295.1	1.42E−05
13.0	216.66	16,579	0.26659	295.1	1.42E−05
13.5	216.66	15,327	0.24646	295.1	1.42E−05
14.0	216.66	14,170	0.22785	295.1	1.42E−05
14.5	216.66	13,101	0.21065	295.1	1.42E−05
15.0	216.66	12,112	0.19475	295.1	1.42E−05
16.0	216.66	10,353	0.16647	295.1	1.42E−05
17.0	216.66	8,849.6	0.1423	295.1	1.42E−05
18.0	216.66	7,565.2	0.12165	295.1	1.42E−05
19.0	216.66	6,467.4	0.10399	295.1	1.42E−05
20.0	216.66	5,529.3	0.08891	295.1	1.42E−05
21.0	216.66	4,728.9	0.07572	295.1	1.42E−05
22.0	216.66	4,047.5	0.06451	295.1	1.42E−05
23.0	216.66	3,466.9	0.05558	295.1	1.42E−05
24.0	216.66	2,955.4	0.04752	295.1	1.42E−05
25.0	216.66	2,527.3	0.04064	295.1	1.42E−05
30.0	231.24	1,185.5	0.01786	295.1	1.49E−05

B-1B Lancer (U.S. Air Force photo).

Appendix C: Airspeed

IAS—indicated airspeed (read from cockpit instrumentation, includes cockpit-instrument error correction)
CAS—calibrated airspeed (indicated airspeed corrected for airspeed-instrumentation position error)
EAS—equivalent airspeed (calibrated airspeed corrected for compressibility effects)
TAS—true airspeed (equivalent airspeed corrected for change in atmospheric density)

$$\mathrm{TAS} = \mathrm{EAS}/\sqrt{(\rho/\rho_0)}$$

$$\mathrm{EAS} = \mathrm{CAS}\sqrt{P/P_0}\left[\frac{(q_c/P + 1)^{0.286} - 1}{(q_c/P_0 + 1)^{0.286} - 1}\right]^{0.5}$$

where

$$q_c = P([1 + 0.2M^2]^{3.5} - 1)$$

Mach number:

$$M = \mathrm{TAS}/a$$

where

a = speed of sound
P_0 = pressure, sea level
ρ_0 = density, sea level

The following are equivalent at 15,000 ft, 30°C day:

$M = 0.428$
TAS = 290 knots
CAS = 215 knots
EAS = 213 knots

F-22 (U. S. Air Force photo).

Appendix D: Airfoil Data

NACA 0006

Stations and ordinates given in percent of airfoil chord]

Upper surface		Lower surface	
Station	Ordinate	Station	Ordinate
0	0	0	0
1.25	.95	1.25	−.95
2.5	1.31	2.5	−1.31
5.0	1.78	5.0	−1.78
7.5	2.10	7.5	−2.10
10	2.34	10	−2.34
15	2.67	15	−2.67
20	2.87	20	−2.87
25	2.97	25	−2.97
30	3.00	30	−3.00
40	2.90	40	−2.90
50	2.65	50	−2.65
60	2.28	60	−2.28
70	1.83	70	−1.83
80	1.31	80	−1.31
90	.72	90	−.72
95	.40	95	−.40
100	(.06)	100	(−.06)
100	0	100	0

L. E. radius: 0.40

NACA 0009

[Stations and ordinates given in percent of airfoil chord]

Upper surface		Lower surface	
Station	Ordinate	Station	Ordinate
0	0	0	0
1.25	1.42	1.25	−1.42
2.5	1.96	2.5	−1.96
5.0	2.67	5.0	−2.67
7.5	3.15	7.5	−3.15
10	3.51	10	−3.51
15	4.01	15	−4.01
20	4.30	20	−4.30
25	4.46	25	−4.46
30	4.50	30	−4.50
40	4.35	40	−4.35
50	3.97	50	−3.97
60	3.42	60	−3.42
70	2.75	70	−2.75
80	1.97	80	−1.97
90	1.09	90	−1.09
95	.60	95	−.60
100	(.10)	100	(−.10)
100	0	100	0

L. E. radius: 0.89

NACA 2415

[Stations and ordinates given in percent of airfoil chord]

Upper surface		Lower surface	
Station	Ordinate	Station	Ordinate
0	-----	0	0
1.25	2.71	1.25	−2.06
2.5	3.71	2.5	−2.86
5.0	5.07	5.0	−3.84
7.5	6.06	7.5	−4.47
10	6.83	10	−4.90
15	7.97	15	−5.42
20	8.70	20	−5.66
25	9.17	25	−5.70
30	9.38	30	−5.62
40	9.25	40	−5.25
50	8.57	50	−4.67
60	7.50	60	−3.90
70	6.10	70	−3.05
80	4.41	80	−2.15
90	2.45	90	−1.17
95	1.34	95	−.68
100	(.16)	100	(−.16)
100	-----	100	0

L. E. radius: 2.48
Slope of radius through L. E.: 0.10

NACA 4415

[Stations and ordinates given in percent of airfoil chord]

Upper surface		Lower surface	
Station	Ordinate	Station	Ordinate
0	-------	0	0
1.25	3.07	1.25	−1 79
2.5	4.17	2.5	−2.48
5.0	5.74	5.0	−3.27
7.5	6.91	7.5	−3.71
10	7.84	10	−3.98
15	9.27	15	−4.18
20	10.25	20	−4.15
25	10.92	25	−3.98
30	11.25	30	−3.75
40	11.25	40	−3.25
50	10.53	50	−2.72
60	9.30	60	−2.14
70	7.63	70	−1.55
80	5.55	80	−1.03
90	3.08	90	−.57
95	1.67	95	−.36
100	(.16)	100	(−.16)
100	-------	100	0

L. E. radius: 2.48
Slope of radius through L. E.: 0.20

NACA 23015

[Stations and ordinates given in percent of airfoil chord]

Upper surface		Lower surface	
Station	Ordinate	Station	Ordinate
0	-----	0	0
1.25	3.34	1.25	−1.54
2.5	4.44	2.5	−2.25
5.0	5.89	5.0	−3.04
7.5	6.90	7.5	−3.61
10	7.64	10	−4.09
15	8.52	15	−4.84
20	8.92	20	−5.41
25	9.08	25	−5.78
30	9.05	30	−5.96
40	8.59	40	−5.92
50	7.74	50	−5.50
60	6.61	60	−4.81
70	5.25	70	−3.91
80	3.73	80	−2.83
90	2.04	90	−1.59
95	1.12	95	−.90
100	(.16)	100	(−.16)
100	-----	100	0

L. E. radius: 2.48
Slope of radius through L. E.: 0.305

NACA 64-006

[Stations and ordinates given in percent of airfoil chord]

Upper surface		Lower surface	
Station	Ordinate	Station	Ordinate
0	0	0	0
.50	.494	.50	−.494
.75	.596	.75	−.596
1.25	.754	1.25	−.754
2.5	1.024	2.5	−1.024
5.0	1.405	5.0	−1.405
7.5	1.692	7.5	−1.692
10	1.928	10	−1.928
15	2.298	15	−2.298
20	2.572	20	−2.572
25	2.772	25	−2.772
30	2.907	30	−2.907
35	2.981	35	−2.981
40	2.995	40	−2.995
45	2.919	45	−2.919
50	2.775	50	−2.775
55	2.575	55	−2.575
60	2.331	60	−2.331
65	2.050	65	−2.050
70	1.740	70	−1.740
75	1.412	75	−1.412
80	1.072	80	−1.072
85	.737	85	−.737
90	.423	90	−.423
95	.157	95	−.157
100	0	100	0

L. E. radius: 0.256

NACA 65(216)-415 $a=0.5$

[Stations and ordinates given in percent of airfoil chord]

Upper surface		Lower surface	
Station	Ordinate	Station	Ordinate
0	0	0	0
.244	1.236	.756	−.960
.469	1.498	1.031	−1.110
.930	1.947	1.570	−1.359
2.121	2.837	2.879	−1.801
4.564	4.175	5.436	−2.411
7.044	5.208	7.956	−2.832
9.540	6.073	10.460	−3.169
14.561	7.465	15.439	−3.673
19.608	8.518	20.392	−4.022
24.669	9.315	25.331	−4.267
29.742	9.900	30.258	−4.428
34.825	10.279	35.175	−4.507
30.916	10.467	40.084	−4.523
45.019	10.438	44.981	−4.446
50.153	10.131	49.847	−4.251
55.263	9.512	54.737	−3.940
60.305	8.645	59.695	−3.521
65.308	7.575	64.692	−2.995
70.281	6.373	69.719	−2.409
75.237	5.152	74.763	−1.848
80.180	3.890	79.820	−1.278
85.117	2.639	84.883	−.723
90.062	1.533	89.938	−.305
95.020	.606	94.980	−.030
100.000	0	100.000	0

L. E. radius: 1.498
Slope of radius through L. E.: 0.233

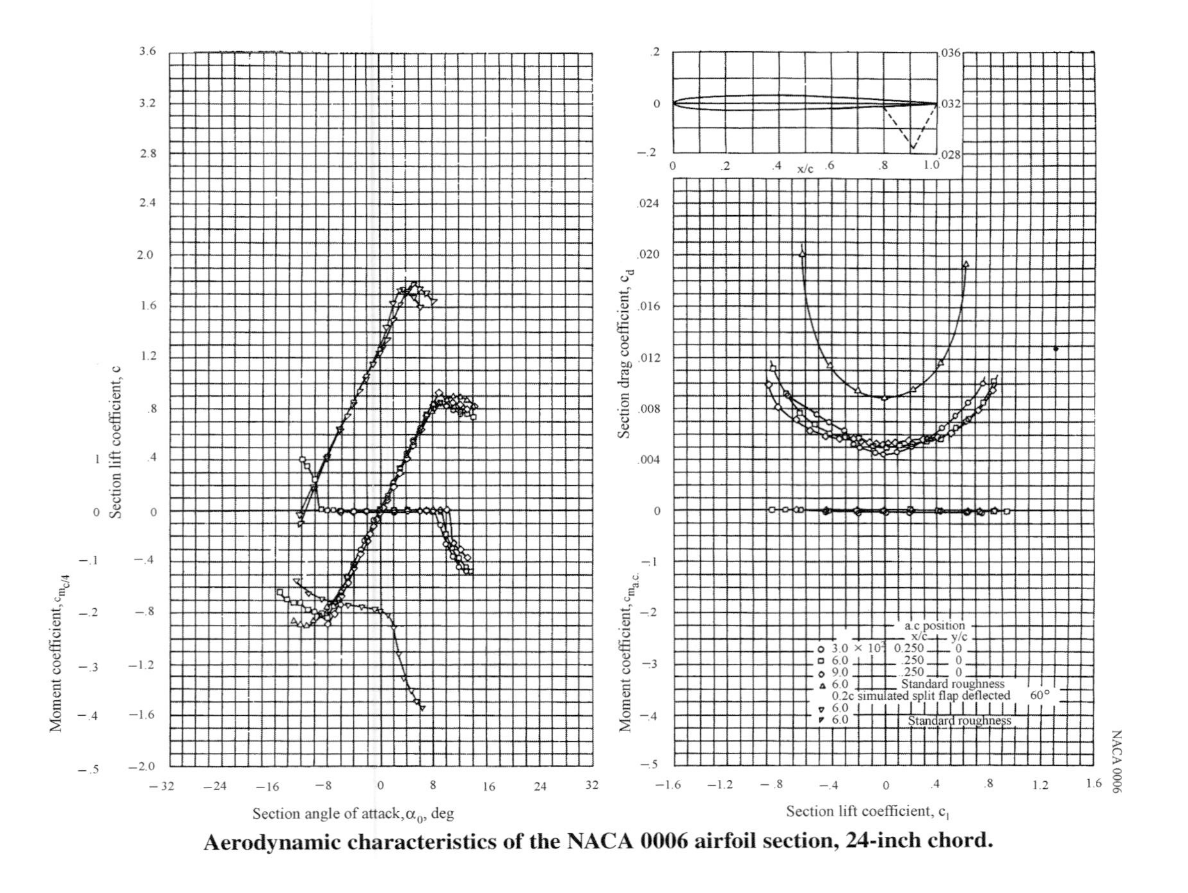

Aerodynamic characteristics of the NACA 0006 airfoil section, 24-inch chord.

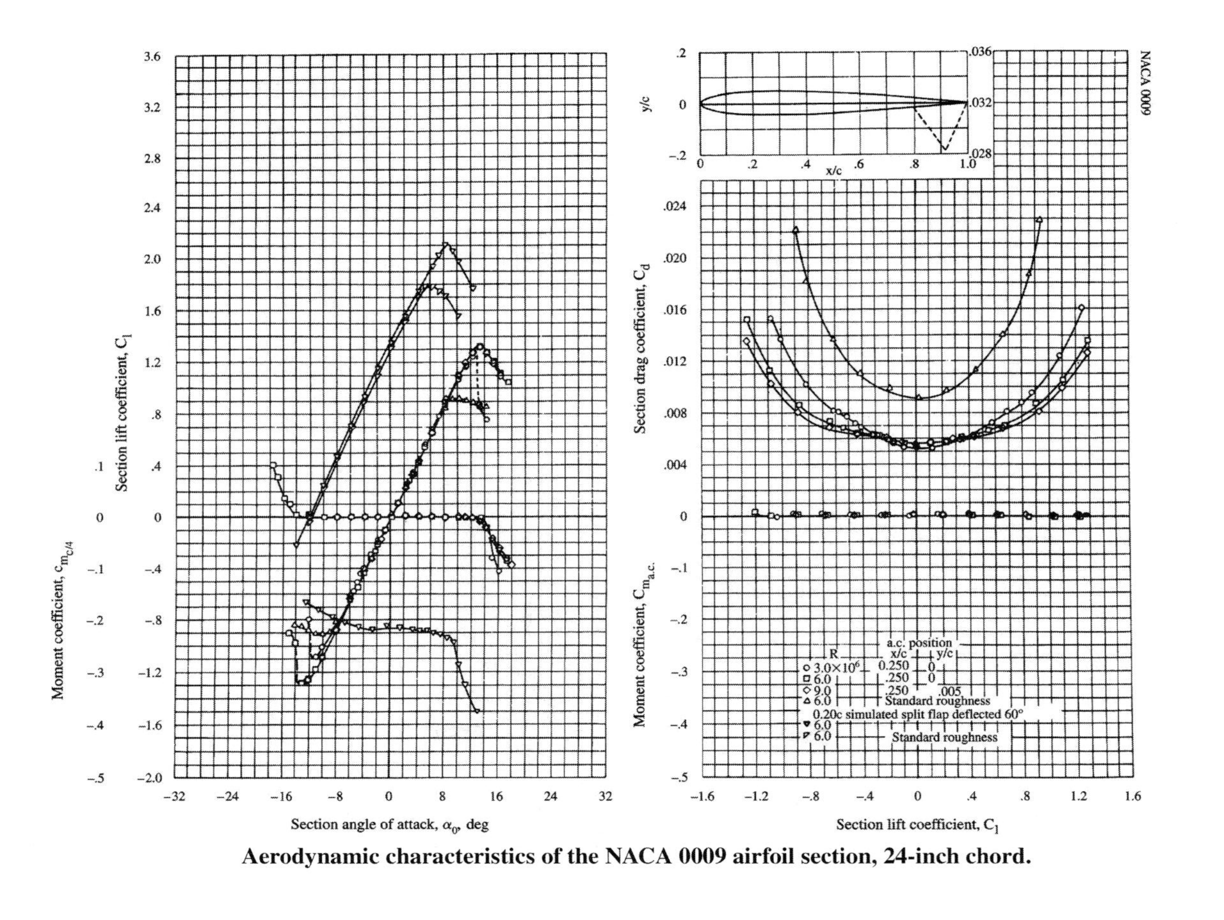

Aerodynamic characteristics of the NACA 0009 airfoil section, 24-inch chord.

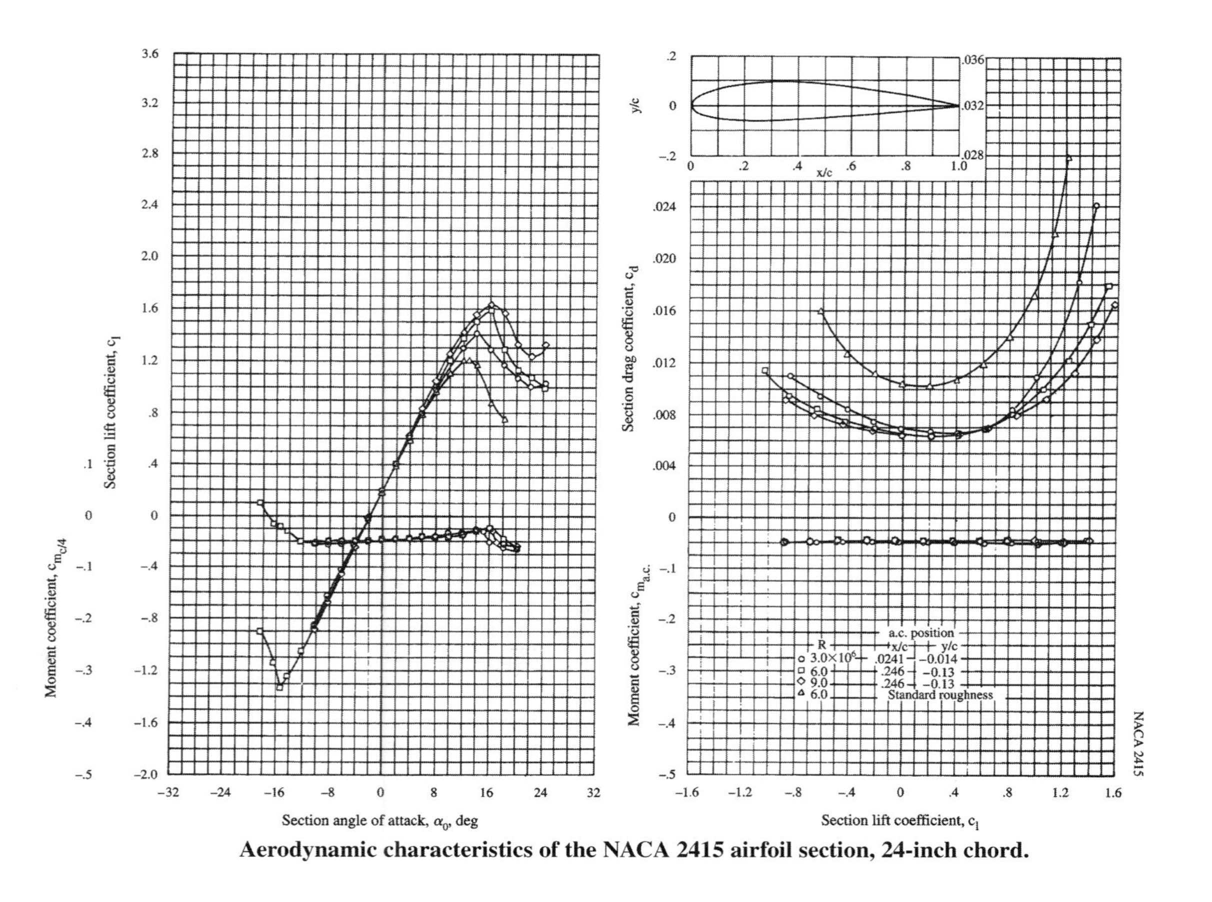

Aerodynamic characteristics of the NACA 2415 airfoil section, 24-inch chord.

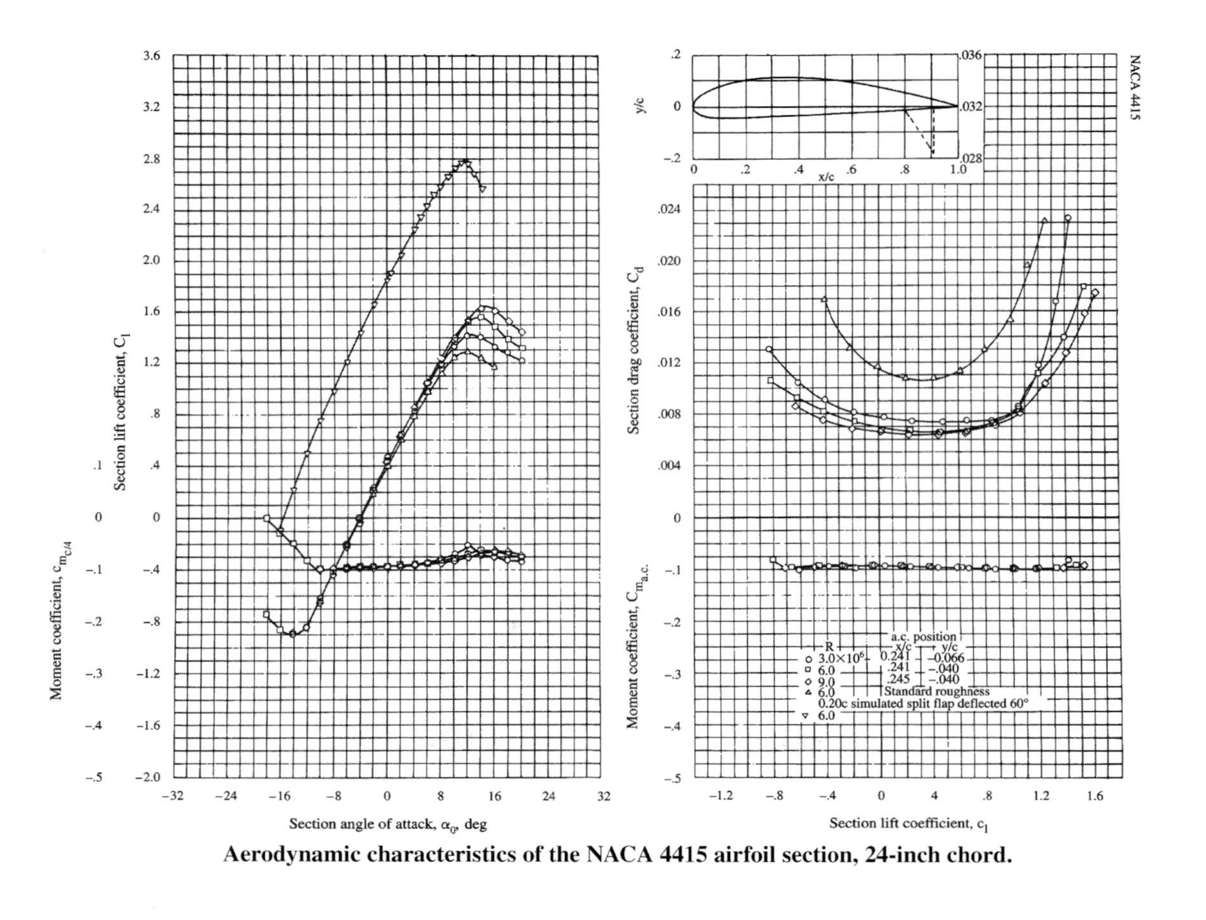

Aerodynamic characteristics of the NACA 4415 airfoil section, 24-inch chord.

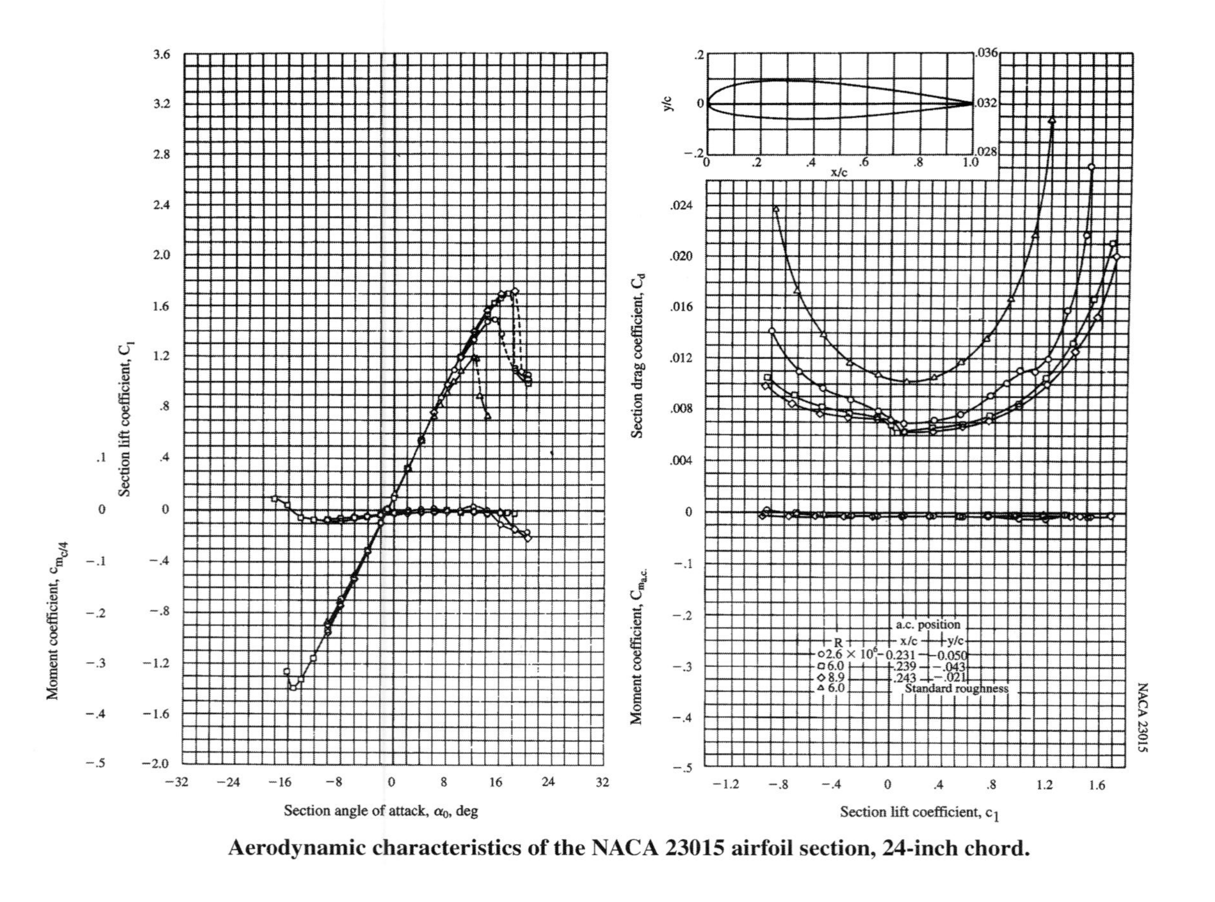

Aerodynamic characteristics of the NACA 23015 airfoil section, 24-inch chord.

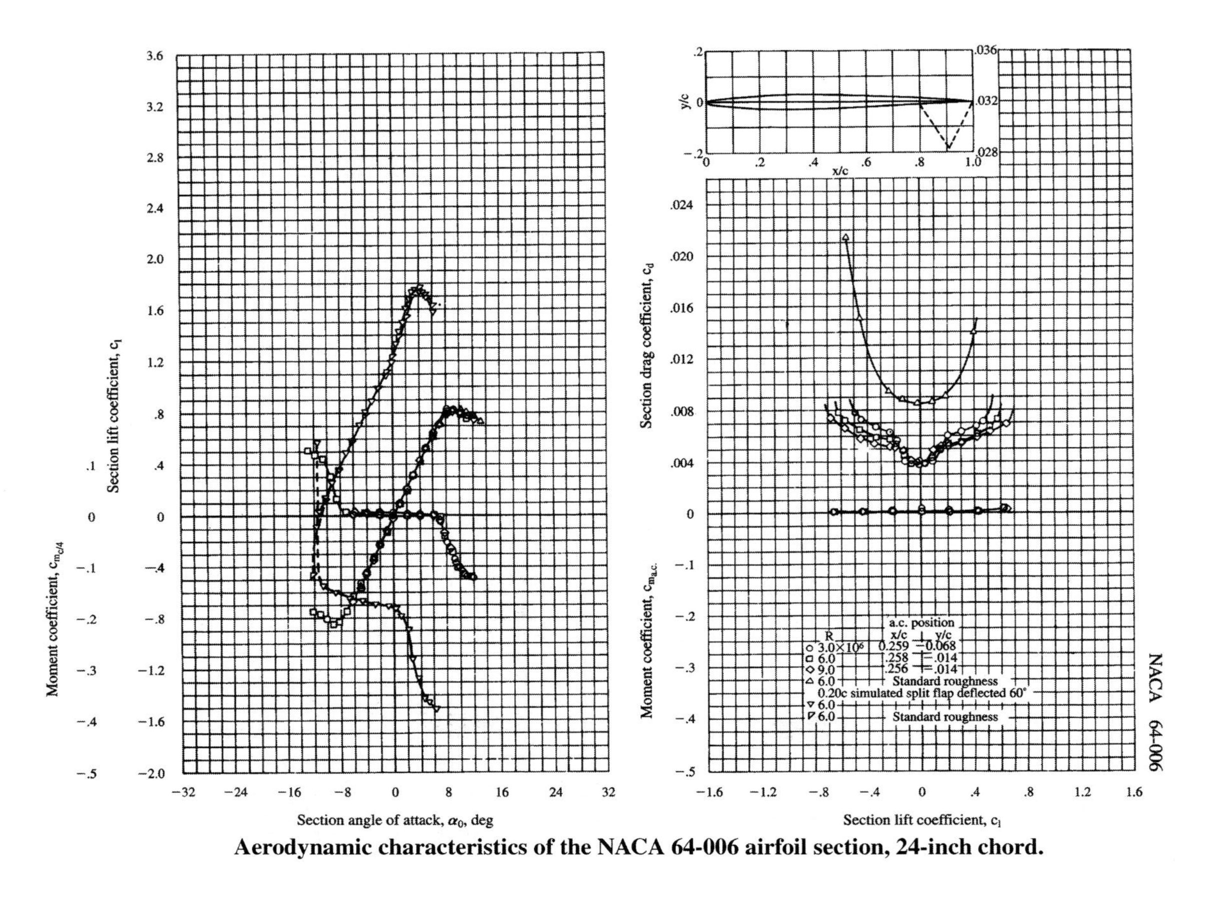

Aerodynamic characteristics of the NACA 64-006 airfoil section, 24-inch chord.

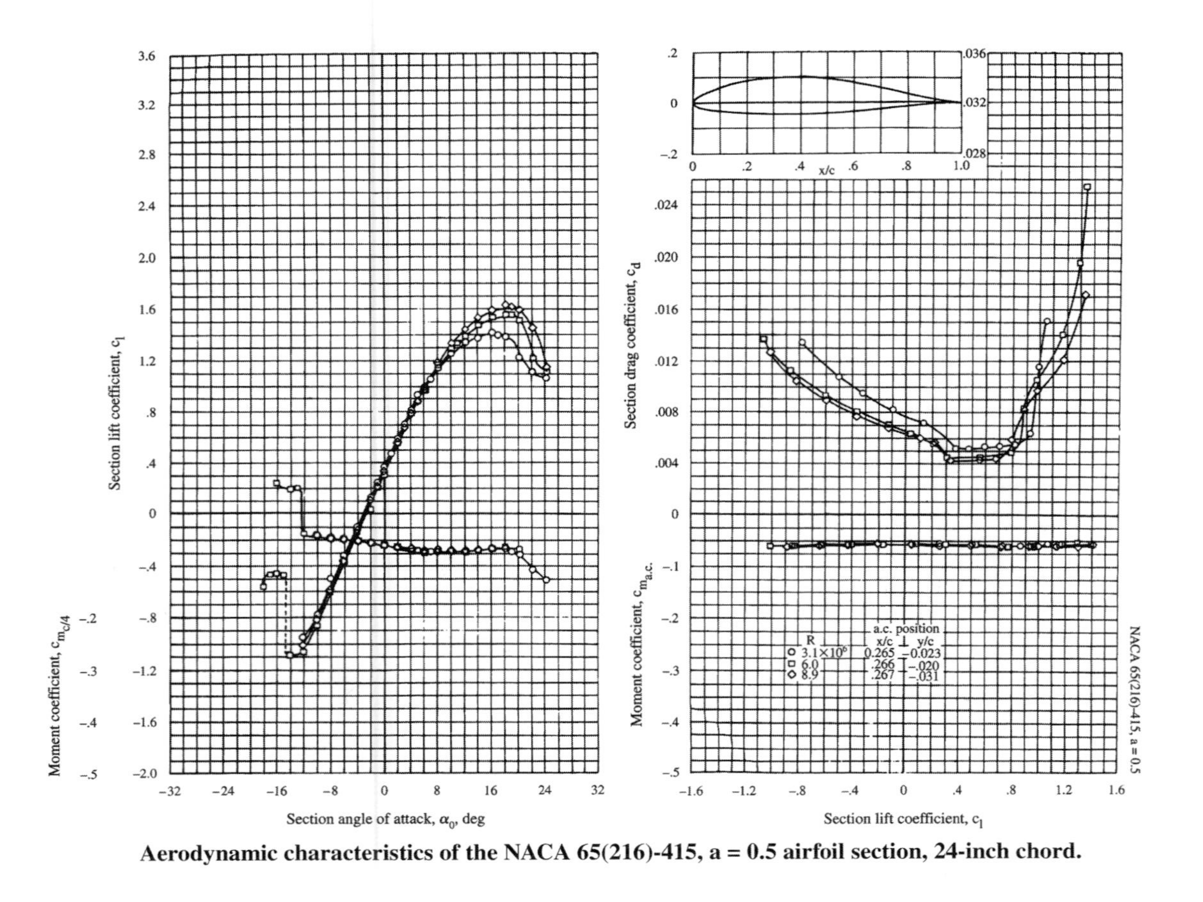

Aerodynamic characteristics of the NACA 65(216)-415, a = 0.5 airfoil section, 24-inch chord.

NLF(1)-0215F Airfoil coordinate ($\delta_f = 0°$)
[c = 60.960 cm (24.000 in.)]

Upper surface		Lower surface	
X/C	*Z/C*	*X/C*	*Z/C*
.00240	.00917	.00000	−.00006
.00909	.01947	.00245	−.00704
.02004	.03027	.01099	−.01211
.03527	.04120	.02592	−.01656
.05469	.05201	.04653	−.02052
.07816	.06250	.07242	−.02399
.10546	.07247	.10324	−.02699
.13635	.08175	.13854	−.02954
.17050	.09019	.17788	−.03166
.20758	.09761	.22073	−.03334
.24720	.10389	.26654	−.03456
.28894	.10887	.31473	−.03531
.33237	.11240	.36468	−.03554
.37702	.11428	.41576	−.03519
.42253	.11427	.46731	−.03415
.46864	.11219	.51867	−.03225
.51524	.10784	.56920	−.02925
.56247	.10147	.61825	−.02441
.61010	.09373	.66662	−.01663
.65752	.08513	.71614	−.00705
.70408	.07603	.76645	.00167
.74914	.06673	.81565	.00804
.79206	.05746	.86198	.01155
.83222	.04844	.90359	.01198
.86902	.03983	.93862	.00990
.90193	.03175	.96588	.00655
.93044	.02428	.98504	.00323
.95409	.01737	.99630	.00086
.97285	.01082	1.00000	.00000
.98710	.00507		
.99658	.00126		
1.00000	.00000		

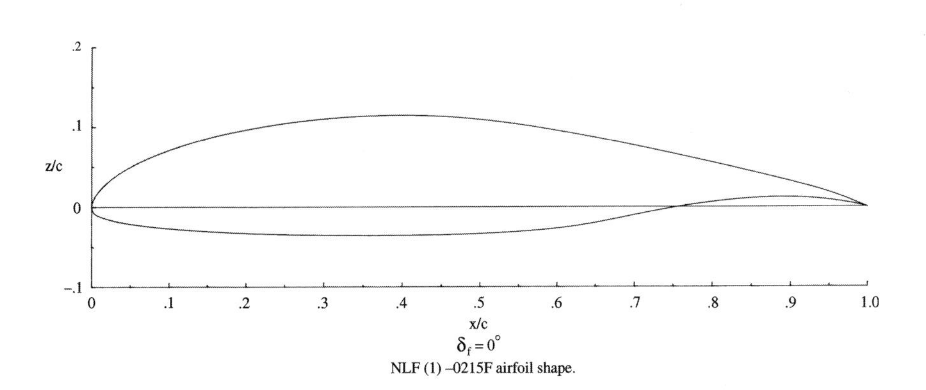

NLF (1) –0215F airfoil shape.

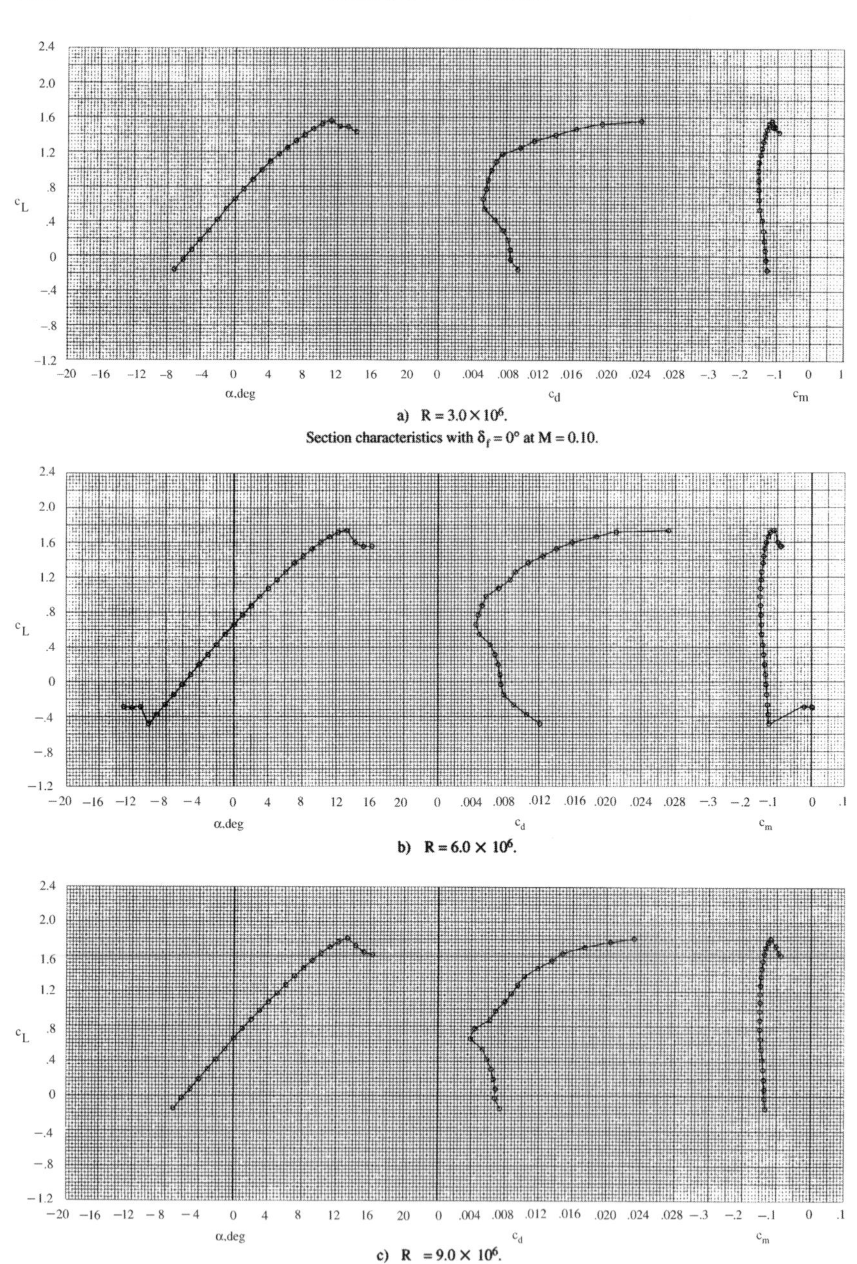

a) $R = 3.0 \times 10^6$.

Section characteristics with $\delta_f = 0°$ at $M = 0.10$.

b) $R = 6.0 \times 10^6$.

c) $R = 9.0 \times 10^6$.

Appendix E: Typical Engine Performance Curves

This appendix contains performance curves for three different types of aircraft engines: an afterburning turbofan, a high-bypass-ratio turbofan, and a turboprop. Although not identical to any currently in production, these engines are representative of advanced technology suitable for use in aircraft design studies. The engines may be scaled using the scaling laws presented in Chapter 10.

These curves were generated using Version 2.0 of the engine-cycle analysis programs ONX and OFFX that are based on the methods contained in the AIAA Education Series textbook, *Aircraft Engine Design*, by Jack D. Mattingly, William H. Heiser, and Daniel H. Daley. Both the textbook and the programs with their user guide are available from the publisher. The author would like to thank Jack Mattingly for preparation of these engine performance curves.

Note: All altitudes are in feet.

E.1 Afterburning Turbofan

Table E.1 Afterburning turbofan characteristics

Characteristic	Value
Sea-level static thrust	30,000 lb {133 kN}
Sea-level static TSFC	1.64 1/hr {46 mg/NS}
Sea-level static airflow	246 lbm/s {112 kg/s}
Bar-engine weight	3,000 lb {1361 kg}
Engine length (including axisymmetric nozzle)	160 in. {407 cm}
Maximum diameter	44 in. {112 cm}
Fan-face diameter	40 in. {102 cm}
Overall pressure ratio	22
Fan pressure ratio	4.3
Bypass ratio	0.41

The following installed engine data reflect these assumptions:

1) Mil. Spec. MIL-E-5008B inlet pressure recovery and inlet duct total pressure ratio of 0.97
2) Power extraction of 320 kW to drive electric generators and auxiliary equipment at all power settings and flight conditions
3) High-pressure bleed airflow at rate of 1.7 lb/s

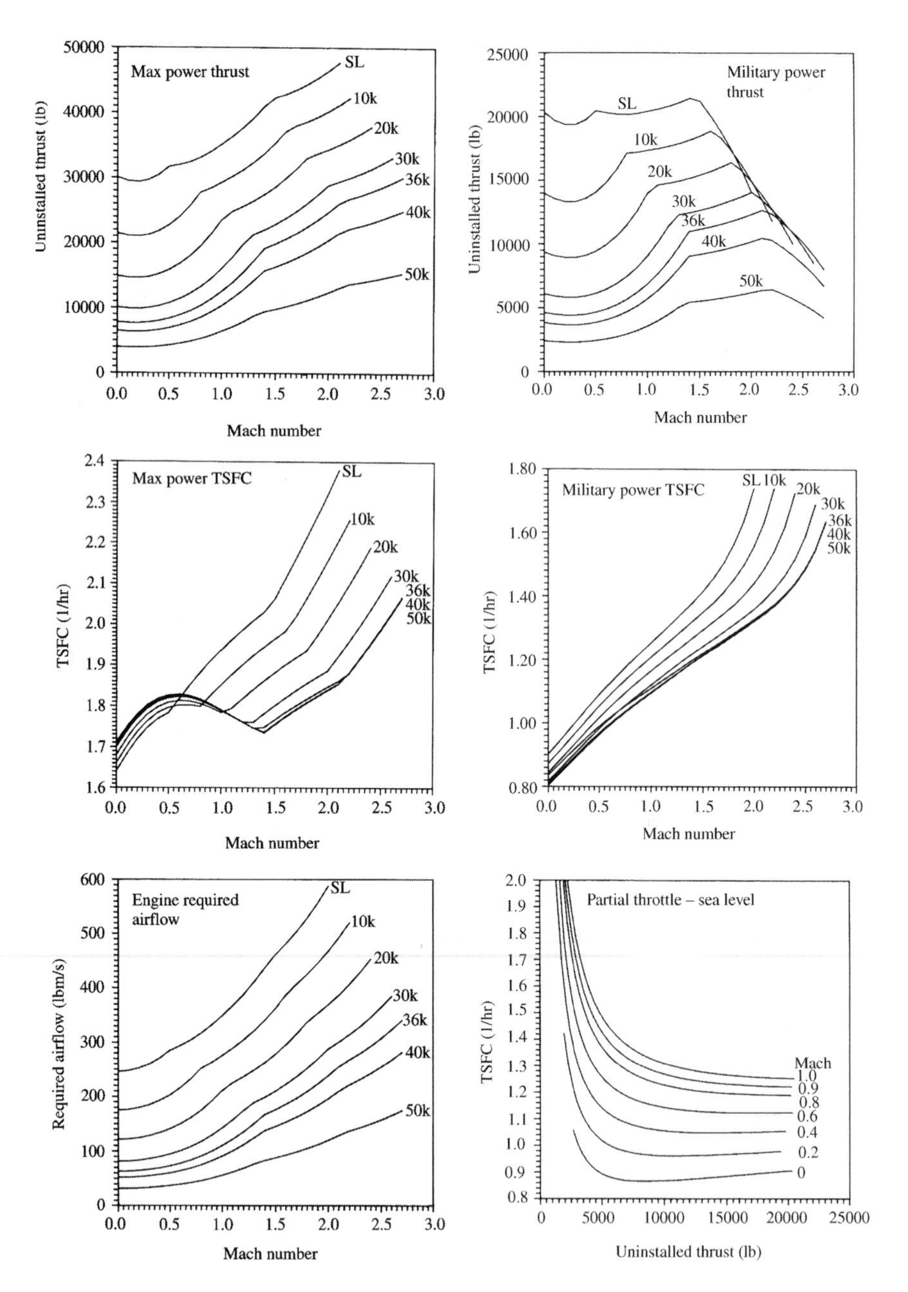

Max power thrust
Uninstalled thrust (lb)
Mach number
SL
10k
20k
30k
36k
40k
50k
Military power thrust
Uninstalled thrust (lb)
Mach number
Max power TSFC
TSFC (1/hr)
Military power TSFC
Engine required airflow
Required airflow (lbm/s)
Partial throttle – sea level
Mach
1.0
0.9
0.8
0.6
0.4
0.2
0
Uninstalled thrust (lb)

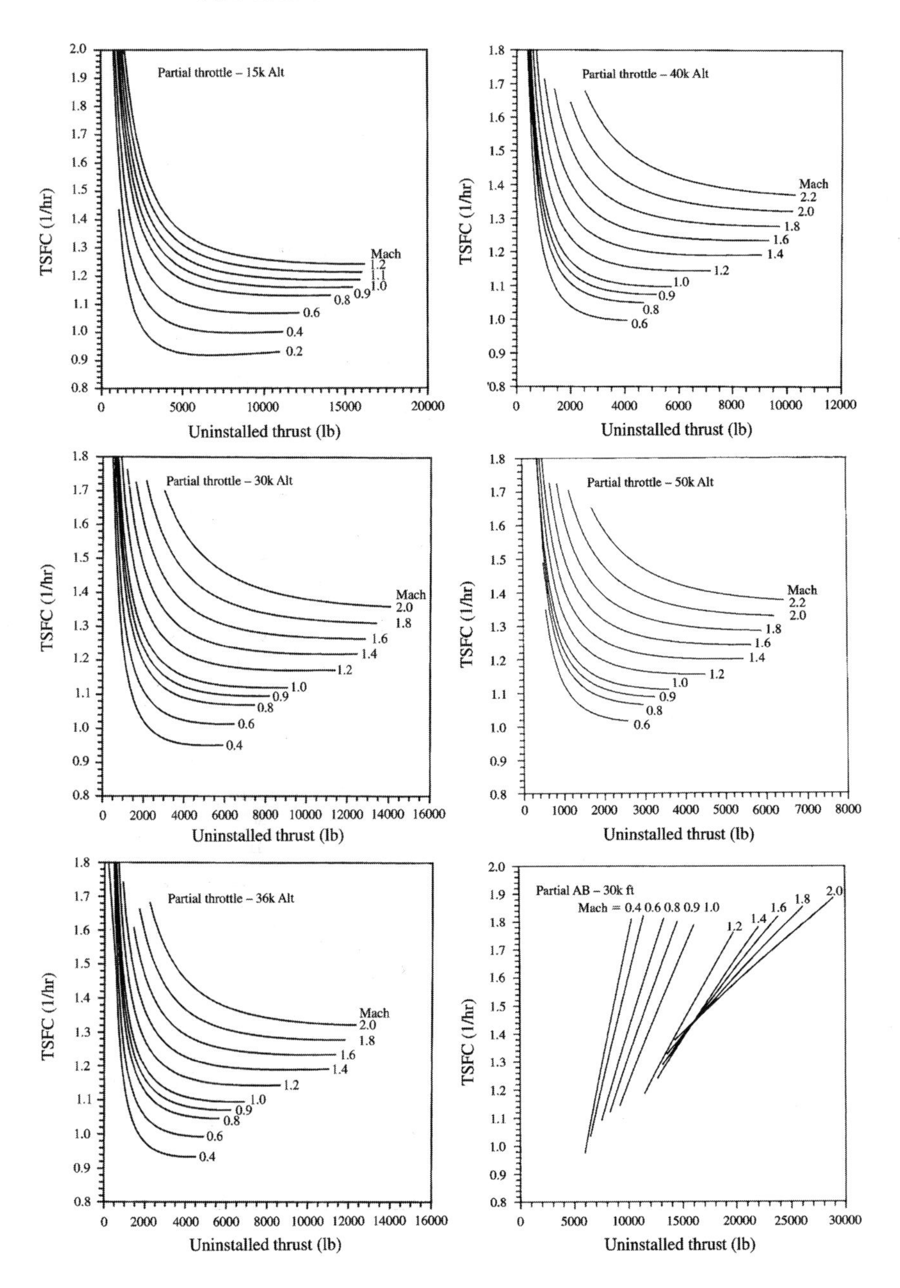
Partial throttle – 15k Alt
TSFC (1/hr)
Uninstalled thrust (lb)
Mach 1.2 1.1 1.0 0.9 0.8 0.6 0.4 0.2
Partial throttle – 40k Alt
Mach 2.2 2.0 1.8 1.6 1.4 1.2 1.0 0.9 0.8 0.6
Partial throttle – 30k Alt
Mach 2.0 1.8 1.6 1.4 1.2 1.0 0.9 0.8 0.6 0.4
Partial throttle – 50k Alt
Mach 2.2 2.0 1.8 1.6 1.4 1.2 1.0 0.9 0.8 0.6
Partial throttle – 36k Alt
Mach 2.0 1.8 1.6 1.4 1.2 1.0 0.9 0.8 0.6 0.4
Partial AB – 30k ft
Mach = 0.4 0.6 0.8 0.9 1.0
1.2 1.4 1.6 1.8 2.0

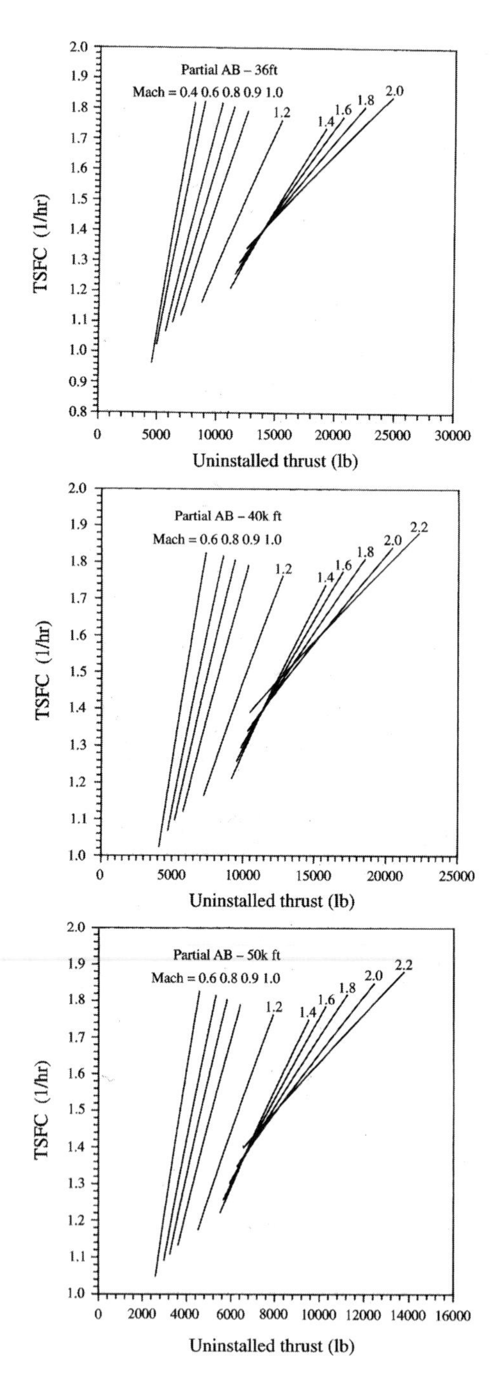
Partial AB – 36ft
Mach = 0.4 0.6 0.8 0.9 1.0
1.2
1.4
1.6
1.8
2.0
TSFC (1/hr)
Uninstalled thrust (lb)
Partial AB – 40k ft
Mach = 0.6 0.8 0.9 1.0
1.2
1.4
1.6
1.8
2.0
2.2
TSFC (1/hr)
Uninstalled thrust (lb)
Partial AB – 50k ft
Mach = 0.6 0.8 0.9 1.0
1.2
1.4
1.6
1.8
2.0
2.2
TSFC (1/hr)
Uninstalled thrust (lb)

E.2 High-Bypass Turbofan

Table E.2 High-bypass turbofan characteristics

Sea-level static thrust	50,000 lb {222 kN}
Sea-level static TSFC	0.40 1/hr {11.3 mg/NS}
Sea-level static airflow	1,680 lb/s {762 kg/s}
Bare-engine weight	7,700 lb {3493 kg}
Engine length	150 in. {381 cm}
Maximum engine diameter	100 in. {254 cm}
Overall pressure ratio	30
Fan pressure ratio	1.6
Bypass ratio	8.0

The following installed-engine data reflect these assumptions:

1) Inlet total pressure ratio of 0.97
2) Power extraction of 650 kW to drive electric generators and auxiliary equipment at all power settings and flight conditions
3) High-pressure bleed airflow at rate of 2.0 lb/s

Maximum rated performance is plotted with dashes.

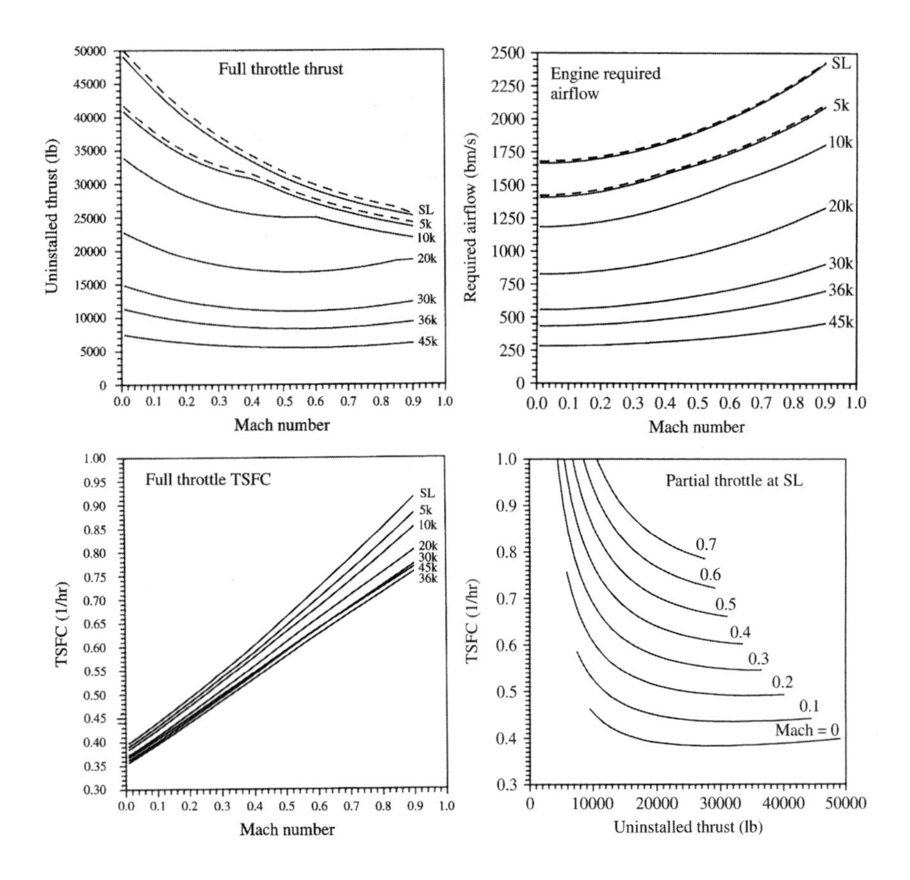

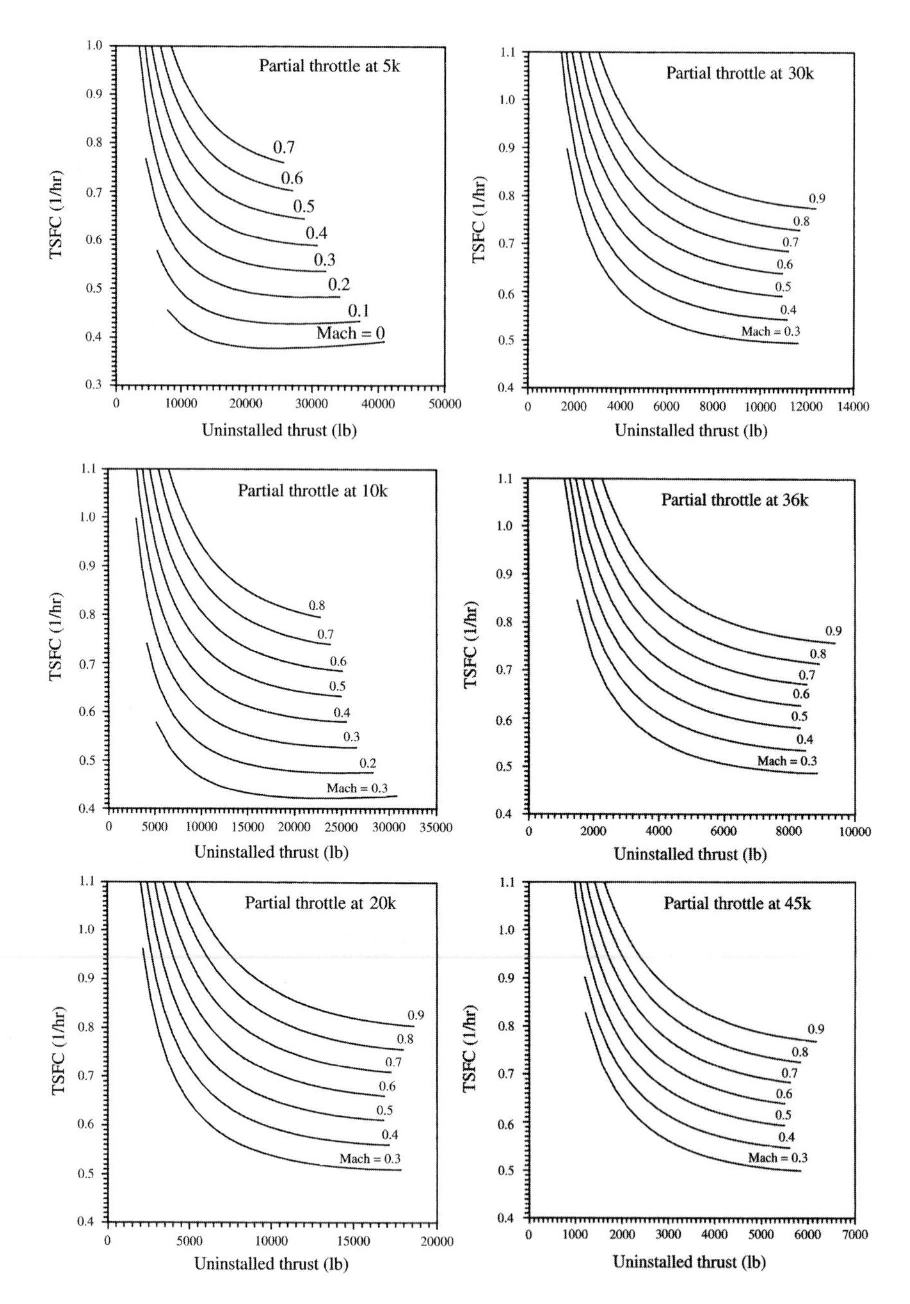

Partial throttle at 5k
TSFC (1/hr)
Uninstalled thrust (lb)
0.7
0.6
0.5
0.4
0.3
0.2
0.1
Mach = 0
Partial throttle at 30k
0.9
0.8
0.7
0.6
0.5
0.4
Mach = 0.3
Partial throttle at 10k
0.8
0.7
0.6
0.5
0.4
0.3
0.2
Mach = 0.3
Partial throttle at 36k
0.9
0.8
0.7
0.6
0.5
0.4
Mach = 0.3
Partial throttle at 20k
0.9
0.8
0.7
0.6
0.5
0.4
Mach = 0.3
Partial throttle at 45k
0.9
0.8
0.7
0.6
0.5
0.4
Mach = 0.3

E.3 Turboprop

Table E.3 Turboprop characteristics

Sea-level static thrust	32,000 lb {142 kN}
Sea-level static power	6,500 hp {4847 kW}
Sea-level static TSFC	0.14 1/hr {4.0 mg/NS}
Sea-level static airflow	42.3 lbm/s {19.2 kg/s}
Bare engine weight (including gear box and propeller)	2,600 lb {1179 kg}
Diameter of four-bladed, two-row propeller	20.5 ft {6.2 m}
Engine length (propeller to exhaust)	200 in. {508 cm}
Engine diameter	46 in. {117 cm}
Overall pressure ratio	30

The following uninstalled-engine data reflect these assumptions:
1) Inlet total pressure ratio of 0.97
2) Power extraction of 54 kW to drive electric generators and auxiliary equipment at all power settings and flight conditions
3) High-pressure bleed airflow at rate of 0.8 lb/s
Maximum rated performance is plotted with dashes.

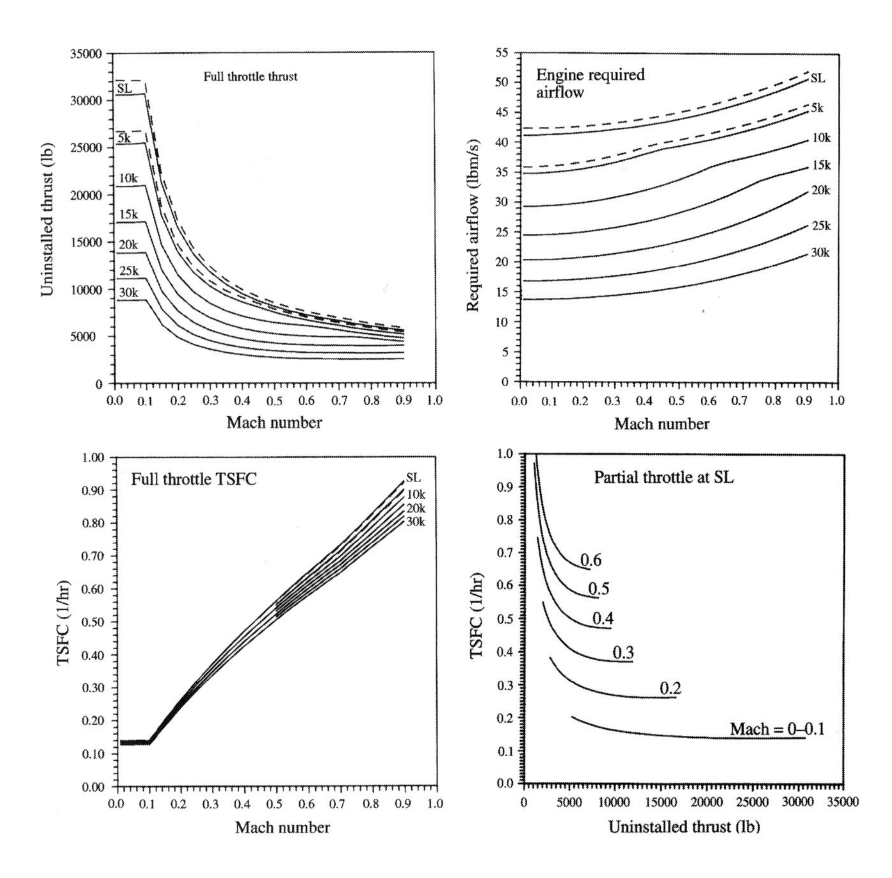

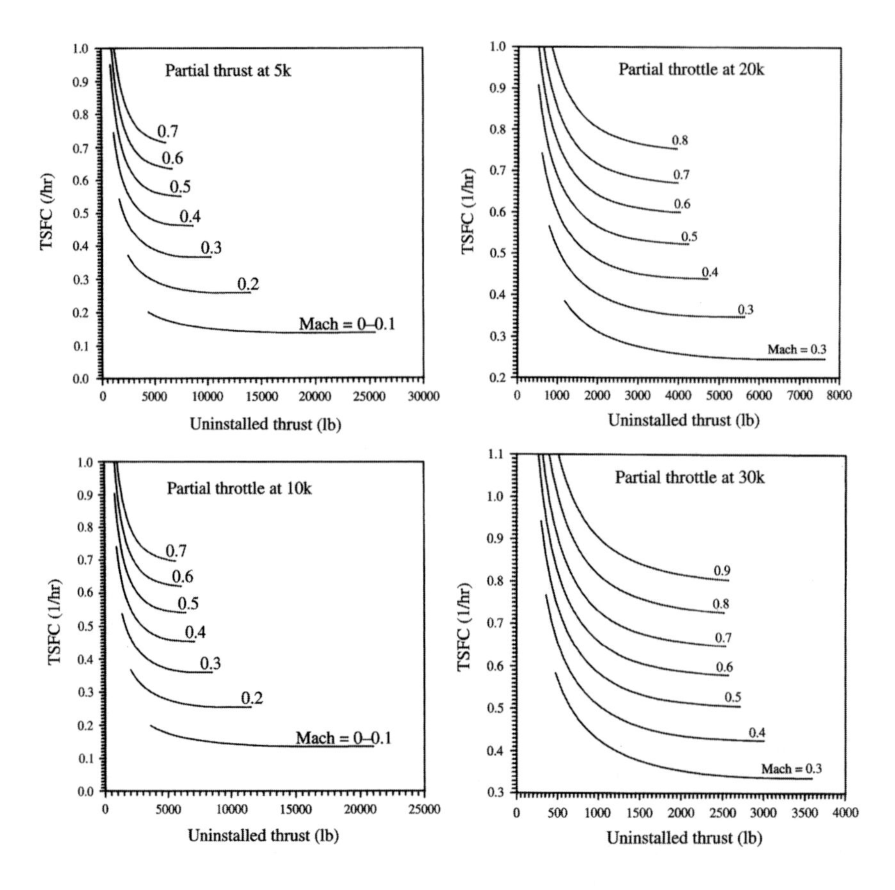
Partial thrust at 5k
TSFC (/hr)
Uninstalled thrust (lb)
0.7
0.6
0.5
0.4
0.3
0.2
Mach = 0–0.1
Partial throttle at 20k
TSFC (1/hr)
Uninstalled thrust (lb)
0.8
0.7
0.6
0.5
0.4
0.3
Mach = 0.3
Partial throttle at 10k
TSFC (1/hr)
Uninstalled thrust (lb)
0.7
0.6
0.5
0.4
0.3
0.2
Mach = 0–0.1
Partial throttle at 30k
TSFC (1/hr)
Uninstalled thrust (lb)
0.9
0.8
0.7
0.6
0.5
0.4
Mach = 0.3

Appendix F: Design Requirements and Specifications

Table F.1 Federal Aviation Regulations (FAR)—Applicability[a]

Category	Various[b]	Normal	Transport
A) Characteristics			
Maximum takeoff weight, lb	≤12,500	≤12,500	—
Number of engines	One or more	Two or more	Two or more
Type of engine	All	Propeller engines only	All
Minimum crew:			
Flight crew	One or more	Two	Two or more
Cabin attendants	None	<20 Pass.: None ≥20 Pass.: One	<10 Pass.: None ≥10 pass.: One or more
Maximum number of occupants	10	11–23	Not restricted
Maximum operating altitude, ft	25,000	25,000	Not restricted
B) FAR Applicability			
Airworthiness standards airplanes	Part 23	Part 23	Part 25
Airworthiness standards engines	Part 33	Part 33	Part 33
Airworthiness standards propellers	Part 35	Part 35	Part 35
Noise standards	Part 36: Prop-Driven, Appendix F		Part 36
General operating and flight rules	Part 91	Part 91	Part 91
Operations:			
Domestic, flag and supplemental comm. operators of large aircraft	—	—	Part 121
Air travel clubs using large aircraft	—	—	Part 123
Air taxi and comm. operators	—	Part 135	—
Agricultural aircraft	Part 137	—	—

[a]After E. Torenbeck (Ref. 23).
[b]Normal, utility, aerobatic, and agricultural.

Table F.2 Takeoff specifications[a]

Item	MIL-C5011A (Military)	FAR Part 23 (Civil)	FAR Part 25 (Commercial)
Velocity	$V_{\mathrm{TO}} \geq 1.1\ V_s$ $V_{\mathrm{CL}} \geq 1.2\ V_s$	$V_{\mathrm{TO}} \geq 1.1\ V_s$ $V_{\mathrm{CL}} \geq 1.2\ V_s$	$V_{\mathrm{TO}} \geq 1.1\ V_s$ $V_{\mathrm{CL}} \geq 1.2\ V_s$
Climb	Gear up: 500 fpm @ S.L. (AEO)[b] 100 fpm @ S.L. (OEI)[c]	Gear up: 300 fpm @ S.L. (AEO)	Gear down: 1/2%@ V_{TO} Gear up: 3% @ V_{CL} (OEI)
Field-length definition	Takeoff distance over 50-ft obstacle	Takeoff distance over 50-ft obstacle	115% of takeoff distance with AEO over 35 ft or balanced field length
Rolling coefficient	$\mu = 0.025$	Not defined	Not defined

[a]After L. Nicolai, Ref. 10.
[b]AEO = all engines operating. [c]OEI = one engine inoperative.

Table F.3 Landing specifications[a]

Item	MIL-C5011A	FAR Part 23	FAR Part 25
Velocity	$V_A \geq 1.2\ V_s$ $V_{\mathrm{TD}} \geq 1.1\ V_s$	$V_A \geq 1.3\ V_s$ $V_{\mathrm{TD}} \geq 1.15\ V_s$	$V_A \geq 1.3\ V_s$ $V_{\mathrm{TD}} \geq 1.15\ V_s$
Field-length definition	Landing distance over 50-ft obstacle	Landing distance over 50-ft obstacle	Landing distance over 50-ft obstacle divided by 0.6
Braking coefficient	$\mu = 0.30$	Not defined	Not defined

[a]After L. Nicolai, Ref. 10.

Table F.4 FAR climb requirements for multi-engine aircraft

Turbine-Engine Aircraft: FAR 25

All segments with one engine stopped, except go-around in landing configuration, which has all engines operating. Engine power or thrust set at "maximum rated," except being "maximum continuous" for third-segment climb. Maximum thrust attained after 8 s from flight idle for go-around. AEO: all engines operating.

Operation	Speed	Flaps	Landing gear	Minimum climb gradient for aircraft with n engines, %: $n=2$	$n=3$	$n=4$
Takeoff climb						
First-segment	LOF[a]	Takeoff	Down	≥ 0	0.3	0.5
Second-segment	V_2[b]	Takeoff	Up	2.4	2.7	3.0
Third-segment	$\geq 1.25\, V_s$[c]	Up	Up	1.2	1.4	1.5
Landing						
Go-around in approach configuration	$\leq 1.5\, V_s$[c]	Approach	Up	2.1	2.4	2.7
Go-around in landing configuration	$\leq 1.3 V_s$[c] AEO	Landing	Down	3.2	3.2	3.2

[a]LOF = liftoff. [b]Climbout speed over 35-ft obstacle. [c]Stall speed in the pertinent condition.

Reciprocating-Engine Aircraft: FAR 25

Power or thrust for operating engines set for takeoff on first and second segments and go-around and for "maximum continuous" during cruise and third segment. One engine windmilling propeller for first and second segments. If plane has automatic feathering, the propeller on an inoperative engine assumed to be feathered. One engine stopped (may be feathered) for third segment and go-around.

Operation	Speed	Flap setting	Landing gear	Minimum steady-climb rate, ft/min
Takeoff climb				
First-segment	V_2[a]	Takeoff	Down	≥ 50
Second-segment	V_2[a]	Takeoff	Up	$\geq 0.046\, V_{s_1}^2$[b]
Third-segment	Best	Up[c]	Up	$\geq \left(0.079 - \dfrac{0.106}{n}\right) V_{s_0}^2$[e,f]
Landing go-around				
(approach configuration)	$\leq 1.5\, V_{s_1}$[b]	Approach[d]	Up	$\geq 0.053 V_{s_0}^2$[e]

[a]V_2 = climbout speed over 35-ft obstacle; out-of-ground effect.
[b]V_{s_1} = stall speed in a specified configuration for reciprocating-engine-powered airplanes, in knots. [c]Or most favorable. [d]But $V_{s_1} \leq 1.1\, V_{s_0}$. [e]$V_{s_0}$ = stall speed in landing configuration for reciprocating-engine-powered airplanes, in knots. [f]At 5000-ft altitude.

Table F.4 FAR climb requirements for multi-engine aircraft (continued)

FAR 23 (Turbine or Reciprocating)

Multi-engine power at maximum continuous except for $W < 6000$ lb.

Aircraft status	Speed	Flaps	Landing gear	Minimum steady-climb rate, ft/min
One engine out (prop feathered)[a]	Most favorable	Most favorable	Up	$\geq 0.027\, V_{s_0}^{2}$ [c]
AEO,[b] $W > 6000$ lb	Most favorable	Takeoff	Up	$\geq$300-ft/min climb gradient $\geq$0.0833 land plane $\geq$0.0667 seaplane
$W < 6000$ lb	Most favorable	Takeoff	Down	$\geq$300 ft/min and $\geq 11.5\, V_{s_0}$ [d]

[a]If $W < 6000$ lb and $V_{s_0} < 61$ knots, there is no engine-out climb requirement.
[b]AEO = all engines operating.
[c]V_{s_0} = stall speed in landing configuration for reciprocating-engine-powered airplanes, in knots at 5000 ft. [d]V_{s_1} = stall speed in a specified configuration for reciprocating-engine-powered airplanes, in knots.

Table F.5 Special carrier suitability requirements (USN)

1) Minimum rate of climb (at design landing weight and approach speed) of 500 ft/min at intermediate thrust (non-AB) with one engine inoperative.
2) Minimum longitudinal acceleration at end of catapult stroke of 0.065 g (at maximum catapult weight; all engines operating).
3) Aircraft to fit on 70- by 52-ft aircraft elevator.
4) Landing gear width not to exceed 22 ft.
5) Folded height of the aircraft not to exceed 18 ft 6 in. Height while folding not to exceed 24 ft 6 in. (18 ft 6 in. desirable).
6) Aircraft maximum weight (loaded and fueled) not to exceed 80,000 lb (elevator limit).
7) Wing span not to exceed 82 ft (64 ft desired).
8) Design landing weight to include high-value stores, empty external fuel tanks, and associated suspension equipment (pylons, ejectors, etc.)

Questions

Following are homework and/or examination questions for each chapter. Additional questions will be posted on the author's Aircraft Conceptual Design Web site, *http://www.aircraftdesign.com.*

Instead of using questions such as these, the instructor may wish to have the students do a design project similar to the design examples of Chapter 21. Properly done, this will test on the design and analysis methods presented throughout the book and give the students a better picture of the design process. A sample RFP is included at the end of this section. The author recommends that the students work individually, each doing an entire design project, rather than working as a team for their first exposure to aircraft design. This ensures that they will experience the entire design process rather than focus on the part of the process that they already understand. The author's *RDS-Student* computer program was specifically developed to reduce the amount of time spent doing design analysis, thus allowing students to do trade studies and carpet plots in a one-semester class.

Author's Note and Chapter 1

1.1 Raymer seems to think that some people make especially good designers, whereas a different type of person is probably better as a design analyst. However, others feel that a well-rounded engineer can and should be equally good at both. Choose and defend either belief. (There is no right answer; your grade depends on demonstrating an understanding of design.)

1.2 True/false? Why? "The ability to use a top-end CAD system is probably the single most important skill required to be successful in conceptual design today."

Chapter 2

2.1 True/false/neither? Why? "Design begins with requirements."

2.2 Describe the impact and process if it is discovered that the wing must be moved during (a) conceptual design, (b) preliminary design, (c) detail design, (d) flight test, and (e) production.

Chapter 3

3.1 Develop an exponential empirical equation to predict the empty weight fraction of the following: (a) homebuilt flying boat, (b) supersonic business jet, (c) stealth fighter, and/or (d) regional jet transport. Use data from *Jane's, Aviation Week*, or similar sources.

3.2 Quickly estimate best and cruise L/D for the designs in Figs. 2.5, 7.1, and 20.6.

3.3 For the ASW design example, conduct a trade study to determine the effect of a change in SFC on sized takeoff weight.

3.4 Resize the ASW design example to a "ferry" mission of 5000 n miles {9260 km}; i.e., delete segments 4 and 5, make range in segment 3 equal to ferry range, resize.

3.5 For the ASW design example, what is the effect on range if an advanced engine can reduce fuel consumption by 20%.

3.6 (advanced) For the ASW design example, calculate the total range obtained if the airplane actually does weigh 50,000 lb {22,680 kg}.

Chapter 4

4.1 For the ASW design example of Chapter 3, what is the lift coefficient at the beginning of cruise? At the end?

4.2 Select wing planform parameters (aspect ratio, taper, etc.) for the ASW design example of Chapter 3. Make reasonable guesses where needed, based on the characteristics of the current aircraft that perform the same mission (S-3).

4.3 If tails are "little wings," why don't they have the same planform parameters as the airplane's wing?

Chapter 5

5.1 Develop an exponential empirical equation to predict the T/W or P/W of the following: (a) homebuilt flying boat, (b) supersonic business jet, (c) stealth fighter, and/or (d) regional jet transport.

5.2 If a piston-propeller aircraft needs an in-flight power loading of 22 to cruise at 16,404 ft {5000 m}, what is the takeoff power loading required to meet cruise requirements if the engine has a supercharger? If it doesn't? Assume $W/W_0 = 0.9$ at the start of cruise. Use the IO-360 and TO-360 engines of Fig. 5.2.

5.3 Select T/W and W/S for the ASW design example of Chapter 3. Make reasonable guesses where needed, based on the characteristics of the current aircraft that perform the same mission (S-3).

Chapter 6

6.1 Resize the ASW design example of Chapter 3 using the methods of Chapter 6, assuming fixed-size engines ($T = 9000$ lb {40 kN} each) and a payload drop at the end of the loiter (2000 lb {907 kg}).

6.2 Estimate horizontal and vertical tail sizes for a twin-turboprop regional transport with wing area of 40 sqm, aspect ratio of 10, taper ratio of 0.5, leading-edge sweep of 5 deg, and tail moment arms both equal to 10 m.

6.3 For the airplane in question 6.2, if the flaps are half-span, what should the aileron areas be?

6.4 (advanced) Why is the volume coefficient method not very applicable to canards? Or is it? Prove with historical and analytical results.

Chapter 7

7.1 (a) Manually construct a conic curve with the following points: $A = (0{,}0)$, $B = (10{,}8)$, $C = (2{,}7)$, and $S = (4{,}6)$. (b) What is the ρ of this curve? Construct other curves with the same A, B, and C, but using (c) $\rho = 0.2$, (d) $\rho = 0.4142$, and (e) $\rho = 0.9$.

7.2 For the aircraft of question 6.2, draw the aircraft top view including wing planform plus horizontal tail and fuselage. Show where the center of gravity should be located assuming that the aircraft is to be an unstable, computer-controlled design. Feel free to make up any parameters or features that are not provided!

7.3 (advanced) Construct a flat-wrap interpolated airfoil at 30% of span for the wing in question 6.2, using the NLF(1)-0215F airfoil (see appendices) with 5 deg of twist. Is the interpolated airfoil still an NLF(1)- 0215F airfoil?

7.4 (advanced) Select and research a commercial CAD program, evaluating it for use in *conceptual* aircraft design.

Chapter 8

8.1 Select and research one particular "aerodynamic fix." How could the problem it is fixing possibly have been avoided in conceptual design? Give examples.

8.2 Identify and discuss the major load paths and wing carrythrough structure for the Boeing Blended Wing Body concept (see Chapter 22.3). Compare vs a "normal" jet transport design.

8.3 Congress and the media have long accused the military of "goldplating" their aircraft. When did this actually happen? Why?

8.4 Design for improved survivability can lead to worse maintenance. Explain.

8.5 Design for improved producibility can lead to worse maintenance. Explain.

Chapter 9

9.1 A transport aircraft is to have seven seats across with two aisles. Using the minimum required values for economy seating (Table 9.1), draw the fuselage cross section and determine the minimum-allowed round fuselage diameter. Draw two larger diameter fuselage cross sections and determine a relationship between fuselage diameter and available cross-section area under the floor for cargo. Assume the floor is 1 ft {31 cm} thick.

9.2 The original North American Aviation A-5 Vigilante used a linear bomb bay in which the bombs were ejected out the rear of the aircraft, between the twin engine nozzles. Discuss the pros and cons of this approach.

Chapter 10

10.1 Use Fig. 10.15 to estimate capture area for (a) the afterburning turbofan of Appendix E.1 with design Mach of 2.1, and (b) the high-bypass turbofan of Appendix E.2 with design Mach of 0.92.

10.2 (advanced) Same as question 10.1 but use detailed methods.

10.3 Select propeller diameter for a 3307-lb {1500-kg} aircraft being designed to reach 300 knots {556 km/hr} at 20,000 ft {6096 m}.

Chapter 11

11.1 Evaluate the down locations of the landing gear of the VTOL Harrier by the design practice presented in Chapter 11. What reasons might have led the designers to this arrangement?

11.2 Evaluate the down locations of the landing gear of the cancelled Lockheed Darkstar UAV by the design practice presented in Chapter 11. What reasons might have led the designers to this arrangement?

11.3 Do a detailed calculation of the landing gear stroke required for the DR-3 example jet design of Chapter 21.

Chapter 12

Wing:		{metric}
Sref-wing	= 279.4	{26.0}
Sexp-wing	= 230.4	{21.4}
A true	= 9.6	
Lambda = Ct/Cr	= 0.6	
Sweep-LE	= 2.0	
t/c average	= 0.088	
Delta Y	= 1.768	
CL-design	= 0.5	
CLmx-airfoil	= 1.9	
X (quarter-chord)	= 12.0	{3.7}

Horizontal Tail:

S-tail	= 70.0	{6.5}
Sexp-tail	= 47.5	{4.4}
A true	= 6.0	
A effective	= 6.0	
Lambda 5Ct/Cr	= 0.57	
Sweep-LE	= 4.6	
t/c average	= 0.08	
Delta Y	= 1.629	
X (quarter-chord)	= 34.0	{10.4}

Fuselage

Swet	= 516.9	{48.0}
length	= 35.5	{10.8}
diam-effctiv	= 6.1	{1.9}
Q (interfer)	= 1.1	
Upsweep deg	= 10.0	
X (nose)	= 0.0	{0}

12.1 For the aircraft just defined, prepare the wing lift curve at Mach 0 and at Mach 0.6. Also graph lift-curve slope vs Mach up to Mach 2.

12.2 Estimate Mdd. Calculate total parasitic drag at Mach 0.01 and Mach 0.6 at sea level using the component buildup method. Compare to result via the equivalent skin-friction method. Which is better? Easier?

12.3 Calculate drag due to lift at $C_L = 0.5$ and $C_L = 0.9$ using the LE suction method. Compare to result via the Oswald method. Which is better? Easier?

12.4 True/false/neither? Why? "CFD should only be used after the external lines of the aircraft are finalized, and is really not suited to conceptual design."

Chapter 13

13.1 Calculate installed thrust and SFC at Mach 0.8, 30,000 ft {9144 m} for the high-bypass turbofan of Appendix E.2 if the inlet pressure recovery is 99% and bleed airflow is increased to 5 lb/s {2.3 kg}.

13.2 If a piston-propeller aircraft needs an in-flight power loading of 17 to meet a climb requirement at 12,000 ft {3658 m}, what is the takeoff power loading required to meet this requirement if the engine has a supercharger? If it doesn't? (assume $W/W_0 = 0.95$ at the start of cruise).

13.3 Calculate propeller thrust for an engine with 300 hp {223.7 kW} having a 6-ft-diam {1.8 m} three-bladed propeller of activity factor 100 and blade design lift coefficient of 0.5, at its design condition of 150 knots {278 km/hr} at 12,000 ft {3658 m}. Also calculate its thrust at 110 knots {204 km/hr} at that altitude if it has variable pitch, and if it doesn't.

Chapter 14

14.1 Describe what an "ideal" material for aircraft structures would be like. How do the most commonly used actual materials compare to this ideal?

14.2 Construct the *V-n* diagram for the as-drawn DR-3 design example in Chapter 21, including gust effects.

14.3 Calculate the maximum wing bending moment for the as-drawn DR-1 design example in Chapter 21 under a 6-*g* turn loading at sea level at 100 knots {185 km/hr}. Calculate and compare the maximum wing bending moment if the designer adds a bracing strut attached at 30% of the span, at a 30-deg angle.

Chapter 15

15.1 A transport aircraft has a wing weight totaling 10% of takeoff gross weight. If a trade study is conducted changing the wing aspect ratio from 8 to 12, by what percent of takeoff gross weight should the fuel weight increase or decrease for the unsized as-drawn aircraft?

15.2 You are the weights engineer, and a designer has come to you with a design concept unlike anything ever done before. Describe different techniques you could follow to attempt to estimate its weights.

15.3 What must you *never* do on a group weights statement?

15.4 (advanced) Develop an analytical equation for the weight of a simplified wing box with spar caps and a single shear web only. Compare to statistical equations.

Chapter 16

16.1 For the aircraft defined for the Chapter 12 questions, calculate the neutral point, pitching moment derivative, and static margin at Mach 0.01 at sea level if (a) $X_{cg} = 13$ ft {4 m} and (b) $X_{cg} = 16$ {4.9 m}.

16.2 Same aircraft, prepare a trim plot and determine the trim deflection at 200 knots {370 km/hr} at sea level for (a) and (b).

16.3 Same aircraft, determine elevator deflection in a steady 3-*g* pull-up at 200 knots {370 km/hr} at sea level for (a) and (b).

16.4 Prepare a Cooper–Harper rating evaluation (explain your reasoning at each decision point) for the worst car/boat/motorcycle/skateboard you ever operated. Extra points for humor or life-saving tips.

Chapter 17

$$C_{D_0} = 0.0250 \quad A = 6 \quad e = 0.80 \quad W/S = 20\,\text{psf}\ \{98\,\text{kg/sqm}\}$$

$$P = 500\,\text{hp}\ \{373\,\text{kW}\} \quad \eta_p = 0.75 \quad c = 0.4\ \text{per hr}\ \{11.3\,\text{mg/Ns}\}$$

$$W = 2000\,\text{lb}\ \{907\,\text{kg}\} \quad W_f = 600\ \{272\}$$

For this propeller-powered aircraft, what is:

17.1 Best cruise speed at 10,000 ft {3048 m}?

17.2 Cruise range at 10,000 ft {3048 m} at 150 knots {278 km/hr}, assuming 100 lb of fuel {45 kg} is used for takeoff and landing?

17.3 Rate of climb at 100 knots {185 km/hr} at sea level and at 10,000 ft {3048 m}?

17.4 Best climb speed at sea level?

17.5 If the engine dies during cruise at 10,000 ft {3048 m}, and you have 100 lb of fuel {45 kg} remaining in the tank, how far can you glide, ignoring prop drag (a)? How long can you stay in the air, trying to restart the engine (b)?

17.6 Describe the minimum fuel-to-climb flight profile for the aircraft analyzed in Fig. 17.15 to reach a high altitude loiter at 40,000 ft {12,192 m} and Mach 0.8.

Chapter 18

18.1 Describe some situations where the minimum weight aircraft would not be the minimum cost aircraft, specifically (a) purchase price, (b) operating cost, and (c) life-cycle cost.

18.2 The Sneaky Aircraft Company stole Boeing's cost data to help make a more accurate cost estimation for their new transport. Why was this a dumb idea (law and ethics aside)?

18.3 Estimate the development and procurement cost of the DR-3 design example in Chapter 21, assuming (a) traditional aluminum construction and (b) substantial use of composites.

Chapter 19

19.1 Develop an empty weight sizing equation for the DR-1 design example in Chapter 21 by estimating the change to empty weight for a 20% increase in takeoff weight, then applying that result to Eq. (19.13).

19.2 Revise the carpet plots of Figs. 19.4 and 19.6 showing a takeoff constraint of 600 and 700 ft. What is the optimal aircraft for each of these relaxed requirements?

Chapter 20

20.1 Revise the ASW design example of Chapter 3 for VTOL using lift engines in bumps on each side of the fuselage. Add a reasonable increase in empty weight fraction to account for the weight of the required lift engines.

Increase S_{wet}/S_{ref} an appropriate amount, estimate the new L/D, and resize the aircraft. Compare the results to the CTOL design.

20.2 The Sikorsky S-58 is an older, single main rotor plus tail-rotor design that was once very common. Estimate the power it requires during cruise at 128 knots, 5000 ft {237 km/hr at 1524 m}, based on the following actual data:

$W = 11{,}867$ lb	{847 kg}
Rotor diameter = 28 ft	{8.5 m}
Total parasitic drag: $D/q = 36.5$ ft^2	{3.4 sqm}

20.3 Estimate for the Sikorsky S-58 the power required to hover, and the autorotate descent speed.

Chapter 21

21.1 Size a helium balloon to carry your own weight, using reasonable assumptions.

21.2 Size a rocket to put yourself into orbit, using reasonable or unreasonable assumptions.

21.3 Go read something by Heinlein, Forward, or Niven. Enjoy, then check their orbital mechanics calculations (number of days to travel, etc...).

Chapter 22

22.1 Analyze the change in sized TOGW for the ASW aircraft example in Chapter 3, if a joined wing is used. Assume a wing weight savings of 28% and a parasitic drag increase of 4%. If these numbers hold up, is the joined wing a good idea for this application?

22.2 Prepare a design sketch of a forward-swept, joined wing, hypersonic asymmetric wing-in-ground effect UAV. Save it to prove you will do anything for a grade.

SAMPLE REQUEST FOR PROPOSAL SUPERSONIC BUSINESS JET RFP

The customer seeks a conceptual design study of a supersonic business jet to take to investors, with intention to develop, test, and produce it as a commercial venture. The following requirements are not firm, and suggestions are welcome (in other words, there just might be some "requirements" given that the students should determine are excessive or unwise!).

- Nonstop supersonic cruise from Paris to New York at Mach 1.8 or greater
- Divert distance of 200 n miles {370 km}
- Eight passengers and crew of two
- 400-lb {181-kg} baggage in cargo hold of 80 cubic ft {2.3 cubic m}
- 5000-ft {1524-m} takeoff balanced field length and landing field length
- 5000-fpm {1524-mpm} rate of climb at sea level
- 140-knots{259-km/hr} approach speed
- 110-knots{204-km/hr} stall speed
- Descend from cruising altitude to 15,000 ft {4572 m} in less than 5 min.
- Meet FAA and ICAO requirements and specifications

Students shall present the following in report form:

Assessment of requirements
Design sketches (at least four)
Rationale for selection
Initial sizing, selection of T/W, W/S, and wing geometry
Design layout
Analysis: Aero, Weights, Propulsion
Sizing and performance
Trade studies: range, payload, cruise speed, etc. . . . (at least one!)
Optional: Carpet plot—T/W and W/S
Optional: Development and production cost
Summary and next step

B-52 (U.S. Air Force photo).

References

[1]Taylor, J., *Jane's All the World Aircraft*, Jane's, London, England, UK, 1976.
[2]Abbott, I., and von Doenhoff, A., *Theory of Wing Sections*, McGraw-Hill, New York, 1949.
[3]Purser, P., and Campbell, J., "Experimental Verification of a Simplified Vee-Tail Theory and Analysis of Available Data on Complete Models with Vee Tails," NACA 823, 1945.
[4]Perkins, C., and Hage, R., *Airplane Performance, Stability, and Control*, Wiley, New York, 1949.
[5]Loffin, L., "Subsonic Aircraft: Evolution and the Matching of Size to Performance," NASA Ref. 1060, 1980.
[6]Sheridan, H., "Aircraft Preliminary Design Methods Used in the Weapons System Analysis Division," Navy Department Weapons Systems Analysis Div., Rept. R-5-62-13, 1962.
[7]Roskam, J., *Methods for Estimating Drag Polars of Subsonic Airplanes*, Published by Author, 1971.
[8]Hoerner, S., *Fluid Dynamic Drag*, Published by Author, 1958.
[9]Wallner, L., "Generalization of Turbojet and Turbine-Propeller Engine Performance in Windmilling Condition," NACA RM-2267, 1951.
[10]Nicolai, L., *Fundamentals of Aircraft Design*, Univ. of Dayton, Dayton, OH, 1975.
[11]Roskam, J., *Airplane Design*, Roskam Aviation and Engineering Corp., Ottawa, KS, 1985.
[12]Diehl, W., *Engineering Aerodynamics*, Ronald, New York, 1928.
[13]Raymer, D., "Supercruise for a STOL Dogfighter," *Aerospace America*, Aug. 1985, pp. 72–75.
[14]Raymer, D., "CDS Grows New Muscles," *Astronautics & Aeronautics*, June 1982, pp. 22–31.
[15]Liming, R., *Practical Analytic Geometry with Applications to Aircraft*, Macmillan, New York, 1944.
[16]Sears, W., "On Projectiles of Minimum Wave Drag," *Quarterly of Applied Mathematics*, Vol. IV, No. 4, Jan. 1947.
[17]Whitcomb, R., "A Study of the Zero-Lift Drag Rise Characteristics of Wing-Body Combinations near the Speed of Sound," NACA RM-L-52H08, 1952.

[18]Ball, R., *Fundamentals of Aircraft Combat Survivability Analysis and Design*, AIAA Education Series, AIAA, New York, 1985.

[19]Stadmore, H., "Radar Cross Section Fundamentals for the Aircraft Designer," AIAA Paper 79-1818, AIAA Aircraft Systems and Technology Meeting, New York, Aug. 1979.

[20]Fuhs, A. E., *Radar Cross Section Lectures*, AIAA, New York, 1984.

[21]Ruck, G., et al., *Radar Cross Section Handbook*, Vols. I and II, Plenum, New York, 1970.

[22]"Anthropometry of Flying Personnel," Wright Air Development Center, TR 52-321, 1954.

[23]Torenbeek, E., *Synthesis of Subsonic Airplane Design*, Delft Univ. Press, Delft, The Netherlands, 1982.

[24]Smyth, S., and Raymer, G., "Aircraft Conceptual Design," Course Notes, Lockheed Corp., 1980.

[25]Seddon, J., and Goldsmith, E. L., *Intake Aerodynamics*, AIAA Education Series, AIAA, New York, 1985.

[26]Crosthwait, E., Kennon, I., and Roland, H., "Preliminary Design Methodology for Air Induction Systems," U.S. Air Force SEG-TR-67-1, Wright-Patterson AFB, OH, 1967.

[27]Ball, W., "Propulsion System Installation Corrections," Air Force Flight Dynamics Lab., AFFDL TR-72-147, Wright-Patterson AFB, OH, 1972.

[28]Stinton, D., *The Design of the Aeroplane*, Granada, London, 1983.

[29]Bingelis, T., *Firewall Forward*, Published by Author, Austin, TX, 1983.

[30]*Aircraft Tire Data*, Goodyear Corp.

[31]Smiley, R., and Horne, W., Mechanical Properties of Pneumatic Tires, NASA Rept. TR-R-64, 1960.

[32]Currey, N., *Aircraft Landing Gear Design: Principles and Practices*, AIAA Education Series, AIAA, Washington, DC, 1988.

[33]Conway, H., *Landing Gear Design*, Chapman & Hall, London, 1958.

[34]Tomkins, F., "Installation *Handbook-Airborne Gas Turbine* Auxiliary Power Units," Garrett Corp., 1983.

[35]Covert, E. E., James, C. R., Kimzey, W. F., Richey, G. K., and Rooney, E. C. (eds.), *Thrust and Drag: Its Prediction and Verification*, Progress in Astronautics and Aeronautics, Vol. 98, AIAA, New York, 1985.

[36]Lowry, J., and Polhamus, E., "A Method for Predicting Lift Increments due to Flap Deflection at Low Angles of Attack in Incompressible Flow," NACA TN-3911, 1957.

[37]Hoak, D., Ellison, D., et al., "USAF DATCOM," Air Force Flight Dynamics Lab., Wright-Patterson AFB, OH.

[38]*Proceedings of Evolution of Aircraft Wing Design Symposium*, AIAA, New York, 1980.

[39]Dillner, B., May, F., and McMasters, J., "Aerodynamic Issues in the Design of High-Lift Systems for Transport Aircraft," *AGARD Fluid Dynamics Panel Symposium*, Brussels, Belgium, May 1984.

[40]Bonner, E., Clever, W., and Dunn, K., "Aerodynamic Preliminary Analysis System II—Part I—Theory," NASA Contractor Rept. 165627, 1981.

[41]Jones, R. T., "Theory of Wing-Body Drag at Supersonic Speeds," NACA Rept. 1284, 1956.

[42]Harris, R., "An Analysis and Correlation of Aircraft Wave Drag," NASA TM-X-947, 1964.

[43]Jumper, E., "Wave Drag Prediction Using a Simplified Supersonic Area Rule," *Journal of Aircraft*, Vol. 20, Oct. 1983, pp. 893–895.

[44]O'Conner, W., "Lift and Drag Prediction in Computer-Aided Design," U.S. Air Force ASD/XR-73-8, 1973.

[45]Cavallo, B., "Subsonic Drag Estimation Methods," U.S. Naval Air Development Center, Rept. NADC-AW-6604, 1966.

[46]Mattingly, J. D., Heiser, W., and Daley, D. H., *Aircraft Engine Design*, AIAA Education Series, AIAA, New York, 1987.

[47]Cawthon, J., Truax, P., and Steenkin, W., "Supersonic Inlet Design and Airframe—Inlet Integration Program (Project Tailor-Mate)," Air Force Flight Dynamics Lab., AFFDL TR-72-124, Wright-Patterson AFB, OH, 1973.

[48]Fraas, A., *Aircraft Power Plants*, McGraw-Hill, New York, 1943.

[49]Lan, C., and Roskam, J., *Airplane Aerodynamics and Performance*, Roskam Aviation and Engineering Corp., Ottawa, KS, 1980.

[50]"Generalized Method of Propeller Performance Estimation," United Aircraft Corp., Hamilton Standard Rept. PDB 6101A, 1963.

[51]Cardinale, S., "Basic Loads—General Information for Designers," Lockheed Corp., 1982.

[52]Corning, G., "Supersonic and Subsonic, CTOL and VTOL, Airplane Design," Published by Author, 1976.

[53]Schrenk, O., "A Simple Approximation Method for Obtaining the Spanwise Lift Distribution," NACA TM-948, 1940.

[54]Peery, D., *Aircraft Structures*, McGraw-Hill, New York, 1950.

[55]Timoshenko, S., *Strength of Materials*, Van Nostrand, New York, 1930.

[56]Neubert, H., and Kiger, R., "Modern Composite Aircraft Technology," *Sport Aviation*, Experimental Aircraft Association, July 1976.

[57]Crossley, F. A., "Aircraft Applications of Titanium: A Review of the Past and Potential for the Future," *Journal of Aircraft*, Vol. 18, Dec. 1981, pp. 993–1002.

[58]Tsai, S., *Composites Design—1985*, Think Composites, Dayton, OH, 1985.

[59]Hoskin, B., and Baker, A. (eds.), *Composite Materials for Aircraft Structures*, AIAA Education Series, AIAA, New York, 1986.

[60]Bruhn, E., *Analysis and Design of Flight Vehicle Structures*, Tri-State Offsett, 1973.

[61]"Military Standardization Handbook—Metallic Materials and Elements for Aerospace Vehicle Structures," Mil-Hdbk-5B, Sept. 1971.

[62]Staton, R., "Statistical Weight Estimation Methods for Fighter/Attack Aircraft," Vought Aircraft, Rept. 2-59320/8R-50475, 1968.

[63]Staton, R., "Cargo/Transport Statistical Weight Estimation Equations," Vought Aircraft, Rept. 2-59320/9R-50549, 1969.

[64]Jackson, A., "Preliminary Design Weight Estimation Program," Aero-Commander Div., Rept. 511-009, 1971.

[65]Toll, T., et al., "Summary of Lateral-Control Research," NACA 868.

[66]Phillips, W., "Appreciation and Prediction of Flying Qualities," NACA 927, 1949.

[67]Roskam, J., "Airplane Flight Dynamics—Part I," Roskam Aviation and Engineering Corp., Ottawa, KS, 1979.

[68]Seckel, E., *Stability and Control of Airplanes and Helicopters*, Academic Press, New York, 1964.

[69]Lockenour, J., "Flying Qualities Considerations in Initial Design," Air Force Flight Dynamics Lab., AFFDL/FGCTM-74-222, Wright-Patterson AFB, OH, 1978.

[70]Hoerner, S., and Borst, H., *Fluid Dynamic Lift*, Hoerner Fluid Dynamics, Bricktown, NJ, 1975.

[71]Tallman, F., *Flying the Old Planes*, Doubleday, Garden City, New York, 1973.

[72]Harper, R. P., Jr., and Cooper, G. E., "Handling Qualities and Pilot Evaluation," *Journal of Guidance, Control, and Dynamics*, Vol. 9, No. 5, 1986, pp. 515–529.

[73]Rhodeside, G., "Investigation of Aircraft Departure Susceptibility Using a Total-G Simulator," AIAA Paper 86-0492, AIAA 24th Aerospace Sciences Meeting, Reno, NV, Jan. 1986.

[74]NACA Rept. TN-1045.

[75]Herbst, W. B., and Krogull, B., "Design for Air Combat," AIAA Paper 72-749, AIAA 4th Aircraft Design, Flight Test, and Operations Meeting, Los Angeles, CA, Aug. 1972.

[76]Skow, A., "Advanced Fighter Agility Metrics," Eidetics International, TR 84-05, Torrance, CA, 1984.

[77]Marx, H., "Comparative Costs of Military Aircraft—Fiction vs Fact," AIAA Paper 83-2565, AIAA Aircraft Design, Systems, and Technology Meeting, Fort Worth, TX, 1983.

[78]Hess, R. W., and Romanoff, H. P., "Aircraft Airframe Cost Estimating Relationships," Rand Corp., Rept. R-3255-AF, Santa Monica, CA, 1987.

[79]Birkler, J. L., Garfinkle, J. B., and Marks, K. E., "Development and Production Cost Estimating Relationships for Aircraft Turbine Engines," Rand Corp., Rept. N-1882-AF, Santa Monica, CA, 1982.

[80]Bengelink, R. L., "The Integration of CFD and Experiment: An Industry Viewpoint," AIAA Paper 88-2043, AIAA 15th Aerodynamic Testing Conf., San Diego, CA, 1988.

[81]Raj, P., Keen, J., and Singer, S., "Application of an Euler Aerodynamic Method to Free-Vortex Flow Simulation," AIAA Paper 88-2517, AIAA 6th Applied Aerodynamics Conf., Williamsburg, VA, 1988.

[82]Miranda, L. R., "Transonics and Fighter Aircraft: Challenges and Opportunities for CFD," NASA Transonic Symposium, NASA Langley Research Center, Hampton, VA, 1988.

[83]Niu, M., *Airframe Structural Design*, Conmilit, Hong Kong, 1988. (Distributed by Technical Book Company, Los Angeles, CA).

[84]Hughes, T., *The Finite Element Method*, Prentice-Hall, Englewood Cliffs, NJ, 1987.

[85]Desai, C., and Abel, J., *Introduction to the Finite Element Method*, Von Nostrand Reinhold, New York, 1971.

[86]Prziemienecki, J., "Finite Element Analysis," (unpublished Notes), 1988.

[87]Kalemaris, S., "Weight Impact of VTOL," SAWE Paper 1326, SAWE, New York, 1979.
[88]Kohlman, D., *Introduction to V/STOL Aircraft*, Iowa State Univ. Press, Ames, Iowa, 1981.
[89]*Proceedings of Special Course on V/STOL Aerodynamics*, Advisory Group for Aerospace Research and Development, AGARDR-710, 1984.
[90]Raymer, D. P., "The Impact of VTOL on the Conceptual Design Process," AIAA Paper 88-4479, AIAA/AHS/ASEE Aircraft Design, Systems, and Operations Meeting, Atlanta, GA, Sept. 1988.
[91]Hoblit, F. M., *Gust Loads on Aircraft: Concepts and Applications*, AIAA Education Series, AIAA, Washington, DC, 1988.
[92]DeMeis, R., "F-117A: First in Stealth," *Aerospace America*, Feb. 1991, pp. 32–42.
[93]Raymer, D., et al., "Supersonic STOVL; The Future is Now," *Aerospace America*, Aug. 1990, pp. 18–22.
[94]Raymer, D., "Post-Stall Maneuver and the Classic Turn Rate Plot," AIAA Paper 91-3170, Sept. 1991.
[95]Raymer, D., "RDS-Professional in Action: Aircraft Design on a Personal Computer," SAE/AIAA Paper 96-5567, Oct. 1996.
[96]Schaufele, R. D., "Applied Aerodynamics at the Douglas Aircraft Company," AIAA Paper 99-0118, Jan. 1999.
[97]Lorell, M. A., and Levaux, H. P., "The Cutting Edge," Rand Corp., MR-939-AF, Santa Monica, CA, 1998.
[98]Aronstein, D. C., "The Development and Application of Aircraft Radar Cross Section Prediction Methodology," SAE/AIAA Paper 965539, Los Angeles, 21 Oct. 1996.
[99]Piccirillo, A. C., "The Have Blue Technology Demonstrator and Radar Cross Section Reduction," SAE/AIAA Paper 965538, Los Angeles, CA, 21 Oct., 1996.
[100]Sweetman, B., *Stealth Bomber*, Airlife Publishing, Shropshire, England, UK, 1989.
[101]Miller, J., *Northrop B-2 Spirit*, Midland Publishing, Leicester, England, UK, 1995.
[102]Strang, W. Z., Tomaro, R. F., and Grismer, M. J., "The Defining Methods of Cobalt: A Parallel, Implicit, Unstructured Euler/Navier-Stokes Flow Solver," AIAA Paper 99-0786, Reno, NV, Jan. 1999.
[103]Rester, S. A., Rogers, C. J., and Hess, R. W., "Advanced Airframe Structural Materials," Rand Corp., R-4016-AF, Santa Monica, CA, 1991.
[104]Rutowski, E. S., "Energy Approach to the General Aircraft Performance Problem," Douglas Aircraft Co., Los Angeles, CA, 1953 (IAS Paper).
[105]Staley, J. T., "New and Emerging Aluminum Products for Cost Effective Airframe and Space Structure," *SAE Journal of Weight Engineering*, Vol. 55, Spring 1996.
[106]Hart-Smith, L., "The Ten Percent Rule for Preliminary Sizing of Fibrous Composite Structures," SAWE Paper 2054, Chula Vista, CA, May 1992.
[107]Jameson, A., "Optimum Aerodynamic Design Using the Navier-Stokes Equations," AIAA Paper 97-0101, Reno, NV, Jan. 1997.
[108]Raymer, D. P., "Multivariable Aircraft Optimization on a Personal Computer," SAE/AIAA Paper 965609, Oct. 1996.

[109]Sobieski, J., "Multidisciplinary Design Optimization," *Advances in Structural Optimization*, Kluwer Academic, The Netherlands, 1995, pp. 483–496.

[110]Crossley, W., Wells, V., and Laananen, D., *The Potential of Genetic Algorithms for Conceptual Design of Rotor Systems*, Overseas Publishers Association, Amsterdam, 1994.

[111]"Air Force Materiel Command Guide on Integrated Product Development," Air Force Materiel Command, Dayton, OH, May 1993.

[112]Johnson, C. L., and Smith, M., "More Than My Share of It All," Smithsonian Institution, Washington, DC, 1985.

[113]Hale, F. J., *Aircraft Performance, Selection, and Design*, Wiley, New York, 1984.

[114]Prouty, R. W., *Helicopter Performance, Stability and Control*, PWS Publishers, 1986.

[115]Saunders, G. H., *Dynamics of Helicopter Flight*, Wiley, New York, 1975.

[116]Raymer, D., and Burnside Clapp, M., "Pathfinder Rocketplane Conceptual Design Study," SAE/AIAA Paper 985596, World Aviation Congress, Anaheim, CA, 1998.

[117]Regan, F. J., *Re-Entry Vehicle Dynamics*, AIAA Education Series, AIAA, New York, 1984.

[118]Biswas, R., and Strawn, R. C., "A New Procedure for Dynamic Adaption of Three-Dimensional Unstructured Grids," *Applied Numerical Mathematics*, Vol. 13, 1994, pp. 437–452.

[119]Griffin, M., and French, J., *Space Vehicle Design*, AIAA Education Series, AIAA, Washington, DC, 1991.

[120]Begin, L., "The Northrop Flying Wing Prototypes," Northrop Corp., Hawthorne, CA.

[121]Nickel, K., and Wohlfahrt, M., *Tailless Aircraft in Theory and Practice*, AIAA Education Series, AIAA, Washington, DC, 1994.

[122]Liebeck, R. H., Page, M. A., and Rawdon, B. K., "Blended-Wing-Body Subsonic Commercial Transport," 36th Aerospace Sciences Meeting and Exhibit, AIAA Paper 98-0438, Reno, NV, Jan. 1998.

[123]Lippisch, A., *The Delta Wing*, Iowa State Univ. Press, Ames, Iowa, 1981.

[124]Wiler, C., and Raymer, D., "Advanced Strategic Aircraft Concepts," AIAA 18th Aerospace Sciences Meeting, Paper 80-0188, Jan. 1980.

[125]Garrison, P., "Boomerang: Technicalities Column," *Flying Magazine*, Vol. 124, No. 2, Feb. 1997, p. 104.

[126]Rutan, E., unpublished e-mail communication to author, 1999.

[127]Wolkovitch, J., "The Joined Wing: An Overview," AIAA 23rd Aerospace Sciences Meeting, AIAA Paper 85-0274, Reno, NV, Jan. 1985.

[128]Huebner, L. D., "Computational Inlet–Fairing Effects and Plume Characterization on a Hypersonic Powered Model," *Journal of Aircraft*, Vol. 32, No. 6, 1995, pp. 1240–1245.

[129]Raymer, D., "Next Generation Attack Fighter: Design Tradeoffs and Notional System Concepts," RAND Corp., MR-595-AF, Santa Monica, CA, 1996.

[130]Raymer, D., *Simplified Aircraft Design for Homebuilders*, Design Dimension Press, Los Angeles, CA, 2003.

[131]Moir, I., and Seabridge, A., *Aircraft Systems: Mechanical, Electrical, and Avionics Subsystems Integration*, AIAA Education Series, AIAA, Reston, VA, 2001.

[132]Zanzig, J., Analytical Methods, Inc., Renton, WA, telecon, 9 Feb. 2005.

[133]Raymer, D., "Dynamic Lift Airship Design Study Final Report (Phase Two)," Conceptual Research Corp., Rept. to Ohio Airships, Inc., Playa del Rey, CA, Jan. 2003.

[134]Raymer, D., "Approximate Method of Deriving Loiter Time from Range," *Journal of Aircraft*, Vol. 41, No. 4, 2004, pp. 938–940.

[135]Raymer, D., "Use of Net Design Volume to Improve Optimization Realism," *Weight Engineering Journal*, Vol. 61, No. 2, 2002.

[136]Raymer, D., "Enhancing Aircraft Conceptual Design Using Multidisciplinary Optimization," Ph.D. Dissertation, Swedish Royal Inst. of Technology, Stockholm, 2002.

[137]Jenkinson, L., Simpkin, P., and Rhodes, D., *Civil Jet Aircraft Design*, AIAA Education Series, AIAA, Reston, VA, 1999.

[138]Raymer, D., and Silva Oliveira, M., "Santos-Dumont: The First Homebuilder," AIAA Paper, July 2003.

[139]Santos-Dumont, A., *My Airships*, University Press of the Pacific, Honolulu, HI, 2002; Reprint of translation edition, *Dans l'Air- In the Air*, Dover, 1973.

[140]Burgess, C., *Airship Design*, Ronald, New York, 1927.

[141]Raymer, D., "Dynamic Lift Airship Design Study Final Report (Phase One)," Conceptual Research Corp., Rept. to Ohio Airships, Inc., Playa del Rey, CA, Dec. 2001.

[142]*Rigid Airship Manual*, USN Bureau of Aeronautics, U.S. Government Printing Office, Washington, DC, 1927.

[143]Ley, W., *Rockets, Missiles and Space Travel*, Viking Press, New York, 1957.

[144]Forward, R., *Indistinguishable from Magic*, Baen Publishing, New York, 1995.

[145]Raymer, D., Sponable, J., Fry, T., et al., "A Rocket-Powered Technology Demonstrator for Responsive Access to Space," AIAA Paper RS3 2005-6006, 2005.

[146]Sutton, G., *Rocket Propulsion Elements*, 5th ed., Wiley, New York, 1986.

[147]Bate, R., et al., *Fundamentals of Astrodynamics*, Dover, New York, 1971.

Index

Supporting Materials

Many of the topics introduced in this book are discussed in more detail in other AIAA publications. For a complete listing of titles in the AIAA Education Series, as well as other AIAA publications, please visit http://www.aiaa.org.